FLORA OF IRAQ

VOLUME SIX

FLORA OF IRAQ

VOLUME SIX

COMPOSITAE

EDITED BY

SHAHINA A. GHAZANFAR
JOHN R. EDMONDSON
AND
D. J. NICHOLAS HIND

With the collaboration of the staff of
the National Herbarium of Iraq of the
Ministry of Agriculture, Baghdad

Published on behalf of the
MINISTRY OF AGRICULTURE
Republic of Iraq
by
ROYAL BOTANIC GARDENS, KEW
2019

Royal Botanic Gardens Kew

First published in 2019 by
Royal Botanic Gardens, Kew,
Richmond, Surrey, TW9 3AB, UK
www.kew.org

ISBN 978-1-84246-685-8
e-ISBN 978-1-84246-686-5

Distributed on behalf of the Royal Botanic Gardens, Kew in North America by
The University of Chicago Press, 1427 East 60th Street, Chicago, IL 60637, USA

British Library Cataloguing in Publication Data
A catalogue record for this book is available from the British Library

Design and page layout: Fakenham Prepress Solutions
Production manager: Georgina Hills

Frontispiece: *Gundelia tournefortii* L., painted by Işik Güner 2013.

Printed in the UK by Bell and Bain Ltd, Glasgow
Printed in the USA by The University of Chicago Press

FSC www.fsc.org MIX Paper from responsible sources FSC® C007785

CONTENTS

FOREWORD

The field of botany in the world has reached a key moment of the century, a lot of money has been invested and many professionals have been carrying out important research. Despite this, we are far from being able to say that we know our natural environment and its diversity well. With this in mind, I hope this volume of *Flora of Iraq* will enrich to a significant extent, not only the field of botany, but will also add to the knowledge of biochemical, medical and other research fields.

The flora of Iraq is particularly rich and diverse in Compositae. The collection of plant specimens of this family started in the 1920's and early taxonomic work on Iraqi Compositae was conducted by foreign experts and Iraqi scientists. Work done on this family, through the collaboration between the Iraqi National Herbarium and Kew Gardens, allowed the completion of the unpublished volume 6. This publication is part of the ongoing work in the *Flora of Iraq* project.

The book in your hands describes around 123 genera and 432 species that represent the diversity of our country. This immense achievement was not accomplished by one person, but by great teamwork, mutual understanding and a great deal of cooperation. Therefore, I would like to express my sincere gratitude to the staff at the Iraqi National Herbarium and Kew Gardens and the public servants at the State Board of Seed Testing and Certification at the Ministry of Agriculture, as well as the Iraqi Seed company for its financial support in this project.

Finally, a special thanks to the honourable Minister of Agriculture Dr Saleh Jabr Al-Hassani and the senior officials.

Mohammed Zain Al-Abdeen Mohammed Raoof
Director General of SBSTC, Iraq National Herbarium
Ministry of Agriculture, Abu Ghraib, Iraq

PREFACE TO VOLUME 6

The present instalment of the Flora of Iraq covers a single family, the Compositae, also known as the Asteraceae. It is our largest angiosperm family in terms of the number of species (123 genera and 433 species). This is the penultimate volume in the Flora series as the monocots (vols. 8 & 9) have already been published.

The editors again wish to express their appreciation for the contributions of 12 external authors of taxonomic accounts to the present volume, in addition to those from Kew staff who were responsible for writing parts of this volume. Since the publication of the previous volume of the Flora of Iraq we have seen further steady progress with the Flora of Iran (in Persian), with the timely appearance of the second part of their Asteraceae treatment. As mentioned previously, a growing number of specimen images and associated metadata are now available to view online from several major herbaria.

The subgeneric and sectional classification of the genera of Compositae has been significantly altered in recent years, thanks to research that incorporates both morphological and molecular data. As the family is well defined and separate from its closest relatives, the impact of the APG project has been minimal, save for repositioning the family earlier in the sequence than traditionally placed. The final volume of the Flora of Iraq will retain the Hutchinson treatment, recognising that some changes in family limits have been made so as to more closely reflect their relationships.

We are again grateful to all the artists who have prepared drawings for this volume, with most genera represented by at least one plate, and again we have been granted permission to reproduce drawings from other Floras to complement those which were specially commissioned for this work. Sadly, Pat Halliday who has contributed several accounts and illustrations for this volume, passed away in May 2019.

We are extremely grateful to the staff of the National Herbarium, Abu Ghraib, Baghdad, especially Shukur Taher Al-Kaisi who has meticulously checked specimen citations against material at their herbarium (BAG). We are also thankful for the support of the Iraq Embassy in London with the transport of copies of the Flora of Iraq to Iraq.

Cliff Townsend, the co-editor of the Flora from its inception up to vol. 4 (plus 8 and 9) and a major contributor of accounts for the present volume, died on 8 January 2018. His many years of service on the Kew herbarium staff and his numerous publications testify to his significant contributions to botanical knowledge, not only in Iraq but around the world.

Shahina A. Ghazanfar
Royal Botanic Gardens, Kew

John R. Edmondson
Hon. Research Associate, Royal Botanic Gardens, Kew

D. J. Nicholas Hind
Royal Botanic Gardens, Kew

LIST OF CONTRIBUTORS

S. I. Ali, Karachi University, Karachi
G. Bromley, Royal Botanic Gardens, Kew
F. G. Davies, Royal Botanic Gardens, Kew
R. Doll, Rostock University, Rostock
J. R. Edmondson, Royal Botanic Gardens, Kew (Honorary Research Associate)
A. Ghafoor, Karachi University, Karachi
S. A. Ghazanfar, Royal Botanic Gardens, Kew
† P. Halliday, Royal Botanic Gardens, Kew
D. J. N. Hind, Royal Botanic Gardens, Kew
H.W. Lack, Free University Berlin & Berlin Botanic Garden, Berlin
† A. Huber-Morath, Basel
C. Jeffrey, Komarov Botanical Institute, St. Petersburg (formerly at R.B.G. Kew)
S.T. Al-Kaisi, Directorate of Seed Testing, Ministry of Agriculture, Baghdad
A. Kästner, Martin Luther University Halle-Wittenberg, Halle (Saale)
I. Mehregan, Islamic Azad University Tehran, Tehran
† H. Meusel, Martin Luther University Halle-Wittenberg, Halle (Saale)
† P. D. Sell, University of Cambridge, Cambridge
† C. C. Townsend, Royal Botanic Gardens, Kew
† G. Wagenitz, Georg-August University, Göttingen

115. COMPOSITAE[1]

Annual or perennial herbs, subshrubs, shrubs, rarely trees or climbers; hermaphrodite, monoecious, or rarely dioecious, many with laticiferous and/or resinous canals; rootstock fibrous, taprooted, or variously tuberous, sometimes woody and xylopodiaceous. Leaves rosettiform or cauline, cauline leaves usually alternate, more rarely opposite or whorled, sessile or petiolate, exstipulate, petiole sometimes auriculate at base, leaf blade entire to variously lobed or divided (pinnate or bipinnate), usually without apical tendrils. Inflorescences in capitula, solitary or few to many, arranged in various forms, rarely in secondary or tertiary aggregations (these with a secondary receptacle and termed synflorescences). Capitula homogamous and discoid or ligulate, or heterogamous and disciform, radiate, or radiant; involucres cylindrical to campanulate or hemispherical; phyllaries (involucral bracts) uniserate to multiseriate; calyculus of outer bracts sometimes present and often differing markedly from phyllaries; receptacle usually flattened, sometimes slightly concave or convex, rarely conical, smooth or alveolate, paleaceous (scaly), paleae lanceolate, oblanceolate, linear-lanceolate, narrowly spatulate, oblong-spatulate, oblong-elliptic, ovate, membranous, coriaceous to papery, or epaleaceous, with bristles, scales, hairs, or naked. Florets hermaphrodite (bisexual or perfect), female (pistillate), or functionally male (functionally staminate), solitary or few to many, central florets in disciform, radiate and radiant capitula termed disc florets, marginal florets termed radiate or radiant when corollas expanded into a variously lobed limb; corollas usually (4-)5-lobed, gamopetalous, tubular, tubular-filiform, regular, bilabiate, radiate or ligulate; stamens (4 or)5, inserted within basal tube; filaments usually free, adnate to proximal part of tube; anthers basifixed or dorsifixed, usually connate into a tube, base caudate or not, apex with (rarely without) a sterile, ovate or lanceolate appendage; pollen grains usually tricolporate, echinate or sometimes lophate or spinulate, often caveate; ovary inferior, unilocular; ovule 1, basal, anatropous; style apically bifid (rarely entire), style arms variously shaped, with or without an apical appendage, hairs, or papillae. Fruit an achene (cypsela); seed erect, endosperm scanty, embryo straight; carpopodium present or absent; pappus consisting of 1 to many rows of scales or bristles (setae), setae often barbellate, rarely plumose, awns, coroniform, cup shaped, variously coloured, persistent, deciduous or caducous or completely absent.

The largest family of dicotyledonous plants, widely distributed in temperate and tropical regions; it comprises, conservatively, four major subfamilies (although sometimes more than 12 are recognized), three of which are represented in Iraq. With more than 1,600 genera, around 123 of which occur in Iraq, and some 25,000 species, of which 433 species occur in Iraq, it includes many economically important crop species such as *Carthamus tinctorius* (safflower – Cardueae), *Helianthus annuus* (sunflower – Heliantheae), *Lactuca sativa* (lettuce – Lactuceae) and the herbs *Artemisia absinthium* (absinthe – Anthemideae) and *A. dracunculus* (tarragon – Anthemideae). Several species are also cultivated as food crops including: *Cichorium intybus* (chicory – Lactuceae), *Cichorium endivia* (endive), *Cynara cardunculus* (globe artichoke, cardoon – Cardueae), *Helianthus tuberosus* (Jerusalem artichoke – Heliantheae), *Scorzonera hispanica* (salsify – Lactuceae), etc. Many species are cultivated as ornamentals and for cut flowers, including *Calendula* (Corn Marigold, Pot Marigold – Calenduleae), *Callistephus* (China Aster – Astereae), *Chrysanthemum* (Florists' Chrysanthemum – Anthemideae), *Cosmos, Dahlia* (Coreopsideae), *Gerbera* (Mutisieae), *Pericallis* (Florists' Cineraria – Senecioneae), *Rudbeckia* (Cone flower – Heliantheae), *Symphyotrichum* (Michaelmas Daisy), *Tagetes* (French and African Marigolds – Tageteae), *Zinnia* (Heliantheae) and others. Many are also considered noxious weeds, including species of *Ambrosia* (Heliantheae), *Carduus* (Cardueae), *Centaurea* (Cardueae), *Chromolaena* (Eupatorieae), *Cirsium* (Cardueae), *Mikania* (Eupatorieae), *Parthenium* (Heliantheae), and *Senecio* (Senecioneae), etc.

Selected general references with relevance to the Compositae

Anderberg, A. A. et al. (contribs). [2006] (2007). Compositae. In: The families and genera of vascular plants (K. Kubitzki, ser. ed.). Flowering plants: Eudicots: Asterales (J. W. Kadereit & C. Jeffrey, vol. eds). Vol. VIII. pp. 61–588.

[1] Family description by D. J. Nicholas Hind

Blakelock, R. A. (1949). The Rustam Herbarium, Iraq. Part II. Systematic List. [Compositae]. Kew Bull. 4(1): 41–65.
Boulos, L. & D. J. N. Hind. (2002). Compositae. In: L. Boulos, Flora of Egypt, Vol. 3 (Verbenaceae – Compositae). pp. 134–320.
Bremer, K. (1994). Asteraceae: Cladistics & Classification. Timber Press, Portland, Oregon.
Brummitt, R. L. (2008). Proposal to add the 1754–1756 German translation of S. Vaillant, *Établissement de nouveaux caractères de trois familles*, 1719–1725, to the "Opera utiqueoppressa". Taxon 57(2): 663.
Davis, P. H. (ed.). (1975). Flora of Turkey and the East Aegean Islands. Vol. 5. pp. [i]–xviii, [i]–890.
Funk, V. A. et al. (eds) (2009). Systematics, Evolution, and Phytogeography of Compositae. pp. [i]–xxxiv, [1]–965.
Greuter, W. (2008). Proposals to conserve the names *Leysera* against *Asteropterus*, *Volutaria* against *Amberboi*, *Rhaponticum* with a conserved type, and *Rhaponticoides* against *Bielzia* (Compositae). Taxon 57(3): 1001.
Greuter, W. & J. McNeill. (2008). Damn Vaillant? A reply to Brummitt in Taxon 57(3): 663. 2008, and some alternative proposals, including adding all Steinwehr's translations in Königl. Akad. Wiss. Paris Phys. Abh. 5–9. 1754–1760 to the "Opera utiqueoppressa". Taxon 57(3) 1015–1016.
Hind, D. J. N. (2007). Plate 603. *Helichrysum orientale.* Compositae. Curtis's Bot. Mag. 24 (4): 251–259.
Tutin, T. G. et al. (eds) 1976). Flora Europaea. Vol. 4. Plantaginaceae to Compositae (and Rubiaceae). [Compositae] pp. 103–410.
Wu, Z. Y. et al. (eds). (2011). Flora of China. Vol. 20–21 (Asteraceae). pp. 1–988.

TERMINOLOGY RELATED TO THE COMPOSITAE[2]

achene (see cypsela) – a single-seeded indehiscent dry fruit with the seed free from the pericarp except at the placenta (the fruit in almost all Compositae; achenes may be pappose (bearing pappus) or epappose (without a pappus) or beaked (with a short or elongated distal seedless part of the achene in some Compositae (e.g. *Taraxacum*). In the tribes Eupatorieae, Heliantheae and Tageteae, achenes are usually blackened by the deposit of phytomelanin; in most Cyanareae some layers of the achene accumulate phytomelanins. Achenes of Compositae, derived from inferior ovaries, are often termed cypselas.

bilabiate corolla – corolla with a 2-lobed adaxial lip and a 3-lobed abaxial lip.

palea (pl. paleae) – receptacular bracts (e.g. in some Gnaphalieae) that individually or collectively subtend central perfect florets within head.

bracts – modified leaves associated with flowers and flower clusters; bracts in Compositae include units of a calyculus, phyllaries (primary and secondary involucral bracts), and paleas (receptacular bracts).

calcarate anther base – base of anther theca prolonged below the insertion of the anther as a spur-like projection.

calyculus – an outer row of differentiated bracts immediately subtending the phyllaries (sometimes described as an outer involucre). (Characteristic of tribes Senecioneae, Heliantheae).

capitulum (pl. capitula) – short, dense indeterminate inflorescence of sessile flowers attached to a common receptacle. Capitula are the primary inflorescence type characteristic of Compositae (generally with an involucre of phyllaries, a receptacle (with or without paleae), and one to many florets of one or more kinds. Capitula may be solitary or in various types of usually determinate secondary inflorescences.

corolla – collective term for petals of a flower.

corolla limb – in a disk floret or ligulate floret the portion of the corolla distal to insertion of anthers, composed of corolla throat and lobes; in a ray floret the more or less expanded distal portion, including the lamina.

corolla tube – cylindric portion of the corolla of a floret; in disk and ligulate florets the portion proximal to insertion of filaments.

coroniform pappus – very short, distinct or connate pappus, collectively forming a crown-like ring.

cypsela – indehiscent, syncarpic, unilocular and monospermic dry fruit derived from an inferior ovary (developmentally the fruit of the Compositae).

dioecious – plant in which all flowers are imperfect, and staminate and pistillate flowers are produced on different individuals.

disciform capitulum – type of heterogamous capitulum bearing perfect or functionally staminate disk florets in the center and peripheral filiform florets (e.g. *Pluchea*, *Pseudognaphalium*), naked florets (e.g. *Cotula coronopifolia*) or functionally neutral florets with reduced corollas (some *Centaurea*); or a homogamous capitulum consisting only of filiform florets. A disciform capitulum superficially resembles a discoid capitulum in lacking ray florets.

---

[2] Adapted from: Roque, N., Keil, D.J. & Susanna, A. (2009). Illustrated glossary of Compositae. In: V.A. Funk, A. Susanna, T.F. Stuessy & R.J. Bayer (eds), *Systematics, Evolution & Biogeography of Compositae*, pp.781–806.

discoid capitulum – homogamous capitulum that contains only disk florets (ray florets always absent).
filiform floret – pistillate floret characterized by a narrowly tubular corolla, sometimes with an unlobed limb or with vestigial lobes.
floret – single flower characteristic of Compositae. Florets have a modified calyx, the pappus, consisting of bristles, scales, and/or awns and is adnate to the distal end of the inferior ovary/ fruit; the corolla is (3–)5-merous, sympetalous, diversely tubular, radial or variously bilateral; androecium is of (3–)5 equal stamens with (usually) distinct, epipetalous filaments and (usually) connate anthers forming a tube; the gynoecium is of 2 connate carpels with a distally 2-branched style and a unilocular inferior ovary containing 1 erect, basal ovule.
heterocarpous – capitulum in which achenes of two or more morphologies are present.
heterogamous capitulum – capitulum with florets of different sex, usually pistillate and bisexual. (Radiate capitula and most disciform capitula are heterogamous. Radiant capitula may be homogamous or heterogamous).
homogamous capitulum – capitulum with florets of the same sex usually all perfect (bisexual) and fertile. Discoid capitula are homogamous.
involucre – disk-like to cup-shaped, ovoid, or cylindric group of bracts in one or more series that collectively subtend or surround the florets of a capitulum (involucres are absent in a few genera (e.g. *Xanthium*).
imperfect floret – floret in which the androecium or gynoecium or both are nonfunctional.
ligulate floret – perfect floret with a proximal tube and a distal, more or less flat, strap-shaped, 5-lobed, bilateral corolla limb (ligule).
monoecious – plant in which all florets are imperfect, and staminate and pistillate flowers are produced on the same individual.
palea (pl. paleae) – small bracts borne on the receptacle subtending all or some of the florets of a capitulum; paleae may be similar in size and/ or texture to phyllaries or more commonly smaller and of membranous to chartaceous texture. (Paleae are always found in the Heliantheae (with only a few exceptions) and in some Anthemideae and Eupatorieae. In some genera, each palea is associated with a floret, and if the palea is removed from the capitulum the floret is also removed. In *Xanthium* all of the paleae of pistillate capitula are fused into a spiny, knobby, or winged bur containing one or two achenes).
pappus – modified calyx consisting of bristles, dry scales, and/or awns, characteristic of the Compositae.
phyllary (pl. phyllaries) – bracts that forms the primary involucre in Compositae. Phyllaries are usually distinct, but in some groups they are connate; they may be arranged in one to many series and range from equal or subequal in length to strongly graduated.
radiant head – heterogamous or homogamous head bearing disk florets in the center and peripheral florets with more or less dilated, radial to strongly bilateral corollas; the peripheral florets may be perfect and fertile, pistillate, or neuter.
radiate head – heterogamous head with disk florets in the center and peripheral ray florets.
ray floret – bilaterally symmetric floret with a (2–)3(–4)-lobed (or some- times unlobed) limb and generally no adaxial lobes. Ray florets are usually pistillate or neuter; neuter ray florets have an ovary (sometimes much reduced) and may or may not have a style. Ray florets are located peripheral to central disk florets in radiate heads.
receptacle – basal part of the head on which phyllaries, florets, and often paleas are inserted. The shape of receptacle may be from weakly concave to flat, convex, conical or cylindric. The surface may be paleate or epaleate, smooth, foveolate or shallowly to deeply alveolate, glabrous, scaly or variously pubescent, bristly, or scaly.
synflorescence – an aggregation of individual capitula onto a secondary receptacle and surrounded by a secondary involucre (e.g. *Sphaeranthus* – Tribe *Inuleae*).

Following Kubitzki (ser. ed.), the family Compositae (Asteraceae) is represented by 16 tribes in Iraq. These include 123 genera and 433 species (including 14 cultivated taxa):

Tribe Mutisieae Cass. [cultivated]
Tribe Cynareae Lam. & DC. (Tribe Cardueae Cass.)
Tribe Gymnarrheneae Panero & V.A.Funk
Tribe Cichorieae Lam. & DC. (syn. Lactuceae Cass.)
Tribe Gundelieae DC. ex Lecoq & Juillet
Tribe Arctotideae Cass.
Tribe Senecioneae Cass.
Tribe Calenduleae Cass.
Tribe Gnaphalieae (Cass.) Lecoq & Juillet
Tribe Astereae Cass.
Tribe Anthemidae Cass.
Tribe Inuleae Cass.

Tribe Heliantheae Cass.
Tribe Coreopsideae Lindl. [cultivated]
Tribe Tageteae Cass.
Tribe Eupatorieae Cass.

KEY TO THE TRIBES OF COMPOSITAE[3]

1. Capitula of 2 kinds, subterranean and subaerial; subaerial capitula congested in the centre of sessile leaf rosette .... **Gymnarrheneae** (p. 146)
   Capitula all aerial, not congested in the centre of sessile leaf rosette 2
2. Capitula with 4 phyllaries and 4 florets .... **Eupatorieae** (p. 434)
   Capitula with > 4 phyllaries and florets 3
3. Leaves and/or phyllaries with pellucid glands or pustules, usually aromatic (cultivated in Iraq) .... **Tageteae** (p. 421)
   Leaves and/or phyllaries without pellucid glands or pustules 4
4. Achenes with phytomelanin layer in pericarp (achenes appearing black) (cultivated in Iraq) 5
   Achenes without phytomelanin layer in pericarp 6
5. Calyculus present; phyllaries with evident elongated oil-glands. .. **Heliantheae** (p. 415)
   Calyculus absent; (cultivated in Iraq) .... **Coreopsideae** (p.)
6. Corollas all strap-shaped, 5-toothed at the apex .... **Cichorieae** (p. 415)
   Corollas not strap-shaped or if some or all strap-shaped then these with 3 or fewer apical teeth 7
7. Phyllaries uniseriate, interlocking by their overlapping margins or partly or wholly connate, with or without a calyculus of tiny external bracts; pappus present 8
   Phyllaries imbricate in 2 or more series, free or connate, if uniseriate then pappus absent, of scales or coroniform, or capitula unisexual or achenes densely villous with long hairs 9
8. Achenes black; phyllaries with evident elongated oil-glands ..... **Heliantheae** (p. 415)
   Achenes not black; phyllaries without elongated oil-glands ..... **Senecioneae** (p. 268)
9. Flowering structure an involucrate synflorescence composed of secondary level capitula of one central fertile single-flowered capitulum subtended by 5–7, single-flowered, primary capitula forming a discrete disseminule .... **Gundelieae** (p.)
   Flowering structure not a synflorescence as above 10
10. Capitula homogamous, all florets tubular and bisexual with 5 relatively long and narrow lobes or all or only the outer florets bilabiate, or if capitula unisexual, then dioecious 11
    Capitula heterogamous, or if homogamous then florets with 5–4–3 relatively short and wide lobes, or phyllaries with scarious appendages or scarious usually brownish margins, or if capitula unisexual then monoecious 14
11. Leaves spiny 12
    Leaves not spiny 13
12. Anthers tailed; receptacle densely setose or capitula with only 1 floret .... **Cynareae** (p. 13)
    Anthers not tailed; receptacle alveolate with fringed alveoli ..... **Arctotideae** (p. 267)
13. Receptacle densely setose; florets all tubular .... **Cynareae** (p. 13)
    Receptacle not setose, but naked or scaly; florets all tubular or the outer or all bilabiate .... **Mutisieae** (p. 13)
14. Phyllaries cartilaginous, hyaline; pappus of a single plumose bristle .... **Gnaphalieae** (p. 284)
    Phyllaries otherwise, not hyaline; pappus otherwise usually of numerous barbellate bristles, scales or coroniform or absent 15

[3] S. A. Ghazanfar – adapted from H. Beentje & S.A. Smith (eds) (2000). *Flora of Tropical East Africa – Compositae (part 1)*. A.A. Balkema.

15. Style branches of bisexual disc florets connate nearly to the tips, with upper part of the style shaft abruptly or gradually thickened below them; phyllaries if connate in lower part then in several series; capitula always bisexual . . . . **Arctotideae** (p. 267)
Style branches of disc florets long to short or absent but the upper part of the style shaft not thickened as above, or capitula unisexual and monoecious; phyllaries if connate uniseriate. 16
16. Receptacle scaly, if not scaly then pappus of scales and/or leaves opposite and/or achenes black and/or capitula unisexual . . . .**Heliantheae** (p. 415)
Receptacle not scaly or if so, then leaves not opposite, achenes not black and capitula not unisexual 17
17. Style branches with a subulate to triangular appendage. . . . .**Astereae** (p. 315)
Style branches without a subulate to triangular appendage, acute to rounded, or truncate and fringed with short hairs or papillae. 18
18. Phyllaries with scarious usually brown often erose margins; style branches truncate and fringed; pappus of hairs absent . . . . . .**Anthemideae** (p. 331)
Phyllaries green and herbaceous or with hyaline margins or scarious membranous appendages; pappus of hairs present, if absent, then style branches not truncate and fringed, or phyllaries uniseriate 19
19. Pappus absent; anthers not tailed. 20
Pappus present; anthers tailed, sometimes minutely so . . . .**Inuleae** (p. 385)
20. Fruits large curved or angular or winged achenes or smooth drupes . . . . **Calenduleae** (p. 279)
Fruits small columnar achenes . . . . **Senecioneae** (p. 268)

KEY TO THE GENERA[4]

1. Florets all ligulate; corollas 5-toothed at apex . . . . GROUP A
Florets not all ligulate; marginal corollas zygomorphic and rayed (limbs with 4 or less teeth) or radiant, or corollas actinomorphic. 2
2. Plant with leaves or phyllaries, or both, with straight or hooked spines . . . . . . GROUP B
Neither leaves nor phyllaries spiny. . . . . GROUP C

GROUP A

1. Plant strongly spiny and thistle-like . . . . 31. **Scolymus**
Plant not strongly spiny and thistle-like 2
2. Corollas pale to bright blue 3
Corollas yellow (often with darker lines) to reddish or purple, never pale to bright blue 5
3. Phyllaries scarious, silvery with a dark midrib extendedin a cusp or arista . . . . 33. **Catananche**
Phyllaries not scarious and silvery, midrib not excurrent in a cusp or arista. 4
4. Inflorescence branches and phyllaries ± glandular-pilose; pappus of scabrid or barbellate, fine white hairs. . . . . .39. **Cicerbita**
Inflorescence branches and phyllaries without glandular hairs; pappus of short, obtuse scales . . . . 32. **Cichorium**
5. Phyllaries uniseriate, or if 2-seriate then these also equal or outer longest. 6
Phyllaries 2- or more seriate, if 2 then subequal inner series much exceeding short outer phyllaries. 8
6. Leaves narrowly linear to narrowly oblong, parallel-veined, often ± grass-like . . . . 57. **Tragopogon**
Leaves not narrowly linear to narrowly oblong and parallel-veined, not grass-like. . . . 7
7. Phyllaries long-hispid, subequal; achenes bipartite, upper part a sterile, basally inflated beak; pappus apical . . . . 52. **Urospermum**
Phyllaries glabrous or floccose, outer often longest; achenes not bipartite; pappus arising well below apex. . . . . .53. **Epilasia**
8. Pappus completely lacking or represented by a short fringe or cup 9

[4] C.C. Townsend – revised by D.J.N. Hind, J.R. Edmondson, S.A. Ghazanfar & S.T. Al-Kaisi

Pappus distinct, represented by long hairs or bristles at least on inner achenes 12
9. Pappus completely lacking 10
Pappus represented in all achenes only by a short fringe or blunt bristles or cup or of several minute, stout, blunt bristles developing into a crown of radiating hooked spines as fruit matures. 11
10. Achenes linear, ± spreading, at least outer more than 5 mm, outer tightly enclosed by involute inner phyllaries; pappus completely absent .... 51. **Rhagadiolus**
Achenes oblong, less than 5 mm long, neither spreading nor enclosed by inner phyllaries; pappus present at least in central achenes .... 37. **Lapsana**
11. Leaves narrowly linear to filiform; achenes curved, furnished with rows of hooked, barbed prickles and tipped with several recurved prickles .... 55. **Koelpinia**
Leaves entire, dentate-pinnatifid or lobed. 12
12. Achenes heteromorphic 13
All achenes similar. 17
13. Outer achenes straight or incurved, terete, subulate, persistent and partly enclosed by persistent phyllaries; central achenes incurved, beaked, with pappus of a crown-like structure or of bristles. .... 46. **Garhadiolus**
Achenes with pappus of plumose bristles 14
14. Achenes with a long, slender beak; pappus of plumose bristles. ... 48. **Helminthotheca**
Achenes without a long slender beak. 15
15. Plant scapose, stem only with small bracts .... 49. **Leontodon**
Plant not scapose, stem with at least one well-developed leaf. 16
16. Achenes with a long, slender beak; phyllaries strongly differentiated, outer series ovate to cordate .... 48. **Helminthotheca**
Achenes at most subrostrate; phyllaries less differentiated, outer not ovate or cordate .... 50. **Picris**
17. Pappus of only inner achenes long, that of outer achenes a short, toothed, membranous cup .... 47. **Hedypnois**
Pappus of inner and outer achenes not differentiated. 18
18. Pappus setae long-plumose at least below 19
Pappus hairs smooth or scabrid 20
19. Plant scapose, stem with only small bracts; leaves deeply pinnatipartite or pinnatisect with narrow lobes. .... 49. **Leontodon**
Plant rarely scapose, and if so then some or all leaves undivided ...... 56. **Scorzonera**
20. Stem ± scapose, with or without bracts, capitula solitary or few. 21
Stem with at least one well-developed leaf; capitula mostly numerous 22
21. Achenes muricate or spinulose above; receptacle naked .... 38. **Taraxacum**
Achenes ribbed but smooth, if spiculate then receptacle paleaceous ...... 35. **Crepis**
22. Median and upper leaves with long-decurrent auricles forming narrow wings along the ± shining, whitish branches .... 40. **Lactuca**
Median and upper leaves not or only slightly decurrent 23
23. At least inner achenes distinctly beaked 24
All achenes truncate, rounded or slightly narrowed above, never distinctly beaked . 29
24. Achenes homomorphous, all beaked. 25
Achenes heteromorphic, central beaked, marginal not. 28
25. Achenes strongly compressed. 26
Achenes not strongly compressed 27
26. Pappus uniseriate of long, copious capillary hairs; no annular rim between beak and pappus-disc in mature fruit .... 40. **Lactuca**
Pappus biseriate, outer row of very short hairs; annular rim present between beak and pappus-disc in mature fruit .... 41. **Steptorhamphus**
27. Capitula ± sessile on twiggy branches; large squamae present at apex of mature achene .... 34. **Chondrilla**
Capitula on slender pedicels; apex of mature achene scabrid, with no squamae .... 40. **Lactuca**
28. Achenes minutely scabrid above, without large squamae around base of beak. .... 35. **Crepis**

Achenes muricate above, with large squamae around base of beak . . . 36. **Heteroderis**
29. Pappus stramineous (straw-coloured) to brownish . . . . 45. **Pilosella**
Pappus pure white 30
30. Many lobes and teeth of leaves white-mucronate 31
Lobes and teeth of leaves not white-mucronate 32
31. Achenes 4-ribbed, the outer dark with ribs with large tubercles . . . . 43. **Reichardia**
Achenes ribbed, angled or winged, scabrid, rugulose or pilose, not with large tubercles . . . . 42. **Launaea**
32. Achenes distinctly compressed; outer phyllaries basally ± incrassate in mature fruit . . . . 44. **Sonchus**
Achenes not compressed; outer phyllaries not basally incrassate in mature fruit . . . . .35. **Crepis**

GROUP B

1. Plant monoecious, with separate male and female capitula; female capitula with 1 or 2 florets, producing a spiny fruit (burr) . . . . 117. **Xanthium**
Plant without separate male and female capitula, if with separate male and female flowers then these occur together on same capitulum 2
2. Phyllaries with hooked spines, forming a burr . . . . 18. **Arctium**
Phyllaries without hooked spines 3
3. Marginal florets rayed 4
Marginal florets not rayed 5
4. Leaves and outer phyllaries spinous (also *Carduus* see 25) . . . . 3. **Atractylis**
Only phyllaries spinous . . . . .103. **Pallenis**
5. Florets in single-flowered capitula, these sometimes aggregated into secondary capitula of several capitula, aggregated into synflorescences (or syncephalia), subtended by small bracts 6
Florets with a common involucre of several rows of phyllaries. 8
6. Capitula single-flowered and aggregated into secondary capitula (eventually becoming disseminules) of up to 8 primary capitula, secondary capitula aggregated into synflorescences . . . . 58. **Gundelia**
Capitula single-flowered with these agregated into synflorescences 7
7. Annual herb with hemispherical synflorescences (or syncephalia); leaves entire and unarmed or with weak spines . . . . 6. **Acantholepis**
Perennial herbs with globose synflorescences (or syncephalia); leaves pinnatifid or pinnatisect, rarely entire, and with strong, rigid spines . . . 5. **Echinops**
8. Inner phyllaries coloured, resembling corollas. 9
Inner phyllaries not coloured. 10
9. Leaves entire, lacking spines, floccose . . . . 10. **Siebera**
Leaves spiny . . . . 2. **Carlina**
10. Leaves not spiny. 11
Leaves spiny 14
11. Pappus absent from all achenes 12
Pappus present, or absent only from marginal achenes 13
12 Leaves not pinnatipartite; branches often rigid, glabrescent to ± arachnoid and shortly glandular; leaves usually spinose; capitula homogamous . . . . 27. **Carthamus**
Leaves pinnatipartite; branches not rigid, often tomentose to scabrous to hirsute with multicellular hairs, sessile glands very frequent, rarely quite glabrous; leaves not spinose (in Iraq); capitula heterogamous. . .28. **Centaurea**
13. Capitula homogamous, all florets hermaphrodite . . . . .26. **Klasea**
Capitula heterogamous, marginal florets neuter . . . . 28. **Centaurea**
14. Achenes densely hairy. 15
Achenes glabrous. 16
15. Style-branches very short and connivent . . . . 3. **Atractylis**
Style-branches elongate, subulate. . . . . .4. **Thevenotia**
16. Pappus of plumose hairs or bristles, at least in part 17
Pappus of scabrid hairs, bristles or scales. 21

17. Receptacle not long-pilose, honeycombed with membranous-fringed pits . . . . . 17. **Onopordum**
Receptacle with long hairs or bristles 18
18. Median and outer phyllaries terminating in a pinnate spine . . . . . 14. **Picnomon**
Median and outer phyllaries terminating in a simple spine 19
19. Receptacle fleshy; capitula very large, 6–10 cm in diameter; phyllaries 8–18 mm wide . . . . . 12. **Cynara**
Receptacle not fleshy; neither capitula nor phyllaries so wide 20
20. Leaves milky-veined on upper surface; pappus of outer florets not plumose; apex of achene with no umbo . . . . . 16. **Notobasis**
Leaves not milky-veined above; pappus of outer florets plumose; apex of achene umbonate . . . . . 13. **Cirsium**
21. Outer pappus of about 10 long bristles, inner pappus of very short bristles . . . . . 28. **Centaurea**
Pappus not so formed, if processes dissimilar in length then outer shortest 22
22. Pappus persistent, of several rows of paleae of differing length, inner bristly above or not . . . . . 27. **Carthamus**
Pappus deciduous, of hairs or bristles, sometimes absent 23
23. Pappus bristles not fused below, rapidly-caducous individually, sometimes absent and achene with a denticulate corona; filaments glabrous . . . . . 19. **Cousinia**
Pappus bristles fused below, deciduous in mature fruit on a basal ring, never absent; filaments hairy or papillose 24
24. Phyllaries with aherbaceous, pectinate-spinous appendage terminating in a stout spine; leaves milky-veined above . . . . . 15. **Silybum**
Phyllaries without such an appendage; leaves not milky-veined above . . . . 11. **Carduus**

GROUP C

1. Capitula radiant with outer florets tubular 2
Capitula either not radiate, or ray-florets ligulate (flat or filiform) 7
2. Phyllaries without a terminal appendage 3
Phyllaries with a distinct terminal appendage 4
3. Phyllaries tapering, acute; outer florets only weakly radiate . . . . . 22. **Crupina**
Phyllaries blunt and rounded with a membranous margin; marginal florets strongly radiate . . . . . 24. **Amberboa**
4. Achene with 3 concentric grooves at apex; appendages of phyllaries pectinate with very fine cilia . . . . . 21. **Zoegea**
Achene not with 3 concentric grooves at apex; appendages of phyllaries either not pectinate or cilia less fine 5
5. Outer radiant floret corollas (cornflower) blue; phyllaries with a black or brown decurrent border . . . . . 28. **Centaurea**
Corollas never blue; phyllaries lacking a decurrent black or brown border 6
6 Corollas yellow; appendages triangular, with 4–8 pairs of cilia, ending in a stiff spinule 2–4 mm long . . . . . 23. **Stizolophus**
Corollas rose-purple, whitish or yellow; appendages consisting of a much longer rigid spine, base ± dilated . . . . . 28. **Centaurea**
7. Leaves conspicuously decurrent to form wings along stem and branches; florets in few-flowered heterogamous capitula subtended by roundish, spinous-mucronate bracts, congested into ± globose, dense, secondary capitula . . . . . 108. **Sphaeranthus**
Leaves rarely decurrent; florets not in few-flowered capitula gathered into dense secondary capitula 8
8. Leaves opposite, at least the lower 9
Leaves all alternate 19
9. Capitula discoid, disciform, or a mixture of radiate or discoid capitula in a congested inflorescence 10
Capitula radiate, discoid capitula very rarely present and inflorescences never congested 14
10. Pappus absent 11

Pappus present. 12
11. Plants densely woolly; receptacle with hooded woolly scales enclosing female florets, indurate at maturity and falling with achenes. . . . . 76. **Micropus**
Plants not woolly; receptacle naked . . . . .115. **Flaveria**
12. Corollas yellow; pappus of 2–4 retrorsely barbed awns . . . . 111. **Bidens**
Corollas not yellow; pappus of scales or scabrid hairs, not retrorsely barbed 13
13. Pappus of awned scales; corollas blue . . . . 122. **Ageratum**
Pappus of barbellate hairs; corollas pink . . . . 123. **Eupatorium**
14. Receptacle naked; leaves and phyllaries gland-dotted . . . . 116. **Tagetes**
Receptacle furnished with scales or bristles; leaves and phyllaries lacking gland dots 15
15. Leaves simple, entire or dentate. 16
Leaves narrowly pinnately divided 18
16. Corollas white . . . . 118. **Eclipta**
Corollas not white 17
17. Ray floret corollas red or purple, sometimes white or yellow, ray corollas marcescent; pappus absent . . . . 121. **Zinnia**
Ray floret corollas usually yellow (except in certain cultivated forms), deciduous; pappus of thin, chaffy scales, rapidly caducous . . . . 119. **Helianthus**
18. Achenes unbeaked, compressed; anther filaments glabrous 112.**Coreopsis**
Achenes beaked, narrowly clavate or fusiform; anther filaments pubescent . . . . .113. **Cosmos**
19. Achenes with no pappus or pappus reduced to a small corona or auricle or a few short bristles up to 1.5 mm long. 20
Achenes with a distinct pappus of hairs, bristles or scales 45
20. Receptacle naked. 21
Receptacle not naked, bearing hairs, paleae or bristles 29
21. Leaves simple, margins entire or remotely dentate; phyllaries 1–2-seriate 22
Leaves ± pinnately divided, if merely dentate then phyllaries multiseriate 24
22. Achenes uniform, up to 2 mm long, compressed with a marginal rib. . . . . . . .81. **Bellis**
Achenes di- or trimorphic, larger than 2 mm 23
23. Ray floret corollas yellow or orange-yellow on both surfaces . . . . .65. **Calendula**
Ray floret corollas yellow above, whitish and ± suffused with purple below . . . . .66. **Dimorphotheca**
24. Achenes dimorphic, marginal differing from those of disc 25
Achenes all similar. 26
25. Capitula appearing discoid, marginal female florets few and inconspicuous. . . . . .95. **Aaronsohnia**
Capitula conspicuously radiate with large rays . . . . 96. **Glebionis**
26. Much-branched bushy herbs with very numerous discoid capitula at most 3 mm in diameter; achenes smooth or almost so . . . . 88. **Artemisia**
Plant not much-branched and bushy, capitula if discoid then 4 mm or more wide; achenes ribbed 27
27. Fruiting receptacle flat or only slightly convex at fruit fall; achenes regularly ribbed . . . . 92. **Tanacetum**
Fruiting receptacle strongly convex to conical at fruit fall; achenes not regularly ribbed 28
28. Achenes strongly 3-ribbed adaxially, with 1–2 reddish-brown glands near apex abaxially . . . . 93. **Tripleurospermum**
Achenes feebly 3–5-ribbed adaxially, eglandular. . . . . .94. **Matricaria**
29. Marginal florets rayed; capitula radiate, rays sometimes few 30
Marginal florets not rayed; capitula discoid or disciform 33
30. Achenes with a broad, shining wing ± 1.75 mm wide . . . . 90. **Anacyclus**
Achenes not winged 31
31. Tall perennials; rays yellow, 2–5 cm long . . . . 120. **Rudbeckia**
Shorter plants; rays white or much smaller and yellow. 32
32. Achenes compressed, with no apical corona; base of disc florets saccate, enveloping tops of achenes; midribs of paleae not reaching apex. . . . . . 89. **Achillea**
Achenes not compressed, with at least traces of an apical corona; base

of disc florets not saccate, midribs of paleae excurrent in a mucro or aristate. . . . . 91. **Anthemis**
33. Leaves pinnately divided 34
Leaves simple, entire or dentate 36
34. Receptacle hairy. . . . . 88. **Artemisia**
Receptacle paleaceous. 35
35. Base of disc florets saccate and enveloping tops of achenes; much-branched low shrubs . . . . . 89. **Achillea**
Base of disc florets not saccate and enveloping tops of achenes; annual to perennial herbs . . . . . 91. **Anthemis**
36. Outer phyllaries connate at base, with foliaceous appendages; leaves spatulate . . . . . 101. **Anvillea**
Outer phyllaries neither connate nor with foliaceous-appendages; leaves not spatulate 37
37. Plants perennials and subshrubs; leaves often ± toothed, if entire then not lanate . . 38
Annual herbs; leaves entire 42
38. At least some outer phyllaries with a greenish apical appendage. . . . . . .27. **Carthamus**
No phyllaries appendiculate 39
39. Stems greenish-tomentose; at least some capitula with larger, leafy outer phyllaries. . . . . .105. **Chrysophthalmum**
Stems subglabrous to sparsely or densely white-tomentose; no outer phyllaries leaf-like 40
40. Leaves entire, dentate or serrate or with sharp divergent teeth; phyllaries not closely appressed, imbricate, chartaceous 41
Leaves regularly crenate-serrulate; phyllaries closely appressed, not chartaceous. . . . . 89. **Achillea**
41. Achenes oblong, somewhat angular, sparsely hairy with conspicuous red resin ducts; pappus of free barbellate capillary bristles in one row . . 109. **Doellia**
Achenes angled or ribbed, glabrous without red resin ducts; pappus of broad flattened brtistles; marginal achenes usually epappose, or all achenes epappose . . . . . 98. **Rhanterium**
42. Leaves linear, glabrescent; achenes cylindrical, spiny, exserted 29.**Dipterocome**
Leaves ± woolly; achenes compressed, included 43
43. Female florets and achenes subtended but not enclosed by paleae . . . . 69. **Filago**
Female florets ± completely enclosed by paleae and falling with achene 44
44. Paleae not indurate at maturity; a few deciduous pappus-bristles present . . . . . .68. **Cymbolaena**
Paleae indurate at maturity; pappus totally lacking . . . . 67. **Bombycilaena**
45. Capitula obviously radiate with rays clearly exceeding disc 46
Capitula discoid, or if radiate then rays inconspicuous, not or scarcely longer than disc 63
46. Ray limbs yellow or orange [note: some plants with yellow rays when fresh may dry reddish; they will be found here] 47
Ray limbs red, pink, blue or purple, not yellow. 55
47. Receptacle setose or paleaceous. 48
Receptacle naked. 49
48. Receptacle with subulate setae; corollas of disc florets purple-tipped. . . 110. **Gaillardia**
Receptacle with concave paleae; corollas of disc florets entirely yellow .102. **Asteriscus**
49. Phyllaries connate for at least half their length into a basal cup 60.**Gazania**
Phyllaries not connate for at least half their length 50
50. Phyllaries 1–2-seriate, inner equal in length or almost so. 51
Phyllaries imbricate, 3-several seriate 52
51. Involucre calyculate; phyllaries erect, laterally coherent, acute to rather blunt . . . . . .63. **Senecio**
Involucre ecalyculate; phyllaries ± spreading, free, acuminate . . . . .61. **Doronicum**
52. Pappus with an outer row of very small scales . . . . . .99. **Pulicaria**
Pappus with no outer row of scales, formed entirely of hairs. 53
53. Disc florets sterile, achenes empty; caespitose plants of high mountains. . . . . .83. **Psychrogeton**

Disc florets fertile; not caespitose high mountain plants 54
54. Plant scapose . . . . 1. **Gerbera**
Plant not scapose 55
55. Achene obovoid; pappus basally connate . . . . 100. **Dittrichia**
Achenes cylindrical or angular, pappus uniseriate, barbellate, mostly free. . . 104. **Inula**
56. Phyllaries ±uniseriate, subequal 56
Phyllaries imbricate, multiseriate, unequal 58
57. Plant densely covered with multicellular glandular hairs . . . . .63. **Senecio**
Plant not covered with multicellular hairs 59
58. Corollas of ray florets yellow; wild . . . . 64. **Iranecio**
Corollas of ray florets purple or blue; cultivated . . . . .62. **Pericallis**
59. Receptacle with long, subulate setae . . . . 110. **Gaillardia**
Receptacle naked or alveolate-fimbrillate 60
60. Plant with dense viscid, glandular hairs . . . . 80. **Lachnophyllum**
Plant not densely furnished with viscid, glandular hairs 61
61. Rays filiform, thread-like . . . . .87. **Erigeron**
Rays flat, not thread-like 62
62. Scapose garden plant . . . . 1. **Gerbera**
Plant not scapose 63
63. Phyllaries arachnoid-pubescent, inner very broadly hyaline at apex . . . . . . .59. **Arctotis**
Phyllaries neither arachnoid nor inner broadly hyaline at apex 64
64. Outer phyllaries leafy, spreading; capitula large, up to 12 cm in diameter. . . 82. **Callistephus**
Outer phyllaries not leafy and spreading; capitula not more than 3 cm diameter. . . 65
65. Receptacle naked. 66
Receptacle not naked, furnished with bristles or scales 85
66. Capitula homogamous, only tubular hermaphrodite florets present 67
Capitula heterogamous, with filiform, minutely ligulate or rarely
tubular female flowers present outside disc 69
67. Phyllaries uniseriate, equal, with an outer calyculus of short, small bracts . .63. **Senecio**
Phyllaries in several unequal series. 68
68. Phyllaries green . . . . .104. **Inula**
Phyllaries scarious, hyaline, whitish, yellow, orange or red. . . . . 72. **Helichrysum**
69. Pappus setae plumose from base 70
Pappus setae simple or barbellate, not plumose from base 71
70. Prostrate annuals, capitula minute, congested into woolly glomerules. 74. **Lasiopogon**
Erect annuals, capitula not woolly, with chartaceous white or coloured
phyllaries. . . . . 79. **Rhodanthe**
71. Capitula with the appearance of fluffy balls from long, white, spreading
hairs of phyllaries; pappus uniseriate with c. 5 hairs . . . . 106. **Pentanema**
Capitula not like fluffy balls; pappus of more than 5 hairs. 72
72. Pappus with an outer row of minute scales outside hairs . . . . .99. **Pulicaria**
Pappus of hairs only 73
73. Shrubs to 1 m tall; leaves broad, entire; corollas pinkish . . . . .107. **Pluchea**
Herbs; leaves broad, toothed; corollas not pinkish 74
74. Phyllaries rather blunt, at least inner with a broad scarious to chartaceous tip 75
Phyllaries narrow, gradually tapering, none with a broad chartaceous or
membranous tip. 79
75. Capitula solitary; phyllaries solid, chartaceous . . . . 77. **Phagnalon**
Capitula in corymbs; phyllaries scarious-margined, white to brown or
brightly coloured 76
76. Female florets few, hermaphrodite numerous; phyllaries brightly coloured. . 72. **Helichrysum**
Female florets numerous, hermaphrodite few; phyllaries with
membranous tips hyaline to brownish 75
77. Low cushion plants; phyllaries with dark brown margins . . . . 71. **Gnaphalium**
Erect plants; phyllaries with hyaline or straw-coloured tips 78
78. Leaves spatulate or oblanceolate, obtuse; capitula in clusters subtended
by leaf-like bracts in a spiciform inflorescence . . . . .70. **Gamochaeta**
Leaves oblanceolate, acute, capitula in rounded clusters without
conspicuous bracts . . . . 78. **Pseudognaphalium**

79. Inner phyllaries narrowly hyaline-margined above; pappus-setae few, <10. . . 77. **Phagnalon**
Inner phyllaries sometimes pale- but never hyaline-margined; pappus hairs numerous, >10 . . . 80
80. Anthers sagittate-caudate at base . . . 81
Anthers obtuse at base. . . 82
81. Corollas pale mauve . . . 109. **Doellia**
Corollas yellow. . . 104. **Inula**
82. Inner florets functionally male, with empty achenes; plants of high mountains. . . 83. **Psychrogeton**
Inner florets hermaphrodite, fertile; not high mountain plants . . . 83
83. Plant either with a ± dense mixed indumentum or stout white marginal hairs on leaves . . . 86. **Conyza**
Plant subglabrous, leaves neither with a ± dense mixed indumentum nor margins with stout white hairs . . . 84
84. Phyllaries linear-oblong, obtuse, 1.75–2.5 mm wide; wild salt-marsh plants; leaves semi-succulent . . . 84. **Tripolium**
Phyllaries linear, acute, not more than 1 mm wide; cultivated or escapes from cultivation; leaves herbaceous . . . 85. **Symphyotrichum**
85. Phyllaries with a ± well-developed apical appendage . . . 86
Phyllaries without apical appendages. . . 89
86. Achenes with 3 concentric grooves at apex; phyllary-appendages pectinate-fimbriate with fine, almost hair-like fimbriae . . . 21. **Zoegea**
Achenes not with 3 concentric grooves at apex; phyllary appendages, if pectinate-fimbriate, less finely divided . . . 87
87. Capitula homogamous, all florets hermaphrodite and fertile . . . **86**
Capitula heterogamous, marginal florets neuter . . . 28. **Centaurea**
88. Phyllaries mostly 4–5-seriate, each with a basal ± veined, appressed cartilaginous portion ± concealed by overlapping papery appendages, these varying much in form from ovate and erose to acuminate . . . 25. **Leuzea**
Phyllaries leaf-like, regularly imbricate, veined, with an appendage usually reduced to an apical slender spine, often reflexed. . . 26. **Klasea**
89. Capitulum a discoid ‘button’ . . . 90
Capitulum not a yellow discoid ‘button’ . . . 91
90. Pappus of numerous short bristles not exceeding the involucre . 105. **Chrysanthemum**
Pappus of 2–3 slender bristles exceeding the involucre . . . 97. **Rhanteriopsis**
91. Leaves pinnatisect, margins with many-barbed glochidiate hairs. . . 22. **Crupina**
Leaves not pinnatisect, or if so margins not glochidiate-pilose . . . 92
92. Acaulescent herbs with capitula crowded in a ± sessile mass; Receptacular scales leathery, exserted, white-tipped . . . 30. **Gymnarrhena**
Plant not acaulescent with crowded capitula, nor with leathery, exserted receptacular scales . . . 91
93. Graceful annuals; leaves entire, filiform, glandular-hairy, viscid; capitula narrow, on slender pedicels. . . 75. **Leysera**
Annuals and perennials; leaves spatulate, elliptic, oblong, linear or pinnatifid, variously hairy, never glandular-hairy and viscid, if densely gland-dotted then leaves ovate-elliptic to obovate with denticulate margins. . . 94
94. Phyllaries scarious above . . . 95
Phyllaries not scarious above . . . 96
95. Pappus hairs plumose at tip; leaves narrow, linear . . . 73. **Ifloga**
Pappus hairs not plumose at tip; leaves spatulate to oblong . . . 69. **Filago**
96. Annuals; pappus paleaceous. . . 97
Perennials; pappus either of hairs or bristles or if scaly then divided into bristles . . . 98
97. Inner phyllaries coloured; pappus of short scales, included; achenes all similar . . . 8. **Xeranthemum**
Inner phyllaries not differing in colour from outer; pappus of long finally exserted scales; achenes dimorphic. . . 9. **Chardinia**
98. Florets filiform, not expanded above, ± equalling the abundant and conspicuous pappus; pappus of undivided bristles or hairs . . . 20. **Jurinea**
Florets expanded above narrow tube, exceeding pappus; pappus of scales deeply divided into scabrous or barbellate bristles . . . 7. **Staehelina**

TRIBE **MUTISIEAE** CASS.

1. **GERBERA** L. *nom.cons.*

Opera Varia: 247 (1758)
Gerbera Cass. in Bull. Soc. Philom. Paris 1817: 34 (1817)

C. C. Townsend

Acaulescent perennial herbs with radical, alternate leaves. Leaves lanceolate to subrotund, entire to toothed or more rarely pinnately lobed, usually pilose or tomentose on lower surface. Capitula solitary on scapes, heterogamous. Involucre hemispherical or campanulate; phyllaries 2–5-seriate, linear to linear-lanceolate. Receptacle flat or slightly convex, somewhat alveolate, naked. Marginal florets female; disc florets usually hermaphrodite. Achenes 5–10-ribbed, tapering above or beaked; pappus of numerous scabrid bristles.

About 70 species in the warmer regions of the Old World.
Gerbera (named in honour of the Silesian-German physician, naturalist and traveler Traugott Gerber, 1710–1743 who became the director of Moscow Botanic Garden.)

Hansen, H.V. 1985. A taxonomic revision of the genus *Gerbera* (Compositae, *Mutisieae*) sections *Gerbera, Parva, Piloselloides* (in Africa) and *Lasiopus*. Opera Bot. 78:1–36.
Hind, D.J.N. 1992. Typification of *Gerbera jamesonii* Adlam (Compositae). Kew Bull. 47 (1): 110.

1. **Gerbera jamesonii** *Adlam* in Gard. Chron. 3: 772, f. 2 (1888).

G. jamesonii Bolus ex Hook.f., Bot. Mag. 115: t. 7087 (1889); Pobedimova in Fl. S.S.S.R. 28: 590 (1963); Husain & Kasim, Cult. Pl. Iraq: 115 (1975); Hansen in Opera Bot. 78: 21 (1985).]

Scapose perennial (15–)25–40(–65) cm tall, vegetative parts softly hairy throughout, leaves densely white-tomentose beneath. Leaves erect, runcinate-pinnatifid, long-petiolate, lamina with margins ± undulate, with broad, acute or obtuse, angular, sinuately-toothed lobes. Capitula solitary or several; involucre hemispherical, 2–4.5 cm in diameter. Ray limbs narrow (1.5–)2.5–4 cm, yellow or orange to orange-red or dark red (white and pink forms occur in cultivation), duller below. Disc floret corolla lobes very short, greenish-yellow or coloured like rays limbs. Achenes puberulous, 6–13 mm; pappus bristles minutely scabrid, off-white.

HAB. & DISTRIB. Cultivated at the Agricultural Research Station; fl. Apr.
LCA: Abu Ghraib, *Sahira* C. 343!

Barberton Daisy (Eng., Am.), Transvaal Daisy (Am.). According to Husain & Kasim this species is cultivated throughout Iraq, but only the above specimen has been seen.

TRIBE **CYNAREAE** LAM. & DC.

2. **CARLINA** L.

Sp. Pl.: 828 (1753); Gen. Pl.: 360 (1754)
Sussana & Garcia-Jacas in Kubitzki, Fam. Gen. Vasc. Pl. 8: 126 [2006] (2007)

H. Meusel & A. Kästner

Perennial or annual herbs with spiny pinnatilobed or pinnatisect leaves. Capitula homogamous, discoid. Involucre campanulate to ± globose, with three types of phyllaries: outer ± foliaceous, pinnatifid, spiny; median linear-lanceolate, erect, not much longer than florets; inner linear, scarious, yellow, straw-coloured or reddish, radiating, much longer than florets; receptacle with each floret surrounded by several scarious or cartilaginous, laciniate paleae. Corollas yellow. Achenes terete, covered with appressed hair; pappus setae plumose, fused and branched at base.

About 30 species in Europe, temperate Asia and N Africa.
Carlina (named after Emperor Charlemagne (c.742–814), who is said to have cured an epidemic in his army by the use of *C. acaulis* L.).

Meusel, H. & Kästner, A. (1990). Monographie der Mediterran-Mitteleuropäischen Compositen: Gattung *Carlina*, Vol. 1. Wien.

Annual herb; inner phyllaries purple or silvery–purple . . . . 1. *C. lanata*
Perennial herb with a woody mostly branched rootstock; inner phyllaries yellow . . . . 2. *C. kurdica*

1. **Carlina lanata** *L.*, Sp. Pl.: 828 (1753); Dinsmore in Post, Fl. Syria, Palest. & Sinai ed. 2, 2: 81 (1933); Zohary in Dep. Agr. Iraz Bull. 31: 152 (1950); Linchevski in Fl. S.S.S.R. 27: 83 (1962); Rawi in Dep. Agr. Iraq Tech. Bull. 14: 108 (1964); Meusel & Kästner in Fl. Turkey 5: 600 (1975); Feinbrun-Dothan, Fl. Palaest. 3: 369 (1978); Rechinger, Fl. Iranica 139a: 102 (1979); Mouterde, Nouv. Fl. Liban et Syrie 3: 445 (1983); Meikle, Fl. Cyprus 2: 945 (1985).

Annual herb, (3–)10–35(–45) cm tall. Stem erect, pubescent, simple or cymosely branched. Leaves lanate to tomentose; basal leaves soft, oblanceolate, pinnatilobed; cauline leaves rigid, semiamplexicaul, 4–7.5 cm, pinnatisect, oblanceolate to broadly ovate, with long squarrose spines. Capitula 1.5–2.5 cm in diameter, surrounded by 3–4 cm long leaves resembling outermost phyllaries. Outer phyllaries lanceolate, squarrose-spinous, exceeding the linear-oblong, purple or silvery-purple inner phyllaries.

HAB. Grazed slope with light *Quercus* woodland, scree; alt. ± 758 m.; fl. Aug.
DISTRIB. Only found once or twice in the extreme N mountain region of Iraq. **MAM**: Sawara Tuka, *Haines* W1597! (Dohuk area, *Zohary*, l.c. – specimen not seen).

Mediterranean Europe from Spain to Turkey, S Bulgaria, Cyprus, Syria, Palestine, Turkey, ?Iran, Transcaucasia, N Africa (Libya to Morocco).

2. **Carlina kurdica** *Meusel & Kästner* in Feddes Repert. 83: 231 (1972). – Type: Iraq, Jabal Khantur near Sharanish, *Rawi* 23290 (K!, holo.); Rechinger, Fl. Iranica 139a: 102 (1979).

C. corymbosa auct. non L., Rawi in Dep. Agr. Iraq Tech. Bull. 14: 108 (1964).

Perennial herb, 15–25(–30) cm tall, partly pubescent, with a woody, branched rootstock. Stems erect, mostly simple, with or without basal leaves. Cauline leaves 4.5–6 cm, semiamplexicaul, ovate, the upper triangular with 6–7 spiny segments, terminal spine 1.5–2 cm. Capitula obconical to broadly obconical, 2–2.5 cm in diameter; outer phyllaries lanceolate, 2.8–3.5(–4) cm, only slightly exceeding the yellow, radiant inner phyllaries. Fig. 1: 1–4.

HAB. Dry fields, dry grassy and stony limestone mountain slopes, open *Quercus* forest; alt. 300–1400 m; fl. Jul.-Sep., fr. Aug.-Sep.
DISTRIB. Widespread in the mountain-forest region of Iraq, just extending to the adjacent dry steppe. **MAM**: Jabal Khantur nr Sharanish, *Rawi* 23290 (type)!, *Rechinger* 10741!; 1 km S of Zakho, *Rawi, Noori & Tikriti* 28970!; Dohuk, *Haines* 1217! **MRO**: Haibat Sultan Dagh, *Rawi & Serhang* 26468!; Shirwan Mazin, *Hadač et al.* 6232!; Sidakan to Diana, *Guest & Husham* 15868!; NW of Haruna, *Gillett* 9804! **MSU**: Darband-i Basian, *Rawi* 22173!; 15km NE of Rania Pushtashan village, *Rawi & Serhang* 24250B (BAG!); **FKI**: 32 km NNE of Kirkuk on road to Koi Sanjaq, *Rawi* 26458!

Endemic.

3. **ATRACTYLIS** L.

Sp. Pl.: 829 (1753); Gen. Pl. ed. 5: 360 (1754)
Bremer, Asteraceae: Cladistics & Classification: 136 (1994); Susanna & Garcia-Jacas in Kubitzki, Fam. Gen. Vasc. Pl. 8: 127 [2006] (2007)

Frances G. Davies

Low-growing prickly herbs. Cauline leaves lobed or toothed with slender spines. Capitula large, subtended by outer foliaceous phyllaries resembling cauline leaves or greatly differing from them, discoid and homogamous or radiant; phyllaries multiseriate, spine-tipped; receptacular scales lanceolate, laciniate-fimbriate at apex, straw-coloured. Florets all alike, or outer female and radiant. Achenes angular, truncate above, silky-lanate; pappus of plumose hairs, fused at base or a little above.

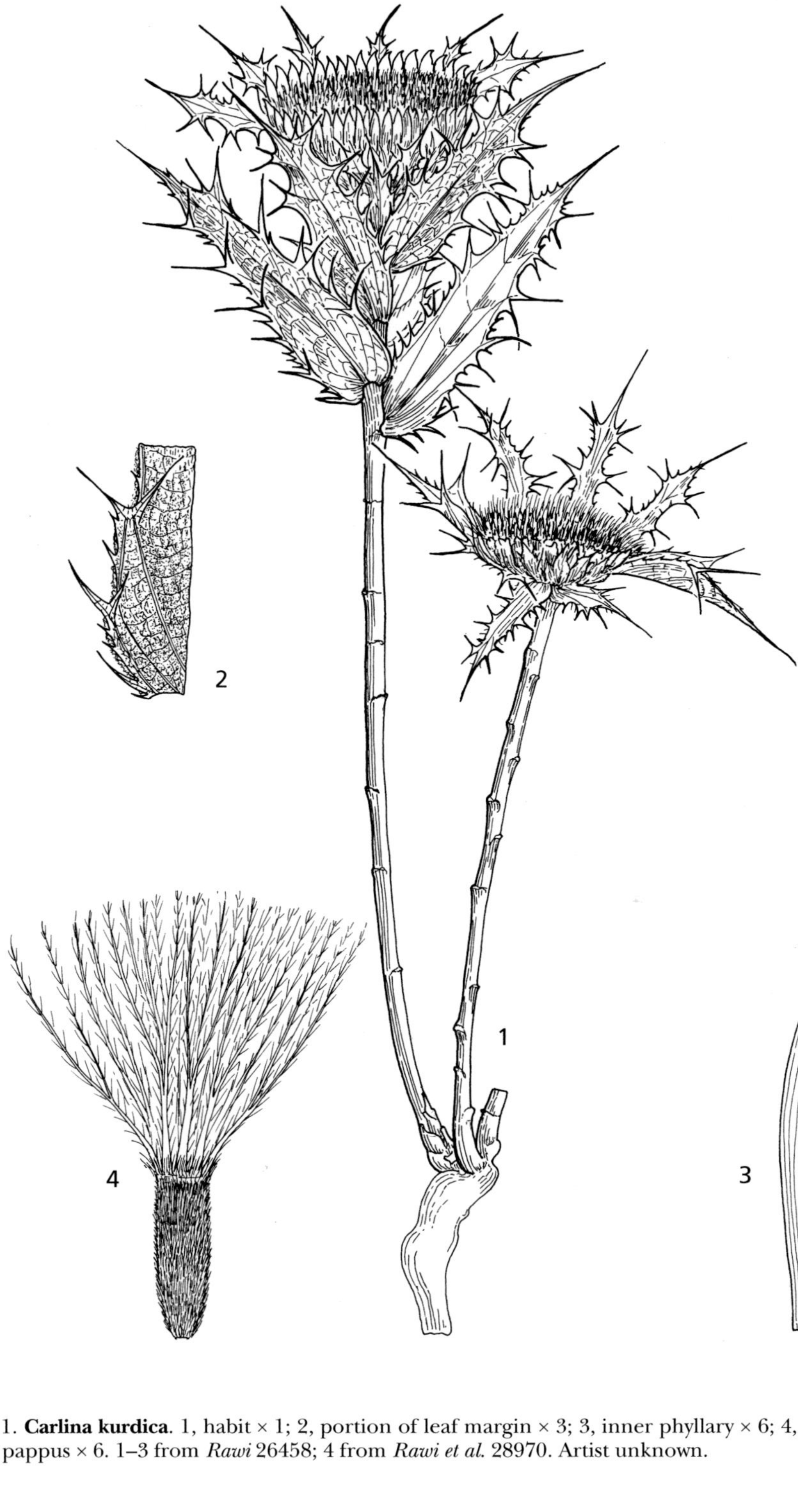

Fig. 1. **Carlina kurdica**. 1, habit × 1; 2, portion of leaf margin × 3; 3, inner phyllary × 6; 4, achene with pappus × 6. 1–3 from *Rawi* 26458; 4 from *Rawi et al.* 28970. Artist unknown.

Around 30 species from S Europe, SW Asia, N Africa and Macaronesia.

Atractylis (from Gr. ἄτρακτος, atraktos, spindle for a thistle-like plant used for making spindles.)

Petit, D. (1987). Analyse cladistique du genre *Atractylis* L. (Compositae, Cardueae). Bull. Soc. Bot. France, Lett. Bot. 134 (2): 165–184.

– (1987). Revision des genres *Atractylis*, *Carlina* et *Chamaeleon* (Compositae, Cardueae) au Maroc.", Bull. Mus. Nation. Hist. Nat., B, Adansonia 9: 407–440.

– (1988). Note nomenclaturale sur les subdivision du genre *Atractylis* L. Bull. Soc. Bot. France Lett. Bot. 135(1): 71–72. (1988).

1. Leaves with spiny teeth of different sizes; foliaceous phyllaries equalling or scarcely exceeding involucre; corollas purple .... 3. *A. cancellata*
   Leaves with distinct triangular lobes tipped by spines, foliaceous phyllaries 1.5–2 × as long as involucre; corollas yellow or reddish yellow 2
2. Marginal florets ligulate; corollas yellow .... 1. *A. carduus*
   All florets tubular; corollas reddish yellow........................... 2. *A. arabica*

1. **Atractylis carduus** (*Forssk.*) *Christ.* in Dansk Bot. Ark. 4 (3): 27 (1922); Täckholm, Stud. Fl. Egypt: 2: 534 (1974); Matthews in Fl. Turkey 5: 603 (1975); Franco in Fl. Europ. 4: 211 (1976); Feinbrun-Dothan, Fl. Palaest. 3: 370 (1978); Collenette, Wild Flow. Saudi Arabia: 174 (1999); Boulos & Hind in Fl. Egypt. 3: 152 (2002); Abdulridha, Taha & Wedad, Ecology and Fl. Basra: 197 (2016).

Centaurea carduus Forssk., Fl. Aegypt.-Arab.: 152 (1775).

Atractylis flava Desf., Fl. Atlant. 2: 254 (1799); Boissier, Fl. Orient. 3: 452 (1875); ; Dinsmore in Post, Fl. Syria, ed. 2, 2: 81 (1933); Dickson, Wild Fls. Kuwait: 22 (1955); Rawi in Dep. Agr. Iraq Tech. Bull. 14: 207 (1964); Rechinger, Fl. Lowland Iraq: 640 (1964).

A. citrina Coss. & Kralik., Bull. Soc. Bot. France 4: 361 (1857).

Perennial herb 10–40 cm tall, simple or branched from base with several simple or branched stems covered with white arachnoid pubescence. Leaves 15–60 × 4–10 mm (excluding spines), each side with 3–5 triangular or oblong lateral lobes ending in spines 0.5–2 mm long, both surfaces arachnoid-hairy. Capitula broadly conical. Outer foliaceous phyllaries 8–12 in a whorl, up to 8 × 1 cm excluding spines, spreading, resembling cauline leaves. Phyllaries obovate to spatulate, firm, acute to acuminate, pale-margined, tomentose, inner 10–2 mm, terminating in a long spine. Ray florets marginal, limb yellow, 1–2 mm, few; inner florets 10–14 mm, lobes 3 mm. Achenes 5–7 mm; pappus dirty white, 8–13 mm. Fig. 2: 1.

HAB. Desert hills and plains, on sandy, sometimes silty and gravelly soils; alt. up to 650 m; fl. & fr. (Apr.-) May-Jun.

DISTRIB. Quite common in the western and southern deserts of Iraq. **DLJ**: in the Jazira by Wadi Thirthar canal, nr Samarra, *C. & E. Guest & Sharif Muhammad Amin Haidar* 18323!; 70 km NNW of Falluja, *Rawi* 20259! **DWD**: 160 km from Ramadi to Rutba, *Rawi* 31309 (BAG!); 45 km NW of Rutba, *Rawi* 21172!; 30–100 km E of Rutba, *Alizzi* 35162!; 180 km W of Ramadi, *Rawi & Noori* 27026!; 145 km W by S of Ramadi, *Rawi* 26839!; Habbaniya, *Rawi & Alizzi* 34461! **DSD**: 15 km SE of Ashuriya, *Guest, Rawi & Rechinger* 19394!; 80 km W of Shabicha, *Chakravarty, Rawi, Khatib & Alizzi* 30071!, *Salman, Fauzi, Hazim & Hamid* 38766!; 2 km N of Khidr al-Ma'i, *Weinert, Fauzi, Addel & Husain* 47639!; Ichrishi, 35 km E by N of Busaira, *Guest & Rawi* 14198!; nr Aidaha, *Guest & Rawi* 19145!; 62 km WNW of Ansab, *Guest & Rawi* 19064!; "Falluja desert" (i.e over the river, W of Falluja), *Haines* W.795!

The colloquial name for this plant, SAMNA, (*Guest & Rawi* 14198) from Ichrishi where it is said to be eaten by camels and sheep. Dickson (1955) also reported that this plant, known locally by the name JALWA, is said to be eaten by camels in Kuwait.

A number of varieties were recognized by Täckholm & Boulos in Egypt (Publ. Cairo Univ. Herb. 5: 24, 1974); these appear not to be present in Iraq.

N Africa, Sinai, Arabian Peninsula, Palestine, Syria.

2. **Atractylis arabica** *Rech.f.* in Ber. Deutsch. Bot. Ges. 72: 281 (1959).– Type: Iraq, 8–15 km NW of Aidaha, *Rechinger* 9414 (W!, holo.); Rawi in Dep. Agr. Iraq Tech. Bull. 14: 207 (1964); Rechinger, Fl. Lowland Iraq: 641 (1964); Rechinger, Fl. Iranica 139a: 103, t. 100 (1980); Collenette, Wild Flow. Saudi Arabia: 173 (1999) [as *memephthae*]; Kilian & Ghazanfar in Fl. Oman 3: 211 (2015).

Fig. 2. **Atractylis carduus**. 1, capitulum with phyllaries × 2. **Atractylis arabica**. 2, habit × 1; 3, capitulum with phyllaries × 2. **Atractylis cancellata**. 4, capitulum with phyllaries × 2. 1–2 from *Rechinger* 19172; 3 from *Rawi* 20847; 4 from *Omar et al.* 37062. Drawn by Moira Rae Carter.

Biennial or perennial herb with several decumbent stems from base, to 30 cm tall, white-arachnoid when young, glabrescent. Leaves sessile, 10–30 × 1.5–6 mm excluding spines, with up to 4 pairs of spinous lobes, spines 0.5–3 mm long, reddish, lamina arachnoid-pubescent when young. Outer foliaceous phyllaries in a ring of 5–6, 18–40 × 1.5–4 mm excluding spines, resembling the stem-leaves. Capitula conical; phyllaries ovate to spatulate, ± thinly pilose or glabrescent, inner 10–14 mm, pale or purplish-margined, tipped by a spine varying from 2 mm long in the outer phyllaries to 6 mm long in the inner. Florets all tubular and hermaphrodite, corollas reddish yellow, 8–10 mm, lobes 2 mm. Achenes 5–6.5 mm; pappus 8–11 mm, dirty white. Fig. 2: 2–3.

HAB. In sandy wadi, on silty flat with low stony ridges and on silty soil in depression on plain; alt. 120–400 m; fl. Mar.-Apr., fr. Apr.-May.
DISTRIB. Occasional in the Southern Desert of Iraq. **DWD**: (exact localities not cited), *Hadač* 4120, 4254, 4279 (fide Rechinger, Fl. Iranica, l.c.). **DSD**: 8–15 km NW of Aidaha, *Guest, Rawi & Rechinger* 19170A!, 19172!; *Rechinger* 9414 (type)!; 40–60 km SSE of Salman, *Guest, Rawi & Rechinger* 18875!; 62 km WNW of Ansab, on Darb el Haj at Saudi border, *Guest, Rawi & Rechinger* 19063!; 20 km W of Nukhaib, *Rawi* 31035!; 99 km E of Shabicha, *Rawi, Khatib & Tikriti* 29276!

Pakistan, Arabian Peninsula (Saudi Arabia, Oman).

3. **Atractylis cancellata** L., Sp. Pl. 830 (1753); Boissier, Fl. Orient. 3: 452 (1875); Nábělek in Publ. Fac. Sci. Univ. Masaryk 52: 30 (1925); Dinsmore in Post, Fl. Syria, ed. 2, 2: 82 (1933); Anthony in Notes Roy. Bot. Gard. Edinburgh 18: 291 (1935); Bornmüller in Beih. Bot. Centralbl. 50: 201 (1939); Zohary in Dep. Agr. Iraq Bull. 31: 152 (1950); Dickson, Wild Fls. Kuwait: 22 (1955); Rawi in Dep. Agr. Iraq Tech. Bull. 14: 107 (1964); Täckholm, Stud. Fl. Egypt ed. 2: 536 (1974); Matthews in Fl. Turkey 5: 604 (1975); Franco in Fl. Europ. 4: 211 (1976); Feinbrun-Dothan, Fl. Palaest. 3: 372 (1978); Rechinger, Fl. Iranica 139a: 104 (1979); Collenette, Wild Flow. Saudi Arabia: 173 (1999); Boulos & Hind in Fl. Egypt. 3: 155 (2002); Kilian & Ghazanfar in Fl. Oman 3: 210 (2015); Abdulridha, Taha & Wedad, Ecology and Fl. Basra: 197 (2016).

Annual herb branched from base and sometimes above, stems ± erect or decumbent, 3–15(–25) mm tall. Stems ± arachnoid, sometimes glabrescent, reddish or brown. Leaves sessile, narrowly oblong-spatulate, 15–50 × 2–8 mm, with spine-tipped teeth, margins with smaller spines; outer phyllaries 15–20, cartilaginous, bipectinate, 7–15 mm, the undivided centre of lamina very narrow. Capitula ± globose. Phyllaries lanceolate to oblong, inner acuminate, 10–14 mm, ± tomentose centrally. Florets tubular; corolla 7 mm, purple. Achenes 4 mm; pappus dirty white, 8–10 mm. Fig. 2: 4.

HAB. In sandy wadis and desert, stony plains and hillsides, by roadsides and in arable land, on sandy and clay soil; alt. up to 850 m; fl. & fr. Mar.-Apr. (-May).
DISTRIB. Occasional, locally frequent, in the mountain, steppe and desert regions of Iraq. **MRO**: Haibat Sultan Dagh, *Rawi, Noori & Kass* 28176! **MSU**: Jarmo, *Helbaek* 1856!, *Haines* W261! **FUJ**, Jabal Makhul, *Rawi & Gillett* 7213! **FNI**: 10 km N of Mosul, *Hossain* 4259! **FPF**: 22 km S of Badra, *Rawi* 18420!; Jabal al-Muwaila, *Guest, Rawi & Rechinger* 17553!; **DWD**: 160–190 km W of Ramadi to Rutba, *Rawi* 31354!; Tall an-Nisr, *Rawi & Noori* 27130! **DSD**: 40 km WNW of Shabicha, *Guest, Rawi & Rechinger* 19309!; 80 km ESE of Busaiya, *Guest & Rawi* 14285!; nr Jabal Sanam, 20 km W by S of Safwan, *Guest, Rawi & Rechinger* 17012! **LEA**: 5 km W of Tib, *Thamer* 47685A! **LCA**: between Baghdad and Falluja, *Omar & Wedad* 47516! **LSM**: Fakka-Laqlaq valley, on Iranian frontier, *Rawi* 25847!; Jabal Hamrin nr. neww dam, *Al-Kaisi* 48681 (BAG!); 14 km Jabal Hamrin S of Saadiya, *Al-Kaisi* 48815 (BAG!); Jabal Sinjar, *Al-Kaisi & Wedad* 52017 (BAG!).

Mediterranean Europe from Portugal to Greece, Mediterranean islands, Turkey, Syria, Arabian Peninsula, Palestine, Iran, N Africa (Egypt to Morocco).

4. **THEVENOTIA** DC.

in Arch. Bot. Par. 2: 331 (1833)

C. C. Townsend

Low spiny annual herbs, ± densely white-tomentose. Leaves rigidly coriaceous, oblong-lanceolate to ovate, prominently veined, strongly spinous. Capitula homogamous, in ± sessile clusters, subtended by involucriform upper leaves; involucre narrowly ovoid; phyllaries multiseriate, becoming narrower and more elongate-spinous inwards, ± white-tomentose; receptacle flat, paleaceous; paleae lanceolate, split into cilia at apex. Florets tubular, hermaphrodite; corollas purple; style-branches elongate, subulate. Achenes subterete, densely appressed-sericeous, hilum basal; pappus of uniseriate, long-plumose setae.

Two species, chiefly from Iran to Pakistan; only one recorded from Iraq.

Thevenotia (after the French explorer Jean de Thévenot, 1633–1667, who died at Mīāneh, Kermanshah province, Iran).

1. **Thevenotia persica** *DC.*, Prodr. 6: 110 (1838); Boissier, Fl. Orient. 3: 454 (1875); Zohary in Dep. Agr. Iraq Bull. 31: 152 (1950); Rechinger, Fl. Lowland Iraq: 642 (1964); Rawi in Dep. Agr. Iraq Tech. Bull. 14: 126 (1964); Rechinger, Fl. Iranica 139a: 90 (1979); Breckle, Hedge & Rafiqpoor, Vasc. Pl. Afghan. Checkl.: 194 (2013).

Atractylis persica (DC.) Boiss., Diagn. Pl. Orient., ser. 1, 6: 110 (1846).

Low rigid annual, 3–10(–20) cm tall, simple or branched from base or above, often with scattered slender spines. Leaves broadly elliptic to ovate or obovate, 2.5–6 × 1.2–3 cm, prominently veined, acute and terminating in a strong spine up to 1 cm long, margins with 2–4 long spines, small spines also frequently present on veins, surfaces ± densely arachnoid-tomentose. Capitula narrowly ovoid, subtended by several smaller, involucriform, still strongly spinous leaves with rather long stiff petioles, solitary or usually clustered; involucre 12–15 × 7–10 mm; phyllaries appressed-imbricate, spinous-tipped, longer and longer-spined inwards, arachnoid-tomentose centrally with stramineous margins. Corollas purple. Pappus 1.5–2 × as long as appressed-sericeous achene.

HAB. Only found once, nearly two centuries ago, on the desert plateau or adjoining foothill regions of Iraq.
DISTRIB. **FUJ/DGA/LEA**: In dry stony places between Mosul and Baghdad, *Olivier* s.n.!

Iran, Afghanistan, Pakistan.

The fact that this plant has never been re-collected from such a well-travelled route, and that Olivier certainly gathered it in Iran, tempts one to consider transposition of labels. However, the plant is widespread in Iran and there seems no reason why it should not occur rarely in Iraq.

5. **ECHINOPS** L.

Sp. Pl.: 814 (1753); Gen. Pl. ed. 5: 356 (1754)
Susanna & Garcia-Jacas in Kubitzki, Fam. Gen. Vasc. Pl. 8: 128 [2006] (2007)

C. C. Townsend

Erect biennial or perennial spinous herbs with 1-several erect, generally angular-sulcate stems, indumentum of tomentose, arachnoid or glandular hairs, rarely absent. Leaves thin to coriaceous, simple and dentate to ± pinnately divided, feebly to very fiercely armed. Capitula 1-flowered, few to very numerous, congested into a synflorescence resembling a spherical 'capitulum', easily falling to leave a ± globose (secondary) 'receptacle' generally subtended by small, reflexed, ± spatulate and dentate bracts remaining from the lowest capitula; (primary) phyllaries multiseriate, outermost often very delicate grading into a brush of hairs, minutely spatulate, intermediate usually broad-clawed with a flabellate or rhomboid limb contracting to a spinous apex, inner more delicate when young, lanceolate to narrowly oblong, generally fimbriate above, ± fused or free at an advanced stage of flowering and fruiting. Florets hermaphrodite, long-tubular with tube often papillose, limb deeply divided, white greenish, blue or red (not in Iraq); anthers caudate at base, basal appendages often bearded. Achenes oblong-cylindrical, ± pilose or more rarely glabrous; pappus short and rather rigid, setae free or more often fused near base.

A genus of possibly about 120 species in Europe, temperate Asia and Africa, but much in need of general revision. Some are widely cultivated in gardens as ornamentals for their beautiful blue capitula.

Echinops is an extremely difficult genus for the taxonomist. As Hedge observed in Fl. Turkey, it is a genus with few reliable taxonomic characters and few clear-cut species. Little short of a monographic treatment will suffice to clarify the Eurasian species, and this is certainly not possible as a preliminary to a regional Flora. The following account is thus to be regarded as provisional only. Material is not easy to collect. Because of their frequently large size and ferocious armament, specimens often consist of merely juvenile capitula and a few leaves. Basal and cauline leaves and mature capitula are necessary for reliable work, and several herbarium sheets may be needed for a single plant. Notes should be made at the time of collection of the position of the largest leaf, and a colour photograph taken with scale. In addition to collecting difficulties, there is a problem with mounting of even well-prepared material since the flowering capitula fall apart and have to be kept in packets. In the descriptions which follow, 'cornigerous' refers to horn-like processes which may or may not be present.

The genus is usually divided by workers in SW Asia into three Sections: *Echinops*, with the inner phyllaries free; *Oligolepis*, with 12–20 phyllaries, the inner being connate and forming a pentagonal coriaceous tube below in fruit; and *Ritropsis* (*Ritrodes*) with 20–30 phyllaries, the inner being loosely connate and forming a membranous tube in fruit. I have not adopted these sections as I am by no means convinced that the line drawn between the last two is as clear-cut as would appear. The species are, however, arranged in the order of these sections in this account. *Echinops* is a genus eminently suitable for study by an Iraqi taxonomist able to observe the species and their variability in the field in as many localities as possible.

Echinops (from εχίνος, echinos, hedgehog or sea-urchin and -όψις, -opsis, resembling).

Feddes Repert. Trautvetter, E.R. (1833). De *Echinope* Genere capita II. Mitavia.
Bunge, A. (1846). Über die Gattung *Echinops*. Bull. Acad. Imp. Sc. Petersb. 6: 390–411.
Bornmüller, J. 1918. Revisionsergebnisse einiger orientalischer und zentralasiatischer Arten der Gattung *Echinops*. Beih. Bot. Centralbl. 36 (2): 200–228.
Kruse, J. & Meusel, H. (1969). Zur Blattfolge und Blattbildung einiger Cynareen-Gattungen. 1. *Echinops*. Feddes Repert. 80: 339–356.
Rechinger, K.H. (1974). Neue orientalische *Echinops*-Arten. Candollea 29: 121–133.
Sánchez-Jiménez & al. (2010). Molecular systematics of Echinops L. (Asteraceae, Cynareae): A phylogeny based on ITS and *trnL-trnF* sequences with emphasis on sectional delimitation. Taxon 59 (3): 698–708.

1. Synflorescences 12–15 mm in diameter . . . . 1. *E. aff. parviflorus*
   Synflorescences 2.5 cm in diameter or more 2
2. Outer and median phyllaries densely glandular-pilose above; leaves softly spinous . . . . 3. *E. inermis*
   Outer and median phyllaries not or very sparingly glandular-pilose above, leaves ± sharply spinous 3
3. Leaves thick and leathery, largest crowded about middle of stem; branches all below terminal 'capitulum' divergent at 90 degrees, clothed with very reduced, bract-like leaves . . . . 10. *E. rectangularis*
   Leaves not so thick, longest not crowded about middle of stem; branches not divergent at right angles with reduced and bract-like leaves 4
4. Involucre in late flower and fruit with innermost phyllaries free to base or almost so. 5
   Involucre in late flower and fruit with innermost phyllaries fused into a tube below . 6
5. Flowering involucre 18–21 mm long; branches several, long, upwardly curving and giving the plant a candelabra-like form; upper leaves pinnatisect . . . . 2. *E. candelabrum*
   Flowering involucre 25–28 mm; synflorescence solitary, or 1–2 ascending branches present; upper leaves dissimilar from lower, lanceolate-oblong, dentate or shallowly lobed . . . . 4. *E. heterophyllus*
6. Pappus long, 2.5–3 mm, dark at base, exserted beyond apical hairs of achene in flower and fruit; leaves broadly toothed to pinnatipartite, densely glandular, glands of very variable length . . . . 8. *E. haussknechtii*
   Pappus short, clearly overtopped by apical hairs of achene in advanced flower and fruit; leaves, if densely covered with glands of variable length, pinnatisect 7
7. Stem and branches with long multicellular hairs, mixed with other glandular hairs of very diverse length or purplish and set in a dense white tomentum; leaves 2–3–pinnatisect . . . . 11. *E. spinosissimus*
   Stem and branches with short or no glandular hairs, if diverse multicellular hairs present then leaves 1–pinnatisect 8
8. Synflorescences not or very inconspicuously horned 9
   Synflorescences conspicuously horned 14

9. Upper surface of leaves clothed with arachnoid indumentum, bipinnatisect with up to 10(-15) pairs of pinnae . . . . 13. *E. chardinii*
Upper surface of leaves totally devoid of arachnoid or floccose indumentum, clothed only with short, evenly distributed glandular hairs, or if arachnoid indumentum is present then leaves only simply pinnately lobed or pinnatifid 10
10. Leaves bi- to subtripinnatisect. . . . . 12. *E. orientale*
Leaves pinnatifid to pinnatisect with dentate-spinous lobes. 11
11. Upper part of stem below synflorescence leafless for 3–8 cm; stem eglandular at least above uppermost leaf (or glands concealed in tomentum), usually ± clothed with very short glands below. 12
Stem with gradually reducing leaves almost or right up to head, glandular also up to head 13
12. Inner phyllaries concrescent to about half their length and finally indurate and fibrillose below . . . . .9. *E. fraudator*
Inner phyllaries loosely joined for less than half their length, not finally indurate and fibrillose . . . . 6. *E. amoenus*
13. Leaves in upper part of stem approaching synflorescence broad-based, lacking a long apical spine . . . . 7. *E. nitens*
Leaves approaching synflorescence narrow, with a long apical spine . . . 5. *E. faucicolus*
14. Lower stem leaves bi- to subtripinnatisect 15
Lower stem leaves pinnatilobed to pinnatisect 16
15. Web of tissue along rhachis of larger stem leaves mostly 7–13 mm wide; upper surface of leaves ± obviously clothed with arachnoid hairs, glands present or not. . . . . 13. *E. chardinii*
Web of tissue along rhachis of larger stem leaves 4–6 mm wide; upper surface of leaves lacking or with rather sparse arachnoid hairs, clad with very short uniform glands . . . . .14. *E. tenuisectus*
16. Horned phyllaries shorter, rarely 2 cm longer than head, often fewer; leaves pinnatilobed to pinnatipartite, web of tissue along midrib at least 5 mm wide . . . . .9. *E. fraudator*
Horned phyllaries numerous, long, when mature 4–5 cm longer than head; leaves pinnatisect, web of tissue along midrib rarely 5 mm wide 15. *E. polyceras*

1. **Echinops** aff. **parviflorus** *Boiss. & Buhse* in Nouv. Mém. Soc. Imp. Naturalistes Moscou 12: 124 (1860); Rechinger, Fl. Iranica 139a: 78 (1979).

Plant up to 35 cm tall. Stem slender, with slender short branches in inflorescence, bluntly and shallowly angular-sulcate, densely lanate and shortly but unequally glandular-pilose. Basal leaves not seen. Median cauline leaves sessile with rather small but very spinous semiamplexicaul auricles, lamina oblong and up to 12 × 8 cm in outline, bipinnatisect with up to 9 pairs of narrowly lanceolate lobes with terminal spines up to 10 mm long; upper cauline and branch leaves reducing, pinnatifid, sharply spinous; all leaves densely white-tomentose below, glandular-papillose above. Synflorescences 12–15 mm in diameter, non-cornigerous; receptacle 1.5 mm; involucre 7–9 mm in flower; brush 1.5. mm long, very sparse; phyllaries 13–15; outer long-spatulate with a distally lacerate limb; median rhomboid-lanceolate with shoulders distally minutely ciliate, becoming lanceolate; inner oblong, free, lacerate, apex blunt; all phyllaries greenish-white. Corollas 6–7 mm, whitish, tube smooth; pappus ± 0.75 mm, much exceeding apical hairs of ovary.

HAB. Rocky waste beside dry stream bed; alt. (recorded once) 1200 m; fl. Oct.
DISTRIB. Only found once in the mountain region of Iraq. **MSU**: Pira Magrun, Zawiyah village, *Haines* s.n.!

True *E. parviflorus* has been otherwise recorded only from Iran.
The above specimen was referred to *E. parviflorus* by Rechinger in *Fl. Iranica* with considerable reservation; he pointed out that although the characters of corolla and involucre agree with *E. parviflorus*, the leaves are glandular-papillose. I share his reservations and the reason for them, but would be highly reluctant to describe a new species from a single specimen on this basis in view of the extremely fragmentary nature of the type in Hb. Boissier.

Haines' gathering appears also to be a mixture, as much larger involucres and corollas are to be found loose in a capsule on the same sheet.

A further specimen certainly related to *E. parviflorus* is *Rawi & Serhang* 23799 from MRO, Kaiwa Rush nr Rania. This has more broadly dissected leaves which are much less fiercely armed than the *Haines* specimen. The inflorescence branches, however, are damaged and only one capitulum is available; this has the involucres distinctly lanate. Population studies and further collections are needed from the above localities.

2. **Echinops candelabrum** *Rech.f.* in Candollea 29: 132 (1974); Rechinger, Fl. Iranica 139a: 74 (1979).

Plant of unknown height, but over 60 cm tall. Stems strongly sulcate below, densely villous, with unequal glandular hairs, clothed about base with a fibrous crown of remains of previous years' leaf-sheaths, above and on several slender upward-curving branches densely whitish-tomentose with scattered glandular hairs. Basal leaves distinctly petiolate, broadly elliptic in outline, to ± 48 × 20 cm including petiole, pinnate below, pinnatisect above with 6–8 pairs of broad, deeply pinnatisect lobes, final segments lanceolate, tipped with short but rigid spines, whitish-tomentose and glandular along veins beneath, thinly arachnoid and ± densely unequally glandular above; median cauline leaves sessile with rounded spinous auricles, strongly spinous with stramineous spines up to about 7 mm; indumentum similar; upper cauline and branch leaves small, pinnatisect, strongly spinous. Synflorescences 4.5–7 cm in diameter with terminal largest; receptacle 4–5 mm; involucre 18–21 mm in flower, those of terminal capitulum largest; brush whitish, 7–10 mm, inner bristles somewhat expanded and toothed above; phyllaries in capitula of terminal head about 20 (of branch capitula fewer); outer with a long, ± pectinate claw, limb ± rhomboidal, ciliate, rigid, mucronate; median lanceolate, long-fimbriate with a slenderly spinous apex; inner free, similar but broader below and shorter; all phyllaries glabrous, whitish, median and inner bluish-violet above. Corollas ± 15 mm, pale blue, tube minutely papillose above; pappus 1.5 mm, dense, brownish, setae connate below, not overlapped by long hairs from apex of young achene.

HAB. Heavily grazed open mountainside and rocky slope; alt. (recorded once) 1515 m; fl. Aug.
DISTRIB. Found only twice in the extreme northern mountain region of Iraq, near the Turkish frontier. **MAM**: Sarsang, *Haines* 1353!; Ser Amadiya, *Haines & Agnew* s.n.!

Endemic.

3. **Echinops inermis** *Boiss. & Hausskn.* in Boissier, Fl. Orient. 3: 436 (1875). – Type. Iraq, Kurdistan or Iran, M. Avroman et Schahu, *Haussknecht* 561 (B, JE, K, P, syn.). Zohary in Dep. Agr. Iraq Bull. 31: 151 (1950); Rechinger, Fl. Iranica 139a: 80 (1979).

Plant 60–100 cm tall. Stem sulcate, simple or with one or two slender ascending branches, both stem and branches terminating in a single head, densely glandular-hairy throughout and upper part whitish-arachnoid. Lower cauline leaves broadly deltoid-ovate to obovate in outline, up to 32 × 17 cm including petiole; lamina pinnatipartite with 3–4 pairs of lobes, lobes lanceolate-oblong, irregularly toothed and softly spinulose, green and glandular-pilose on upper surface, greyish- or whitish-arachnoid below; petiole long, with scattered spinous bract-like lobes; median cauline leaves similar but sessile with base spinous-auriculate; upper cauline leaves rapidly reducing, simple, dentate, spinulose. Synflorescences 3.5–4 cm in diameter, non-cornigerous; receptacle ± 5 mm; involucre 16–22 mm long in flower, to 26 mm in fruit, about twice as long as brush, brush formed from scabrid-ciliate, upwardly slightly dilated, flattened bristles; phyllaries numerous, 20 or more, all free; outer flabellate at apex, toothed and mucronate; median broadly flabellate and abruptly contracted to a short rigid point, with marginal antrorsely scabrid teeth; inner lanceolate, spine-tipped, with long forwardly-directed marginal cilia, blue-tipped in flower; outer and median phyllaries densely glandular-pilose above. Corolla 10–14 mm, blue, tube glabrous with a few minute papillae at apex; pappus ± 2 mm, dense, setae barbellate, free, not overlapped by any apical achene hairs.

HAB. Dry stony slopes and in clay depression in *Quercus* forest; alt. 1060–1360 m; fl. Jul.-Aug.
DISTRIB. Apparently rare in the mountain region of Iraq. **MAM**: Khantur mt. NE Zakho, *Rawi* 23405 (BAG!); above Sarsang, *Haines* s.n.!, *Rechinger* 11908!; Khantur, *Rawi* 23388! **MRO**: Baradost Dagh, *Field*

& Lazar 874! **MSU**: Qopi Qaradagh, *Haines* 1080!; "M. Avroman et Schahu", *Haussknecht* 561 (syntype)!; Qara Dagh beyond Jafaran, *Eig, Zohary & Feinbrun* s.n.

W Iran.

4. **Echinops heterophyllus** *P.H.Davis* in Notes Roy. Bot. Gard. Edinburgh. 22: 72, f. 1, II (1958); Rawi in Dep. Agr. Iraq Tech. Bull. 14: 114 (1964); Hedge in Fl. Turkey 5: 616 (1975); Rechinger, Fl. Iranica 139a: 67 (1979).

[*E.* aff. *tournefortii* auct. non Ledeb.: Blakelock in Kew Bull. 4: 53 (1949); Rawi in Dep. Agr. Iraq Tech. Bull. 14: 114 (1964)].

Plant (15–)20–68 cm tall. Stem sulcate, stout, simple or with one or two slender, ascending branches, both stem and branches terminating in a single synflorescence, eglandular, densely white-tomentose above, thinly so or glabrescent towards base, with a basal collar of fibrous remains of previous years' leaf sheaths. Basal and lower cauline leaves ovate or deltoid-ovate in outline, up to 38 × 18 cm, bipinnatisect with 5–9 pairs of deeply divided, strongly spinous lobes, spines up to 8 mm long; petioles with a large, expanded, strongly spinous basal sheath, free part rapidly becoming very short; median cauline leaves strongly differentiated, sessile with a spinous-auriculate base, simple, lanceolate-oblong, spinous-dentate with shallow lobes and shorter spines between lobes; upper leaves rapidly reducing, similar in form to median; all leaves green above, sparsely arachnoid, ± glandular or not, densely tomentose on lower surface. Synflorescences 7–9 mm in diameter when flowering, non-cornigerous; receptacle 12 mm; involucre 25–28 mm long in flower, to 30–35 mm in fruit; brush 10–12 mm, of rather few, scabrid, slightly flattened bristles; phyllaries about 20, all free, glabrous; outer linear with a small, lanceolate, ± membranous-ciliate tip; median rhomboidally expanded above with tip extended into a long, ciliate spine; inner lanceolate and tipped with a long, ciliate spine; median and inner phyllaries bluish-tipped in flower. Corollas 18–20 mm, blue, tube glabrous, sharply pentagonal; pappus *c.* 2 mm, dense, setae ± flattened and fused near base, about ½ overlapped by hairs of achene apex.

HAB. Rocky and stony mountain slopes; alt. 1900–2525 m; fl. Aug.
DISTRIB. Rare in the thorn-cushion zone of the mountain region of Iraq, only occurring on the country's highest mountain. **MAM**: Khantur mountain, NE of Zakho, *Rawi* 23405 (BAG!); **MRO**: Algurd (Helgord) Dagh, *Guest* 2937!, *Gillett* 12471!; above Nowanda, *Rechinger* 11888!

Turkey (SE Anatolia), W Iran.

5. **Echinops faucicolus** *Rech.f.* in Candollea 29: 130 (1974); Rechinger, Fl. Iranica 139a: 57 (1979).

Plant with numerous stems 40–70 cm tall. Stem angular-sulcate with a few short, ascending branches above, slightly arachnoid upwards, densely clad with very short, pale glands. Leaves somewhat coriaceous, basal not seen, lower and median stem leaves sessile with distinct broadly toothed auricles, teeth tipped with fine spines; lamina elliptic in outline, up to 24 × 10 cm, pinnatifid with up to 9 pairs of pinnae or in largest leaves lower pinnae bifid with deflexed lower lobes, all lobes deltoid-lanceolate and toothed with stramineous spines up to 15 mm long; upper leaves pinnate with strongly spinous lobes; all leaves ± densely, evenly shortly glandular above, densely white-tomentose beneath. Synflorescences 4–5 cm in diameter, not or inconspicuously cornigerous; receptacle 5 mm; involucre 15–17 mm, pure white and fine; phyllaries 15–20; outer long-spatulate, lamina ± rhomboid, antically denticulate, with minute subsessile glands; median obovate with narrow, antically denticulate shoulders abruptly contracted to a long spine; inner subacute or blunt, ciliate above, lightly fused to about the middle; median and outer phyllaries tipped with violet-blue. Corolla not seen. Pappus *c.* 1 mm, dense, fused below, completely overlapped by apical achene hairs.

HAB. Rocky hillside, crevices in rocks; alt. 500–700 m; fl. & fr. Aug.-Sep.
DISTRIB. Only in one restricted area in the mountain region of Iraq. **MRO**: Bekhal falls near Rowanduz, *Guest & Husham* 15914!; in gorges between Rowanduz and Barsarini, *Rechinger* 11267!

Of the two specimens cited, the first is in fruit and the second has only tight buds, so that the corolla remains unknown.

Endemic.

6. **Echinops amoenus** *Rech.f.*, Fl. Iranica 139: 58 (1979).

Plant 50–70 cm tall. Stems angular-sulcate, erect, rather slender, simple or rarely branched, ± glandular-papillose below especially in grooves, sparsely glandular and arachnoid above, finally ± glabrescent. Basal and lower cauline leaves not seen. Median cauline leaves sessile with ± clasping, roundish, strongly spinous auricles, lamina elliptic in outline, up to 16 × 8 cm, pinnatipartite with 5–10 pairs of narrowly triangular lobes with terminal spines up to 7 mm long, median pairs further lobed or toothed; upper leaves rapidly reducing; all leaves coriaceous, upper surface ± densely clad with sessile or subsessile glands, lower densely white-tomentose. Synflorescences *c.* 6 cm in diameter when flowering, non-cornigerous; receptacle 5 mm; involucre 20–24 mm in flower, to 28 mm in fruit; brush 8–9 mm, of slightly flattened bristles; phyllaries ± 20; outer long-spatulate, lamina with rounded margins, slightly mucronate, strongly white-arachnoid, ciliate; median obovate, becoming oblanceolate, with a blue spinous apex, blue-ciliate at ± rounded shoulders and above, upper margins densely arachnoid; inner lanceolate-oblong, fimbriate at apex, loosely connate for less than half their length. Corolla 24–26 mm, blue, tube glabrous or papillose. Pappus *c.* 1 mm, dense, fused below, just overlapped by apical achene hairs (in one or two flowers seen these have been sparse but still overlapping pappus).

HAB. On dry hills, alt. not recorded; fl. Jul.
DISTRIB. Only known from two localities in the extreme N of Iraq not far from the Turkish frontier. **MAM**: Dagh al Radzijan (Darhawzan?), between Zakho and Sharanish, *Rechinger* 10732!; 20 km NE of Zakho, *Rawi* 23171!

Endemic.

7. **Echinops nitens** *Bornm.*, Feddes Repert. Spec. Nov. Regni Veg. 13: 7 (1913) & Bornmüller in Beih. Bot. Centralbl. 60B: 199 (1939); Rechinger, Fl. Lowland Iraq: 637 (1964) & Fl. Iranica 139a: 46 (1979).

Plant 40–60 cm tall. Stem angular and sulcate, erect or somewhat curved, stout to slender, simple or with a few slender branches, densely foliose, densely white-tomentose at least above, below and where the tomentum is ± evanescent numerous subsessile or sessile glands visible. Basal and lowest cauline leaves similar, shortly petiolate, otherwise similar to the sessile, spinous-auriculate median stem leaves, elliptic to ovate-elliptic in outline, up to 17 × 9 cm, pinnatifid to pinnatipartite with 4–6 pairs of broadly to narrowly triangular lobes with terminal spines up to 10 mm long and smaller marginal spines; upper leaves similar, only rapidly reducing towards apex; all leaves coriaceous, green, sometimes shining and ± sessile-glandular on upper surface, very densely white-tomentose below. Synflorescences 4–6 cm in diameter when flowering, non-cornigerous; receptacle 3–4 mm; involucre 14–19 mm, long in flower, to 25 mm in fruit, brush 5–6(–9) mm, of rather few slightly flattened bristles; phyllaries *c.* 15; outer long-spatulate with a ciliate, shortly aristate, glabrous or ± arachnoid limb; median lanceolate, antrosely ciliate, spine-tipped; inner lanceolate-oblong, fimbriate at apex, connate for about ½ of their length; median and inner phyllaries bluish at apex in flower. Corollas 12–15 mm, blue, tube shortly papillose. Pappus 2–2.5 mm, dense, setae ± fused at base, completely overlapped by long hairs from achene apex.

HAB. Rocky limestone slopes of gorge and in rock crevices, in *Quercus* scrub; alt. 680–800 m; fl. Jun.-Aug.
DISTRIB. Confined to the NE forest zone of Iraq, but there it is apparently locally common. **MRO**: Rowanduz gorge (Gali Ali Beg), *Guest* 2987!, *Hadač* 5772!, *Agnew, Hadač & Haines* s.n.!, *Rechinger* 11260!, *Gillett* 9441!, *Bornm.* 1468 p.p. & 4269; Shirwan Mazin, *Hadač et al.* 5861!; between Zawita and Hupa, *Hadač et al.* 6183!

Endemic.

8. **Echinops haussknechtii** *Boiss.*, Fl. Orient. 3: 433 (1875); Rechinger, Fl. Iranica 139a: 33 (1979).

E. armatus Boissier. & Hausskn. in Boissier, Fl. Orient. 3: 436 (1875); Zohary in Dep. Agr. Iraq Bull. 31: 151 (1950); Rawi in Dep. Agr. Iraq Tech. Bull. 14: 114 (1964); Rechinger, Fl. Iranica 139a: 43 (1979), quoad var. *armatus*.

Plant up to 60 cm tall. Stems slender to rather stout, with long, divergent or ascending branches above, angular-sulcate, ± lanate and densely clad with very unequal, ± dark-tipped glandular hairs. Basal leaves not seen. Lower cauline leaves shortly petiolate, petiole triangular in section, lamina obovate-oblong, to 35 × 20 cm in outline, pinnatifid to pinnatipartite with up to 7 pairs of broadly triangular lobes with terminal spines up to 10 mm long; median cauline leaves similar but sessile with rounded, amplexicaul, strongly spiny auricles; upper cauline and branch leaves reducing, ovate or oblong, sharply spinous but not at all pinnately divided; all leaves arachnoid-tomentose on lower surface, above sparsely arachnoid and densely clad with long, yellowish glandular hairs. Synflorescences 5–7 cm in diameter when flowering, non-cornigerous; receptacle 7–8 mm; involucre 20–22 mm in flower; brush 11–18 mm, whitish; phyllaries 13–18; outer long-spatulate with a distally denticulate, minutely glandular-ciliate limb; median rhomboid-lanceolate, entire or somewhat lacerate, shoulders minutely glandular-ciliate; inner forming a deeply grooved tube below for about half their length; all phyllaries greenish to pale silvery-blue above. Corollas 18–22 mm, pale blue to whitish, tube minutely papillose. Pappus 2.75–3 mm, dirty white, dark below, much longer than apical hairs of achene. Fig. 4: 1–2.

HAB. Hillside, steppe and dry roadside banks, altitude not recorded in any of the specimens; fl. & fr. May-Jun.
DISTRIB. Occasionally found in the foothill region of Iraq, perhaps extending to the mountains. **MSU/FKI**?: between Kirkuk and Sulaimaniya, *Haussknecht* 562! **FKI**: 3 km from Kirkuk on way to Sulaimaniya, *Karim, Hamid & Jasim* 40484!; Kirkuk, *Haines* s.n.! **FAR**: Arbil, *Agnew, Hadač & Haines* 5753!

W Iran.

Clearly differs in both indumentum and pappus length from the Iranian and Iraqi material named by Rechinger in Fl. Iranica as "E. armatus var. cryptadenius" and "E. armatus var. papillosus".

9. **Echinops fraudator** *C.C.Towns.* in Kew Bull. 43 (1): 113 (1988). –Type. Iraq, Baba Gurgur, *Guest* 4009 (K! holo.; BAG, iso.)

E. cephalotes auct. non DC.: Anthony in Notes Roy. Bot. Gard. Edinburgh 18: 291 (1935).
E. armatus Boiss. & Hausskn. var. *cryptadenus* Rech. f. in Fl. Iranica 139a: 44 (1979). –Type. Iraq, 10 km E of Mandali, *Rechinger* 9638 (W).
E. armatus Boiss. & Hausskn. var. *papillosus* Rech. f. in Fl. Iranica 139a: 45 (1979). –Type. Iraq, prope Kirkuk, *Rechinger* 9968 (W).

Plant 15–60 cm tall. Stems simple or with a few long, ascending branches from near base, slender to rather stout, angular-sulcate below, very variable in indumentum from white-tomentose and apparently eglandular by concealment to ± shortly stipitate-glandular or with sessile glands throughout, often tomentose above and shortly glandular at least in grooves below. Leaves elliptic or oblong-elliptic, lower shortly petiolate and median with rounded, fiercely spinous, otherwise similar, to 26 × 14 cm, pinnatilobed to pinnatifid with 5–7 pairs of broadly triangular lobes simple or deeply dentate at base or throughout, teeth also strongly armed; upper stem and branch leaves rapidly reducing, strongly spinous; all leaves white-tomentose on lower surface, conspicuously glandular along veins or not, arachnoid to glabrous above with very short and papillose, or long and unequal, glands. Synflorescences cornigerous (spines sometimes small), 5–7 cm in diameter excluding spines, which extend up a further 2 cm; receptacle 5–8 mm; involucre 15–22 mm in flower, up to 25 mm in fruit, brush whitish, 7–11 mm; phyllaries 16–19(–23) mm, outer long-spatulate, with a ciliate, mucronate, rhomboidal or flabellate limb, glabrous or somewhat arachnoid at apex; median lanceolate, spinous to very strongly spinous, pectinate along margins; inner rather blunt, fimbriate at apex, fused to about half-way, fibrillating below; all phyllaries usually bluish above, sometimes not. Corollas 16–20 mm, white or pale blue, tube densely minutely papillose, occasionally also densely lanate. Pappus 2 mm, dense, brownish, completely overlapped by long hairs at top of achene. Fig. 3: 1–7.

HAB. Roadside, waste ground, cultivated fields, on a sandstone bluff, by a river in conglomerate hills, on hillsides, on alluvium, sandy or stony soil on clay gravel; alt. 110–400 m; fl. May-Jun., fr. July.
DISTRIB. Apparently frequent in the foothill regions of Iraq, rare in the adjacent mountains and plains. **MSU**: NE of Darband-i Khan, *Poore* 572! **FUJ**: Shargat, Rao s.n.! **FAR**: Makhmur Dagh, *Gillett* 11246! **FKI**: Kirkuk, *Rechinger* 9968!, *Hadač* 1267!; Baba Gurgur, *Guest* 4009!; nr Tang, *Uvarov* 24!; Mirjani (Injana?), *Sutherland* 42!; Injana, *Rawi* 22872! **FPF**: Naft Khana, *Rogers* 0321!; Koma Sang, *Rawi* 20590!, *Hadač* 4596!; Inaiza, *Haines* s.n.!; 10 km E of Mandali,, *Rechinger* 9638!; 4 km E of Badra, *Haddad* 9954!; 30 km SE of Badra, *Rechinger* 9168! **FPF/LCA**: 75 km N of Amara, *Hazim & Noori* 30643! **FPF/LEA**: Jabal

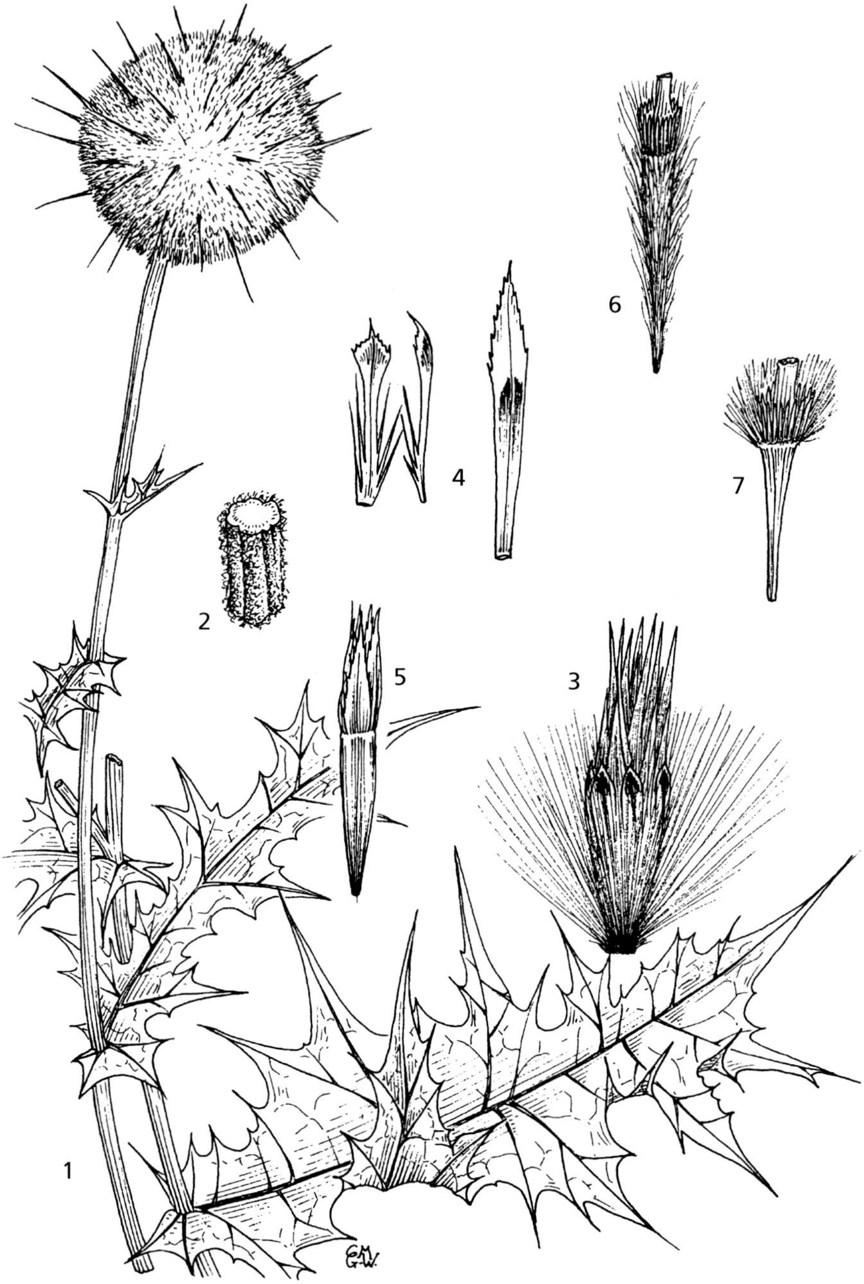

Fig. 3. **Echinops fraudator**. 1, habit × 2/3; 2, part of stem enlarged to show glandular hairs; 3, involucre and brush ×2; 4, outer and median phyllaries × 2; 5, inner (coalescent) phyllaries × 2; 6, young achene with pappus (some pappus hairs removed) × 4; 7, young achene with pappus × 4. 1, 3–6 from *Guest* 4009, 2 from *Rawi* 22872; 7 from *Rechinger* 9638. Drawn by Christine M. Grey-Wilson.

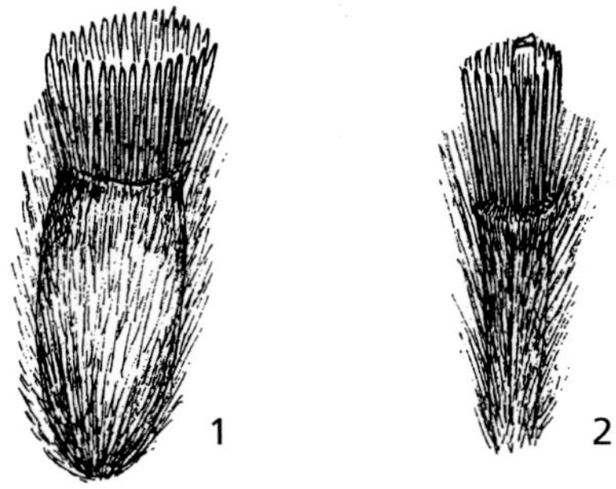

Fig. 4. **Echinops haussknechtii**. 1, 2, young achenes with pappus × 4. 1, 2 from *Haussknecht* 563a. Drawn by Christine M. Grey-Wilson.

Hamrin by road from Ba'quba to Khanaqin, *Hadač et al.* 4660! LEA: Kani Sakht, *Ani & Bharucha* in Barkley 7655! Shahraban, *Haines* s.n.!; 17 km E of Zurbatya *Al-Kaisi & Al-Khayat* 50678 (BAG!); 30 km NE Mandali *Al-Kaisi & Al-Khayat* 50780 (BAG!).

W Iran.

10. **Echinops rectangularis** *Rech. f.* in Ber. Deutsch. Bot. Ges. 72: 280 (1959). Type – Iraq, Erbil, in valle inter Hayat et Haji Omran, viii. 1957, *Rechinger* 11276 (E, M, W, syn.); Rawi in Dep. Agr. Iraq Tech. Bull. 14: 114 (1964); Rechinger, Fl. Iranica 139a: 61 (1979).

Plant up to 45 cm tall. Stem stout, branched towards apex, branches short, divergent at 90 degrees with numerous very small leaves, both stem and branches densely white-lanate, eglandular, bluntly and shallowly angular-sulcate. Basal leaves not seen. Cauline leaves coriaceous, densely crowded in middle part of stem, oblong in outline, pinnatipartite with up to 4 pairs of broadly lanceolate lobes or merely dentate with deltoid teeth, teeth of lobes terminating in rigid stramineous spines up to 8 mm long; all leaves densely white-tomentose below, moderately or densely so above, especially along veins which thus appear whiter. Synflorescences 7–8 cm in diameter, non-cornigerous; receptacle 7–8 mm; involucre about 15 mm in flower; brush 8–11 mm, whitish with hairs brownish at base; phyllaries 20–22; outer slenderly spatulate with a distally lacerate limb; median rhomboid-lanceolate with shoulders denticulate or dentate; inner lanceolate-oblong, fimbriate at apex, connate for less than half their length; all phyllaries green, or slightly bluish above. Corollas pale blue, 20–22 mm, tube minutely papillose. Pappus 1.25 mm, exceeded by apical hairs of ovary.

HAB. Upland valley and mountainside; alt. 1000–1400 m; fl. Jul.-Aug
DISTRIB. Only found twice in the mountain region of Iraq. MRO: between Rayat and Haji Umran, *Rechinger* 11276!; 47 km N of Rowanduz, Darband village, *Rawi* 24260!

Endemic.

11. **Echinops spinosissimus** *Turra* in Farsetia Nov. Pl. Gen.: 13 (1765); Meikle, Fl. Cyprus 2: 939 (1985); .

E. viscosus DC., Prodr. 6: 525 (1838); Boissier, Fl. Orient. 3: 429 (1875); Nábělek in Publ. Fac. Sci. Univ. Masaryk 52: 27 (1925); Dinsmore in Post, Fl. Syria, Palest. & Sinai ed. 2, 2: 75 (1933); Guest in Dep. Agr. Iraq Bull. 27: 30 (1933); Blakelock in Kew Bull. 4: 53 (1949); Zohary in Dep. Agr. Iraq Bull. 31: 151 (1950); Rawi, ibid. Tech. Bull. 14: 114 (1964); Hedge in Fl. Turkey 5: 618 (1975).

Plant 0.5–1.5(–2) m tall. Stems rarely simple, usually branched above, stout, angular-sulcate, densely white-tomentose to floccose or glabrescent, thinly to ± densely clad with unequal, multicellular glandular hairs. Basal and lowest cauline leaves shortly petiolate, elliptic to lanceolate or oblong-elliptic in outline, up to 30 × 15 cm, 2–3-pinnatisect with up to 8 pairs of pinnae (often with further pairs ± reduced to spines below this); pinnules narrowly triangular with strong to rather weak terminal and lateral spines, margins recurved; median cauline leaves sessile with rounded, strongly spinose auricles, otherwise similar; upper leaves reducing and finally bract-like, strongly spinous; all leaves green and densely clothed on upper surface with unequal multicellular glandular hairs, lower surface white-tomentose. Synflorescences 2.5–7(–10) cm in diameter, cornigerous or not; receptacle 4–7 mm; involucre 15–35 (–40) mm in flower, 18–38(–45) mm in fruit; brush whitish, 5–12(–24) mm; phyllaries 20–24, outer long-spatulate, with a distinctly ciliate, mucronate, brown-spotted limb; median oblanceolate with blunt, ciliate shoulders, apex ± spinose; inner linear-lanceolate, ± acuminate, lightly fused to about ⅓, fimbriate; all phyllaries whitish dorsally, often with a greenish submarginal vitta below, with or without blue pigmentation, glabrous. Corollas 18–25 mm, blue to whitish, tube minutely papillose. Pappus 1.5–2 mm, dense, brownish to off-white, completely overtopped by hairs at top of achene.

DISTRIB. (of species). Sicily, Balkans, Greece, Crete, Aegean Is., Cyprus, Turkey, Lebanon, Syria, Palestine, Iran.

subsp. **spinosissimus**

E. glandulosus E.Weiss in Berh. zool.-bot. Ges. Wien 18: 433 (1868).
E. viscosus subsp. *glandulosus* (E.Weiss) Rech.f. in Fl. Aegaea: 641 (1943).
E. syriacus Boiss., Diagn. Pl. Orient. ser. 1, 10: 84 (1849); Rechinger in Fl. Iranica 139a: 62 (1979).
E. bicolor Nábělek in Publ. Fac. Sci. Univ. Masaryk 52: 28 (1928); Hedge in Fl. Turkey 5: 622 (1975); Rawi in Dep. Agr. Iraq Tech. Bull. 14: 114 (1964).

HAB. Rocky places on limestone hills, and silt at the base of such a hill; alt. 800–1100 m; fl. Jul.-Oct.
DISTRIB. Only in the northernmost mountain region of Iraq, not far from the Turkish frontier. **MAM**: Mar Ya'qub, *Nábělek* 3650!, 3654! (syntypes); Gali Dohuk, *Barkley & Ishaq* 9025!; between Sandul and Zawita, *Alkas* 18631!; valley above Sharanish, *Rechinger* 11507!; Atrush, Guest 3646! **MSU**: Darband-i Khan, *Haines* s.n.!

Although Rechinger in *Fl. Iranica* sets out the differences between *E. bithynicus* and *E. syriacus*, he did not make clear the differences which he considered as separating *E. syriacus* and *E. spinosissimus*, to which latter (as *E. viscosus*) Boissier himself subordinated *E. syriacus*. My view is that Boissier's reduction was wholly justified.

Rechinger did not deal with *E. bicolor* Nábělek for *Fl. Iranica*, being presumably under the impression that it was described from Turkey. Hedge, in Fl. Turkey, correctly surmises that it was described from Iraq, the type locality being Mar Ya'qub, where Nábělek also recorded "E. viscosus". Hedge stated that *E. bicolor* is "closely related to *E. phaeocephalus*, and the two taxa are possibly conspecific". However, *E. phaeocephalus* falls within sect. *Oligolepis*, with 12–20 phyllaries connate in the basal half, whereas *E. bicolor* was described with 20–22 phyllaries; this I have confirmed from the type. The most striking feature of *E. bicolor* is the sparse, very short brush, much more scanty than I have seen in any other specimen of *E. spinosissimus*. This alone could scarcely be taken as establishing a new species, however, and in all other ways the type of *E. bicolor* differs in no respect from *E. spinosissimus*, including degree of fusion of the inner phyllaries, if capitula at the same stage are examined.

subsp. **bithynicus** (*Boiss.*) *Kožuharov* in Bot. J. Linn. Soc. 71: 41 (1975).

E. bithynicus Boiss., Diagn. Pl. Orient. ser. 1, 6: 100 (1845); Rechinger, Fl. Iranica 139a: 61 (1979).
E. viscosus subsp. *bithynicus* (Boiss.) Rech.f. in Fl. Aegaea: 641 (1943); Hedge in Fl. Turkey 5: 619 (1975).

Stems densely white-tomentose with scattered purplish multicellular glandular hairs, the cells usually conspicuously lighter and darker in colour.

HAB. (of subsp.). Mountain slopes and ravines, and on cultivated ground. **MRO**: Rowanduz, *Regel* 84; Salah ad-Din, *Noori & Hamid* 41172!; **MSU**: 19 km W of Sulaimaniya, *Rechinger* 12490!

Crete, Aegean Is., Turkey, W Iran.

12. **Echinops orientalis** *Trautv.* in Echin. Gen. Cap. II: 22, f. 8 (1833); Bobrov in Fl. S.S.S.R. 27: 25 (1962); Hedge in Fl. Turkey 5: 620 (1975); Rechinger, Fl. Iranica 139a: 52 (1979).

Plant 30–100(–150) cm tall. Stem angular-subsulcate, rather slender, with a few long, ascending branches, somewhat white-tomentose above when young, this indumentum very rapidly evanescent or ± persistent, when evanescent the stem and branches brownish-stramineous and ± evenly shortly glandular as in lower parts; branches naked above and with a few scattered, spinous, reduced and bract-like leaves. Lower cauline leaves slenderly subpetiolate with a few slender spines about junction with stem, lamina up to 40 × 20 cm, elliptic or oblong-elliptic in outline, bi- to subtripinnatisect with up to 10(–12) pairs of pinnae; pinnae lanceolate, lowest entire or bifid with lower lobe backwardly directed; all lobes toothed, teeth terminating in sharp stramineous spines up to 10 mm long; web of tissue along rhachis (4–)7–12(–15) mm wide; median cauline leaves sessile with rounded, small but strongly spinous auricles, passing into pinnate, strongly spinous-toothed upper stem leaves, all leaves usually free of arachnoid hairs above, lower surface densely white-tomentose, glandular hairs often present on veins. Synflorescences not cornigerous, (4–)5–7 cm in diameter; receptacle 5–11 mm; involucre 20–26 mm in flower, 26–30 mm in fruit; brush whitish, 8–15 (–20) mm; phyllaries 19–27; outer long-spatulate with a ± rhomboidal, entire or denticulate limb; median expanded-obovate, rather abruptly or more gradually contracted to ciliate spine, rather bluntly to more sharply shouldered with shoulders distally ciliate, often bluish above, inner acute to obtuse, fimbriate at apex, fused to about halfway, often bluish-tipped. Corollas 20–25 mm, blue to whitish, tube papillose. Pappus 1–1.5 mm, dark and fused below, completely overtopped by apical hairs of achene.

HAB. Rocky mountain slopes, among limestone rocks in *Quercus* forest, by a roadside and on clay plains; alt. 1200–2500 m; fl. Jun.-Sep.
DISTRIB. Scattered in the mountain region of Iraq, rare in the foothills. **MAM**: above Basingera (Barzur?), N of Zakho, *Rechinger* 11529! **MRO**: between Haji Umran and Rayat, *Rechinger* 11837!; road between Bardanas and the Qandil range, *Rawi & Serhang* 24612!; Musaiyif Nawanda, nr Rost, *Guest & Husham* 15809!; Qerra Qaw valley, N of Pushtashan, *Rawi & Serhang* (BAG!); Rayat, *Guest* 13122 (BAG!). **FNI**: 5 km from Eski Kellek on road to Mosul, *Omar, Qaisi & Wedad* 49626!

Turkey, Iran, Transcaucasia, C Asia (Turkmenia).

A specimen from **LSM**: Amara, *Rawi & Haddad* 25751! is close to this plant in vegetative characters, but the capitula are very young. In view of this solitary occurrence in the Mesopotamian plain and the unsatisfactory state of the gathering it is best left un-named until it is collected again.

13. **Echinops chardinii** *Boiss. & Buhse* in Nouv. Mém. Soc. Imp. Naturalistes Moscou 12: 124 (1860); Rechinger, Fl. Iranica 139a: 53 (1979).

E. mosulensis Rech.f. in Fl. Iranica 139a: 54 (1979). Type: Iraq, 18 km E of Dohuk, *Rechinger* 11978 (W, holo.; G!, K, iso.).

Plant 0.4–1 m or more tall. Stem angular-sulcate, erect, rather slender to stout, with long ascending branches or more rarely 1-headed in small individuals, ± densely white-tomentose when young, becoming glabrescent with age or tomentum sometimes persistent, ± eglandular or numerous small glands revealed with evanescence of indumentum; branches naked or with scattered and very reduced bract-like leaves above. Lower cauline leaves slenderly subpetiolate with a few slender spines about junction with stem, lamina up to 45 × 24 cm, oblong or elliptic-oblong in outline, bipinnatisect with up to 10(–15) pairs of pinnae; pinnae lanceolate, lowest often bifid with a backwardly directed lower lobe, all lobes toothed, teeth terminating in sharp stramineous spines up to 10 mm long but often less; web of tissue along rhachis usually 7–13(–18) mm wide (sometimes narrower in small plants); median stem leaves sessile with rounded, small but strongly spinous auricle, passing into pinnate, strongly and sharply toothed upper cauline leaves; all leaves densely to rather sparsely arachnoid on upper surface, almost eglandular to moderately clothed with rather short, uniform glands often well hidden in the indumentum, lower surface densely white- or greyish-tomentose. Synflorescences distinctly cornigerous or phyllary-spines merely obviously unequal, 5–7.5 cm in diameter excluding the cornigerous spines, which project up to a further 1.5 cm; receptacle 6–7 mm; involucre 20–25 mm in flower, 24–30 mm in fruit; brush whitish, 10–17 mm; phyllaries (19–)23–26; outer long-spatulate with a subentire to distally denticulate, rhomboidal to flabellate limb; median expanded-obovate, contracted to the denticulate-fimbriate spine with sharp shoulders which are distally denticulate-ciliate, green or bluish-tinged above; inner acute to obtuse, fimbriate at apex, fused to about halfway. Corollas blue, 20–22 mm, tube papillose; pappus 1–1.5 mm, fused and darkened below, completely overtopped by apical hairs of developing achene.

HAB. Dry hills on rocky limestone, stony ground, sandy and silty ridges and hill bases, also along channels in former cultivation and in grassland in degraded *Quercus* forest; alt. 300–1200 m; fl. Jun.-Aug.
DISTRIB. Widespread in the forest zone of the mountain region of Iraq, occasional in the foothills, encountered once in the Lower Mesopotamian plain. **MAM**: 18 km E of Dohuk, *Rechinger* 11978 (type of *E. mosulensis*)!; Dohuk, *id.* 10622!; 3 km E of Agra, *Chapman* 26187! **MRO**: nr Batas, *Guest* 3014!; Walash nr Rowanduz, *Thesiger* 324; **MRO/MSU**: between Dukan and Mirza Rustam, *Rechinger* 10987! **MSU**: Qaranjir, *Rawi* 21603!; ridge NE of Darband-i Khan, *Poore* 554!; Jarmo, *Helbaek* 1969!; Sulaimaniya, *Field & Lazar* 978! **MSU/FKI**: 10–18 km E of Kirkuk, *Rechinger* 9993! **FUJ**: Hammam Ali, *Hossain* 4313!; 49 km W of Mosul, *Rawi* 22935! **FNI**: 42 km NE of Mindan bridge, *Chapman* 26214!; ? **FPF**: 25 km N of Bibar, *Rawi, Husham & Noori* 29435!; **LCA**: Daltawa (Khalis), *Guest* 2469!

W Iran.

HAISHARAN; according to a note on *Guest* 2469 this plant is eaten by animals and people when young, and dried for fodder.

I can make little of the characters used by Rechinger in Fl. Iranica to separate *E. mosulensis* from *E. chardinii.* The key characters which he uses are the cornigerous capitula of *E. mosulensis* (non-cornigerous in *E. chardinii*) and bullate leaves of *E. mosulensis* (non-bullate in *E. chardinii*). I cannot comprehend the latter even in material named by Rechinger himself and would not consider it very significant in the absence of other characters; and since his notes on *E. mosulensis* state that occasionally the horn-formation is almost suppressed or is reduced to a strongly dissimilar length of the phyllaries, this character seems also of dubious worth. The type specimen of *E. chardinii* (G-Boiss.) has so much the character of most Iraqi "mosulensis" that I have little hesitation in relegating it to synonymy here.

14. **Echinops tenuisectus** *Rech.f.* in Anz. Österr. Akad. Wiss. Wien, math.-nat. Kl. 101: 16 (1964). Type – Iraq, Diyala. Mandali, v. 1957, *Rechinger* 9714 (W, holo.); Rechinger, Fl. Lowland Iraq: 637 (1964) and Fl. Iranica 139a: 63 (1979).

Plant 40–80 cm tall. Stem angular-sulcate, erect, slender to rather stout, simple or with a few ascending branches, densely clothed with a close white indumentum when young, this soon evanescent, leaving stem brownish and ± densely clad with minute, subpapillose glands. Basal and lowest cauline leaves not seen. Lower and median cauline leaves sessile with a ring of large, sharp spines up to 3 cm long but scarcely perceptible auricles at base, lamina up to 27 × 14 cm, elliptic or oblong-elliptic in outline, subtripinnatisect with up to 9 pairs of pinnae; pinnae narrowly lanceolate, teeth terminating in rigid stramineous spines up to 15 mm long web of tissue along rachis 3–6 mm wide; upper leaves much reduced, pinnate, strongly spinous; all leaves green, green or sparingly floccose on upper surface and ± densely clad with very short, uniform glands, lower surface densely and (especially when young) very contrastingly white-tomentose. Synflorescences very cornigerous, 5–7 cm in diameter excluding cornigerous spines which project a further 10 mm; receptacle 6–7 mm; involucre 20–25 mm in flower, not seen in fruit; brush whitish, *c.* 10 mm (young); phyllaries 18–23; outer long-spatulate with a distally dentate, sometimes very minute ± flabellate limb; median expanded-obovate, contracted to ciliate spine with sharp shoulders which are distally denticulate-ciliate; inner acute to obtuse, fimbriate at apex, fused to about halfway. Corolla and pappus not seen (but pappus almost certainly, as with plant's close relatives, overtopped by the achene-hairs).

HAB. Mountain slopes and rocky places, grassland in degraded *Quercus* forest, cultivated fields, uncultivated fields on dry hills; alt. 212–1790 m; fl. May-Aug.
DISTRIB. Scattered in the mountain and foothill regions of Iraq, with one record just within the Lower Mesopotamian region. **MAM**: Ser Amadiya, *Guest* 4978!; mountain N of Zakho, *Rechinger* 10738! **MSU**: Azmira Dagh nr Sulaimaniya, *Hadač* 1859!; 60 km between Chamchamal & Sulaimaniya, *Joseph Al-Kas* 18850 (BAG!). **FKI/MSU**: 30 km from Kirkuk to Sulaimaniya, *F. Karim, H. Hamid & M. Jasim* 40534 (BAG!). **FKI**: Tuz Khurmatu, *Guest* 1416! **FPF**: Badra, *Rechinger* 9200! **LEA**: Shahraban (Muqdadiya), *id.* 9714!

Endemic.

Two other gatherings have been provisionally referred to this species: MAM, Zawita, *Guest* 4585!; MRO, Rowanduz gorge, *Guest* 9440! Both of these have narrow leaf divisions, but are much less distinctly cornigerous, the leaves are somewhat less divided and less fiercely armed. The second is particularly suspect. I am decidedly uneasy about the *specific* distinctness of both *E. tenuisectus* and *E. chardinii* from *E. orientalis,* but for the moment can only recognise them lest information be lost. This unease is increased by the scatter of specimens of the three species over much the same area in Iraq. Even closely related species in Iraq tend to occupy different areas. The bulk of sect. *Ritropsis* is a very difficult assembly, and is in need of intensive study in the field.

15. **Echinops polyceras** *Boiss.*, Diagn. Pl. Orient. ser. 1, 10: 85 (1849); Zohary in Dep. Agr. Iraq Bull. 31: 151 (1950); Feinbrun-Dothan, Fl. Palaest. 3: 363 (1978).

E. lasioclinius Boiss., Diagn. Pl. Orient. ser. 1, 10: 89 (1849).
E. blancheanus Boissier, Fl. Orient. 3: 430 (1875); Dinsmore in Post, Fl. Syria, Palest. & Sinai ed. 2, 2: 76 (1933); Blakelock in Kew Bull. 4: 53 (1949); Rawi in Dep. Agr. Iraq Tech. Bull. 14: 114 (1964); Rechinger, Fl. Lowland Iraq: 636 (1964).

Plant 40–60(–100) cm tall. Stem simple or with few branches, ± stout, sulcate, angular below, ± densely white-arachnoid and eglandular above, often somewhat to distinctly glabrescent below with few to numerous short, pale, glandular hairs. Basal and lowest cauline leaves congested, shortly petiolate, elliptic to oblong-elliptic in outline, up to 30 × 16 cm, pinnatipartite or pinnatisect into 5–8 pairs of narrowly triangular lobes with strong, stramineous terminal spines up to 12 mm long, lobes shallowly to deeply dentate or some bifid, teeth also strongly spinous; median cauline leaves sessile, spinous-auriculate, becoming more deeply and narrowly lobed; upper leaves much-reduced, spinous-bract like above; all leaves moderately arachnoid on upper surface, ± white-veined with denser tomentum, densely white-tomentose below, eglandular, margins ± revolute, veins often narrowly white above. Synflorescences considerably cornigerous, 4.5–5 cm in diameter excluding these spines, which project to 2–4(–5) cm further; receptacle 4–8 mm; involucre 15–16 mm long in flower, to 20 mm in fruit; brush pure white, 7–10 mm, very dense, obconical, of slightly flattened bristles; phyllaries 19–24; outer long-spatulate with a distally ciliate, aristate, glabrescent limb; median obovate with blunt, ciliate shoulders, apex produced into a distinct to extremely long spine; inner obtuse to subacute, fimbriate, connate to about middle; all phyllaries pale green, white-margined, glabrous or slightly arachnoid when young. Corollas 12–14 mm, white to pale mauve, tube shortly papillose or with a few long whitish hairs. Pappus 1.5 mm, bristles fused into a corona below, completely overtopped by apical hairs of achene. Fig. 5: 1–4.

HAB. In rocky and stony limestone desert, featureless degraded *haswa* plain, sandy and clay depressions; alt. 45–350 m; fl. Apr.-May.
DISTRIB. Scattered in the desert regions of Iraq, rare in the Lower Mesopotamian Plain and foothills. **FUJ/FKI**: Fatha, *Calder* 2076. **DWD**: 230 km NW of Ramadi, *Rawi* 20934!; 100 km from Rutba to Ramadi, *Botany staff* 41822!; 2 km E of Rutba, *Rechinger* 12810!; ± 80 km SW of Karbala, *Guest, Rawi & Rechinger* 19470! **DSD**: 25 km SE of Salman, *Rawi* 14902!; 4 km W of Busaiya, *Haddad* 9576!; 4 km SW of Umaiyid, *Guest, Rawi & Rechinger* 18722!; Kraa, 51 km N of Neutral Zone, *Rawi, Khatib & Tikriti* 29190! **LCA**: Daltawa (Khalis), *Guest* 2469!; Falluja desert, *Hadač* 1795!

Syria, Lebanon, Palestine.

One strange-looking specimen with narrowly dissected leaves, *Alizzi* 35206 from Ghadaf (no doubt the Wadi), 100 km S of Rutba, may belong here. The specimen is poor and immature, the area is that of *E. polyceras*, from which only the finely divided leaves separate it; even in this character, however, it is approached by *Rawi* 14902. The appearance of the plant closely resembles that of *E. paradoxus* Rech. f., an Afghan endemic.

E. descendens Hand.-Mazz. has been listed by Zohary and Rawi, and Rechinger deals with it in his *Fl. Lowland Iraq*, but the localities from which Handel-Mazzetti described it are beyond the Iraqi frontier in Syria.

6. **ACANTHOLEPIS** Less.

Linnaea 6: 88 (1831)
Bremer, Asteraceae: Cladistics & Classification: 135 (1994); Susanna & Garcia-Jacas in Kubitzki, Fam. Gen. Vasc. Pl. 8: 128 [2006] (2007).

Frances G. Davies

Annual herbs with slender spines and much arachnoid pubescence. Leaves simple and entire or with spinulose margins. Synflorecences hemispherical, usually solitary or several, of 1-flowered capitula, subtended by a number of bract-like leaves. Primary capitula

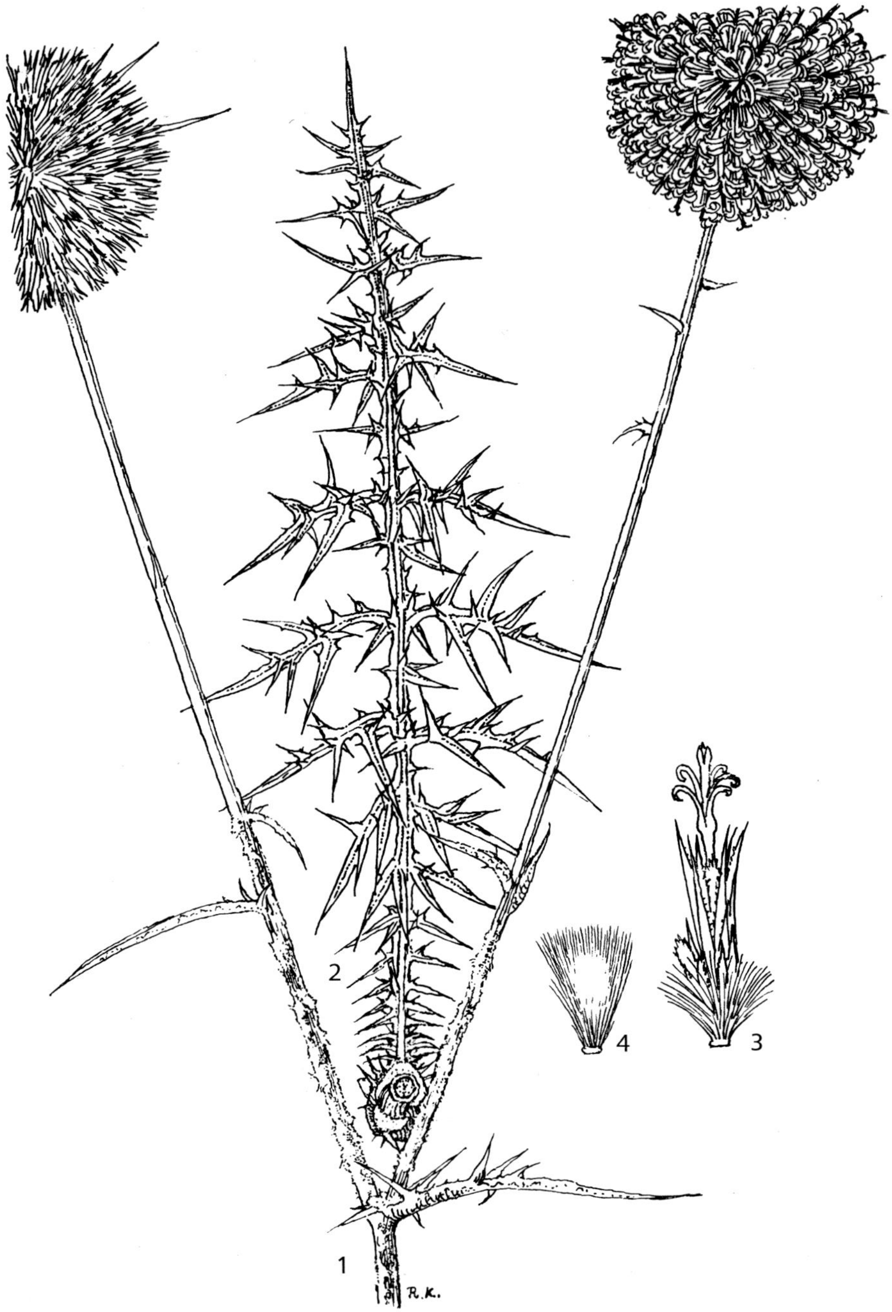

Fig. 5. **Echinops polyceras**. 1, habit × 1; 2, cauline leaf × 1/2; 3, floret x 2; 4, achene × 10. From Naomi Feinbrun-Dothan, *Flora Palaestina*, Part three: Ericacaceae to Compositae, Plates, Jerusalem: The Israel Academy of Sciences & Humanities, 1977, pl. 610. © The Israel Academy of Sciences and Humanities. Reproduced with permission.

Fig. 6. **Acantholepis orientalis**. 1, habit × 1; 2, fruiting cluster × 3; 3, phyllary × 6; 4, floret × 12. 1 from *Rawi* 2113; 2–4 from *Weinert et al.* 47618. Artist unknown.

with a multiseriate involucre; outermost phyllaries reduced to a brush of hyaline scales or hairs, middle phyllaries firm, shortly setose, inner phyllaries tipped with a long seta and innermost closely surrounding a single floret; receptacle small, epaleaceous. Florets hermaphrodite, corollas pink, regularly and deeply 5-partite with only a short tube. Achenes oblong, densely setuliferous; pappus a ring of barbellate uniseriate scales which are shortly connate at base.

A genus of one species differing scarcely from *Echinops* and considered synonymous in the account of *Echinops* in the Flora of China (2011). However, Garnatje et al. (2005) merely pointed out that *Acantholepis*, and *Echinops nanus* were basal to other species of *Echinops* studies; they drew no conclusion as to the continued validity of *Acantholepis*.

Acantholepis (from Gr. ἄκανθα, *akantha*, spine and λέπις, *lepis*, a scale).

1. **Acantholepis orientalis** *Less.* in Linnaea 6: 88 (1831); Boissier, Fl. Orient. 3: 442 (1875); Dinsmore in Post, Fl. Syria, ed. 2, 2: 77 (1933); Dickson, Wild Fls. Kuwait: 11 (1955); Linchevsky in Fl. S.S.S.R. 27: 54 (1962); Rawi in Dep. Agr. Iraq Tech. Bull. 14: 104 (1964); Kupicha in Fl. Turkey 5: 622 (1975); Feinbrun-Dothan, Fl. Palaest. 3: 365 (1978); Al-Mayah, Al-Edani & Al-Asadi, Ecology & Fl. Basra: 189 (2016).

Annual herb (5–)10–15(–30) cm tall, stems white, lanate, slender, leafy, simple or few-branched. Leaves oblong-linear, 1.5–6 × 0.2–0.8(–1.4) cm, acuminate, margins revolute, both surfaces greyish arachnoid-pubescent, lower more densely so, slender, 2–3 mm, long straw-coloured spines arising submarginally. Primary capitula subtended by leafy bracts as long as or longer than capitulum; phyllaries multiseriate, outermost reduced to a dense brush of hair-like scales, intermediate and inner pectinate-ciliate and densely lanate, intemediate shortly setose-tipped with a pale hair, inner terminating in a long brownish setose point, lamina of phyllaries linear-lanceolate, 6–8 mm, awn 4–5 mm. Achenes obovoid, *c.* 6 mm; pappus 5–7 mm, of ± free barbellate scales. Fig. 6: 1–4.

HAB.: In the desert, on sandy soil, gravel, and, in a depression, on silty soil; alt. (100–)300–600 m; fl. & fr. Apr.(-Jun.).
DISTRIB. Occasional in the desert region of Iraq: **DWD**: Tal al Nisr, 48 km NE of Rutba, *Rawi* 31211!; 40 km ESE of Rutba, *Rawi* 31312!; 20 km WNW of Rutba, *Rawi* 21113!; 130 km N of Nukhaib, *Rawi* 31118!; 20 km W of Nukhaib, *Rawi* 30990!; 15 km SE of Ashuriya, *c.* 50 km W of Shabicha, *Guest, Rawi & Rechinger* 19358!; **DSD**: 2 km N of Khadhar al-Ma'i, *Weinert, Fawzi, Adil & Hussain* 47618!; Jabal Sanam, *Guest & Rawi* 14390A!; nr. Saudi frontier, *Rawi* 31156! **LCA**: "Falluja desert" (near Falluja, but on opposite side of R. Euphrates), *Haines* W797!; 20 km SW of Baghdad to Falluja, *S.Omar, Janan, Sahira & K. Hamad* 52732 (BAG!).

Jordan, Syria, Turkey, Caucasia, Kuwait, Iran, Pakistan, Afghanistan, C Asia (Turkmenia to Tajikistan).

7. **STAEHELINA** L.

Sp. Pl.: 840 (1753); Gen. Pl. ed. 5: 364 (1754)
Hirtellina Cass. in Dict. Sci. Nat., ed. 2, 50: 441 (1827)

Frances G. Davies

Perennial dwarf shrubs or subshrubs, often with woody rootstocks. Leaves alternate, sessile, gland-dotted, simple, unarmed. Inflorescence a lax corymb of 2–5 branches. Capitula homogamous; involucres longer than broad; phyllaries multiseriate, not petaloid; receptacle flat, sparsely paleaceous. Florets homomorphic, 1–few; corollas pink, limb spilt deeply into 5 lobes; anthers caudate at base. Achenes homomorphic, cylindrical, densely off-white villous or glabrous; pappus a uniseriate ring of many barbellate hairs, cohering into flattened setae at base.

A genus of around 10 species occurring in both the Mediterranean region and southern Africa; one species endemic to Iraq.

Stahaehelina (named in honour of Benedikt Stähelin (1695–1750), a Swiss botanist, student of Vaillant and von Haller and known to Linnaeus.)

1. **Staehelina kurdica** *Merxm. & Rech.f.*, Ber. Deutsch. Bot. Ges. 72: 282 (1959). –Type. Iraq, Kurdistan, between Hora and Baibu, Raikan district, *Thesiger* 1327 (BM!, holo.); Rechinger, Fl. Iranica 139a: 100 (1979).

Woody-based herb with many rather strictly erect unbranched glabrous but densely scabrid-glandular stems up to 45 cm tall. Leaves sessile, ovate-elliptic to obovate, larger and more densely clustered above stem base, 30–50 × 12–21 mm, margins denticulate, base slightly clasping stem, lamina on both surfaces densely gland-dotted and bright green, rounded at base, sessile. Inflorescence with 2–3(–5) capitula in a lax corymb, pedicels 1–5 cm long, bearing minute ovate bracts which in turn pass into phyllaries. Involucre ellipsoidal in bud, becoming cylindrical in flower, ± 27 mm long; phyllaries (5–)7–8-seriate, hard, cartilaginous, broadly ovate, outer surface pale, glabrous, margins pilose, tips brown and mucronate; receptacle flat. Corolla tube ± 6 mm, limb scarcely wider, ± 9 mm long, lobes 2.5–5 mm; corollas pale purple. Achenes 5 mm, densely white-sericeous; pappus 12 mm, with about 20 bundles of 5 or 6 bristles coherent for ⅓ of their length into setae, off-white.

HAB. Rock fissures on cliffs, limestone outcrops; alt. ± 1200 m; fl. & fr. Jun.-Aug.
DISTRIB. Rare and local in the northern sector of the mountain region of Iraq. **MAM**: between Hora and Baibu, Raikan district, *Thesiger* 1327 (type); Jabal Khantur, nr Sharanish N of Zakho, *Rechinger* 10781!; between Dohuk and Amadiya, above Sarsang, *Rechinger* 11905!; Sarsang, *Haines* 1242 (E), *Agnew* in Haines s.n.

Endemic.

8. **XERANTHEMUM** L.

Sp. Pl.: 857 (1753); Gen. Pl., ed. 5: 369 (1754)

Frances G. Davies

Annual herbs with greyish stems and entire grey-green tomentose leaves. Capitula heterogamous, radiant, fusiform or ovoid in bud, later campanulate; involucre multiseriate; outer series of phyllaries scarious, ovate, inner series conspicuous, petaloid, pink or white, linear and acute; receptacle paleaceous. Outer florets unequally bilabiate, filiform, corollas usually pinkish, sterile. Disc florets hermaphrodite, regular, 5-lobed, corollas yellow; style long, very shortly bifid. Achenes angular, truncate above, narrowed to base, densely pilose; pappus of 5 or 10–15 scales gradually or abruptly contracted upwards into barbellate or ciliate, equal or unequal awns, light brown.

About 5 species in the Mediterranean region and SW Asia; 3 species in Iraq.

Xeranthemum (from the Gr. ξηρός, xeros, dry, ἄνθος, anthos, flower – from the husk-like nature of the phyllaries).

1. Outer and middle phyllaries with a green, herbaceous, tomentose centre, not mucronate; pappus-scales 7–15 .... 1. *X. cylindraceum*
   Outer and middle phyllaries entirely hyaline, mucronate; pappus-scales 5 2
2. Involucre ovoid to subglobose in flower; pappus-scales at most slightly longer than achene .... 2. *X. annuum*
   Involucre fusiform in flower; pappus-scales about twice as long as achene... 3. *X. longipapposum*

1. **Xeranthemum cylindraceum** *Sm.* in Sibthorp & Smith, Prodr. Fl. Graec. 2: 172 (1813); Boissier, Fl. Orient. 3: 464 (1875); Linchevsky in Fl. S.S.S.R. 27: 66 (1962); Rawi in Dep. Agr. Iraq Tech. Bull. 14: 127 (1964); Kupicha in Fl. Turkey 5: 606 (1975); Feinbrun-Dothan, Fl. Palaest. 3: 366 (1978); Rechinger, Fl. Iranica 139a: 96 (1979); Taifour & El-Oqlah, Pl. Jordan Ann. Checkl.: 49 (2017).

[*X. foetidum* auct. non Moench.: Dinsmore in Post, Fl. Syria, Palest. & Sinai ed. 2, 2: 78 (1933).]

Plant 6–65 cm tall with angled, widely branching stems. Leaves sessile, oblong to linear, 18–30(–40) × 3–8(–12) mm, smaller towards stem tips, minutely apiculate, both surfaces sparsely lanate. Capitula ± 15-flowered; involucre 15–21 × 8–10 mm, fusiform in bud; outer phyllaries 4-seriate, broadly ovate, rounded, yellowish-brown, centre herbaceous and lanate; inner phyllaries1–2-seriate, lanceolate, acute, scarious, pink. Sterile outer florets ± 9 mm long; fertile inner florets ± 5 mm. Achenes obconical, 5–6 × 1–2.5 mm, dark brown, with appressed white hairs; pappus scales very unequal, 1.5–6 mm, contracted 1–1.5 mm above base into brown barbellate scales. Fig. 7: 4–5.

HAB. Invariably in *Quercetum*, on dry slopes, boulder scree and grassland; alt. 1250–1360 m; fl. May-Jun.

DISTRIB. A rare plant of the mountain region of Iraq. **MAM**: 13 km from Amadia towards Sarsang, *Sharif & Hamad* 50350!; Ispindari Saddle, Sawara Tuka, *Chapman* 26317! **MRO**, Kani Mazu Shirin, *Agnew, Hadač & Haines* 2007!; Sefin Kuh nr Shaqlawa, *Bornmüller* 1410; between Shahidan and Pushtashan, *Rechinger* 11723 p.p. **MSU**: NE of Darband-i Khan, *Poore* in Kew 540!; Atrush- Rabetki, *Chapman* 9341 (BAG!).

S Europe from Portugal and France to Bulgaria and Czech Republic, eastwards to Crimea, Syria, Turkey, Lebanon, Palestine, Caucasia and Iran.

2. **Xeranthemum annuum** *L.*, Sp. Pl.: 857 (1753); Boissier, Fl. Orient. 3: 444 (1875); Dinsmore in Post, Fl. Syria, Palest. & Sinai ed. 2, 2: 78 (1933); Linchevsky in Fl. S.S.S.R. 27: 62 (1962); Kupicha in Fl. Turkey 5: 604 (1978).

X. squarrosum Boiss., Diagn. Pl. Orient., ser. 1, 6: 101 (1845); Boissier, Fl. Orient. 3: 444 (1875); Nábělek in Publ. Fac. Sci. Univ. Masaryk 52: 29 (1925); Dinsmore in Post, Fl. Syria, Palest. & Sinai ed. 2, 2: 78 (1933); Guest in Dep. Agr. Iraq Bull. 27: 108 (1933); Zohary, ibid. 31: 152 (1950); Linchevsky in Fl. S.S.S.R. 27: 63 (1962); Rawi in Dep. Agr. Iraq Tech. Bull. 14: 127 (1964); Rechinger, Fl. Iranica 139a: 93 (1979).

X. squarrosum var. *unicolor* Boiss., Diagn. Pl. Orient., ser. 1, 6: 101 (1845); Boissier, Fl. Orient. 3: 444 (1875); Bornmüller in Beih. Bot. Centralbl. 60B: 200 (1940); Blakelock in Kew Bull. 4: 65 (1949); Rawi in Dep. Agr. Iraq Tech. Bull. 14: 127 (1964).

[*X. longepapposum* auct. non Fisch. & C.A. Mey.: Guest in Dep. Agr. Iraq Bull. 27: 108 (1933).]

Plant 10–60(–75) cm tall, simple or widely branched with angled stems. Leaves sessile, elliptic, oblong or narrowly oblong, 5–40 × 1.5–8 mm, apiculate, sometimes drying revolute, both surfaces sparsely grey-lanate. Capitula 30–40– (or more-) flowered; involucre ovoid in bud, 15–25 × 7–15 mm in flower; outer phyllaries 4–5-seriate, ovate, rounded, prominently apiculate, scarious throughout. Inner phyllaries linear, bright pink or white, sharply acute, equalling or longer than lower part of involucre. Sterile outer florets 9–11 mm; fertile inner florets ± 5 mm. Achenes obconical, 3–5 × 6 × 1–2 mm, with appressed white hairs; pappus scales 5, contracted above base into very unequal, 2–5 mm awns. Fig. 7: 1–3.

HAB. On open grassy or rocky mountain slopes or valleys, frequently in degraded *Quercus* forest, in rock crevices, on rocky clay or grey sandstone; alt. 750–1970 m; fl. Jun.-Sep.

DISTRIB. Widespread, locally abundant, in the E mountain regions of Iraq, increasing in frequency northwards. **MAM**: Zawita, *Guest* 3691!, 4729!; Dohuk, *Rawi* 8643!; Sarsang, *Haines* 567!; Khantur, *Rawi* 23497!; Jabal Bekher, *Rawi* 23006!; 8 km S of Zakho, *Rechinger* 10695. **MRO**: between Haji Umran and Rayat, *Gillett* 9473!; between Karoukh mt. and Dargala, *Noori & Kass* 27751!; Jabal Saffen (?), Shaqlawa, *Omar, Sahira, Karim & Hamid* 38261!; Chia-i Mandali, *Guest* 2762!; mt. Qandil between Shahidan and Pushtashan, *Rechinger* 11723 p.p.; Kuh Sefin, *Bornmüller* 1408. **MSU**: 7 km W of Tawila, Balcha village, *Rawi* 22368!; Zalu (?) mountain, *Rawi, Husham & Noori* 29395!; mt. Hauraman (Avroman), Sosaka, *Rechinger* 10162.

C & E Europe from Austria eastwards to S Russia including Crimea, Turkey, Lebanon, Syria, Iran, Caucasia, C Asia (Turkmenia). Cultivated as an ornamental and occasionally escaped or naturalized elsewhere.

According to Husain & Kasim (1975), this species is occasionally cultivated in Iraq. Strangely for such an attractive plant, no Kurdish or Arabic vernacular name has been recorded.

Some authors consider the Asian specimens of this species to be distinctive in their smaller capitula and in detailed pappus characters. The Iraq plants are slender and few-flowered and would fall within *X. squarrosum* Boiss. However, this degree of splitting is not believed to be warranted by the present author.

3. **Xeranthemum longipapposum** *Fisch. & C.A. Mey.* in Nouv. Mém. Soc. Imp. Naturalistes Moscou 4: 337 (1835); Boissier, Fl. Orient. 3: 445 (1875); Handel-Mazzetti in Ann. Naturh. Mus. Wien 27: 440 (1913); Dinsmore in Post, Fl. Syria, Palest. & Sinai ed. 2, 2: 78 (1933); Zohary in Dep. Agr. Iraq Bull. 31: 152 (1950); Linchevsky in Fl. S.S.S.R. 27: 65 (1962); Rawi in Dep. Agr. Iraq Tech. Bull. 14: 127 (1964); Kupicha in Fl. Turkey 5: 606 (1975); Rechingher, Fl. Iranica 139a: 95 (1979); Breckle, Hedge & Rafiqpoor, Vasc. Pl. Afghan. Checkl.: 196 (2013).

Plant 5–35 cm tall with angular stems, divaricately branched from base upwards, or rarely only above middle. Leaves sessile, linear to linear-lanceolate, 6–50 × 1.5–4 mm, reducing upwards, acute or apiculate, margins strongly revolute, both surfaces sparsely to considerably lanate. Capitula 10–15-flowered; involucre 15–20 × 5–8 mm, fusiform in bud; outer phyllaries in 5–6 rows, broadly ovate, predominantly apiculate, scarious throughout.

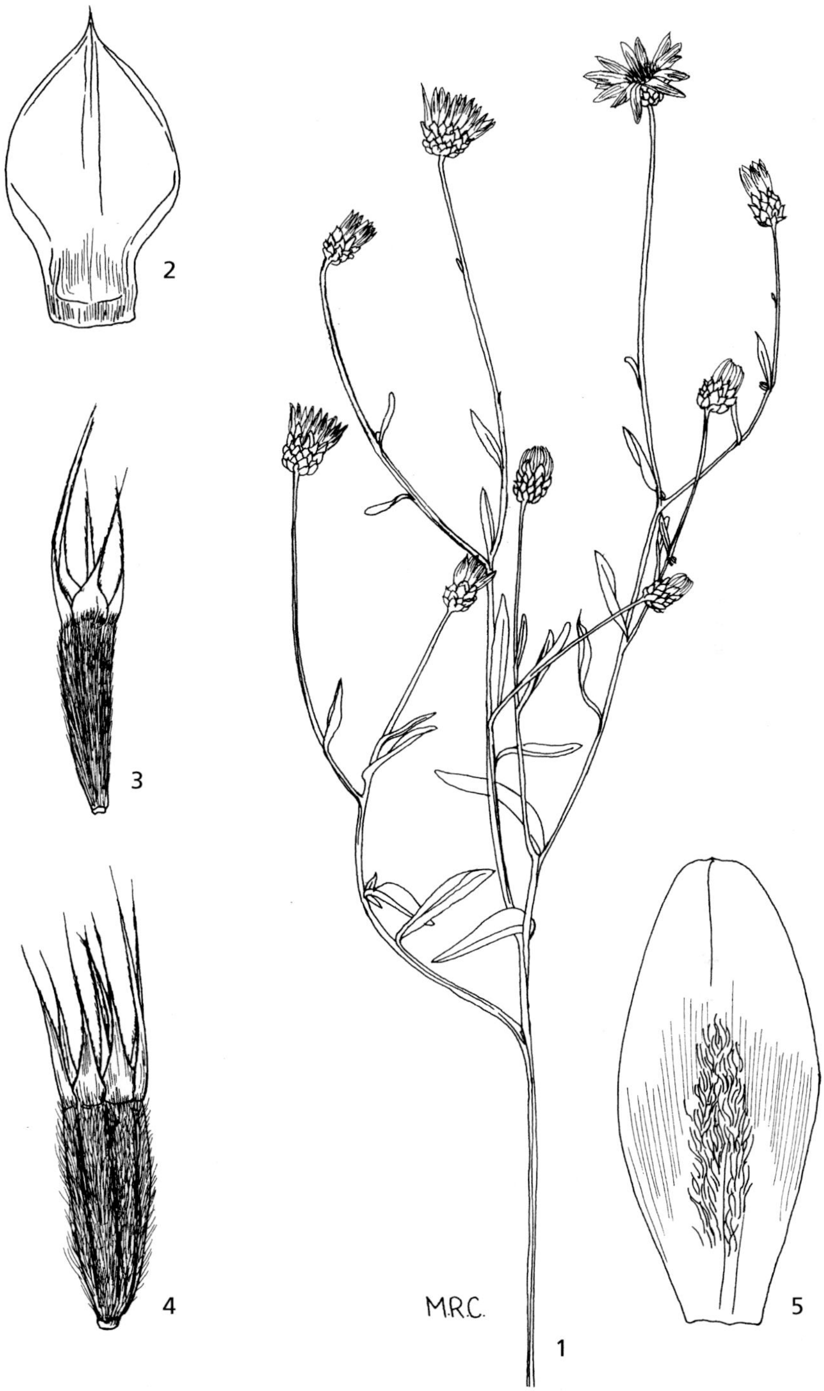

Fig. 7. **Xeranthemum annuum**. 1, habit × ½; 2, median phyllary × 10; 3, achene and pappus × 8. **Xeranthemum cylindraceum**. 4, achene and pappus × 6; 5, medium phyllary × 10. 1–2 from *Rechinger* 10695; 3 from *Gillett* 9473; 4–5 from *Chapman* 26317. Drawn by Moira Rae Carter.

Inner phyllaries linear, bright pink, shorter than lower part of involucre. Sterile outer florets ± 8 mm, few; fertile inner florets ± 5 mm. Achenes obconical, 5–6 mm, with appressed white hairs; pappus scales 5, contracted above base into subequal, ± 10 mm awns.

HAB. Stony fields and mountain slopes; alt. 600–1300 m.
DISTRIB. Very rare, not seen recently, in the mountain region of Iraq. **MJS:** Jabal Sinjar, *Handel-Mazzetti* 1489. **MRO**: Rowanduz district, *Zohary* s.n.:

Turkey, Palestine, Lebanon, Transcaucasia, Iran, Afghanistan, C Asia.

I was informed by the authorities at Jerusalem that although Zohary stated in his publication that a specimen from Rowanduz district was represented in herb HUJ, no specimen was to be found there in the covers of any species of the genus.

9. **CHARDINIA** Desf.

Mém. Mus. Hist. Nat. Paris 3: 455 (1817)

Frances G. Davies

Annual herbs with entire greyish-green leaves and widely diverging branches. Capitula terminal, solitary, fusiform in bud, cylindrical in flower, broadly obconical in fruit, heterogamous, disciform. Phyllaries multiseriate imbricate, scarious in fruit. Receptacle with scales transitional from inner phyllaries, narrowing inwards. Marginal florets female, 3–5-lobed, (sometimes lacking), central florets hermaphrodite; corollas narrow, shortly 5-lobed. Achenes of female florets glabrous, winged, compressed, with 1 central rib, those of hermaphrodite florets cylindrical, many-ribbed, white-pilose especially in lower part; pappus (reduced to a denticulate corona on female florets) of 8–10 lanceolate scales, emergent from the fruiting involucre.

A monotypic genus.

Chardinia (named in honour of Jean-Baptiste Chardin, 1643–1713, noted traveller in Iran and author of *The travels of Sir John Chardin* (1711).)

1. **Chardinia orientalis** (*L.*) *Kuntze* in Acta Horti Petrop. 10: 201 (1887); Anthony in Notes Roy. Bot. Gard. Edinburgh 18: 292 (1935); Linchevsky in Fl. S.S.S.R. 27: 69 (1962); Rawi in Dep. Agr. Iraq Tech. Bull. 14: 110 (1964); Kupicha in Fl. Turkey 5: 608 (1975); Feinbrun-Dothan, Fl. Palaest. 3: 367 (1978); Rechinger, Fl. Iranica 139a: 98 (1979); Gabrielian & Fragman-Sapir, Flow. Transcaucasus & Adj. Areas: 68 (2008); Taifour & El-Oqlah, Pl. Jordan Ann. Checkl.: 39 (2017).

Xeranthemum annuum L. var. *orientale* L., Sp. Pl.: 858 (1753).
X. orientale Miller, Gard. Dict. ed. 8, no. 3 (1768).
Chardinia xeranthemoides Desf., Mém. Mus. Hist. Nat. Paris 3: 455 (1817); Boissier, Fl. Orient. 3: 446 (1875); Guest in Dep. Agr. Iraq Bull. 27: 21 (1933); Zohary in Dep. Agr. Iraq Bull. 31: 152 (1950).

Annual, 8–30 cm tall, simple or divaricately branched. Leaves alternate, lowest narrowed with petioloid bases, remainder sessile, narrowly elliptic to linear, very variable in size, up to 30 × 10 mm, acute, entire, sparsely tomentose and green above, densely tomentose and canescent beneath. Capitula terminal on stems and branches on slender, 1.5–12 cm, canescent pedicels. Outer phyllaries broadly ovate, brown or greenish centrally, margins broadly hyaline; intermediate phyllaries broadly obovate, apiculate to acute; inner phyllaries oblanceolate to elliptic, slightly elongating in fruit, 12–18 mm; receptacular scales whitish, narrowing inwards, trifid to simple. Female florets 1–2 or rarely absent, corolla 5.5 mm, lobes 1.2 mm. Hermaphrodite florets about 8, corolla 9.5 mm, lobes 0.3 mm. Achenes of female florets ± 4 mm long, with a central rib on one side and 2 marginal toothed wings; achenes of hermaphrodite florets 7–9 mm long, many-veined, stramineous, densely pilose near base, more sparingly above; pappus 10–15 mm. Fig. 8: 1–3.

HAB. Chiefly on dry stony or grassy slopes and hilltops, sometimes in coppiced *Quercus* forest, on dry banks or roadsides, as a cornfield weed, also noted from a *Juglans* grove by a stream, on clay, loam, gravel or rocky limestone ground; alt. 220–1900 m; fl. & fr. Mar.-Jun.

Fig. 8. **Chardinia orientalis**. 1, habit × 2/3; 2, capitulum × 4; 3, achene × 3¼. 1 from *Bornmüller* 1412. Drawn by Moira Rae Carter.

DISTRIB. Very common in the mountains, occasional in the foothills and on the plains. **MJS**: 36 km E of Karsi, *Barkley & Majid Ibrahim* 8012!; Jabal Sinjar, *Qaisi & Widad* 52131!, *Qaisi & Hamad* 49147! **MAM**: Dohuk, *Guest* 2322!; 8 km S of Dohuk, *Botany staff* 43369!; nr Zawita, *Guest* 2197!; Sharifa nr Amadiya, *O. Polunin* 5132!; Matina, *Rawi* 8682!; Aqra, *Qaimaqam of Aqra* 3085!; between Sarsang and Amadiya, *Karim, Hamid & Jasim* 40938!; nr Zakho, *Karim, Dabbagh & Hamad* 44896!; Chiya-i Gara, *Dabbagh & Jasim* 46934!; Sawara Tuka, *E. Chapman* 26366!; Zawita gorge, *Guest* 13290! **MRO**: Kuh Sefin above Shaqlawa, *Bornmüller* 1412!, *Gillett* 8096!; 31 km N of Kirkuk on the way to Koi Sanjaq, *Rawi, Noori & Alkas* 27998!; Rowanduz gorge, *Guest* 2143!; 19 km SW of Salah ad-Din, *Barkley* 5630!; Salah ad-Din, *Mooney* 4255!, *Chapman* 119931; Shaqlawa, *Haines* W.701!; 17 km NW of Rania, *Rawi, Noori & Alkas* 28612!; Gali Warta, 30 km NW by N of Rania, *id.* 28818!; Haibat Sultan Dagh, *id.* 28209!; Galala, *Alkas* 18589!; Sefin Dagh, *Gillett* 8096! **MSU**: Jarmo, *Helbaek* 463!; mt. Avroman nr Tawila, *Rechinger* 10320!; Abu Abida, nr Halabja, *Omar, Karim, Hamza & Hamid* 37449!; between Kirkuk and Chemchamal, *Rawi & Gillett* 7561!; Chemchamal, *Rogers* 0194!; 60 km from Chemchamal on the way to Sulaimaniya, *Alkas* 185451; Kamarspa, between Halabja and Tawila, *Rawi* 22253!; Dara Tri, *id.* 22009!; mt. Avroman N of Halabja, *id.* 22072!; Penjwin, *id.* 8831! **FUJ**: Qaiyara, *Bayliss* 43!; Mosul, *Lazar* 3385!; Jabal Makhul, *Rawi & Gillett* 7229! **FKI**: Kani Domlan hills, *Guest* 4332!; Kifri, *Rawi & Gillett* 7422! **FKI/FPF**: Jabal Hamrin, nr Table Mountain, *Guest* 1902!; Pilkhana, *Rawi* 12811!

BANISH IUK (Kur.), GIA GUR (Kur.)

Palestine, Jordan, Syria, Turkey, Caucasia, Iran, Afghanistan, Pakistan, C Asia (Turkmenia to Sinkiang).

10. **SIEBERA** J.Gay *nom. cons.*

Mém. Soc. Hist. Nat., Paris 3: 344 (1827)
Susanna & Garcia-Jacas in Kubitzki, Fam. Gen. Vasc. Pl. 8: 131 [2006] (2007)

Frances G. Davies

Annual greyish, lanate herbs with widely diverging branches and simple leaves, the uppermost subtending capitula as an involucel. Capitula solitary or aggregated, pedunculate; phyllaries multiseriate, the outer spine-tipped from a broad base, the upper portion of the inner acute, pink, petaloid; receptacle paleaceous, scales setiform, brown. Florets of 2 kinds, outer sterile and limb bilabiate with linear, subspinous segments, inner hermaphrodite, soft, narrowly tubular. Achenes angular, pubescent; pappus (of inner florets only) of 7–10 scales, ciliate midrib extending as a plumose or barbellate awn.

A genus of 2 species distributed over SW Asia from Syria to Turkmenia; both occur in Iraq.

Siebera (named in honour of Franz Sieber (1789–1844), Czech botanist and traveler).

Plant mostly 15–30 cm tall; uppermost leaves hardly differing from cauline ones, not forming a distinct involucel; pappus scales without brown spots, awns barbellate .... 1. *S. pungens*
Plant 3–12 cm tall; uppermost leaves much broader and more densely hairy than cauline, forming a distinct involucel; pappus scales with brown spots, awns plumose .... 2. *S. nana*

1. **Siebera pungens** (*Lam.*) *J.Gay* in Mém. Soc. Hist. Nat. Paris 3: 343 (1827); Boissier, Fl. Orient. 3: 447 (1875); Handel-Mazzetti in Ann. Naturh. Mus. Wien 27: 440 (1913); Nábělek in Publ. Fac. Sci. Univ. Masaryk 52: 30 (1925); Dinsmore in Post, Fl. Syria, Palest. & Sinai ed. 2, 2: 79; Zohary in Dep. Agr. Iraq Bull. 31: 152 (1950); Rawi in Dep. Agr. Iraq Tech. Bull. 14: 125 (1964); Rechinger, Fl. Lowland Iraq: 639 (1964); Grierson & Kupicha in Fl. Turkey 5: 607 (1975); Rechinger, Fl. Iranica 139a: 97 (1979).

Xeranthemum pungens Lam., Encycl. [J. Lamarck et al.] 3: 236 (1789).

Herb up to 30 cm tall with diverging, lanate branches from base or above. Leaves sessile, elliptic, 15–40 × 4–10 mm, acute and apiculate, entire, margins usually ± revolute, upper surface canescent, lower white-lanate, uppermost leaves not or scarcely forming a distinct involucre around capitulum. Capitula without differentiated pedicels, solitary, subtended by, but not overtopped by, involucel leaves; involucre 22–45 × 7–14 mm; outer phyllaries in about 4 regularly imbricate rows, central part of ovate base lanate, border brownish-scarious, apex obtuse with an apiculate or spiny tip, spines of inner series c. 10–20 mm; inner phyllaries 20–40 mm, linear, brown below, bright pink above, coloured part about half total length or more. receptacular scales linear, 9–12 mm. Hermaphrodite florets c. 8–10 mm, corolla lobes c. 3–4 mm. Achenes 6–10 mm, shortly pilose; pappus c. 9–12 mm, scales extending into 5 barbellate awns. Fig. 9: 1–3.

Fig. 9. **Siebera pungens**. 1, habit × 1; 2, achene and pappus × 4; 3, tip of pappus × 10. **Siebera nana**. 4, habit × 1; 5, tip of pappus × 10. 1–3 from Turkish material; 4–5 from *Rawi* 23438. Drawn by Moira Rae Carter.

HAB. Very rare in Iraq; only found a few times in the lower forest zone and upper steppe zone, not far from the Turkish frontier; clay rocky mountain side; alt ± 850 m; fl. & fr. Jun.-Jul.
DISTRIB. **MAM**: Sandur, *Qaisi & Khalid* 46118!; **MJS**: Jabal Sinjar, *Handel-Mazzetti* 1355 & 1466; **FNI**: below Mar Ya'qub, *Nábělek* 3980.

Syria, Lebanon, Turkey, Iran, Turkmenia.

2. **Siebera nana** (*DC.*) *Bornm.* in Beih. Bot. Centralbl. 60B: 201 (1939); Grierson & Kupicha in Fl. Turkey 5: 607 (1975); Rechinger, Fl. Iranica 139a: 97 (1979); Breckle, Hedge

& Rafiqpoor, Vasc. Pl. Afghan. Checkl.: 188 (2013); Taifour & El-Oqlah, Pl. Jordan Ann. Checkl: 47 (2017).

S. pungens (Lam.) J.Gay var. *nana* DC., Prodr. 6: 531 (1838).

Herb 3–12 cm tall, simple or with lanate branches from base or above. Median leaves narrowly oblanceolate to narrowly elliptic, 10–40 × 1–6 mm, acute, entire, margins ± revolute, both surfaces ± lanuginose. Capitula 1–4, densely clustered, partly concealed by involucel leaves which are 10–17 × 6–10 mm, broadly ovate, obtuse and densely lanate; involucre 15–25 × 5–7 mm, outer phyllaries lanate over their entire surface, margins whitish, base ovate, with soft spines 4–7 mm; inner phyllaries 16–24 mm, linear, upper half pink. Receptacular scales linear, 8–10 mm. Hermaphrodite florets 7–8 mm long, lobes 2–3 mm. Achenes 6–9 mm; pappus 11–13 mm, brownish-spotted throughout the scales and plumose awns. Fig. 10: 4–5.

HAB. On rocky mountain slopes among *Quercus* and *Pistacia* trees, on a sandstone hillside below the tree line; alt. 400–1500 m; fl. & fr. May-Jul.
DISTRIB. Rather rare in Iraq. **MAM**: Khantur, NE of Zakho, *Rawi* 23438!, 23776!; Ispindari, above Sawara Tuka, *E. Chapman* 26311!; Kani Mazi and Matina mt., Summit of Barugha mt., *Al-Khayat, Jawhar & Abdullah* 58000 (BAG!). **MSU**: Jarmo, *Helbaek* 1855!, *Haines* W. 262! **FKI**: 7 km NE of Kirkuk, *Rawi, Noori & Kass* 27909!

Syria, Turkey, Iran, Afghanistan.

11. **CARDUUS** L.

Sp. Pl.: 820 (1753); Gen. Pl. ed. 5: 358 (1754)

C. C. Townsend

Annual, biennial or perennial herbs. Stem glabrous to densely pilose. Basal leaves forming a rosette, entire or incised, spinous-margined; cauline leaves spinous, commonly decurrent to form spinous wings down the stem, entire or sinuate to bipinnatisect, surfaces glabrous to hairy but never spinulose-strigose. Capitula homogamous, ± globose to cylindrical, solitary and long-pedunculate to clustered and sessile. Phyllaries multiseriate, erect to reflexed, all but innermost spine-tipped, glabrous to ± floccose or lanate. Receptacle clothed with long, bristly hairs. Florets regular or zygomorphic, corollas purple to pink or rarely white, bisexual. Achenes oblong-ovate, compressed, glabrous, smooth or variably striate, apex with a circumferential rim and central umbo; setae multiseriate, of scabrous hairs united at base and falling complete.

About 80 species in the temperate and subtemperate regions of the Old World, and at high altitudes in tropical Africa; 3 species in Iraq.

The thistles of this genus differ from *Cirsium* by the shortly scabrid (not plumose) pappus setae.

Carduus (from Gr. καρδυυς, karduus, the name used by Dioscorides for this plant). Plumeless Thistle (Eng.).

Kazmi, S.M.A. (1964). Revision der Gattung *Carduus* (Compositae), Teil 1 in Mitt Bot. Staatssamml München 5: 139–198 (1963); Teil 2, ibid. 5: 279–550 (1964).

1. Capitula globose to broadly ovoid, scarcely longer than wide; median phyllaries long and narrow, gradually tapering from base to apex with straight sides . . . . 1. *C. getulus*
   Capitula oblong, distinctly longer than broad; median phyllaries not gradually tapering with straight sides, either ± gradually rounded or constricted above a broad base to a narrow apex 2
2. At least some median phyllaries as long as innermost phyllaries, narrow, the subulate green tips with a strong stramineous midrib and border, each ± equalling width of green surface between them . . . . 2. *C. argentatus*
   Median phyllaries all shorter than innermost, green tips wider than midrib and border, usually distinctly narrower than the green surface between them . . . . 3. *C. pycnocephalus*

1. **Carduus getulus** *Pomel*, Nouv. Mat. Fl. Atlant. 2: 275 (1875); Kazmi in Mitt. Bot. Staatssamml München 5: 431 (1964); Täckholm, Stud. Fl. Egypt, ed. 2: 532 (1974); Feinbrun-Dothan, Fl. Palaest. 3: 376 (1978); Rechinger, Fl. Iranica 139a: 225 (1979); Boulos & Hind in Fl. Egypt. 3: 149 (2002); Abdulridha, Taha & Wedad, Ecol. & Fl. Basra: 199 (2016).

Annual herb, 5–35 cm, varying from acaulescent to branched from base with rather few sparingly branched stems; stem and leaves green, sparingly floccose, stem strongly and irregularly spinous-winged from decurrent leaves. Basal and lower cauline leaves dentate to deeply pinnately lobed, lobes broadly oblong to triangular, tipped with fine spines, margins coarsely to finely spinous–denticulate, very variable in size, up to 20 × 8 cm but usually less; upper leaves rapidly reducing, narrower and with narrower lobes and teeth, finally bract-like. Capitula globose to broadly ovoid, 1.5–2 cm long, solitary or in clusters of 2–5, on a short (to 1 cm) densely white-lanate peduncle, or sessile. Phyllaries glabrescent to floccose, erect and imbricate, linear-lanceolate, outer with a strong midrib and excurrent stramineous spine, inner 12–15 mm, rather feebly 3-veined, weakly spinous or submuticous, pale-margined and often with a pale tip. Corollas with subequal lobes, ± 10 mm, purple. Achenes 3–4 mm; pappus 8–10 mm, pure or dirty white. Fig. 10: 6.

HAB. Sandy desert soils; alt. up to 100(–350) m; fl. & fr. (Feb.-)Mar.-Apr.
DISTRIB. Occasional, mainly in the Southern Desert of Iraq. **FPF**: Jabal Hamrin, between Ba'quba and Khanaqin, *Hadač* 560; **DLJ**: 70 km NW of Falluja, nr Wadi Thirthar, *Chakravarty & Rawi* 30413! **DSD**: 15 km SE of Ashuriya, *Guest, Rawi & Rechinger* 19364!; Um Qasr, *id.* 16906!; 44 km S of Zubair, *Thamer* 47422!; Batin, c. 65 km SW of Basra, *Guest et al.* 17083!; Jabal Sanam, *Chakravarty, Rawi, Khatib & Tikriti* 29892. **LBA**: 70 km E of Zubair, *Thamer* 47450!; Al-Shabi near Al-Aujah, *Al-Kaisi, K. Hamad & H.Hamid* 48417 (BAG!); Al-Samah, *Al-Kaisi, K. Hamad & H.Hamid* 48471 (BAG!); 10 km to Um-Qasir, *Al-Kaisi, Ghina & Saif* 58411 (BAG!).

Lebanon, Palestine, Jordan, Egypt, Iran, N Africa. Perhaps adventive in Turkey.

2. **Carduus argentatus** *L.*, Mant. Alt.: 280 (1771); Boissier, Fl. Orient. 3: 522 (1875); Dinsmore in Post, Fl. Syria, ed. 2, 2: 88 (1933); Anthony in Notes Roy. Bot. Gard. Edinburgh 18: 291 (1935); Rawi in Dep. Agr. Iraq Tech. Bull. 14: 107 (1964); Kazmi in Mitt. Bot. Staatssamml München 5: 453 (1964); Täckholm, Stud. Fl. Egypt ed. 2: 534 (1974); Davis in Fl. Turkey 5: 432 (1975); Franco in Fl. Europ. 4: 230 (1976); Feinbrun-Dothan, Fl. Palaest. 3: 376 (1978); Boulos & Hind in Fl. Egypt. 3: 149 (2002).

Annual with a slender taproot, 14–80 cm, simple or with generally a few long, ascending branches in lower half or sometimes above, stem shortly irregularly and sometimes interruptedly spinous-winged from the decurrent leaves, ± white-lanate. Basal and lower stem leaves deeply or shallowly pinnatifid with lobes broadly triangular, dentate and spinulose, green and sparingly pilose on upper surface, ± white-tomentose below, variable in size, up to 17 × 6 cm; upper leaves rapidly reducing, usually becoming narrower with narrow lobes and more slender teeth, finally bract-like. Capitula solitary or in clusters of 2–3(–5), oblong, 16–22 × 8–12 mm, generally on long naked white-tomentose pedicels 5–17 cm long, but in shortly pedunculate groups in the most robust forms; phyllaries green or ± white-floccose, outer short, broad-based, the slender spinose tip with thickened margins and midrib; intermediate phyllaries similar but the slender upper part very long, subulate, spine-tipped, the longest shorter than to exceeding the broader, not or very weakly spinose, feebly 3-veined inner phyllaries. Corolla 9–12 mm long, purple with equal lobes. Achenes 4–6 mm long; pappus 10–14 mm, dirty white.

subsp. *argentatus*, with broader intermediate phyllaries shorter than or rarely sub-equalling the inner ones, occurs in the Aegean Is., Syria, Lebanon, Palestine and Turkey, but not in Iraq, where the species is represented by:

subsp. **acicularis** (*Bertol.*) *Meikle* in Fl. Cyprus 2: 949 (1985). Fig. 10: 7.

C. acicularis Bertol. in Ann. Stor. Nat. Bologna 1: 274 (1829); Kazmi in Mitt. bot. Staatssamml München 5: 454 (1964); Davis in Fl. Turkey 5: 434 (1975); Rechinger, Fl. Iranica 139a: 229 (1980).

Intermediate phyllaries with the subulate upper part very narrow, the longest equalling or exceeding the broader inner phyllaries.

HAB. Irrigated gardens; alt. 750–840 m; fl. & fr. May-Jun.
DISTRIB. Occasional in the E sector of the lower mountains in Iraq. **MSU**: Jarmo, *Helbaek* 2178!; between Chemchemal and Darband i Basian, *Haussknecht* s.n.; 55 km NW of Sulaimaniya on road to Dukan, *Rawi* 21780; 60 km between Chemchamal & Sulaimaniya, *Josef & Al-Kass* 18542 (BAG!).

Europe (Mediterranean region from S France to the Balkans (including Bulgaria), Cyprus, Syria, Lebanon, Palestine, Egypt, N Africa (Libya).

The differences between *C. argentatus* and *C. acicularis* Bertol., regarded by some authors including Kazmi (l.c.) as distinct species, seem to me to be purely of degree and some specimens, especially from Cyprus, might reasonably be assigned to either. Separation with confidence is possible in the great majority of cases, and so subspecific rank seems appropriate.

3. **Carduus pycnocephalus** *L.*, Sp. Pl. ed. 2: 1153 (1763); Boissier, Fl. Orient. 3: 520 (1875); Handel-Mazzetti in Ann. Naturh. Mus. Wien 27: 44 (1913); Guest in Dep. Agr. Iraq Bull. 27: 19 (1933); Bornmüller in Beih. Bot. Centralbl. 60B: 266 (1939); Blakelock in Kew Bull. 4: 45 (1949); Zohary in Dep. Agr. Iraq Bull. 31: 153 (1950); Rawi, Dep. Agr. Iraq Tech. Bull. 14: 45 (1964); Rechinger, Fl. Lowland Iraq: 644 (1964); Kazmi in Mitt. bot. Staatssamml München 5: 444 (1964); Täckholm, Stud. Fl. Egypt ed. 2: 534 (1974); Davis in Fl. Turkey 5: 435 (1978); Franco in Fl. Europ. 4: 251 (1976); Boulos & Hind in Fl. Egypt 3: 149 (2002); Abdulridha, Taha & Wedad, Ecol. & Fl. Basra: 199 (2016).

C. australis L.f., Suppl. Pl.: 348 (1782).

Annual with a rather slender taproot, 5–75(–100) cm, simple or sparingly branched with long, ascending branches from near base upwards. Stem prominently irregularly winged from the decurrent leaves, wings sharply and often rigidly spinose, green to floccose. Basal and lower stem leaves pinnatilobed to pinnatipartite, lobes broadly triangular-oblong with broadly triangular secondary lobes, dentate and spinose or spinulose, usually green and sparingly pilose on upper surface, lower surface similar to ± densely white-tomentose; upper leaves similar but smaller and often narrower. Capitula oblong, terminal in groups of 3–5, more rarely solitary, 12–20 × 8–15 mm, on short naked, white-tomentose pedicels up to 2.5 cm or stems narrowly spinous-winged right up to capitula; lateral sessile capitula sometimes present on stem. Phyllaries green to ± densely floccose; outer phyllaries short, strongly spinose; intermediate phyllaries distinctly constricted above the broad base to a strongly veined, narrower spinose tip to very indistinctly constricted and more shortly veined, less spinose tip; innermost phyllaries not or weakly spinose, feebly 3-veined, sometimes purplish at apices. Corollas 11–16 mm, purple to pink, with equal lobes. Achene 4–6 mm long; pappus 10–15 mm, dirty white. Fig. 10: 1–5.

subsp. PYCNOCEPHALUS has the median phyllaries obviously constricted between the broad base and the narrow apical part, and the latter is strongly veined; the involucre is glabrous or sparingly floccose, long (1.7–2.3 cm) and the outer and especially median phyllaries divergent; the corolla is 13 mm or more in length.
It occurs in S & SE Europe and the Mediterranean region, but does not occur in Iraq, where the following two subspecies are found:

subsp. **albidus** (*M.Bieb.*) *Kazmi* in Mitt. Bot. Staatssamml. München 5: 446 (1964).

Carduus albidus M.Bieb, Fl. Taur.-Cauc. 2: 269 (1808).

Median phyllaries obviously constricted between the broad base and the narrow green apical part, scarcely divergent, vein very strong, reaching almost or quite to base of apical part, the margins of which are convex and appear somewhat thickened; involucre 12–16 mm, distinctly floccose; inner phyllaries green throughout.

HAB. Mountain slopes on limestone in destroyed *Quercus* forest, on a clay plain, in an irrigated orchard pasture, by a wadi in gypsaceous hills, along ditches in a field, a weed in gardens and waste ground on the alluvial plain: alt. to 1350 m; fl. & fr. (Feb.-) Mar.-Apr. (-Jun.).
DISTRIB. Quite common in the lower mountains and steppe, a frequent weed in fields gardens and waste ground on the alluvial plains. **MAM**: Zakho valley, *Guest* 2266! **MRO**: Shaqlawa, *Haines* W.693!; Baradost, *Erdtman & Goedemans* 15636!; 12 km N of Haibat Sultan Dagh, *Rawi, Noori & Kass* 28383!

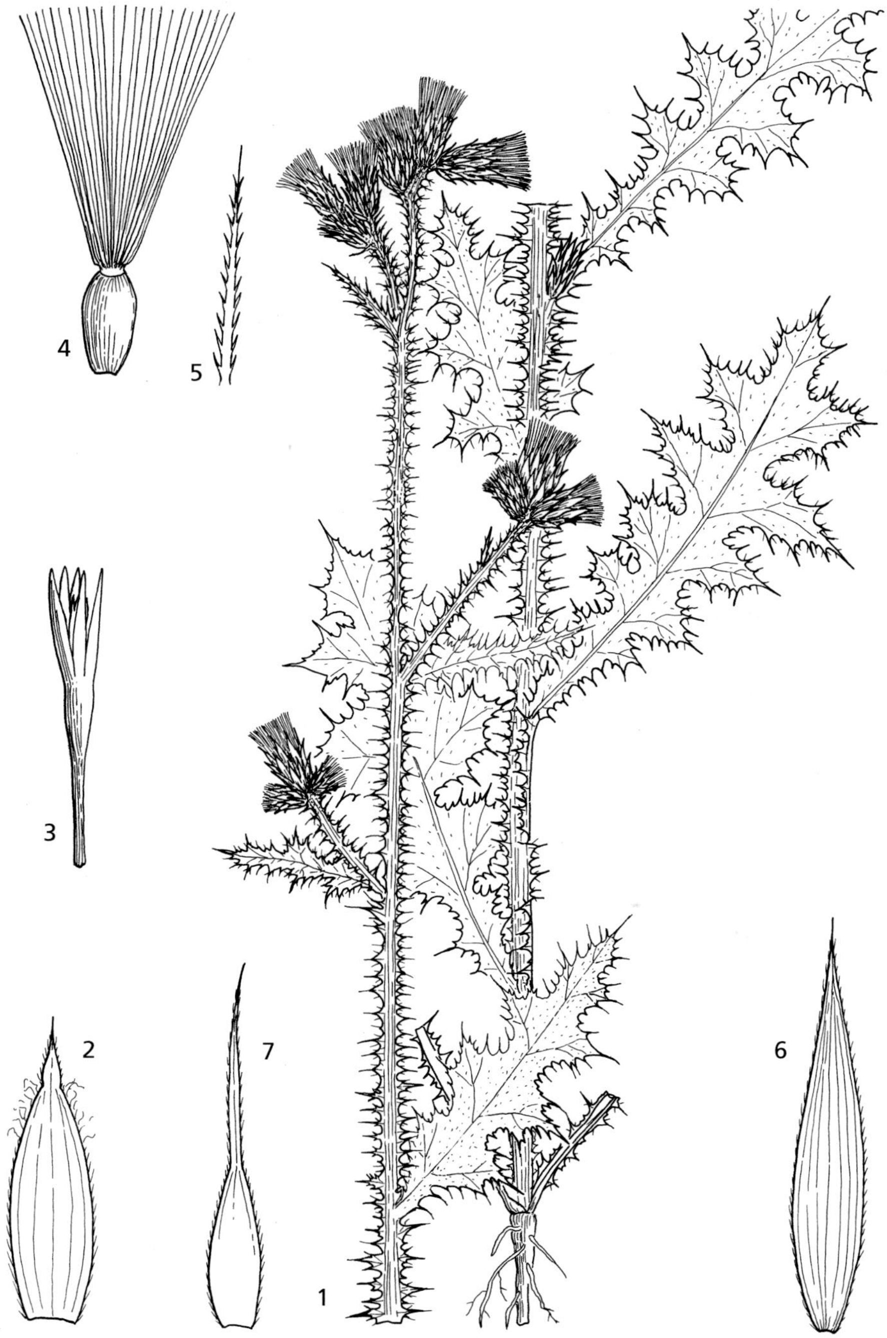

Fig. 10. **Carduus pycnocephalus** subsp. **arabicus**. 1, habit × 1; 2, median phyllary × 6; 3, floret; 4, achene × 6; 5, single pappus enlarged × 9. **Carduus getulus**. 6, median phyllary × 6. **Carduus argentatus** subsp. **acicularis**. 7, median phyllary × 6. 1–4 from *Rawi et al.* 28289; 5 from *Thamer* 47450; 6 from *Hausskecht* s.n. Drawn by Christine M. Grey-Wilson.

MSU: Jarmo, *Helbaek* 583!, 1279!; Tainal, *Rawi & Gillett* 11644!; Tawila, *Rawi* 22300! **FKI**: Tuz, *Guest* 1415!; Kifri, *Rawi & Gillett* 7414! **FPF**: Pilkhana, *Rawi* 12859!; Mandali, *Guest* 899!; 15 km SE of Bagsaiya, *Rawi & Haddad* 25361! **LCA**: Karrada Mariam, *Haines* W.183!; Baghdad, *Colvill* s.n., *Paranjpye* s.n. (comm. Graham), *Guest* 1952!; Rustamiya, *Lazar* 1146!; Za'faraniya, *Omar & Sahira* 37817!; Hilla, *Paranjpye* s.n. (comm. Graham). **LEA**, 10 km E of Ba'quba, *Rogers* 046!, *Haddad & Barkley* 7305!; Kut al-Imara, *Gillett* 6567!; Sharhani well, 15 km N of Tib, *Husain al-Ani* 39857! **LSM**: Amara, *Watson* s.n. (comm. Sharples); Musaida, nr Amara, *Field & Lazar* 76!

My concept of this subspecies agrees with that of Davis in *Fl. Turkey* rather than that of Rechinger in *Fl. Iranica*. All the Iraqi specimens cited by the latter as subsp. *pycnocephalus* which I have seen I have referred to subsp. *albidus*.

subsp. **arabicus** (*Jacq. ex Murr.*) *Nyman*, Consp. Fl. Eur.: 414 (1879).

C. australis L.f., Suppl. Plant.: 348 (1781).
C. arabicus Jacq. ex Murr., Syst. Veg. ed. 14: 724 (1784).
C. marmoratus Boiss. & Heldr. in Boissier, Diagn. Fl. Orient., ser. 2, 3: 43 (1856).
C. arabicus Jacq. ex Murr. ssp. *marmoratus* (Boiss. ex Heldr.) Kazmi, Mitt. Bot. Staatssamml. München 5: 450 (1964).
C. pycnocephalus subsp. *cinereus* (M.Bieb) P.H.Davis in Notes Roy. Bot. Gard. Edinburgh 33: 429 (1975), saltem p.p.

HAB. On sandy, silty, gravelly and stony soils; alt. up to ± 550 m; fl. & fr. (Feb.-) Mar.-Apr.(-Jun.).
DISTRIB. Occasional in the mountain, forest and steppe regions, common in the desert region of Iraq. **MAM**, Dohuk, *Guest* 1321!; Sarsang, *Alkas* 18686!; **MRO**: Chinarok, N of Haibat Sultan Dagh, *Rawi, Noori & Kass* 28289! **MSU**: Qarachitan, *Gillett* 7967!; Sulaimaniya, *Gowan* 2407! **FUJ/FNI**: nr Mosul, *Barkley & Haddad* 5496! **FNI**: Nimrud, *Helbaek* 960!; nr Nineveh, *Guest* 1488! **FAR**: 30 km W of Arbil on Mosul road, nr Eski Kellek, *Shahwani* 25170! **FPF**: Jabal Hamrin, nr Table Mt., *Guest* 1899!; Chilat, nr Iranian frontier, *Rawi & Haddad* 25704!; 16 km S of Badra, *Rawi* 18191!; Garmashaya, 77 km NE of Kut, *Rawi, Khatib & Alizzi* 31936! **DLJ**: 70 km N of Rawa, *id.* 31920!; between Rawa and Haditha, Omar, *Qaisi et al.* 45156!; 3 km E of Hit, *Rawi & Gillett* 6814! **DWD**: Tal an-Nisr, 50 km N of Rutba, *Rawi & Noori* 27128!; 260 km W by N of Ramadi, *Rawi* 20940!; Wadi Hauran, nr Khan Baghdadi, *Omar, Qaisi, Hamad & Hamid* 44574!; T.1, *id.* 45060!; between Nahiya and Qaim, *id.* 44404!; 12 km SE of Hit, in Wadi Al Ghadhaf (between Rutba & Nuqaib), *Rawi et al.* 14743!; 8 km E of Shithatha, *Rawi & Gillett* 6479!; **DSD**: Wagaga, 11 km E of Salman, *Botany Staff* 42104!; 30 km W by N of Busaiya, *Guest, Rawi & Long* 14160!; **LCA**: Sudur, *Jack Isaac & Buthaina Makki* 347!; Babylon, *Reuter* 3557! **LEA**: Fakka-Laqlaq valley, E of Amara, *Rawi* 25826! **LSM**: 70 km E of Amara, N of Rashida well, *Husain al-Ani* 39777!; 30 km S of Amara, *Hazim & Noori* 30652!

SE Europe (from Corsica to Balkans), SW Asia (Turkey, Iran).

CHAUBAZAR (Kurd.), *Gillett* 7697; KALOR (Kurd.-Sulaimaniya), *Gowan* 2407; eaten by all grazing animals in spring; GUL AGHĀN (Kurd.–Dohuk), *Guest* 1321, also widely called KASŪB, a general name used for thistly plants by Arabs.

It will be seen that my position on *C. pycnocephalus* is midway between that of Davis in *Fl. Turkey* and Rechinger in *Fl. Iranica*. I cannot maintain *C. australis* (*C. arabicus*) as a distinct species, as is done by the latter; but Davis's recognition of five subspecies within *C. pycnocephalus* seems to me to be unnecessarily complex. Having examined Linnaean specimen 966/11, regarded by Davis as the type of *C. australis*, I cannot separate it from certain Iraqi material such as *Rawi* 18191, and thus must include it within the more southerly material at least of "subsp. *cinereus* (M.Bieb.) Davis".

C. PYCNOCEPHALUS subsp. BREVIPHYLLARIUS P.H.Davis in Notes Roy. Bot. Gard. Edinburgh 33: 412 (1975).

Vein of median phyllaries weaker, rarely reaching base of the broader green apical portion, unconstricted or only a slight constriction present between this and the basal part, margin plane, pale and membranous, not appearing thickened; involucre 10–17 mm, glabrous or sparingly floccose; inner phyllaries not infrequently purplish at tip.

Not recorded so far from Iraq but likely to occur there.

S & SE Europe (from Corsica and Italy to Greece and Serbia), SW Asia (Syria, Lebanon, Palestine, Egypt, Turkey, Iran, Afghanistan).

12. **CYNARA** L.

Sp. Pl.: 827 (1753); Gen. Pl. ed. 5: 359 (1754)
Sussana & Garcia-Jacas in Kubitzki, Fam. Gen. Vasc. Pl. 8: 132 [2006] (2007)

C. C. Townsend

Robust perennial herbs with stout, sparingly branched, winged or unwinged stems. Leaves large, pinnatisect or pinnate, with a winged rachis, spiny or rarely unarmed. Capitula very large, homogamous, solitary or in few-headed corymbs; involucre ovoid or globose; phyllaries multiseriate, thick, outer coriaceous, with no central costa, spinose or not; receptacle fleshy, densely setose. Corollas blue, purplish-blue or whitish, tubular with an obliquely 5-lobed limb; filaments free, glandular. Achenes oblong-obovoid, compressed and ± 4-angled, hilum basal; pappus multiseriate, setae long, stout, plumose, connate at base into a deciduous ring.

About 8 species in the Mediterranean region and SW Asia; two species in Iraq.

(*Cynara*, either from the Gr. κυνάρα, kynara, canine; or from the Lat. cinara, a plant native to the island of Cinara (now Zinara) in the Aegean.)

Wiklund, A. (1992) The genus *Cynara* L. (Asteraceae – Cardueae), Bot. J. Linn. Soc. 190 (1): 75–123.

Leaves spineless; phyllaries erect or spreading; receptacular bristles 16–50 mm . . . . 1. *C. cardunculus*
Leaves strongly spiny; outer phyllaries ± squarrose-recurved; receptacular bristles 35–70 mm . . . . 2. *C. aurantiaca*

1. **Cynara cardunculus** *L.*, Sp. Pl.: 827 (1753).

C. scolymus L., Sp. Pl.: 827 (1753); Dinsmore in Post, Fl. Syria, Palest. & Sinai ed. 2, 2: 94 (1933); Guest in Dep. Agr. Iraq Bull. 27: 27 (1935); Kupicha in Fl. Turkey 5: 323 (1975); Franco in Fl. Europ. 4: 249 (1976); Feinbrun-Dothan, Fl. Palaest. 3: 381 (1978).
C. hortensis Mill., Gard. Dict., ed. 8 (1768).
C. sylvestris Lam., Encycl. [J. Lamarck et al.] 1: 277 (1783).
C. sylvestris Lam. var. *spinosus* Lam., Encycl. [J. Lamarck et al.] 1: 277 (1783).
C. scolymus L. var. *major* Brot., Fl. Lusit. 1: 338 (1804).
C. scolymus L. var. *minor* Brot., Fl. Lusit. 1: 338 (1804).
C. cardunculus L. var. *sylvestris* Brot., Fl. Lusit. 1: 339 (1804).
C. cardunculus L. var. *hortensis* Brot., Fl. Lusit. 1: 339 (1804).
C. horrida Ait., Hort. Kew. 3: 148 (1811).
C. cardunculus L. var. *inermis* ?Hort., Le bon Jardinier: 38 (1818).
C. corsica Viv., Fl. Libyc. Spec.: 68 (1824).
C. spinosissima C. Presl, Delic. Prag.: 109 (1822).
C. cardunculus L. *β altilis* DC., Prodr. 6: 620 (1838). Type: not stated. De Candolle referred to having seen only live material – '(v. v.)'.
C. cardunculus L. var. *sativa* Moris, Fl. Sardoa: 460 (1840-1843).
C. scolymus L. var. *pungens* Vis., Fl. Dalmat. 2: 46 (1847).
C. scolymus L. var. *mutica* Vis., Fl. Dalmat. 2: 46 (1847).
C. cardunculus L. var. *scolymus* (L.) Fiori in Fl. Anal. Ital. (Fiori & Beguinot) 3: 381 (1904).

Plant mostly 0.9–1.5 m tall, sometimes more. Lower leaves large, broadly oblong in outline, up to 75 × 35 cm, green on upper surface and greyish-arachnoid below, pinnatisect with ovate or lanceolate, mostly coarsely toothed but sometimes entire segments, teeth rather blunt and mucronate. Capitula depressed-globose, 6–7 × 7–10 mm; phyllaries imbricate at base (which in lower phyllaries is fleshy), appressed or lower ± spreading, with a deltoid-ovate to lanceolate, slightly concave, coriaceous apical appendage which is truncate to cuspidate at apex; innermost phyllaries with a narrower membranous appendage. Corollas violet-blue or blue, exceeding inner phyllaries. Achene obovate, 3.7–7 mm, grey brown to straw coloured, usually with dark brown and sometimes also whitish dots, faintly 2–4 ribbed (mature achenes not seen – descript. from Wiklund, 1992, p. 116.); pappus 2–3 mm, dirty white to brownish.

Occasionally cultivated as a vegetable in Iraq. **LCA**: Abu Ghraib, *Omar* 38927! **LBA**: Basrah, *Al-Izzi & S. Omar* s.n. (BAG!).

Artichoke (Eng.), KHARSHŪF, ANGÎNĀR (Turk.). The young flower-capitula are cooked and eaten for the fleshy bracts and receptacle. They may be baked, fried, boiled, stuffed or served with various sauces, and when very young are sometimes eaten raw.

Poorly known in the wild state, but widely cultivated particularly in the Mediterranean region; the cultivated derivative has usually been treated as the separate species *C. scolymus* – Cardoon and Globe Artichoke.

2. **Cynara auranitica** *Post* in Bull. Herb. Boiss. 1: 400 (1893); Dinsmore in Post, Fl. Syria, Palest. & Sinai ed. 2, 2: 95 (1933).

C. syriaca auctt. non Boiss., Diagn. Pl. Orient. ser. 1, 10: 94 (1849); Boissier, Fl. Orient. 3: 558 (1875); Feinbrun-Dothan, Fl. Palaest. 3: 381 (1978); Rechinger, Fl. Iranica 139a: 282 (1979).
C. kurdica Hand.-Mazz. in Ann. Naturh. Mus. Wien 27: 445 (1913). –Type. Mesopotamien-Expedition des naturwissenschaftl. Kurdistan, *Handel-Mazzetti* 1949 (W, holo.; WU, iso.). Nábělek in Publ. Fac. Sci. Univ. Masaryk 52: 34 (1925); Blakelock in Kew Bull. 4: 52 (1949).

Plant to 1–2 m tall. Stems sulcate-striate, glabrous throughout or somewhat arachnoid above, corymbosely branched above with up to 6 capitula. Lower leaves large, 30–60(–75) × 20–35(–50) cm, broadly oblong in outline, pinnatisect to pinnate with (4–)5–6 pairs of opposite or alternate, distant segments bearing narrowly triangular strongly spinous teeth, green above, greyish-arachnoid beneath; petiole of basal leaves long, spinous. Upper leaves rapidly reducing with narrower segments, uppermost very reduced, spiny and bract-like. Capitula ovoid, 7–10 × 6.5–10 cm including the spreading phyllaries; pedicels long, naked or with small bract-like leaves; phyllaries imbricate at base, innermost erect and less firm, narrow, median commonly patent and outermost squarrose-reflexed; all coriaceous and spinescent, at least outer frequently purplish, appendages not demarcated. Corollas bright or violet-blue, exceeding inner phyllaries. Achene not seen, reported to be ± 9 mm; pappus 4–4.5 cm, dirty white or pale buff. Fig. 11: 1–5.

HAB. In hillside pastures and fields, in cultivated fields and along roadsides; alt. 35–750 m; fl. Jun.-Jul. DISTRIB. Apparently uncommon in the mountain and upper steppe region of Iraq. **MAM**: 51 km NW of Mosul near Filda police post, *Rawi* 22943! **MSU**: Chemchemal, *Haines* W1513!; along Sulaimaniya to Dukan road, 1979, *Hossain* s.n.!; Susa, *F. Karim* 39277 (BAG!). **FNI**: Ain Sifni, *Guest* 4040!

SE Turkey, coastal and SW Syria and S Lebanon, Palestine, W Iran.

13. **CIRSIUM** Mill.

Gard. Dict. ed. 4, 1 (1754)
Susanna & Garcia-Jacas in Kubitzki, Fam. Gen. Vasc. Pl. 8: 132 [2006] (2007)

C. C. Townsend

Biennial or perennial, very rarely annual herbs, with stems winged from the decurrent leaves or not. Leaves simple or more commonly pinnately lobed or divided, spinous with rather weak to extremely strong spines, often ± cottony-arachnoid at least on lower surface, upper with or without setae or spinules. Capitula solitary or in corymbs, racemes or dense clusters, homogamous, hermaphrodite or rarely dioecious; phyllaries multiseriate, closely imbricate at least below, glabrous to ± floccose, outer and median ± spine-tipped with spines erect to squarrose, margins entire to spinulose or fimbriate; receptacle long-pilose. Corollas all tubular, usually pink to purple, sometimes white or yellow, limb usually divided to about halfway but sometimes deeper. Achenes smooth, glabrous, compressed to subterete; pappus multiseriate, connate into a ring at base, usually deciduous, innermost hairs often only scabrid at apex and ± clavate.

Probably about 250 species in the northern temperate and subtemperate regions.
Cirsium (from Gr. Κιρσιον, *kirsion*, a kind of thistle, in turn from Κιρσος, *kirsos*, 'a swollen vein or welt', because thistles were often used as a remedy against such things).

1. Plant with a deep, creeping underground root system; flowers functionally unisexual, the female capitula with abortive anthers, the male capitula with generally abortive ovaries; corolla limb divided almost or quite to the base; pappus accrescent in fruit, soon much longer than the corolla . . . . SECT. CEPHALONOPLOS

Fig. 11. **Cynara auranitica**. 1, basal leaf × 1; 2, capitulum × 1; 3, floret × 3; 4, achene with pappus × 3; 5, pappus × 9. 1, 4–5 from *Guest* 4040; 2–3 from *Hossain* s.n. Drawn by Christine M. Grey-Wilson.

Rootstock not creeping; with adventitious shoots; flowers hermaphrodite; corolla limb divided to about the middle; pappus not accrescent to become much longer than corolla 2
2. Leaves densely or sometimes very sparsely clothed with setae or spinules . . . . SECT. EPITRACHYS
Leaves without setae or spinules on the upper surface . . . . SECT. CIRSIUM

1. Leaves neither strigulose-setose nor spinulose on upper surface 2
Leaves strigulose-setose or spinulose on upper surface 7
2. Capitula on very long peduncle-like branches with no or only reduced bract-like leaves, or capitula up to 3 on shorter peduncles at top of such branches; leaves elliptic to oblanceolate, serrate with broad-based long-aristate but scarcely spinose teeth . . . . 8. *C. canum*
Capitula not borne on very long, naked branches; leaves ± pinnate, spinose 3
3. Capitula subtended by at least one pinnately lobed bract-like upper leaf with strong yellowish spines . . . . 2. *C. hygrophyllum*
Capitula not subtended by any bract-like, pinnate upper leaves, if bracts present at top of pedicels resembling the phyllaries and pass into them 4
4. Leaves not or only very shortly decurrent along the stem 5
Leaves considerably decurrent along the stem, sometimes forming a wing complete to the next leaf down 6
5. Fruiting pappus greatly exceeding corolla; corolla limb divided to base or almost so . . . . 13. *C. arvense*
Fruiting pappus not or scarcely exceeding corolla; corolla divided to about halfway . . . . 9. *C. pubigerum*
6. Lower stem leaves subentire to pinnately lobed, but rarely lobed to more than halfway; corolla finally 13–17(–19) mm, lobes 4.5–5(-5.5) mm; pappus (9–)10–12(–13) mm . . . . 10. *C. elodes*
Lower stem leaves pinnately lobed to ½ to ¾; corolla finally 17–20 mm, lobes 5–8 mm; pappus 12–15 mm . . . . 11. *C. libanoticum*
7. Stem leaves distinctly decurrent, wings indicated by at least rows of spines 7. *C. vulgare*
Stem leaves not at all decurrent, stem wingless 8
8. Outer and median phyllaries with flattened stramineous spinules up to 1 mm long about the apex just below the spine . . . . 1. *C. ciliatum*
Outer and median phyllaries smooth or very minutely hirtellous or scabrid at apex, without flattened stramineous spinules 9
9. Setae of upper surface of leaves very sparsely scattered; capitulum (excluding the long, ± squamose terminal spines of the outer and median phyllaries) 2.5–4 cm in diameter . . . . 5. *C. karduchorum*
Setae of upper surface of leaves numerous to dense; capitulum (excluding the erect to squarrose spines) less than 2.5 cm in diameter 10
10. Involucre small, obconical to shortly cylindrical, 8–10 mm in diameter; stems and lower surface of leaves densely white-tomentose . . . . 3. *C. haussknechtii*
Involucre subglobose or broadly ovoid, if cylindrical and as narrow as 10 mm then stems and lower leaf surface ± glabrous 11
11. Capitula numerous at ends of stem and branches, densely congested into roundish clusters; median phyllaries 18–25 mm including slender, 5–10 mm apical spine . . . . 4. *C. sorocephalum*
Capitula solitary or few at ends of stem and branches, not densely congested; median phyllaries up to c. 15 mm 12
12. Lower and median stem leaves with 10–12 pairs of triangular lobes, lobes (excluding apical spine) to c. 15 mm long . . . . 6. *C. pseudobracteosum*
Lower and median stem leaves with up to 8 pairs of deltoid to narrowly oblong-lanceolate lobes, largest lobes (excluding spine) to at least 2 cm and attaining 4–5 cm . . . . 2. *C. lappaceum*

SECT. EPITRACHYS DC. (INCL. SECT. *Pseudepitrachys* PETRAK).

1. **Cirsium ciliatum** (*Murr.*) *Moench*, Meth. Pl.: 227 (1802); Boissier, Fl. Orient. 3: 527 (1875); Charadze in Fl. S.S.S.R. 28: 130 (1963); Petrak in Fl. Iranica 139a: 248 (1979).

Carduus ciliatus Murr., Comment. Goetting.: 35 (1784).
Cnicus ciliatus (Murr.) Willd., Sp. Pl. ed. 4, 3 (3): 1667 (1803).

Erect biennial or perennial herb, (50–)90–120(–150) cm tall; rootstock not seen. Stems striate-sulcate, angular below, pale with green furrows, sparsely to moderately arachnoid, unwinged, with few to numerous slender leafy branches which are frequently again shortly branched, each division bearing a single large erect capitulum. Basal leaves withered or broken at time of flowering. Lower and median cauline leaves oblong in outline, not at all decurrent, 8–26 × 5–10 cm excluding spines, deeply pinnatilobed with 4–8 pairs of simple or commonly bifid lateral lobes, lobes linear or linear-lanceolate, 20–50 × 4–7 mm, terminating in a stout 3.5–10 mm stramineous spine, upper surface and margins densely spinulose-setose but otherwise glabrous, lower surface whitish-arachnoid and ± tomentose. Upper leaves reducing, auriculate and semi-amplexicaul, less deeply divided but remaining strongly armed. Capitula solitary on branches and branchlets, rarely much congested, pedicels with numerous reduced upper leaves but these not obviously involucriform below the capitula. Involucre depressed-globose to broadly ovoid, 20–35(–40) mm in diameter, not arachnoid; phyllaries mostly 10–15-seriate, pale and smooth below, purplish above and ± scabrid dorsally below this purple area, apex strongly stramineous-fimbriate along margins and spinous-aristate; outer and median phyllaries with a broader, lanceolate or lanceolate-ovate squarrose apex, median sometimes merely divergent, 12–20 mm long with a 2–4 mm stramineous apical spine; inner phyllaries larger and more tapering, 22–35 mm, apex ± erect. Corolla purple or pinkish-purple, 28–38 mm, tube larger than limb, which is divided to ⅔ into 5–10 mm segments. Achenes (3.5–)4.5–5.5 mm, light brown, smooth; pappus 20–28 mm, dirty white. Fig. 12: 4.

HAB. Mountain forest, in damp spots in the *Astragalus* zone; alt. ± 1700 m; fl. & fr.
DISTRIB. Iranian frontier in the central sector of the mountain forest region; only found once. **MRO**: Haji Umran, *Gillett* 12430!

Turkey, Caucasus, Iran.

The above specimen consists of two pieces: one a leafy stem, the other a small piece of immature partial inflorescence. I am unsure whether the two belong together. The partial inflorescence is certainly *C. ciliatum*, but the leafy stem may possibly be another species, such as *C. haussknechtii*. Nor do I know if this is the specimen from Haji Umran cited by Petrak in *Fl. Iranica* as *Gillett* s.n.; it bears no Petrak determination slip.

The two subspecies recognised within this species in both Fl. Iranica and Fl. Turkey, subsp. *ciliatum* and subsp. *szovitsii* (C. Koch) Petrak, seem to be of doubtful value. The Iraqi material is in any event too young to place at this level.

2. **Cirsium lappaceum** (*M. Bieb.*) *Fisch.* in Cat. Hort. Gor. ed. 2: 35 (1812); Boissier, Fl. Orient. 3: 531 (1875); Dinsmore in Post, Fl. Syria, Palest. & Sinai ed. 2, 2: 89 (1933); Zohary in Dep. Agr. Iraq Bull. 31: 153 (1950); Charadze in Fl. S.S.S.R. 28: 101 (1963); Rawi in Dep. Agr. Iraq Tech. Bull. 14: 111 (1964); Petrak in Fl. Iranica 139a: 253 (1979).

Carduus lappaceus M.Bieb., Beschr. Länd. Kasp.: 193 (1800).
Cnicus lappaceus (M.Bieb.) M.Bieb., Fl. Taur.-Cauc. 2: 277 (1808).
Cirsium penicillatum C.Koch in Linnaea 17: 41 (1842).

Erect perennial herb, 30–100 cm tall, with several stems from a tough rootstock. Stems striate-sulcate and sometimes ± angular, thinly arachnoid to white-tomentose, unwinged, branched, branches short and slender, forming a dense raceme, or longer to form a lax panicle, branches with rather few, narrow leaves. Basal leaves withered at time of flowering. Lower and median cauline leaves sessile, oblong or broadly elliptic in outline, not at all decurrent, 10–18 × 3.5–8 cm excluding spines, deeply pinnatilobed with (3–)4–8 pairs of simple or bifid lateral lobes, lobes deltoid to linear-lanceolate or narrowly oblong-lanceolate, 5–55 × 4–12 mm (the terminal lobe often longer and larger), terminating in a rigid 5–22 mm stramineous apical spine, upper surface densely spinulose-setose and thinly to (more rarely) moderately arachnoid, lower surface ± densely tomentose. Upper leaves reducing, semi-amplexicaul, with fewer and narrower divisions but remaining strongly armed. Capitula solitary or in groups of 2–3, subtended at base by up to 5 narrow involucral leaves of variable size; involucre globose or ovoid-globose, 12–22 mm in diameter, moderately to densely arachnoid especially around tips of phyllaries; phyllaries 7–10-seriate, pale and smooth below, dark green along prominent midrib above; outer and median phyllaries lanceolate, median 8–12 mm long with a reflexed or ± squarrose, 1–2 mm, stramineous apical spine; inner phyllaries linear-lanceolate, 15–20 mm, often coloured and rarely feebly spinose at apex. Corollas whitish to pale mauve or purplish, (14–)17–21 mm, tube ± equalling limb, which is divided to halfway into 3–4 mm segments. Achenes 5 mm, pale or greyish-brown, smooth; pappus (10–)12–15 mm, dirty white.

subsp. LAPPACEUM, with long slender branches and branchlets and whitish flowers, occurs in Turkey and Transcaucasia. The species is represented in Iraq by the most widespread subspecies:

subsp. **anatolicum** *Petrak* in Tr. Tifl. Bot. Sada 12 (1): 12 (1912). Fig. 12: 5–6.

C. anatolicum (Petrak) Grossh., Fl. Kavk. 4: 182 (1934).
[*C. haussknechtii* auct. non Boiss., Petrak in Fl. Iranica 139a: 252 (1979), p.p.]

At least branchlets short, the capitula in a raceme; flowers pale mauve to pinkish-purple; capitula commonly smaller than in subsp. *lappaceum*.

HAB. On stony and rocky mountain slopes; alt. 1700–3000 m; fl. & fr. Aug.-Sep.
DISTRIB. Scattered in the central area of the mountain region of Iraq, and locally abundant. **MRO**: Hisar-i Rost, *Guest & Husham* (*Alizzi*) 15813!; Haji Umran, *id.* 15897!, 15898!, *Gillett* 12437, *Rechinger* 11311!

Turkey, Transcaucasia, Iran.

The limited Iraqi material referred to this subspecies is somewhat variable. *Guest & Husham* 15898 and *Gillett* 12437 are typical, but *Guest & Husham* 15897 from the same locality has the leaves more arachnoid above and florets and pappus short for the species; the population at Haji Umran would repay further study. The material from Hisar-i Rost has leaves less tomentose than usual beneath and not erect as in the rather similar *C. hakkaricum* Davis & Parris. I have thus assigned all these plants to the rather variable subsp. *anatolicum*.

Both *Guest & Husham* 15897 & 15898 bear the note that the plant is "eaten by all animals"; CHIFAR (Kur.) (*Guest & Husham* 15813)

3. **Cirsium haussknechtii** *Boiss.*, Fl. Orient. 3: 535 (1875); Petrak in Fl. Iranica 139a: 252 (1979).

C. straussianum Hausskn. in Mitt. Thür. Bot. Ver. 9: 18 (1891).

Erect perennial herb, 0.9–1.2 m tall, with several stems from a rather woody rootstock. Stems striate-sulcate and somewhat angular, white-tomentose with arachnoid hairs, unwinged, with numerous short and slender (the lowest sometimes longer) branches with few, narrow reduced leaves. Basal leaves withered or broken at time of flowering. Lower and median cauline leaves sessile, broadly or more narrowly oblong in outline, not at all decurrent, 12–28 × 3–13 cm excluding spines, deeply pinnatilobed with 4–8 pairs of simple or bifid lateral lobes, lobes deltoid-lanceolate to narrowly oblong, 10–55 × 6–13 mm (the terminal lobe often longer and larger), terminating in a stout 6–14 mm stramineous spine, upper surface and margins densely spinulose-setose and moderately arachnoid, lower surface densely whitish-tomentose. Upper leaves reducing, becoming semi-amplexicaul, more narrowly divided but still strongly armed. Capitula rarely solitary, usually in racemose clusters of 3–7 at ends of short branches, sessile or very shortly pedunculate, subtended at base by 1–3 involucriform narrow leaves of variable size; involucre obconical to shortly cylindrical, 8–10 mm in diameter, moderately to densely arachnoid especially around tips of phyllaries; phyllaries 6–9-seriate, closely imbricate, pale and smooth below, darkish green by prominent midrib above; outer and median phyllaries deltoid-lanceolate, median 5–8 mm long with a short, erect or slightly divergent 0.5–1 mm stramineous apical spine; inner phyllaries linear-lanceolate, 14–16 mm, coloured and not or only feebly spinose at tip. Corolla pale pink or pale purplish, 12–15 mm, tube subequalling limb, which is divided to halfway into 2.5–4 mm segments. Achenes ± 4 mm, pale brown speckled with darker streaks, smooth; pappus 7–12 mm, dirty white.

HAB. Only found in two localities in the mountain region of Iraq; alt. 1700–1820 m; fl. Aug.
DISTRIB. **MAM**: Ser Amadiya, on top of the ridge, *Guest* 4985!; ibid., open rocky valley leading to the summit, *Agnew & Haines* 2074!; **MRO**: Haji Umran, *Gillett* 15897, 15898.

Turkey (SE Anatolia), W. Iran.

Strangely, neither of the above Gillett specimens have been located at Kew, as Gillett's gatherings usually are, and I have not seen them. Petrak saw them for *Fl. Iranica*, and there seems no reason to doubt his identification.

Although Davis & Parris [Fl. Turkey 5: 392 (1975)] described the upper surface of the leaves in this species as "spinose-strigose … otherwise glabrous" it is usually ± arachnoid – and in the isotype material at Kew (the most extreme seen) whitish-tomentose.

4. **Cirsium sorocephalum** *Fisch. & C.A.Mey.*, Ind. Sem. Hort. Petrop. 33 (1835); Boissier, Fl. Orient. 3: 526 (1875); Charadze in Fl. S.S.S.R. 28: 91 (1963).

C. congestum Fisch. & C.A.Mey. ex DC., Prodr. 6: 641 (1838); Boissier, Fl. Orient. 3: 525 (1875); Charadze in Fl. S.S.S.R. 28: 90 (1963); Rawi in Dep. Agr. Iraq Tech. Bull. 14: 111 (1964); Petrak in Fl. Iranica 139a: 238 (1979).
C. congestum var. *sorocepalum* (Fisch. & C.A.Mey.) Petrak in Fl. Iranica 139a: 239 (1979).

Plant erect, probably perennial, (0.4–)0.9–1.5(2.3) m tall. Stems striate-sulcate, glabrous or very thinly arachnoid, angular below, whitish with grooves green, simple in shorter forms, more usually with long ± divergent branches, unwinged. Basal leaves apparently absent at time of flowering. Lower cauline leaves narrowed below, oblong to elliptic-oblong in outline, c. 5.5–43 × 1–13 cm excluding spines, pinnatilobed with the 4–11 pairs of deltoid to oblong lobes simple (in small forms) or up to 5-fid (in large forms), the tapering teeth excurrent in strong 5–17 mm stramineous spines, margins with smaller spines and spinules, upper surface with numerous rather unequal spinules, lower surface glabrous or rarely sparingly arachnoid. Median and upper leaves sessile and semi-amplexicaul, strongly armed, not at all decurrent, reducing upwards with mostly simple or bifid teeth, the larger passing into the lower stem-leaves. Capitula congested into terminal clusters of (3–)5–10 at ends of stem and branches, clusters subtended and interspersed by reduced, narrow, bract-like leaves with strong, spreading spines. Involucre oblong to broadly cylindrical, 10–22 mm in diameter, glabrous; phyllaries mostly 6–9-seriate, pale and smooth at base, darker and somewhat grooved on either side of midrib above; outer and median phyllaries somewhat broader and more rapidly narrowed, median 18–25 mm including the erect or arcuate-divergent, slender, 5–10 mm stramineous spine; innermost phyllaries similar or commonly more gradually tapering into a somwhat longer spine; margins of all phyllaries minutely spinulose-scabrid. Corolla pale pink or mauve, 18–25 mm, tube subequalling limb, which is divided to up to ¾. Achenes 5.5–7 mm, brown, somewhat shining; pappus 20–25 mm, dirty white.

HAB. Rocky mountain slopes and scree, ascending to the *Astragalus* zone; alt. 1000–1820 m; fl. Jul.-Aug., fr. Aug.-Oct.
DISTRIB. Widespread in the central, less so in the southern mountain regions of Iraq. **MRO**: Musaiyif Nawanda, *Guest & Husham* (*Alizzi*) 15807!; Baski Hawaran, *Rawi & Serhang* 23962!; between Rayat and Haji Umran, *Rechinger* 11785!; Haji Umran, *Qaisi, Karim & Hamad* 49799!; Serin, *Rawi & Serhang* 24010A!; Qurnago, *Serhang & Rawi* 26599! **MSU**: Sarsang, *Haines* 1849!

Turkey, Transcaucasia, Iran, Afghanistan, C Asia (Turkmenia).

Even with material named by Petrak before me I am not able sufficiently to comprehend the differences in phyllary shape said by him to characterize var. *sorocephala*; and of the eight specimens which he cites he confesses that two are near to the typical plant. I see no useful practical advantage in recognising this form at varietal rank, much less as a species as is done by Kharadze in Fl. S.S.S.R.

5. **Cirsium karduchorum** *Petrak* in Anz. Öst. Akad. Wiss., Math.-Nat. Kl. 101: 435 (1964); Davis & Parris in Fl. Turkey 5: 394 (1975); Petrak in Fl. Iranica 139a: 243 (1979).

Erect perennial herb, 75–150 cm tall, with several stems arising from base. Stems glabrous, striate-sulcate with shallow grooves, brownish, branched from about middle

Fig. 12. **Cirsium karduchorum**. 1, habit leaf × 1; 2, flowering stem × 1; 3, floret × 3. **Cirsium ciliatum**. 4, median phyllaries × 6. **Cirsium lappaceum** subsp. **anatolicum**. 5, portion of leaf × 1; 6, leaf surface × 9. 1 from *Rechinger* 11330; 2–3 from *Gillett* 12333; 4 from *Gillett* 12430; 5–6 from *Gillett* 1243. Drawn by Christine M. Grey-Wilson.

or above with long, ascending or shorter, more divergent branches, unwinged. Basal leaves not seen. Lower cauline leaves narrowed below, broadly oblong to oblong-ovate in outline, 23–35 × 12–18 cm excluding spines in the few such leaves seen, pinnatilobed to ⅔ with the 5–7 pairs of deltoid 2–4-fid lobes, the short broad teeth excurrent in a strong stramineous or pale brown 8–12 mm spine, margins with spines of varying size. Median and upper leaves rapidly reducing in size, sessile and broadly amplexicaul, very strongly armed with spines 5–20 mm long, not or only very shortly decurrent. All leaves glabrous on both surfaces or very sparingly pilose above when young, with few scattered spinules above. Capitula few, solitary on stem and (sometimes short) branches, usually subtended by one or a few narrow, spinous-margined bract-like upper leaves. Involucre broadly campanulate to subglobose, 2.5–4 mm in diameter, glabrous; phyllaries mostly 12–15-seriate, pale and smooth at base, darker about apex and base of terminal spine; outer and median phyllaries abruptly narrowed from a lanceolate-oblong base to a narrow, ± squarrose, apex excurrent in a fine but rigid, stramineous spine, median 17–27 mm including the 7–12 mm spine; inner phyllaries linear, more gradually narrowed, innermost 25–40 mm with apical spine flexuose or divergent; margins of all phyllaries minutely ciliate-denticulate. Corolla purple, 25–30 mm, lobes 5–8 mm, about ½ length of limb, which subequals tube. Mature achenes not seen; pappus about 20 mm, dirty white. Fig. 12: 1–3.

HAB. Mountain slopes and rocky hillsides up to the *Astragalus* zone; alt. 1900–2900 m; fl. & fr. Aug.-Sep.
DISTRIB. Apparently rare, only collected three times in the central area of the mountain region of Iraq. **MRO**: Algurd (Helgord) Dagh, *Gillett* 12333!; Rayat, *Guest* 13121!; Kani Sawaran range, *Rawi & Serhang* 24672!

Turkey (SE Anatolia).

Two good Iraqi specimens of *Cirsium* which I have found difficult to place are the following: **MRO**: Algurd (Helgord) Dagh, *Gillett* 9508!; Rayat, *Guest* 12121A! The latter was not cited by Petrak, but was determined by him as *C. karduchorum* with the qualifying note that it is a small-headed form approaching *C. strigosum*. The former was cited unequivocally as *C. strigosum* by Petrak in *Fl. Iranica* 139a: 257 (1979), though his *determinavit* label on the sheet is qualified by the remark that it differs from the type by the somewhat smaller, narrower capitula and the arachnoid indumentum on the lower surface of the leaves.

These two specimens are virtually identical. The capitula are indeed similar to those of *C. karduchorum* on a small scale, with their finely tipped, needle-like phyllaries of which the median to outer are patent to distinctly squarrose. They also share with *C. karduchorum* the broadly auriculate-based upper leaves of that species; the leaf-surfaces of the median and lower leaves, however, are as densely strigulose as those of *C. lappaceum*, and the segments almost as narrow. It is striking that *C. karduchorum* is found both on Helgord Dagh and at Rayat – two of its only three localities – and the two collections cited above may well be hybrids between *C. karduchorum* and *C. lappaceum* subsp. *anatolicum*. The latter species, admittedly, has not yet been recorded from either locality, but there is no doubt that the thistles are under-collected in Iraq. In my judgement, the broadly auriculate stem leaves, very fine and patent outer phyllaries and arachnoid lower leaf surfaces isolate these specimens from *C. strigosum*, which I cannot therefore recognise as an Iraqi species. *Cirsium* is a genus prone to hybridisation; pollen taken from *Gillett* 9508 and examined in Kew's palynology department proved to have 30% sterility, and 3% for *Guest & Husham* 15898 (*C. lappaceum* subsp. *anatolicum*) and 12% for *Rawi & Serhang* 24672 (*C. karduchorum*), which seems to support the possibility of hybrid status for these two plants. No pollen was found on *Guest* 13121A, which is at a much earlier state of development of the capitula.

6. **Cirsium pseudobracteosum** *Davis & Parris* in Notes Roy. Bot. Gard. Edinburgh 33: 419 (1975) & Fl. Turkey 5: 395 (1975).

[*C. bracteosum* auct. non DC., Petrak in Fl. Iranica 139a: 245 (1979), p.p.]

Perennial herb with numerous stems from base, 50–90 cm tall. Stems glabrous or sparingly arachnoid, striate-sulcate, whitish with shallow greenish grooves, sparingly branched above with slender, ascending or shorter, thicker branches, unwinged. Basal leaves narrowed to base, linear in outline, 20 × 2.5–3 cm excluding spines, pinnatilobed to about ¾ with 10–12 pairs of simple to trifid, triangular lobes, lobes excurrent in a rigid stramineous spine

c. 5–10 mm long, spinules on upper surface very few. Median and lower cauline leaves sessile, not decurrent, subamplexicaul, venation very prominent below, progressively shorter than basal leaves and with fewer lobes, glabrous or somewhat arachnoid on lower surface of lamina and midrib and also on upper surface, where the spinules are rather sparse and rarely exceed 1 mm long, terminal spines of the patent or slightly forward-directed mostly simple or bifid (sometimes trifid) lobes up to 15 mm. Upper leaves considerably reducing but very strongly armed. Capitula solitary or in clusters of up to 5, subtended and usually considerably exceeded by several narrow, strongly spinous-margined bract-like upper leaves. Involucre subglobose, 1.5–2 mm in diameter, glabrous to thinly arachnoid; phyllaries mostly 5–7-seriate, olive-green to brownish and smooth below, darker towards apex; outer and median phyllaries narrowly oblong-lanceolate, rather quickly narrowed to a fine, erect or divergent spine, median 9–14 mm including 3–15 mm spine; innermost phyllaries linear, more gradually narrowed, 16–20 mm with a slender flexuose point; margins of phyllaries minutely ciliate-denticulate. Corollas purple to pinkish-lilac or white, 12–18 mm, lobes 3–5 mm, about ½ length of limb, which subequals the tube. Achenes about 5 mm, brown, rather shining; pappus 10–13 mm, dirty white.

HAB. By a roadside and on overgrazed slopes in the *Astragalus* zone; alt. 2500–2800 m; fl. & fr. Aug.-Sep.
DISTRIB. Apparently local in the central area of the mountain region of Iraq. **MRO**: Ser Kurawa, *Gillett* 9769!; Algurd (Helgord) Dagh, *Rechinger* 11468, 11863; road between Bardanas and the Qandil range, *Rawi & Serhang* 24613!; Pushtashan, *Rechinger* 15492!; Qandil range nr Germasur lake, *Rechinger* 11117!

Turkey, Iran.

Differs from *C. bracteosum* in its leaves with more numerous and thus more crowded leaf segments, with less wide sinuses between which are rounded rather than square, the upper pair of segments being less decurrent in appearance and the terminal smaller; the upper leaf surface is more sparingly spinulose than in *C. bracteosum* with the spinules rarely 1 mm.

7. **Cirsium vulgare** (*Savi*) *Ten.*, Fl. Nap. 5: 209 (1835–6); Charadze in Fl. S.S.S.R. 28: 135 (1963); Davis & Parris in Fl. Turkey 5: 397 (1975); Petrak in Fl. Iranica 139a: 259 (1979); Meikle, Fl. Cyprus 2: 954 (1985).

Carduus lanceolatus L., Sp. Pl.: 821 (1753).
Cirsum lanceolatum (L.) Scop., Fl. Carn. ed. 2, 2: 130 (1772); Boissier, Fl. Orient. 3: 538 (1875); Dinsmore in Post, Fl. Syria, Palest. & Sinai ed. 2, 2: 90 (1933); Zohary in Dep. Agr. Iraq Bull. 31: 153 (1950); Rawi, ibid. Tech. Bull. 14: 111 (1964), non Hill, Herb. Brit. 1: 80 (1769).
Carduus vulgaris Savi, Fl. Pis. 2: 241 (1798).
Cirsium lanceolatum var. *hypoleucum* DC., Prodr. 6: 636 (1838).

Erect biennial herb, (30–)50–150(–180) cm, with a long tough rootstock. Stems striate-sulcate and usually ± angular, pale with green furrows, sparingly or usually moderately arachnoid, variably winged, branched from about the middle upwards with short or more commonly rather long and slender branches. Basal leaves mostly ± withered at time of flowering, oblong to lanceolate-ovate in outline, narrowed and shortly petiolate below, 15–30 × 8–15 cm excluding spines, deeply pinnatilobed with 3–6 pairs of deltoid to deltoid-lanceolate mostly 2(–3)-fid lateral lobes, lobes and teeth terminating in strong, mostly 5–15 mm, stramineous spines; upper surface with dense setae of variable length, mostly to 3 mm, but with scattered or numerous spinous ones at or near margins; lower surface moderately arachnoid to tomentose. Lower and median cauline leaves similar but reducing in size upwards and sessile, often with a larger terminal lobe, distinctly but variably decurrent, sometimes fairly shortly so or with lower part of wing indicated only by a few spines, sometimes (even on the same plant) the wing complete along the internodes. Upper leaves considerably reducing, sessile but not semi-amplexicaul, very strongly armed with spines up to 20 mm. Capitula erect, solitary or 2–3 on branches and branchlets, only congested when young, occasionally with one or two linear bract-like leaves below but not genuinely involucrate. Involucre ovoid or subglobose, 2.5–4(–4.5) cm in diameter, usually moderately arachnoid; phyllaries 8–11-seriate, pale olivaceous below, darker green at base of stramineous terminal spine; outer and median phyllaries narrowly lanceolate, gradually tapering to a ± spreading but not squarrose apex, median 15–20 mm including the 2.5–5 mm spine; inner phyllaries narrowly linear-lanceolate, 28–35 mm, tip erect and not or only feebly spinose. Corollas purple or pinkish-purple, rarely white, 25–34 mm, limb much shorter than tube, divided to halfway with 5–7 mm lobes. Achenes 3.5–4.5 mm, yellowish or brownish with darker markings, pappus 19–30 mm, dirty white.

HAB. On mountain slopes, by streams in shade of *Quercus* forest or *Juglans*, along irrigation ditches or at valley edges; alt. 700–1150 m; fl. Jul.-Aug., fr. Aug.-Sep.
DISTRIB. Occasional in the northern and central mountain regions of Iraq. **MAM**: Sikrun near Sarsang, *Haines* 1586!; Sarsang, *Haines* 1599!; 25 km from Zakho on way to Kani Masi, *Botany staff* 43831! **MRO**: Pushtashan, *Rawi* 23883!, *Rawi & Serhang* 26586!

SECTION **CIRSIUM**

8. **Cirsium canum** (*L.*) *All.*, Fl. Pedem. 1: 151 1785); Boissier, Fl. Orient. 3: 541 (1875); Charadze in Fl. S.S.S.R. 28: 169 (1963); Davis & Parris in Fl. Turkey 5: 400 (1975); Petrak in Fl. Iranica 139a: 272 (1979).

Carduus canus L., Mant. Pl.: 105 (1767).

Perennial herb, 40–200 cm tall, with fleshy fusiform-cylindrical clustered roots. Stems striate-sulcate, partially winged, more deeply grooved below, ± cottony-arachnoid, moderately leafy below, simple or usually with several long, slender branches above bearing a solitary capitulum at apex and clothed with a few bract-like leaves or none, or capitula up to 3 on shorter pedicels at tops of such branches. Basal leaves attenuate to base, 14–38 × 1.5–5 cm, narrowly elliptic to oblanceolate, acuminate at apex, serrate with broad-base patent or ascending teeth with an apical arista up to 8 mm long with smaller teeth and cilia between, subglabrous to ± arachnoid on both surfaces. Median cauline leaves similar in form but sessile and long-decurrent, also subglabrous to arachnoid, major teeth sometimes longer and narrower and occasionally bifid. Upper leaves rapidly reducing to the bract-like leaves of the branches. Capitula not subtended by leaves. Involucre broadly campanulate to subglobose, glabrescent or sparingly floccose especially along phyllary margins, 18–25 mm in diameter; phyllaries mostly 7–9-seriate, olive-green below with a black apical vitta; outer phyllaries broadly ovate-lanceolate, shortly spinous-mucronate, 3–5 mm, tip purplish-black; median lanceolate-ovate, 7–9 mm, also dark-tipped and shortly mucronate; innermost phyllaries linear-lanceolate with a flexuose to reflexed, membranous, purplish tip, 16–18 mm. Corollas purple, 18–25 mm, limb divided to more than halfway into very narrow, 6–9 mm segments. Achenes ± 4 mm, brownish, somewhat shining; pappus 13–16 mm, dirty white.

HAB. In damp shaded area; only found, but said to be "common"; alt. ± 560 m; fl. & fr.
DISTRIB. **MSU**: 5 km from Sulaimaniya to Sayed Sadiq, *Omar, Qaisi & Wedad* 49375!

C & S Europe from Poland and Austria to C Italy and S Russia, Turkey, Iran, Transcaucasia, Siberia.

9. **Cirsium pubigerum** (*Desf.*) *DC.*, Prodr. 6: 650 (1838); Boissier, Fl. Orient. 3: 544 (1875); Davis & Parris in Fl. Turkey 5: 403 (1975).

Cnicus pubigerus Desf., Cat. Hort. Par. ed. 3: 399 (1829).
Cirsium depilatum Boiss. et Bal. in Boissier, Fl. Orient. 3: 543 (1875); Charadze in Fl. S.S.S.R. 28: 185 (1963).

Perennial herb 1–1.25 m tall. Stems striate-sulcate, angular, glabrous or more commonly thinly arachnoid, leaves not or only very shortly decurrent along stem, not forming a distinct ring. Branches few to numerous, long and slender, leaves not at all decurrent. Basal leaves (only two seen) gradually tapering into a narrowly winged petiole, oblanceolate, pinnatilobed to less than half with broad, mostly deeply bifid lobes, or lobed to ¾ with up to 4-fid lobes, teeth ending in stout stramineous spines up to 10 mm long, margins with smaller spines. Median cauline leaves broadly lanceolate to oblong-elliptic in outline, 7–18 × 2.5–8 cm, shallowly sinuate-lobed to deeply pinnatifid, lobes simple to 4-fid, broadly triangular, glabrous or with scattered hairs particularly along primary venation, narrowly tapering teeth ending in strong stramineous spines up to 8 mm long, sinuses between lobes rounded to broad and square. Upper cauline and branch leaves shorter and smaller with a narrower lamina but strongly armed. Capitula commonly but not invariably subtended by at least one sessile, pinnatipartite, involucral leaf differentiated from phyllaries; involucre hemispherical, 15–20 mm in diameter, glabrous or sparingly arachnoid, especially around margins of tips of phyllaries; phyllaries 5–7-seriate, with a long, narrow, black apical vitta usually especially clear in the median phyllaries; outer phyllaries broadly lanceolate-ovate, 3–4 mm with a spinous mucro 1–2 mm long; median phyllaries lanceolate-ovate, 6–7 mm with a mucro 1–3 mm long; innermost phyllaries linear, 15–18 mm with a firm, flexuous, generally purplish tip. Corolla purplish-pink, 18–22 mm, with lobes 5–7 mm and about half length of limb. Achenes 2.5–4 mm, pale to dark brown, somewhat shining; pappus 14–17 mm, dirty white.

Distrrib. (of species): Turkey, Iran, Transcaucasia.

Cnicus pubigerus was described from cultivated material of unknown origin. The type specimen, which must be regarded as representing var. PUBIGERUM, is, according to Davis & Parris [Fl. Turkey 4: 405 (1975)] a plant with a very lax inflorescence with few, large, long-pedunculate capitula, and is unusually arachnoid. In Iraq the species is represented by:

var. **spinosum** *Petrak* in Mitt. Thür. Bot. Ges. 2: 19 (1960).

C. iraqense Petrak in Anz. Öst. Akad. Wiss., Math.-Nat. Kl. 109: 172 (1972); Petrak in Fl. Iranica 139a: 273 (1979).

Capitula of moderate size, 15–18 mm in diameter; leaves pinnatifid to ½-¾ with rather narrow, acute lobes, glabrous or usually somewhat arachnoid on lower surface, especially around primary venation.

HAB. On hill slopes, often under trees and by streams, on stony or rocky clay soil, locally frequent; alt. 700–2800 m; fl. & fr. Jul.-Sep.
DISTRIB. Occasional in the central mountain region of Iraq. **MRO**: Chiya-i Mandali, *Guest* 2736!; Algurd (Helgord) Dagh, *Guest* 1584!, *Rechinger* 11864; Pushtashan, *Rawi* 23882!, *Rechinger* 11080.

Turkey (SE Anatolia), W Iran.

In *Fl. Turkey*, Davis & Parris record subsp. *caniforme* Petrak from Iraq also, but I have not seen any Iraqi material which I would refer to this subspecies, nor any Iraqi specimen so named by them.

10. **Cirsium elodes** *M.Bieb.*, Fl. Taur.-Cauc. 3: 555 (1819); Boissier, Fl. Orient. 3: 547 (1875); Zohary in Dep. Agr. Iraq Bull. 31: 153 (1950); Charadze in Fl. S.S.S.R. 28: 191 (1963); Rawi in Dep. Agr. Iraq Tech. Bull. 14: 111 (1964); Davis & Parris in Fl. Turkey 5: 407 (1975); Petrak in Fl. Iranica 139a: 271 (1979).

Perennial herb, 35–110 cm, with clustered, slender or somewhat incrassate roots. Stem striate-sulcate, slightly to ± densely arachnoid-tomentose, leaves decurrent along stem to form ± continuous sinuosely lobed wings, wings spinulose or with strong yellowish spines up to 8 mm long, narrower towards base of stem and sometimes reduced to a narrow spinous border above often indicated by spinous lobes. Branches numerous, long and ascending with very narrow spinous-lobed wings and sometimes small leaves, sometimes again branched, bearing solitary or usually clusters of 2–5 sessile or shortly pedunculate capitula at apex. Basal leaves variable in size, oblanceolate to narrowly elliptic, ± long-attenuate at base, entire and denticulate to pinnatilobed almost to axis with broadly toothed, spine-tipped lobes and usually smaller marginal spines, densely floccose on both surfaces to sparsely arachnoid (though usually more densely so beneath). Median cauline leaves narrowly- to oblong-elliptic or narrowly oblong-ovate, subentire and denticulate or pinnatilobed (very rare to as much as halfway to axis), 8.5–19 × 2.5–4.5 cm, armature and indumentum similar to that of basal leaves. Upper cauline leaves with lamina much reduced but the yellowish spines less so. Capitula not subtended by sessile involucral leaves strongly dissimilar to phyllaries. Involucre hemispherical, rounded at base; phyllaries mostly 6–7(–8)-seriate, all with a blackish terminal vitta, glabrous or thinly arachnoid especially marginally and towards apex, 8–12 mm in diameter; outer phyllaries ovate-lanceolate, 2–4 mm with a spinous mucro up to 2 mm long; median phyllaries narrowly ovate-lanceolate, 5–6 mm with a 0.5–2 mm mucro; innermost phyllaries linear-lanceolate, (11–)12–15(–17) mm, with a ± membranous, often purplish, ± undulate tip. Corollas mauve to purplish-pink, (13–)14–17(–19) mm with lobes 4.5–5.5 mm, about half length of limb. Achenes 3 mm, compressed, stramineous, oblong, smooth; pappus finally (9–)10–12(–13) mm, dirty white.

HAB. On limestone mountains, on slopes or in *Quercus* forest, by streams; alt 610–1100 m; fl. Jun.-Aug., fr. Sep.-Oct.
DISTRIB. Occasional in the N, C and probably the S mountain regions of Iraq. MAM: Sarsang, *Barkley* 9051, *Anders* 2339. MRO: Rowanduz gorge, *Guest* 454!, *Gillett* 12419!; below Rowanduz, *Rechinger* 11253; Pushtashan, *Serhang & Rawi* 26499!; Pushtashan, NE of Rania, lower slope of Qandil Range, *Rawi & Serhang* 26499 (BAG!); Haji Omran, *Gillett* 12446, *Rawi* 24943!; 20 km NNE of Kirkuk – on road to Koisanjaq, *Rawi* 26451A (BAG!). MSU: (or Iran): "in montibus calcareis Avroman et Schahu", *Haussknecht* s.n.!

Turkey (E Anatolia), W Iran, Transcaucasia.

11. **Cirsium libanoticum** *DC.*, Prodr. 6: 647 (1838); Boissier, Fl. Orient. 3: 548 (1875); Dinsmore in Post Fl. Syria, Palest. & Sinai ed. 2, 2: 90 (1933); Rawi in Dep. Agr. Iraq Tech. Bull. 14: 111 (1964); Davis & Parris in Fl. Turkey 5: 408 (1975).

C. apiculatum DC., Prodr. 6: 642 (1838).
C. kotschyanum Boiss., Diagn. Pl. Orient. ser. 1, 6: 103 (1846).

Robust perennial herb, 1–3 m. Stems striate-sulcate, angular, glabrous to moderately arachnoid, leaves practically continuously decurrent to form a complete strongly spinose wing along internodes, though wing is sinuate and sometimes reduced to a narrow margin or row of spines. Branches numerous, long and ascending, strongly armed in a manner similar to stem. Basal leaves gradually tapering to a distinct petiole, narrowly elliptic in outline, sinuate-lobed to less than ½ with broad, toothed lobes, teeth ending in spines up to 5 mm long, margins with smaller spines. Median cauline leaves narrowly elliptic or elliptic-oblong in outline, pinnatilobed to ½ to ¾, 6–30 × 3–11 cm, lobes triangular to broadly lanceolate, simple to trifid, lobes and teeth terminating in strong spines to 14 mm long, sinuses between lobes broad and square to narrower and more rounded, leaf surfaces glabrous or very sparingly arachnoid to sparingly arachnoid above and whitish-subtomentose below. Upper cauline and branch leaves shorter and smaller with a narrower lamina, but remaining strongly and stoutly spinose. Capitula not subtended by sessile involucral leaves strongly dissimilar to the phyllaries. Involucre hemispherical or somewhat more narrowed at base, 7–10 mm in diameter excluding spines, glabrous or thinly arachnoid, or more densely so along phyllary margins; phyllaries 6–8-seriate, all with a black terminal vitta; outer phyllaries ovate to broadly ovate-lanceolate, 5–7 mm with a spinous mucro 1–3 mm long which is commonly divergent or deflexed in median and upper phyllaries; innermost phyllaries linear, 13–18 mm, with a ± membranous, often purplish, ± undulate tip. Corolla purplish-pink, 17–20 mm with lobes 4.5–7 mm, about half length of limb. Achenes 3 mm, stramineous, smooth; pappus finally (12–)13–15 mm, dirty white.

DISTRIB. (of species): Turkey, Syria, Lebanon, Iran, C Asia (Turkmenia).

subsp. LIBANOTICUM, which occurs in Syria, Lebanon and western Iran, is characterised by having the leaves (including the underside) glabrous or very thinly arachnoid, and the wings and leaves very strongly armed with spines to 10 mm or more. In Iraq the species is represented by:

subsp. **arachnoideum** *Davis & Parris* in Notes Roy. Bot. Gard. Edinburgh 33: 419 (1975).

[*C. alatum* non (S.G.Gmelin) Bobrov: Petrak in Fl. Iranica 139a: 269 (1979), saltem quoad sp. *Haines* 421]
[*C. uliginosum* non M.Bieb.: Petrak, ibid.: 275 (1979)]

Leaves moderately arachnoid to whitish-subtomentose on lower surface, wings and leaves less strongly armed (in Iraq the spines only to 6(–8) mm in length). Capitula somewhat less broadly rounded at base.

HAB. (of subsp.). Streamsides and mountain slopes, ascending to the *Astragalus* zone; alt. 800–1900 m; fl. & fr. Jul.-Sep.
DISTRIB. (of subsp.). Occasional in the N & C mountain regions of Iraq, are in the adjacent foothills, frequency not noted by any collector. MAM: nr Gara, *Kotschy* 421!; Sarsang, *Haines* 421!; Bel-Kaif, ± 70 km NE of Mosul, *Hossain* s.n.! MRO: Haji Umran, *Gillett* 12446!, *Rawi* 24943!; Kani Sawaran range, near Alana village, *Rawi & Serhang* 24669! FAR: Khaznah, *Haley* 243.

Turkey, Syria.

Although *Haines* 421 was cited by Petrak as *C. alatum* in *Fl. Iranica*, the Kew sheet of this number bears his determination as *C. apiculatum* DC. (a synonym of *C. libanoticum*), which is what it seems to be. Davis has also determined it as his subsp. *subarachnoideum*.

Kotschy 421 was distributed by Hohenacker as "C. uliginosum M.B. var. longepedunculata C.H. Schultz, Bip. ms.". It was cited by Petrak (the only specimen which he does cite) as *C. uliginosum*, and P.H. Davis has also named the Kew sheet of this gathering as *C. uliginosum*. In spite of having to contradict these authorities, I believe this gathering – which is very big and lax – to be a form of *C. libanoticum* subsp. *arachnoideum*, probably drawn up by growing in rank vegetation. The wings formed by the decurrent leaves are complete along the internodes even though narrow and sometime present only as a row of spines. In *C. uliginosum*, to judge from more recent Caucasian material, the leaves are more shortly decurrent, rarely to half the length of the stem internodes. Petrak in fact uses the character of an incomplete wing in his key to *Cirsium*, and because of it *Kotschy* 421 keys out to *C. libanoticum*, not *C. uliginosum*. The corolla length in *Kotschy* 421 is on the long side for *C. libanoticum*, but *Rawi & Serhang* 24669 is equally long (20 mm). The inner phyllaries of *Kotschy* 421 also attain 19 mm, while *C. libanoticum* usually has these processes at 13–18 mm, but this can scarcely be of great significance.

12. **Cirsium hygrophilum** *Boiss.*, Diagn. Pl. Orient. ser. 1, 10: 89 (1849); Boissier, Fl. Orient. 3: 549 (1872); Charadze in Fl. S.S.S.R. 28: 187 (1963) – indexed as *C. hydrophylum*; Petrak in Fl. Iranica 139a: 274 (1979).

Probably perennial, stature not known but almost certainly at least 1 m. Stems striate-sulcate, angular at least below, moderately to rather densely clothed with multicellular arachnoid hairs, leaves rarely sufficiently decurrent along stem to form a complete wing along internodes, wing strongly yellowish-spinose. Branches slender, ascending, sometimes again branched, terminated by (1–)2–4 sessile or shortly petiolate capitula, leaves similar to those of upper stem. Basal leaves not seen. Median stem leaves narrowly elliptic or elliptic-oblong in outline, 10–20 × 6–10 cm, deeply pinnatilobed to about ⅔, lobes narrowly triangular and very acute, often further forked, lobes and secondary lobes terminating in very long, stout, yellowish spines up to 15 mm long, margins with small spines, thinly or more densely clad with arachnoid hairs on midrib and primary venation of lower surface (more rarely on surface of lamina also), glabrous or thinly hairy above. Upper stem and branch leaves shorter and smaller with a narrower lamina but equally strong spines rendered more conspicuous by the lamina-reduction, very shortly decurrent. Capitula subtended by at least one, usually two or more reduced leaves with a narrow lamina and long, strong yellow spines. Involucre hemispherical, rounded at base; phyllaries mostly 5–7(–8)-seriate, all with a long, black terminal vitta, glabrous or thinly arachnoid dorsally, more densely so along margins of phyllaries, 12–14 mm in diameter excluding spines; outer phyllaries ovate-lanceolate, 2–4 mm with a spinous mucro 2–4 mm long; median phyllaries ovate-lanceolate, 5–8 mm with a 2–3 mm mucro; innermost phyllaries linear-lanceolate, 16–18 mm with a ± membranous, often purplish, ± undulate tip. Corollas purplish-pink, 18–21 mm with lobes 6–7 mm and about ⅔ length of limb. Mature achenes not seen; pappus 14–16 mm, dirty white.

HAB. On a mountain slope?; alt. ± 900 m.
DISTRIB. Recorded only once in Iraq. **MAM**: Sarsang, *Haines* s.n.

Although Haines's herbarium is at Edinburgh (E), with many duplicates at Kew, the above specimen has not been found in either. It was cited by Petrak in his *Fl. Iranica* account, and since other material so named by him has proved correct there seems no reason to suspect the record.

SECTION **CEPHALONOPLOS** (*Necker*) *DC.*

13. **Cirsium arvense** (*L.*) *Scop.*, Fl. Carn. ed. 2, 2: 126 (1772); Boissier, Fl. Orient. 3: 552 (1875); Dinsmore in Post, Fl. Syria, Palest. & Sinai ed. 2, 2: 91 (1933); Zohary in Dep. Agr. Iraq Bull. 31: 154 (1950); Charadze in Fl. S.S.S.R. 28: 213 (1963); Rawi in Dep. Agr. Iraq Tech. Bull. 14: 111 (1964); Rechinger, Fl. Lowland Iraq: 645 (1964); Davis & Parris in Fl. Turkey 5: 410 (1975); Petrak in Fl. Iranica 139a: 277 (1979).

Serratula arvensis L., Sp. Pl. ed. 1: 820 (1753).
S. setosa Willd., Sp. Pl. ed. 4, 3: 1664 (18030.
Cirsium setosum (Willd.) M.Bieb., Fl. Taur.-Cauc. 3: 560 (1819); Charadze in Fl. S.S.S.R. 28: 210 (1963).
C. arvense var. *setosum* (M.Bieb.) Hoh. in Bull. Soc. Imp. Nat. Mosc. 11: 282 (1838), repr. Enum. Pl. Talysch: 52 (1838).

Perennial herb, (20–)30–120(–150) cm, with an initial long taproot producing an underground system of slender, widely creeping, whitish roots sending up numerous sterile and flowering shoots. Stems striate, angular at least below, glabrous or (especially above) floccose, branched mostly above middle, usually with slender, leafy branches; leaves not much decurrent along stem, which is thus not spinous-winged. Basal leaves occasionally somewhat congested but not rosette-forming. Lower and median cauline leaves narrowly to broadly elliptic or elliptic-oblong in outline, very variable in size but mostly 5–17 × 1.5–7 cm, entire to pinnatisect, when entire spinous-ciliate, when divided with broadly triangular strongly spinous lobes (spines to 5 mm), with margins spinous-ciliate between; median and upper leaves diminishing in size, median sometimes shortly decurrent; all leaves glabrous on both sides to sparsely floccose above and densely white-tomentose on lower surface. Capitula not involucrate with leafy bracts, several to numerous, solitary or ± clustered on branches. Involucre hemispherical to broadly ovoid, 10–17 mm in diameter, glabrous to sparingly floccose; phyllaries 5–7–seriate, with a black terminal vitta; outer phyllaries ovate, 2–3.5 mm, with a spinous mucro 0.5–1 mm long; median phyllaries 7–10 mm, lanceolate-ovate, similarly mucronate-tipped; innermost phyllaries linear-lanceolate, 15–26 mm, they and all rows from median inwards with a ± membranous, frequently purplish undulate tip. Corollas purplish, 12–20 mm long with lobes 3–5 mm, tube long and narrow. Achenes 2.5–3.5(–4) mm, oblong, smooth, pale to dark brown; pappus (of mature fruit) 20–25 m, dirty white.

DISTRIB. (of species). Practically throughout Europe, Cyprus (probably introduced), Turkey, Caucasia, Iran, Afghanistan; introduced into N America and elsewhere.

var. ARVENSE, with leaves ± glabrous above and below and usually ± pinnately lobed or sinuate-pinnatifid, is not represented in Iraq. The only specimen of this species seen which may of Iraqi origin is referable to:

var. **vestitum** *Wimmer & Grabowski*, Fl. Siles. 2, 2: 92 (1829).

C. incanum (S.G.Gmel.) Fisch. ex M.Bieb., Fl. Taur.-Cauc. 2: 267 (1808); Charadze in Fl. S.S.S.R. 28: 211 (1963).
C. arvense var. *incanum* (S.G.Gmel.) Ledeb., Fl. Ross. 2: 735 (18460; Petrak in Fl. Iranica 139a: 279 (1979).
C. arvense var. *hypoleucum* Boiss., nomen in sched. Kotschy Iter cilic.-kurd. 1958, no. 290.
C. arvense subsp. *vestitum* (Wimmer & Grabowski) Petrak in sched. (1913); Davis & Parris in Fl. Turkey 5: 410 (1975).
C. arvense subsp. *incanum* (S.G.Gmel.) Iljin, Fl. Yugo-Vost. S.S.S.R. 6: 408 (1936).

Leaves densely white-tomentose on lower surface, margins denticulate or sinuate, rarely pinnatifid.

HAB. & DISTRIB. (of var.). Only one specimen has been seen which might possibly be from Iraq: Kotschy's "Pl. Mesopot. Kurdistan & Mossul" no. 337, or which could equally well be from Turkey or Syria.

Distributed through much of the native area of the species but predominantly southern and eastern.

This species was also recorded from Iraq by Anthony in Notes Roy. Bot. Gard. Edinburgh 18: 291 (1935) from Amara, but such a southerly record cannot be accepted in the absence of a voucher specimen.

Davis [Fl. Turkey 5: 395 (1975)] stated that *Cirsium aduncum* Fisch. & C.A.Mey. ex DC. subsp. *aduncum* occurs in northern Iraq. Petrak does not record it from Iraq in *Fl. Iranica*, and I have not seen a specimen. There is a specimen at Kew from "Nestorian Mountains & Gawar", 1857, *Capt. Garden* and determined by Petrak as "*Cirsium haussknechtii* nahert sich dem *C. aduncum* F. et M. ". Someone has written "Iraq" in pencil by the locality, perhaps believing Gawar to be Gawara (MRO). However, Garden's collection was made in Asia Minor – see Kew Bull. 1901: 25 (1901).

Petrak [Fl. Iranica 139a: 277 (1979)] mentioned *Cirsium rhizocephalum* C.A.Mey. for northern Iraq, but he cites no specimen and I have not seen one.

14. **PICNOMON** Adans., *nom. cons.*

Fam. Pl. 2 : 116 (1763)
Susanna & Garcia-Jacas in Kubitzki, Fam. Gen. Vasc. Pl. 8: 132 [2006] (2007)
Acasma Vaill., (1754), nom. rej.

Frances G. Davies

Spiny herb with continuously winged stems. Leaves shallowly lobed with prominent straw-like spines. Capitula solitary or in corymbs, surrounded by bract-like leaves; involucre ovoid to cylindrical, densely cottony; phyllaries multiseriate; receptacle long-pilose. Florets all similar, tubular, corollas pinkish-purple or whitish, unequally 5–lobed, gradually expanded from base without clear division into tube and limb. Achenes ellipsoid to narrowly obovoid, compressed, light brown, shining, with an apical pit and central umbo, hilum slightly eccentric; pappus of several series of off-white plumose bristles, connate at base, deciduous.

A monotypic genus.

Picnomon (from the Gr. name of a plant now known as *Picomon acarna*).

1. **Picnomon acarna** (*L.*) *Cass.*, Dict. Sci. Nat. ed. 2, 40: 188 (1826); Blakelock in Kew Bull. 4: 60 (1949); Kupicha in Fl. Turkey 5: 412 (1975); Feinbrun-Dothan, Fl. Palaest. 3: 380 (1978); Petrak in Fl. Iranica 139a: 283 (1979); Breckle, Hedge, Podlech & Rafiqpoor, Vasc. Pl. Afghan. checklist: 180 (2013); Taifour & El-Oqlah, Pl. Jordan Ann. Checkl.: 45 (2017).

Carduus acarna L., Sp. Pl.: 820 (1753).
Cnicus acarna (L.) L., Sp. Pl. ed. 2: 1158 (1762).
Cirsium acarna (L.) Moench, Meth. Pl. Suppl.: 226 (1802); Handel-Mazzetti in Ann. Naturh. Mus. Wien 27: 444 (1913); Dinsmore in Post, Fl. Syria, Palest. & Sinai ed. 2,2: 91 (1933); Anthony in Notes Roy. Bot. Gard. Edinburgh 18: 291 (1935); Bornmüller. in Beih. Bot. Centralbl. 60B: 207 (1939); Zohary in Dep. Agr. Iraq Bull. 31: 154 (1950); Tamamshian in Fl. S.S.S.R. 28: 218 (1963); Rawi in Dep. Agr. Iraq Tech. Bull. 14: 111 (1964); Rechinger. Fl. Lowland Iraq: 645 (1964).

Annual herb 0.3–0.6 m. Stems whitish-arachnoid, all except basal part winged with decurrent leaf bases, wings regularly ciliate-spinous or spinulose. Leaves oblong-linear, sessile, shallowly lobed, lobes terminating in a 5–15 mm yellow spine, those at centre of stem 60–140 × 6–12 mm excluding lobes, margins ciliate-spiny, both surfaces densely greyish arachnoid-hairy, lower surface conspicuously reticulate with an intramarginal longitudinal vein. Capitula solitary or in ± dense corymbs, surrounded and exceeded by a ring of 6–8 bract-like leaves; involucre 20–30 × 8–10 mm, of many rows of closely imbricate phyllaries; phyllaries multiseriate, gradate, imbricate, densely arachnoid, linear-oblong, outermost spinous-mucronate, median tipped with yellow divergent pectinate spines, innermost acute, linear, membranous-tipped, apex reddish. Corollas 16–24 mm, two longer lobes 4–6 mm, two shorter 2.5–4 mm. Achenes 5–6 mm, pappus 15–20 mm.

HAB. In cultivated fields and on hillsides, by roadsides and in *Quercus* forest, on light soils and on rocky ground; alt. ± 20–1700(–1800) m; fl. & fr. Jul.-Aug. in the hills, Jun.-Jul. on the plains.
DISTRIB. Quite common in the lower mountains, occasional in the steppe and desert regions. **MAM**: Jabal Bekher, nr Zakho, *Rawi, Noori & Tikriti* 28968!; Khantur, nr Zakho, *id.* 28968!; Ser Amadiya, above Gali Mazurka, *Guest* 4996!; between Zawita and Atrush, *Mosharraf Hussain* 629!; Badi, 10 km ENE of Dohuk, *id.* 4421!; Sarsang, *Haines* 425. **MRO**: Gali Ali Beg, *Gillett* 12415!; between Arbil and Rowanduz, *Bornm.* 1458!; Rust, *Guest & Husham (Alizzi)* 15846!; Qurnago, N of Pushtashan, *Rawi & Serhang* 26602!; between Pushtashan and Surad, *id.* 26529!; Hab as-Sultan Dagh, *id.* 26769!; Chinaruk, on road to Qandil, *id.* 26471!; Haji Umran, *Rechinger* 11843. **MSU**: Sulaimaniya, *Guest* 12922!; Penjwin, *id.* 12934!; **MRO/FAR**: between Arbil and Shaqlawa, *Rawi, Chakraverty, Alizzi & Noori* 19682! **FPF**: Kani Baz, *Rawi & Serhang* 24855! **FAR**: Arbil, *Hadač et al.* 5747. **FNI**: 0.5 km NE of Mindan bridge, *E. Chapman* 26209! **FPA/LEA**: 78 km NE of Fakka, *Hazim* 30656! **LCA**: Karrada, Baghdad, *H.J. Paranjpye* s.n.! (comm. Graham). **LCA**: Hilla, *id.* s.n.! **LSM**: 30 km S of Amara, *Hazim & Noori* 30651! **LSM/LBA**: 45 km E of Kut towards Al Amara, *Rechinger* 9697!

Mediterranean Europe (Portugal to S Russia and Romania), Cyprus, Palestine, Syria, Lebanon, Turkey, Iran, Caucasia, Afghanistan, C Asia, N Africa (Algeria).

15. **SILYBUM** Adans.

Fam. Fl. 116 (1763)
Susanna & Garcia-Jacas in Kubitzki, Fam. Gen. Vasc. Pl. 8: 133 [2006] (2007)

C. C. Townsend

Robust biennial thistle with a basal rosette, stems unwinged. Leaves lobed, spiny. Capitula solitary, homogamous; involucre globose; phyllaries in several series, outer foliaceous, spiny with terminal spine very strong, innermost lanceolate, entire. Receptacle fleshy, densely setose. Corolla purple (rarely white), tube very slender, twice as long as the deeply and obliquely 5-lobed limb; filaments connate, glandular. Achenes obovate-ellipsoid, laterally compressed, hilum slightly eccentric at base, apical umbo surrounded by a raised rim; pappus of several rows of off-white, flattened barbellate hairs which are connate into a deciduous ring at base.

Two species in the Mediterranean region, SW Asia and E to Afghanistan; one species in Iraq.

Silybum (from σιληβον, *silybon*, applied by Dioscorides to various thistle-like plants and mentioned by Pliny as a remedy for jaundice).

1. **Silybum marianum** (*L.*) *Gaertn.*, Fruct. Sem. Pl. 2: 378 (1791); Boissier, Fl. Orient. 3: 556 (1875); Handel-Mazzetti in Ann. Naturh. Mus. Wien 27: 497 (1913); Nábělek in Publ. Fac. Sci. Univ. Masaryk 52: 94 (1925); Dinsmore in Post, Fl. Syria, Palest. & Sinai ed. 2, 2: 93 (1933); Guest in Dep. Agr. Iraq Bull. 27: 94 (1933); Anthony in Notes Roy. Bot. Gard. Edinburgh 18: 291 (1935); Blakelock in Kew Bull. 4: 64 (1949); Zohary in Dep. Agr. Iraq Bull. 31: 154 (1950); Rawi in Dep. Agr. Iraq Tech. Bull. 14: 125 (1964); Rechinger, Fl. Lowland Iraq: 646 (1964); Hassawi, Taminy & Husham al-Izzi in Dep. Agr. Iraq Tech. Bull. 167: 129 (1968); Kupicha in Fl. Turkey 5: 369 (1975); Franco in Fl. Europ. 4: 249 (1976); & in Fl. Iranica 139b: 281 (1977); Feinbrun-Dothan Fl. Palaest. 3: 382 (1978); Collenette, Wild Flow. Saudi Arabia: 218 (1999); Boulos & Hind in Fl. Egypt 3: 158 (2002); Breckle, Hedge & Rafiqpoor, Vasc. Pl. Afghan. Checkl.: 188 (2013); Haloob & Al- Kaisi, Illustr. Fl. Lowland Iraq 1(1): 56 (2016); Taifour & El-Oqlah, Pl. Jordan Ann. Checkl.: 47 (2017).

Carduus marianus L., Sp. Pl.: 823 (1753).

Stem (25–)40–150(–200) cm, juicy and rather weak, simple or with a few long branches, slightly arachnoid or glabrous. Leaves sessile, rather thin in texture, oblong in outline, pinnatifid with broad, triangular-ovate, spinous lobes, basal leaves up to 50 × 25 cm but usually less, mostly withered at time of flowering; lower cauline leaves similar but leaves rapidly reducing upwards, all amplexicaul with spinous-auriculate bases and veined and mottled with white. Capitula moderate to large, 3.5–12 cm in diameter including spines of involucre, on a finally long peduncle which is naked or bears a few spiny bract-like leaves; outer and median phyllaries 8–15 mm with a broad subrotund to ovate base and a large pectinate-spinous appendage terminating in a long, strong and in outermost phyllaries ± reflexed spine; inner phyllaries lanceolate with a small appendage. Corolla exceeding inner phyllaries. Achenes 6–7 × 3 mm, obscurely striate, mottled with black and brown, about half length of pappus. Fig. 13: 1–6.

HAB. Mostly a weed of cultivation in fields, orchards and gardens and along irrigation channels, common on waste ground; also occurring in a *Phoenix* plantation, on loam in wadis and on disturbed ground in semi-desert; alt. 2–40 m, recorded, probably extends higher; fl. & fr. Mar.-Jun.(-Oct.).
DISTRIB. Quite common in the Lower Mesopotamian Region of Iraq, occasional in the Iranian foothills. **FUJ**: Mosul, *Musharraf Husain* 2010! **FPF**: 10 km N of Jalaula, *Barkley* 7699!; Khanaqin, *Rogers* 0367! **DLJ**: nr Balad, *C. & E. Guest & Sharif Muhammad Amin Haidar* 18343! **LCA**: Baghdad, *Hausskn.* s.n.!, *Colvill* s.n., *Paranjpye* (comm. Graham) s.n., *Guest* 1953!; nr Baghdad, *Lazar* 500!; Abu Ghraib, *Alizzi* 34149!, *Hashim & Hasan* 34123!, *Omar, Janan & Sahira Abdul* 36963!, *Alizzi & Omar* 34258!; Za'faraniya, *Janan & Sabah* 41356! **LEA**: Ba'quba, *Rogers* 047! **LSM**: Musaida, nr Amara, *Field & Lazar* 74!; 40 km from Amara to Basra, *Khatib & Alizzi* 42623! **LBA**: 10 km S of Basra, *Rawi* 25911!

Fig. 13. **Silybum marianum**. 1, habit × 1; 2, leaf × 1; 3, outer phyllary × 2; 4, inner phyllary × 2; 5, achene and pappus × 2; 6, top of achene × 6. 1 from *Maitland* 117 (Palestine material); 2 from *Omar et al.* 36963; 3–6 from *Guest & Haidar* 18343. Drawn by Moira Rae Carter.

Apparently native in the Mediterranean region, SW Europe and SW Asia east to Afghanistan and Pakistan; casual or naturalised in much of rest of Europe and elsewhere in the world, e.g. Australia, tropical Africa and S America. A variety with white corollas has been recognized in Palestine and Egypt.

GULAGHĀN, (Kurd.-Iraq); Milk Thistle (Eng.).

16. **NOTOBASIS** Cass.

Dict. Sci. Nat. 35: 170 (1825)
Susanna & Garcia-Jacas in Kubitzki, Fam. Gen. Vasc. Pl. 8: 134 [2006] (2007)

Frances G. Davies

Thistly annual herbs with stem not winged. Leaves simple or usually pinnately lobed, veins ending in marginal spines, lamina somewhat white-marbled. Inflorescence a group of medium-sized homogamous capitula, each surrounded and overtopped by spiny, leafy bracts. Florets all tubular, corollas purple, unequally 5-lobed; receptacle setose. Achenes obovoid, compressed, smooth, dark brown; hilum a narrow slit; pappus cream, that of marginal florets barbellate, that of inner florets plumose.

A monotypic genus.

Notobasis (from Gr. νωτος, *notos,* dorsal and βασις, *basis,* base, because its achenes are attached at the base of their dorsal side.

1. **Notobasis syriaca** (*L.*) *Cass.*, Dict. Sci. Nat., ed. 2, ed. 2, 25: 225 (1822) & 35: 170 (1825). Boissier, Fl. Orient. 3: 553 (1875); Nábělek in Publ. Fac. Sci. Univ. Masaryk 52 : 34 (1925); Guest in Dep. Agr. Iraq Bull. 27: 67 (1933); Bornmüller in Beih. Bot. Centralbl. 60B: 207 (1940); Rawi in Dep. Agr. Iraq Tech. Bull. 14: 120 (1964); Kupicha in Fl. Turkey 5: 419 (1975); Werner in Fl. Europ. 4: 242 (1976); Feinbrun-Dothan, Fl. Palaest. 3: 377 (1978); Petrak in Fl. Iranica 139a: 286 (1979); Boulos & Hind in Fl. Egypt. 3: 150 (2002); Kilian & Ghazanfar in Fl. Oman 3: 217 (2015); Taifour & El-Oqlah, Pl. Jordan Ann. Checkl.: 45 (2017).

Carduus syriacus L., Sp. Pl.: 823 (1753).
Cirsium syriacum (L.) Gaertn., Fruct. Sem. Pl. 2: 383, t. 163 f. 2 (1791); Anthony in Notes Roy. Bot. Gard. Edinburgh 18: 291 (1935).
Cnicus syriacus Roth in Usteri, Neue Ann. Bot. 3(8): 10 (1794).

Little-branched herb to 70 cm tall. Stem fleshy, arachnoid, green, pale below. Leaves sessile, ± auriculate, expanded at base, elliptic to oblanceolate, lower to 15 × 7 cm, decreasing in size upwards, shallowly lobed, glabrous or sparsely arachnoid-pubescent or scabrid, veins very prominent on underside and excurrent as weak pale spines at lamina margins but strong spines on auricle, upper surface white-marbled along veins. Inflorescence of 1-several groups of 1–7 sessile or very shortly pedunculate capitula 18–25 mm in diameter, excluding spines. Involucre campanulate; outermost phyllaries leaf-like, up to 7 × 4 cm, lamina reduced to a narrow band along prominent veins, which end in strong spines; intermediate phyllaries cartilaginous with a terminal green spine-tipped appendage; inner phyllaries acuminate, with soft spine at apex. Corolla 15–17 mm long, limb 7–10 mm, 2 long lobes 5–7 mm, 3 short lobes 2.5–3.5 mm; filaments long-papillose. Achenes 5–7 × 3.5–5 mm, dark brown; pappus 12–14 mm. Fig. 14: 1–4.

HAB. Rocky mountain slopes, hill slopes on clay with humus, hill wadis, roadsides, ditches in cornfield; alt. 105–2730 m; fl. & fr. Apr.–Jun.
DISTRIB. Quite common in lower mountains and steppe region of Iraq. **MAM**: nr Dohuk, *Guest* 2228!; Amadiya, *Botany staff* 43403! **MRO**: Algurd Dagh, *Guest* 2958!; Hab as-Sultan Dagh, *Rawi, Noori & Kass* 28385! **FKI/MSU**: between Koi Sanjak and Kirkuk, *id.* 28062! **MSU**:, 55 km NW of Sulaimaniya, on Dukan highway, *Rawi* 21790!; Jarmo, *Helbaek* 1277! **FUJ**: Shargat (Shirqat), *Y.R. Rao* (comm. Graham) s.n.!; Mosul, *Emberger, Guest, Long, Schwan & Yusuf Serkahia* 15421! **FNI**: Nimrud, *Helbaek* 959! **FAR**: nr Arbil, *Bornmüller* 1416; **FAR/FKI**: between Arbil and Altun Kupri, *Nábělek* 3643! **FKI**: Tuz Khurmatu, *Guest* 1417! **FKI/MSU**: 30 km from Kirkuk, on road to Sulaimaniya, *Karim, Hamid & Jasim* 40533! **FPF**: 2 km N of Sa'diya, *Qaisi* 47240!; 20 km from Sa'diya along Khanaqin road, *H. Hamid* 46616!

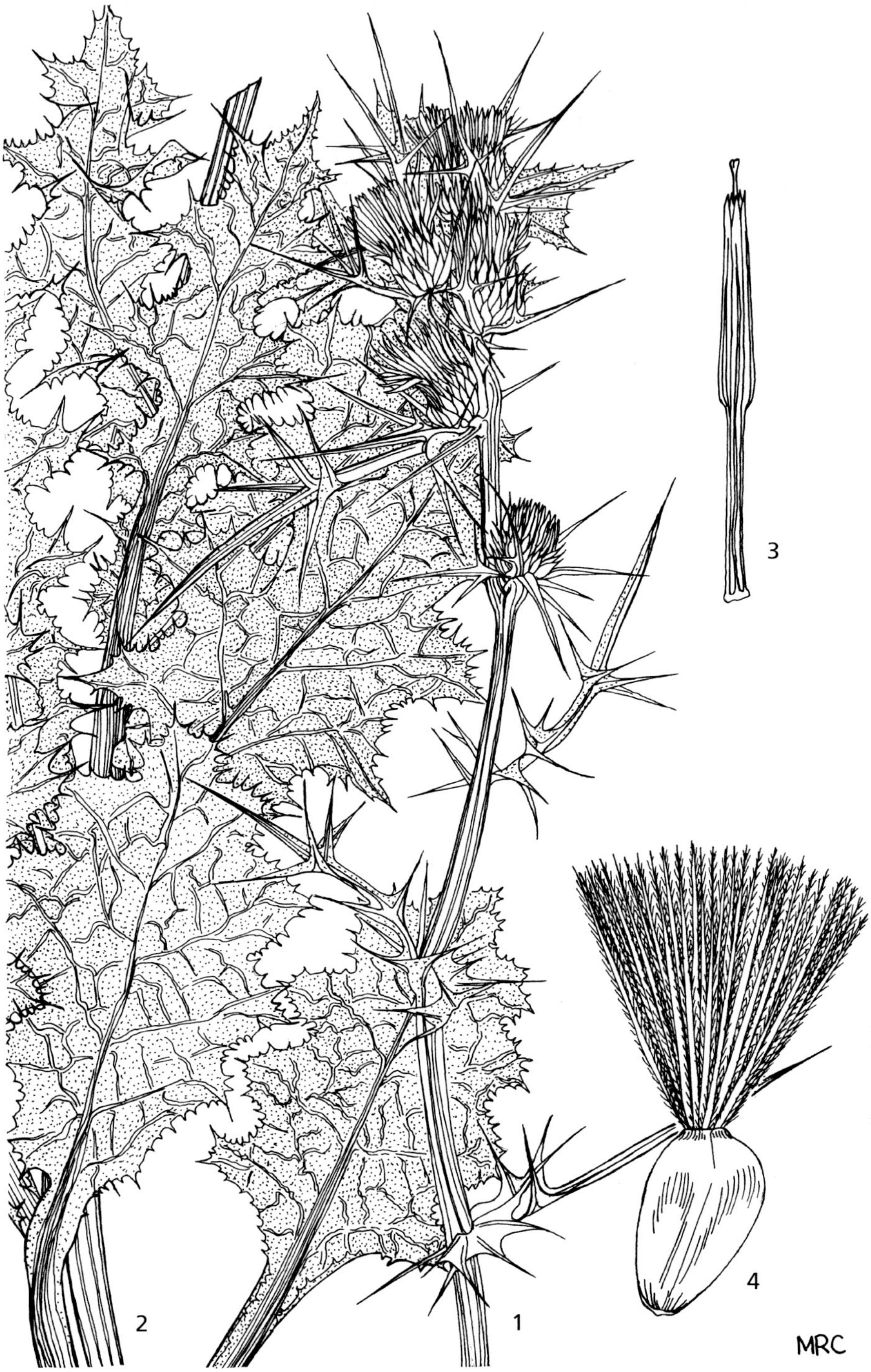

Fig. 14. **Notobasis syriaca**. 1, habit × ⅔; 2, leaf × ⅔; 3, floret × 6; 4, achene and pappus × 6. 1–2 from *Helbaek* 1277; 3–4 from *Emberger et al.* 15421. Drawn by Moira Rae Carter.

Spain, Crete, Aegean Is., Cyprus, Egypt, Jordan, Palestine, Syria, Iran, Azerbaijan.

Syrian Thistle (Eng.).

17. **ONOPORDUM** L.

Sp. Pl.: 827 (1753); Gen. Pl. ed. 5: 359 (1754); Bremer, Asteraceae: Cladistics & Classification: 146 (1994); Susanna & Garcia-Jacas in Kubitzki, Fam. Gen. Vasc. Pl. 8: 134 [2006] (2007)

C. C. Townsend

Annual or perennial herbs with ± arachnoid indumentum, stems distinctly winged from spinously decurrent leaves. Leaves simple and dentate to variously pinnatifid or pinnatisect. Capitula large, homogamous, solitary to corymbose; involucre globose to hemispherical; phyllaries multiseriate, linear to lanceolate, imbricate below with broad to narrow and very spinous apex more loosely appressed to ± reflexed; receptacle honeycomb-like, achene pits with toothed, membranous margins. Florets all tubular, corollas deep to paler purple or rarely white, limb ± unequally lobed, lobes linear. Achenes tetragonal, ± laterally compressed, transversely broadly rugose and with fine longitudinal veins, hilum basal; pappus setae multiseriate, connate into a ring at base and deciduous as a unit, scabrid, barbellate or plumose.

Probably about 50 species in Europe, SW & C Asia and N Africa.

Onopordum (from the Gr. όνος, *onos*, an ass, πέρδομαι, *perdomai*, to noisily expel wind: name of a thistle in Dioscorides, so called from its effect when eaten by donkeys).

A genus in need of more and better collections for the herbarium and also of field study. As with all such large plants, too many collections consist of "tops" or branches with no indication of the height of the plant, flower colour, probable duration and habit. Although the lowest basal leaves have often dried and shrivelled at the time of flowering, it is usually possible to collect or at least determine the shape of the inner basal leaves, and this should always be done, noting variation within each population. At present such leaves are so rarely represented in collections that it is scarcely possible even to decide whether they yield reliable taxonomic characters. Scaled colour photographs could also be extremely useful. Careful and competent revisions within geographical, especially phytogeographical, areas would build towards the much-needed but at present scarcely practicable general revision of the genus if adequate voucher material was preserved.

Eig, A. (1942). Revision of the *Onopordon* species of Palestine, Syria and adjacent countries. Pal. Bot., Jerusalem Series 2: 185–199.

1. Capitula with all the phyllaries erect and ± imbricate 2
   Capitula with apical half or more of at least median phyllaries patent or reflexed . . . 3
2. All phyllaries very narrow, linear, green, at most 2 mm wide below. . . . 4. *O. sirsangense*
   Phyllaries linear-lanceolate, outer and median ± white-floccose, 3–4 mm wide below . . . . 2. *O. leptolepis*
3. Reflexed part of median phyllaries wide, broadly triangular-acuminate, rigid but with only a short apical spine. . . . . 1. *O. illyricum*
   Reflexed part of median phyllaries narrow and subulate above, ± flat ventrally and convex dorsally, tapering into a long-spinous apex. 4
4. All phyllaries narrow and shorter, median mostly 15–25 mm long with base 2–3 mm wide, gradually tapering above. . . . . 3. *O. acanthium*
   Phyllaries broader and longer, median rarely as short as 25 mm and then abruptly narrowed above imbricate base 5
5. Outer phyllaries densely whitish-arachnoid on dorsal surface; median phyllaries ± gradually narrowed upwards, or slightly more abruptly so at about halfway (including spine); plants taller, up to c. 1.5 m. . .5. *O. heteracanthum*
   Outer phyllaries densely floccose along margins, dorsal surface glabrescent; median phyllaries abruptly narrowed above basal ⅓ (including spine); plants not attaining 1 m 6

6. Biennial; pappus setae plumose, longest side hairs 0.8–1 mm long; leaves ± pinnatilobed with broad segments . . . . 6. *O. ambiguum*
Perennial; pappus setae barbellate, longest side hairs ± 0.4 mm; leaves pinnatisect with narrow, distant segments . . . . 7. *O. sarrafii*

1. **Onopordum illyricum** *L.*, Sp. Pl.: 827 (1753); Boissier, Fl. Orient. 3: 560 (1875); Nábělek in Publ. Fac. Sci. Univ. Masaryk 52: 35 (1925); Dinsmore in Post, Fl. Syria, Palest. & Sinai ed. 2, 2: 95 (1933); Zohary in Dep. Agr. Iraq Bull. 31: 154 (1950); Rawi in Dep. Agr. Iraq Tech. Bull. 14: 120 (1964); Danin in Fl. Turkey 5: 361 (1975);

O. carduchorum Bornm. & Beauv. in Bull. Soc. Bot. Gen. ser. 2, 6: 150, f. 4 (1914); Danin in Fl. Turkey 5: 362 (1975); Rechinger, Fl. Iranica 139a: 162 (1979).

[*O. tauricum* Willd. subsp. *corymbosum* auct. non (Wilk.) Rouy, Nábělek in Publ. Fac. Sci. Univ. Masaryk 52: 35 (1925); Zohary in Dep. Agr. Iraq Bull. 31: 154 (1950); Rawi, ibid. Tech. Bull. 14: 120 (1964)].

Biennial herb with a stout stem 1–1.5(–2) m tall, green and rather thinly arachnoid-hairy to densely white-tomentose, with few to numerous ascending branches shorter than to considerably exceeding apical capitulum, continuously winged with a sinuous wing 2–8 mm wide with narrowly or more usually broadly triangular teeth up to 8 mm long terminating in rigid spines up to 7 mm long interspersed with shorter, more shortly spinous teeth, parts of wings sometimes only feebly sinuous and with slender teeth. Basal leaves broadly lanceolate or lanceolate-oblong in outline, pinnatilobed, narrowed and subpetiolate below, usually withered at time of flowering. Lower and median cauline leaves similar or somewhat narrower, sessile, 18–38 × 4–14 cm, pinnatilobed to commonly somewhat more than, but sometimes less than, ½, lobes broadly to narrowly triangular, spinous-toothed, terminal spines of lobes up to 7 mm. Upper leaves rapidly diminishing and narrowing, remaining strongly spinous. Capitula solitary or up to 3 together on stem and branches, hemispherical to ovoid, 3.2–6.5 cm in diameter excluding spreading part of phyllaries; phyllaries 6–8-seriate, broad, often ± purplish-tinged above, outer and median with triangular upper part reflexed, phyllaries margins minutely subscabrid-ciliate and occasionally somewhat arachnoid-floccose; median phyllaries lanceolate-oblong, c. 16–24 mm with base 5–7 mm broad, rather gradually to more abruptly narrowed to a 2–4 mm, stramineous apical spine; innermost phyllaries linear-lanceolate, 35–40 mm, gradually narrowed to a similar apical spine. Corolla 30–42 mm, purple, either 4 lobes equal and one longer at base, or of 2 ± dissimilar pairs and one longer, or otherwise variable, shorter lobes of deeply divided limb 7–12 mm, longer lobes 12–16 mm. Achenes 4.5–6.5 mm, dark brown to blackish; pappus setae 12–15(–17) mm, shortly plumose with longest barbs ± 0.4 mm, dirty white.

HAB.: On slopes of limestone hills, usually dry and rocky, sometimes in *Quercus* forest and scrub, also on disturbed ground under cliffs, in cultivated stony steppe and plantations and on waste ground; alt. 250–1200 m; fl. May–Jul., fr. Jul.–Aug.

DISTRIB. Widespread in the mountain region and adjacent foothills of Iraq. **MJS**: Karsi, *Gillett* 10873! **MAM**: Amadiya, *Rawi* 8673!; Sarsang, *Haines* 1198!; 20 km NE of Zakho, *Rawi* 23167!; between Maltai and Tel Kef, N of Mosul, *Nábělek* 3561. **MRO**: Saran, *Noori & Kass* 27278! **MSU**: Kani Spi, *Rechinger* 12353!, *Rawi* 22387!; Darband-i Khan, *Haines* 1531!, 1707!; 53 km NW of Sulaimaniya, *Rawi* 21812! **MSU/FKI**: 10–18 km E of Kirkuk, *Rechinger* 9988! **FUJ**: Nineva plantation, Mosul, *Chapman* 25400! **FNI**: E bank of R Zab nr Eski Kellek, *Gillett* 8187! **FKI**: 10–15 km N of Kirkuk, *Rawi* 21559A! **FPF**: Al Hashima, 9 km NE of Al Fateha police station, *Al-Kaisi, A. Haloob, H. Adnan & Riyadh* 58878 (BAG!).

Mediterranean Europe from Portugal to Bulgaria, Aegean Is., Cyprus, Turkey, W Iran.

KAUK (*Gillett* 10873).

The original description of *O. carduchorum*, together with the account by Danin (Fl. Turkey 5: 360, 1975) made much of the division of the corolla to separate that species from *O. illyricum*, and Rechinger (Fl. Iranica 139a: 162, 1979) also described the corolla limb of *O. carduchorum* as "in lacinias valde inaequales fissus" (divided into distinctly unequal segments). I find this character, as in many "thistles", very variable. The degree of abruptness of the narrowing of the phyllaries is also variable: smaller capitula, for example, tend to have more abruptly narrowed phyllaries. Iraqi material could be matched in this character with European *O. illyricum*. Asiatic material of *O. illyricum* (including that cited as this species in Fl. Turkey) is less white-tomentose than European, and the pappus tends to be longer; but I cannot feel able to give *O. carduchorum* any formal taxonomic rank. It is interesting to note that there is a specimen from Spain at Kew (*Verdcourt* 4317 from Valdepenzo) named as *O. carduchorum* by Danin, but Spain is not cited in the distribution of the species in Fl. Turkey.

2. **Onopordum leptolepis** *DC.*, Prodr. 6: 619 (1838); Boissier, Fl. Orient. 3: 564 (1875); Tamamshyan in Fl. S.S.S.R. 28: 237 (1963); Rechinger, Fl. Iranica 139a: 163 (1979); Breckle, Hedge, Podlech & Rafiqpoor, Vasc. Pl. Afghan. Checklist: 179 (2013).

Biennial herb with a stout stem 30–120 cm tall, moderately to densely greyish– or white–tomentose throughout, simple or with few ascending branches, continuously winged with a sinuous wing 2–5 mm wide with narrowly to broadly deltoid teeth up to 8 mm long with terminal spines also to 8 mm long, or wing almost straight with shorter and more slender spines. Basal leaves narrowly oblong-lanceolate, narrowed and subpetiolate below, mostly withered at time of flowering. Lower and median cauline leaves similar but somewhat narrower, of thick texture, venation prominent at least on lower surface, 4–25 × 1.5–8 cm, pinnatifid to ½ or rather less with broad or narrow triangular teeth terminated by strong yellowish or brownish spines up to 10 mm long, margins with smaller spines of variable length. Upper leaves rapidly diminishing and narrowing, uppermost simple and bract-like with a long terminal spine. Capitula solitary on stem and branches, hemispherical to ovoid, (3.5–)4.5–6 cm in diameter excluding spines, ± densely white-floccose at least about outer and median phyllaries; phyllaries c. 7–8–seriate, linear–lanceolate, gradually tapering to apex, all erect and ± imbricate, very shortly scabrid along margins, glabrous and purplish or stramineous above; median phyllaries 18–32 mm, base ± 4 mm broad, with a rigid, indistinctly demarcated 2–4 mm spine; innermost phyllaries 35–50 mm, more feebly spinous at apex. Corolla 28–40 mm long, pale to deep bluish–purple, limb much shorter than tube and deeply divided into 7–9 mm lobes. Achenes 5.5–6.5 mm, dark brown; pappus setae 20–24 mm, shortly plumose, white.

HAB. Only found once in Iraq, on a dry hill slope; alt. ± 1700 m; fl. Aug.
DISTRIB. **MRO**: Haji Umran, *Rechinger* 11327!

Iran, Transcaucasia, Afghanistan, Pakistan, C Asia (widespread).

3. **Onopordum acanthium** *L.*, Sp. Pl.: 827 (1753); Boissier, Fl. Orient. 3: 559 (1875); Zohary in Dep. Agr. Iraq Bull. 31: 154 (1950); Tamamshyan in Fl. S.S.S.R. 28: 231 (1963); Rawi, ibid. Tech. Bull. 14: 120 (1964); Danin in Fl. Turkey 5: 368 (1977); Rechinger, Fl. Iranica 139a: 163 (1979).

O. stenostegium Boissier, Fl. Orient. 3: 563 (1875).

Biennial herb with a stout stem (0.3–)0.5–1.5 m tall, with a stout taproot, green and rather sparsely to rather densely whitish-arachnoid, sparsely or more usually ± repeatedly corymbose-branched above, continuously winged with a sinuous wing 2–8 mm wide with mostly broadly and shallowly triangular lobes to 5 mm long, strong midribs excurrent in stramineous spines 2–6(–8) mm long interspersed with shorter, more weakly spinous teeth. Basal leaves broadly elliptic to broadly oblong, narrowed to a short petiole, ± withered at time of flowering. Lower cauline leaves similar, sessile, 17–40 × 10–24 cm, sinuate-lobed with broad, irregularly toothed lobes terminating in strong spines mostly up to 5 mm long but sometimes longer, margins with spines of variable length and strength. Upper leaves gradually diminishing and narrowing, remaining strongly spinous. Capitula solitary or up to 3 ± approximate, subglobose, 2.3–5 cm in diameter excluding spines; phyllaries 7–10-seriate, narrow, outer and median with upper part patent or ± reflexed, stramineous below, green to purplish towards apex, glabrous to ± arachnoid, margins minutely subscabrid-ciliate; median phyllaries lanceolate-subulate, 15–25 mm long with base 2–3 mm broad, gradually narrowed to a (2–)3–6 mm long apical spine; innermost phyllaries similar but longer, 25–40 mm. Corolla 20–32 mm long, purple (rarely white), linear lobes 5–9 mm. Achenes 4–6 mm long, dark brown to blackish; pappus setae 7–10 mm long, shortly barbellate, dirty white.

HAB. Rarely occurring in the mountain region of Iraq; alt. 540–1500 m.
DISTRIB. **MAM**: Sarsang, *Haines* s.n. **MSU**: Penjwin, *Haussknecht*! **FNI**: Alqosh, *Socin* s.n. (1871)!

C & S Europe from Portugal to Greece, Romania & C Russia, Turkey, Caucasia, N & W Iran, Afghanistan, Pakistan, Turkmenia, C Asia, E & W Siberia, N Africa (Algeria, introduced?). Introduced in N Europe, N America etc.

I have examined the type specimen of *O. stenostegium* Boiss. which was collected at FNI: "Elgosh" (Alqosh) by a Dr Socin in 1871. This is a poor, badly pressed specimen, with phyllaries and pappus lying loose in a capsule, no trace of a capitulum remaining. The pappus is shortly barbellate and up to 8 mm long; the median phyllaries are 20 × 2.5 mm at the base, and gradually taper to a 4 mm spine; the outer phyllaries were clearly recurved. All these characters are those of *O. acanthium*, and the only discrepancy from this species is the weakly armed stem. I have, however, noticed such reduction in stem armature on many thistles which have sent up secondary growth following cutting or grazing, and in my opinion *O. stenostegium* is such a secondary growth of *O. acanthium*.

4. **Onopordum sirsangense** *Rech.f.* in Fl. Iranica 139a: 164 (1979). – Type. Iraq, Mosul: Sarsang, 9.viii.1959, *Haines* 1585 (E, holo.); Tuzlacı in Fl. Turkey 11 (Suppl. 2): 161 (2000).

Biennial herb to 70 cm tall, whitish-tomentose below on stem and young leaves, green and thinly arachnoid above, with a few ascending branches, continuously winged with a widely sinuous wing 5–10 mm wide, teeth wide and shallow with short and rather weak terminal spines not exceeding 2 mm. Basal leaves of first year plant densely white-tomentose, but soon more loosely arachnoid, broadly elliptical, narrowed and shortly petiolate below, withered at flowering time. Lower and median cauline leaves thin in texture, broadly oblong–elliptic, 7–15 × 4–7 cm, simple and compound-dentate, larger teeth broad and shallow, thin terminal spine 2–5 mm, margins with shorter spines of variable length. Upper leaves gradually decreasing but scarcely proportionately narrower, very broadly decurrent. Capitula solitary on stem and branches, ovoid to depressed-globose, 2.8–3.5 cm in diameter, green, not at all floccose; phyllaries 6–8-seriate, linear, gradually tapering to apex, all erect and imbricate, very shortly scabrid along margins and on dorsal surface; median phyllaries 12–15 mm long, with a 1–2 mm stramineous apical spine; innermost phyllaries ± 25 mm long with a larger slender apical spine. Corolla 25–28 mm, apparently purple, limb subequalling tube and clothed with minute, sessile, yellowish glands, deeply divided with 7–9 mm lobes. Achenes 5–5.5 mm, dark brown to blackish; pappus setae short, 10–12 mm, shortly barbellate, white.

HAB. Only found once in the mountain region of Iraq, on a stony bank below a cliff in a gorge, alt. ± 1200 m. fl. Aug.
DISTRIB. **MAM**: Sarsang, *Haines* 1585! (type)

Turkey (SE Anatolia). A species in need of investigation in the field; it may possibly be a shade form of *O. acanthium*. The type locality is only ± 35 km from that of *O. stenostegium*.

5. **Onopordum heteracanthum** *C.A.Mey.*, Verz. Pfl. Cauc.: 68 (1831); Boissier, Fl. Orient. 3: 563 (1875); Bornmüller in Beih. Bot. Centralbl. 60B: 208 (1939); Zohary in Dep. Agr. Iraq Bull. 31: 154 (1950); Tamamshyan in Fl. S.S.S.R. 28: 239 (1963); Rechinger, Fl. Iranica 139a: 159 (1979); Taifour & El-Oqlah, Pl. Jordan Ann. Checkl.: 45 (2017).

O. canum Eig in Pal. Journ. Bot. J. Ser. 2: 193 (1942). – Type. Makatu, near Mandali, *Guest* 892 (K, holo.); Zohary in Dep. Agr. Iraq Bull. 31: 154 (1950); Rawi, ibid. Tech. Bull. 14: 120 (1964); Rechinger, Fl. Lowland Iraq: 647 (1964); Rechinger, Fl. Iranica 139a: 161 (1979); Tuzlacı in Fl. Turkey 11 (Suppl. 2): 160 (2000), sphalm. 'heteraeanthum'; Haloob & Al-Kaisi, Illustr. Fl. Lowland Iraq 1(1): 51 (2016).

Biennial herb with a stout solid stem 65–150 cm tall, moderately to densely canescent or white-tomentose with few to many long ascending branches, continuously winged with a sinuous wing 3–12 mm wide with broadly deltoid teeth up to 16 mm long with rigid terminal spines 4–10 mm long, or wing slightly sinuous or almost straight with shorter and more slender spines. Basal leaves oblong-elliptic, 25–50 × 6–17 cm, narrowed below, mostly withered at time of flowering, shallowly and broadly pinnately lobed to pinnatipartite, with broad or narrow deltoid teeth with rigid spines up to 10 mm long, margins with spines of variable length, veins prominent on lower surface. Lower and median cauline leaves similar but somewhat narrower; upper cauline and branch leaves rapidly diminishing, uppermost bract-like with a terminal spine. Capitula solitary on stem and branches, hemispherical,

4–6 cm in diameter excluding spines, ± densely cobwebby-floccose with white hairs below; phyllaries 7–8-seriate; outer phyllaries lanceolate to ovate-lanceolate, appressed or more rarely shortly spinous tips deflexed; median phyllaries 45–65 mm, gradually tapering to apex from 4–7 mm broad base or occasionally somewhat more abruptly so about middle, 25–40 mm spinous apex ± sharply patent or reflexed, commonly purple with a pale tip, scabrid along margins and usually on upper inner surface, glabrous or scabrid and sometimes with sessile glands dorsally; inner phyllaries 30–45 mm, narrowly lanceolate, erect, innermost feebly or sometimes not at all spinous. Corolla 30–40 mm, purple, limb shorter than tube and subequally to equally divided into 7–10 mm lobes. Achenes 5–6 mm; pappus setae 15–20(–22) mm, shortly plumose with longest barbs c. 0.4 mm, dirty white.

HAB. On dry sandy and clay hillsides, in cultivated places and patches in the desert, and as a weed along irrigation channels; alt. 130–780 m; fl. Apr.-Jun., fr. Jun.-Aug.
DISTRIB. Scattered in the foothills and lower parts of the mountain regions of Iraq, also occurring in the Lower Mesopotamian plain and rarely in the desert. **MAM**: 51 km NW of Mosul near Filda police station, *Rawi* 22942!; **MSU**: 10 km from Sulaimaniya towards Saiyid Sadiq, *Omar, Qaisi & Wedad* 49417!; Jarmo, *Helbaek* 1954!; **MSU/FKI**: 10–18 km E of Kirkuk, *Rechinger* 9989! **FUJ**: ± 25 km S of Mosul on Hammam Alil road, *Hossain* 4310!; Kirrana village, 49 km W of Mosul, *Rawi* 22934! **FNI**: 52 km N of Mosul on way to Dohuk, *Rechinger* 10618! **FKI** 10–15 km N of Kirkuk, *Rawi* 21559! **FPF**: Lajama police station, 25 km SE of Mandali, *Rawi* 20723!; Makatu, nr Mandali, *Guest* 892 (type of *O. canum*)! **DLJ**: Wadi Shuraimiya (Schreimije") near Tikrit, *Handel-Mazzetti* 1013! **DGA**: Adhaim, *Robertson* 102! **DSD**: 69 km from Ansat on way to Samah, *Fawzi, Hazim & Hamid* 38881! **LEA**: Fakka-Laqlaq valley, *Rawi* 25852! **LCA**: Shifta railway station, 15 km NE of Baghdad, *Anderson* 200! **DWD**: 40 km W Ramadi to Rutba, *Al-Kaisi & K. Hamad* 48943 (BAG!).

Turkey (SE Anatolia), Iran, Transcaucasia.

O. heteracanthum var. *anisacanthum* (Boiss.) Boiss. was recorded by Handel-Mazzetti (Ann. Natur. Mus. Wien 27: 447, 1913) from "Steinige Steppe (Kalk und Lava) bei Skerije am Dschebel Sindschar, Tell Kokeb am Chabur, ober Urfa und bei Dschülman, in steppis lapidosis". I have seen Handel-Mazzetti's specimen, no. 1855 (W) and it is localised as "Mesopotamia, in altiplanitie ad septentrionem urbis Urfa ad Euphratem versus, Dschülman, in steppis lapidosis". This locality is in Turkey. No specimen from Jabal Sinjar is apparently extant, and so the identity of the plant which Handel-Mazzetti saw there is in doubt. Probably it was *O. heteracanthum*, but to judge from No. 1855 not *O. anisacanthum* Boiss.

After some hesitation I have reduced *O. canum* Eig to synonymy with this species, since of all the characters used to separate it by Eig himself and by Rechinger in Fl. Iranica, only the broader outer phyllaries and the somewhat more abruptly tapering intermediate phyllaries seemed to be in any way reliable, and even here transitional forms occur, as indeed Rechinger's remarks on *O. canum* imply. For example, Eig states that *O. canum* has a longer pappus with longer side hairs, giving the pappus length of *O. canum* as 25–27 mm. The holotype of *O. canum* (*Guest* 892) has one capitulum in good flower, and I find the pappus to be only about 20 mm, and comparing barb-length with that of Caucasian *O. heteracanthum* I see no difference whatsoever. Another Iraqi specimen (*Robertson* 102) with broad outer phyllaries and more constricted median ones, which would no doubt be referable to *O. canum* on these grounds, has a 22 mm pappus but little of the facies of the type. Eig goes so far as to describe the pappus hairs of *O. canum* as "ca. 0.06 mm thick", but as a few inner pappus hairs in *Onopordum* species are commonly longer than and up to 3 × as thick as the outer it is doubtful which are intended, and the character is unhelpful.

6. **Onopordum ambiguum** *Fresen.* in Mus. Senckenb. 1: 85 (1834); Boissier, Fl. Orient. 3: 562 (1875); Dinsmore in Post, Fl. Syria, Palest. & Sinai ed. 2, 2: 96 (1933), p.p.; Täckholm, Stud. Fl. Egypt ed. 2: 538 (1974); Feinbrun-Dothan, Fl. Palaest. 3: 385 (1978); Boulos & Hind in Fl. Egypt. 3: 161 (2002); Taifour & El-Oqlah, Pl. Jordan Ann. Checkl.: 45 (2017).

O. ambiguum var. *horridum* Post, Fl. Pal.: 457 (1896).

Biennial herb with a stout stem 35–60 cm tall, moderately to densely canescent or white-tomentose with few to many divergent-ascending branches, continuously winged with a sinuous wing 3–10 mm wide with broadly or more narrowly deltoid teeth up to 15 mm long with stout terminal spines 0.5–1(–20) mm. Basal and lower cauline leaves similar (the basal

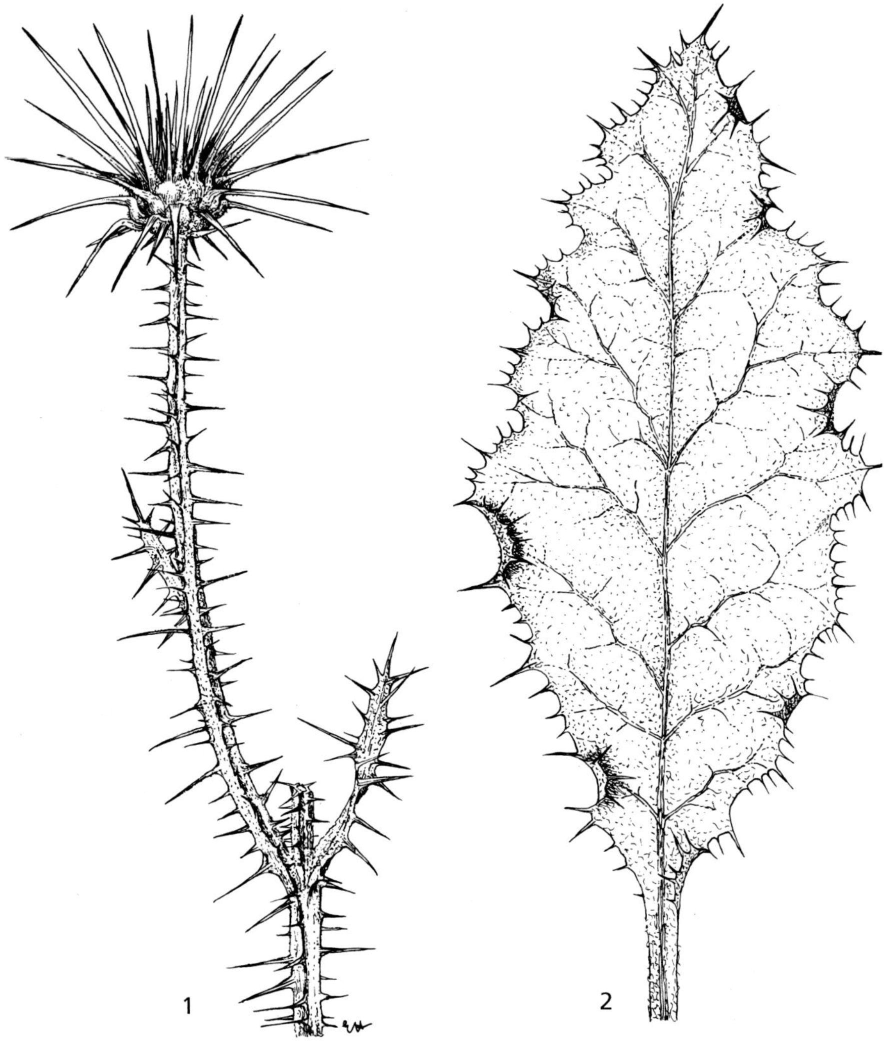

Fig. 15. **Onopordum ambiguum**. 1, habit flowering branch × ½; 2, leaf × 1. Drawn by Ruth Koppel. From Naomi Feinbrun-Dothan, *Flora Palaestina*, Part three: Ericacaceae to Compositae, Plates, Jerusalem: The Israel Academy of Sciences & Humanities, 1977, pl. 648. © The Israel Academy of Sciences and Humanities. Reproduced with permission.

mostly withered at flowering time), narrowly to broadly oblong-elliptic, 8–38 × 3.5–14 cm, shallowly and broadly pinnatifid to pinnatipartite to ⅔ with broad or narrower deltoid teeth with rigid spines similar to those of wings, margins with spines of variable length, veins narrowly prominent on lower surface. Median cauline leaves similar but reducing and somewhat narrower; upper stem and branch leaves rapidly reducing, uppermost bract-like and spinous. Capitula solitary on stem and branches, hemispherical, 4–6 cm in diameter, excluding spines, moderately to densely floccose along edges of phyllaries, sometimes thinly to moderately so on their dorsal surfaces, phyllaries 5–7-seriate, outer phyllaries short, (6–)10–20 mm, rather slender spinous point ± deflexed; median phyllaries (20–)35–60 mm, abruptly narrowed above 6.5–8(–10) mm wide basal ⅓ into (12–)23–45 mm, stramineous or ± purplish, reflexed spinous apex, scabrid along margins, smooth or occasionally somewhat scabrid on midrib ventrally, sometimes with sessile glands dorsally;

inner phyllaries 20–38 mm, narrowly lanceolate, erect, innermost feebly spinous. Corolla (25–)29–36 mm, purple, limb shorter than tube and ± unequally divided into 6–9 mm lobes. Achenes 5–6 mm, blackish; pappus setae (16–)18–25 mm, long–plumose with longest side barbs 0.8–1 mm, dirty white. Fig. 15: 1–2.

HAB. In the desert, on plains or in open sandy/gravelly or clay hollows, along roadsides and near a river; alt. 40–710 m; fl. & fr. Apr.-Jun.(-Sep.).
DISTRIB. Apparently widespread and locally common in the Western Desert of Iraq and the adjoining Lower Mesopotamian plain, rare in the Eastern Desert region. **DWD**: Rutba, *Field & Lazar* 133; 5 km S of H.3, *Rawi* 21215!; 160–190 km NW of Ramadi on road to Rutba, *Rawi* 31320!; 55 km W of Ramadi, *Alizzi* 35128!; 10 km S of Ana, *Alizzi & Husain* 34002! **DGA**: 50 km SE of Samarra, *Rawi & Shahwani* 25469! **LCA**: Falluja desert, *Haines* 152!; Hillah, *Graham* s.n.!

Jordan, Egypt, Sinai, Saudi Arabia, Yemen.

7. **Onopordum sarrafii** *C.C.Towns.* in Kew Bull. 42 (2): 439 (1987). –Type. Iraq, Lower Jazira, 2 May 1978, *Hossain* s.n.! (K!, holo.).

Perennial herb 50–85 cm tall, ± whitish or greyish-white with arachnoid tomentum, with numerous slender ascending branches, stem also rather slender and solid, winged with wing up to 4 mm wide but above frequently very narrow and on branches often reduced to a thin green rim or row of spines and ± interrupted, teeth broad-based and narrowly acuminate, up to 8 mm, stramineous terminal spine also to 8 mm, rigid. Basal leaves withered at flowering time, 25 × 5–6 cm, pinnatipartite (according to Al-Sarraf). Larger cauline leaves narrowly lanceolate-elliptic, 12–30 × 4.5–8 cm, pinnatisect with broadly to narrowly lanceolate lobes terminated by sharp spines up to 8 cm long, margins with shorter spines of variable length. Upper leaves rapidly diminishing in length and width, uppermost bract-like and spinous. Capitula solitary on stem and branches, hemispherical to depressed-globose, 3–3.5 cm in diameter excluding spines, densely white-floccose only around top of peduncle, otherwise only so along edges of phyllaries, dorsally very thinly so or glabrescent; phyllaries 7–9-seriate; outer phyllaries 8–13 mm, spinous point ± deflexed; median phyllaries 26–38 mm, abruptly narrowed above 5–7 mm broad basal ⅓ into 15–25 mm reflexed spinous apex, which is stramineous or purplish about centre, scabrid along margins and scabrid at least along upper part of midrib ventrally; inner phyllaries 35–40 mm, narrowly lanceolate, erect, innermost shortly and more feebly spinose. Corolla ± 30 mm, purple, limb shorter than tube and ± unequally divided into 7–9 mm segments. Mature achenes not seen; pappus setae 14–16 mm, shortly plumose with longest barbs 0.4 mm, white.

HAB. Sand dunes, occurring only in a limited area in the Lower Jazira; alt. ± 115 mm; fl. Apr.-May.
DISTRIB. **FUJ**: 3 km S of Baiji along Mosul-Baghdad road, 2 May 1978, *Hossain* s.n.! (type); 6 km S of Baiji along same road, 18 Apr. 1979, *Hossain* s.n.!

This species was distinguished, written up and given the name of "O. arenarium Hossain & Al-Sarraf" by Al-Sarraf (1979) in "A taxonomic revision of the Iraqi species of *Onopordum* Linn. (Compositae)", an M.Sc. thesis from the University of Mosul, of which there is a copy in the Kew library. It should be noted that the plant referred to in this work as *O. acanthium* is *O. heteracanthum*. The name "O. arenarium" as applied to the Iraqi plant was never validly published, and in any event would be antedated by *O. arenarium* Pomel.

18. **ARCTIUM** *L.*

Sp. Pl. 816 (1753); Gen. Pl. ed. 5: 357 (1754)
Bremer, Asteraceae: Cladistics & Classification: 143 (1994); Susanna & Garcia-Jacas in Kubitzki, Fam. Gen. Vasc. Pl. 8: 136 [2006] (2007).

C. C. Townsend

Robust biennial herbs with a long, tough taproot, much-branched, branches divergent. Leaves petiolate, alternate, unarmed, simple, cordate at base. Inflorescence racemose or cymose; capitula homogamous; involucre somewhat depressed-globose to shortly ovoid; phyllaries narrow, subulate, multiseriate, appressed at base, outer and median spreading above with rigidly uncinate tips, innermost erect, hooked or (not in Iraq) apices acuminate; receptacle flat, densely bristly. Florets hermaphrodite, tubular. Corollas pale lilac to purple, rarely white, glabrous or shortly glandular, limb rather abruptly expanded. Achenes oblong-obovoid, compressed, striate; pappus short, deciduous, of several rows of free, scabrid hairs.

About 11 species in temperate Eurasia.

Arctium, from αρκτιον, *arktion,* a name in Pliny, also referred to as *Arcturus* or – in the English translation by P. Holland – "clot-burr".

Arènes, J. (1950). Monographie du genre *Arctium.* Bull. Jard. Bot. Brux. 20: 73–156.

Susanna, A., Garcia Jacas, N., Vilatersana R. & Garnatje, T. (2003); Generic boundaries and evolution of characters in the *Arctium* group: a nuclear and chloroplast DNA analysis. Collect. Bot. 26: 101–118.

Inflorescence corymbose; inner phyllaries about equalling the florets at anthesis; expanded limb of corolla distinctly shorter than filiform tube . . . 1. *A. lappa*
Inflorescence of capitula disposed in racemes; at least the inner florets exceeding the inner phyllaries at anthesis; expanded limb of corolla shorter than or equalling tube . . . . 2. *A. minus*

1. **Arctium lappa** *L.*, Sp. Pl. ed. 1: 816 (1753), emend. Mill., Gard. Dict. ed. 8, no. 1 (1768); Yuzepchuk & Sergievskaya in Fl. S.S.S.R. 27: 97 (1962); Rawi in Dep. Agr. Iraq Tech. Bull. 14: 106 (1964); Rawi & Chakravarty in Dep. Agr. Iraq Tech. Bull. 15: 15 (1964); Rechinger, Fl. Iranica 139a: 107 (1979); Meikle, Fl. Cyprus 2: 947 (1985); Zhu Shi & Greuter in Fl. China 20–21: 153 (2011); Breckle, Hedge & Rafiqpoor, Vasc. Pl. Afghan: 148 (2013).

Lappa vulgaris Hill, Veg. Syst. 4: 28, t. 25 f. 1 (1762).
Arctium majus (Gaertn.) Bernh., Syst. Verz. Pflanz. Erfurt.: 154 (1800), nom. illeg. superfl.
Lappa major Gaertn., Fruct. Sem. Pl. 2: 379, t. 162 f. 3 (1791), nom. illeg. superfl.; Boissier, Fl. Orient. 3: 457 (1875).
Arctium vulgare (Hill) Druce in Ann. Scot. Nat. Hist. no. 60: 222 (1906); Dinsmore in Post, Fl. Syria, ed. 2, 2: 84 (1933).
A. lappa subsp. *major* (Gaertn.) Arènes in Bull. Jard. Bot. Brux. 20: 75 (1950).
[*A. minus* (non Hill.) Bernh.; Kupicha in Fl. Turkey 5: 355 (1975), pro parte.]

Erect, widely branching biennial mostly 75–150 cm tall; stems ± angular-furrowed, subglabrous or usually thinly floccose-arachnoid, often ± purplish, pithy. Leaves cordate-ovate, obtuse to subacute, with shallow, broad teeth which are apiculate from percurrent veins or subentire, green and more sparsely hirtellous above, grey-green and tomentellous beneath with prominent veins, lamina in basal leaves up to 50 cm long and somewhat less in width, cauline leaves reducing in size; petiole of basal leaves up to 30+ cm, canaliculate on adaxial surface. Inflorescence a lax to rather dense corymb. Capitula usually rather few, 2.5–4 cm in diameter including spines, depressed-globose in flower, on slender pedicels; phyllaries multiseriate, dense, subulate, prominently 1-veined, longest 1–2 cm long, glabrous to shortly ciliate, occasionally arachnoid. Florets glabrous, subequalling phyllaries, at anthesis tube 6–8 mm, slender, longer than limb 4–6 mm. Corolla purple. Achenes compressed-obovate, 6–7.25 × 2.5–3.25 mm, brownish, sometimes dark-marbled, striate; pappus early deciduous, hairs unequal, longest 2–3.25 mm. Fig. 16: 1–3.

HAB. In shady places, in orchards under *Malus* (apple) or *Populus* trees or on mountains under *Quercus* or *Juglans,* usually near streams or in humid areas; alt. 700–1550 m; fl. & fr. Jun.-Sep.
DISTRIB. Widespread in the mountain regions of Iraq. **MAM**: Baidaho village 5 km from Kani Masi, *Botany Staff* 43906!; Zawita, *Guest* 3759! **MRO**: Chia-i Mandali, *id.* 2658!; Haji Umran, *Dabbagh & Hamid* 46297!, *Qaisi, Karim & Hamad* 49788!; Rayat, *Qaisi, Karim & Hamad* 49804!; Pushtashan, *Rawi, Uthman & Serhang* 26495!; lower slopes of Qandil range NE of Rania, *Rawi & Serhang* 18236!; **MSU**: Avroman, *Rawi, Chakrawarty, Noori & Alizzi* 19801!; Saiyid Sadiq, Penjwin, *Haines* s.n.!

Almost throughout Europe, Cyprus, Lebanon, Turkey, Caucasia, Iran, Afghanistan, C & N Asia eastwards to China. Introduced in N America and elsewhere.

WAISAR (Kurd.), *Rawi, Uthman & Serhang* 26495; 'URQAH (Kurd.), *Guest* 3759.

2. **Arctium minus** (*Hill.*) *Bernh.*, Syst. Verz. Pfl.: 134 (1800); Yuzepchuk & Sergievskaya in Fl. S.S.S.R. 27: 100 (1962); Kupicha in Fl. Turkey 5: 355 (1975), pro parte; Rechinger, Fl. Iranica 139a: 197 (1979); Breckle, Hedge & Rafiqpoor, Vasc. Pl. Afghan: 148 (2013).

Fig. 16. **Arctium lappa**. 1, large lower leaf × ⅔; 2, inflorescence × ⅔; 3, capitulum × 2. **Arctium minus**. 4, inflorescence × ⅔; 3, capitulum × 2. 1–3 from *Guest* 3759; 2–3 from *Guest* 2658; 4–5 from *Milne-Redhead* 6500 (British material). Artist unknown.

Lappa minor Hill, Veg. Syst. 4: 28, t. 25 f. 3 (1762); Boissier, Fl. Orient. 3: 457 (1875).

Erect biennial mostly 60–120 cm tall with divergent-erect or divergent branches; stems ± angular-furrowed, ± floccose-arachnoid or subglabrescent, pithy. Leaves similar to those of *A. lappa* but sometimes with teeth and lower usually distinctly longer than wide. Inflorescence rather narrow, racemes terminal on the stem and branches, pedicels slender, narrowly divergent; capitula numerous. Capitula 2–3 cm in diameter including spines, globose-ovoid; phyllaries multiseriate, dense, subulate, prominently 1-veined, longest 1–1.8 cm, ciliate to arachnoid, rarely glabrous. Corollas purple, glabrous, at least central exceeding phyllaries at anthesis, tube (3–)4.5–5(–6) mm, equalling or shorter than the (4–)4.5–5.5(–6) mm limb. Achenes compressed-obovate, 5–6.5(–7) mm, brownish and often dark-marmorate, striate; pappus soon deciduous, longest of unequal hairs 2–3.5 mm. Fig. 16: 4–5.

HAB. Shady places beneath trees and in a gully; alt. (only once recorded) 1000–1200 m; fl. & fr. Jul.-Sep. DISTRIB. Apparently rare in the mountain region of Iraq. **MAM**: Sarsang, *Haines* s.n.! **MRO**: Naprdan, *Hadač* 5376; Pushtashan, *Rechinger* 11029.

Almost throughout Europe, Turkey, Caucasus?, Iran, Afghanistan, N Africa (Algeria, Morocco), Macaronesia (Madeira). Introduced in Australasia, N tropical and temperate S America.

19. **COUSINIA** Cass.

Dict. Sci. Nat. ed. 2, 47: 503 (1827)
Susanna & Garcia-Jacas in Kubitzki, Fam. Gen. Vasc. Pl. 8: 136 [2006] (2007)
Auchera DC., Prodr. 6: 557 (1838)

A. Huber-Morath
Revised by I. Mehregan

Biennial or perennial spiny herbs, often with woody stock, subdichotomously branched, corymbose above, ± many-headed, mostly arachnoid-woolly or tomentose and often glabrescent, rarely glabrous. Leaves often leathery, prickly or spiny, dentate, lobed, pinnatifid or pinnatipartite, often decurrent and thereby stems winged. Capitula homogamous, discoid; involucre cylindrical to globose; phyllaries multiseriate, rigid, imbricate, with or without an appendage, bearing a prickle or spine at tip. Receptacle densely aristate-paleaceous, with smooth or barbellate bristles. Florets hermaphrodite; corollas yellow, pink or purple, rarely white. Achenes glabrous, obovate, ± compressed, apex rounded or truncate, often with a ± distinctly denticulate margin; pappus bristles scabrous, free, strongly caducous.

Some 600 species in SW & C Asia, mainly in the Irano-Turanian region. There is a centre of diversity in Iran and Afghanistan (about 355 species), and the genus extends westwards to Anatolia, Syria, Lebanon and Palestine; 11 species in Iraq.

Cousinia [possibly derived from the vernacular French word for a sticky burr or *cousin*, defined as "plantes dont les fruits chargé des aspérités s'accrochent aux habits des passans" (plants whose spiny fruits attach themselves to the clothes of passers-by); Audouin 1823, t. 4 p. 624].

Huber-Morath, A. (1972). Die anatolischen Arten der Gattung *Cousinia* Cass. in Ber. Schweiz. Bot. Ges. 82 (3): 223–268.
- (1975). *Cousinia* in Davis (ed.) Fl. Turkey 5: 329–353.
Mehregan, I. & Kadereit, J. W. (2008). Taxonomic revision of *Cousinia* sect. *Cynaroideae* (Asteraceae, Cardueae). Willdenowia 38: 293^362.
Mehregan, I. (2011). Notes on the taxonomy of *Cousinia* sect. Haussknechtianae (Asteraceae; Cardueae). Iran J. Bot. 17 (2): 137–149.
Rechinger, K.H. (1972). *Cousinia* in Fl. Iranica 90: 1–329.
Winkler, C. (1893). Synopsis specierum generis *Cousiniae* Cass, in Acta Horti Petrop. 12, 2: 181–286.
- 1897. Mantissa synopsis specierum *Cousiniae* Cass. in Acta Horti Petrop. 14, 2: 187–243.

1. Cauline leaves not decurrent, amplexicaul nor adnate to stem 2
   Cauline leaves decurrent, amplexicaul or adnate to stem 3
2. Capitula sessile, ± 30-flowered; incolucre broadly ovoid, 2–3 cm in diameter including spines .... 2. *C. haussknechtii*

Capitula pedunculate, 4–10-flowered; involucre oblong-cylindrical, 0.8–1 cm in diameter including spines . . . . 1. *C. stencephala*
3. Stems mostly winged; phyllaries abruptly contracted from an ovate-lanceolate base into a keeled linear-lanceolate spreading spine . . . . . . 11. *C. wesheni*
Stems usually not winged; phyllaries not forming a keeled spine 4
4. Perennial, not monocarpic; phyllaries linear-subluate, narrowed into long stramineous spines . . . . 3. *C. rawanduzensis*
Biennials or perennials, monocarpic; phyllaries abrubptly enlarged from an adpressed constricted base in to a ± large appendage 5
5. Corollas yellow 6
Corollas pink or purple 7
6. Stems densely arachnoid-floccose; phyllaries rigid, with spreading or subrecurved, lanceolate, carinate, longitudinally folded appendage; receptacle bristles glabrous. corollas white . . . . 7. *C. algurdina*
Stems arachnoid-tomentose or glabrescent; phyllaries with prominent yellow midrib (middle ones), their free part abruptly expanded above into a usually flat, erect, deltoid or deltoid-sagittate appendage; receptacular bristles rough; corollas yellow . . . . 9. *C. pergamacea*
7. Appendages of median phyllaries with entire margins 8
Appendages of median phyllaries with lateral prickles or spines 9
8. Appendages 4–6 mm broad at base, broadly triangular, suddenly contracting into a 5 mm long spine . . . . .5. *C. kopi-karadaghensis*
Appendages 10 mm broad at base, triangular-acuminate, spine 3(–5) mm,10. *C. inflata*
9. Middle phyllaries densely imbricate; appendage attenuate into a short spine at apex with 3–8 prickles on both sides; receptacular bristles rough; corollas 28–35 mm, purple . . . . 8. *C. macrolepis*
Middle phyllaries ± spreading, spreading-incurved, appressed-incurved or -erect or imbricate; appendage attenuate into a short or long spine with 2–4(–6) spines or prickles on both sides; corolla 18–30 mm, reddish-pink, pink or pale-purple 10
10. Plant, densely arachnoid-tomentose, tomentose or ± glabrescent; leaves tomentose on both sides, or tomentose beneath and glabrescent above; phyllaries appendage 4–10 mm broad, attenuate into a long spine at apex, with 2–4(–6) spines on both sides. receptacular bristles usually rough . . . . .6. *C. odontolepis*
Plant arachnoid- or canescent-tomentose to glabrous; leaves, arachnoid- or canescent-tomentose on both surfaces, ± glabrescent above or rarely glabrescent on both sides; phyllaries appendage 7–15 mm broad, usually attenuated into a long or short spine at apex, with 2–5 or rarely more tiny spines or prickles on both sides; receptacular bristles rough or smooth. . . . . 4. *C. aintabensis*

SECTION **STENOCEPHALAE** *BUNGE.*

1. **Cousinia stenocephala** *Boiss.*, Diagn. Pl. Orient. ser. 1, 6: 115 (1846). – Type: Iraq, Pesh Khabur to Bawart, Kotschy 400 (W!). Boissier, Fl. Orient. 3: 486 (1875); Handel-Mazzetti in Ann. Naturh. Mus. Wien 27: 51 (1913); Nábělek in Publ. Fac. Sci. Univ. Masaryk 52: 31 (1925); Anthony in Notes Roy. Bot. Gard. Edinburgh 18: 291 (1935); Bornmüller. in Beih. Bot. Centralbl. 60B: 202 (1939); Blakelock in Kew Bull. 4: 51 (1949); Zohary in Dep. Agr. Iraq Bull. 31: 152 (1950); Rawi in ibid. Tech. Bull. 14: 112 (1964); Huber-Morath in Fl. Turkey 5: 347 (1975); Rechinger, Fl. Iranica 90: 168 (1979).

Perennial, 15–60 cm, dichotomously branched from base, loosely corymbose. Stems slender, terete, arachnoid-woolly, soon glabrescent and shining above, densely white-tomentose below, conspicuously reticulate-veined; basal leaves oblong, 6–19 × 3–7 cm, gradually narrowed into a 1–7 cm long petiole, irregularly short-lobate and dentate with spines 1–5 mm long; cauline leaves gradually smaller, not decurrent, with few regularly-prickled teeth. Capitula 5–50 or more, 4–10-flowered; involucre oblong-cylindrical, 0.8–1 cm in diameter including spines; phyllaries 12–30, white-tomentose, linear to linear-lanceolate, 10 mm long, with glabrous, reflexed, hooked prickles not more than 2 mm long; receptacle bristles smooth. Corollas

Fig. 17. **Cousinia stenocephala**. 1, habit × ⅔. **Cousinia macrolepis**. 2, capitulum × ⅔. **Cousinia wesheni**. 3, capitulum × 3. 1 from *Rawi* 22933A; 2 from *Wheeler Haines* W2042; 3 from *Rawi* 52664. Drawn by Moira Rae Carter.

yellow or reddish-suffused, 10–12 mm. Achenes oblong-obovate, light brown, 3–4 × 1.5–2.2 mm long, truncate, apex scarcely marginate and denticulate. Fig. 18: 1.

HAB. Stony hillsides and steppic grassland on conglomerate and sandy clay, a weed in fields; alt. (150–)300–900 m; fl. & fr. Jun.-Jul.
DISTRIB. Occasional in the lower forest zone of Iraq, common in the steppe region. **MJS**: Jaddala, 17 km W of Balad Sinjar, *Handel-Mazzetti* 1549. **MAM**: Atrush, *Guest* 3643!, *Anders* 1589; Dohuk, *Rechinger* 10621! **MRO**: Baba Chichak, between Arbil and Rowanduz, *Bornmüller* 1401! **MSU**: Jarmo, *Helbaek* 1926!; Sarchinar, *Hadač* 4781; 65 km SW of Sulaimaniya, *Rawi* 21817!; between Surdash and Dukan, *Rechinger* 10122. **MSU/FKI**: between Kirkuk and Chemchamal, *Hadač* 119; **FUJ**: Balad Sinjar, *Guest* 4101!; Karya Bat Khan, nr Balad Sinjar, *Field & Lazar* 611!; between Balad Sinjar and Tal Afar, *Guest* 4143!, 4192!; Mir Qasim, *Field & Lazar* 553; Hamaida, *Bornmüller* 1322; Shargat (Ashur), *Bornmüller* 1032; 60 km S of Balad Sinjar, *Alizzi & Husain* 33926! **FNI**: Maltai, between Tal Kaif and Mar Ya'qub, *Nábělek* 3907; between Pesh Khabur and Bawart, *Kotschy* 400! (type); NE of Mindan bridge, *E. Chapman* 26211!; **FAR**: Jabal Qara Choq, *Gillett* 4245; Arbil, *Guest* 30021!; between Arbil and Kirkuk, *Thesiger* 472; **FKI**: Altun Kupri, *Guest* 4029; 40 km W of Kirkuk on road to Altun Kupri, *Alkas* 18568!; Kani Domlan hills, nr Kirkuk, *Guest* 4248!; 2 km N of Kirkuk, *Rawi* 21552!; **FPF**: 10 km E of Mandali, *Rechinger* 9643!; Koma Sang, nr Mandali, *Rawi* 20591!, 20659!

Syria, SW Turkey, W Iran.

Guest (1933) noted the Kurdish name GUFGŪF for *Cousinia*, probably a vernacular name for this species in the plains around Arbil.

SECTION **HAUSSKNECHTIANAE** *RECH.F.*

2. **Cousinia haussknechtii** *C.Winkler*, Acta Horti Petrop. 12: 238 (1893). –Type: Avroman [et Schahu], *Haussknecht* 569 (G!, holo.); Rechinger, Fl. Iranica 90: 187 (1972).

C. serotina Rech.f. in Anz. Öst. Akad. Wiss., Math.-Nat. Kl. 101 (1964).

Perennial, 50–110 cm, Stems erect, with strict branches, striate-furrowed, whitish, finely appressed-pubescent, finally glabrous. Leaves yellowish-green above, appressed white-tomentose below; basal leaves with petiole up to 8 cm long, lamina thick, rigid, narrowly lanceolate, up to 20 × 3.5 cm, deeply irregularly sinuate-lobate, lobes at right angles to midrib, veins prominent, prolonged into strong spines; cauline leaves smaller, subsessile or sessile, not decurrent, leaves of flowering branches much reduced. Capitula ± 30-flowered, sessile in strictly erect racemes, densely yellowish or brownish glandular branches. Involucre broadly ovoid, 2–3 cm in diameter including spines; phyllaries ± 100, subulate, erect-spreading, stramineous, densely clothed with yellowish glands, arachnoid-lanate in lower part; receptacle bristles smooth. Corollas purple or pink, 12–15 mm. Achenes greyish-brown, thick, compressed, 5.5 × 3 mm, indistinctly striate lengthwise, apex truncate. Fig. 17: 2–3.

HAB. On rocky mountain slopes; alt. 1600–2400 m; fl. & fr. Jul.-Sep.
DISTRIB. Rare in Iraq, only found a few times in the SE sector of the thorn-cushion zone. **MSU**: Pira Magrun, *Hadač, Haines & Walid al-Hashimi* W.1855!; *Eig & Durdavani* 505; Avroman [et Schahu], *Haussknecht* 569 G-BOIS (type)!

Iran.

3. **Cousinia rawanduzensis** *Mehregan*, Iran. J. Bot. 17, 2: 144 (2014). –Type. Iraq, mons Helgurd ad confines Persiae, (36°40'N, 44°50'E), in valle supra pagum Nowanda, ± 2000–2600 m, *Rechinger* 11347 (W!, holo.).

C. leptolepis auct. non Bornm. & Gauba, Rech.f. Rechinger, Fl. Iranica 90: 111 (1972) p. p.

Perennial, up to 100 cm, corymbosely and paniculately branched above, many-headed. Stems slender, erect, whitish, glabrous or minutely glandular, finely striped. Leaves green, coriaceous, glabrous, minutely glandular-punctate, conspicuously white-veined, prickly; basal leaves with petiole 3–4 cm, lanceolate, up to 35 × 10 cm, pinnatilobed, with triangular irregular segments and stramineous 5–10 mm long terminal spines; cauline leaves similar but ovate or triangular, attenuate at base, sessile, shortly adnate-decurrent. Capitula 30–50-flowered on 1–5 cm long peduncles. Involucre ovate, 1–2 cm in diameter including spines; phyllaries 90–100, arachnoid and often glabrescent, linear-subulate, narrowed into long stramineous spines, lower and median ± spreading or recurved. Receptacle bristles smooth. Corollas pink, 15–20 mm. Achenes brown with darker spots, linear-oblong, 4.5–6 × 1–3 mm.

HAB. High rocky mountain slopes in the thorn-cushion zone, sometimes overgrazed; alt. 1850–2500 m; fl. & fr. Jul.-Sep.
DISTRIB. Quite common in the central sector of the mountain region of Iraq, occasional in the eastern sector. **MRO**: Algurd (Helgord) Dagh, *Gillett* 12470!; Mt. Helgurd, *Rechinger* 11347 W (type)!; Mt. Qandil above Pushtashan, *Rechinger* 11109!; Hisar-i Rost, *Guest & Husham* (*Alizzi*) 15814!; Chiya-i Mandau, nr Walza, *Guest* 2735!; on the track from Bardanas to Qandil, *Rawi & Serhang* 24620!; mt. Serin near road to Qandil, *Rawi & Serhang* 24010!

Note. Herbarium material belonging to *C. rawanduzensis* were formerly misidentified by Rechinger (1972) as *C. leptolepis*, which is a narrow endemic from C Elburz in N Iran.

N Iraq; endemic.

SECT. **CYNAROIDEAE** *BUNGE*.

4. **Cousinia aintabensis** *Boiss. & Hausskn.* in Boissier, Fl. Orient. 3: 503 (1875); Mehregan & Kadereit in Willdenowia 38: 304 (2008).

Arctium aintabense (Boiss. & Hausskn.) Kuntze, Revis. Gen. Pl. 1: 307 (1891).
Cousinia arbelensis C.Winkl. & Bornm. in Bull. Herb. Boissier 3: 566 & 567, t. 14 (1895). –Type. Iraq, habitat in Kurdistaniae terra Arbela, ad radices montis Kuh-Sefin prope pagum Shaklava, 1200 m, 4.6.1893, *Bornmüller 1403b* (B-100088397, lecto.!; B!, G!, isolecto.; designated by Mehregan, 2008); Bornm. in Beih. Bot. Centralbl. 60B: 205 (1939); Zohary in Dep. Agr. Iraq Bull. 31: 153 (1950); Rawi, ibid. Tech. Bull. 14: 112 (1964); Hub.-Mor. in Fl. Turkey 5: 349 (1975); Rechinger, Fl. Iranica 90: 236 (1972).
C. arbelensis var. *pinnata* C.Winkl. & Bornm. in Bull. Herb. Boissier 3: 568, t. 14 (1895). –Type. Iraq, habitat in Kurdistaniae terra Arbela, ad radices montis Kuh-Sefin prope pagum Shaklava, 1000 m, 30.5.1893, *Bornmüller* 1403 (B-100088396!, lecto.; JE!, K!, P!, W!, WU!, isolecto., designated by Mehreganin in Willd. 38: 304 (2008)).
C. handelii Bornm., Öst. Bot. Zeitschr. 62: 187 (1912). –Type. Iraq, Mespotamia, in lapidosis (substrato calcareo) montium Dschebel Sindschar supra oppidum Sindschar, c. 600–700 m, 8.6.1910, *Handel-Mazzetti* 1359 (W!, holo.; B!, WU!, iso.); Handel-Mazzetti in Ann. Naturh. Mus. Wien 27: 442 (1913); Zohary in Dep. Agr. Iraq Bull. 31: 153 (1950); Rawi, ibid.: 112 (1964).
C. sinjarensis Rech.f. in Publ. Cairo Univ. Herb. 7/8: 289 (1977). –Type. Iraq, west side of Jabal Sinjar, 800 m, rocky slope, ravine, 30.5.1968, *Anders* 2172 (W!, holo.; Iraq, west side of Jabal Sinjar, 600–1200 m, 12.5.1969, *Anders* 2689 (W!, paratype).

Plant up to 60 cm tall, arachnoid- or canescent-tomentose to glabrous. Stems branched from the base. Leaves herbaceous or leathery, arachnoid- or canescent-tomentose on both surfaces, ± glabrescent above or rarely glabrescent on both sides; basal leaves up to 25 × 10 cm, lyrate, pinnatisect or pinnately oblong-lanceolate, spiny-lobed or with spiny-dentate margin; cauline leaves gradually smaller and less divided towards the apex, ovate, ovate-lanceolate to linear-lanceolate, spiny-lobed to -dentate, usually long-decurrent to form winged stems. Capitula 3–7 cm broad with appendages, arachnoid-tomentose, ± glabrescent or glabrous; phyllaries 40–100; middle ones ± spreading to spreading-incurved or rarely bent downward, their free part abruptly expanded above into usually concave, keeled, ovate, long-deltoid, deltoid- or rhomboid-lanceolate appendage; appendage 15–30 mm long, 7–15 mm broad, usually attenuated into a long or short spine at apex, with 2–5 or rarely more tiny spines or prickles on both sides. Receptacular bristles rough or smooth. Flowers (30–)50–100; corolla reddish-pink, pink, purple or straw-coloured, 18–30 mm long; anthers reddish-pink, pink, purple or pale-purple. Achenes 4–5 mm long.

HAB. Lowlands of northern Iraq, on stony slopes and in oak forests; alt. 320–1400 m; fl. May-Jun.
DISTRIB. **MAM**: Mosul, Atrush-Rabetki, *Chapman* 9338!; between Atrush village and pine woodland, *Hossain* 4268!; 30 km from Mosul to Aqra, *Qaisi* 49718!; **MRO**: Shaqlawa, *Haines* 2049! *Agnew & al.* W.2049!; Jabal Safin, *Sakira* 38252!, *Rawi* 9078!; Shaikh Adi nr Ayn Sifni, 13.6.1934, *Field & Lazar* 712!; Pirmum Dagh, *Gillett* 8024!

Turkey, (?) N Syria.

5. **Cousinia kopi-karadaghensis** *Rech.f.*, Fl. Iranica 90: 249 (1972). – Type. Iraq, Qopi Qaradagh, rocky slope, 1500 m, 17.6.1959, *Haines* W.1545 (E!, holo.; K!, W!, iso.). Mehregan & Kadereit in Willdenowia 38: 307 (2008).

Biennial or perennial, monocarpic, 20–60 cm, finely appressed tomentose, soon glabrescent. Stem divaricately branched from base, with few 1-headed branches. Basal leaves herbaceous, with petiole up to 2 cm long, lamina undivided, oblong-ovate, up to 6 × 4 cm, entire, obsoletely prickly or lyrate-pinnatisect with up to 5 pairs of small, unequal triangular or oblong-lanceolate lateral segments and an enlarged ovate or oblong-lanceolate terminal lobe to 10 × 4 cm; cauline leaves soon smaller, undivided, oblong to lanceolate, sinuate-spinulose, decurrent into long and narrow wings. Capitula 50–70-flowered. Involucre globose, 2.5–4 cm in diameter, including spines; phyllaries 120–130, densely imbricate, arachnoid, glabrescent, with coriaceous, broadly triangular appendages which suddenly contract into a spine up to 5 mm long, hastate, 4–6 mm broad at base, entire at margin; receptacle bristles scabrous. Corollas purple, about 22 mm. Achenes 4 × 1.5 mm, striped lengthwise, apex truncate and obsoletely denticulate.

HAB. Mountainside in rich *Quercus* forest on steep rocky slopes; alt. 600–1400 m; fl. & fr. May-Jun.
DISTRIB. Occasional in the SE to southern sector of the forest zone of Iraq. MSU: Qopi Qaradagh, *Haines* W. 1545 (type)!; Kopi Qaradagh, 5 km SW of Kanitacht, *Eig & Feinbrun* 507 (W!); Qaradagh Mt, *Makki* 525!; 16 km NW of Darband-i Khan, *Rawi, Husham (Alizzi) & Noori* 29443!; Darband-i Khan, *Haines* W.1635!

Endemic to NE Iraq.

6. **Cousinia odontolepis** *DC.*, Prodr. 6: 556 (1838). –Type: Iraq, in deserto Assyriae, 1837, *Aucher-Eloy* 3496 (G-DC!, lecto[5].; P!, K!, isolecto., designated by Mehregan in Willdenowia 38: 312 (2008); Boissier, Fl. Orient. 3: 509 (1875); C. Winkler, Synopsis 276 (1892); Zohary in Dep. Agr. Iraq Bull. 31: 153 (1950); Rawi, ibid. Tech. Bull. 14: 112 (1964); Rechinger, Fl. Iranica 90: 229 (1972); Mehregan & Kadereit in Willdenowia 38: 312 (2008).

Arctium odontolepis (DC.) Kuntze, Revis. Gen. Pl. 1: 308 (1891).

Plant up to 50 cm high, densely arachnoid-tomentose, tomentose or ± glabrescent. Stems branched from the base or higher. Leaves leathery to herbaceous, tomentose on both sides, or tomentose beneath and glabrescent above; basal leaves up to 25 × 10 cm including spines, usually lyrate, sometimes lanceolate or oblanceolate, dentate to deeply pinnatisect or pinnatipartite, with spiny-dentate lobes; stem leaves sessile, gradually smaller and less divided towards the apex, ovate to oblong-lanceolate, spiny-lobed, cuneate at base, decurrent to form winged stems. Capitula 2.5–7 cm broad with appendages, usually arachnoid-tomentose or ± glabrescent; phyllaries 50–120; middle ones recurved, ± spreading, spreading-incurved, appressed-incurved or -erect or imbricate, their free part usually abruptly expanded above into a usually keeled, concave or flat, rhomboid, deltoid, rhomboid- or deltoid-lanceolate or lanceolate appendage; appendage 10–30 mm long, 4–10 mm broad, attenuate into a long spine at apex, arachnoid-tomentose, tomentose or glabrescent, with 2–4(–6) spines on both sides. Receptacular bristles usually rough. Flowers (20–)40–120; corolla purple, reddish-pink or flavescent, 18–30 mm long; anthers concolorous. Achenes 4–5 mm long. (Description from Mehregan & Kadereit l.c.).

Two subspecies in Iraq:

Uppermost leaves not enclosing the involucre; phyllaries usually sparsely arachnoid-tomentose or glabrescent, with ± spreading, recurved or inflated-incurved appendages (Iraq, provinces Mosul and Sulaymaniyah) . . . . b. subsp. *odontolepis*
Uppermost leaves enclosing the involucre; phyllaries usually densely arachnoid-tomentose, with recurved appendages (Iraq, Erbil, Qandil Mts and E of Rawanduz) . . . . a. subsp. *kurdica*

a. subsp. **odontolepis** Fig. 18: 4.

Cousinia cymbolepis Boiss., Diagn. Pl. Orient. ser. 1: 119 (1846). –Type. Iraq, in regionibus superioribus montis Gara Kuristaniae, 6.8.1841, *Kotschy* 380 (lecto., G-BOIS!, lecto.; B!, E!, G!, JE!, K!, M!, P, isolecto. [also lectotype of *C. schultziana*]!, W!; designated by Mehregan in Willdenowia 38: 312 (2008).

[5] As noted before by Rechinger (1972), Aucher-Eloy's material clearly cannot have been collected "in deserto Assyriae" because the species grows in mountainous regions of N and NE Iraq and not in deserts (Mehrangan l.c.).

C. baueri Bornm. & Nábělek in Österr. Bot. Z. 63: 61. 1913. – Type: Iraq, Assyria, in derelictis horti coenobii Mâr-Jakub prope Mosul, c. 1000 m, 7.6.1910. *Nábelek* 3903 (BRA, photo!).
C. odontolepis var. *leiolepis* Bornm. in Repert. Spec. Nov. Regni Veg. 46: 282. 1939. – Type. Jebel Baykhair near Zakho, 15.-18.6.1934, *Field & Lazar* 798 (W!, lecto.; P!, K!, G! isolecto. designated by Mehregan in Willdenowia 38: 312 (2008).
C. leatherdalei Rech. f. in Österr. Bot. Z. 100: 462. 1953. –Type. Iraq, Mosul, Sarsang, 8 km W of Amadia, 3200 ft., 4.6.1947, *Leatherdale* 101 (BM-000927947!, holo.; W, iso. [fragment]!).
C. mazu-shirinensis Rech. f., Fl. Iranica 90: 226. 1972. – Type. Iraq, Kurdistan, Arbil, Kani Mazu Shirin, 5000 ft., 20.6.1961, *Wheeler-Haines* 2046 (holo.; E!; K!, W!, iso.).
C. acanthophysa Rech. f., Fl. Iranica 90: 227. 1972. –Type. Iraq, Kurdistan, Mosul, ad confines Turciae prov. Hakkari, in ditione pagi Sharanish, in montibus calc. a Zakho septentrionem versus, in declivibus saxosis (Tonschiefer) montis Zawita, 1400 m, 4.-9.7.1957, *Rechinger* 10922 (W! holo; B!, E!, K!, M!, iso.).

Very variable. Uppermost leaves not enclosing the involucre. Phyllaries usually sparsely arachnoid-tomentose or glabrescent, with tightly imbricate, appressed-incurved, ± spreading, spreading-erect, recurved or inflated-incurved appendages.

HAB. Mountains of N and NE Iraq, on stony slopes and in oak forests; alt. 800–1400 m; fl. (May-) Jun.-Aug.
DISTRIB. Endemic to the mountains of N and NE Iraq. **MAM**: Mosul: Galli Zawita, NE of Zakho, 1400 m, 8.7.1957, *Al Rawi* 23563; Zawita gorge, Aqra-Dinart, *Ghafran* 26137; prov. Hakkari, in ditione oppidi Zakho, pass between Dagh al Radzjiem and Sharanish, *Rechinger* 10796!; Dohuk to Amadiya, *Rechinger* 11552! 21 km E. of Dohuk to Amadiya, at Zawita, *Rechinger* 11549 (E, W); 103 km N. of Mosul towards Zakho, *Rechinger* 10652!; Jabal Khantur nr Sharanish, *Rechinger* 12049!; Ispindari, above Suwaratuka, 1250 m, *Chapman* 26328! Sarsang, *Haines* W. 581!. W. 1333! 23 km NE of Zakho, 25 km S of Sharanish, *Rawi* 23189; 5 km S of Zakho, *Rawi* 23092; Bikher Mt near Zakho, *Rawi* 23049 & 22984; Badi, *Omar & al.* 49639; Dori village, 5 km E Kani Masi, *Omar & Kaisi* 45432; Kergedraija, *Rawi* 9140; Khantur Mt, NE of Zakho, *Rawi* 23478 & 23403; Seramadia, *Guest* 4981; Seramadia, 13.8.1961, *Agnew & Haines* W. 2107!; Sulaf, *Haines* W. 2062; Sundur, near Dohuk, *Haines* W. 1570! Zawita gorge, *Guest* 3715!; M. Potine (Botin) 10 km NNW of Shirwan Mazin (Mazu Shirin), *Haines s.n.*!; Zawita nr Shirwan Mazin (Mazu Shirin), *Haines* 2147!; on the mountain NNW of Seri Hasan Beg, *Guest* 2907!; **MRO**: Berd Agha Gin Village, c. 18 km NW of Rania, *Rawi et al.* 28691. **MSU**: Sulaymaniyah: Asme, *Haines* W. 2107!; 10 km from Sulaimaniyah to Choart, *Noori & Khamidi* 41254; Azmir, 860 m, 15.5.1971, *Omar et al.* 38041; Gweija Dagh above Sulaimaniya, *Rawi & Gillett* 11715.

Study of material of *Cousinia odontolepis* subsp. *odontolepis, C. cymbolepis* (≡*C. schultziana*), *C. baueri, C. leatherdalei, C. mazu-shirinensis* and *C. acanthophysa* revealed that all these are very variable but connected by intermediate forms. None of them can be circumscribed as separate species (Mehregan & Kadereit op.cit.).

b. **Cousinia odontolepis** subsp. **kurdica** (*C.Winkl. & Bornm.*) *Mehregan*, Willdenowia 38: 315 (2008).

C. kurdica C.Winkl. & Bornm. in Bull. Herb. Boissier 3: 564 & 567. 1895. – Type. Iraq, Arbil (Erbil), in terrae Riwandous monte Sakri-Sakran altitudine 13-1700 m, 16.1893, *Bornmüller* 1404 (B-100093376!, lecto.; B!, G!, isolect., designated by Mehregan in Willdenowia 38: 315 (2008).
C. carduchorum C.Winkl. & Bornm. [as "*minor*"] in Bull. Herb. Boissier 3: 566 & 568 (1895). –Type. Iraq, in Kurdistaniae terra Riwandous, in monte Händarin, 12-1300 m, 28.6.1893, *Bornmüller* 1406 (B-100088384!, lecto.; isolectotypes: B ["1406b"], G!, JE!, isolecto., designated by Mehregan in Willdenowia 38: 315 (2008).
C. carduchorum var. *major* C.Winkl. & Bornm. [as "*major*"] in Bull. Herb. Boissier 3: 569. (1895). – Type. Iraq, in monte Sakri-Sakran ditionis Riwandous, 1700 m, 23.6.1893, *Bornmüller 1405* (B-100088383!, holo.; fragment in JE!).
C. qandilica Rech.f., Fl. Iranica 90: 223 (1972). – Type. Iraq, Arbil, Montes Qandil, ad confines Persiae, in quercetis saxosis inter Shahidan & Pushtashan, substr. calc., 1200 m, 28.7.1957, *Rechinger* 11003 (W!, holo.; M!, iso.) (for detailed notes on synonomy see Mehregan & Kadereit op.cit. 315).

Uppermost leaves enclosing the involucre. Phyllaries usually densely arachnoid-tomentose, with recurved appendages.

HAB. Mountains, on stony slopes and in oak forests; alt. 1000–2600; fl. Jun.-Aug.
DISTRIB. Endemic to N Iraq. **MRO**: Kermasur lake, Qandil range, *Rawi & Serhang* 24137; Mt. Helgurd (Algurd) above Nowanda, *Rechinger* 11348!; Qandil mts., below Pushtashan towards Shahidan, *Rechinger* 11022!; Baaki Hawaran Mt, *Rawi* 23960; Koorak Mt, N of Shahidan Project, *Rawi & Serhang* 23800.

Cousinia odontolepis subsp. *kurdica, C. carduchorum* and *C. qandilica,* all distributed in NE Iraq (Qandil mountain range and neighbouring area), are connected by intermediate forms. Subspecies *kurdica* is similar to subsp. *odontolepis* and differs mainly in having a densely arachnoid-tomentose indumentum.

Endemic.

7. **Cousinia algurdina** *Rech.f.* in Anz. Öst. Akad. Wiss., Math.-Nat. Kl. 101: 344 (1964). –Type. Iraq, Erbil, Mons Helgurd (Algurd) ad confines Persiae, in valle supra pagum Nowanda, 2600–3000 m, 10.-14.8.1957, *Rechinger* 11465-II (W-1972-01633!, lecto.; B!, E!, K!, M!, W!, isolecto., designated by Mehregan in op.cit.); Rawi in Dep. Agr. Iraq Tech. Bull. 14: 112 (1964); Hub.-Mor. in Fl. Iranica 122: 222(1977); Mehregan & Kadereit in Willdenowia 38: 317 (2008).

C. cynaroides (M.Bieb.) C.A.Mey. var. *arlgirdensis* Blakelock in Kew Bull. 4: 50 (1949); Rawi in Dep. Agr. Iraq Tech. Bull. 14: 112 (1964). –Type. Iraq, Sulaymaniyah, Al Gird Dagh, 2700 m, 24.7.1932, *Guest* 2951 (lK [specimen with five capitula]!, lecto.; K!, isolecto.; designated by Mehregan in Willdenowia 38: 317 (2008).

Biennial or perennial, monocarpic, 20–50 cm, densely arachnoid-floccose. Stems erect, robust, divaricately branched with single-headed branches. Basal leaves rosulate, with petiole up to 4 cm, lamina oblong- to linear-lanceolate, 8–15 × 3–4 cm, deeply irregularly sinuate-lobate, lobes with prominent veins prolonged into strong spines up to 10 mm long; lower cauline leaves similar but subsessile or sessile and then decurrent into wings up to 2 cm long; upper cauline leaves smaller and less incised, triangular. Capitula more than 100-flowered. Involucre depressed-globose, 5–7 mm in diameter including spines; phyllaries 40–60, coriaceous, rigid, with spreading or subrecurved, lanceolate, carinate, longitudinally folded appendage, 5–8 mm broad at base, prolonged into a straight spine up to 10 mm; receptacle bristles glabrous. Corollas white, 23–26 mm. Achenes not known. Fig. 17: 1.

HAB. Rocky alpine summit, on an east-facing mountain slope and southern slope; alt. 2400–3500 m; fl. & fr. Jul.-Aug.
DISTRIB. Occasional in the central sector of the thorn-cushion zone and alpine regions of Iraq. **MRO**: Algurd (Helgord) Dagh, *Guest* 2951!; in valley above Nawanda, *Rechinger* 11465; Qandil range, *Rawi & Serhang* 24042!; Malakh (Gomasur), *Rawi* 2401!; Montes Qandil ad confines Persiae, *Rechinger* 11116!

Although *Cousinia algurdina* along with *C. odontolepis* subsp. *kurdica* are distributed in the Qandil range in N Iraq, they are not sympatric. *C. algurdina* grows mainly in the alpine zone, whereas *C. odontolepis* subsp. *kurdica* grows in the subalpine zone or lower. *C. algurdina* is more similar to *C. canescens* but differs by appendages with spines on both sides and a yellowish indumentum especially in its upper part.

Endemic to the alpine zone of Qandil range, close to the border with Iran and likely to occur in W Iran.

8. **Cousinia macrolepis** *Boiss. & Hausskn.* in Boissier, Fl. Orient. 3: 505 (1875); Winkler, Synopsis 380 (1892); Zohary in Dep. Agr. Iraq Bul. 31: 153 (1950); Rawi, ibid. Tech. Bull. 14: 112 (1964); Rechinger, Fl. Iranica 90: 223 (1972); Mehregan & Kadereit in Willdenowia 38: 325 (2008).

Arctium macrolepis (Boiss. & Hausskn.) Kuntze, Revis. Gen. Pl. 1: 308 (1891). – Type. Iraq, Sulaymaniya, in monte Pir Omar Gudrun Kurdistaniae Persicae, 6.1867, *Haussknecht* 570 (G-BOIS [sheet with tree labels designated as *C. macrolepis,* lecto. designated by Mehregan & Kadereit Willdenowia 38: 325 (2008), B, G, JE, W, isolecto.)

Cousinia gigantosphaera Rech.f. in Anz. Öst. Akad. Wiss., Math.-Nat. Kl. 101: 345 (1964). –Type. Iraq, Kurdistan, Asma near Sulaimaniya, slope above Casino, poor oak scrub, 13.7.1961, *Haines s.n.* (E, holo.; W, iso.); Rechinger, Fl. Iranica 90: 225 (1972).

Plant up to 50 cm high, densely arachnoid-tomentose or ± glabrescent. Stems divaricately branched from base or higher, densely leafy to apex. Leaves leathery, tomentose on both sides; basal leaves up to 25 × 15 cm including spines, obovate, oblong-lanceolate to lanceolate, dentate to deeply pinnatisect with spiny-dentate lobes; stem leaves sessile, gradually smaller and less divided towards the apex, ovate-elliptic to broadly lanceolate,

Fig. 18. **Cousinia algurdina**. 1, habit × 1. **Cousinia haussknechtii**. 2, leaf (upper surface) × ⅔; 3, inflorescence × ⅔. **Cousinia odontolepis** subsp. **odontolepis**. 4, capitulum × ⅔. 1 from *Rawi & Serhang* 24501; 2–3 from *Haines* W1855; 4 from *Omar & Kaisi* 45432. Drawn by Moira Rae Carter.

spiny-lobed, decurrent to form winged stems, upper leaves enclosing lower part of the involucres. Capitula 4–8 cm broad with appendages, usually arachnoid-tomentose. Phyllaries 80–100 or more; outer ones similar to uppermost leaves but gradually smaller; middle ones densely imbricate, their free part abruptly expanded above into a usually keeled, concave or apically recurved, deltoid, rhomboid- or sagittate-lanceolate appendage; appendage 15–25 mm long, 8–15 mm broad, attenuate into a short spine at apex, arachnoid-tomentose, with 3–8 prickles on both sides. Receptacular bristles rough. Flowers 70–160; corolla purple (dirty white in some herbarium material), 28–35 mm long; anthers concolorous. Achenes 4–5.5 mm long. (Description from Mehregan & Kadereit op. cit.). Fig. 18: 2.

HAB. Hillside with oak scrub, and in oak forests; alt. 950–3500 m; fl. & fr. May-Jun.
DISTRIB. NE Iraq. **MRO**: Haibat Sultan Dagh (Mt), N of Koi Sanjaq, *Rawi et al.* 28264!; Pira Magrun, *Haines* W. 2042!, 1860!; Talan region near Dokan, *Karim* 39293; Azmir (Asme) Mt., *Karim* 39320!.

Cousinia macrolepis and *C. gigantosphaera*, both described from around Sulaimaniya, are characteristic in having uppermost leaves enclosing the leaf-like outer phyllaries, and differ from each other mainly in the shape of the phyllaries (deltoid versus rhomboid or sagittate-lanceolate). The two taxa are connected by intermediate forms and cannot be maintained as separate species.

Iran.

9. **Cousinia pergamacea** *Boiss. & Hausskn.* in Boissier, Fl. Orient. 3: 513 (1875); Winkler, Synopsis 272 (1892); Zohary in Dep. Agr. Iraq Bull. 31: 153 (1950); Rawi, Tech. Bull. 14: 112 (1964); Rechinger, Fl. Iranica 90: 246 (1972); Mehregan & Kadereit in Willdenowia 38: 330 (2008).

Arctium pergamaceum (Boiss. & Hausskn.) Kuntze, Revis. Gen. Pl. 1: 308 (1891).
Cousinia wheeler-hainesii Rech.f. in Anz. Öst. Akad. Wiss., Math.-Nat. Kl. 101: 353 (1964); Rechinger, Fl. Iranica 90: 246 (1972).

Plant up to 80 cm high, arachnoid-tomentose or glabrescent. Stems branched from base or higher. Leaves leathery or herbaceous, tomentose on both sides or glabrescent above and tomentose to ± glabrescent beneath; basal leaves up to 40 × 15 cm including spines, obovate, ovate, lanceolate to oblong-lanceolate, undivided to deeply sinuate-lobate with spiny-dentate lobes; stem leaves sessile, rarely distinctly attenuate at base, gradually smaller and less divided towards the apex, oblong-ovate to lanceolate, spiny-lobed, usually long-decurrent to form winged stems. Capitula 1.5–5 cm broad with appendages, glabrous, sparsely arachnoid or glabrescent. Phyllaries 50–140; middle ones imbricate, with prominent yellow midrib, their free part abruptly expanded above into a usually flat, erect, deltoid or deltoid-sagittate appendage, at least two times broader than base; appendage 5–17 mm long, 5–15 mm broad, attenuate into a spine at apex, usually glabrous, usually without or rarely with very tiny appressed prickles on both sides. Receptacular bristles rough. Flowers 40–110; corollas yellow, 20–27 mm long; anthers rose or yellow. Achenes 4–6 mm long.

HAB. Stony slopes and mountain sides, usually in oak forests; alt. 1300–1600 m; fl. & fr. Jun.-Aug.
DISTRIB. NE Iraq. **MRO**: Penjwin, Malakawa pass, *Rechinger* 10429! 12238!; Mela Kowa (on Sulaymaniya-Penjwin highway), *Rawi* 22433!

Cousinia pergamacea subsp. *sardashtensis* (Rech. f.) Mehregan, distributed in southern parts of Iranian Azerbaijan, differs from subsp. *pergamacea* only in having spines on both sides of the phyllary appendages. It is endemic to W Iran.

W Iran.

10. **Cousinia inflata** *Boiss. & Hausskn.* in Boissier, Fl. Orient. 3: 512 (1875). –Type. Avroman [& Shahu, in Iran], *Haussknecht* s.n. Winkler, Synopsis 273 (1892); Zohary in Dep. Agr. Iraq Bull. 31: 153 (1950); Rawi, Tech. Bull. 14: 112 (1964); Rechinger in Fl. Iranica 90: 249 (1972); Mehregan & Kadereit in Willdenowia 38: 305 (2008).

C. straussii Hausskn. & C.Winkler in Acta Horti Petrop. 14: 235 (1895); Rechinger, Fl. Iranica 90: 242 (1972).

Biennial or perennial, monocarpic, 20–40 cm, appressed arachnoid-tomentose, soon glabrescent. Stem divaricately branched from base, with few 1-headed branches. Basal leaves membranous, prickly denticulate, with petiole up to 6 cm long, lamina undivided, broadly ovate, acuminate, up to 15 × 7 cm, or lyrately pinnatisect, oblong, with 2–5 pairs of small lateral segments and an enlarged terminal segment to 15 × 9 cm; cauline leaves soon smaller, oblong and ovate, undivided, subsinuate-spinose or prickly denticulate, long decurrent into broad cuneate wings. Capitula many-flowered; involucre depressed-globose, 3–4.5 cm in diameter including spines; phyllaries 100 or more, densely imbricate, with coriaceous, glabrous, reflexed, horizontally or erecto-patent, triangular-acuminate appendages up to 10 mm broad, entire at margin, with a terminal spine up to 3(–5) mm long; receptacle bristles scabrous. Corollas purple, *c.* 22 mm. Achenes 6 × 2 mm, sub-prismatic, striped lengthwise, apex obsoletely margined.

HAB. On mountain sides; alt. (1350–)1700–2450 m; fl. & fr. Jun.
DISTRIB. Occasional in the SE sector of the thorn-cushion zone of Iraq. **MSU**: Hawara Birza, nr Surdash, *Rawi, Noori & Kass* 29467!; Malakawa, *Rawi, Husham* (*Alizzi*) *& Noori* 29507!; Kamarspa, Rawi 22179!; Tawila, *Rawi* 21947!, *Rechinger* 10215!; Mt. Hawraman (Avroman), *Haussknecht* s.n. (type).

Endemic, unless it occurs on Mt. Shahu in W Iran.

SECT. **PUGIONIFERAE** BUNGE

11. **Cousinia wesheni** *Post*, Pl. Postianae 2: 13 (1891); Winkler, Synopsis 235 (1892); Dinsmore in Post, Fl. Syria, Palest. & Sinai 2, 2: 86 (1933); Rechinger, Fl. Lowland Iraq: 643 (1964); Rechinger, Fl. Iranica 90: 280 (1972); Huber-Morath in Fl. Turkey 5: 353 (1975).

C. chaborasica Bornm. & Hand.-Mazz., Öst. Bot. Zeitschr. 62: 183 (1912); Handel-Mazzetti in Ann. Naturh. Mus. Wien 27: 451 (1913); Zohary in Dep. Agr. Iraq Bull. 31: 152 (1950).

Biennial or perennial, 25–50 cm, densely white woolly-tomentose, glabrescent above. Stem branching from base, paniculately corymbose above. Basal leaves with petiole 2–4 cm long, lamina narrowly or broadly oblong-lanceolate, 20–25 × 5–10 cm, pinnatisect with 5–8 pairs of broadly triangular spinose lateral segments, spines up to 10 mm long; cauline leaves soon smaller, prickly dentate, uppermost triangular, appressed to capitula, all decurrent along the whole internode into broad, spiny wings. Capitula 15-numerous, 30–40-flowered; involucre ovoid to subglobose, 4–7 cm in diameter including spines; phyllaries 25–35, abruptly contracted from an ovate-lanceolate base into a keeled linear-lanceolate spreading 1–2 cm long spine; receptacle bristles smooth. Corollas purple, 12–15 mm. Achenes oblong-obpyramidate, 3–3.5 × 3 mm, indistinctly striped lengthwise, apex truncate and obtusely denticulate. Fig. 18: 3.

HAB. Stony hills and steppic plains sometimes at foot of lower mountains; up to alt. 600–700 m; fl. & fr. May-Jul.
DISTRIB. Occasional in the steppe region of Iraq. **MSU**: Jarmo, *Helbaek* 1954. **FNI**: between Khabur river and Bara, Jabal Sinjar, *Handel-Mazzettii* 1588. **FKI**: 5 km SW of Daquq, *Rawi* 22855!; Kifri, *Rechinger* 9966! **DGA**: 4 km E of Samarra, *Rawi* 20335!, *Rechinger* 9554.

Syria, Turkey (SE Anatolia).

TAXA DOUBTFULLY RECORDED FROM IRAQ

Cousinia lucida *DC.*, Prodr. 6: 553 (1838). – Type. "In deserto Assyriae", *Aucher-Eloy* 3500. Boissier, Fl. Orient. 3: 469 (1875); Winkler, Synopsis 218 (1892); Zohary in Dep. Agr. Iraq Bull. 31: 162 (1950); Rawi in ibid. Tech. Bull. 14: 112 (1964); Rechinger, Fl. Iranica 90: 150 (1979).

Perennial, glabrous, 15–25 cm, corymbosely branched above. Stems slender, erect, whitish, finely striped. Leaves green, coriaceous, conspicuously reticulate-veined, shining, ± regularly lobate-dentate, with spines 2–5 mm long; cauline leaves congested, gradually smaller, lower ovate, 3–5 × 2–3 cm, decurrent into broad wings 1–2 cm long, uppermost enveloping capitula. Capitula 5–6, 4–5-flowered, sessile or subsessile. Involucre cylindrical, 5–8 mm in diameter including spines; phyllaries 15–18, linear-lanceolate, obtusely carinate, up to 10 mm long, with erect or spreading prickles 2–3 mm. Receptacle bristles smooth. Corollas yellow, 10 mm. Achenes obovate, brown, striped lengthwise, apical margins subdenticulate.

Only found once, with locality unclear, or possibly only occurring in W Iran. ?**FPF/LEA**, "in deserto Assyriae", *Aucher-Eloy* 3500 (type!).

Cousinia boissieri *Buhse* in Boiss. et Buhse, Nouv. Mém. Soc. Imp. Nat. Mosc. 12 (18): 1207 (1860); Boissier, Fl. Orient. 3: 499 (1875); Winkler, Synopsis 196 (1892); Rechinger, Fl. Iranica 139a: 179 (1979); Huber-Morath in Fl. Turkey 5 : 347 (1975).

C. layardii Ball & Barbey, Lausanne (1890); Winkler, Synopsis 196 (1892).

Biennial or perennial, 30–90 cm. Stems erect, glabrous, ivory-white, many-branched, not winged, forming a broad, 15–25-headed panicle. Basal leaves 4–6, in a rosette, coriaceous, densely veined, loosely arachnoid-woolly and soon glabrescent above, white-tomentose below, obovate to suborbicular, 6–14 × 4–10 cm, shortly or indistinctly petiolate, with short lobes, densely spiny-dentate; cauline leaves abruptly smaller and bract-like, lower ones broadly lanceolate to orbicular, auriculate, upper ones small with very long spines. Capitula 25–60-flowered. Involucre broadly ovoid to globose, 1.5–2.5 cm in diameter including spines; phyllaries 40–60, soon glabrescent and becoming bright green, oblong to lanceolate, suddenly constricted into a glabrous 3–6 mm long yellow spine; receptacle bristles finely denticulate above. Corollas yellow, sometimes reddish-suffused, 12–15 mm. Achenes obovate to obpyramidal, brown, 4–5 × 2–2.5 mm long, with white, undulate longitudinal ribs, apex shortly obtusely denticulate.

Known from only one very old gathering: "Mountains of Kurdistan, Mahmudiya", *Layard* s.n. (type of *C. layardii*)! The location of this place in the mountains is unknown. The somewhat brief reference to the place of publication of this species reflects the fact that the name was published at Lausanne in an 8-page pamphlet with "*Cousinia layardi*" as the title.

20. **JURINEA** Cass.

in Bull. Sci. Philom. Par. 1821: 140 (1821)
Susanna & Garcia-Jacas in Kubitzki, Fam. Gen. Vasc. Pl. 8: 137 [2006] (2007)
Outreya Jaub. & Spach, Ill. Pl. Orient. 1: 131, t. 68 (1843); *Jurinella* Jaub. & Spach, Ill. Pl. Orient. 2: 103 (1847)

C. C. Townsend

Perennial herbs or subshrubs, stems simple or frequently branched. Leaves simple and entire to pinnatilobed or pinnatisect, unarmed, frequently ± tomentose especially on lower surface. Inflorescence of a single capitulum to corymbosely branched. Capitula homogamous, discoid; involucre cylindrical to campanulate or globose; phyllaries linear to linear-lanceolate, very numerous, multiseriate, closely imbricate or outer and median with tips ± spreading, acute and ± spinose-tipped to rather blunt and foliose; receptacle flat or slightly convex, clothed with simple or dentate paleae usually subequalling achenes. Florets hermaphrodite, tubular, corollas mostly pink to purplish, rarely white to darker purple, gradually widened above and deeply 5-lobed. Achenes shortly cylindrical to obpyramidal, usually tetragonous, with a jagged, very short, membranous corona around apical margin; pappus long, multiseriate, deciduous or persistent, barbellate or plumose.

Around 100 species in the N temperate regions of the Old World, chiefly in W & C Asia. Four species in Iraq.

Jurinea (named in honour of Louis Jurine, 1751–1819, a Swiss zoologist).

Iljin, M.M. (1963). De genere *Jurinella* Jaub. et Sp. Not. Syst. (Leningrad) 22: 285–287.

1. Innermost pappus hairs not longer and coarser; pappus deciduous; dwarf caespitose mountain plants . . . . 4. *J. moschus*
   Innermost pappus hairs longer and coarser; pappus persistent; not dwarf caespitose mountain plants 2
2. Pappus hairs conspicuous, thistle-like, almost concealing florets . . . . . 3. *J. carduiformis*
   Pappus hairs inconspicuous, florets distinctly exserted 3
3. Phyllaries all closely appressed-imbricate; leaves with margins revolute, lower surface densely white-tomentose; upper leaves very narrow, not ± amplexicaul . . . . 1. *J. mesopotamica*
   Outer and median phyllaries ± recurved or squarrose; leaf-margins not revolute, lower surface not white-tomentose; upper leaves ± amplexicaul . . . . 2. *J. inuloides*

1. **Jurinea mesopotamica** *Handel-Mazzetti* in Ann. Naturh. Mus. Wien 27: 443, t. 17 f. 9, t. 18 f. 1 (1913).– Type. Iraq, Jabel Sinjar District, Wadi Shilu and a slope towards Bara, Jabal Sinjar, *Handel-Mazzetti* 1557 (B, W, syn.). Zohary in Dep. Agr. Iraq Bull. 31: 154 (1950); Rawi, ibid. Tech. Bull. 14: 118 (1964); Danin & Davis in Fl. Turkey 5: 441 (1977).

Perennial herb 30–45 cm tall, base with dense tufts of woolly hairs among scaly remains of previous years' leaves. Stems several, wiry, densely foliose, striate-angular, ± white-lanate, with slender suberect branches from about middle. Basal leaves absent at time of flowering; median cauline leaves 6–8(–11) × 0.5–0.8(–1.3) cm, upper and branch leaves much reducing and narrow; all leaves with strongly revolute margins, white-tomentose on lower surface, scabridulous above. Capitula solitary on stem and branches; involucre ovoid-cylindrical, 2.5–3 × 1–1.2 cm, narrowed to base and obconical in fruit; phyllaries in 6 series, all imbricate, tomentose dorsally, lanceolate to linear-lanceolate inwards. Florets 30–35 mm, lobes ± 6 mm, corollas rose-pink. Ripe achenes not seen; pappus brownish-red, shortly plumose, 15–17 mm with a few central hairs longer.

HAB. & DISTRIB. Lime-rich loam in a wadi and on a hill slope; very rare, only found once in the Sinjar region; fl. Jun. **MJS**: Wadi Shilu and a slope towards Bara, Jabal Sinjar, *Handel-Mazzetti* 1557 (type); Jabal Sinjar, *Kamel Khalaf* 31 (36514).

Turkey (SE Anatolia), Syria (Jabal Abd-al Aziz).

2. **Jurinea inuloides** *Boiss. & Hausskn.* in Boissier, Fl. Orient. 3: 581 (1875). –Type. Mt Avroman, *Haussknecht* 576 (K, G, P, W; syn.).

Zohary in Dep. Agr. Iraq Bull. 31: 154 (1950); Rawi, ibid. Tech. Bull. 14: 118 (1964).

Perennial herb 30–50 cm tall, base with dense tufts of woolly hairs among scaly remains of previous years' leaves, these occurring also in axils of low branches where present. Stem branched from near base or only in upper half, striate to somewhat angular, glabrous or sometimes tomentellous above. Basal leaves absent at time of flowering; median cauline leaves 4–10 × 0.75–1.6 cm, acute to apiculate, ± amplexicaul at base, minutely scabrid; branch leaves rapidly reducing. Capitula solitary on stem and branches; involucre campanulate, 2.5–3 × 1.8–2 cm; phyllaries in about 6 series, tomentose centrally, inner linear-aristate and erect, outer lanceolate, 7–8 mm and ± squarrose, median transitional. Florets 28–30 mm, lobes ± 6 mm, colour of corollas unknown but probably rose to purple. Ripe achenes not seen; pappus stramineous with plumose hairs, 14–17 mm with a few central hairs distinctly longer.

HAB. & DISTRIB. On mountain rocks at 1820–2420 m, if indeed it occurs in Iraq; fl. Jun.: ? **MSU**: Avroman mts., *Haussknecht* 576 (type!).

Iran?

This is one of the many species localized by Haussknecht as gathered on "Avroman et Schahu", and could have been gathered in Iran rather than Iraq. However, several such species have been found on Iraqi mountains and *J. inuloides* is included in view of the uncertainty.

3. **Jurinea carduiformis** (*Jaub. & Spach*) *Boiss.*, Diagn. Pl. Orient. ser. 1, 6: 111 (1846); Boissier, Fl. Orient. 3: 577 (1875); Handel-Mazzetti in Ann. Naturh. Mus. Wien 27: 444 (1913); Nábělek in Publ. Fac. Sci. Univ. Masaryk 52: 36 (1925); Bornmüller in Beih. Bot. Centralbl. 60B: 208 (1939); Zohary in Dep. Agr. Iraq Bull. 31: 154 (1950); Rawi in Dep. Agr. Iraq Tech. Bull. 14: 18 (1964).

Outreya carduiformis Jaub. & Spach, Ill. Pl. Orient. 1: 131, t. 68 (1843); Rechinger in Fl. Iranica 139a: 210 (1979); Breckle, Hedge & Rafiqpoor, Vasc. Pl. Afghan: 179 (2013).

Perennial herb 6–13(–30) cm tall with a slender but tough rootstock clad with a ± thick crown of remains of previous years' leaves. Stems mostly several, branched from near base upwards, slender, striate, densely white-lanate. Basal leaves withered at time of flowering; cauline leaves decurrent, scarcely reducing above, 9–25 × 0.8–6 cm, narrowly linear and entire with flexuous margins or usually narrowly and remotely pinnatisect with segments up to 6 mm wide with entire or sinuate-undulate margins, sinuses between segments rather square, with a narrow entire or undulate portion of leaf-lamina present all along rhachis; upper leaf surface green and more sparsely arachnoid, lower surface densely white-lanate. Capitula solitary on stem and branches, broadly campanulate, 2.2–3.5 × 1.5–2.8 cm, involucrate with upper leaves; phyllaries 6–8-seriate, linear-lanceolate, ± densely white-tomentose dorsally, outer squarrose-recurved, innermost erect, median ± spreading. Florets 25–37 mm, limb 7–9 mm, segments 2–4 mm, tube filiform. Achenes ± 5 mm, body black when mature with transverse whitish scales; pappus long and conspicuous, (2.2–)2.8–4 cm, dirty white.

HAB. Usually on rocky and stony mountain slopes, also in a wadi, noted on gypsum, clay and silt; alt. 600–850 m; fl. & fr. May-Aug.
DISTRIB. Widespread in the lower mountain region, also in the foothills and once found on the adjacent plain. **MJS**: Jabal Sinjar, *Gillett* 11147!, *Qaisi & Hamad* 49165! **MRO**: Khalakan, *Omar, Qaisi & Wedad* 49460!; Haibat Sultan Dagh, *Rawi & Serhang* 26467!; Sefin Dagh, *Bornmüller* 1463!; Baba Chichak, *Bornmüller* 1464; between Der Harir and Kani Watman, *Nábělek* 3563. **MSU**: Jarmo, *Haines* 292!, *Helbaek* 1857! **MSU/FKI**: between Kirkuk and Qara Anjir, *Omar & Karim* 37869! **FKI**: 3 km along the road from Kirkuk to Sulaimaniya, *Kasim, Hamid & Jasim* 40487! **LEA**: Fakka, 70 km E of Amara, *Rawi* 25783!

Iran, Afghanistan, Pakistan,Tadzhikistan, Turkmenistan.

4. **Jurinea moschus** (*Habl.*) *Bobrov* in Bot. Zhurn. 43: 1545 (1958).

Centaurea moschus Habl., Neue nord. Beytr. 4: 54 (1783).
Serratula depressa Stev. in Mem. Soc. Imp. Nat. Mosc. 4: 61 (1813).
Jurinea depressa (Stev.) C.A. Mey., Verz. Pfl. Cauc.: 67 (1831); Boissier, Fl. Orient. 3: 583 (1875).
Jurinella aucheri Jaub. & Spach, Ill. Pl. Orient. 2: 101, t. 183 (1847).
Jurinea depressa var. *pinnatisecta* Boissier, Fl. Orient. 3: 583 (1875).
Jurinella moschus (Habl.) Bobrov in Bot. Zhurn. 43: 1534 (1958); Iljin in Fl. S.S.S.R. 27: 705 (1962); Danin & Davis in Fl. Turkey 5: 450 (1975); Rechinger, Fl. Iranica 139a: 212 (1979).
Jurinella moschus subsp. *pinnatisecta* (Boiss.) Danin & Davis in Notes Roy. Bot. Gard. Edinburgh 33: 432 (1975); Rechinger, Fl. Iranica 139a: 213 (1979).

Densely caespitose perennial herb with a tough, usually stout rootstock densely crowned with scaly remains of previous years' leaves. Leaves forming rosettes, simple and entire to pinnatisect with a large cordate-ovate to elliptic 17–50 × 1.2–3 cm terminal leaflet and 1–3 pairs of much smaller, remote leaflets up to 5 mm long, or (not in Iraq) pinnatisect with up to 8 pairs of ± equal, dentate to pinnatifid leaflets; all leaves thinly to moderately arachnoid (sometimes glabrescent) above and densely white-lanate below, with distinct petioles 1.5–4 cm. Capitula solitary and sessile in basal rosette or up to 8 congested together, sessile or on short pedicels up to 2 cm long; involucre subglobose, 15–30 × 15–40 mm; phyllaries 4-seriate, aristate, dorsally tomentose, stramineous below with green tips, inner lanceolate and erect, outer broadly lanceolate and ± reflexed. Florets 18–28 mm, corollas whitish to lilac, limb 8–10 m, lobes 5–6 mm. Achenes cylindrical-tetragonous, 5–6 mm, black when ripe, finely longitudinally striate; pappus 8–12(–16) mm, white or purplish upwards, with or without a "skirt" of shorter hairs 2.5–3 mm long, deflexed onto the achene.

HAB. Rocky summits and slopes of high mountains; alt. 3000–3500 m; fl. Jul.-Aug., fr. Aug.-Sep.
DISTRIB. Confined to the extreme N. mountain region of Iraq along the Iranian frontier. **MRO**: Qandil range NE of Rania, *Rawi & Serhang* 18278!, 24479!, 26778!; summit of the Qandil mts., *Rechinger* 11150!

Turkey (C & E Anatolia), NW Iran, Transcaucasia.

The "skirt" of short outer pappus hairs appears to be constant in Iraqi material (which is from a very limited area) but present or absent elsewhere.

21. **ZOEGEA** L.

Mant. Pl. 1: 15 (1767)
Susanna & Garcia-Jacas in Kubitzki, Fam. Gen. Vasc. Pl. 8: 139 [2006] (2007).

C. C. Townsend[6]

Annual, usually much-branched, slender herbs. Leaves simple to lyrate-pinnatisect below, simple and linear above, glabrous to scabrid-hirtellous. Capitula heterogamous,

[6] It should be acknowledged here that about half of the initial determinations for the treatment of this genus, with a preliminary rough draft, were performed by Dr Frances G. Davies. The taxonomic opinions, and especially the notulae, are however the responsibility of the author alone.

radiant, numerous, solitary at ends of stem and branches; involucre globose in bud, finally hemispherical; phyllaries multiseriate, outer with long-pectinate appendages with fimbriae finely drawn-out, inner longer, shining, stramineous to purplish, fimbriate or denticulate at about apex; receptacle setose. Outer florets neuter, radiant, ligulate, lobes short to rather long, inner florets hermaphrodite, tubular, corollas whitish to pink-tinged or bright to orange-yellow. Achenes compressed-ovoid, smooth, glabrous and pilose, with concentric furrows at apex; hilum lateral; pappus of several series of outer scales, longer median minutely scabrid or barbellate bristles and again shorter, usually darker, bristles or scales within.

A small genus of 3 species distributed from SE Mediterranean to SW & C Asia; all found in Iraq.

Zoegea – named in honour of Johan Zoega, 1742–1788, a student of Linnaeus who investigated the flora of Iceland.

1. Phyllary appendages terminating in a long, usually purplish-black arista much longer than lateral fimbriae; innermost rows of pappus setose; corollas white or pinkish . . . . 3. *Z. purpurea*
   Phyllary appendages terminating in a short arista no longer than lateral fimbriae; innermost rows of pappus of short scales; corollas bright yellow or orange-yellow 2
2. Rays not or slightly exceeding involucre, 7–8 mm; achenes 4(–5) mm, initially with long fine hairs above; pappus with longest bristles 7–9mm. . 1. *Z. crinita*
   Rays up to 1.5 × as long as involucre, 9–14 mm; achenes 2.5–3(–3.75) mm, initially puberulous; pappus with longest bristles (3.5–)4–5(–6.5) mm . . . . 2. *Z. leptaurea*

1. **Zoegea crinita** *Boiss.*, Diagn. Pl. Orient., ser. 2, 3: 63 (1856); Boissier, Fl. Orient. 3: 698 (1875); Anthony in Notes Roy. Bot. Gard. Edinburgh 18: 291 (1935); Rawi in Dep. Agr. Iraq Tech. Bull. 14: 127 (1964); Rechinger, Fl. Lowland Iraq: 648 (1964) & Fl. Iranica 139b: 424 (1980).

Z. baldschuanica C.Winkl. in Trudy Imp. S.-Peterburgsk. Bot. Sada 9: 426 (1886). Cherepanov in Fl. S.S.S.R. 28 : 248 (1963).
Z. crinita subsp. *baldschuanica* (C.Winkl.) Rech.f. in Fl. Iranica 139b: 425 (1980).

Plant 20–60 cm tall, vegetative parts glabrous or shortly whitish scabrid-hirtellous, branched from near base upwards or only in upper half, simple when depauperate, branches long, slender and divaricate. Basal and lower cauline leaves petiolate, simple to lyrate-pinnatisect with up to 4 pairs of narrow or very reduced and roundish lateral lobes and one large terminal elliptic lobe, variable in size, up to 12 × 1.8 cm, upper cauline and branch leaves narrowed at base, lamina linear to linear-oblanceolate or narrowly elliptic, rapidly reducing. Capitula on slender, glabrous or scabridulous pedicels 1–5 cm; involucre 11–14 mm in diameter excluding setae, hemispherical; outer phyllaries striate below, finely pectinate with 8–12 pairs of fimbriae extended as stramineous hairs, longest 5–7 mm, intermediate phyllaries similar, more striate and coloured below, glabrous to ± floccose or glandular-pilose dorsally, becoming transitional to outer phyllaries, inner phyllaries 15–20 mm, pale-tipped, shortly fimbriate at extreme tip. Marginal florets 7–8 mm, not or slightly exceeding involucre, corollas golden-yellow. Achenes smooth, sparsely softly pilose with long hairs above, 4–5 mm, white or stramineous; pappus white with a short, 1.5 mm, central blackish-purple or brownish tuft of scales, longest bristles 7–9 mm.

HAB. Rocky mountain; alt. 1200–1400 m; fl. & fr. Jun.-Jul.?
DISTRIB. Very rare (only found in one locality) in the mountain region of Iraq. **MSU**: mt. Avroman nr Tawila, *Rechinger* 12444!

N, W & S Iran, Afghanistan.

This and the following plant are extremely close, and I vacillated much before recognizing them as distinct species. Some characters used by Boissier to distinguish them are very unconvincing, for example the length of the cilia of the phyllary-appendages. In *Z. leptaurea* these are at their shortest in Palestine and Saudi Arabia in the west of its range and the longer ones of *Z. crinita* might well have represented the eastern end of cilinal variation in the already notably variable *Z. leptaurea.* Other characters, including those used here, are equally comparative of dimension. The overlap of achene and pappus character is, however, not as great as the dimensions given in the descriptions might make it appear. In my experience, if the pappus is at the long extreme in *Z. leptaurea* the achene is short and vice versa, so that using combined achene + pappus length, confusion between the two species is normally unlikely. The fact that the tip of the inner phyllaries is rather blunt and shortly denticulate in the type of *Z. crinita* (as in the western forms of *Z. leptaurea*) rather than laciniate-fimbriate (as predominantly in the eastern forms of the latter) seems another good reason for maintaining *Z. crinita* as distinct. However, comparatively few specimens have mature achenes, and unfortunately the specimen cited above is the one true intermediate which I have seen. It has the large achenes and pappus and short rays of *Z. crinita,* but lacerate tips to the inner phyllaries: ± as in *Z. leptaurea* "subsp. mesopotamica". Another specimen from the same locality, collected by Rechinger (12443!) on the same day, has 14 mm rays and I would name it as *Z. leptaurea.* The status of the two species thus remains uncertain to me or, if they prove distinct, the presence of *Z. crinita* in Iraq needs further confirmation.

2. **Zoegea leptaurea** *L.*, Mant. Pl. 1: 117 (1767); Boissier, Fl. Orient. 3: 697 (1875); Handel-Mazzetti in Ann. Naturh. Mus. Wien 27: 447 (1913); Nábělek in Publ. Fac. Sci. Univ. Masaryk 52: 46 (1925); Dinsmore in Post, Fl. Syria, ed. 2, 2: 117 (1933); Guest in Dep. Agr. Iraq Bull. 27: 110 (1933); Anthony in Notes Roy. Bot. Gard. Edinburgh 18: 291 (1935); Blakelock in Kew Bull. 4: 65 (1949); Zohary in Dep. Agr. Iraq Bull. 31: 157 (1950); Rawi in Dep. Agr. Iraq Tech. Bull. 14: 127 (1964); Rechinger, Fl. Lowland Iraq : 649 (1964); Kupicha in Fl. Turkey 5: 585 (1975); Rechinger, Fl. Iranica 139b : 422 (1980).

Centaurea calendulacea Lam., Encycl. 1: 668 (1785).
Zoegea mianensis Bunge ex Boissier, Fl. Orient. 3: 697 (1875).
Z. mesopotamica Cherep. in Not. Syst. Herb. Inst. Bot. Acad. Sci. URSS 19: 451 (1959); Rechinger, Fl. Lowland Iraq: 648 (1964).
Z. leptaurea subsp. *mesopotamica* (Cherep.) Rech.f. in Fl. Iranica 139b: 423 (1980).

Plant 16–75 cm tall, whitish scabrid-hirtellous throughout in its vegetative parts, branched from near base or occasionally only above with long, slender, divaricate branches, rarely (when depauperate) simple. Basal and lower cauline leaves petiolate, sinuate-lobed to lyrate-pinnatisect with up to 4 pairs of narrow lateral lobes and a larger elliptic terminal one, variable in size, up to 8 × 1.5 cm, upper cauline and branch leaves broadly to narrowly linear, rapidly reducing, narrowed at base. Capitula on slender, scabrid pedicels 1–9 cm; involucre hemispherical, 8–12 mm in diameter excluding setae; outer phyllaries with base striate, terminal appendages elegantly finely pectinate with 7–11 pairs of spreading fimbriae extended as long brownish hairs, longest (2–)3.5–5 mm long, median phyllaries transitional, inner phyllaries 14–18 mm with a whitish, shining, finely lacerate, denticulate or fimbriate apex. Marginal florets radiant, golden or orange-yellow, rays 9–14 mm, exceeding involucre. Achenes puberulent, smooth, 2.5–3(–3.75) mm, white or brownish to almost black; pappus white with a short, (less than 1 mm), central tuft blackish-purple to greenish-black, longest bristles (3.5–)4–5(–6.5) mm. Fig. 19: 1–9.

HAB. Coppiced *Quercus aegilops* areas on hillsides, stony hill slopes, *Pistacia* steppe, silty depressions on desert plateau, cultivated and fallow fields, roadsides, etc., chiefly on stony and sandy clay; alt. 340–1250 m; fl. May-Jul.

DISTRIB. Abundant in the mountain region of Iraq, quite frequent in the foothills and upper plain, scarce in the desert and lower plain. MAM, Bekma gorge, *Gillett* 8222!; **MAM**: Aqra, *Rawi* 11330!; Ba'idra, nr Shaikhan, *Salim Effendi* 2592!; Dohuk Gorge, *Chapman* 26268!; Aradin, *Qaisi & Hamad* 46062!; 84 km NW of Mosul, *Rawi* 22950! **MRO**: 31 km N of Kirkuk on road to Koi Sanjaq, *Rawi, Noori & Kass* 28000!; Shaqlawa, *Omar, Sahira, Karim & Hamid* 38285!; Haibat Sultan Dagh, *Rawi, Noori & Kass* 28285!, 28233!; between Shaqlawa and Harir, *Omar* 37625!; Sefin Dagh, Shaqlawa, *Bornm.* 1415! **MSU**: 4 km E of Qaranjir, *Rawi* 21705!; Jarmo, *Helbaek* 1718!, 1859!; 10 km N of Darband-i Khan, *Omar, Qaisi & Wedad* 49363!; 13 km SW of Sulaimaniya, *Rawi* 21705!; 20 km NW of Sulaimaniya, *Rawi* 21732! **MJS**: Jabal Sinjar, *Gillett* 11096!, *Qaisi, Khayat & Karim* 50860!; 47 km E of Karsi, *Bharucha & Ani* 8800! **FUJ**: Mir Qasim, between Balad Sinjar and Tell Afar, *Field & Lazar* 549!; Jabal Khatchra, *id.* 638!; Hammam Ali, *Anders* 980! **FNI**: Hinnis, nr Ain Sifni, *Guest* 4050!; 5 km from Eski Kellek on road to Mosul, *Omar, Qaisi & Wedad* 49628! **FAR**: Makhmur, *Gillett* 11221! **FKI**: above Bawa Nur, *Poore* 550! **FKI/MSU**: from Kirkuk

Fig. 19. **Zoegea leptaurea**. 1, habit × 2/3; 2, marginal floret × 6; 3, inner floret × 9; 4, achene and pappus × 12; 5, anthers × 18; 6, style × 18; 7, outer phyllary × 6; 8, median phyllary × 6; 9, inner phyllary × 9. All from *Bharucha & Ani* 8800. Drawn by M. Millar Watt.

to Sulaimaniya, *Rawi* 21520! **FPF**: Al-Hashimah, between Badra and Mandali, *Qaisi & Khayat* 50728!; Koma Sang police station, *Rawi* 20614! **DSD**: 12 km WNW of Ansab, *Guest, Rawi & Rechinger* 18983! **LCA**: Abu Ghraib, *Sahira* 38594!

Syria, Lebanon, Turkey, W Iran, Afghanistan.

FUTÎR (*Radhi* 3841, "Arbil Province"), SAIFRAH (*Mudir of Tal Kaif* 3188, Tal Kaif), GŪLÎL ZĀRIK (Kurd., *Salim Effendi* 2592).

Notes on the above three sheets refer to this species as being "grazed by animals, "useful for sheep grazing" and "fodder plant" respectively.

Rechinger in *Fl. Iranica* refers the Iraqi material to subsp. *mesopotamica* (Cherep.) Rech.f. This is distinguished from subsp. *leptaurea* by ray length and small differences in the form of the phyllaries. However, I do not find the separation convincing and both forms and intermediates occur freely in Turkey, from which country only subsp. *leptaurea* is recorded in *Fl. Iranica.* Similarly, although Iraqi material is fairly uniform in having long-fimbriate apices to the inner phyllaries, the differences in this respect between, for example, *Rawi* 20614 and *Qaisi & Khayat* 50728, cited above, and *Rechinger* 9633, cited in *Fl. Iranica,* all from the Mandali region, seem to me to be greater than between the first two and "subsp. leptaurea". Thus no subspecies are recognized here. When published, Linnaeus 'changed' the spelling of the generic name (in Specierum Editionis II) to '*Zoegaea*' (p. 117) from that originally published (in the Generum Editionis VI), and in the index.

3. **Zoegea purpurea** *Fresed.* in Mus. Senckenberg. 1: 86, t. 5 (1834); Boissier, Fl Orient. 3: 698 (1875); Dinsmore in Post, Fl. Syria, ed. 2, 2: 118 (1933); Cherepanov in Fl. S.S.S.R. 28 : 247 (1963); Rechinger, Fl. Lowland Iraq: 649 (1964); Täckholm, Stud. Fl. Egypt, ed. 2: 538 (1974); Feinbrun-Dothan, Fl. Palaest. 3: 387 (1978); Rechinger, Fl. Iranica 139b: 425 (1980); Collenette, Wild Flow. Saudi Arabia: 225 (1999); Boulos & Hind in Fl. Egypt. 3: 161 (2002); Kilian & Ghazanfar in Fl. Oman 3: 219 (2015); Taifour & El-Oqlah, Pl. Jordan Ann. Checkl.: 49 (2017).

Z. aristata DC., Prodr. 6: 562 (1838).

Plant (10–)15–30(–40) cm tall, whitish scabrid-hirtellous throughout in its vegetative parts, branched from near base or occasionally only above, branches long, slender, divaricate. Basal and lower cauline leaves petiolate, lamina entire and scarcely even sinuate-denticulate to lyrate-pinnatisect with up to 4 pairs of narrow lateral lobes and a larger elliptic terminal one, variable in size, up to 7 × 1 cm, petiolate; upper (or all) cauline and branch leaves linear, rapidly reducing above, narrowed at base. Capitula on slender scabrid pedicels 1–9 cm long; involucre hemispherical, 7–12 mm indiameter excluding setae; outer and outer intermediate phyllaries striate, terminal appendages finely pectinate with 6–10 pairs of spreading fimbriae extended as long whitish or purplish hairs, apex prolonged into a fine awn up to ± 7 mm long, awn and centre and apex of appendage usually dark purplish, tips of inner intermediate phyllaries lacerate and with a long terminal awn, frequently purplish, inner phyllaries 10–14 mm, pale or purplish-tipped, muticous. Corollas white or pink-tinged, marginal florets very narrowly ligulate and scarcely exceeding involucre; inner florets tubular, ± equalling involucre. Achenes smooth, glabrous, 2 mm, white; pappus 5–6 mm, white, central rows of short, dirty white bristles.

HAB. Stony places in the desert, rocky soil; alt. 260–340 m; fl. & fr. Apr.-Jun.
DISTRIB. An apparently very local plant in the Southern and Western Deserts. **DWD**: Fatha depression, 70 km SW of Nukhaib, *Rawi* 31068!; Jal al-Rawaq, 110 km SE of Salman, *Rechinger* 9340. **DSD**: 27–46 km WNW of Ansab, *id.* 9366; 12 km WNW of Ansab, *id.* 13807, *Guest, Rawi & Rechinger* 18983!

Palestine, Jordan, Sinai, Egypt, Saudi Arabia, Oman, Iran, Afghanistan, Pakistan, C Asia (Turkmenia to Tajikistan).

The following colloquial names have been noted for this plant: KULIL ZARIK (*Salim* 2592, Kurd.), FUTÎR (Kurd. "grazed by all animals", *Radhi*); SAIFRAH "grazed by sheep", *Mudur of Tel Kaif* 3188).

22. **CRUPINA** (Pers.) DC.

Ann. Mus. Hist. Nat. Paris 16: 157 (1810)
Susanna & Garcia-Jacas in Kubitzki, Fam. Gen. Vasc. Pl. 8: 139 [2006] (2007)
Centaurea sect. *Crupina* Pers., Syn. Pl. 2: 481 (1807)

C. C. Townsend

Slender erect annual herbs. Leaves alternate, mostly narrowly and deeply pinnatisect, clothed at margins with multicellular many-barbed glochidiate hairs. Inflorescences laxly corymbose. Capitula heterogamous, radiant; involucre narrow; phyllaries in several rows, papery, unarmed; receptacle flat, bristly. Outer florets sterile. Disc florets hermaphrodite; corolla purple to pink, limb deeply regularly 5-lobed, tube pilose. Achenes smooth, velutinous above, compressed to terete and shortly cylindrical, hilum basal or lateral, truncate at apex with an annular rim surrounded by pappus; pappus of several rows of outer barbellate pale to dark bristles and an inner series of short dark scales.

Three species in Iraq, the Mediterranean region and W Asia.

Crupina (from the Low German *krupen*, to creep – alluding to the ability of the seed to 'creep' forward using its rigid, barbellate pappus; for this reason, mature seeds do not fall from the involucre in dried herbarium material of the genus.)

1. Achenes not at all compressed, bullet-shaped and blunt with a round, apical, totally transverse hilum; hairs of tube of florets strongly shortly barbellate at cell walls . . . . 3. *C. vulgaris*
   Achenes compressed and ± keeled at least towards apex, hilum lateral or oblique; hairs of tube of florets smooth or with minute projections at cell walls . 2
2. Hairs of tube of disc-florets smooth, usually numerous and long; hilum of achene lateral, narrow, often wedge-shaped . . . . 1. *C. crupinastrum*
   Hairs of tube of disc-florets with minute projections at cell walls, less numerous and usually shorter; hilum of achene oblique, ± broadly ellipsoidal . . . . 2. *C. intermedia*

1. **Crupina crupinastrum** (*Moris*) *Vis.*, Fl. Dalm. 2: 42 (1849); Boissier, Fl. Orient. 3: 699 (1875); Nábělek in Publ. Fac. Sci. Univ. Masaryk 52: 46 (1935); Dinsmore in Post, Fl. Syria, Palest. & Sinai ed. 2, 2: 118 (1933); Anthony in Notes Roy. Bot. Gard. Edinburgh 18: 291 (1935); Bornmüller in Beih. Bot. Centralbl. 60B: 215 (1939); Zohary in Dep. Agr. Iraq Bull. 31: 167 (1950); Cherepanov in Fl. S.S.S.R. 28: 253 (1963); Rawi in Dep. Agr. Iraq Tech. Bull. 14: 113 (1964); Täckholm, Stud. Fl. Egypt ed. 2: 538 (1974); Kupicha in Fl. Turkey 5: 587 (1975); Feinbrun-Dothan, Fl. Palaest. 3: 388 (1975); Rechinger, Fl. Iranica 139b: 449 (1980); Meikle, Fl. Cyprus 2: 966 (1985); Boulos & Hind in Fl. Egypt 3: 162 (2002); Taifour & El-Oqlah, Pl. Jordan Ann. Checkl.: 40 (2017).

Centaurea crupina L. var. *maculata* Pers., Syn. Pl. 2: 488 (1807).
C. crupinastrum Moris, Enum. Sem. Regi Horti Bot. Taur. 1842: 12 (? late 1842/?early 1843); Bot. Zeit. 1(10): 168 (March 1843); Fl. Sardoa 2 : 443 (?1840-43)[7].

Slender, wiry, erect annual 30–60 cm (shorter if depauperate), with long, slender, divergent-erect branches from below the middle or sometimes simple below inflorescence, glabrous or thinly pilose, stem and branches angular, sulcate. Leaves mainly in lower part of plant; basal leaves bluntly pinnatisect with a larger, usually pinnatifid terminal lobe, frequently withered by flowering time; cauline leaves deeply and narrowly pinnatisect with linear, irregularly serrate-dentate lobes, oblong in outline, lower 2–8 × 0.8–3.8 cm, petiolate, upper more narrowly divided, smaller, sessile; all with midrib ± villous below, margins with short many-barbed glochidiate hairs. Inflorescence lax, capitula on long, leafless,

[7] Notes: It is not at all clear which was the earliest publication – the *Enumeratio* is dated 1842, but this may merely be the year it covers, although it may well be early in 1843 but before the Bot. Zeit. paper – a full description is provided. The *Botanische Zeitung* paper merely repeats Moris's description from the *Enumeratio*. The accurate date of publication of *Flora Sardoa* is unknown as the title page merely gives '1840-1843', but most probably early in 1843 but after the *Enumeratio*; a full description is provided and reference to the *Enumeratio*.

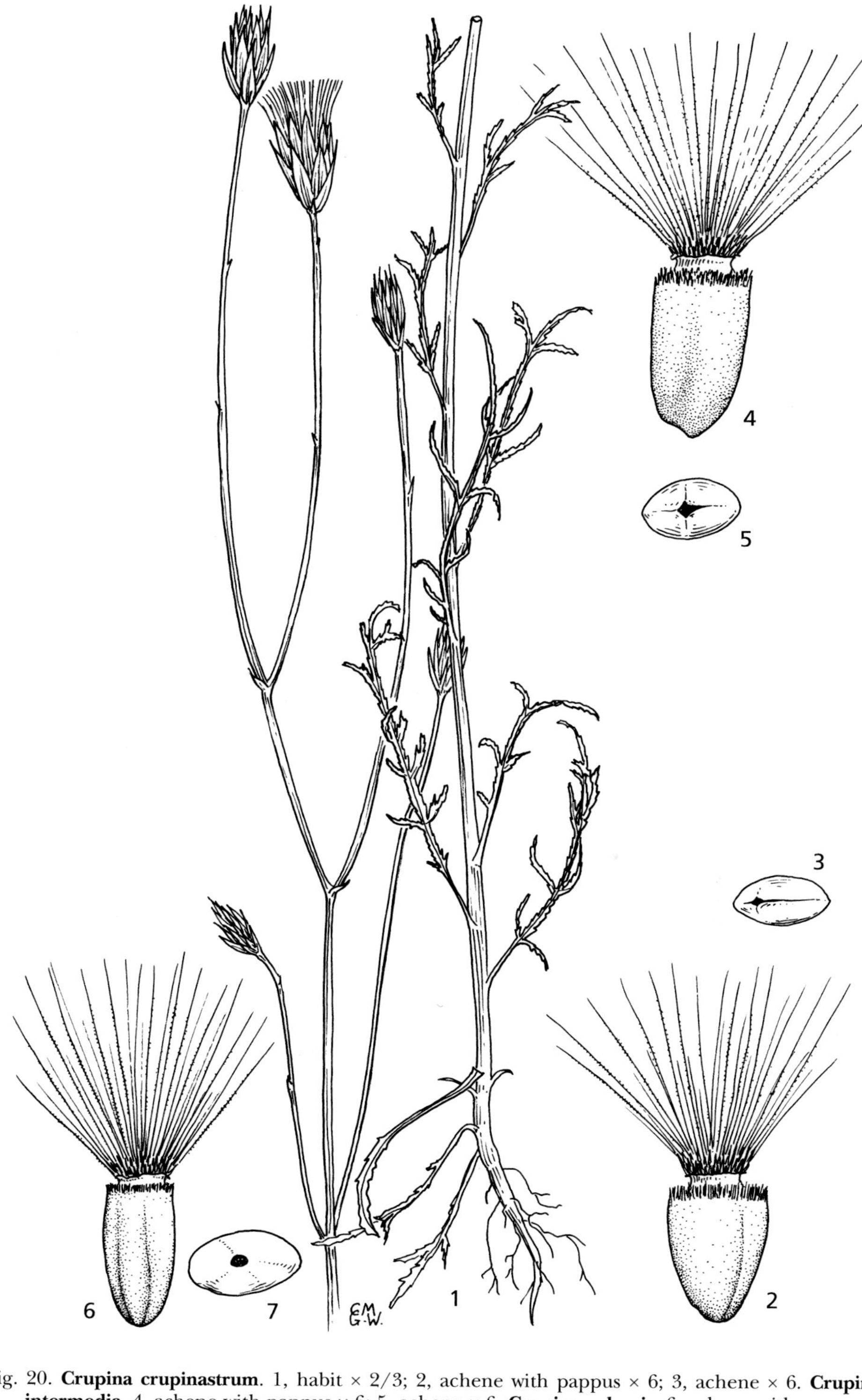

Fig. 20. **Crupina crupinastrum**. 1, habit × 2/3; 2, achene with pappus × 6; 3, achene × 6. **Crupina intermedia**. 4, achene with pappus × 6; 5, achene × 6. **Crupina vulgaris**. 6, achene with pappus × 6; 7, achene × 6. 1 from *Guest* 15417; 2–3 from *Field & Lazar* 663; 4–5 from *Sentensis* 910; 6–7 from *Rawi* 21935. Drawn by Christine M. Grey-Wilson.

pedunculiform branches up to 15(–20) cm long; involucre fusiform in bud, narrowly obconical and 15–22 × 8–12 mm at anthesis; phyllaries minutely glandular dorsally, often purplish-tinged, scarious-margined, outer ovate, 2–4 mm long, inner lanceolate, 15–20 mm. Corollas bright purple, well exceeding inflorescence, marginal slightly radiant, inner with divergent segments. Achenes distinctly compressed and keeled, 3.5–6 × 2–3 mm, apex brownish-velutinous, base glabrous with a narrow, lateral, often cuneiform hilum; pappus often dark brown to blackish, longest bristles 4–8 mm. Fig. 20: 1–3.

HAB. Commonly on hill slopes of stony or rocky clay, also on steep, sometimes unstable, limestone slopes, in *Quercus* forest, on calcareous sandstone and on conglomerate, on silty soil, and in a wadi in the mountains; alt. 600–1400 m; fl. Apr.-Jun., fr. May-Jul.
DSTRIB. Common in the mountain region of Iraq, scarce in the foothills. **MJS**: Jabal Sinjar, *Gillett* 10973!, *Qaisi, Khayat & Karim* 50886!; Karsi, *Qaisi & Wedad* 52252! **MAM**: Kani Masi, *Hamad & Fadhil* 45637!; nr Zakho, *Karim, Dabbagh & Hamad* 44902!; Dohuk, *Guest* 15417!; Jabal Bekher, *Rawi* 8457!, *Field & Lazar* 828!; Sawar Tuka, *E. Chapman* 26321! **MRO**: above Salahaddin, *Barkley* 5679!; Haibat Sultan Dagh, *Rawi, Noori & Kass* 28173!, 28230!; Saran village, *Nuri, Kass & Serhang* 27285!; between Haji Umran and the Iranian frontier, *Alkas* 18597! **MSU**: Tashija, *Omar & Karim* 47062!; Jarmo, *Helbaek* 1188!; 7 km & 10 km W of Tawila, *Rawi* 22360!, 22125!; Halabja, *Rawi* 8826!; mt. Goizha (Gweija), *Mooney* 4330! **FUJ**: Jabal Khatchra, *Field & Lazar* 663! **FKI**: 31 km N of Kirkuk on road to Koi Sanjaq nr Suwan village, *Rawi, Noori & Kass* 27993!

Mediterranean Europe and islands from Spain to Turkey, Syria, Lebanon, Palestine, Egypt, Transcaucasia, Iran, N Africa (Libya to Algeria).

2. **Crupina intermedia** (*Mutel*) *Walp.*, Ann. Bot. Syst. 1: 447 (1849); Kupicha in Fl. Turkey 5: 588 (1975); Rechinger, Fl. Iranica 139b: 450 (1980).

Centaurea intermedia Mutel in Rev. Bot. 1: 400 (1846).
Crupina vulgaris Cass. var. *β intermedia* (Mutel) Batt. in Fl. Algér (Battandier et al.) 1(3): 492 (1889).

Differs from *C. crupinastrum*, which it superficially closely resembles, in the characters given in the key. Intermediate in character, as the name suggests, between *C. crupinastrum* and *C. vulgaris*, it is tempting to suppose that it arose as a fertile hybrid between the two as suggested by Le Vaillant. (Rév. Gén. Bot. 77: 111–124. 1970). Fig. 20: 4–5

HAB. Dry stony and earthy hillsides and limestone rocks; alt. 700–1800 m; fl. Apr.-May, fr. May-Jun.
DISTRIB. Scattered in the eastern mountain region of Iraq. **MRO**: mt. Hawara Blinda, NE of Haji Umran, *Kass & Noori* 27814! **MSU**: Jarmo, *Haines* 327!; Qaranjir, *Rawi & Gillett* 7536!; 20 km NW of Sulaimaniya on the Dukan highway, *Rawi* 21735!; Darband-i Basian, *Rawi* 22849!, *Rechinger* 10575!

Turkey, Iran, Algeria.

3. **Crupina vulgaris** *Cass.*, Dict. Sci. Nat. 12: 68 (1818); Boissier, Fl. Orient. 3: 699 (1875); Post (revised Dinsm.), Fl. Syria, Palest. & Sinai ed. 2, 2: 118 (1933); Cherepanov in Fl. S.S.S.R. 28: 254 (1963); Kupicha in Fl. Turkey 5: 587 (1975); Rechinger, Fl. Iranica 139b: 448 (1980).

Centaurea crupina L., Sp. Pl.: 909 (1753).
Serratula crupina (L.) Vill., Hist. Pl. Dauphiné (Villars) 3(1): 38 (1788).

Differs from the two preceding species in the characters given in the key. No material has been seen from Iraq. Rechinger (Fl. Iranica 139b: 449, 1980) records it from "Sinjar", but cites no specimen and believes (pers. comm. 3 Sept. 1985) that this record could well have been based on an incorrectly named specimen in another institution. *C. vulgaris* is a species with a more northerly distribution and not very likely to occur in Iraq. Fig. 20: 6–7.

S, C & Mediterranean Europe to S Ukraine and Crimea, Turkey, Caucasus, Iran, C Asia (Turkmenia to Xinjiang), N Africa (Libya to Morocco).

23. **STIZOLOPHUS** Cass.

Dict. Sci. Nat. ed. 2, 44: 36 (1826)
Susanna & Garcia-Jacas in Kubitzki, Fam. Gen. Vasc. Pl. 8: 140 [2006] (2007)

Centaurea sect. *Stizolophus* (Cass.) DC.
Centaurea sect. *Balsamitophyllum* Boiss.

G. Wagenitz

Annuals, branched from base or above, with few to numerous medium-sized capitula. Leaves undivided or lower lyrate, scabrid. Inflorescences of solitary capitula or of lax corymbs. Capitula heterogamous, globose; involucre ovoid; phyllary appendages triangular, scarcely decurrent, with numerous regular cilia and ending in a slender spinule; receptacle setose. Outer florets radiant, sterile, corollas yellow, with staminodes; central florets perfect, styles very long. Achenes medium-sized, glabrous, base with rugulose elaiosome; pappus biseriate, deciduous, outer series with long pinnules, inner series short, pinnules few.

Two species distributed in arid regions of SW & C Asia.

Stizolophus (from στιζο, *stizo*, speckled, λοφος, *lophos*, hill, perhaps an example of one of Cassini's meaningless but plausible generic names).

Wagenitz, G. (2006). A revision of *Centaurea* (Compositae-Cardueae) in the flora of Iraq. Rostaniha 7. (suppl. 2): 343–394.

Appendages cartilaginous, c. 3 mm broad at base excluding cilia, with 8–12 cilia on each side; achenes 5–6 mm long .... 1. *S. balsamita*
Appendages not cartilaginous, only 1 mm broad at base, with 4–6 cilia on each side; achenes 3–3.5 mm long .... 2. *S. balsamitoides*

1. **Stizolophus balsamita** (*Lam.*) *C. Koch* in Linnaea 24: 423 (1851); Takhtajan in Takhtajan & Fedorov, Fl. Erevana: 329 (1945).

Centaurea balsamita Lam., Encycl [J. Lamarck et al.] 1: 667 (1785); Boissier, Fl. Orient. 3: 679 (1875); Handel-Mazzetti in Ann. Naturh. Mus. Wien 27: 451 (1913); Náběiek in Publ. Fac. Sci. Univ. Masaryk 52: 42 (1925); Dinsmore in Post, Fl. Syria, Palest. & Sinai ed. 2, 2: 111 (1933); Anthony in Notes Roy. Bot. Gard. Edinburgh 18: 291 (1935); Wagenitz in Fl. Turkey 5: 485 (1975) & in Fl. Iranica 139B: 331 (1980); Breckle, Hedge, Podlech & Rafiqpoor, Vasc. Pl. Afghan. Checkl.: 154 (2013).
Stizolophys balsamitaefolium Cass., Dict. Sci. Nat., ed. 2. [F. Cuvier] 51: 49, 50 (1827), nom. illegit., based on *C. balsamita* Lam.

Annual with an erect stem 30–80(–120) cm tall. with several long simple branches in upper half, rarely simple. Stem and branches straw-coloured, glabrescent. Leaves scabrous with very short hairs, entire or denticulate (rarely lower with a few coarse teeth at base), lower leaves oblong, withered at flowering time, median and upper leaves gradually smaller and narrower, ending in a yellow mucro 1–3 mm. Involucre ovoid with truncate base, 15–25 × 12–24 mm; phyllaries multiseriate, lower part nearly concealed by appendages; appendages cartilaginous, appressed, yellowish to light brown, triangular, 2–3.5 mm long and broad excluding cilia, scarcely decurrent, with 7–12 cilia 2–3 mm long on each side, gradually narrowed into a spinule 2–4 mm, innermost phyllaries narrowed into a slender often reddish tip. Marginal florets scarcely radiant, with staminodes; corollas yellow. Achenes 4–5.5 mm; pappus 2–5 mm, scabrous to barbellate, inner series short.

DISTRIB. (of species) Syria, Turkey, Iran, Afghanistan, Transcaucasia, C Asia (Turkmenia to Kyrgyzia).

subsp. **balsamita**. Involucre 18–25 × (15–)17–24 mm; appendages (without the cilia) 2.5–3.5 × 2.5–3.5 mm with 8–12 cilia on each side and a terminal spinule 3–4 mm. Achenes 4.5–5.5 mm; pappus (3.5–)4–5 mm.

HAB. Among *Quercus* forest and scrub; alt. 1600–1700 m; fl. & fr. Jul.-Aug.
DISTRIB. In the eastern section of the forest zone of Iraq (and also in the irrigated tree plantation near Mosul). **MSU**: Qopi Qaradagh, *Haines* 2051!; Qara Dagh, *Buthaina Makki* 497! **FKI**: Nineveh plantation, N of Mosul, *Anders* 1581!

This subspecies occurs throughout the range of the species. A second subspecies occurs in southern and eastern Iran.

2. **Stizolophus balsamitoides** (*Post*) *Soják* in Novit. Bot. Prag. 1962: 48 (1962).

Centaurea balsamitoides Post, Pl. Postiane. 2: 15 (1891); Wagenitz in Fl. Iranica 139B: 332 (1980).
C. stramenticia Hand.-Mazz. in Ann. Naturh. Mus. Wien 27: 448 (1913).

Annual, stem erect, 20–60 cm tall, branched from near base or in middle part, branches simple or with several capitula. Stem and branches ivory-white, shining, nearly glabrous. Leaves firm, puberulent and covered with sessile glands, lower leaves withered at flowering-time, lyrate (or sometimes undivided?), median leave lanceolate to linear-lanceolate, 3–5 mm broad, upper leaves only 1–2 mm broad, very small. Capitula solitary at ends of branches; involucre ovoid, 13–17 × 6–9 mm; phyllaries multiseriate, imbricate, with slightly elevated longitudinal ribs, finely tomentose, appendages small, narrowly triangular (c. 1 mm broad at base excluding cilia), straw-coloured, with 4–6 pairs of distinct cilia 1.5–2 mm long and a terminal mucro 2–3.5 mm. Marginal florets not radiant, with staminodes; corollas sulphur-yellow. Achenes 3–3.5 mm long, greyish, at base with a cartilaginous yellowish swelling at hilum; pappus (3.5–)4–5 mm long, inner series short.

HAB. On gravelly hilltops, stony hillsides, gypsiferous steppe, clay slopes; alt. 100–400 m; fl. & fr. Mar.-May(-Jun.).
DISTRIB. In the dry steppe region of Iraq. **FUJ**: between Lake Khatanya (in Syria) and Jabal Sinjar, *Handel-Mazzetti* 1590!; between R. Khabur and Jabal Sinjar, *id.* 1609 (type of *C. stramenticia*). **FKI/FPF**: between Tikrit and Mandali, *Sutherland* s.n.! **FPF**: Jabal Hamrin, nr. Muqdadiya (Shahraban), *Haines* 1396!; Jabal Hamrin on the Khanaqin road, *Agnew* 1301!, *Hadač* 1709!; Jabal Hamrin, *Sutherland* 51!; Quraitu, nr Khanaqin, *Rawi* 5697!; Kani Mari nr Khanaqin, *Hadač, Haines & Hashimi* 4648! **DLJ**: Umm al Matin, 74 km NW of Baiji-Haditha road, *Chakravarty, Rawi, Khatib & Alizzi* 32040!

Syria.

S. balsamitoides is closely allied to *S. coronopifolius* (Lam.) Cass. from Turkey (C & NE Anatolia) and Transcaucasia, in which it was included later by Post himself. The capitula are however smaller in *S. balsamitoides*, the pappus is longer and present in all achenes, while in *S. coronopifolius* there is a distinct heterocarpy with outer achenes epappose. Further investigations will show if the last named character is a really constant one.

24. **AMBERBOA** (Pers.) Less.

Syn. Gen. Comp. : 8 (1932), *nom. cons.*
Susanna & Garcia-Jacas in Kubitzki, Fam. Gen. Vasc. Pl. 8: 142 [2006] (2007).

Centaurea subgen. *Amberboa* Pers., Syn. Pl. [Persoon] 1: 481 (1805)
Centaurea Sect. *Amberboa* (Pers.) Pers., Sp. Pl. 5 : 316 (1821)

C. C. Townsend

Annual (or biennial) herbs, stem usually branched. Leaves simple to lyrate or pinnatipartite, subglabrous. Capitula rather few and relatively large, heterogamous, radiant; involucre ovoid, firm; phyllaries multiseriate, all unarmed, obtuse, closely imbricate; receptacle setose. Marginal florets sterile. Disc florets hermaphrodite and fertile; all corollas rose-pink to yellow. Achenes laterally compressed, villous, with a slender rim around apex and a swollen margin around deep lateral hilum; pappus of numerous multiseriate narrow scabrid scales.

6 species in eastern Europe, S W & C Asia; one species, *A. moschata*, is widely cultivated and is present in Iraq.

Amberboa (said to be derived from a French word, *amberboi*, which has not been traced, originating from the medieval Latin *ambar* or *amber*, for the well-known fossil resin; this has a musky odour similar to that of *A. moschata*, the best-known species of the genus).

1. **Amberboa moschata** (*L.*) *DC.*, Prodr. 6: 560 (1837); Boissier, Fl. Orient. [Boissier] 3: 605 (1875); Tzvelev in Fl. S.S.S.R. 28: 325 (1963); Wagenitz in Davis, Fl. Turkey 5: 461 (1975) Shi & Martins in Fl. China 20–21: 186 (2011); Abdulridha, Taha & Wedad, Ecol. & Fl. Basra: 191 (2016).

Centaurea moschata L., Sp. Pl.: 909 (1753).

Annual or biennial, (12–)20–50(–70) cm tall, branched from base and frequently also above, stems glabrous or ± floccose. Leaves subglabrous, lower petiolate, lanceolate to elliptic, 4–6 × 1.2–1.5 cm, dentate to lyrate, median short-petiolate, pinnatifid to pinnatipartite with linear to oblong, frequently sharply dentate segments, uppermost leaves ± sessile much reduced. Capitula pedunculate, pedicels very variable in length but up to *c.* 20 cm; involucre 10–18 mm in diameter; outer phyllaries roundish with narrow membranous borders, inner oblong, apical appendage membranous, ovate to orbicular. Corollas rose-pink to yellow, rarely white, marginal strongly radiant, lacerate at tip.

HAB. & DISTRIB. Cultivated in a garden; fl. Apr. **LBA**: Basra, *Alizzi & Omar* 35109!

Turkey, Transcaucasia, NW Iran; widely cultivated.

Sweet Sultan (Engl.). A favourite garden plant decorative for its handsome capitula and with a musky odour.

25. **LEUZEA** DC.

Fl. Franc. (DC. & Lamarck), ed. 3, 4: 109 (1815), nom. nov. pro *Rhacoma* Adans. (1763) non L. (1759) [= Crossopetalum P. Browne (1756) (CELASTRACEAE)]
Rhaponticum Vaill., Königl. Akad. Wissensch. Paris Phys. Abh. 5: 177 (1754), nom. inval., non Ludw. (1757) (=*Centaurea* L.), nec Adans. (1763), nom. illeg.; Dittrich in Davis, Fl. Turkey 5: 460 (1975) (as *Rhaponticum* 'Hill'); Susanna & Garcia-Jacas in Kubitzki, Fam. Gen. Vasc. Pl. 8: 143 [2006] (2007); *Stemmacantha* Cass., Bull. Sci. Soc. Philom. Paris 1817: 12 (1817); Bremer, Asteraceae: Cladistics & Classification: 155 (1994). *Acroptilon* Cass., Dict. Sci. Nat., ed. 2, 50: 464 (1827).

D. J. Nicholas Hind[8]

Bushy perennial herbs with numerous many-branched green and leafy stems from a woody rhizomatous stock. Leaves sessile, simple, entire or dentate, unarmed. Inflorescence subcorymbose or of solitary capitula. Capitula medium-sized to large, ovoid, homogamous and discoid; phyllaries mostly 4–5-seriate, each with a basal ± veined, appressed cartilaginous portion ± concealed by overlapping papery appendages, these varying much in form from ovate and erose to acuminate; receptacle flat, white-setose. Florets tubular, well-exserted from involucre, corollas purple or yellow, rarely white, corolla-limb twice as wide as tube, lobed to about half-way, corolla lobes narrow. Achenes obovoid, compressed, glabrous, faintly ribbed, hilum basal to sublateral; pappus persistant, deciduous or caducous, of several rows of hairs, outer from ¼ to ½ as long as inner, flattened, barbellate, plumose towards apex, white.

A genus of about 26 species widespread in the mountains of Europe and Asia, with one doubtfully native species from Australia. One species, *L. repens* ('Russian knapweed') is considered a noxious weed in N America (USA and Canada) and is present in the Flora area.

Leuzea was named after de Candolle's friend, the botanist, librarian, astrologer, hypnotist, physician, alchemist, etc., Joseph Phillipe François Deleuze (1753–1835), from Sisteron, Alpes-de-Haute-Province, France.

This account originally was drafted under the generic name *Acroptilon*. Subsequent research has shown that the concept of *Acroptilon* is congeneric with that of '*Rhaponticum* Vaill.' sensu auctt. Unfortunately, Vaillant's names were published long before the start date of the binomial classification accepted in the Code, and the work that many authors use to propose their validity is a German translation of Vaillant's. Several authors discount their validity (e.g. Dittrich 1984, Hind 2007, Brummitt 2008, Susanna 2009), as does the current author. Hidalgo et al. (2006), using DNA sequencing data have shown that *Acroptilon*, *Leuzea* and '*Rhaponticum* Vaill.' are congeneric and as consequence Hidalgo provided a new combination for *Acroptilon repens* as he considered that '*Rhaponticum* Vaill.' was the valid generic name for the complex. Although Dittrich (1984) had earlier transferred most species to *Stemmacantha*, Hidalgo et al.'s results have shown that a broader approach might have been taken. Whilst Holub (1973) had probably anticipated future problems, and made a number of combinations in *Leuzea*, the transfer of *Acroptilon repens* had not been made; this was rectified by Hind (2019).

Bremer, K. (1994). Asteraceae: Cladistics & Classification: 153.
M. Dittrich, M. (1984). Neukombinationen in der Gattung *Stemmacantha* Cass. (Compositae) mit Bemerkungen zur Typisierung einiger ihrer Arten. Candollea 39: 45–49.
Holub, J. 1973. Contributions to the taxonomy and nomenclature of *Leuzea* DC. and *Rhaponticum* auct. *Folia Geobot. Phytotax., Praha* 8 (4): 377–395.

---

[8] Based on F. G. Davies's account of *Acroptilon* Cass.

Hidalgo, O., Garcia-Jacas, N., Garnatje, T. & Susanna, A. (2006). Phylogeny of *Rhaponticum* (Asteraceae, Cardueae-Centaureinae) and related genera inferred from nuclear and chloroplast DNA sequence data: taxonomic and biogeographic implications. Ann. Bot. (Oxford) 97(5): 705–714.

1. **Leuzea repens** (*L.*) *D.J.N.Hind*, Kew Bull. 74(2): 20 (2019).

Centaurea repens L., Sp. Pl. ed. 2: 1293 (1763).
Centaurea picris Pall. ex Willd., Sp. Pl. ed. 4, 3: 2302 (1803).
Serratula picris (Pall. ex Willd.) M. Bieb., Fl. Taur.-Caucas. 3: 546 (Dec. 1819 or early 1820).
Acroptilon obtusifolium Cass., Dict. Sci. Nat., ed. 2, 50: 465 (1827).
A. subdentatum Cass., Dict. Sci. Nat., ed. 2, 50: 465 (1827).
A. serratum Cass., Dict. Sci. Nat., ed. 2, 50: 466 (1827).
A. angustifolium Cass., Dict. Sci. Nat., ed. 2, 50: 466 (1827).
Carduus picris (Pall.) ex Willd.) Sweet, Hort. Brit. [Sweet], ed. 2: 281 (1830).
Acroptilon picris (Pall. ex Willd.) C.A. Mey., Verz. Pfl. Cauc.: 67 (1831); Boissier, Fl. Orient. 3: 612 (1875); Rawi in Dep. Agr. Iraq Tech. Bull. 14: 105 (1964).
Acroptilon picris (Pall. ex Wild.) DC., Prodr. 6: 662 (1838), comb. superfl.
A. picris (Pall. ex Willd.) C. A. Mey. var. α *obtusifolia* (Cass.) DC., Prodr. 6: 662 (1838).
A. picris (Pall. ex Willd.) C. A. Mey. var. β *angustifolia* (Cass.) DC., Prodr. 6: 662 (1838).
A. picris (Pall. ex Willd.) C. A. Mey. var. γ *subdentata* (Cass.) DC., Prodr. 6: 662 (1838).
A. picris (Pall. ex Willd.) C. A. Mey. var. δ *oxylepis* DC., Prodr. 6: 662 (1838).
A. picris (Pall. ex Willd.) C. A. Mey. var. ε *serrata* (Cass.) DC., Prodr. 6: 663 (1838).
A. repens (L.) DC., Prodr. 6: 663 (1838); Cherepanov in Fl. S.S.S.R. 28: 345 (1963); Kupicha in Davis, Fl. Turkey 5: 462 (1975); J. Dostál in Fl. Europ. 4: 254 (1976); Fl. Iranica 139B: 308 (1980).
A. australe Iljin in Bot. Mater. Gerb. Bot. Inst. Komarova Akad. Nauk S.S.S.R. 7: 59 (1937); Cherepanov in Fl. S.S.S.R. 28: 344 (1963).
A. repens (L.) DC. f. *cylindrica* Bornm., Bieh. Bot. Centralbl., Abt. B., 60: 209 (1940), nom. inval.
A. repens (L.) DC. f. *macrocephala* Bornm., Bieh. Bot. Centralbl., Abt. B., 60: 209 (1940), nom. inval.
A. repens (L.) DC. f. *macrolepis* Bornm., Bieh. Bot. Centralbl., Abt. B., 60: 209 (1940), nom. inval.
A. repens (L.) DC. f. *tincta* Bornm., Bieh. Bot. Centralbl., Abt. B., 60: 209 (1940), nom. inval.
A. repens subsp. *australe* (Iljin) Rech. f., Fl. Iranica 139B:309 (1980).
Rhaponticum repens (L.) Hidalgo, Ann. Bot. (Oxford) 97(5): 714 (2006); Shi Zhu & Martins in Fl. China 20–21: 178 (2011).

Stems 10–40(–60) cm tall, branched throughout, arachnose at least when young. Leaves many, oblong to lanceolate or linear, very variable in size, up to 7 × 2 cm, apex acute, mucronate, margins entire or often with a few teeth towards base, surfaces with depressed glandular dots, glabrescent, ± scabrid especially towards margins, lower surface somewhat more arachnose than upper, green; involucre very variable in size, mostly 10–16 × 5–10 mm; appendages of outer phyllaries pilose to ± glabrous, varying greatly from acute to eroded, ovate or depressed-ovate, those of inner phyllaries pilose, acuminate. Corolla glabrous throughout, ± 1.5 × as long as involucre, limb about half length of tube. Achenes 3–4 mm, cream to brown; pappus setae 8–10 mm. Fig. 21: 1–6.

HAB. In the mountains; alt. ± 1700 m; fl. & fr. Apr.–Aug.
DISTRIB. Very rare in Iraq; found only once, near the Iranian frontier. **MRO**: Haji Umran, *Rechinger* 11312!

S Russia, Turkey, Caucasia, Iran, Afghanistan, Pakistan, Kashmir, C Asia (Turkmenia to Uzbekistan), W Mongolia, China (Sinkiang), W Siberia. Introduced and widespread as a noxious weed in N America and now recognized as a weed in Argentina.

Iljin distinguished *Acroptilon australe* based on the shape of the appendages of the median and outer phyllaries (acute, not obtuse) and the fact that they conceal the cartilaginous bases of succeeding phyllaries. Rechinger has, as indicated in the synonymy above, reduced *A. australe* to subspecific rank. However, even at this rank he recognised that the only known Iraqi gathering is intermediate between the two subspecies, and there is no point in attempting to separate them here. Hidalgo et al. (2006) abstained from making a new combination (under *Rhaponticum*) as they considered its position unresolved. Bornmüller's names are all be invalid as they were apparently published with brief German 'diagnoses' at a date when Latin was the prerequisite for descriptions of new taxa.

Fig. 21. **Leuzea repens**. 1, habit × 2/3; 2, inner phyllary × 5; 3, median phyllary × 5; 4, outer phyllary × 5; fruit and pappus × 5; 6, floret × 6. All from *Davis* 13222 (Turkey). Drawn by Moira Rae Carter.

26. **KLASEA** Cass.

Dict. Sci. Nat. 35: 173 (1825); Susanna & Garcia-Jacas in Kubitzki, Fam. Gen. Vasc. Pl. 8: 144 [2006] (2007)

Shahina A. Ghazanfar[9]

Perennial unarmed herbs. Leaves simple, entire or serrate or sometimes pinnatisect, eglandular, glabrous or with lanate or villous pubescence. Inflorescence of terminal, solitary or corymbosely grouped capitula, homogamous. Phyllaries (involucre bracts) leaf-like, regularly imbricate, veined, with an appendage usually reduced to an apical slender spine, often reflexed. Florets all alike, narrowly tubular, limb ± equal to tube and deeply 5-lobed, corollas purplish, pink or yellow. Achenes obovoid, compressed, truncate, ribbed, glabrous, hilum almost basal; pappus of many coarse barbellate or barbellate-plumose hairs, persistent, double with the inner row raised above the outer.

Sixty-five species, Europe, Mediterranean region, SW Asia and northern Africa; 4 species in Iraq.

Greuter, W. (2003). The Euro+Med treatment of *Cardueae* (*Compositae*) – generic concepts and required new names. Willdenowia 33: 49–61.

Martins, L. & Hellwig, F.H. (2005). Systematic position of the genera *Serratula* and *Klasea* within Centaureinae (Cardueae, Asteraceae) inferred from ETS and ITS sequence data and new combinations in *Klasea*. Taxon 54(3): 632–638.

1. Leaf margins entire, leaves obtuse and amplexicaul . . . . 1. *K. cerinthifolia*
   Leaf margins toothed or lobed 2
2. Capitula in groups of 5–25; basal leaves regularly saw-toothed . . . . 2. *K. grandifolia*
   Capitula solitary or 2–3 together; basal leaves crenate or irregularly toothed 3
3. Leaves lyrate, margins crenate; phyllaries acute . . . . 3. *K. oligocephala*
   Leaves pinnatisect, margins irregularly toothed; phyllaries spine-tipped . . 4. *K. radiata*

1. **Klasea cerinthifolia** (*Sm.*) *Greuter & Wagenitz* in Willdenowia 33(1): 58 (2003).

Serratula cerinthifolia (Sm.) Boiss., Fl. Orient. 3: 585 (1875); Nábělek in Publ. Fac. Sci. Univ. Masaryk 52: 36 (1925); Dinsmore in Post, Fl. Syria, Palest. & Sinai ed. 2, 2: 98 (1933); Guest in Dep. Agr. Iraq Bull. 27: 90 (1933); Blakelock in Kew Bull. 4: 63 (1949); Zohary in Dep. Agr. Iraq Bull. 31: 154 (1950); Rawi in Dep. Agr. Iraq Tech. Bull. 14: 125 (1964); Matthews in Fl. Turkey 5: 453 (1975); Rechinger, Fl. Iranica 139b: 288 (1980).
Centaurea cerinthifolia Sm., Prodr. Fl. Graec. 2: 197 (1813); Boissier, Fl. Orient. 3: 589 (1875); Bornmüller in Beih. Bot. Centralbl. 60B: 209 (1939).
[*Serratula behen* sensu Lam., Tabl. Encycl. 3: 242 (1823), quoad descr., non L. (1753); Handel-Mazzetti in Ann. Naturh. Mus. Wien 27: 147 (1913).]
Schumeria cerinthifolia (Sm.) Iljin in Not. Syst. (Leningrad) 20: 362 (1960).

Stem branched, 15–65 cm, angular, white or green-striated, leafy, glabrous. Petiole of basal leaves up to 10 cm, lamina oblong-elliptic, up to 12 × 4 cm, simple or pinnatifid with unequal oblong obtuse or acute lateral lobes, much smaller than terminal lobe, glabrous. Cauline leaves sessile and amplexicaul, ovate to oblong, median 3–8 × 0.5–2.5 cm, obtuse, occasionally mucronate, entire, glabrous, uppermost hardly to considerably reduced in size. Capitula solitary at ends of branches; involucre at maturity 20–30 × 6–14 mm, with a rounded base, buds ovoid-fusiform; phyllaries scarious, outer ovate, subacute, papillose above, glabrous or sparingly floccose, inner linear, faintly striated, base and middle glabrous, tip papillose, ciliate or lanate apically, hair sometimes tufted. Corollas 25–30 mm, lobes of limb ± 6 mm, yellow. Achenes ± 7 mm, many-ribbed, whitish; pappus subplumose, 16 mm long.

HAB. Mountain slopes in *Quercus* and *Pinus* forest, occasionally on the upper foothills; alt. (300–)700–1000(–2000) m; fl. & fr. May-Jul.
DISTRIB. Common in the western and central zones of the forest region of Iraq, but very rare in the eastern sector and in the steppe. **MAM**: Jabal Khair, nr Zakho, *Rawi* 23052!; Ain Nuni, 15 km NW of Amadia, *Nábělek* 3363!, 3921!; Gali Mazurka below Ser Amadiya, *Dabbagh, Qaisi & Hamid* 46027!; Zawita, *Guest* 4901!; Dairalok, *Qaisi & Widad* 45870!; Atrush, *Guest* 3637!; Aqra, *E. Chapman* 26159! **MRO**: Salah

[9] Based on C.C. Townsend's account of *Serratula*.

ad-Din, *Qasim* 37536!; above Shaqlawa, *Bornmüller* 1459!; Chinaruk, Omar, *Qaisi & Widad* 49475!; Hab as-Sultan Dagh, *Rawi, Noori & Kass* 28262!; Qura, NE of Rania, *Rawi & Serhang* 23824!; Sakri-Sakran, *Bornm.* 4060; **MSU**: Jarmo, *Helbaek* 1895!; Darband-i Basian, *Rawi* 22823!; Pira Magrun, *Haussknecht* s.n. **FNI**: Maltai, between Tel Kaif and Mar Ya'qub, *Nábělek* 3663. **MJS**, Jabal Sinjar, *Handel-Mazzetti* 1473, *Mudir of Sinjar* 3117!, *Qaisi & Hamad* 53742! **MJS/FNI**: "Muwasal Tiatan Muzule Nuwar", *Field & Lazar* 447!;

Mediterranean Europe, Cyprus, Syria, Lebanon, Palestine, Turkey, Iran.

Guest noted that this plant was on sale in the market at Arbil and was sometimes used as a substitute for tea.

2. **Klasea grandifolia** (*P.H.Davis*) *Greuter & Wagenitz* in Willdenowia 33(1): 58 (2003).

Serratula grandiflora P.H. Davis in Notes Roy. Bot. Gard. Edinburgh 33: 293 (1974); Rechinger, Fl. Iranica 139b: 298 (1980).
[*S. coriacea* auct. non Fisch. & C.A.Mey.: Rawi in Dep. Agr. Iraq Tech. Bull. 14: 125 (1964)].

Erect, robust herb about 1.5 m tall, branched above. Stems woody, angular, green-striated, glabrous. Basal leaves with lamina 29–33 × 14–19 cm, long-petiolate, simple or lyrate basally with sometimes a few small lobes on petiole, main lamina elliptic, obtuse, base cordate, margin regularly saw-toothed, both surfaces glabrous apart from an initial lanate pubescence on veins. Cauline leaves large and petiolate below, rapidly reducing above, simple, uppermost sessile and entire. Inflorescence a corymb of many capitula, pedicels 0.3–1.5 cm. Involucre ovoid, urceolate with slightly spreading inner phyllaries, 15–20 × 6–8 mm; phyllaries 4–5-seriate, rather few (c. 40), outer and median lanceolate, acute, shorty spine-tipped, coriaceous, green, apically darker-striated, shortly pilose, feebly floccose or ± glabrous; inner phyllaries with slightly spreading, scarious, stramineous, simple, lanceolate appendages up to 10 mm. Corollas exserted from involucre, purplish-pink, 15–18 mm, lobes of limb 4–5 mm. Achenes oblong, 8–9 × ± 2.8 mm, brown, faintly striated; pappus c. 8 mm, subplumose, stramineous. Fig. 22: 1–7.

HAB. In *Quercus* woodland on a mountain slope facing N.; alt. ± 1350 m ; fl. & fr. Jun-Jul.
DISTRIB. Very rare in Iraq; only found in one locality in the SE sector of the forest zone. **MSU**: Qopi Qara Dagh, *Haines* 1534!, *Agnew* 492, *Poore* 542!, *Hadač* 6269.

W Iran.

3. **Klasea oligocephala** (*DC.*) *Greuter & Wagenitz* in Willdenowia 33(1): 58 (2003).

Serratula oligocephala DC., Prodr. 6: 669 (1837); Boissier, Fl. Orient. 3: 589 (1875); Nábělek in Publ. Fac. Sci. Univ. Masaryk 52: 37 (1925); Dinsmore in Post, Fl. Syria, Palest. & Sinai ed. 2, 2: 98 (1933); Guest in Dep. Agr. Iraq Bull. 27: 90 (1933); Blakelock in Kew Bull. 4: 63 (1949); Zohary in Dep. Agr. Iraq Bull. 31: 155 (1950); Rawi in Dep. Agr. Iraq Tech. Bull. 14: 125 (1964); Davis & Kupicha in Fl. Turkey 5: 458 (1975); Rechinger, Fl. Iranica 139b: 294 (1980).

Rhizomatous herb about 30–70 cm tall. Stem slender, sulcate-striate, white-woolly at base and at first throughout, unbranched, leafy only in about lower part or with a few very reduced leaves above. Basal leaves petiolate, petiole 3–10 cm, often floccose when young, lamina lyrately lobed with 1–2 pairs of small basal lobes, main lamina ovate, 5–18 × 2–11 cm, obtuse, base cordate, truncate or cuneate, margins deeply (sometimes compoundly) crenate-dentate, both surfaces at first floccose, soon glabrescent. Stem leaves few, rapidly decreasing, simple or pinnatisect. Capitula solitary or few, long-pedunculate; pedicels 2–10 cm, without or with minute bract-like leaves. Involucre hemispherical, 18–25 × 18–25 mm. Outer phyllaries green or brown, lanate, glabrescent, gradually narrowed to a sharply acute point, with or without striations, dark near apex, sparingly short-hairy and papillose; inner phyllaries with brown or stramineous scarious acuminate appendages, indumentum similar. Corollas exceeding involucre, purple, 15 mm, lobes 4 mm. Achenes dark brown, 6–7 × 3 mm, smooth; pappus 7–8 mm, of barbellate flattened hairs.

HAB. Rocky mountain slopes, on limestone, often in *Pinus* and *Quercus* woods; alt. (500–)1000–2000 m; fl. & fr. Apr.-Jul.
DISTRIB. Quite common in the NW sector of the forest zone of Iraq. **MAM**: Zakho pass, *Guest* 2264!; Khantur, ± 23 km NE of Zakho, *Rawi* 23309!; Sharanish, NE of Zakho, *Rawi* 23208!; Ain Nuni, *Nábělek* 3663; Zawita, *Guest* 4592!, *Emberger, Guest, Long, Schwan & Yusuf* (*Sarkahia*) 15387!; Zawita gorge, *Guest*

Fig. 22. **Klasea grandifolia**. 1, habit × 2/3; 2, leaf × 1; 3, leaf base × 1; 4, phyllary × 3; 5, anthers enlarged; 6, style enlarged; 7, achene and pappus × 6. All from *Haines* 1534. Drawn by M. Millar Watt.

2206!; Gara Dagh, *Dabbagh & Jasim* 46977!; Gara Dagh near Sarsang, *E. Chapman* 26421!; Sarsang, *Haines* W.1169!; Pl. Mesopot., Kurdistan & Mussul", *Kotschy* 316.

Syria, Turkey (E & S Anatolia).

4. **Klasea radiata** (*Waldst. & Kit.*) *Á.Löve & D.Löve*, Bot. Not. 114: 43 (1961) subsp. **biebersteiniana** (*Grossh.*) *Greuter* in Willdenowia 35(2): 236 (2005).

Carduus radiatus Waldst. & Kit., Pl. Rar. Hung.: t. 2 (1800).
Serratula radiata (Waldst. & Kit.) M.Bieb., Fl. Taur.-Cauc. 3: 545 (1819).
Serratula radiata M.Bieb. subsp. *biebersteiniana* Iljin ex Grossh. Fl. Kavkaza [Grossheim] 4: 194 (1934).
S. biebersteiniana (Iljin ex Grossh.) Takht., Atl. Fl. Yerevana 323 (1945); Borissova in Fl. S.S.S.R. 28: 275 (1963).
Klasea biebersteiniana (Grossh.) Hidalgo, Ann. Bot. (Oxford) 97(5): 714 (2006).

Rhizomatous herb 30–60(–100) cm tall, scabrid. Stems ridged, slender, with a few long, slender branches. Basal leaves (only the sheathing bases remaining at flowering time) lyrate-pinnatisect to pinnatisect, up to 40 × 12 cm, petiolate, with 6–8 (rarely more) pairs of linear to oblong, acute, subentire to toothed lobes, terminal lobe slightly to considerably larger both surfaces residually lanate, lower surface with a conspicuous pale network of veins; upper leaves either similar in form (but more shortly petiolate) or uppermost simple and narrow. Capitula solitary on leafy branches or sometimes in lax clusters; involucre campanulate, broadly ovoid to hemispherical, 15–25 mm in diameter; phyllaries closely imbricate, outer with an ovate base ± abruptly narrowed to a short mucro or subaristate with a straight or deflexed dark spine 1.5–3 mm., inner with a linear to lanceolate-mucronate, scarious, stramineous apical appendage. Corollas purplish, 20–22 mm, lobes narrow, 7–8 mm, exceeding involucre. Achenes 5 mm long, brownish; pappus hairs 6–8 mm, smooth.

HAB. On barren denuded hills; alt. ± 450 m; fl. Apr.
DISTRIB. Very rare in Iraq; found only once in the moist steppe zone. **FKI**: nr Kirkuk, *Guest et al.* 5097!

Turkey, Iran, Caucasia.

This seems to be rather a weak subspecies.

Serratula haussknechtii Boiss. was cited for Iraq by Zohary in Dep. Agr. Iraq Bull. 31: 155 (1950) on the basis of Boissier's record in Fl. Orient. 3: 589 (1875) from "montibus calcareis Avroman et Schahu" – a citation repeated by Rawi in Dep. Agr. Iraq Tech. Bull. 14: 125 (1964). However, Mt. Shahu is entirely in Iran, as is the eastern side of Mt. Avroman. Although many botanists have collected on the Iraqi side of Mt. Avroman, none has as yet detected the presence of *S. haussknechtii*. It may possibly occur, as may a further little-known species of the Iran/Iraq frontier area, *S. melanochila* Boiss.

27. **CARTHAMUS** L.

Sp. Pl.: 830 (1753); Gen. Pl.: 361 (1754)
Susanna & Garcia-Jacas in Kubitzki, Fam. Gen. Vasc. Pl. 8: 144 [2006] (2007)
Kentrophyllum Neck. ex DC., Ann. Mus. Natl. Hist. Nat. 16: 158 (1810)

C. C. Townsend

Annual herbs or subshrubs (two W European species), corymbosely branched with rigid, often whitish stems and branches, glabrescent to ± arachnoid and shortly glandular. Leaves coriaceous, usually spinose, dentate to pinnatisect, rarely simple and entire, upper sessile and ± auriculate or amplexicaul. Capitula homogamous, many-flowered. Involucre subglobose to broadly or more narrowly ovoid, ± arachnoid and glandular. Phyllaries multiseriate, outer and median leafy or with a leafy or fimbriate appendage, inner lanceolate and ± striate. Corolla bluish to orange or bright to pale yellow, rarely whitish, limb deeply divided. Filaments bearded, with scattered hairs or glabrous. Achenes glabrous, turbinate, mostly bluntly to distinctly 4-angled, the marginal rugose and almost always lacking a pappus, the inner smooth except usually about the apex, with a pappus of scales or barbellate to shortly plumose bristles.

About 14 species in the Mediterranean region and SW & C Asia; 7 species in Iraq.

Carthamus (from the Arab. name *Qurtūm* for safflower, *C. tinctorius*, q.v.).

Vilatersana, R., Garnatje, T., Susanna, A. & Garcia-Jacas, N. (2005). Taxonomic problems in Carthamus (Asteraceae): RAPD markers and sectional classification. Bot. J. Linn. Soc. 147 (3): 375–383.

González, L. (1990). Acerca de la clasificación natural del género *Carthamus* L., s. l. Anales del Jardín Botánico de Madrid 47: 11–34.

1. Leaves without or with very short marginal spines or teeth, these when present less than 2 mm long . . . . 1. *C. tinctorius*
   Leaves with marginal teeth or spines more than 2 mm long 2
2. Corollas purple, rarely pink or whitish. 3
   Corollas bright to pale yellow, orange or orange-red 4
3. Median phyllaries with a lacerate, lanceolate to rhomboid, frequently purple-streaked appendage present between innermost lanceolate and outer foliose-appendiculate phyllaries . . . . 3. *C. dentatus*
   Median phyllaries with a lacerate, fimbriate appendage (occasionally absent in some capitula) between the innermost lanceolate and the outer foliose-appendiculate phyllaries . . . . 2. *C. lanatus*
4. Outer phyllaries foliose-appendiculate, passing into inner lanceolate phyllaries without intermediates of different form . . . . 4. *C. glaucus*
   Iintermediate phyllaries transitional from outer to inner, with gradually reducing foliaceous appendages and marginal teeth, spinulose at apex. 5
5. Achenes of central florets with a very short, caducous pappus much shorter than achene body . . . . 5. *C. oxyacantha*
   Achenes of central florets with a well-developed pappus, the largest bristles clearly longer than achene body 6
6. Inner foliose-appendiculate phyllaries equalling or exceeding florets, with 4–9 stramineous to yellowish spines along each margin . . . . 6. *C. persicus*
   Inner foliose-appendiculate phyllaries shorter than florets, with 10–13 yellowish spines along each margin . . . . .7. *C. curdicus*

1. **Carthamus tinctorius** *L.*, Sp. Pl.: 830 (1753); Boissier, Fl. Orient. 3: 709 (1875); Handel-Mazzetti in Ann. Naturh. Mus. Wien 27: 452 (1913); Nábělek in Publ. Fac. Sci. Univ. Masaryk 52: 47 (1925); Dinsmore in Post, Fl. Syria, Palest. & Sinai ed. 2, 2: 122 (1933); Guest in Dep. Agr. Iraq Bull. 27: 19 (1933); Dickson, Wild Fls. Kuwait: 30 (1955); Shostakovski in Fl. S.S.S.R. 28: 586 (1963); Hanelt in Feddes Repert. 67: 90 (1963); Kupicha in Fl. Turkey 5: 327 (1975); Husain & Kasim in Cult. Pl. Iraq: 476 (1975); Chakravarty in Plant Wealth Iraq 1: 103 (1976); Feinbrun-Dothan, Fl. Palaest. 3: 403 (1978); Rechinger, Fl. Iranica 139B: 433 (1980); Meikle, Fl. Cyprus 2: 978 (1985); Breckle, Hedge & Rafiqpoor, Vasc. Pl. Afghan: 146 (2013); Kilian & Ghazanfar in Fl. Oman 3: 212 (2015); Abdulridha, Taha & Wedad, Ecol. & Fl. Basra: 201 (2016).

Annual herb 0.25–1.25 m tall. Stem unbranched below, with slender ascending branches above, shining white, glabrous, striate to angular-sulcate. Basal and cauline leaves similar, glabrous or sparsely glandular, lanceolate-oblong to narrowly ovate or elliptic, 2–11 × 0.8–2 cm, sessile, acute with a spinulose tip, margins entire or regularly serrate with shortly spinulose teeth. Involucre subglobose, shortly glandular, thinly arachnoid or not, 1.8–3 cm in diameter, solitary or in lax corymbs, foliaceous outer phyllaries 4-seriate with base appressed and appendages ovate, acute with a spinulose tip, margins entire or spinulose; median and inner phyllaries from ovate-appendiculate to lanceolate inwards, stramineous below, prominently veined. Corollas orange to orange-red. Achenes turbinate, quadrangular, 7–8 × 4–5 mm; pappus absent. Fig. 23: 5.

HAB. & DISTRIB. Cultivated, or as a relic of, or escape from cultivation; grown mainly on the alluvial plain, but also in the mountain region and perhaps elsewhere; alt. 35–2300 m; fl. & fr. Mar.–Jun. on the plains, Jun.–Aug. in the mountains. **MRO**: Qandil range, NE of Rania, *Rawi & Serhang* 26818!; Haji Umran, *Rawi, Uthman & Serhang* 26070! **LCA**: cultivated at the College of Agriculture, Abu Ghraib, *Barkley & Noori Mohamed* 4989!; Abu Ghraib, *Alizzi* 34141!, 34143!, *Sahira* C684!; Baghdad, at Za'faraniya Horticultural Station from seed obtained from Palestine, *Aftan al-Rawi* 13944! **LBA**: Basra, in garden, *Omar, Karim & Hamid* C180!

Cultivated in the Mediterranean region, W & C Asia and Africa since ancient times; not known outside cultivation, and of uncertain origin.

Bastard Saffron, Safflower, ZA'FARĀN, QURTŪM used as colouring for food and other purposes. Formerly the strip of land on the left of the R. Tigris was known as Za'faraniya because of bastard saffron cultivated there; but this land has now been swallowed up by the southward extension of the city of Baghdad.

2. **Carthamus lanatus** *L.*, Pl.: 830 (1753); Boissier, Fl. Orient. 3: 706 (1875); Handel-Mazzetti in Ann. Naturh. Mus. Wien 27: 452 (1913); Dinsmore in Post, Fl. Syria, Palest. & Sinai ed. 2, 2: 120 (1933); Guest in Dep. Agr. Iraq Bull. 27: 19 (1933); Blakelock in Kew Bull. 4: 45 (1949); Zohary in Dep. Agr. Iraq Bull. 31: 157 (1950); Shostakovski in Fl. S.S.S.R. 28: 583 (1963); Hanelt in Feddes Repert. 67: 135 (1963); Rawi in Dep. Agr. Iraq Tech. Bull. 14: 108 (1964); Rechinger, Fl. Lowland Iraq: 660 (1964); Kupicha in Fl. Turkey 5: 592 (1975); Rechinger, Fl. Iranica 139B: 436 (1980); Meikle, Fl. Cyprus 2: 436 (1985); Collenette, Wild Flow. Saudi Arabia: 176 (1999); Boulos & Hind in Fl. Egypt 3: 178 (2002); Breckle, Hedge & Rafiqpoor, Vasc. Pl. Afghan: 154 (2013).

Annual herb 20–90 cm tall. Stem stramineous to brownish, simple or usually divaricately branched above middle, sparsely to densely furnished with shaggy arachnoid hairs. Basal and lower cauline leaves 2.5–14 × 1–4 cm, oblong- or ovate-lanceolate, sessile and ± amplexicaul, veins prominent, green to greyish-arachnoid and minutely glandular, spinous-pinnatifid or sometimes pinnatisect with 3–8 pairs of lobes; upper lobes narrowly oblong with 1 or more pairs of secondary lobes, proximal lobes linear; upper cauline leaves less deeply lobed. Involucre ovoid, ± arachnoid and shortly glandular; outer phyllaries foliaceous to foliaceous-appendiculate, 25–50 × 6–12 mm with 3–5 pairs of spines or short spinous lobes; innermost phyllaries 15–25 mm, lanceolate, entire; intermediate phyllaries with a lanceolate to narrowly ovate fimbriate appendage, or these occasionally absent in some capitula. Corollas bright to pale yellow. Achenes white to greyish, turbinate, quadrangular, 5–6 × 4–5 mm; pappus whitish to brown, variably developed, longest bristles to 15 mm.

DISTRIB. (of species): S Europe and the Mediterranean region generally, SW & C Asia. Introduced in tropical and S.Africa, Australia, N & S America.

subsp. LANATUS, with green leaves, bright yellow flowers and the inner pappus-scales very short, is not found in Iraq. In Iraq the species is represented by:

subsp. **turkestanicus** (*Popov*) *Hanelt* in Feddes Repert. Spec. Nov. Regni Veg. 67: 148 (1963); Rechinger, Fl. Iranica 139B: 436 (1980).

C. turkestanicus Popov in R. Uzb. Gos. Univ. ser. Biol. 27: 36 (1941).

Leaves canescent. Corollas pale yellow. Inner pappus-scales longer, half as long as to subequalling the intermediate phyllaries.

HAB. Hillside, and as a weed of cultivation and broken ground between fields of *Triticum*; alt. 100–850 m; fl. & fr. May-Sep.
DISTRIB. Scattered in the mountains, foothills and central alluvial plain of Iraq. **MSU**: Jarmo, *Haines* 240! **FPF**: Sa'diya, 31 km SW of Khanaqin, *Qaisi* 47208!; Koma Sang, 7 km W of Mandali, *Rawi* 20607!; Mandali, *Guest* 893! **LCA**: Daltawa (Khalis), *Guest* 2468! **FPF**: near Zurbatia, *A. Haloob & Saif* 58698 (BAG!).

Guest (1933) gives the colloquial Arabic name of this plant as DAM AL-HUSSAIN (Hussain's Blood), and states that if a phyllary is torn off and held with the wounded surface against the hand it leaves a mark resembling an old bloodstain.

3. **Carthamus dentatus** *Vahl*, Symb. Bot. 1: 69, t. 17 (1790); Boissier, Fl. Orient. 3: 708 (1875); Dinsmore in Post, Fl. Syria, Palest. & Sinai ed. 2, 2: 121 (1933); Blakelock in Kew Bull. 4: 45 (1945); Hanelt in Feddes Repert. 67: 92 (1963); Rawi in Dep. Agr. Iraq Tech. Bull. 14: 108 (1964); Kupicha in Fl. Turkey 5: 592 (1975); Rechinger, Fl. Iranica 139B: 435 (1980); Meikle, Fl. Cyprus 2: 979 (1985).

Annual herb 30–60(–100) cm tall. Stems whitish to pale brown, ± villous-arachnoid to glabrescent, shortly glandular, branched mostly above middle. Leaves sessile, ovate to linear-lanceolate, (1–)2–6 × 0.8–1.5 cm, acuminate, greyish arachnoid-pubescent on both surfaces and shortly glandular, underside prominently veined and somewhat bullate, lowermost pinnatifid to pinnatisect, remainder toothed or pinnatifid with 3–7 pairs of teeth or lobes, spiny. Involucre ovoid, tapering above, ± arachnoid, densely shortly pubescent and glandular; outer phyllaries foliaceous, toothed, not totally concealing the inner, linear to lanceolate, 2.5–5.5 × 0.4–0.8 cm, exceeding florets; innermost phyllaries lanceolate, cartilaginous, gradually tapering; intermediate phyllaries with a lanceolate to rhomboid, lacerate-fimbriate, frequently purple-streaked apical appendage; both median and inner phyllaries with a slender terminal spine. Corollas purple. Achenes turbinate, bluntly quadrangular, whitish, 4–6 × 4–5 mm; pappus shining-stramineous, longest (median) bristles 12–20 mm long, or all bristles of this length. Fig. 23: 6–7.

HAB. Rare in Iraq, only found once in the mountain region, not far from the Turkish frontier; alt. ± 925 m; fl. Oct.
DISTRIB. **MAM**: nr Zawita, 925 m, *Guest* 1657!

Aegean Is., Cyprus, Crete, Syria, Turkey, W Iran.

4. **Carthamus glaucus** *M.Bieb.*, Tabl. Prov. Casp.: 118 (1798) emend. Hanelt in Feddes Repert. 67: 105 (1963); Boissier, Fl. Orient. 3: 707 (1875); Handel-Mazzetti in Ann. Naturh. Mus. Wien 27: 452 (1913); Nábělek in Publ. Fac. Sci. Univ. Masaryk 52: 46 (1925); Anthony in Notes Roy. Bot. Gard. Edinburgh. 18: 292 (1935); Blakelock in Kew Bull. 4: 45 (1949); Zohary in Dep. Agr. Iraq Bull. 31: 67 (1950); Shostakovski in Fl. S.S.S.R. 28: 584 (1963); Rawi in Dep. Agr. Iraw Tech. Bull. 14: 108 (1964); Rechinger, Fl. Lowland Iraq: 660 (1964); Kupicha in Fl. Turkey 5: 593 (1975); Feinbrun-Dothan, Fl. Palaest. 3: 403 (1978); Rechinger, Fl. Iranica 139B: 435 (1980); Mouterde, Nouv. Fl. Liban et Syrie 3: 498 (1983); Boulos & Hind in Fl. Egypt. 3: 179 (2002).

Kentrophyllum alexandrinum Boiss. & Heldr. in Boissier, Diagn. Pl. Orient., ser. 2, 6: 115 (1859).
Carthamus glaucus M.Bieb. var. *alexandrinus* (Boiss. & Heldr.) Boiss., Fl. Orient. 3: 707 (1875).
Carthamus alexandrinus (Boiss. & Heldr.) Bornm., Verh. Zool.-Bot. Ges. Wien 48: 605 (1898).
Carthamus glaucus M.Bieb. subsp. *alexandrinus* (Boiss. & Heldr.) Hanelt, Feddes Repert. 67: 113 (1963).

Annual herb 15–70 cm tall. Stems whitish to pale stramineous, ± densely arachnoid or glabrescent, shortly glandular, branched from about the middle or at top only. Leaves sessile, ± amplexicaul, lanceolate to ovate-lanceolate, 1.5–5 × 0.4–1.2 cm, acuminate, mostly sparingly arachnoid, shortly pubescent and glandular, prominently veined, lower pinnatifid to pinnatisect with 3–6 pairs of teeth or lobes, spiny. Involucre ovoid, ± arachnoid, densely shortly pubescent or glandular; outer phyllaries foliaceous, 15–30 × 4–6 mm, linear to lanceolate, subequalling or exceeding florets; innermost phyllaries linear, acuminate; intermediate phyllaries transitional from outer to inner, with gradually reducing foliaceous appendages and marginal teeth, spinulose at apex. Corollas purple or rarely pink or whitish. Achenes turbinate or rarely ovoid, slightly angled, 5 × 4 mm, whitish; pappus stramineous or brownish, longest setae to 4 mm.

HAB. Stony mountain slopes, fields and hillsides in the *Quercus* zone, rocky places near streams, dry grassy roadsides; alt. 300–1600 m; fl. & fr. Jun.-Sep.
DISTRIB. Frequent in the lower mountain region, occasional in the steppe region of Iraq. **MJS**: Jabal Sinjar, *Handel-Mazzetti* 1497! **MAM**: Atrush, *Guest* 3639!; Bekher, nr Zakho, *Rawi, Noori & Tikriti* 28966!; Mar Ya'qub, *Nábělek* 3718; Sarsang, *Haines* 408! **MRO**: between Arbil and Rowanduz, *Bornmüller* 1461!; Gali Ali Beg, *Omar, Sahira, Karim & Hamid* 38327!; 3 km S of Rayat, *Rawi & Serhang* 26791!; Pushtashan, *id.* 23843!; Batas, *Guest* 3013!; 2 km below Darband, *Gillett* 9455! **MSU**: 16 km NW of Darbandikhan, *Rawi, Husham & Noori* 29440!; Zalam, nr Khormal, *id.* 29417! **FNI**: 1 km NE of Mindan, *E. Chapman* 26202! **FPF**: Jabal Hamrin, *Sutherland* 102, 103. **LEA**: Shahraban (Muqdadiya), *Rechinger* 13374!

Crimea, Turkey, Caucasia, W Iran.

5. **Carthamus oxyacanthus** *M.Bieb.*, Tabl. Prov. Casp.: 108 (1798); Boissier, Fl. Orient. 3: 709 (1875); Guest in Dep. Agr. Iraq Bull. 27: 19 (1933); Anthony in Notes Roy. Bot. Gard. Edinburgh 18: 292 (1935); Bornmüller in Beih. Bot. Centralbl. 60B: 215 (1939); Blakelock in Kew Bull. 4: 45 (1949); Zohary in Dep. Agr. Iraq Bull. 31: 157 (1950); Dickson, Wild Fls. Kuwait: 30 (1955); Shostakovski in Fl. S.S.S.R. 28: 584 (1963); Hanelt in Feddes Repert.

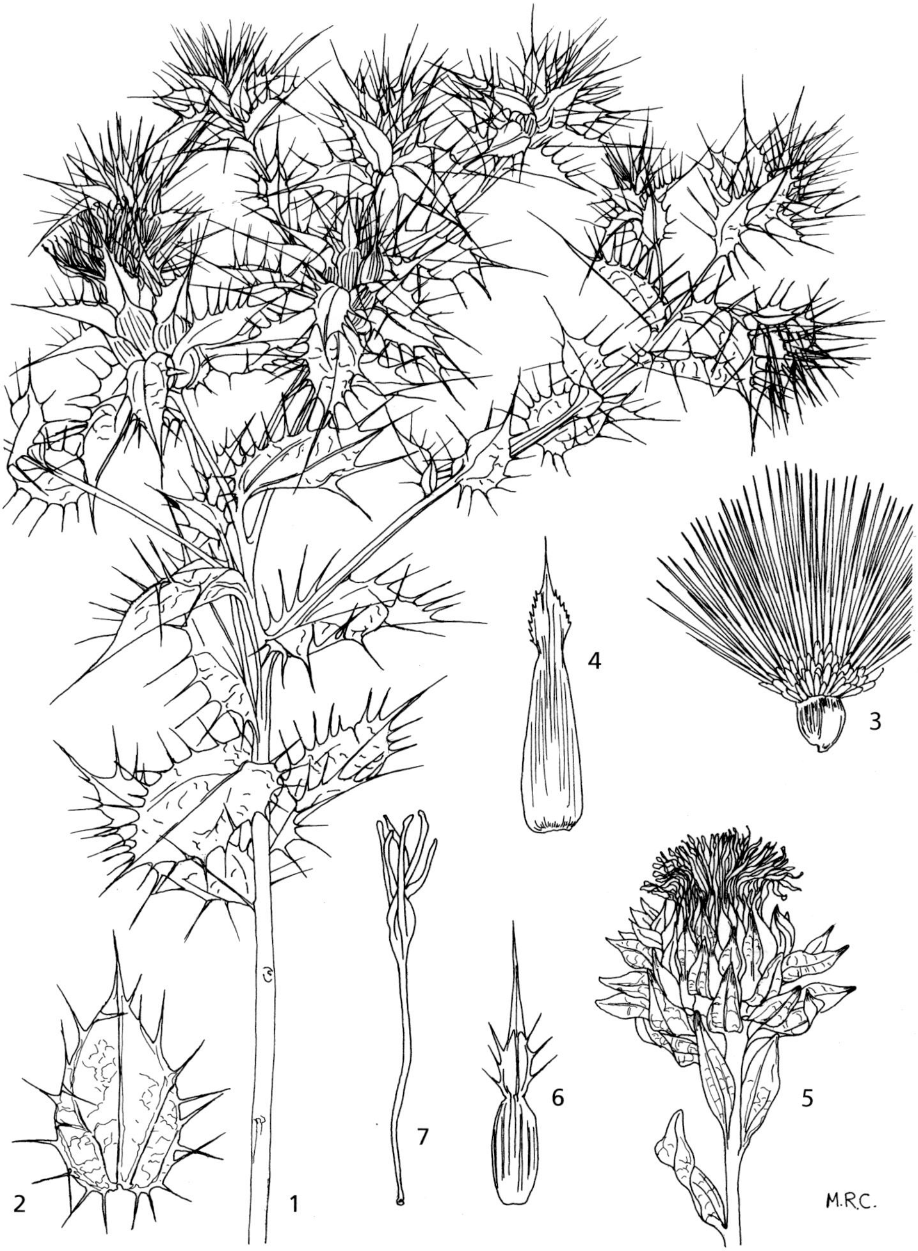

Fig. 23. **Carthamus oxyacanthus**. 1, habit × 2/3; 2, leaf × 2; 3, floret × 3; 4, outer phyllary × 2. **Carthamus tinctorius**. 5, capitulum × 1; **Carthamus dentatus**. 6, intermediate phyllary × 2; 7, achene × 2. 1–4 from *Gillett* 6551; 5 from *Barkley & Noori Mohammed* 4989; 6–7 from *Guest* 1657. Drawn by Moira Rae Carter.

67: 83 (1963); Rawi in Dep. Agr. Iraq Tech. Bull. 14: 108 (1964); Rechinger, Fl. Lowland Iraq: 661 (1964); Hasawi, Tammimi & Alizzi in Dep. Agr. Iraq Tech. Bull. 167: 112 (1968); Rechinger, Fl. Iranica 139B: 433 (1980); Mouterde, Nouv. Fl. Liban et Syrie 3: 496 (1983); Collenette, Wild Flow. Saudi Arabia: 176 (1999); Breckle, Hedge & Rafiqpoor, Vasc. Pl. Afghan: 154 (2013); Abdulridha, Taha & Wedad, Ecology and Fl. Basra: 201 (2016).

Annual herb 20–50(–70) cm tall. Stems whitish, usually considerably divaricately corymbose-branched from middle or below, glabrous or sparsely glandular below, increasingly glandular and often arachnoid above. Basal and lower stem leaves to 23 × 6 cm, irregularly narrowly pinnatisect with 4–5 pairs of spinose lobes or lowest almost unarmed, all narrowed and subpetiolate with a short expanded basal sheath. Upper cauline and branch leaves oblong or lanceolate-oblong, sessile, ± amplexicaul, very variable in size, with mostly 6–9 pairs of teeth terminating in long, yellow (rarely reddish) spines up to 10 mm. Involucre ovoid, densely shortly glandular and often ± arachnoid; outer phyllaries foliaceous, patent, 15–50 × 4–14 mm, margins and tip with long, yellow spines, exceeding florets; innermost phyllaries lanceolate, tip feebly spinose but otherwise unarmed, strongly ribbed; intermediate phyllaries with gradually reducing foliose appendages, then entire and quickly narrowed. Corollas bright yellow. Achenes obovoid, only bluntly quadrangular, 4–6 × 2–4 mm, shining, white to chestnut; pappus usually absent, rarely very short. Fig. 23: 1–4.

HAB. Stony hillsides, by streams in the hills, on limestone slopes and silty plain, also as a weed of (especially irrigated) cultivation, on clay, sandy clay and conglomerate; alt. 30–920 m; fl. & fr. May-Oct.

DISTRIB. Quite common on the lower mountains, steppe and lower Mesopotamian plain. **MAM**: Badi, *Guest* 4416!; Atrush, *Guest* 3644! **MRO**: Diana, nr Rowanduz, *Guest* 2988! **MSU**: Jarmo, *Helbaek* 1978!; Sarchinar, nr Sulaimaniya, *Fedde* 5354!; Sulaimaniya, *Guest* 12923! **FNI**: Gerwona nr Ain Sifni, *Field & Lazar* 722!; Bashiqa, *Tikriti* 16386! **FAR**: Ankawa, nr Arbil, *Bornmüller* 1462; Bawanur, *Poore* 552! **FPF**: Sa'diya, *Qaisi* 47207!; Ab-e-Naft, W of Mandali, *Guest* 776!; Lajama, *Rawi* 20724! **DLJ**: 12 km NW of Falluja, *Rawi & Shahwani* 25464! **DSD**: 78 km E by S of Salman, *Guest, Rawi & Long* 14128! **LCA**: Bani Sa'd, 25 km N of Baghdad, *B. Anderson* 1995!; Abu Ghraib, *Gillett* 6551!; Aqqur Kuf, *Sahira & Ali* 39493! **LEA**: 78 km W of Fakka, *Hazim* 30668! **LSM**: 30 km S of Amara, *Hazim & Noori* 30649!

Transcaucasia, Iran, Afghanistan, Pakistan, C Asia.

6. **Carthamus persicus** *Willd.*, Sp. Pl. ed. 4, 3: 1707 (1803); Hanelt in Feddes Repert. 67: 71 (1963); Kupicha in Fl. Turkey 5: 591 (1975); Feinbrun-Dothan, Fl. Palaest. 3: 403 (1978); Rechinger, Fl. Iranica 139B: 143 (1980).

C. palaestinus Eig in Feddes Repert. 63: 48 (1931), nomen, ex Rech. f. in Anz. Math.–Nat. Kl. Öst. Akad. Wiss. 101: 17 (1964); Hanelt in Feddes Repert. 67: 77 (1963).

Annual herb 20–60(–90) cm tall. Stems white, branched from near base or in upper half only, and branches glabrous or sparingly floccose and/or glandular approaching the capitula. Basal and lower stem leaves pinnatifid to narrowly pinnatisect with up to 6 pairs of narrow, toothed lobes and irregular shorter lobes, oblong and 4–9 × 1.2–3.5 cm in outline. Upper cauline leaves sessile, auriculate, oblong to lanceolate- or ovate-oblong, acuminate, dentate to pinnatisect with stramineous spines up to 6 cm long, glabrous or feebly arachnoid or glandular, 3–5 × 0.8–1.5(–2) cm; branch leaves rapidly diminishing. Involucre ovoid, thinly to densely arachnoid and glandular, particularly the median and inner phyllaries; outer phyllaries resembling the upper branch leaves, subequalling to slightly exceeding florets, patent, 15–40 × 3–7 mm, with 4–9 stramineous to yellowish spines on each margin; inner phyllaries lanceolate, tip feebly spinous but otherwise unarmed, moderately ribbed; intermediate phyllaries with reducing leafy appendages, then entire and quickly narrowed above. Corollas bright yellow. Achenes turbinate-quadrangular, 4–5 mm, slightly longer than broad, whitish or marmorate with a rugose apical rim; pappus stramineous or brownish, longer bristles 7–11 mm, barbellate.

HAB. On low hills, by roadsides, in wadis and depressions in the subdesert, on clay, sand, gravel or silt; alt. 220–710 m; fl. & fr. May-Jun.

DISTRIB. Occasional in the desert, rare on the alluvial plain of Iraq. **DWD**: Al-Masad depression, *Alizzi & Omar* 36183!; 5 km S of H.3, *Rawi* 21211!; 160–190 km NW of Ramadi on road to Rutba, *Rawi* 31322!; 91 km W of Ramadi on road to Rutba, *Rechinger* 12685!; 46 km W of Ramadi on road to Rutba, *Guest, Knowles & Rawi* 25005! **LEA**: 70 km N of Amara, *Hazim & Noori* 30625!

Hanelt, l.c., postulates that *C. palaestinus* Eig ex Rech.f. may be a hybrid between *C. persicus* and *C. tinctorius*. I cannot, however, but agree with Feinbrun-Dothan l.c., in regarding it as referable to the very variable *C. persicus*.

7. **Carthamus curdicus** *Hanelt* in Feddes Repert. 67: 82 (1963); Rechinger, Fl. Iranica 139B: 432 (1980).

Annual herb 40–60 cm high. Stems whitish, with a few ascending slender branches above middle, glabrous. Basal leaves not seen. Lower stem leaves pinnatisect with 5 pairs of narrow, weakly spinous segments, 8–10 × 2–2.5 cm; upper stem leaves sessile, auriculate, finely spinous-dentate with yellow spines, 4–6 × 0.8–2 mm, rapidly diminishing to the small, narrow branch leaves; all leaves glabrous. Involucre ovoid, median and inner phyllaries densely glandular, ± arachnoid; outer phyllaries resembling upper branch leaves, linear-lanceolate, 15–20 × 2–3 mm, with 10–13 yellowish spines on each margin, much shorter than florets; inner phyllaries lanceolate, tip feebly spinose but otherwise unarmed, strongly ribbed; intermediate phyllaries with reducing leafy appendages, then entire and quickly narrowed above. Corollas bright yellow. Achenes turbinate, ± sharply angular, 5.5 × 4.5 mm, whitish with a brown upper margin; pappus stramineous, darker below, longest bristles 7–8.5 mm, barbellate.

HAB. Very rare; two collections from dry hills in the northernmost mountain region and one in the foothills; alt. (once recorded) 550 m; fl. Jul.
DISTRIB. **MAM**: 20 km NE of Zakho, *Rawi* 23169!; Dagh al Radziem between Zakho and Sharanish, *Rechinger* 10731! **FPF**: N of Kani Masi, *Hadač* 4653!

Turkey.

Hanelt suggests that this plant may be a hybrid between *C. persicus* and *C. gypsicolus* Iljin, though the latter species is hitherto unknown outside Transcaucasia and C Asia. This, and the apparent fertility of *C. curdicus*, seems to make this suggestion rather improbable. At least one gathering has been made in Turkey (*Davis & Hedge* 31021 from Pertek in Tunceli province, E Anatolia) which is indistinguishable from the Iraqi material. The range of variation shown by Turkish material of *C. persicus* makes it much more likely that *C. curdicus* will eventually be reduced to synonymy under that variable species.

28. **CENTAUREA** L.

Sp. Pl.: 909 (1753); Gen Pl.: 389 (1754)
Susanna & Garcia-Jacas in Kubitzki, Fam. Gen. Vasc. Pl. 8: 146 [2006] (2007)

G. Wagenitz

Annual, biennial or perennial (sometimes suffrutescent) herbs, rarely shrubs. Stem and leaves often tomentose to scabrous to hirsute with multicellular hairs, sessile glands very frequent, rarely quite glabrous. Leaves alternate, sometimes nearly all radical, very variable, but (in Iraq) never spiny; often pinnatifid, pinnatisect or lyrate, sometimes decurrent. Capitula nearly always heterogamous with neuter marginal and hermaphrodite central florets; involucre very variable in size and form, nearly globose or hemispherical to ovoid or fusiform; phyllaries multiseriate, imbricate, ± rigid and leathery, nearly always with a scarious, straw-textured or coriaceous appendage of extremely multifarious form, entire or fringed to ciliate, orbicular, lanceolate or triangular, often ending in a mucro or spine, sometimes the appendage consisting of a mere mucro or spine with or without lateral spinelets; receptacle with numerous smooth bristles. Corollas pink, purple, blue, yellow or whitish; marginal florets neuter, sometimes with staminodes, funnel-shaped with 5–8 or more lobes (higher numbers found in corollas of radiant florets) or nearly filiform and very inconspicuous with 4–5 linear lobes; central florets hermaphrodite, tubular. Achenes usually glabrous and smooth when ripe, ± laterally compressed, apex rounded or truncate, hilum lateral near base, sometimes with an elaiosome; pappus multiseriate, with unequal scabrous, barbellate or plumose bristles, gradually elongated towards centre, innermost row usually shorter, or pappus sometimes absent.

About 600 species mainly in the Mediterranean region, S W & C Asia, but extending to N Europe and tropical Africa, a few geographically and taxonomically isolated species in N & S America. Some species occur nearly worldwide as weeds.

Iraq is not rich in species of this genus if compared with the countries adjacent in the

N and W Iran and Turkey. More than half of the 43 native species recognised here are restricted to the mountainous area in the north and west and the Jabal Sinjar. The limited number of species makes determination of the Iraqi species of this taxonomically difficult genus much easier than in the countries mentioned above. Only in sect. *Acrocentron* problems in delimitation and determination of species are met with and the account is still provisional. The species of this section are limited to the area near the frontiers of Turkey and Iran and the material available is still insufficient. Collectors should try to gather well-developed whole plants including the underground parts and the basal leaves.

Centaurea (from Lat. *centum*, a hundred, *aureus*, golden).

Garcia-Jacas, N., Uysal, T., Romashchenko, K., Suárez-Santiago, V.N., Ertuğrul, K. and Susanna, A. (2006). *Centaurea* revisited: a molecular survey of the *Jacea* group. Ann. Bot., 98(4): 741–753.

Greuter, W., Wagenitz, G., Agababian, M. and Hellwig, F.H. (2001). (1509) Proposal to conserve the name *Centaurea* (Compositae) with a conserved type. Taxon 50(4): 1201–1205.

Wagenitz, G. (1955). Pollenmorphologie und Systematik in der Gattung *Centaurea* L. s. 1. Fl. Allg. Bot. Zeit., 142(2): 213–279.

Wagenitz, G. (1986). *Centaurea* in South-West Asia: patterns of distribution and diversity. Proceedings of the Royal Society of Edinburgh. Section B. Biological Sciences 89: 11–21.

Wagenitz, G. (2006). A revision of *Centaurea* (Compositae – Cardueae) in the flora of Iraq. Rostaniha 7 (suppl. 2): 343–394.

In the key and the descriptions, statements concerning the appendages refer to the middle row of the phyllaries if not stated otherwise. These show the diagnostic features best. Measurements given for the diameter of the involucre exclude all patent parts (especially the spines).

KEY TO GROUPS

1. Median (and upper) leaf bases distinctly decurrent 2
   Leaf bases not decurrent 3
2. Plants annual, with small capitula, involucre up 10–20 mm in diameter . . . . . GROUP A
   Plants perennial or biennial, if biennial then with much larger capitula . . . . . GROUP B
3. Plants annual or sometimes biennial . . . . GROUP C
   Plants perennial . . . . GROUP D

GROUP A

1. Decurrent wings of the leaves entire 2
   Decurrent wings dentate 4
2. Florets blue 3
   Florets never blue . . . . 22. *C. solstitialis*
3. Heads 10–20 mm long; involucre 10–15 mm in diameter; intermediate row of pappus shorter than the achene . . . . 44. *C. cyanus*
   Heads 20–25 mm long; involucre 15–20 mm in diameter; intermediate row of pappus longer than the achene . . . . 42. *C. depressa*
4. Stems and leaves ± cobwebby with long glossy hairs; achenes 2.2–2.8 mm long, hilum nearly 1/3 of its length . . . . 22. *C. solstitialis*
   Stems and leaves with very short stiff hairs; achenes 3–3.5 mm long, hilum only about 1/4 of its length . . . . 24. *C. pseudosinaica*

GROUP B

1. Phyllaries without an appendage, or with a minute mucro up to 1.5 mm long 2
   Phyllaries with a distinct appendage 4
2. Stem and leaves with crisped hairs . . . . 17. *C. polypodiifolia*
   Stem and leaves nearly glabrous 3
3. Basal leaves lyrate with very large terminal segment and few lateral segments; involucre 10–18 mm broad; inner pappus setae distinctly shorter than the preceding . . . . 18. *C. behen*
   Basal leaves pinnatisect with several pairs of lateral lobes; involucre 5–8 (–10) mm broad; inner pappus setae about as long as the preceding. 21. *C. koeieana*
4. Appendage a simple spine (sometimes with a pair of minute lateral spinules) 5

Appendage not a simple spine 7
5. Involucre 6–9 mm in diameter . . . . 20. *C. rigida*
Involucre more than 14 mm diameter 6
6. Capitula on short peduncles forming a raceme; pappus 10–15 mm; inner row long . . . . 16. *C. amadensis*
Capitula on long divaricate branches; pappus 8–9 mm long, inner row shorter . . . . 19. *C. alveicola*
7. Appendage ending or gradually narrowing to a long or short spine; appendage ciliate with numerous or cilia distant 8
Appendage scarious; ciliate, cilia 1.5–3.5 mm long, brown at base, silvery in upper part . . . . 42. *C. triumfettii*
8. Basal part of phyllaries totally covered by appendages 9
Basal part of phyllaries not totally covered by appendages 10
9. Capitula 4–5 cm in diameter, on pedicels 10–15 cm long; terminal spine 4–15 mm long . . . . 12. *C. regia*
Capitula 3–4 cm in diameter, on very short pedicels; terminal spine only 1–3(–4) mm long . . . . 13. *C. imperialis*
10. Corollas yellow . . . . 15. *C. gudrunensis*
Corollas rose-purple or whitish 11
11. Biennial plant with several capitula in a raceme; terminal spine of appendages distinct, longer than cilia . . . . 14. *C. gigantea*
Perennial plant with 1–3 capitula: terminal spine of appendages scarcely distinct from cilia . . . . 11. *C. handelii*

GROUP C

1. Outer phyllaries with a green foliaceous appendage, median ending in a simple spine . . . . 31. *C. bruguieriana*
Phyllaries not with green foliaceous appendage 2
2. Corollas yellow 3
Corollas rose-purple or whitish 5
3. Phyllaries with a broad hyaline border; spines without lateral spinules or with one pair near base . . . . 29. *C. hyalolepis*
Phyllaries without a hyaline border; spines with 2–4 pairs of lateral spinules often going up to middle of spine 4
4. Phyllaries with a fine, appressed, mealy tomentum; achene 2.3–2.5 mm long; corollas remaining yellow when dry . . . . 25. *C. mesopotamica*
Phyllaries sparsely arachnoid (cobwebby); achene 2.8–3.2 mm long; corollas drying bluish . . . . 26. *C. microcnicus*
5. Appendage a rigid spine with 1–3 pairs of lateral spinules near base . . . . . 28. *C. iberica*
Appendage triangular with 5–10 pairs of cilia and ending in a slender spinule 6
6. Main stem very short, central capitulum nearly sessile; achene 2.5 mm long, pappus 3–4 mm long . . . . 6. *C. ammocyanus*
Main stem elongated; achene 1.8–2.3 mm long, pappus 2–3 mm long . . . . 7. *C. laxa*

GROUP D

1. Appendage a firm spine, simple or with 1–2 pairs of 2 short spinules at or near base . 2
Appendage spiny or not, but always with several pairs of lateral cilia or teeth or pinnatilobed; pappus absent 3
2. Stem erect, median leaves entire or with few teeth; pappus 7–8 mm long . . . . 25. *C. dethesiana*
Stem ascending or prostrate, median leaves pinnatisect . . . . 30. *C. postii*
3. Corollas yellow 4
Corollas purplish or whitish 5
4. Appendage very small, narrowly triangular with few cilia; pappus 1–2 mm long . . . . 9. *C. rhizantha*
Appendage large, with numerous irregular cilia; pappus 15–20 mm long 10. *C. aucheri*
5. Plant copiously branched with numerous small capitula; involucre 3–10 mm in diameter 6

Plant with few branches with one large capitulum; involucre more than 15 mm broad 10

6. Appendage nearly round with irregularly dentate or ciliate margins; involucre 8–10 mm in diameter . . . . 8. *C. foveolata*

Appendage triangular with distinct regular cilia; involucre 3–6 mm in diameter 7

7. Appendage ending in a very short mucro shorter than or as long as the lateral cilia . 8

Appendage ending in a spinule slightly to distinctly longer than the lateral cilia 9

8. Capitula crowded in fascicles at end of branches; upper leaves lanceolate or oblong . . . . 1. *C. aggregata*

Capitula solitary at end of branches; upper leaves linear . . . . 2. *C. fusiformis*

9. Involucre 4–6 mm broad; capitula with 10–20 hermaphrodite florets . . 5. *C. singarensis*

Involucre 3–4 mm broad; capitula with 5–8 hermaphrodite floret . . . . . . 4. *C. intricata*

10. Appendage with short spine or mucro, this only up to 12–8 mm long in median phyllaries 11

Appendage with a distinctly longer spine 12

11. Appendages totally concealing basal part of phyllaries; involucre c. 25 mm in diameter . . . . 37. *C. hadacii*

Appendages not totally concealing basal part of phyllaries; involucre 15–20 mm in diameter . . . . 35. *C longipedunculata*

12. Plants branched from near base . . . . 33. *C. urvillei*

Plants unbranched or with few branches in upper part 13

13. Basal and lower leaves usually lyrate; appendages triangular, very gradually narrowed into the spine . . . . 34. *C. ochrocephala*

Basal and lower leaves interruptedly pinnatisect to sublyrate; appendages with triangular basal part, but rather abruptly narrowed into the spine 14

14. Pappus 10–15 mm long; innermost phyllaries with very broad rounded appendage 8–15 mm wide . . . . 38. *C. luristanica*

Pappus 5–10 mm long; appendages of innermost phyllaries, usually not as wide as in the preceding species 15

15. Stem branched from near base . . . . 40. *C. irritans*

Stem simple or branched in or above middle (branching from base may occur in damaged plants) 16

16. Involucre tomentose . . . . 39. *C. davisii*

Involucre glabrous 17

17. Leaves lyrate, nearly glabrous . . . . 36. *C. persica*

Leaves interruptedly pinnatisect to sublyrate, ± hirsute, rarely subglabrous . . . . 33. *C. elegantissima*

SECT. **ACROLOPHUS** (CASS.) DC.

1. **Centaurea aggregata** *Fisch. & C.A.Mey.* ex *DC.*, Prodr. 6: 585 (1838); Boissier, Fl. Orient. 3: 652 (1875); Dinsmore in Post, Fl. Syria, Palest. & Sinai ed. 2, 2: 107 (1933); Blakelock in Kew Bull. 4: 46 (1949); Klokov in Fl. S.S.S.R. 28: 537 (1963); Wagenitz in Fl. Turkey 5: 494 (1975) & Fl. Iranica 139B: 334 (1980).

C. virgata Lam. subsp. *aggregata* (DC.) Gugler in Ann. Hist. Nat. Mus. Hung. 6: 248 (1908).
C. pauciflora C. Koch in Linnaea 24: 433 (1851).
C. chrysolopha Boiss. et Kotschy in Boiss., Diagn. Pl. Orient. ser. 2, 3: 75 (1856).
C. amani Post in J. Linn. Soc., Bot. 24: 434 (1888); Bull. Herb. Boiss. 1: 23 (1893).
C. aggregata DC. var. *argyrea* Bornm. in Beih. Bot. Centralbl. 60B: 210 (1939). –Type: Iraq, Kurdistan, Rowandaz, 21/06/1893, *Bornmüller* 1472 (B!).
Acosta aggregata (DC.) Soják in Čas. Nár. Muz., Odd. Přír. 140: 133 (1972).

subsp. ***aggregata***

Perennial with woody base. Stem erect, (30–)40–75 cm tall, branched in upper part. Leaves thinly arachnoid-tomentose when young, glabrescent, scabrous; lower and median leaves lyrate with a large lanceolate terminal segment 8–20 cm broad and 2–5 pairs of much smaller linear-lanceolate lateral segments; upper leaves entire, lanceolate or oblong, sometimes with a pair of lobes at base. Capitula numerous, crowded in fascicles of 3–5(–10) at ends of branches. Involucre subcylindrical, 10–13 × 3.5–5 mm; phyllaries multiseriate, longitudinally veined, appendages small, only partly concealing basal part, brown, triangular, slightly spreading, with 6–8(–9) cilia 1.5–2 mm long on each side and ending in a very short mucro 0.5 mm. Corollas rose-purple, marginal florets scarcely radiant, few; central florets hermaphrodite, 10–12. Achenes 2.8–3 mm; pappus 2.5–3.5 mm, inner series short.

HAB. In the mountains, on rocky slopes; alt. 1000–2150 m; fl. & fr. Jun.-Jul.(-Aug.).
DISTRIB. Apparently widespread in the mountain region of Iraq: **MAM**: Khantur, *Rawi* 23370!; between Sharanish and Dagh al Rajiem nr Zakho, *Rechinger* 10808! **MRO**: Algurd (Helgurd), nr Nawanda, *Guest* 2809!; Kani Rash, in Qurnaqo valley, NE of Pushtashan, *Serhang & Rawi* 26653!; Rowanduz, mt. Handren, *Bornmüller* 1472 (type of *C. aggregata* var. *argyrea*); Handren, *Bornmüller* 1471!; Saran, on Karokh, SE of Handren, *Noori & Alkas* 27274! **MSU**: Pira Magrun, *anon.* 5807!; Malakawa, *Rawi, Alizzi & Noori* 29475!; Zalam, on Avroman, *id.* 29398!; Kajan, nr Penjwin, *Rawi* 22665!; Tawila, *Rawi* 21913!; 7 km W of Tawila, *Rawi* 22334!; 5 km S of Tawila, *Rawi* 22075!; Pira Magrun, *Thesiger* 1122!

Turkey, W & NW Iran; Transcaucasia.

A variable species. Only the subsp. *aggregata*, described above, occurs in Iraq. It occurs throughout the range of the species.

2. **Centaurea fusiformis** *Blakelock* in Kew Bull. 4(1): 47 (1949); Wagenitz in Fl. Iranica 139B: 335 (1980).

Perennial with a woody base and several erect stems, (30–)40–50 cm tall, branched from below middle with ascending branches, each with several shorter one-headed branchlets. Stem and branches angular, thinly tomentose on the faces, lanate at base. Basal and lower cauline leaves withered by flowering time, petiolate, lanceolate-spatulate, undivided and indistinctly dentate or lyrate; median leaves linear-spatulate, 1–3 mm broad, nearly glabrous, punctate with sessile glands; upper cauline leaves linear, very small, only 3–5 mm long, uppermost similar to phyllaries. Capitula solitary on the branches. Involucre nearly cylindrical, but narrowed at base with unusually gradual transition into the peduncle, 11–13 × 3–4 mm; phyllaries multiseriate, indistinctly veined in lower part with small appendages; appendages appressed, light brown, with 5–6 cilia 1 mm long on each side and ending in a short mucro (0.5–1 mm). Marginal florets not radiant, only 4–6 hermaphrodite florets; corollas probably purplish (pale when dry). Achenes 3.5 mm; pappus 1.5–2.5 mm.

HAB. Mostly in *Pinus* forest on rocky limestone ledges; alt. 450–1150 m; fl. & fr. (Jul.-) Aug.-Oct.
DISTRIB. Uncommon in the mountain region of Iraq. **MAM**: Dohuk, *Guest* 1589!; Sarsang, *Feinbrun & Schwarz* s.n.!; Zawita, *Guest* 3729!, *Rechinger* 11534!; Jabal Bekher, *Rawi* 23061!

Endemic. Only known from a limited area near the Turkish frontier.

3. **Centaurea virgata** *Lam.*. Encycl. [J. Lamarck et al.] 1: 670 (1785); Boissier, Fl. Orient. 3: 651 (1875); Handel-Mazzetti in Ann. Naturh. Mus. Wien 27: 59 (1913); Wagenitz in Willdenowia 6: 482 (1975), Fl. Turkey 5: 496 (1975) & Fl. Iranica 139b: 336 (1980).

Acosta virgata (Lam.) Holub in Folia Geobot. Phytotax. 7: 314 (1972).

Perennial with a woody base and several stems. Stems (20–)30–70 cm tall, ± erect, repeatedly and divaricately branched from base or median part. Leaves thinly arachnoid-tomentose, basal and lower petiolate, pinnatipartite or partly bi-pinnatipartite with linear segments, rarely lyrate, often withered at flowering time; median leaves pinnatipartite with few distant segments 1–2(–3) mm broad, upper linear, undivided. Capitula small, solitary or often in pairs at end of numerous branches, forming a panicle, usually easily deciduous when fruiting. Involucre nearly fusiform,7–9 × 3–4 mm broad; phyllaries multiseriate, basal part with elevated longitudinal veins, not totally concealed by the appendages. Appendages spreading or reflexed, triangular, straw-coloured, but at least the upper ones usually with a purple spot, ciliate, cilia 5–10 on each side, 1–1.5(–2) mm long, terminal spinule ± hooked and 1–2 mm long, rarely only 0.5–1 mm and straight. Marginal florets very few, scarcely radiant; hermaphrodite florets 5–8 florets; corollas rose-purple. Achenes 3–3.8 mm; pappus 1–2(–3.5) mm, rarely absent.

subsp. *virgata*, with 7–10 pairs of cilia on the appendages and a very short mucro of 0.5–1 mm, occurs sporadically in NW & Inner Anatolia. It is unknown from Iraq, where the species is represented by:

subsp. **squarrosa** *Gugler*, Ann. Hist. Nat. Mus. Hung. 6: 248 (1908); Wagenitz in Fl. Iranica 139b: 336 (1980).

C. squarrosa Willd., Sp. Pl ed. 4, 3(3): 2319 (1803), nom. illegit.
C. squarrosa Roth (1800); Klokov in Fl. S.S.S.R. 28: 535 (1963).
C. squarrosa Willd. var. *rubella* Sch.Bip. in Pl. Alepp. Kurd. Moss. 303 (1843).
C. squarrosa Willd. var. *colensis* C.Koch in Linnaea 24: 434 (1851).
Acosta squarrosa (Willd.) Soják in Čas. Nár. Muz., Odd. Přír. 140: 134 (1972).
C. virgata Lam. var. *squarrosa* (Willd.) Boiss., Fl. Orient. 3: 651 (1875); Dinsmore in Post, Fl. Syria, Palest. & Sinai ed. 2, 2: 106 (1933); Blakelock in Kew Bull. 4: 49 (1949).
C. virgata Lam. variant *A*, Wagenitz in Willdenowia 6: 483 (1972).

Cilia 5–8 on each margin, 1–1.5 mm; terminal spinule hooked, (1–)1.5–2 mm; pappus 1–2 mm.

HAB. On rocky slopes and stony plateaux, in *Pinus* and *Quercus* forest; alt. (700–)1000–1500(–2000) m; fl. (May-)Jun.-Jul.(-Aug.).
DISTRIB. Very common in the mountain region of Iraq. **MAM**: 23 km NE of Zakho, 25 km S of Sharanish, *Rawi* 23188!; Daimka, *Omar & Dabbagh* 45382!; Maya, 20 km E of Kani Mazi, *Hamid & Farouk?* 4580!; Tarawanish, 10 km E of Kani Masi, *Qaisi, Hamid & Hamad* 45713!; Sarsang, *Barkley & Abbas al-'Ani* 905!, *Haines* s.n.!; Gali Mazurka, nr Amadiya, Dabbagh, *Qaisi & Hamid* 45996!, 43871!; Jabal Bekher, *Field & Lazar* 809!; Zawita, *Guest* 4581!; Qara Dagh, *Kotschy* 303!; Atrush, *Guest* 3633!; between Atrush and Rabetki, *G.W. Chapman* 9349!; nr Dinarti, *E. Chapman* 26145! **MRO**: nr Salah ad-Din, *Barkley* 3371!; Sefin Dagh, *Rawi* 9076!; Shaqlawa, *Gillett* 11291!, *Omar, Sahira, Karim & Hamid* 38258!; between Walash and Walza, *Guest* 2671!; Kholan, between Darband and Haji Umran, *Rawi* 13772!; Haji Umran, *Rawi & Serhang* 24931!; Chinaruk, *id.* 25356! **MSU**, Jarmo, *Haines* W. 274!; Tasluja, 20 km NW of Sulaimaniya, *Rawi* 22765!; 13 km N of Kani Spi, *Rawi* 22420!; Malakawa (Mela Kowa), *Rawi* 22437!; 11 km N of Penjwin, *Rawi* 22911!; Penjwin, *Guest* 12988!; Zewiya, Pira Magrun Sulaimaniya, *Botany staff* 58934 (BAG!). **MJS**: Jabal Sinjar, *Al Khayat & F. Karim* 50887 (BAG!).

Bulgaria, Syria, Lebanon, Turkey, Iran, Afghanistan, Pakistan, C Asia (Turkmenia).

A species widespread in the mountainous part of SW Asia. In subsp. *squarrosa* the capitula remain closed at fruiting time, are easily deciduous and can be dispersed epizoochorously by means of the hooked spinules of the involucre. The plants found in Iraq are relatively uniform as to the characters of the involucre and may thus all be assigned to subsp. *squarrosa*. They are however remarkably variable in the shape of the basal leaves which are lyrate in some gatherings, a character absent from the greater part of the area of the subspecies.

4. **Centaurea intricata** *Boiss.*, Diagn. Pl. Orient. ser. 1, 6: 130 (1845); Wagenitz in Fl. Iranica 139b: 338 (1980).

Perennial suffrutescent plant, (10–)20–40 cm tall. Stems woody, the old dead ones subspinescent, divaricately branched from near base, ultimate branches delicate, all appressed grey-tomentose. Leaves thinly tomentose, glabrescent, lower withered by flowering-time, pinnatipartite with 2–4 pairs of linear-lanceolate, entire or dentate segments, more rarely basal leaves pinnatilobed or entire; median leaves narrowly linear-lanceolate, 10–15 × 1–1.5 mm, entire or with few teeth; upper cauline leaves minute, filiform, Capitula solitary at ends of branches, often easily deciduous. Involucre nearly fusiform, 8–10 × 3–3.5 mm; phyllaries multiseriate, inner ones striate; appendages small, light brown, 0.6–0.8 mm in diameter at base excluding cilia, with 3–8 pairs of cilia 0.5–2.5 mm. Corollas pale purplish (whitish when dry), marginal sterile florets few, hermaphrodite florets 5–8, inconspicuous. Achenes ± 3 mm; pappus 1–1.5 mm.

HAB. On conglomerate hills, and steppe plains; alt. 150–800 m; fl. & fr. May-Jun.
DISTRIB. In one area near the centre of the steppe region of Iraq and in another near the frontier of Iran. **FAR**: 7 km E of Mahmur, *Gillett* 112118! **FPF**: Koma Sang, nr Mandali, *Rawi* 20655!, *Hadač, Haines & Walid al-Hashimi* 4604!; 10 km E of Mandali, *Rechinger* 9655!; Makatu, 16 km W by N of Mandali, *Hadač, Haines & Walid al-Hashimi* 4629!; 10 km E of Mandali, *Rechinger* 9655!

Iran.

The specimens from Iraq belong to subsp. *intricata*, with easily deciduous capitula and appendages, ending in a slender mucro 1.5–3(–4) mm long, with 5–8 pairs of cilia, 1.5–2.5 mm. The Iraqi specimens differ slightly from the type by the shorter cilia (1–2 mm) and mucro (1.5–2.5 mm).

5. **Centaurea singarensis** *Boiss. & Hausskn.* in Boissier, Fl. Orient. 3: 654 (1875). –Syntypes: Inter Syriaco-Armenicum, 05/1867, Jabal Sinjar, *Haussknecht* s.n.; Mt. Singarae, *Haussknecht* 582 (K!, JE).

Perennial suffrutescent plant, 25–50 cm tall. Stems divaricately branched from near base, greyish, appressed-arachnoid. Leaves thinly tomentose, lower sometimes glabrescent; basal leaves petiolate, lyrate, up to 11 cm long with two pairs of lanceolate lateral segments and a much larger terminal segment of similar shape; median and upper cauline leaves sessile, linear, uppermost very small, 3–5 × 1 mm. Capitula solitary at tips of branches, easily deciduous. Involucre oblong, 12–13 × 4–6 mm. Appendages of phyllaries rather small, only partly concealing basal part of phyllaries, narrowly triangular (excluding cilia scarcely 1 mm in diameter at base), central part reddish-brown, with 5–8 whitish cilia 2 mm long on each side and ending in a spinule 2–4 mm. Marginal sterile florets few and inconspicuous (or sometimes even lacking?), fertile florets 10–20, hermaphrodite; corollas rose-purple. Achenes ± 3 mm; pappus 3–3.5 mm, inner series much shorter. Fig. 24: 1–4.

HAB. In rocky clefts on a limestone mountain; alt. 250–1450 m.
DISTR. **MJS/FUJ**: Jabal Sinjar, *Haussknecht* s.n., & *Haussknecht* 582 (syntypes), *Gillett* 11084!, *Anders* 1989!; **FUJ**: Jabal Khatchra, between Balad Sinjar and Tal Afar, *Field & Lazar* 637!

A rare local endemic, allied to *C. intricata* but with distinctly larger capitula.

SECT. **AMMOCYANUS** BOISS.

6. **Centaurea ammocyanus** *Boiss.*, Diagn. Pl. Orient. ser. 1, 10: 109 (1849); Boissier, Fl. Orient. 3: 640 (1875); Handel-Mazzetti in Ann. Naturh. Mus. Wien 27: 448 (1913); Nábělek in Publ. Fac. Sci. Univ. Masaryk 62: 28 (1925); Dinsmore in Post, Fl. Syria, Palest. & Sinai ed. 2, 2: 106 (1933); Blakelock in Kew Bull. 4: 46 (1949); Zohary in Dep. Agr. Iraq Bull. 31: 153 (1950); Rechinger, Fl. Lowland Iraq: 653 (1964); Feinbrun-Dothan, Fl. Palaest. 3: 393 (1978); Wagenitz in Fl. Iranica 139b: 340 (1980); Mouterde, Nouv. Fl. Liban et Syrie 3: 475 (1983); Collenette, Wild Flow. Saudi Arabia: 171 (1999); Boulos & Hind in Fl. Egypt 3: 165 (2002); Taifour & El-Oqlah, Pl. Jordan Ann. Checkl.: (2017).

Ammocyanus arabicus Dostál in Acta Bot. Acad. Sci. Hung. 19: 78 (1973).

Annual, slightly araneose-tomentose, primary stem usually very short, 1–3 cm long, several prostrate or ascending branches 10–30 cm long arising below the terminal capitulum, often further branched. Basal leaves spatulate to lanceolate in outline, undivided, pinnatilobed or sublyrate; median leaves lanceolate, entire or with few lobes or teeth; uppermost leaves linear-lanceolate, inserted directly below capitulum and usually longer than involucre. Involucre ovoid, 10–12(–13) × 5–7(–8) mm; phyllaries in several series, green, inner often with a purple tinge in upper part, glabrous, only partly concealed by the appendages; appendages triangular, brown or blackish brown, decurrent, ciliate, cilia 5–7(–8) on each side, 1.5–2.5 mm long, appendage ending in a slender mucro 2–4 mm. Marginal florets slightly radiant; corollas rose-purple, Achenes 2.5 mm long, lateral hilum bearded; pappus 3–4 mm.

HAB. Silty desert soils, sandy gravel and stony plains; alt. 200–650 m.
DISTRIB. Common in the Western Desert of Iraq, also quite common in the Southern Desert. **FUJ**: on Jabal Makhul, nr Shargat, *Guest* 13446! **FKI/DLJ**: between Salahiya (Kifri) and Khan Baghdadi, *Rawi & Gillett* 6871! **DLJ**: Umm al-Matin, *Chakravarty, Rawi, Alizzi & Khatib* 31886!; between Nahiya and Ana, *Handel-Mazzetti* 774!; Shbaichan, nr Ana, *Rawi* 5690!; 80 km NW of Rawa, *Chakravarty et al.* 31927!; 36 km NE of Haditha, *id.* 31855!; 45 km NW of Haditha, *Alizzi & Husain* 34019!; 12 km NW of Rawa, *Omar, Qaisi, Hamad & Hamid* 44456!; 30 km NE of K 3, *Rawi & Gillett* 7115!; Jazira desert, *Edmonds* 3796!; **DWD**: Amrah, 120 km NW of Rutba, *Omar* 34246!; 60 km N of Rutba, *Rawi & Khatib* 32388!; 12 km N of Rutba, *Rawi* 21069!; Faithat al-Masad, *Janan al-Mukhtar* 33329!; 19 km NE of Rutba, *Rawi* 31177!; 12 km N of T.1, *Chakravarty et al.* 31741!; 100 km E of Rutba, *Omar* 34324!; 260 & 210 km NW of Ramadi, *Rawi* 20971!, 20914!; 55 km W of Ramadi on Rutba highway, *id.* 20865!; 30 & 20 km W of Nukhaib, *Rawi* 14756!, 31001!; 12 km W of Ukhaidhir, *Rawi* 30854! **DWD/DSD**: Wadi al-Khirr, 15 km N of Lussuf, *Guest, Rawi et al.* 19447! **DSD**: 15 km SE of Ashuriya, 50 km WNW of Shabicha, *id.* 19351!, *Rechinger* 9470; 50 km N of Salman, *Rawi* 14864!; 5 & 20 km S of Ansab, *Omar & Hamid* 36609!, 36641!; 5 & 20 km S of Ansab, nr Saudi frontier, *Rawi* 19017!; **LSM**: 10 km W of Kubaisa, *Omar, Qaisi et al.* 44534!

Syria, Palestine, Jordan, Egypt (Sinai), Saudi Arabia, N Iran.

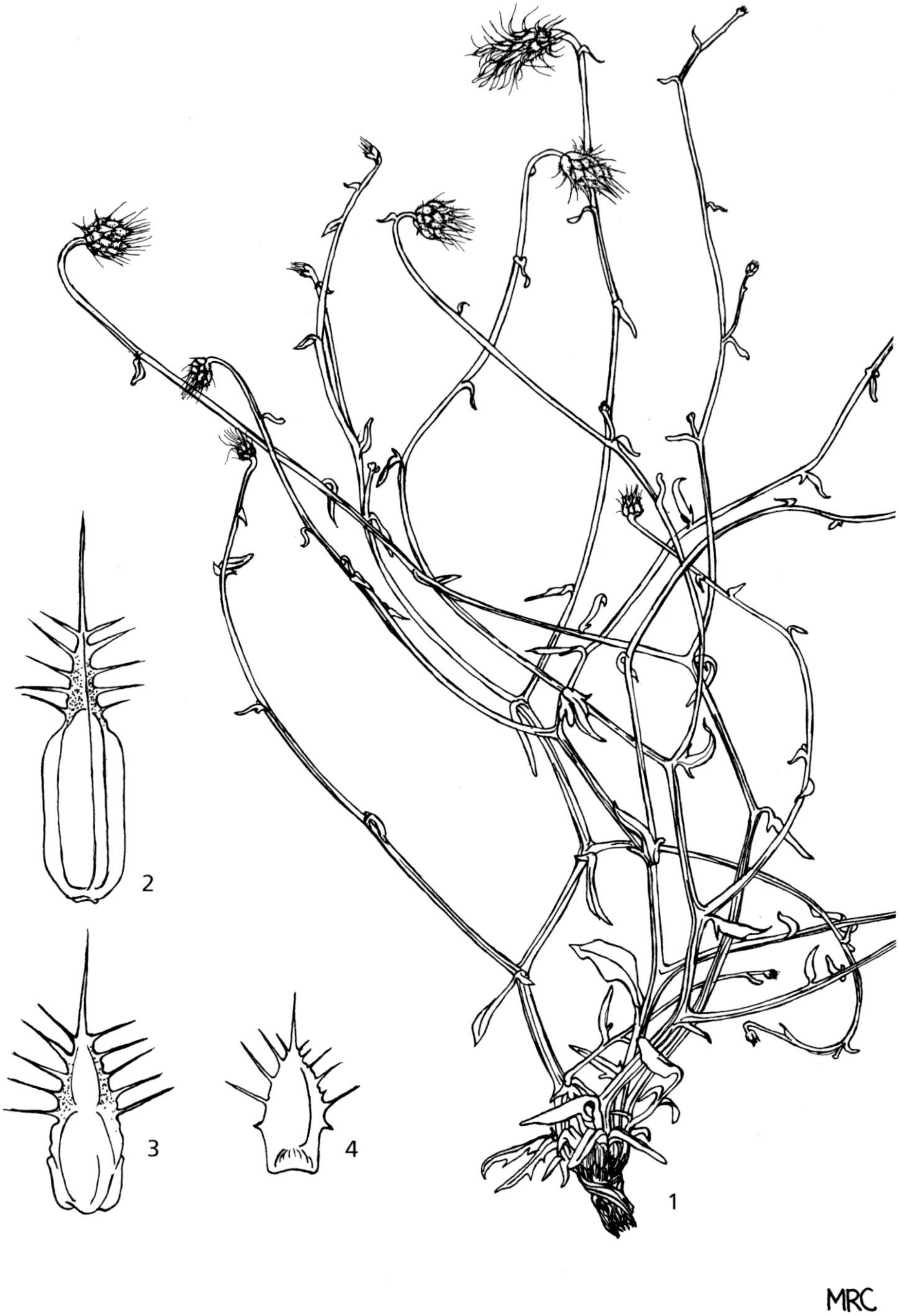

Fig. 24. **Centaurea singarensis**. 1, habit × 1; 2, inner phyllary × 6; 3, median phyllary × 6; 4, outer phyllary × 6. 1 from *Gillett* 11084; 2–4 from *Field & Lazar* 637. Drawn by Moira Rae Carter. Magnifications given here already reduced by 2/3.

7. **Centaurea laxa** *Boiss. & Hauskn.* in Boissier, Fl. Orient 3: 640 (1875); Handel-Mazzetti in Ann. Naturh. Mus. Wien 27: 488 (1913); Dinsmore in Post, Fl. Syria, Palest. & Sinai ed. 2, 2: 106 (1933); Zohary in Dep. Agr. Iraq Bull. 31: 155 (1950); Rawi, Dep. Agr. Iraq. Tech. Bull. 14: 109 (1964); Rechinger, Fl. Lowland Iraq: 653 (1964); Wagenitz in Fl. Turkey 5: 500 (1975); Mouterde, Nouv. Fl. Liban et Syrie 3: 476 (1983).

Annual, 10–30 cm tall, plant branched from near base, erect main stem often overtopped by upper branches but always well developed. Basal and lower cauline leaves lanceolate and undivided or pinnatipartite; median usually lyrate with 2–3 pairs of segments or lobes near base; upper linear-lanceolate with 2 pairs of small teeth at base or entire. Involucre ovoid, 10–11 × 5–6(–7) mm; phyllaries multiseriate, green, inner often with a purple tinge, glabrous, only partly concealed by the appendages; appendages triangular, brown or blackish brown, decurrent, ciliate, cilia 6–8(–10) on each side, 1.5–2.5 mm long, ending in a slender mucro 2–4(–5) mm. Marginal florets moderately radiant; corollas rose-purple. Achenes 1.8–2.5 mm; pappus 2–3 mm.

HAB. On calcareous-gypsiferous soil; alt. 200–700 m: fl. & fr. Mar.-Apr.
DISTRIB. Occasional in the western lower steppe region of Iraq. **MAM**: Sardariya, 15 km SW of Aqra, *Memerian* 10793! **FUJ**: between Jabal Makhul and Khanuqa, *Handel-Mazzetti* 1053!; between Wadi Safa and Hadhr, *Handel-Mazzetti* 1107!; Qaiyara, *Bayliss* 72!; Qal'a Sharqat, *Maresch* 108 in Handel-Mazzetti; 53 km W of Balad Sinjar, *Eig & Zohary* s.n.!; Jabal Makhul (nr. Baiji), *Guest* 13191 (BAG!); **DWD**: 30m NE of K3, *Rawi & Gillett* 7115!

Syria, Turkey.

C. laxa is closely related to C. *ammocyanus*. It is mainly distinguished from this species by the well-developed main axis and the slightly smaller achenes and shorter pappus. The distribution is more northern than that of C *ammocyanus*.

SECT. **PHALOLEPIS** (CASS.) DC.

8. **Centaurea foveolata** *Blakelock* in Kew Bull. 4(1): 46 (1949). –Type: Iraq, Amadia, Kurdistan, Mazurka George above Sulaf, 1500 m, 2/8/1933, *Guest* 3773 (K!, holo.); Rawi in Dep Agr. Iraq Tech. Bull. 14: 109 (1964); Wagenitz in Fl. Iranica 139B: 341 (1980).

Perennials with a woody base and several stems, 35–55 cm tall. Stems nearly glabrous, branched in upper part with several branches with 1–3 capitula. Leaves nearly glabrous, punctate with sessile glands, very slightly arachnoid-pilose on upper side; basal and lower leaves withered at flowering time, petiolate, lyrate-pinnatifid, 5 cm long, with 1–2 pairs of linear lateral segments and a lanceolate larger terminal segment; median leaves lanceolate to linear-lanceolate, mostly with 1–2 pairs of linear lobes; upper leaves linear, 1–2 mm. Involucre ovoid, 12–14 × 8–10 mm; phyllaries multiseriate, indistinctly veined in upper part, totally concealed by their appendages. Appendages nearly orbicular, not decurrent, ± 4 mm broad, the triangular central part firm, brownish, margins hyaline, whitish, irregularly denticulate to ciliate and ± lacerate, appendage ending in a mucro 1.5–3.5 mm. Marginal florets very inconspicuous (or radiant florets lacking?); corollas rose-purple, Achenes 2.5–3 mm; pappus 3.5–5 mm, inner series short (sometimes as long as the others ?).

HAB. In the mountains, by a millstream, in cracks in a cliff; alt. 1000–1500 m; fl. & fr. Jul.-Aug.
DISTRIB. At the western end of the mountain region. **MAM**: between Dohuk and Amadiya, ± 1000 m, *Rechinger* 11627!; near Amadiya, at Sulaf, *Guest* 3769!; Gali Mazurka, above Sulaf, *Guest* 3773! (type), *Agnew & Haines* s.n.!

Endemic.

SECT. **RHIZOCALATHIUM** TZVELEV

9. **Centaurea rhizantha** *C.A.Mey.*, Verz. Pfl. Cauc.: 64 (1831); Tsvelev in Fl. S.S.S.R. 28: 567 (1963); Wagenitz in Fl. Turkey 5: 519 (1975) & Fl. Iranica 139B: 349 (1980).

[*C. sessilis* auctt. non Willd., Boissier, Fl. Orient. 3: 676 (1875); Dinsmore in Post, Fl. Syria, Palest. & Sinai ed. 2, 2: 111 (1933); Blakelock in Kew Bull. 4: 48 (1949).]
C. glaucescens Fisch. & C.A.Mey. in Ann. Sci. Nat.. Bot. ser. 4, 1: 31 (1854).
C. grossheimii Sosn. in Bot. Zhurn. 34: 288 (1949).

C. sessilis Willd. var. *pinnatiloba* Rech.f. in Symb. Bot. Upsal. 11 (5): 33 (1952).
C. rhizanthoides Tzvelev in Fl. S.S.S.R. 28: 620 (1963).

Perennial with a creeping and branched woody rhizome. Capitulum sessile or nearly sessile in a rosette of leaves, rarely short-stalked. Leaves laxly hirsute with long articulate hairs, variable in form, often lyrate with a large triangular or lanceolate terminal segment and 1–3(–5) pairs of linear-lanceolate lateral segments, (in ours) leaves sublyrate or pinnatipartite with terminal segment scarcely longer than lateral; involucre ovoid, 15–20 × 10–20 mm; phyllaries multiseriate, coriaceous, glabrous. Appendages small, straw-coloured or light brown, erect or reflexed, narrowly triangular, only in small part concealing base of phyllaries, with 1–5 cilia, 1–3 mm long on each side, terminal spinule of about the same length. Marginal florets inconspicuous, not radiant; corollas yellow. Achenes 5–6 mm; pappus 1–2 mm.

HAB. In the upper part of the thorn-cushion zone, on a rocky plateau and rocky slopes; alt. (2200–)3000–3800 m; fl. & fr. Jul.-Aug.
DISTRIB. Rare and local, in the highest mountain in Iraq near the summit, adjacent to the Iranian frontier. **MRO**: Algurd Dagh, *Guest & Ludlow-Hewitt* 2881!, *Gillett* 9597!, *Rechinger* 11419!; Algurd Dagh, E side, *Rawi & Serhang* 24739!; S part of mt. Karokh, *Kass & Serhang* 27478!

Algurd Dagh is the southernmost station for the species. These populations differ from most others by the pinnatipartite, not distinctly lyrate leaves.

Turkey (E Anatolia), N, W & E Iran, Turkmenia.

SECT. **PHAEOPAPPUS** (DC.) O.HOFFM.

10. **Centaurea aucheri** (*DC.*) *Wagenitz* in Bot. Jahrb. Syst. 82: 180 (1963).

Tomanthea aucheri DC., Prodr. 6: 564 (1838).
Phaeopappus aucheri (DC.) Heimerl in Stapf, Denkschr. Akad. Wiss. Wien, Math.-Nat. Kl. 50 (2): 65 (1885).
Phaeopappus leuzeoides Boiss., Diagn. Pl. Orient. ser. 1, 6: 124 (1845).
Phaeopappus depressus Boiss., Fl. Orient. Suppl.: 311 (1888).
Tomanthea depressa (Boiss.) Czerep. in Not. Syst. (Leningrad) 20: 480 (1960).

subsp. **aucheri**

Perennial herb with branched woody rhizome and a very short erect or decumbent-ascending stem up to 10 cm, stem shorter than or about as long as basal leaves, simple or with few branches. Leaves ± densely floccose-tomentose, glabrescent; basal leaves petiolate, very variable. often partly undivided and partly lyrate or even pinnatisect; the few cauline leaves usually undivided, lanceolate or oblong. Involucre ovoid, 25–30 × 18–25 mm broad; phyllaries multiseriate, lower part coriaceous, glabrous and only partly concealed by the appendages; appendages triangular or nearly orbicular, light brown or straw-coloured with a brighter margin, rather irregularly ciliate in upper part, with dentate, slightly decurrent auricles towards base and ending in a spinule 2–5 mm. Marginal florets inconspicuous with thread-like lobes; corollas yellow with a purple anther-tube, Achenes 6–8 mm, brown, tetragonous in cross-section with a denticulate upper margin; pappus 15–20 mm, brown, scabrous, innermost series longest.

HAB. Dry slopes; alt. 1700–2000 m; fl. & fr. Jun.-Jul.
DISTRIB. Only found in one locality near the Iranian frontier. **MRO**: Kudu, nr Haji Umran, *Rawi* 9177!; Haji Umran, *Wiltshire* s.n. (in herb. Haines!), *Rechinger* 11295!; Shaikh Shaikhan valley, *Rawi* 24863!

Turkey (E Anatolia), N & W Iran, Transcaucasia.

Only subsp. *aucheri*, described above, occurs in Iraq. It is distributed throughout most of the range of the species.

11. **Centaurea handelii** *Wagenitz* in Bot. Jahrb. Syst. 82: 184 (1963). –Type. Jabal Sinjar, near the foot of the mountain, *Haussknecht* 595 (JE!, holo.)

Phaeopappus haussknechtii Boissier, Fl. Orient. 3: 599 (1875). – Type: Iraq, May 1867, Iter Syriaco-Armeniacum, *Haussknechti* s.n. (holo W; iso BM, JE, K). Rawi, in Dep. Agr. Tech. Bull. 14: 121 (1964).
Tomanthea haussknechtii (Boiss.) Takht. in Trudy Molod. Nauchn. Rabotn. Armen. fil. Akad. Nauk S.S.S.R. 1939: 243 (1939); Cherepanov in Not. Syst. (Leningrad) 20: 480 (1960).

Fig. 25. **Centaurea handelii**. 1, habit × 1; 2, inner phyllary × 3; 3, median phyllary × 3; 4, outer phyllary × 3. 1 from *Rawi* 22667; 3–4 from *Davis* 45661 (Turkish material). Drawn by Moira Rae Carter.

Perennial, 10–30 cm, tall. Simple or often with one or two simple branches in lower part. Stem and leaves ± densely greyish floccose-tomentose. Basal and lower cauline leaves petiolate, ovate-oblong in outline, pinnatisect with 3–6 pairs of lanceolate segments (the upper decurrent along rachis), rarely part of the lower leaves undivided and lanceolate; median and upper leaves strongly decurrent, median often pinnatisect in their terminal part, undivided in lower half, sometimes undivided, resembling upper ones, which surpass the capitulum. Involucre nearly orbicular with a truncate base, 25–30 mm broad; phyllaries multiseriate, coriaceous, glabrous and smooth; appendages not totally concealing basal part of phyllaries, broadly triangular, not decurrent, brown, with 3–7 cilia 2–4 mm long on each side, ending in a spinule scarcely longer or stronger than the cilia. Marginal florets inconspicuous; corollas purple. Mature achenes unknown; pappus 14–19 mm long, whitish or brown, bristles scabrous, innermost not differing from the others. Fig. 25: 1–4.

HAB. In the mountains, on limestone slopes, serpentine rocks etc.; alt. (600–)1400–1800 m; fl. & fr. (Apr.-) May-Aug.
DISTRIB. Local in the mountain region of Iraq. **MSU**: Penjwin, *Rawi* 12234!, *Rechinger* 12255!; Kajan mts., nr Penjwin, *Rawi* 22667! **MJS**: Jabal Sinjar, near the foot of the mountain, *Haussknecht* 595 (type!); Tschil Miran on Jabal Sinjar, *Handel-Mazzetti* 1512!; Rashid on Jabal Sinjar, *Anders* 2682!; *Qaisi & Hamad* 49142; *Omar, Khayat & Qaisi* 52563!; *Thesiger* 538!

Endemic.

SECT. **CYNAROIDES** BOISS. EX WALP.

12. **Centaurea regia** *Boiss.*, Diagn. Pl. Orient. ser. 1, 6: 135 (1845); Boissier, Fl. Orient. 3: 670 (1875); Handel-Mazzetti in Ann. Naturh. Mus. Wien 27: 450 (1913); Bornmüller in Beih. Bot. Centralbl. 60B: 211 (1939); Zohary in Dep. Agr. Iraq Bull. 31: 156 (1950); Wagenitz in Willdenowia 2: 477 (1960) & Fl. Iranica 139B: 365 (1980); Mouterde, Nouv. Fl. Liban et Syrie 3: 480 (1983).

Cynaroides regia (Boiss.) Dostál in Acta Bot. Acad. Sci. Hung. 19: 77 (1973).

Biennial with erect robust stem, 40–110 cm tall, sparsely pilose, branched in upper part. Leaves stiff, papery when dry, scabrous with short hairs; basal leaves with a long petiole, ovate to cordate or more rarely lyrate, very large, with a petiole 25 cm long; lower cauline leaves broadly lanceolate, narrowed into a broad petiole. entire or lobed in basal part; median cauline leaves broadly lanceolate or oblong, sessile and decurrent, up to 20 × 7 cm; upper leaves lanceolate, much smaller, shortly decurrent. Capitula several in a raceme with pedicels 10–15 cm long, sometimes additional longer branches present near middle of stem. Involucre nearly globose, 40–50 mm in diameter; phyllaries multiseriate, coriaceous, totally concealed by their appendages. Appendages large, ovate or broadly triangular, not decurrent, whitish, straw-coloured or brown, ciliate with numerous cilia (3–)4–5 mm long and ending in a spine of variable length (see subspecies). Marginal florets scarcely radiant; corollas reddish purple. Achenes 6–7.5 mm; pappus 10–15 mm, inner series ± 2.5 mm, of narrow scales.

Distribution of species: Syria. Turkey, Iran.

A magnificent plant with the largest capitula known in the genus; it justifies the name given by Boissier.
Two subspecies in Iraq:

Spines of median phyllaries (8–)9–14 mm; appendages of median phyllaries usually whitish or straw-coloured (rarely brown) . . . . a. subsp. *regia*
Spines of median phyllaries (3–)4–8 mm long; appendages brownish . . . . b. subsp. *cyanarocephala*

subsp. **regia**

HAB. (of subsp.). Occasional on the lower slopes of the mountains and on drier hills of the steppe; alt. 350–700 m; fl. & fr. May-Jul.
DISTRIB. **MAM**: 5 km S of Zakho, *Rawi & Rechinger* 16690!; Olaka, *Rawi* 8769!; nr the Turkish frontier between Mosul and Zakho, *Rechinger* 10632!; 20 km from Aqra to Mosul, *Al-Kaisi* 49732 (BAG!). **FKI**: nr Kirkuk, *Haines* s.n.!; S of "Bueer", *Anders* 2083!

Syria, W Iran.

subsp. **cyanarocephala** (*Wagenitz*) *Wagenitz,* Fl. Iranica 139B: 365 (1980).

C. cyanarocephala Wagenitz in Willdenowia 2: 477 (1960).
C. cardunculus Boiss., Diagn. Pl. Orient. ser. 1, 6: 135 (1845), nom. illegit., non Pallas (1771); Boissier, Fl. Orient. 3: 670 (1875); Blakelock in Kew Bull. 4: 46 (1949).

HAB. Rocky mountainsides and gorges, dry hills; alt. 600–1300 m; fl. & fr. May-Jun.
DISTRIB. Quite common in the eastern sector of the mountain sector of Iraq, rare in the central sector. **MRO**: Rowanduz gorge, *Emberger, Guest, Long et al.* 15477!; Razanok, 15 km E of Rowanduz, *Anders* 1419!; between Rowanduz and Harir, *Bornmüller* 1488! **MSU**: Chemchemal, *Rechinger* 10059!; Qaradagh, *Haines* W1184!; 4 km NW of the junction leading to Dukan, *Haines* W12843!; Darband-i Bazian, *Haines* W10578!; Nalparais, nr Penjwin, *Hadač et al.* 4855!; between Khurmal and Halabja, *Haines* W5076!

Turkey (SE Anatolia).

One specimen from **FUJ** Jabal Khatchra nr Balad Sinjar on the way to Tal Afar, *Field & Lazar* 623!, is intermediate between the two subspecies.

13. **Centaurea imperialis** *Hausskn.* ex *Bornm.* in Beih. Bot. Centralbl. 20(2): 168 (1906); Wagenitz in Willdenowia 2: 478 (1960) & Fl. Iranica 139B: 366 (1980).

Biennial plant with a thick turnip-like root, erect, 50–125 cm tall, hirsute with long hairs especially in lower part, branched only in upper part. Leaves densely hirsute to nearly tomentose; the rarely collected basal leaves petiolate, ovate-subcordate, large (± 22 cm long); lower cauline leaves often withered at flowering time. broadly lanceolate, gradually narrowed at base or constricted into a broad petiole; median and upper leaves lanceolate, decurrent. Capitula 5–12, sessile or nearly so and forming a spike, or on short pedicels. Involucre nearly globose, 30–40 mm in diameter; phyllaries in many series, their basal part totally concealed by the appendages. Appendages nearly orbicular or broadly ovate, straw-coloured or whitish, 10–12 mm in diameter without the cilia, cilia 10–15 on each side, 1–4 mm long, terminal spinule 1–3 mm. Marginal florets scarcely radiant, shorter than central ones; corollas reddish or purple. Achenes 5–6.5 mm; pappus 5–9 mm, inner row much shorter.

HAB. Subalpine plains, alt. 1480–1500 m.
DISTRIB. Rare in Iraq. **MRO**: Haji Umran, *Guest & Husham (Alizzi)* 15904!, *Rawi & Serhang* 24933!; NE of Rayat, *Zohary* s.n.! Mt. Avroman, *Haussknecht* s.n.

W Iran.

14. **Centaurea gigantea** *Sch.Bip.* ex *Boiss.*, Fl. Orient. 3: 672 (1875). –Type: Iraq, Kurdistan, nr. Mosul [in apricis lapidosis ad pagum Gara Kurdistaniae prope Mossul], 1 viii 1841, *Kotschy* 349 (holo. G!, iso. GOET! K! LAU! W!). Blakelock in Kew Bull. 4: 48 (1949); Zohary in Dep. Agr. Iraq Bull. 31: 156 (1956); Rawi in Dep. Agr. Iraq. Tech. Bull. 14: 109 (1964); Wagenitz in Fl. Turkey 5: 533 (1975) & Fl. Iranica 139B: 367 (1980).

C. chaldaeorum Nábělek in Publ. Fac. Sci. Univ. Masaryk 52: 40 (1925).
[*C. tomentella* auct. non Hand.-Mazz., Blakelock in Kew Bull. 4: 49 (1949)].

Biennial plant with a thick turnip-like root. Stem erect, (30–)60–180 cm tall, hirsute with long hairs especially in lower part, branched only in upper part. Leaves hirsute to ± tomentose, with long appressed hairs especially conspicuous along the whitish midrib. Basal leaves cordate or oblong-cordate with a long petiole; lower cauline leaves broadly lanceolate or lanceolate, with a short petiole; median and upper leaves lanceolate, decurrent. Capitula in a raceme, pedicels short in upper part, gradually longer towards base of raceme, lowest sometimes with 2–4 capitula. Involucre subglobose to inversely cone-shaped, with a truncate base, gradually narrowed towards apex, 30–35 × (20–)25–30 mm; phyllaries multiseriate, firm, finely tomentose, ± glabrescent. Appendages only partly concealing basal part of phyllaries, firm, coriaceous, straw-coloured or brownish, ± patent, with a narrow triangular base 2.5–4 mm in diameter. without the cilia, gradually narrowed into a 8–12 mm long terminal spine, with 4–8 distant cilia (2–)3–5 mm long on each side. Marginal florets scarcely radiant; corollas pale purplish, whitish when dry. Achenes 5.5–6(–8) mm; pappus (6–)8–10 mm, inner series much shorter.

HAB. Dry and grassy places on lower hillside; locally frequent in open *Quercus* forest, on limestone soils; alt. 750–1500 m; fl. & fr. Jun.-Aug.
DISTRIB. Quite common on the lower mountains in Iraq. **MAM**: Gara 14 km SSW of Amadiya, *Kotschy* 349!; 19 km E of Dohuk, *Rechinger* 45976!; Mazurka gorge by Amadiya, *Guest* 3781!, *Dabbagh, Qaisi & Hamid* 45976!; Sarsang, *Haines* W1244!; **MRO**: Baradost, nr Diana, *Field & Lazar* 873!; between Shahidan and Pushtashan, *Rechinger* 11013!; Karoukh, N of Shahidan, *Rawi & Serhang* 23801!; between Dargala and Karoukh, *Rawi, Noori & Kass* 27734!; Chiya-i Mandau, nr Walash, *Guest* 2673!; **MSU**: on Dukan highway, 20 km NW of Sulaimaniya, *Rawi* 21720!; on the Sulaimaniya–Qaradagh road, *Haines* W1314!

SE Turkey.

15. **Centaurea gudrunensis** *Boiss. & Hausskn.* in Boissier, Fl. Orient. 3: 673 (1875). – Type: Iraq, Pira Magrun [Pir Omar Gudrun], 06. 1867, *Haussknecht* s.n. & *Haussknecht* 585 (lecto. K!, G!, JE).

Perennial 25–50 cm with a woody base, branched nearly from base or from middle, branches with several capitula, forming a lax panicle. Stem and branches whitish, subglabrous or with sparse, long, articulate hairs. Leaves rather rigid when dry, scabrous with short hairs (only visible with hand-lens); basal and lower cauline leaves petiolate, lanceolate in outline, ± 30 cm long, pinnatipartite to sublyrate with 3–5 pairs of linear-lanceolate lateral segments (sometimes a few additional smaller ones in the intervals), segments entire or with a single coarse tooth; median and upper leaves oblong-lanceolate to linear-lanceolate, entire, decurrent. Involucre 15–20 × 10–15 mm, cup-shaped; phyllaries multiseriate, coriaceous, smooth, nearly glabrous; appendages not totally concealing basal part of phyllaries, straw-coloured, triangular, not decurrent, 8–13 × *c.* 2 mm excluding cilia, gradually narrowed into a spine 4–6 mm long, on each side with 3–4 distant cilia 2–3 mm. Appendages of innermost phyllaries rounded with a denticulate margin. Corollas yellow. Achene 4.5–5.5 mm; pappus 4–6 mm, inner series as long as or even slightly longer than the others.

HAB. Dry waste foothills, grassy places, *Quercus* woods; alt. 1000–1200 m.
DSTRIB. Occasional in the neighbourhood of Sulaimaniya district in Iraq. **MSU**: Pira Magrun, *Haussknecht* s.n.! & *Haussknecht* 585 (types), *anon.* 5208!; nr Sulaimaniya, *Haines* W.1313!; between Surdash and Shadala, *Zohary & Feinbrun* s.n.!

A rare narrow endemic, so far only known from four gatherings in the area of Sulaimaniyah.

SECT. **PARAPHYSIS** (DC.) WAGENITZ

16. **Centaurea amadanensis** *Sch.Bip.* in Linnaea 19: 326 (1847), incl. var. *gymnoclada* (Jaub. & Spach) Wagenitz in Bot. Jahrb. Syst. 82: 191 (1963) & Fl. Iranica 139B: 371 (1980).

Amberboa decurrens DC., Prodr. 6: 560 (1838).
Phaeopappus gymnocladus Jaub. & Spach, Ill. Pl. Or. 3: 15 (1847).
P. decurrens (DC.) Boiss., Fl. Orient. 3: 600 (1875).
P. decurrens (DC.) Boiss. var. *gymnocladus* (Jaub. & Spach) Bornm. in Beih. Bot. Centralbl. 32 (2): 413(1914).
Tomanthea decurrens (DC.) Takht. in Trudy Molod. Nauchn. Rabotn. Armen. Fil. A.N. SSSR 1939: 243 (1939).
T. gymnoclada (Jaub. & Spach) Takht., in Trudy Molod. Nauchn. Rabotn. Armen. Fil. A.N. SSSR 1939: 243 (1939).
Phaeopappus decurrens (DC.) Boiss. var. *anoplos* Bornm. in Köie, Beitr. Fl. Südwest Irans 1: 30 (1945); Danish Sci Invest. Iran 4: 30 (1945).
Paraphysis decurrens (DC.) Dostál in Act. Bot. Acad. Sci. Hung. 19: 76 (1973).

Biennial plant with a thick turnip-like root. Stem stout, erect, 50–100(–180?) cm tall, branched only above, hirsute in lower part, in upper part lanate when young, later glabrescent. Leaves papery when dry, sparsely pilose, with whitish elevated veins; basal and lower cauline leaves long-petiolate, undivided, broadly ovate to nearly triangular to lyrate with 1–2 pairs of lateral segments and a very large terminal segment, 12–15 cm broad; median leaves oblong-spatulate to broadly lanceolate, decurrent with broad wings; upper leaves lanceolate or linear, shortly or scarcely decurrent. Capitula (2–)5–15, arranged in a raceme, upper nearly sessile or with short pedicels, lower with a peduncle to 5(–20) cm. Involucre ovoid or obconical, 25–35 × 22–30 mm; phyllaries multiseriate, acute, coriaceous, tomentose (later glabrescent), terminated by a spine 5–20 mm long (var. *amadanensis*) or ending in a mucro only 0.6–2 mm long ("var. *gymnoclada*"). Marginal florets few, shorter than central ones, very inconspicuous; corollas yellowish white, anther tube reddish pink. Achenes 6–8 mm; pappus 10–15 mm, setae of innermost series slightly longer than others.

HAB. Calcareous mountain slopes; 925–1885 m; fl. & fr. May-Nov.
DISTRIB. In the south-eastern mountains of Iraq. **MSU**: Mt Hawraman [Avroman] above Darimar, *Gillett* 11844!; above Khurmal, *Hadač* 5057!; Ahmad Awa village, 924 m, 5/xi/2017, *Ahmad* 11-765 (SUFA)[10]; Darimar mountain, 2,048m, 29/v/2012, *Ahmad* 12-738 (SUFA); Rangin mountain, 1,886 m, 8/vi/2012, *Ahmad* 12-930 (SUFA); Dalane mountain, 1,728m, 9/vi/2012, *Ahmad* 12-1090 (SUFA).

W Iran.

The following specimens from **MRO** seem to represent a new species allied to *C. amadanensis* and the Iranian *C. nemecii* Nábělek but are inadequate for description. The involucre of this material is 4–5 cm in diameter, the appendage a spine 10–22 mm long and broadened at base, pappus (only known from *Rawi & Serhang* 23816) 10 mm long with a short inner row. The plant needs re-collecting with florets and mature fruits. **MRO**: Barsa rini gorge, *Emberger, Guest et al.* 15485!; Gali Warta *c.* 30 km NW by N of Rania, *Rawi, Noori & Kass* 28857!; Koorak (Kurek), N of Shahidan Project, *Rawi & Serhang* 23816!

SECT. **MICROLOPHUS** (CASS.) DC.

17. **Centaurea polypodiifolia** *Boiss.*, Diagn. Pl. Orient., ser. 1, 6: 126 (1845) Fl. Orient. 3: 683 (1875); Tsvelev in Fl. S.S.S.R. 28: 563 (1963); Wagenitz in Fl. Turkey 5: 538 (19750 & Fl. Iranica 139B: 373 (1980).

C. euphratica Boiss., Diagn. Pl. Orient., ser. 1, 125 (1845).
C. pseudobehen Boiss., Diagn. Pl. Orient., ser. 1, 6: 126 (1845).
C. szovitsiana Boiss., Fl. Orient. 3: 683 (1875).
Microlophus polypodiifolius (Boiss.) Agajanov, Fl. Azerb. 8: 443 (1961).
C. polypodiifolia Boiss. var. *pseudobehen* (Boiss.) Wagenitz in Notes Roy. Bot. Gard. Edinburgh 33: 229 (1974) & Fl. Turkey 5: 539 (1975).
C. polypodiifolia Boiss. var. *szovitsiana* (Boiss.) Wagenitz in Notes Roy. Bot. Gard. Edinburgh 33: 229 (1974) & Fl. Turkey 5: 539 (1975).

Biennial or perennial plant with thickened taproot. Stem 25–75 cm tall, hirsute with articulate hairs, profusely branched from about the middle, branches often with several capitula forming a lax corymb. Leaves firm, with prominent veins, especially the lower sparsely hirsute, the others glabrescent; basal and lower leaves petiolate, basal pinnatifid with 8(–10) pairs of linear-lanceolate or narrowly triangular acute segments, sometimes lobed near base: lower leaves pinnatilobed; median ovate-oblong, broadly decurrent; upper leaves ovate to lanceolate, shortly decurrent or sessile, sometimes distinctly enveloping the involucre. Involucre 16–20 × 8–12 mm, nearly obconical (contracted towards apex); phyllaries multiseriate, coriaceous. Appendage very small, a mucro 0.5–1(–2) mm long, deciduous. Marginal florets inconspicuous, not radiant; corollas sulphur-yellow. Achenes 4–5 mm; pappus 4(–8) mm, inner series very short.

HAB. Riverside, open *Quercus* forest; alt. ± 1000 m.
DISTRIB. Rare, so far known from only three Iraqi collections. **MRO**: Koorak nr Shahidan, *Rawi & Serhang* 18247!; between Shahidan and Pushtashan, *Rechinger* 11010! **FNI**: on the bank of the R. Tigris, nr Bawart, S of Zakho, *Kotschy* 172/266!

E Anatolia, NW Iran, Transcaucasia.

The synonymy and the description given above refer to the typical var. *polypodiifolia* which is the only form of the species represented in Iraq, although the broad upper leaves of the few Iraqi specimens seen are not quite typical.

18. **Centaurea behen** *L.*, Sp. Pl.: 914 (1753); Boissier, Fl. Orient. 3: 682 (1875); Nábělek in Publ. Fac. Sci. Univ. Masaryk 52: 43 (1925); Dinsmore in Post, Fl. Syria, Palest. & Sinai ed. 2, 2: 113 (1933); Guest in Dep. Agr. Iraq Bull. 27: 20 (1933); Bornmüller in Beih. Bot. Centrabl. 60B: 213 (1939); Arènes in Bull. Soc. Bot. Fr. 94: 156 (1947); Blakelock in Kew Bull. 4: 46 (1949); Zohary in Dep. Agr. Iraq Bull. 31: 156 (1950); Tsvelev in Fl. S.S.S.R. 28:

---

[10] Herbarium of the Sulaimani University, Faculty of Agriculture (SUFA).

561 (1963); Rawi in Dep. Agr. Iraq Tech. Bull. 14: 108 (1964); Wagenitz in Fl. Turkey 5: 538 (1975); Feinbrun-Dothan, Fl. Palaest. 3: 396 (1978); Wagenitz in Fl. Iranica 139B: 373 (1980); Gabrielian & Fragman-Sapir, Flow. Transcaucasus & Adj. Areas: 60 (2008).

C. alata Lam., Encycl. [J. Lamarck et al.] 1: 665 (1785).
Serratula behen (L.) Lam., Tabl. Encycl. 3: ;242 (1823), excl. descr.
Piptoceras behen (L.) Cass., Dict. Sci. Nat. ed. 2, 54: 482 (1829).
Centaurium behen (sic.) (L.) C. Koch in Linnaea 24: 418 (1851).
Microlophus behen (L.) Takht. in Takhtadzhyan & Fedorov, Fl. Erevana: 326 (1946).

Perennial, stem erect, glabrous, 60–150 cm tall, branched in the upper part, branches overtopping main axis. Leaves firm, with prominent veins, appearing glabrous (very short hairs usually present). Basal and lower cauline leaves petiolate, very large (up to 55 × 18 cm), usually lyrate with 1–3 pairs of retrorse, lanceolate or narrowly triangular segments and a much larger broadly lanceolate or oblong-triangular terminal segment with a truncate base, or basal leaves rarely entire, broadly lanceolate; median cauline leaves oblong or broadly lanceolate, sometimes pinnatilobed, decurrent; upper leaves much smaller, lanceolate to nearly ovate, decurrent or scarcely amplexicaul. Capitula solitary at ends of long branches. Involucre ovoid, strongly narrowed towards apex, 18–26 × 10–18 mm; phyllaries multiseriate, imbricate, coriaceous, glabrous, acute, ending in a very short deciduous mucro 0.2–0.8 mm. Marginal florets inconspicuous with thread-like segments; corollas yellow, numerous. Achenes 4.5–6 mm; pappus 5–7(–8) mm, inner series short (1.6–2 mm).

HAB. On lower mountain slopes and foothills; alt. (250–)400–1700 m; fl. & fr. (Apr.-)Jun.-Aug.
DISTRIB. Quite common in Iraq on the W & C mountains, occasional on the lower hills: **MAM**: "Pl. Mesopot., Kurdist. & Mossul", *Kotschy* 399!; Jabal Bekher, 90 km NW of Mosul nr Zakho, *Rawi* 22968!; Chinaruk, on Haibat Sultan Dagh, *Omar, Qaisi & Widad* 49468!; Sara near Kani Kawan spring, *Karoukh, Noori & Kass* 27281!; Rayat, *Guest* 13042!; 2 km below Rayat, *Gillett* 9625!; Koi Sanjak, *Omar* 37611! **MRO/FAR**: Beharka, nr Arbil, *Bornmüller* 1476! **MRO**: Chiya-i Mandau, *Guest* 2764!; Ser Kurawa, 10 km from Rania, *Omar, Sahira, Karim & Hamid* 38235!; Haji Umran, *Rawi & Serhang* 24270!, 24947!; Koorak project, N of Shahidan, *id.* 23802!; between Gundar Zhor and Darband, *Gillett* 12399!; Umarawa, *Qaisi & Hamad* 43542! **MSU**: Darband-i Basian, *Rawi* 22806!; Tasluja, *id.* 22764!; Qaradagh, *Haines* W. 1183! **MJS**: Jabal Qilat, between Ain Tellawi and Balad Sinjar, *Field & Lazar* 500!; Jaball Sinjar, *Qaisi, Khayat & F. Karim* 50866! **FUJ**: 35 km S of Khanaq, *Eig & Zohary* s.n.!; Bahirah, 35 km NW of Mosul, *id.* s.n.!

Palestine, Lebanon, Turkey, Iran, Transcaucasia.

According to Guest (Dep. Agr. Iraq Bull. 27, 20), this species is known locally by the Kurdish name of KAKHORT and is eaten by camels and sheep in times of scarcity, though generally untouched when other food is abundant. The tips of the young leaves are also eaten by the people and the woody rootstock is dug up and used for fuel.

19. **Centaurea alveicola** *Rech.f.* in Ber. Deutsch. Bot. Ges. 72: 283 (1959). –Type. Between Mandali and Badra, *Rechinger* 9681 (holotype W); Rechinger, FI. Lowland Iraq: 6:55 (1964); Wagenitz in Fl. Iranica 139B: 376 (1980).

Perennial, up to 150 cm tall. nearly glabrous in all parts. Stem pale green with whitish streaks, branching from middle with divaricate branches. Leaves firm, coriaceous, pale green, becoming reddish or purple when withering, glabrous but with minute sessile glands, with prominent veins. Lower leaves withered at flowering time, very variable, large (up to 30 cm long), ovate-elliptic and undivided or lyrate with a large terminal segment and up to four pairs of small oblong lateral segments, segments crenate-dentate at margins; median leaves cordate with an amplexicaul base; uppermost small, linear, acute, sessile, not decurrent. Capitula solitary, on long pedicels. Involucre ovoid, rounded at base, 24–28 × 14–16 mm; phyllaries in several series, coriaceous, glabrous or thinly tomentose, terminated by a yellowish or reddish spine 6–17 mm long (in median phyllaries); inner phyllaries with a scarious unarmed appendage. Spine simple or with a pair of short lateral spinules. Marginal florets shorter than central ones, with threadlike segments; corollas pale yellow. Achenes ± 4 mm; pappus 8–9 mm, brownish, inner series 3–4 mm. Fig. 26: 1–6.

HAB. On steppic hills, sometimes gravelly or even a little rocky: alt. 100–300 m: fl. & fr. Apr.-May (-Jun.).
DISTRIB. Occasional, mostly along the frontier of Iran from Khanaqin to Badra. **FPF**: Kani Mazi 20 km S of Khanaqin, *Hadač, Haines & Hashim* 4645!; Naft Khana, 35 km SE of Khanaqin, *id.* 2054!; between Mandali and Badra, *Rechinger* 9681 (type)!; 35 km NE of Mandali, *Qaisi & Khayat* 50805!; 28 km from

Fig. 26. **Centaurea alveicola**. 1, habit × 1; 2, leaf × 1; 3, inner phyllary × 3; 4, median phyllary × 3; 5, outer phyllary × 3; 6, achene with pappus ×7. All from *Hadač et al.* 2054. Drawn by Moira Rae Carter.

Mandali, *Qaisi & Yahya* 45230!; 55 km SE of Mandali, *Rawi* 20743! **LEA**: Jabal Hamrin nr Shahraban (Muqdadiya), *Haines* 1397!

Endemic (but likely to occur also in W Iran).

20. **Centaurea rigida** *Banks & Sol.* in Russell, Nat. Hist. Aleppo ed. 2, 2: 263 (1794); Blakelock in Kew Bull. 4: 48 (1949); Rechinger, Fl. Lowland Iraq: 656 (1964); Wagenitz in Fl. Turkey 5: 540 (1975); Feinbrun-Dothan, Fl. Palaest. 3: 396 (1978); Wagenitz in Fl. Iranica 139B: 376 (1980); Taifour & El-Oqlah, Pl. Jordan Ann. Checkl.: 38 (2017).

C. *russelliana* Buek, Index ad DC., Prodr. 2: VI (1840). nom. illeg. (based on C. *rigida* Banks & Sol.).
C. *myriocephala* Sch.Bip. ex Boiss., Fl. Orient. 3: 682 (1875).; Dinsmore in Post, Fl. Syria, Palest. & Sinai ed. 2, 2: 112 (1933); Anth. in Notes Roy. Bot. Gard. Edinburgh 13: 291 (1935).
C. *schizophylla* Nábělek in Publ. Fac. Sci. Univ. Masaryk 52: 42 (1925).
C. *myriocephala* var. *schizophylla* (Nábělek) Nábělek, op. cit. 105: 47 (1929).
C. *rigida* f. *schizophylla* (Nábělek) Blakelock in Kew Bull. 4: 4 (1949).
C. *rigida* var. *schizophylla* (Nábělek) Wagenitz in Fl. Iranica 139B: 377 (1980).
C. *myriocephala* Sch.Bip. ex Boiss. var. *erythracantha* Bornm. in Feddes Repert. 24: 372 (1928).
C. *rigida* var. *erythracoantha* (Bornm.) Blakelock in Kew Bull. 4: 48 (1949).

Perennial (sometimes behaving as biennial) nearly glabrous plant with a stout thickened root with several stems, with fibrous remains of withered petioles at base. Stem 30–90 cm, divaricately and repeatedly branched from near base, forming sometimes nearly globose bushes. Leaves rigid, coriaceous; basal and lower cauline leaves petiolate, very large (up to 50 cm long), usually lyrate with a large oblong to triangular terminal segment (not rarely hastate at base) and 1–4+ pairs of lanceolate much smaller lateral segments or (in "var. *schizophylla*") basal leaves pinnatipartite with 8–10 pairs of linear-lanceolate segments, segments irregularly dentate; median leaves pinnatipartite or undivided, dentate, shortly decurrent; upper leaves entire, gradually narrowed from a broad half-clasping base to nearly linear. Capitula numerous. Involucre ovoid, 13–18 × 6–9 mm; phyllaries multiseriate, coriaceous, floccose-tomentose, glabrescent, median ending in an erect or patent spine 5–15 mm long, usually yellow (reddish in "var. *erythracantha* (Bornmüller) Blakelock"), spines sometimes with minute lateral spinules. Marginal florets inconspicuous, not radiant; corollas sulphur-yellow, Achenes 4–5 mm; pappus 4–5 mm, scabrous, inner series short.

HAB. On rocky limestone soil, steppic clay, as a weed in corn and barley field, etc.; alt. up to 900 (–600) m; fl. & fr. May-Aug.(-Sep.).
DISTRIB. Occasional in the foothills and steppe region of Iraq and on the alluvial plain. **MAM**: nr Sandur, *Haines* s.n.!; between Dohuk and Zakho, *Rechinger* 12172!; between Dohuk and Amadiya, *id.* 11497!; between Sulaf & Sarsang, *S. Omar* 37733 (BAG!); Ain Sifni, *Guest* 2581 (BAG!). **MRO**: Shaqlawa on Sefin Dagh, *Bornmüller* 1479!; Haibat Sultan Dagh, N. of Koi Sanjaq, *Rawi, Noori & Kass* 28268 (BAG!); Erbil, *Gillet* 8003!; 4 km S of Taktak Br., *Rawi, Noori & Kass*, 28085 (BAG!). **MSU**: Jarmo, *Haines* W. 326!**FUJ**: Hamam Alil, *Anders* 979!; Mir Qasim, *Field & Lazar* 560! **MJS**: 6 km E of Jabal Sinjar, *Gillet* 11067! **FAR**: Arbil province, *Regel* 75! **FKI**: Daquq, *Imad & Abdul-Jafar* 50109!; 1–18 km E of Kirkuk, *Rechinger* 9997!; 10 km N of Tuz to Kirkuk, *Gillet*, 8003, *Botany staff* 42946 (BAG!). **FPF/LEA**: nr Muqdadiya, *Haines* s.n.! **FPF**: Mandali, *Hadač, Haines & Hashimi* 4579!; 25 km SW of Badra, *Al-Kaisi, Thamer & Salah* 51362 (BAG!); Mendali, *Guest*, 881, id. 1715! **DGA**: Jabal Hamrin, 12 km E of Samarra, *Rechinger* 9564a!; 13 km E of Samarra, *Rawi*, 20364 (BAG!). **LCA**: Khan Bani Sa'd, *Hadač & Agnew* 4561!; 20 km NE of Baghdad, *Rawi* 12563 (BAG!).

TAHLISHK (Kur.) (*Guest* 2581)

Palestine, Jordan, Syria, Turkey.

C. rigida is rather variable especially concerning the length of the spines and the form of the basal leaves. According to Handel-Mazzetti (Ann. Naturh. Mus. Wien 25: 65, 1914) *C. rigida* can behave as a tumbleweed.

21. **Centaurea koeieana** *Bornm.* ex *Koeie*, Beitr. Fl. Südwest Irans 1: 31 (1945); Wagenitz in Fl. Iranica 139B: 377 (1980).

Perennial herb with woody basal parts, 25–50 cm tall, base of stem with fibrous remnants of petioles. Stems erect, sparsely arachnoid, branching repeatedly from near base. Leaves firm, especially beneath with veins elevated and scabrous with very short hairs, glabrescent; basal and lower cauline leaves petiolate, to 30 cm long, pinnatipartite, oblong-lanceolate in outline, with several pairs of linear to narrowly triangular segments, segments dentate or lobate, terminal one much larger, triangular; median cauline leaves sessile, half-clasping and ± decurrent, undivided, irregularly dentate or entire, gradually narrowed from base, 10–2 × 1.5–2 cm (in basal part); upper smaller, linear-lanceolate, entire, shortly decurrent. Capitula numerous, in a lax corymb. Involucre 15–18 × 5–8(–10) mm, narrowed towards top; phyllaries few-seriate, coriaceous, smooth, nearly glabrous, ending in a minute mucro usually only 0.2–0.4 mm (rarely up to 1.5 mm). Florets with few outer sterile ones with threadlike segments; corollas sulphur-yellow. Achenes 5–6 mm; pappus 4–6 mm, inner series as long as intermediate setae.

HAB. Stony hillsides; alt. 1300–1400 m., fl. ?Jun.
DISTRIB. Occasional in the SE mountain region of Iraq: **MSU**: Nalparaiz, *Hadač et al.* 4844!; Mt. Avroman, at Tawila near the Iranian frontier, *Rechinger* 10293!, *Rawi* 21965!; 10 km W of Tawila, *id.* 22142!; Zalam Mt., *Rawi, Husham & Noori* 10293!; 12 km E of Chemchemal, *Rawi & Gillett* 11618!

Iran.

Allied to C. *rigida,* but distinguished by the very short mucro of the phyllaries (only 0.2–0.4 mm, 1–1.5 mm in the specimen collected by *Hadač et al.* 4844) and the long inner pappus.

SECT. **MESOCENTRON** (CASS.) DC.

22. **Centaurea solstitialis** *L.*, Sp. Pl. ed. 1: 917 (1753); Boissier, Fl. Orient. 3: 685 (1875); Handel-Mazzetti in Ann. Naturh. Mus. Wien 27: 452 (1913); Nábělek in Publ. Fac. Sci. Univ. Masaryk 52: 45 (1925); Dinsmore in Post, Fl. Syria, Palest. & Sinai ed. 2, 2: 114 (1933); Anthony in Notes Roy. Bot. Gard. Edinburgh 18: 292 (1935); Blakelock in Kew Bull. 4: 49 (1949); Regel in Candollea 16: 238 (1958); Zohary in Dep. Agr. Iraq Bull. 31: 156 (1950); Cherepanov in Fl. S.S.S.R. 28: 571 (1963); Rechinger, Fl. Lowland Iraq: 656 (1964); Wagenitz in Fl. Turkey 5: 542 (1975) & in Fl. Iranica 139B: 378 (1980); Mouterde, Nouv. Fl. Liban et Syrie 3: 485 (1983); Boulos & Hind in Fl. Egypt 3: 166 (2002); Taifour & El-Oqlah, Pl. Jordan Ann. Checkl.: 39 (2017).

Calcitrapa solstitialis (L.) Lam., Fl. Franç. 2: 34 (1778).
Centaurea adami Willd., Sp. Pl. ed. 4, 3: 2310 (1803).
Cyanus solstitialis (L.) J. & C. Presl Fl. Cechica: 179 (1819).
Triplocentron solstitialis (L.) Fourr., Ann. Soc. Linn. Lyon N.S. 17: 97 (1869).
Leucantha solstitialis (L.) Á.Löve & D.Löve, Bot. Notis. 114: 44 (1961).
Centaurea solstitialis var. *adami* (Willd.) Heuff. in Verh. Zool.-Bol. Ges. Wien 8: 144 (1858).
Centaurea solstitialis subsp. *adami* (Willd.) Nyman, Consp. Fl. Eur.: 430 (1879).

Annual, 15–80 cm, branched from near base or middle part, branches simple in smaller plants, partly branched with several capitula in more robust ones. Stems and leaves appressed-tomentose. Basal and lower leaves (mostly withered at flowering time) lyrate to pinnatipartite, with 3–4 pairs of lanceolate lateral segments and a usually larger lanceolate to triangular terminal segment; median and upper cauline leaves lanceolate to linear-lanceolate, lobed or toothed to entire, widely decurrent into narrow entire wings; uppermost leaves linear. Capitula solitary at ends of long to moderately short branches. Involucre oblong-ovoid, 13–16 × 8–12 mm; phyllaries multiseriate, arachnoid-tomentose; appendage in outer phyllaries 3–5 short spinules of nearly equal length, in median a straw-coloured patent spine 15–25 mm long, with 2–3 short spinules (2–4 mm) on each side near base. Marginal florets not radiant, with threadlike segments; corollas yellow. Achenes 2–3 mm long, dimorphic; marginal dull, blackish, epappose, central glossy, greyish to brown with white pappus 3–4(–5) mm long, innermost series short, white.

HAB. Lower slopes of mountains and foothills of N Iraq. alt. 450–1000 m; fl. Apr.-Jun.
DISTRIB. Occasional on the lower mountain slopes and in the steppe region. **MAM**: 5 km S of Zakho, *Rawi* 23093; Tarawanish, 10 km E . of Kani Masi, *Al-Kaisi, K. Hamad & H. Hamid* 45712 (BAG!); Bikair mountain near Zakho, *Rawi* 23093!; near Dinarta, *Chapman,* 26236!; Sarsang, *Eig & Feinbrun* s.n.! **MSU**: Tasluja, *Hadač* 1946!; 5 km from Penjwin, *M. Noori & K. Hamad,* 41247! **MRO**: Galalah, *S. Omar, Sahira, F. Karim, H. Hamid* 38423! **FNI**: Gerwona, nr Ain Sifni, *Field & Lazar* 723 (BAG!). **DWD**: 40 km E.of Rutba, *S. Omar, Al-Kaisi & K. Hamad* 43964!

W, S & E Europe, Syria, Lebanon, Turkey, Iran, S Russia, Caucasus, C Asia (Turkmenia to Tajikistan). Introduced as a weed in N America and elsewhere; Yellow Star-Thistle (Eng.).

Only subsp. *solstitialis,* described above, occurs in Iraq. It is found nearly throughout the range of the species.

23. **Centaurea sinaica** *DC.*, Prodr. 6: 592 (1838); Dinsmore in Post, Fl. Syria, Palest. & Sinai ed. 2, 2: 114 (1933); Rechinger, Fl. Lowland Iraq: 657 (1964); Täckholm, Stud. Fl. Egypt ed. 2: 541 (1974); Feinbrun-Dothan, Fl. Palaest. 3: 398 (1978); Mouterde, Nouv. Fl. Liban et Syrie 3: 485 (1983); Collenette, Wild Flow. Saudi Arabia: 178 (1999); Boulos & Hind in Fl. Egypt 3: 168 (2002); Taifour & El-Oqlah, Pl. Jordan Ann. Checkl.: 38 (2017).

C. arabica Velen., Sitzungsber. Königl. Böhm. Ges. 1911, xi. 10 (1912).
C. pseudosinaica Mouterde, Saussurea 3: 19 (1973), nom. illeg.

Annual. Primary stem usually very short, branching as in C. *pseudosinaica.* Stem and branches villous with long aciculate hairs (conspicuously glistening if viewed under a strong lens). Leaves sparsely hirsute, slightly arachnoid especially when young. Basal leaves petiolate, lanceolate in outline, dentate, pinnatilobed or pinnatisect (sometimes with lobate lateral segments); median and upper cauline leaves lanceolate, dentate, decurrent with dentate wings. Capitula subtendd by upper leaves, shortly pedunculate. Involucre oblong, 1–14 × 9–11 mm in diameter, slightly arachnoid when young; phyllaries multiseriate, coriaceous, outer and median ending in a yellow spine, in median 12–20(-30) mm long, with (2–)3–4 pairs of lateral spinules 2.5–5 mm long in lower part. Marginal florets inconspicuous, shorter than central ones; corollas yellow or flesh coloured. Achenes 2.2–3 mm long, lateral hilum about 1/3 length of achene; pappus (2–)3–3.5(–4) mm long, inner series short.

HAB. Sandy, clay and gravelly soil in subdesert, rocky low hills, rarely as a weed; alt. (150–)300–700 m; fl. & fr. (Mar.-)Apr.-May (-Jun.)
DISTRIB. Quite common in the western desert of Iraq. **DWD**: 20 km W of Rutba, *Rawi* 14690!; 50 km W of H.3, *Rawi* 31158!; 10 km SW of Rutba, *id.* 21072!; 15 km S of Rutba, *Khatib & Hazim* 32446!; 18 km S of Rutba, *Rawi* 14641!; 40 km SE of Rutba, *Omar* 34208!; between Qaim and H.2, *Omar & Hamid* 36654!; 63 km W of Ramadi, *Hadač* 4306!; 70 km W of Nukhaib, *Rawi* 31399!; Wadi Hauran 11 km W of Rutba dam, *S.Omar, H.Abbas, Al-Kaisi & K. Hamad* 54586!

Egypt (Sinai), Jordan, Palestine, Syria, Arabia (N Saudi Arabia).

24. **Centaurea pseudosinaica** *Cherepanov* in Not. Syst. (Leningrad) 20: 392 (1960); Wagenitz in Fl. Iranica 139B: 380 (1980); Collenette, Wild Flow. Saudi Arabia: 171 (1999); Kilian & Ghazanfar in Fl. Oman 3: 213 (2015); Taifour & El-Oqlah, Pl. Jordan Ann. Checkl.: 38 (2017).

Annual. Primary stem usually very short (terminal capitulum on a peduncle 1–5 cm long, rarely longer), branched from base with several procumbent or ascending branches up to 25 cm long or longer with several capitula. Stem, branches and leaves ± densely hairy with short stiff hairs, not arachnoid. Basal leaves petiolate, lanceolate in outline, dentate, lyrate or pinnatilobed with 2–5 pairs of lateral segments; median and upper leaves decurrent with denticulate wings, lanceolate or linear-lanceolate, median coarsely dentate or denticulate and similar to upper ones. Involucre oblong, 12–15 × 6–10 mm, slightly arachnoid when young, glabrescent; phyllaries multiseriate, coriaceous, outer and median ending in a yellow spine (10–)15–20 mm long in the median leaves, with 1–2 (rarely 3) pairs of lateral spinules 3–5 mm; marginal florets inconspicuous, shorter than the central ones; corollas yellow. Achenes 3–3.5 mm long with a lateral hilum.

HAB. In the desert on silty and sandy or clay depressions or plains, on gravelly soil by a ravine, and as a weed; alt. up to ± 250 m; fl. & fr. Apr.-May.
DISTRIB. Quite common in the desert region of Iraq. **DGA**: 50 km SE of Samarra, *Rawi & Shahwani* 25484! **DLJ**: 85 km NNW of Falluja, *Rawi* 20302!; **DWD**: 12 km W of Ukhaidhir, *id.* 30865!; 35 km E of Nukhaib, *Guest, Rawi & G. Long* 14046!, *Hadač, Agnew et al.* 4200!; 32 km W of Karbala, *M.L. Grant* 15687!; **DWD/DSD**: Faithat al Qaiyara, 65 km W of Lussuf, *Guest, Rawi & Rechinger* 19462!; **DSD**: Ichrishi, *Guest & Rawi* 14193!

Arabia (Kuwait, Oman, Saudi Arabia, Qatar, UAE, Yemen), Jordan, S Iran.

Only subsp. *pseudosinaica,* described above, is present in Iraq.

IMRAR (*Guest & Rawi* 14193); Dickson notes that the common name of this plant is CHALIBA, colloquial Arabic for KALB, and when camels eat it their milk is said to be bitter (*Dickson* 468).

25. **Centaurea mesopotamica** *Bornm.* in Beih. Bot. Centralbl. 20, 2: 170 (1906); Handel-Mazzetti in Ann. Naturh. Mus. Wien 27: 451 (1913); Rechinger, Fl. Lowland Iraq: 655 (1964); Collenette, Wild Flow. Saudi Arabia: 177 (1999).

C. musilii Velen., Sitz.-Ber. Böhm. Ges. Wiss., Math.-Nat. Kl. 1911, XI: 9 (1912).

Annual with a very short main axis (terminal capitulum sessile or with a peduncle to 4 cm tall) and several prostrate to ascending basal branches. Branches 5–30(–35) cm long, simple or branched with 2–4(rarely more) capitula. Leaves thinly tomentose; basal leaves petiolate, lanceolate, mostly dentate, more rarely lyrate or pinnatilobed; median and upper leaves sessile, ± lanceolate, dentate or entire. Involucre ovoid, (11–)12–14(–15) × 8–10(–12) mm broad; phyllaries multiseriate, with a fine mealy tomentum, later glabrescent; appendages of median phyllaries a spine (6–)10–20(–25) mm long with 2–3(–4) pairs of lateral spinules 2–4 mm long in lower part. Marginal florets shorter than central; corollas yellow. Branches of style spreading. Achenes 2–2.5 mm long, with a small lateral hilum scarcely ¼ of length of achene; pappus 2.5–4 mm, inner series very short.

HAB. Generally in sandy places: alt. up to 650 m: fl. & fr. Apr.-May(-Jun.).
DISTRIB. Quite common in the Desert Region of Iraq, occasional in the alluvial plain. **FUJ**: nr K2 pipeline, 50 km from Baghdad, *Agnew, Hadač & Haines* s.n.! **DLJ**: by Wadi Shuraimiya nr Tikrit, 150 km NW of Baghdad, *Handel-Mazzettii* 1020!; 18 km W of Khamran, *Alizzi & Husain* 33789!; W of Baiji, *Anders* 1932! **DWD**: 40 km E of Rutba, *Barkley & Palmatier* 2270!; 5 km W of Rutba, *Rawi & Noori* 27063!; 5 km S of Rutba, *Rawi* 21243!; 45 km S of Rutba, *id.* 21259!; 260 km W of Ramadi, *id.* 20986!; 63 km W of Ramadi, *Hadač* 4307!; Ramadi, *Graham* 45!; 12 km W of Ukhaidhir, *Rawi* 30863!; 15 km SE of Ashuriya (50 km WNW of Shabicha), *Guest, Rawi et al.* 19350!; 10 km SE of Busaiya, *Rawi* 26008! **DSD**: Shaiba (Shu'aiba), 3–6 km NW of Zubair, *Watson* s.n.. **LEA**: Sharaban (Muqdadiya), *Agnew* 9737!, *Haines* 1306!

Jordan, Syria, Arabia (Kuwait, Saudi Arabia). Easily to be distinguished from similar species (e.g. *C. microcnicus)* by the type of indumentum on the phyllaries, which is a fine mealy tomentum easily wiped off. According to Dickson the common name of this plant in Kuwait is CHALBA.

26. **Centaurea microcnicus** *Reese & Samuelsson* in Rech. f., Ark. Bot. ser. 2, 1(5): 459 (1959); Mouterde, Nouv. Fl. Liban et Syrie 3: 485 (1983).

Annual with very short main axis (terminal capitulum nearly sessile) and several prostrate to ascending branches. Branches 5–25(–35) cm long, simple or branched. Leaves with sessile glands and sparse long hairs. Basal leaves petiolate, oblong-lanceolate and dentate, lobate or even interruptedly pinnatipartite; median and upper leaves sessile, sometimes slightly auriculate, lanceolate and dentate (or median pinnatilobed). Capitula solitary at ends of branches. Involucre 13–14 × 6–8 mm; phyllaries multiseriate, sparsely arachnoid; appendage of median phyllaries a spine 7–15(–20) mm, in lower part with 2–4 pairs of lateral spinules 2–3.5 mm. Marginal florets shorter than central, scarcely radiant; corollas yellowish, drying bluish-green. Branches of style not spreading. Achene 2.8–3.2 mm long, lateral hilum large, ⅓ length of achene; pappus 1.5–3 mm long, inner series very short.

HAB. In a sandy clay depression, on a stony hillside: alt. 300–450 m (only recorded on the label of Rawi's gatherings); fl. & fr. Mar.-May.
DISTRIB. Occasional in the steppe and desert regions of Iraq. **FUJ**: 5 km S of Hadhr, *Alizzi & Husain* 33857!; **DLJ**: 10 km S of Manaif, *Omar & H. Hamid* 36587!; 5 km N of Manaif, *Al-Kaisi & K. Hamad* 49022 (BAG!);48 km NW of the Baiji-Haditha track, *Chakrawarty, Rawi, Khatib & Alizzi* 31851!; 74 km NW of Baiji-Haditha track, *id.* 31884!; **DWD**: 95 km NE of Rutba, *Rawi & Khatib* 32339!; Wadi Hauran, *Chakrawarty et al.* 31628!

Syria.

SECT. **CORETHROPSIS** DC.

27. **Centaurea delbesiana** *Arènes* in Notul. Syst. (Paris) 14: 187 (1951); Rechiger, Fl. Lowland Iraq: 658 (1964); Mouterde, Nouv. Fl. Liban et Syrie 3: 478 (1983).

Perennial suffrutescent plant with several erect stems, 25–35 cm tall. Stems appressed-tomentose, grey, usually with a few divaricate branches with one capitulum each. Leaves floccose-tomentose; basal leaves usually withered at flowering time, lanceolate and entire or pinnatipartite with two pairs of linear-lanceolate segments; median cauline leaves sessile, narrowly lanceolate, entire or with few teeth; upper leaves very small, linear. Involucre oblong, 18–19 × 10–12 mm; phyllaries multiseriate, thinly appressed-tomentose, inner glabrescent, purple-suffused; phyllaries (except innermost) ending in a yellowish spine up to 9–11 mm long, usually with 1–2 pairs of short spinules 2–3 mm long at base. Marginal shorter than central florets, with thread-like segments; corollas yellow. Achenes unknown; pappus 7–8 mm, inner series not shorter.

HAB. On a sandy clay hillside, on a rocky limestone hill: alt. 500–600 m: fl. & fr. Jun.
DISTRIB. Very rare in Iraq. **DWD**: 260 km W of Ramadi, *Rawi* 20977!, *Rechinger* 9846!; 40 km E of Rutba, *Omar, Qaisi & K. Hamad* 43964!; 45 km S of Rutba, *Rawi* 21260!

Syria.

This species is only known from the type gathering in Northern Syria (Halabja) and three localities in Iraq. Arènes assigned the species to Sect. *Seridia,* but the typical species of this section are very different. He further on speculated about a hybrid origin from *Centaurea cheiracantha* and *Jurinea stahaelina* DC., although admitting that this hypothesis seems rather improbable due to the fact that the first named species is not known from the area. Of course the idea of a hybrid between two rather distantly related genera is quite absurd. Perhaps *C. delbesiana* can best be allocated to the section *Corethropsis* DC. including *C. scoparia* Sieb. ex Spreng. and *C. schimperi* DC.

SECT. **CALCITRAPA** DC.

28. **Centaurea iberica** *Trev. ex Spreng.*, Syst. Veg. ed. 16, 3: 406 (1826); Boissier, Fl. Orient. 3: 690 (1875); Nábělek in Publ. Fac. Sci. Univ. Masaryk 52: 46 (1925); Dinsmore in Post, Fl. Syria, Palest. & Sinai ed. 2, 2: 115 (1933); Guest in Dep. Agr. Iraq Bull. 27: 20 (1933); Anthony in Notes Roy. Bot. Gard. Edinburgh 18: 291 (1935); Blakelock in Kew Bull. 4: 46 (1949); Regel in Candollea 16: 239 (1958); Cherepanov in Fl. S.S.S.R. 28: 561 (1963); Rechinger, Fl. Lowland Iraq: 657 (1964); Wagenitz in Fl. Turkey 5: 543 (1975); Feinbrun-Dothan, Fl. Palaest. 3: 399 (1978); Wagenitz in Fl. Iranica 139B: 381 (1980); Mouterde, Nouv. Fl. Liban et Syrie 3: 487 (1983); Abdulridha, Taha & Wedad, Ecology and Fl. Basra: 203 (2016); Taifour & El-Oqlah, Pl. Jordan Ann. Checkl.: 38 (2017).

Calcitrapa iberica (Spreng.) Schur., Enum. PI. Transs.: 409 (1866).
C. iberica Spreng. var. *brevispina* Freyn & Sint. in Österr. Bot. Z. 44: 259 (1894).
Centaurea pallescens Delile var. *iberica* (Spreng.) Gugler & Thell. in Ann. Hist. Nat. Mus. Hung. 6: 209 (1908).
Leucantha iberica (Spreng.) Á.Löve & D.Löve, Bot. Notis. 114: 44 (1961).

Annual or biennial, 20–80 cm tall, strongly and divaricately branched in upper part with distinct overtopping of the relative main axis by two or three branches arising a short distance below the terminal capitulum. Leaves sparsely hirsute; lower leaves petiolate, pinnatifid, pinnatipartite or lyrate with up to 6 pairs of lanceolate, denticulate (or rarely pinnatilobed) segments; stem-leaves sessile, with fewer segments but in other respects similar; upper leaves pinnatipartite with 1–2 pairs of segments or undivided, denticulate. Involucre ovoid, 13–18 × 9–12(–15) mm broad; phyllaries multiseriate, glabrous, coriaceous, outer and median with a narrow whitish hyaline margin and ending in a spine; spine in median phyllaries rigid, stramineous or brownish, 10–30 mm long, at base on each side with (1–)2–3 spinules 2–4 mm long; inner phyllaries with a rounded nearly entire scarious brownish appendage. Marginal florets moderately radiant; corollas pink to violet or whitish. Achenes 3–4 mm; pappus 1–2.5(–3) mm, inner series short.

HAB. Along the margins of fields and dry water channels; generally below 150 m (exceptionally up to alt. c. 2000 m); fl. & fr. Apr.-May (Jun.-Aug.).
DISTRIB. Widespread except in the desert regions. **MAM**: Jabal Bekher nr Zakho, *Field & Lazar* 810! **MRO**: nr Iranian frontier above Pushtashan, *Rechinger* 11809!; Seri Hasan Beg, *Guest* 2906! **MSU**: Susa, *F.*

Karim 39254! **FKI**: Kirkuk, *Guest* 1985! **FPF**: Mandali, *id.* 767!, 1804! **LCA**: Baghdad, *Paranjpye* s.n. (comm. Graham), *Rogers* 0304!; nr Falluja, *Guest* 964!; Babylon, *Rogers* 0297!; Hilla, *Paranjpye* s.n.!; nr Daltawa (Khalis), *Guest* 2467!, *Lazar* 406! **LSM**: nr Amara, *Field & Lazar* 60!; 40 km from Amara, on the road to Basra, *Khatib & Alizzi* 32622! **LBA**: between Basra and Fao, by Shatt al-Arab, *Alizzi & Omar* 35078!

SE Europe (Greece to Romania and Crimea), Jordan, Palestine, Syria, Lebanon, Turkey, Iran, Transcaucasia, Afghanistan, Pakistan, Kashmir, C Asia.

KASSŪB, occasionally MURRAIR (bitter-wort) and CHALBA (dog), according to Guest.

29. **Centaurea hyalolepis** *Boiss.*, Diagn. Pl. Orient. ser. 1, 6: 133 (1845); Anthony in Notes Roy. Bot. Gard. Edinburgh 18: 292 (1935); Rechinger, Fl. Lowland Iraq: 658 (1964); Wagenitz in Fl. Turkey 5: 545 (1975); Feinbrun-Dothan, Fl. Palaest. 3: 399 (1978); Wagenitz in Fl. Iranica 139B: 383 (1980); Taifour & El-Oqlah, Pl. Jordan Ann. Checkl.: 38 (2017)

C. *pallescens* Delile var. *hyalolepis* (Boiss.) Boiss., Fl. Orient. 3: 69 (1875).
C. *pallescens* Delile f. *hyalolepis* (Boiss.) Gugler, Ann. Hist. Nat. Mus. Hung. 6: 209 (1908).
Calcitrapa hyalolepis (Boiss.) Holub in Preslia 46: 227 (1974).

Annual to biennial, 30–80(–100) cm tall. Stem repeatedly and divaricately branched from near base with overtopping of the relative main axis as described above for *C. iberica*. Branches whitish with sparse crisp hairs. Basal leaves withered by flowering-time, pinnatisect to lyrate with several pairs of lanceolate toothed or basally lobed segments; median cauline leaves similar or undivided, toothed; upper leaves lanceolate to nearly triangular, sessile with a semiamplexicaul base, with few coarse teeth, subtending the capitula. Involucre ovoid to nearly globose, 11–13 × 8–10 mm; phyllaries multiseriate, glabrous, with a hyaline margin forming broad auricles at base of spines; spines in median phyllaries 15–25(–30) mm long, with 1–2 pairs of very short (1–1.5 mm) spinules at base, sometimes lateral spinules altogether lacking. Corollas yellow. Achenes 2.2–2.5(-3) mm; pappus 2.5–3 mm.

HAB. Sandy clay plain and wadis, riverine thicket, waste, fallow and cultivated land, especially along irrigation channels; alt. 50–450 m; & fl. & fr. Apr.-Jun.
DISTRIB. Quite common in the upper plains and foothills region of Iraq; also in the desert and lower Mesopotamian plain. **FUJ**: Bat Khan, nr Balad Sinjar, *Field & Lazar* 612!; between Ba'aj and Balad Sinjar, *Ani* 8787! **FNI**: between Mosul and Nineveh, *E. Chapman* 25371!; Nineveh, *Helbaek* 907!; *Barkley & Ani* 8979 p.p.! **FAR**: Ankawa, nr Arbil, *Bornmüller* 1423!; Arbil, *Gillett* 8014! **FKI**: Qara Hasan, *Karim, Hamid & Jasim* 40464!; Jabal Hamrin, between Baghdad & Kirkuk, *Bornmüller* 1424! **FPF**: Mandali, *Guest* 903!, *Haddad* 99981! **DLJ**: nr Balad, *C. & E. Guest and Sharif Muhammad Amim Haidar* 18342! & *Alizzi & Husain* 33712! **LCA**: Dukhail, 30 km N of Baghdad, *B. Anderson* 1986!; Daltawa (Khalis), *Rogers* 0333!; Baghdad, *Rogers* 048!, *Haines* 790!; **LEA**: Ba'quba, *Paranjpye* (comm. Graham) s.n.!; Shahraban (Muqdadiya), *Haines* 307! **LSM**: Fakka, 70 km E of Amara, *Rawi & Haddad* 25797!

Cyprus, Palestine, Jordan, Lebanon, Syria, Turkey, S Iran.

SECT. **TETRAMORPHAEA** (DC.) BOISS.

30. **Centaurea postii** *Boiss.*, Fl. Orient. 3: 688 (1875); Dinsmore in Post, Fl. Syria, Palest. & Sinai ed. 2, 2: 103 (1933); Mouterde, Nouv. Fl. Liban et Syrie 3: 479 (1983); Boulos & Hind in Fl. Egypt. 3: 166 (2002); Taifour & El-Oqlah, Pl. Jordan Ann. Checkl.: 38 (2017).

Perennial, sometimes densely tufted with numerous sterile rosettes and ascending or prostate stems 5–50 cm long with several capitula. The main stem usually overtopped by one or two lateral branches. Stem and leaves with ± thick appressed greyish-white indumentum. Basal to median leaves pinnatipartite with 4–6 pairs of linear segments ending in a short firm mucro, upper leaves linear-lanceolate, each margin with 1–3 teeth. Involucre ovate, tomentose, 13–15 mm × 8–9 mm. Appendage of median phyllaries a firm yellowish spine 15–25 mm long, simple or with 1–2 pairs of spinules near base. Corollas white with a creamy anther-tube (or sometimes pinkish red-coloured?). Achenes 4–4.5 mm long with large lateral hilum with a prominent tooth-like elaiosome, without pappus.

DISTRIB. Egypt (Sinai), Syria, Palestine, Lebanon, Jordan.

No material from Iraq has been seen. Zohary (1939: 251) mentions a specimen from "Harrat el Rajil, between Azrak and Rutba" but it has not been located in HUJ and it is not clear if the locality is in Iraq. An occurrence in the area of the flora does seem quite possible.

Boissier placed this species, known to him only from the type collection, in the sect. *Mesocentron* and describes the corollas as pink. The material seen by us has yellow corollas when dry, but according to field-notes by Townsend on material from eastern Jordan they are white with a creamy anther-tube. Both colours may occur (as in *C. sinaica*) or Boissier may have been in error. *C. postii* with non-decurrent leaves and achenes without pappus might be better placed in sect. *Calcitrapa*, but the perennial habit is unusual here too.

31. **Centaurea bruguieriana** (*DC.*) *Hand.-Mazz.* in Ann. Naturh. Mus. Wien 27: 451 (1913); Nábělek in Publ. Fac. Sci. Univ. Masaryk 52: 45 (1925); Wagenitz in Fl. Iranica 139B: 340 (1980).

Tetramorphaea bruguieriana DC., Prodr. 6: 609 (1838).
Centaurea phyllocephala Boiss., Diagn. Pl. Orient. ser. 1, 6: 134 (1845); Boissier, Fl. Orient. 3: 684 (1875), nom. illegit.

Annual, 10–50 cm tall, in larger plants profusely branched from near base with divaricate and intricate branches often overtopping their relative main axis, whole plant often forming a nearly globose bush; small plants with few branches near middle. Stem and branches ivory-white, hirsute with articulate hairs. Leaves sparsely hirsute, white-ciliate at margins; lower leaves petiolate, lyrate-pinnatilobed or undivided, lanceolate; median oblong, sessile and half-clasping; upper smaller, oblong-cordate, often several crowded at base of involucre. Involucre (excluding appendage) 10–11 × 5–6 mm; phyllaries multiseriate, greenish, glabrous, outer with large foliaceous appendages very similar to uppermost leaves, ciliate at margins; median phyllaries with a simple spine 15–20 mm long; inner with small scarious appendages. Marginal florets slightly radiant; corollas pale purplish. Achenes 2 mm; pappus 2.5–3 mm. Fig. 27: 1–6.

DISTRIB. (of species): Syria, Turkey, Iran, Transcaucasia, Afghanistan, Pakistan, C Asia (Turkmenia to Kazakhstan).

subsp. **bruguieriana**

C. bruguieriana (DC.) Hand.-Mazz. var. *spinosissima* Bornm. in Beih. Bot. Centralbl. 60B: 214 (1939).

Foliaceous appendages subcordate, 5–6 mm broad, with cilia 1–3 mm long at margins.

HAB. (of subsp.): On stony or gravelly or sandy soil, sometimes rather saline, also as a weed of irrigated cultivation or roadsides: alt. 40–320 m; fl. May-Jul., fr. Jun.-Aug.
DISTRIB. (of subsp.): Very common in the steppe region of Iraq, also occasionally found in the desert region. **MJS**: Jabal Qilat, between Ain Tellawi and Balad Sinjar, *Field & Lazar* 421! **MSU**: Jarmo, *Helbaek* 1891! **FUJ**: on the road from Baiji to Mosul, *Field & Lazar* 387! **FNI**: Pesh Khabur, *Kotschy* 411!; 0.5 km NW of Mindan Bridge, *E. Chapman* 26210! **FAR**: Mahmur Dagh, on road to Dibaga, *Gillett* 11204!; 8 km NE of Arbil, *id.* 8346!; Ankowa, *Bornmüller* 1422! (type of *C. bruguierana* var. *spinosissima*); **FKI**: Altun Kupri desert, *Haussknecht* s.n.!; 40 km from Kirkuk to Altun Kupri, *Alkas* 18569!; 2 km N of Kirkuk, *Rawi* 21553! **FKI/DGA**: Adhaim, S of cultivated area, *Robertson* 103! **FKI**: nr Saud Ali Khan on W bank of R. Diyala, *Poore* 571! **FPF**: 2 km N of Sa'diya, *Qaisi* 47229!; Naft Khana, *Rogers* 0323!; R. Ab-i Naft near Mandali, *Guest* 902!; Koma Sang, nr Mandali, *Rawi* 20671!; Lajama police post, 25 km SE of Mandali, *id.* 20728!; Chilat police post nr Iranian frontier, *Rawi & Haddad* 25705! **DLJ**: 10 km W of Tikrit, *Alizzi & Husain* 33737!; 32 km NW of Falluja, *Rawi & Shahwani* 25470! **DSD**: Ain Halwa, *Salman, Kasim, Nuri, Hamid & Kadhim* 40122!; 30 km W by N of Busaiya, *Guest, Rawi & Long* 14175!; 60 km SW by S of Zubair, *Guest & Rawi* 14312! **LCA**: Daltawa (Khalis), *Rogers* 0329!; Khan Bani Sa'd, 25 km N of Baghdad, *B. Anderson* 1999!; Falluja desert (E. of Falluja), *Haines* 794!; Abu Ghraib, Samir *Monim* 16493!; 20 km W of Abu Ghraib, *Ani & Haddad* 9931! 30 km N of Diwaniya, *Thamir* 47660! **LEA**: 75 km N of Amara, *Hazim & Noori* 30631! **LBA/DSD**: 52 km NE of Safwan, *Rawi* 25986!

Turkey, Syria, Iran.

A few gatherings are conspicuous by their apparently yellow corollas; in one case noted by the collector [**FUJ**: Qaiyarah, Mosul, *Bayliss* 138!]. It is not impossible that this is due to a hybridization with the yellow-flowered *C. hyalolepis*. Further observations are necessary.

subsp. **belangeriana** (*DC.*) *Bornm.*, in Beih. Bot. Centralbl. 60B: 214 (1939); Wagenitz in Fl. Iranica 139B: 385 (1980).

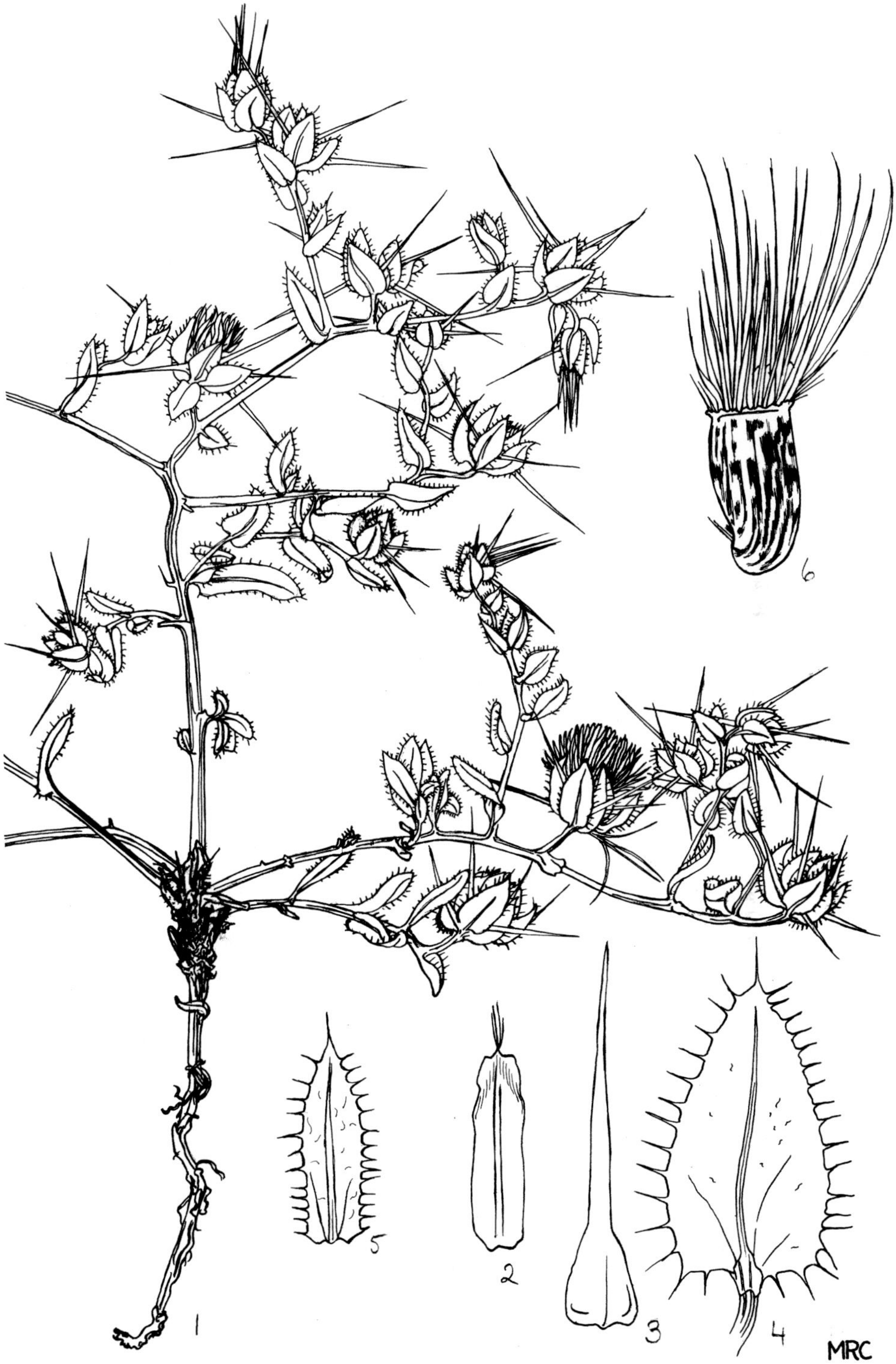

Fig. 27. **Centaurea bruguieriana**. 1, habit × 1; 2, inner phyllary × 5; 3, median phyllary × 5; 4, outer phyllary × 5; 5, outer phyllary × 5; 6, achene with pappus × 20. 1 *from Guest, Rawi & Long* 14175; 2–4 from *Rechinger* s.n.; 5 from *Bornmüller* 1422. Drawn by Moira Rae Carter.

Tetramorphaea belangeriana DC., Prodr. 6: 609 (1838).
Centaurea belangeriana (DC.) Heimerl in Denkschr. Akad. Wiss. Wien, Math.-Nat. Kl. 50: 66 (1885); Cherepanov in Fl. S.S.S.R. 28: 578 (1963).
C. *phyllocephala* Boiss. var. *persica* Boiss., Diagn. Pl. Orient., ser. 1, 6: 134 (1845).

Foliaceous appendages oblong, 4 mm broad, with very short cilia (0.3–0.8 mm).

DISTRIB. (of subsp.): Upper Jazira district, very rare. **FUJ**: Qaiyara, *Bayliss* 139; near the ruins of Hatra (Hadhr), *Bharucha & Majid Brahim* 8648. The former is recorded as having yellow florets.
A few specimens are intermediate between the subspecies, with cilia 0.5–1.5 mm long: **MRO**: Arbil to Salahuddin, *Haines* s.n.! **DSD**: 60 km SW by S of Zubair, *Guest & Rawi* 14312!

Iran, Transcaucasia, Afghanistan, Pakistan, C Asia.

SECT. **ACROCENTRON** (CASS.) DC.

32. **Centaurea urvillei** *DC.*, Prodr. 6: 592 (1838); Boissier, Fl. Orient. 3: 665 (1875); Wagenitz in Willdenowia 6: 488 (1972) & Fl. Turkey 5: 550 (1975) & Fl. Iranica 139B: 388 (1980).

Columbada urvillei (DC.) Holub in Preslia 45: 144 (1973).

Biennial or short-lived perennial herb with thickened main root. Stem 5–35 cm, often shorter than basal leaves, simple or branched from near base. Leaves slightly arachnoid to distinctly tomentose; lower and median leaves lyrate with a triangular, rhombic or oblong-ovate terminal segment and few to numerous lanceolate, oblong or lyrate lateral segments, often interspersed with small lobes. Involucre 20–40 × 15–40 mm, ovoid to globose; phyllaries multiseriate; appendages ± concealing basal part of phyllaries, very variable in form and colour of basal part, length of cilia and of terminal spine (10–40 mm long). Marginal florets slightly radiant; corollas pinkish red to purplish or whitish. Achenes 4–7 mm; pappus (5–)8–14 mm. Receptacle bristles 15–20 mm.

DISTRIB. (of species): Turkey, Iran, Iraq.

A very variable species, especially in Turkey, with several subspecies which are not easy to delimit. Subsp. URVILLEI, which occurs mainly in Outer Anatolia, is lacking in Iraq, where the species is represented by:

subsp. **deinacantha** (*Boiss. & Hausskn.*) *Wagenitz* in Fl. Iranica 139B: 388 (1980). –Type. Pira Magrun, *Haussknecht* 584, & *Haussknecht* s.n.

C. deinacantha Boiss. & Hausskn. in Boiss., Fl. Orient. 3: 665 (1875).
? *C. urvillei* DC. var. *longicaulis* Parsa in Kew Bull. 2: 201 (1948).

Stem stout, sulcate in lower part, 20–35 cm tall, branched from near base with several one-headed branches about as long as main stem. Lower leaves petiolate, lyrate with 2–4 pairs of lanceolate or oblong lateral segments (often additional small segments between the larger ones). Involucre 30–40 × 30–35 mm; appendages from a broad triangular straw-coloured or brown base gradually narrowed into a spine, with spine in median phyllaries 20–30(-40) × 6–9 mm at base excluding cilia, cilia 5–7 mm. Achenes 6–7 mm long; pappus 8–14 mm, inner series short.

HAB. (of subsp.): In the mountains, often in clefts in limestone rock; alt.700–1200 m: fl. & fr. Jun.-Jul. (-Aug.).
DISTRIB. (of subsp.). In the central section of the mountain region. **MRO**: Navanda, *Hadač* 2506!; Kani Mazi Shirin, *Agnew, Hadač & Haines* s.n.!; Rowanduz, *id.* 5776!; Rowanduz, Nalpridan, *Hadač & Kader* 5320!; between Rowanduz and Bersorini, *Bornmüller* 1487!, *Rechinger* 11264!; Kewa Rash, *Rawi & Serhang* 237861. **MSU**: Pira Magrun, *Haussknecht* 584 and s.n. (type)!, *Haines* s.n.!

Iran.

33. **Centaurea elegantissima** *Bornm.* in Feddes Repert. 24: 369 (1928). – Type. Sakri Sakran, nr Rowanduz, 1500 m, 1892–93, *Bornmüller* 1489 (B, holo).

Colymbada elegantissima (Bornmüller) Holub in Folia Geobot Phytotax. 7: 315 (1972).

Biennial (or perennial?), 30–50+ cm tall. Stem erect, ± hirsute at base, glabrous above, simple or with 1–2 branches in upper part. Leaves thin, sparsely to moderately densely hirsute with articulate hairs or subglabrous; basal and lower cauline leaves with a short petiole, interruptedly pinnatisect to sublyrate, with 4–5 pairs of larger and several smaller segments, segments very variable, oblong to obovate (rounded at apex) in outline, with several coarse teeth or lobes, terminal segment often somewhat larger; median leaves similar, but smaller and sessile, sometimes more distinctly lyrate; upper leaves pinnatilobed or entire with few teeth or lobes near base. Involucre nearly globose, 30 × 30–35 mm; phyllaries multiseriate, coriaceous, glabrous; appendages concealing a large part of basal part of phyllaries, triangular, brown or blackish-brown, decurrent, rather abruptly narrowed into a 6–15 mm long spine, regularly ciliate, cilia 5–6 mm long; innermost phyllary nearly circular, unarmed, cucullate, regularly ciliate. Corollas rose-purple (according to the collectors). Mature achenes unknown; pappus 7–9 mm, inner series 2.5 mm.

HAB. On stony and rocky mountain slopes; alt. 900–1650 m; fl. & fr. May-Jun.
DISTRIB. Only found in the central sector of the mountain region, mostly near the Iranian frontier. **MRO**: Sakri Sakran, nr Rowanduz, *Bornmüller* 1489 (type)!; Razanok, *Cuckney* (in Guest) 4848!; Saran, nr Kani Kawan (Kun) spring, Karokh, *Noori & Alkas* 21316!; between Saran and Kilki, *Alkas, Noori & Serhang* 27328!; Berd Agha Gin, 18 km NW of Rania, *Rawi, Noori & Alkas* 28868!; nr Dargala, c. 35 km NW by N of Rania, *id.* 28888!

Endemic.

Most collections are incomplete, and better material is urgently needed to give a clear picture of the variability of this species. As far as our knowledge goes, the large interruptedly pinnatisect basal leaves with rounded segments and the large capitula with regularly ciliate appendages and moderately long spine are characteristic.

34. **Centaurea ochrocephala** *Wagenitz* in Fl. Iranica 139b: 389 (1980).

Perennial. Stem 30–65 cm tall, sparsely hirsute in lower part, nearly glabrous above, with few long one-headed branches near middle or in upper part. Leaves with sparse articulate hairs especially on lower surface, nearly glabrous above, upper leaves ± glabrous; lower leaves often withered at flowering time, petiolate, pinnatisect or lyrate, with 3–4 pairs of entire or denticulate, lanceolate or oblong lateral segments; median leaves lyrate, sessile with a large lanceolate terminal segment and 3–4 pairs of linear-lanceolate segments or lobes; upper leaves undivided, dentate or with small lobes at base. Involucre nearly globose, (20–)22–25 × (20–)25–30 mm; phyllaries multiseriate, basal part totally or nearly totally concealed by their appendages. Appendages straw-coloured, coriaceous, in median phyllaries narrowly triangular, shortly decurrent, very gradually narrowed into a spine, 20–30(-35) mm long, regularly ciliate in lower part, 5–7 mm in diameter at base excluding the 6–9 mm cilia; appendages of innermost phyllaries rounded, ciliate, without a spine. Corollas pinkish red to purple. Achenes 6–6.5 mm; pappus 7–9 mm, inner series 2–3 mm.

HAB. In the mountains on limestone; alt. 1550–1800 m: fl. & fr. Jun.-Jul.
DISTRIB. Very rare in the northern sector of the mountain region. **MAM**: mt. Zawita nr Sharanish, *Rechinger* 10935!; Daimka, 35 km NE of Zakho, *Omar & Dabbagh* 45396! **MSU**: Zalam, *Rawi, Husham & Noori* 29379!

NW Iran. Typical specimens of this species are only known from a small area in Iranian Azerbaijan; the plants gathered by Rechinger (10935) are not yet flowering, but seem to fit best here. The specimens from Zalam mountain differ especially by the brownish appendages with short cilia.

35. **Centaurea longipedunculata** *Sch.Bip.* ex *Boiss.*, Fl. Orient. 3: 668 (1875). –Type. Amadiya district, Gara village, near Sarsang, *Kotschy* 374 (JE, WAG, HAL, P, syntypes)

Colymbada longipedunculata (Boiss.) Holub, Folia Geobot. Phytotax. 7: 316 (1972).

Perennial. Stem 80–90 cm tall, glabrous, with few long one- or few headed branches in upper part. Leaves glabrous or nearly glabrous; basal and lower cauline leaves petiolate, large (basal with petiole 40–50 cm long), lyrate or pinnatisect, terminal segment (if leaves are lyrate) broadly lanceolate, up to 7 cm broad, repand-dentate; lateral segments 3–4 (in pinnatisect leaves up to 7) on each side, oblong-lanceolate, indistinctly dentate, 10–20(–25) cm broad, upper ones in particular distinctly decurrent along rhachis: median cauline leaves similar to lower, but sessile and smaller with only 1–3 pairs of segments;

uppermost leaves undivided, margin entire or dentate. Involucre oblong, 22–25 × 14–18 mm broad; phyllaries multiseriate, greenish, their lower part for the most part concealed by the appendages; appendages straw-coloured, triangular, decurrent, ending in a short spine 3–4 mm long, ciliate, cilia 3–6 mm long, often undulate. Corollas pinkish-red-purple. Achenes 5.5 mm; pappus 7 mm.

HAB. In fissures in calcareous rock in *Quercus* forest: alt. 1000–1400 m: fl. & fr. Jul.-Aug.
DISTRIB. Rare in Iraq, only found in two localities in the NW corner of the forest zone not far from the Turkish frontier. **MAM**: Daimka, 35 km NE of Zakho, *Omar & Dabbagh* 45385!; Gara village, nr Sarsang, *Kotschy* 374 (type); Sarsang, *Haines* s.n.!; above Sarsang, between Dohuk and Amadiya, *Rechinger* 11663!

Endemic.

Closely allied to *C. persica* and mainly differing by the very short terminal spine.

36. **Centaurea persica** *Boiss.*, Diagn. Pl. Orient., ser. 1, 6: 132 (1845).

Colymbada persica (Boiss.) Holub in Folia Geobot. Phytotax. 7: 316 (1972).

Perennial. Stem (25–)30–50(-90) cm tall, at base often with firm remains of petioles, basal part sometimes with long hairs, remaining part or whole stem glabrous, simple or with a few one-headed branches above middle. Leaves firm, subglabrous (scabrid on margin and sometimes with sparse long hairs especially on veins of lower surface); basal and lower leaves petiolate, lyrate, terminal segment rhombic or lanceolate, entire or remotely dentate, lateral segments 2–4(–7) on each side, lanceolate or linear-lanceolate; median leaves similar but sessile and with only 1–2 pairs of lateral segments, sometimes only lobate; cauline leaves few, uppermost undivided. Involucre oblong or subglobose. 20–25(–27) × 16–23 mm; phyllaries multiseriate, not much of lower part covered by the appendages; appendages straw-coloured (scarcely differing in colour from the phyllaries), central triangular part 4–5 mm in diameter at base, extending into narrow decurrent margins and rather abruptly constricted into spreading spine, appendages in median phyllaries 10–20(–25) mm long including spine, ciliate, cilia (3–)4–5(–6) mm long; appendages of inner phyllaries nearly orbicular, without a spine, ciliate or lacerate. Corollas pinkish-red-purple, marginal not radiant. Achenes 5–7 mm; pappus (5–)7–9 mm, inner series ± 2 mm.

DISTRIB. (of species): E Turkey, W Iran.

var. **persica.**
Nearly always at least with some hairs at base of stem or at margins and veins of leaves. Plant scarcely more than 60 cm tall. Basal leaves up to 20 cm long with 2–4(–5) segments on each side. Appendages very variable in length. Spines straight.

HAB. On mountainsides, sometimes by streams; alt. (200–)800–2000 m; fl. & fr. Jun.-Aug.
DISTRIB. Locally common in the central mountain region of Iraq. **MRO**: Algurd, *Guest* 2849!, *Gillett* 9532!, 12465!; between Karokh Mt. and Dargala, *Rawi, Kass & Noori* 27732!; Handren to the Iranian frontier, *Bornmüller* 1486!; Siewaka, at the foot of mt. Karokh, *Kass & Noori* 27562!

E Turkey, W Iran.

var. **multijuga** *Bornm.* in Beih. Bot. Centralbl. 60B: 211 (1939). – Type. Handren, *Bornmüller* 1492 (B, holo).
Totally glabrous, up to 90 cm tall. Basal leaves large (up to 35 cm with petiole), with 3–7 segments on each side. Appendages (12–)15–20 mm long, spines slightly curved.

HAB. & DISTRIB. Local, in the high mountains on rocky places and in ravines; alt. 1300–2600 m; fl. & fr. Jun.-Aug.: **MRO**: Handren, *Bornmüller* 1492 (type)!; above Nawanda, *Rechinger* 11392!; Qandil range, above Pushtashan, *Rechinger* 11812!

Endemic.

37. **Centaurea hadacii** *Wagenitz* in Kew Bull. 40: 793 (1985). –Type. Rowanduz district, above the village of Zaytah, 19 June 1961, *Agnew, Hadač & Haines* 5945 (PR, holo & iso).
Biennial or perennial plant with a tap-root. Stem erect, 45–60 cm tall, subglabrous, with few short or long simple branches in upper part. Leaves firm, appearing glabrous but with sparse short hairs near and at margins and along veins; basal (and lower cauline?) leaves

lyrate, large, up to 30 cm long including the long petiole, terminal segment distinctly larger, lanceolate, lateral segments 3–5 on each side, oblong-lanceolate (the lower much smaller), all remotely dentate; median leaves lyrate-pinnatilobed, sessile; upper much smaller, lanceolate, with 1–2 lobes on each side at base. Involucre nearly glabrous, 22–25 × 25–26 mm broad; phyllaries multiseriate, their lower part totally covered by the appendages. Appendages coriaceous, straw-coloured or light brown, broadly triangular (excluding the cilia 6–7 mm, including the cilia up to 15 mm broad), shortly decurrent, with 14–16 cilia 4–7 mm long on each side, ending in a scarcely 1 mm long, often very indistinct mucro. Marginal florets not radiant; corollas rose-purple. Achenes 5–6 mm; pappus 6–7 mm long, scabrous, inner series 2–3 mm.

HAB. In the thorn-cushion zone of the mountain region; alt. 2140 m; fl. & fr. Jun.-Aug.
DISTRIB. Very rare on igneous rocks and limestone slopes: **MRO**: above the village of Zaytah in the Rowanduz district, *Agnew, Hadač & Haines* 5945 (type)!; above Ari on Ser Kurawa, *Gillett* 9713!

Endemic.

This species is remarkably similar to C. *persica* in vegetative characters, but differs strikingly in the totally unarmed large appendages which completely cover the basal part of the phyllaries.

38. **Centaurea luristanica** *Rech.f.*, in Anz. Math.-Nat. Kl. Österr. Akad. Wiss. 88: 265 (1961); Wagenitz in Fl. Iranica 139B: 397 (1980).

Perennial. 20–50 cm tall, simple or with few branches, ± hirsute at least in lower part. Leaves sparsely hirsute to subglabrous: basal leaves pinnatisect or lyrate, with 4–8 pairs of oblong to obovate, mostly irregularly dentate lateral segments, smaller segments sometimes present between larger ones, terminal segment lanceolate to nearly obovate; median and upper leaves lyrate to pinnatilobed with 1–3 pairs of segments or lobes. Involucre oblong to cup-shaped, rounded at base, 25–33 × 18–30 mm; phyllary appendages with a triangular brown base, gradually narrowed into a slender spine of variable length, margins whitish, ciliate, decurrent; appendage with spine 10–35 mm long, cilia 4–5(–8) mm long; innermost phyllaries with very broad rounded appendages, 8–15 mm broad, rather regularly ciliate. Marginal florets not radiant; corollas rose-purple. Achenes 5.5–6.5 mm, densely hairy when young; pappus 10–15 mm, inner series much shorter.

HAB. On steep limestone slopes; alt. 1100–1400 m; fl. & fr. Apr.-May.
DISTRIB. Very rare in the mountain region of Iraq. **MSU**: on the Avroman range above Darimar, *Gillett* 11834!; above Khurmal (on the same range of mountains), *Hadač* 5070!

Iran.

It is only with some hesitation that I have included in C. *luristanica* these two collections from Avroman, which differ in some respects from each other and the type. *C. luristanica* is only known from a few collections. Characteristic features seem to be the triangular brown basal part of the appendages, the very broad ciliate appendages of the inner phyllaries and the long pappus.

39. **Centaurea davisii** *Wagenitz* in Willdenowia 6: 497 (1972) & Fl. Turkey 5: 554 (1975) & Fl. Iranica 139B: 394 (1980).

Colymbada davisii (Wagenitz) Holub in Preslia 45: 144 (1973).

Very probably perennial, but underground parts still unknown. Stem 50 cm tall or more, ascending, stout, nearly glabrous, with 1–2 short branches each with one capitulum in upper part. Leaves glabrous or nearly glabrous, basal and lower petiolate, long (up to 30 cm including petiole), interruptedly pinnatisect with 8–10 pairs of larger, lanceolate, coarsely dentate or pinnatilobate segments and several smaller leaflets; median and upper leaves gradually smaller, sessile, pinnatisect or pinnatifid with fewer segments. Involucre oblong-ovoid, truncate at base, 25–30 × 20–22 mm; phyllaries multiseriate, floccose-tomentose, greenish, inner ones with purple tinge; appendages only partly concealing basal portion of phyllaries, spreading or slightly recurved, brown, gradually narrowed into a spine from a triangular decurrent base, 15–20 mm long including spine, 3–5 mm in diameter. at base excluding the 4–5 mm cilia. Marginal florets scarcely radiant;

corollas rose-purple. Achene 5–5.5 mm; pappus 8–10 mm, inner series short, elements scale-like.

HAB. Among rocks with a slaty substratum; alt. ± 1500 m; fl. & fr. July.
DISTRIB. Very rare in Iraq; only found by the Turkish frontier, NE of Zakho. **MAM**: Zawita Mt., nr Sharanish, *Rawi* 23651!, *Rechinger* 10902a!

SE Turkey.

40. **Centaurea irritans** *Wagenitz* in Rechinger, Fl. Iranica 139B: 390 (1980).

Perennial, sometimes with several stems from base, 20–40(–60) cm tall. Stem hirsute with articulate hairs in basal part and with white wool in axils of leaves, glabrous in upper part; a few long, simple branches present in lower half. Leaves with sparse to dense, long, articulate hairs especially on lower side. Basal leaves mostly withered at flowering time, petiolate, interruptedly pinnatipartite with distant, lanceolate, dentate segments of unequal size; median leaves similar but sessile or sublyrate; only a few distant very small leaves in upper half of stem and on branches. Involucre oblong, 20–25 × 15–18 mm broad; phyllaries multiseriate, slightly tomentose, often with a purple tinge, their lower part not totally concealed by the appendages; appendages straw-coloured to brownish, triangular, decurrent, gradually narrowed into a spine, with spine (in median phyllaries) 15–25 mm long, 3–5 mm in diameter at base excluding the 3–4(–5) mm cilia; appendages of innermost phyllaries scarious, cucullate, irregularly ciliate, unarmed. Marginal florets scarcely radiant; corollas rose-purple. Achenes ± 5 mm long (only known from one gathering); pappus 8–11 mm, inner series 2.5–3 mm.

HAB. In *Quercus* forest on a steep slope, in crevices in limestone; alt. 1300–1400 m; fl. & fr. Apr.-Jun.
DISTRIB. Rare in the mountain region of Iraq, and only found on or near one mountain. **MSU**: Amoret, nr Qaradagh, *Haines* 1155!; Qara Dagh, *Poore* 436!, *Poore & Hadač* 5235!

Iran.

41. **Centaurea benedicta** (*L.*) *L.*, Sp. Pl.: 826 (1753); Greuter in Euro+Med Plantbase (2006); Garcia-Jacas et al. in Ann. Bot. 98(4): 741–753 (2006); Zhu Shi & Ludwig Martins in Fl. China 20–21: 191 (2011).

Cnicus benedictus L., Sp. Pl.: 826 (1753); Boissier, Fl. Orient. 3: 705 (1875); Nábělek in Publ. Fac. Sci. Univ. Masaryk 52: 46 (1925); Dinsmore in Post, Fl. Syria, Palest. & Sinai ed. 2, 2: 110 (1933); Anthony in Notes Roy. Bot. Gard. Edinburgh 18: 292 (1935); Bornmüller in Beih. Bot. Centralbl. 60B: 215 (1939); Zohary in Dep. Agr. Iraq Bull. 31: 157 (1950); Rawi, ibid. Tech. Bull. 15: 29 (1964); Rechinger, Fl. Lowland Iraq: 662 (1964); Kupicha in Fl. Turkey 5: 589 (1975); Feinbrun-Dothan, Fl. Palaest. 3: 405 (1978); Rechinger, Fl. Iranica 139b: 428 (1980); Collenette, Wild Flow. Saudi Arabia: 181 (1999).

Annual herb (3–)10–40 cm tall, often branched below a capitulum terminating a short primary stem at ground level. Branches ribbed, pale, shaggy-pubescent. Basal leaves petiolate, petiole winged; cauline sessile, 4–16 × 0.5–2 cm, base amplexicaul and shortly decurrent, elliptic, pinnatilobed with 3–5 pairs of oblong-lanceolate, acute, toothed lobes, sparsely hairy on both sides, sometimes sessile-glandular, veins prominent and excurrent as thin spines to 0.5 mm long but sometimes scarcely developed. Upper leaves exceeding the capitula. Outer phyllaries leaf-like, prickly, overtopping and transitional to inner; middle phyllaries shallowly lobed, firm, ovate-lanceolate, glabrescent; inner phyllaries up to 3 cm long, ± hyaline-margined, tip pectinate or erose or (in innermost) entire and acute. Corolla yellow, 18–22 mm, limb slightly shorter than tube and unequally 5-lobed, with 2 lobes 4 mm long and 3 lobes 1.5–2 mm. Filaments papillose. Achenes (4.5–)7–9 × 2–2.5 mm, ovoid-cylindrical, truncate, with 20 strong pale ribs; hilum large. Outer pappus hairs to 10 mm, inner pappus 1.5 mm, yellow or brown.

S Europe from Portugal to the Balkans, Sardinia, Crete, Cyprus, Aegean Is., Arabia, Palestine, Syria, Lebanon, Turkey, Iran, Afghanistan, Pakistan, C Asia (widespread); introduced elsewhere in the Old World.

The species is separable into two well-marked varieties, both of which occur in Iraq.

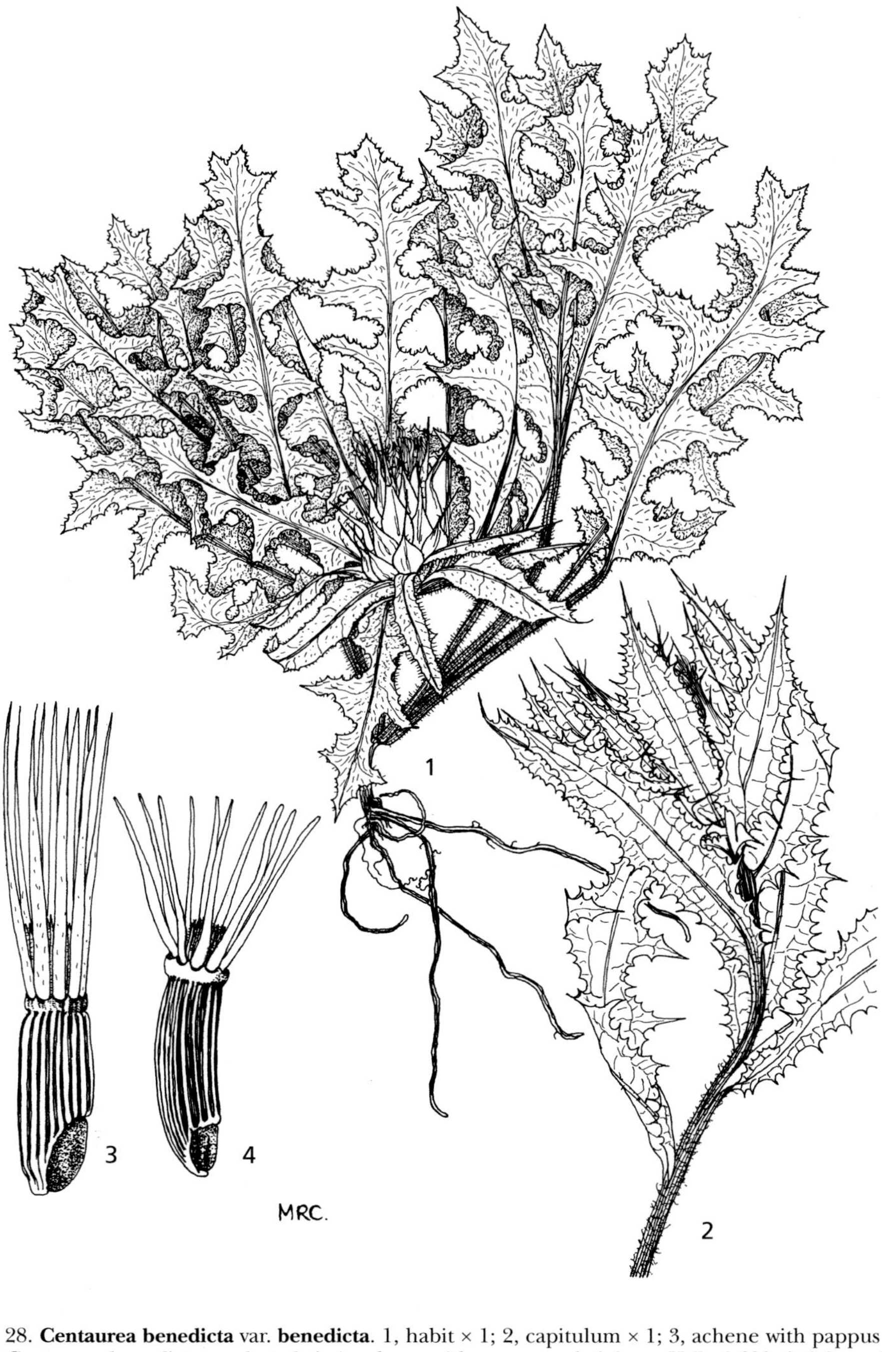

Fig. 28. **Centaurea benedicta** var. **benedicta**. 1, habit × 1; 2, capitulum × 1; 3, achene with pappus × 5. **Centaurea benedicta** var. **kotschyi**. 4, achene with pappus × 6. 1 from *Helbaek* 639; 2–3 from *Rawi* 22838; 4 from *Rawi* et al. 28568. Drawn by Moira Rae Carter.

Achenes 7.5–9 mm long; longest pappus setae equalling or longer than achene, 9–11 mm . . . . var. *benedicta*

Achenes 4.5–8 mm (in Iraqi material seen 6.5–7 mm); longest pappus setae shorter than achene, 3–6 mm (in Iraqi material 4.5–5 mm). . . . . . . var. *kotschyi* var. **benedicta** Fig. 28: 1–3.

Cnicus benedictus L. var. *microcephalus* Boiss, Fl. Orient. 3: 705 (1875).

HAB. On lower mountain slopes and higher steppic uplands, often on disturbed and sometimes damp ground; alt. (120–)350–900(–1200) m; fl. & fr. (Mar.-) Apr.-Jun.
DISTRIB. Quite common in the lower forest zone and moist steppe zone of Iraq, occasional in the dry steppe zone. **MAM**: 5 km S of Zakho, *Rawi* 23088!; Amadiya, *Anders* 2264; **MRO**: Hab as-Sultan Dagh, *Rawi, Noori & Kass* 28193!, *Omar, Karim, Hamza & Hamid* 37199!; Kani Kawan spring, Saran, Karokh, *Noori & Kass* 27315!; between the slope of Karoukh mountain and Dargala village, *Rawi, Noori & Kass* 27724!. **MSU**: Darband-i Basian, *Rawi* 22838!; Jarmo, *Helbaek* 639!, *Haines* W. 342!; Qara Dagh village, *Gillett* 7962! **FUJ**: Qaiyara, *Bayliss* 107! **FKI**: N of Kifri, *Poore* 376!; Kirkuk, *Graham* 783!

Found throughout the range of the species.

var. **kotschyi** (Boiss.) Ghaz. **comb. nov.** Fig. 28: 4.

Cnicus benedictus L. var. *kotschyi* Boiss., Fl. Orient. 3: 706 (1875); Kupicha in Fl. Turkey 5: 589 (1975); Fl. Iranica 139b: 429 (1980).

HAB. Recorded from wet waste land and fields on clay; alt. 120–700 m.
DISTRIB. Apparently uncommon in the steppe region of Iraq. **MRO**: Dolako (Sardkah?) village, 11 km NW of Rania, *Rawi, Noori & Kass* 28568! **MSU**: nr Chemchemal, *Rogers* 0238! **FPF**: 3 km N of Sa'diya, *Qaisi* 46333!

The distribution of this variety is incompletely known, but in SW Asia it seems scattered through most of the range of the species.

Blessed Thistle (Eng.), sometimes called Holy Thistle, as mentioned by Wren (1956), who describes its medicinal properties as tonic, diaphoretic and emmengogue; it is also said to be useful in dyspepsia, intermittent fevers and loss of appetite, as well as in breaking up colds and in menstrual derangements due to colds.

SECT. **CYANUS** (MILL.) DC.

42. **Centaurea triumfettii** *All.*, Auct. Syn. Hort. Taur.: 16 (1773) & Fl. Pedem. 1: 158 (1785); Dinsmore in Post, Fl. Syria, Palest. & Sinai ed. 2, 3: 103 (1933); Bornmüller in Beih. Bot. Centralbl. 60B: 210 (1939); Wagenitz in Fl. Turkey 5: 576 (1975) & Fl. Iranica 139B: 414 (1980).

Cyanus variegata Lam., Encycl. [J. Lamarck et al.] 1: 668 (1785), nom. illeg.
Centaurea axillaris Willd., Sp. Pl. ed. 4, 3 (3): 2290 (1803), nom. illeg.; Boissier, Fl. Orient. 3: 636 (1875).
C. cana auct. vix Sm., Prodr. Fl. Graec. 2: 198 (1813).
C. triumfettii (All.) Á.Löve & D.Löve, Bot. Notis. 114: 44 (1961)

Perennial, with a slender long creeping rhizome. Stem erect, 10–40(–50) cm tall, arising from centre of a rosette (leaves of rosette often withered at flowering time), unbranched or with a few simple branches in upper part. Stem and leaves ± densely floccose-tomentose; leaves lanceolate to linear-lanceolate, lower narrowed into a petiole, median and upper sessile and often shortly decurrent. Capitula solitary at ends of branches; involucre cup-shaped, 15–22 × 9–18 mm. Phyllaries in several series, greenish with decurrent scarious appendage. Appendage light brown to blackish-brown, ciliate, cilia 1.5–3.5 mm long, brown at base, silvery in upper part. Marginal florets strongly radiant, blue or more rarely rose-purple, central purple-violet, anther-tube slightly curved. Achenes 4–6 mm; pappus 1–2.5 mm.

HAB. Rocky and grassy mountain slopes, open *Quercus* forest and scrub, also on a wall and in a vineyard; alt. 600–3000 m; fl. & fr. (Mar.-)May-Jul.
DISTRIB. On the mountains in the thorn-cushion and alpine zones. **MAM**: Matina, *Rawi* 8677!; Khantur, *Rawi* 8553!, *Rechinger* 12069!; Amadiya, *Guest* 12!; Gali Zawita, *Rawi* 23632!; Sarsang, *Barkley* 8120!; Chiya-i Gara, *Dabbagh & Jasim* 46933! **MRO**: Jabal Baradost, *Field & Lazar* 915!; *Karouk, Kass & Noori* 27431! **MSU**: Qandil range, E of Qala Diza, *Thesiger* 1150!; Qopi Qaradagh, *Hadač et al.* 5150!, 5216!, 6295! **MJS**: summit of Jabal Sinjar, *Anders* 1994!, *Omar, Khayat & Qaisi* 52524!; Karsi, *Gillett* 11012!, *Anders* 577!!

C & S Europe, Crimea, N Africa, Lebanon, Turkey, N & W Iran, Caucasus.

All attempts to subdivide this very polymorphic species have so far been unsatisfactory. In Iraq the species is confined to the mountainous area but shows a considerable variation even there. Some plants described by me (in *Fl. Iranica*) as "Group D" from Iraq are rather conspicuous but there are transitional forms and I doubt whether formal taxonomic recognition is advisable at this stage of our knowledge. The situation is made more difficult by the fact that a considerable part of the material seen has been collected without the basal and underground parts, which may be of importance.

43. **Centaurea depressa** *M.Bieb.*, Fl. Taur.-Cauc. 2: 346 (1808); Boissier, Fl. Orient. 3: 635 (1875); Dinsmore in Post, Fl. Syria, Palest. & Sinai ed. 2, 2: 105 (1933); Anthony in Notes Roy. Bot. Gard. Edinburgh 18: 291 (1935); Cherepanov in Fl. S.S.S.R. 28: 415 (1963); Rechinger, Fl. Lowland Iraq: 653 (1964); Wagenitz in Fl. Turkey 5: 580 (1975) & Fl. Iranica 139b: 416 (1980).

Cyanus depressa (M.Bieb.) Soják in Čas. Národ. Muz., Odd. Přír. 140: 131 (1972).
Centaurea anatolica Griseb., Spic. Fl. Rumel. 2: 234 (1844).
C. depressa var. *amasiana* Bornm. in Magyar Bot. Lap. 4: 260 (1905).

Annual, 10–60 cm, main stem erect, often with several ascending branches from near base, with one to several capitula. Branches and leaves greyish floccose-tomentose. Lower leaves oblong or oblong-spatulate, basally narrowed into a broad petiole, entire or lyrate with a large terminal segment and 2 or 3 pairs of lateral lobes; median and upper leaves sessile, gradually smaller, oblong to linear-lanceolate. Involucre (13–)14–18 × 8–12(–13) mm, ovoid to cup-shaped; phyllaries multiseriate, glabrous, greenish (the innermost often violet-tinged), with a triangular appendage widely decurrent along margin, appendage brown or blackish with silvery teeth 1.5–2(–3) mm. Marginal florets blue (cornflower-blue), sterile, distinctly radiant with 6–8 unequal, rather broad lobes; central florets violet with a steel-blue, strongly curved anther-tube. Achenes 4.5–5.5(–6) mm long with large bearded hilum; pappus 5–8(–9) mm, longer than the achne with the inner series short.

HAB. On a silty plain; alt. ± 550 m: fl. & fr. Apr.
DISTRIB. Only one established record from Iraq, in the desert region. **DWD**: 32 km W of Rutba, *Rawi* 31142! Jabal Harim, *Sunderland* 50 (in Fl Lowland Iraq)

SE Europe (Bulgaria, Greece), Crimea, Lebanon, Syria, Turkey, Iran, Transcaucasia, Pakistan, C Asia, Nepal, W Tibet (introduced to Spain).

This species, widespread in most countries of the Nearer East, is remarkably rare in Iraq. Only one collection from this area has been seen by the author and one other (from Jabal Hamrin) is recorded by Anthony, whose determinations are not very reliable. As the species is a common weed in some adjacent countries a casual introduction seems possible.

44. **Centaurea cyanus** *L.*, Sp. Pl.: 911 (1753); Boissier, Fl. Orient. 3: 634 (1875); Dinsmore in Post, Fl. Syria, Palest. & Sinai ed. 2, 2: 105 (1933); Guest in Dep. Agr. Iraq Bull. 27: 20 (1933); Blakelock in Kew Bull. 4: 46 (1949); Cherepanov in Fl. S.S.S.R. 28: 416 (1963); Rechinger, Fl. Lowland Iraq: 652 (1964); Wagenitz in Fl. Turkey 5: 581 (1975) & Fl. Iranica 139B: 418 (1980).

Cyanus segetum Hill, Veg. Syst. 4: 29 (1762); Herb. Brit. 1: 82 (1769).

Annual, 20–60 cm or more tall, sparingly or profusely branched from near base with 1- to several- headed ascending branches. Stem and leaves thinly floccose-tomentose, leaves glabrescent below Lower leaves usually withered at flowering-time, lyrate or lanceolate; median and upper leaves linear-lanceolate to nearly filiform, median usually with 1–3 long subulate teeth in basal part, upper entire. Involucre 10–16 × 5–10 mm broad, oblong, funnel-shaped when fruiting; phyllaries multiseriate, greenish, innermost often violet-tinged, appendage a narrow light brown to blackish brown border divided into teeth 0.5–1 mm. Florets similar to those of *C. depressa* but marginal florets with narrower lobes. Achenes 3–4 mm; pappus 2–3 mm, shorter than achene.

HAB. & DISTRIB. Only found once in Iraq. **MRO**: Mergasur, damp ground at roadside near stream overflow, *Agnew, Hadač & Haines* s.n.!

Quite commonly cultivated in gardens as an ornamental in Iraq, e.g. **LCA**: Baghdad, *Guest*

186!; Zaffaraniya, *Guest & Rawi* 14551!; Zaffaraniya, *Shakir Sabir* 16663! **LSM**: Amara, *Alizzi & S. Omar* 34915!

No doubt the Mergasur plant was adventive.

As a weed in most of Europe, Siberia, C Asia, N America and elsewhere, in natural habitats in Italy, Greece, W Turkey, Caucasus.

Cornflower or Blue-bottle (Eng.). According to Wren (1950) it used to be considered as a medicinal plant, but is now only very occasionally used as a tonic and stimulant. It was formerly a very common weed in cereal fields in Britain; but crop spraying has largely eliminated it, and it is now rare.

DOUBTFUL RECORDS

Centaurea graeca *Boiss. & Spruner.* According to Anthony (1935: 291) this was collected at Ramadi. As *C. graeca* (=*C. attica* Nyman) is a plant from Greece not occurring in the Near East, this is certainly a misidentification: probably of *C. ammocyanus*, which is frequent in the western desert of Iraq and is superficially somewhat similar.

Centaurea spectabilis (*DC.*) *Sch.Bip.* The record by *Shehbaz, Khakani & Mayah* 1983 (Candollea 38: 352) "12 km from Sinjar to Kursi" is within the area of *C. handelii* and the specimen, which unfortunately was not available on loan, probably belongs to that species. *C. spectabilis*, which occurs in E Anatolia and Transcaucasia, is in the same section but can easily be distinguished by the leaves, which are not decurrent or only slightly so.

29. **DIPTEROCOME** Fisch. & C.A.Mey.

Ind. Sem. Horti Petrop. 1: 26 (1835)
Jeffrey in Kubitzki, Fam. Gen. Vasc. Pl. 8: 147 [2006] (2007)
Jaubertia Spach in Ill. Pl. Orient. 3: 131 (1849) non Guillemin in Ann. Sci. Nat. sér. 2, 16: 60 (1841)

Patricia Halliday

Dwarf, erect or spreading annual, much-branched from base, with simple, sessile, narrowly linear, somewhat fleshy entire basal and cauline leaves. Capitula small, heterogamous, campanulate, few-flowered, sessile, axillary and terminal; involucre cylindrical-campanulate, phyllaries 3–4-seriate, erect, imbricate. Receptacle flat, scaly. Florets all tubular, glabrous; ray-florets female, fertile, ± bilabiate, uniseriate; style of ray florets shortly bifid, arms compressed, spreading, spatulate; disc florets functionally male, regular, shortly 5-dentate; style of disc florets imperfect, subdivided, thickened, papillose; filaments fused, free for a short distance below anthers; anthers united, appendiculate, truncate at base, not caudate. Achenes of ray-florets cylindrical, straight to arcuate, spiny on dorsal surface, with 2 unequal, apical, recurved, flat, spiny horns, lower horn longest and ± equalling length of achene; pappus persistent, consisting of a few flattened bristles.

A monotypic genus, superficially resembling *Koelpinia.*

Dipterocome, from Gr. δίπτερω-, diptero-, two-winged, and κόμε, kome, mane or possibly κόρνα, korna, horn.

1. **Dipterocome pusilla** *Fisch. & C.A.Mey.*, Ind. Sem. Horti Petrop. 1: 26 (1835); Boissier, Fl. Orient. 3: 420 (1875); Handel-Mazzetti in Ann. Naturh. Mus. Wien 27: 437 (1913); Nábělek in Publ. Fac. Sci. Univ. Masaryk 52: 27 (1925); Dinsmore in Post, Fl. Syria, Palest. & Sinai ed. 2, 2: 74 (1933); Anthony in Notes Roy. Bot. Gard. Edinburgh 18: 291 (1935); Blakelock in Kew Bull. 4: 52 (1949); Zohary in Dep. Agr. Iraq Bull.. 31: 154 (1950); Rawi in Dep. Agr. Iraq Tech. Bull. 14: 114 (1964); Rechinger, Fl. Lowland Iraq: 635 (1964); Breckle, Hedge, Podlech & Rafiqpoor, Vasc. Pl. Afghan. checklist: 170 (2013).

Koelpinia sessilis Boiss., Diagn. Pl. Orient. ser. 1, 11: 34 (1849), sphalm. 'Kaelpinia'.
Jaubertia koelpinioides Spach in Jaub. & Spach, Ill. Pl. Orient. 3: 131, t. 189 (1849) non Guillemin.

Annual herb to 12(–20) cm tall, branched from base and dichotomously branched in upper part, at first arachnoid, later ± glabrescent. Leaves linear, opposite, or sometimes in

Fig. 29. **Dipterocome pusilla**. 1, habit × 2/3; 2, phyllary × 12; 3, receptacle scale × 12; 4, ray floret × 12; 5, style of ray floret × 21; 6, disc floret × 18; 7, anthers (opened) × 36; 8, style of disc floret × 36; 9, achene × 9. All from *Rawi* 12778. Drawn by Christine M. Grey-Wilson.

whorls of 3, 30(–56) × 2(–6) mm, subacute, scarcely narrowed to base, with a single vein. Capitula 2–3(–5) mm in diameter, with about 7 ray florets and 4 disc florets; phyllaries 3 × 1 mm, oblong-lanceolate, obtuse, with scarious margins in upper half, outer surface of upper half arachnoid; receptacle scales lanceolate, 2 × 0.75 mm, shortly narrowed below, with narrow scarious margins, glabrous. Ray florets 1.5 mm long, with notched upper lip;

disc florets 2–2.5 mm, less than 1 mm in diameter, narrowly funnel-shaped; anthers 1 mm long, united, appendiculate, base truncate, not caudate; filaments united into a tube, free for a short distance below anthers. Achenes of ray florets 6 × 1 mm; pappus of ray florets 1 mm long, of disc florets less than 0.5 mm. Fig. 29: 1–9.

HAB. Low hill slopes, muddy flats, plains, edges of cornfields, dry steppe on gypsaceous hills, on loamy or sandy soil; alt. 60–300 m; fl. & fr. Mar.-May.
DISTRIB. Occasional, locally frequent, in the southern foothills, the Lower Jazira district and the central part of the Lower Mesopotamian plain. **FKI**: SE of Tuz Khurmatu, *Rawi & Gillett* 10313! **FPF**: flats at the base of Jabal Hamrin, *Haines* W1171!; Pilkhana, 10 km E of Khanaqin, *Rawi* 12778!, 12844!; Qara Los, nr Mandali, *id.* 12633!; Tursak, nr Mandali, *Guest* 1739! **DLJ**: Raisha hill nr Shibachan, *Chakravarty, Rawi, Khatib & Alizzi* 32033! **LCA**: 65 km N of Falluja, *Rawi* 3024!

Palestine, Syria, Iran, Afghanistan.

Rechinger in Fl. Lowland Iraq 634 (1964)] observed that the achenes of this species may be occasionally dimorphic; I have seen no evidence of this in Iraqi material.

TRIBE **GYMNARRHENEAE** PANERO & V.A.FUNK

30. **GYMNARRHENA** Desf.

Mém. Mus. Hist. Nat. 4: 1, t. 1 (1818); Jeffrey in Kubitzki, Fam. Gen. Vasc. Pl. 8: 147 [2006] (2007)
Cryptadia Endl., Gen. Pl. Suppl. 1: 1381 (1841)

J. R. Edmondson

Dwarf glabrous annual, ephemeral, rosulate, acaulous, tap-rooted. Leaves in a basal whorl, subsessile. Inflorescences amphicarpic (some capitula buried, others aerial), disciform. Aerial capitula grouped into a glomerule, each with numerous mainly peripheral functionally female florets and a few central hermaphrodite, functionally male, florets; involucral bracts multiseriate, broadly ovate, receptacular scales leathery, apices whitish, cucullate. Aerial achenes with pappus of scabrid setae and 7–9 paleae; subterranean achenes epappose.

The genus is now sometimes recognized as belonging to the monogeneric subfamily, the Gymnarrheneae. One or two species, one doubtfully distinct species (*G. balansae* Coss. & Kralik) from N Africa and the second occurring in Iraq and across the whole of the Saharo-Sindian region.

Due to the distribution of male florets in the capitula, the reduced aerial synflorescences appear to contain several small groups of male florets each originating from a homogamous capitulum, and the single female florets derive from another. Subterranean achenes germinate underground; aerial achenes are wind-dispersed.

Koller, D. & Roth, N. (1964). Studies on the ecological and physiological significance of amphicarpy in *Gymnarrhena micrantha*. Amer. J. Bot. 51: 26–35.
E. Zamski, E., Ucko, O. & Koller, D. (1978). The mechanism of root contraction in *Gymnarrhena micrantha*, a desert plant. New Phytol. 95: 29–35.
V. A. Funk, V.A. & Fragman-Sapir, O. (2009). Gymnarrheneae (Gymnarrhenoideae) in Funk et al. (eds), Syst. Evol. Biogeog. Compositae: 327–332.

1. **Gymnarrhena micrantha** *Desf.*, Mem. Mus. Par. 4: 2, t. 1 (1818); de Candolle, Prodr. 5: 374 (1836); Rawi in Dep. Agr. Iraq Tech. Bull. 14: 116 (1964); Rechinger, Fl. Lowland Iraq: 601 (1964); Feinbrun-Dothan, Fl. Palaest. 3: 310 (1978); Rechinger, Fl. Iranica 145: 134 (1980); Chaudhary, Fl. Kingdom Saudi Arabia 2(3): 140 (2000); Boulos & Hind in Fl. Egypt 3: 201 (2002); Qaiser & Rubina in Fl. Pakistan 210: 93, f. 20 F–K (2003); Kilian & Ghazanfar in Fl. Oman 3: 278 (2015); Breckle, Hedge & Rafiqpoor, Vasc. Pl. Afghan: 174 (2013); Haloob & Al- Kaisi, Illustr. Fl. Lowland Iraq 1(1): 44 (2016); Taifour & El-Oqlah, Pl. Jordan Ann. Checkl.: 42 (2017).

Cryptadia euphratensis Lindl. in Chesney, Narr. Euphr. Exped. 441 (1868).

Rosette leaves linear to narrowly lanceolate, 30–60 × 10–15 mm, acute to attenuate, truncate and membranous at base. Inflorescence a cluster of subsessile capitula and with

Fig. 30. **Gymnarrhena micrantha**. 1, habit × 1; 2, capitulum × 6; 3, outer phyllary × 9; 4, marginal floret × 9; 5, styleof outer floret × 9; 6, inner floret × 9; 7, style of inner floret × 36; 8, anther × 36; 9, central floret displayed to show pappus × 18. 1 from *Chakravarty et al.* 31678; 2–9 from *Wheler Haines* W166. Drawn by Christine M. Grey-Wilson.

cleistogamous subterranean capitula. Aerial capitula several, grouped in centre of rosette, 5–8 mm in diameter with several whorls of chartaceous or reddish somewhat shiny phyllaries, heterogamous, apices spiny. Florets functionally staminate or pistillate, very shortly pedicellate, with 3- or 4- lobed corollas; anthers ecaudate and lacking an apical appendage; aerial achenes very small, 7–8 mm, ovoid, villous, pappus scaly or absent. Subterranean capitula homogamous, subtended by blackish phyllaries, remaining buried in fruit; florets all female; achenes flattened, large. Fig. 30: 1–9.

HAB. Sandy, loamy, alluvial and clay/gravel soil, haswa plains, sand dunes with gypsum; alt. 10–560 m; fl. Mar.-Apr.

DISTRIB. Widespread in the deserts of lowland Iraq, extending into the Lower Jazira and the eastern foothills. **FKI**: 25 km SE of Buksaya, Khazima police station, *Rawi & Haddad* 25637! **FPF**: 22 km S of Badra, *Rawi* 18424!; Koma Sang station, *Rawi* 12615!; 4 km N of Saadiya, *Qaisi* 42574!; Tib al-Sharhani, *Qaisi & Khayat* 50589!; N of Mansuria, *Barkley, Ani & Mati* 6843! **DLJ**: between Baghdad and Mosul, *Olivier & Bruguiere* (type); 15 km W by NW of Falluja, *Rawi* 30152!; Wadi Thirthar canal nr Samarra, *C. & E.R. Guest & Haider* 18333!; nr Rawa, *Rawi & Gillett* 7093!; 20 km W of Tikrit, *Alizzi & Husain* 33762!; nr Baj, 450 km from Baghdad, *Rawi & Hamada* 33579! **DWD**: Shithatha, *Rawi & Gillett* 6487!; 8 km E of Shithatha, *Rawi & Gillett* 6464!; 15 km N of Ramadi, *Rawi & Gillett* 6774!; 140 km W of Ramadi, *Rawi* 26836!; 50 km WSW of Karbala, *Rawi* 30813!; Nukhaib, Rawi 14794!; 70 km W of Nukhaib, *Rawi* 31034!; 35 km from Ramadi towards Rutba, *Jenan al-Mokhtar* 33381!; 125 km NE of Rutba, *Chakravarty, Rawi, Khatib & Allizzi* 31567!; 20 km E of Ana to Qaim, Omar, Qaisi, *Hamad & Hamid* 44419!; Wadi al-Ghadhaf, *Rawi* 14838!; 12 km from T1 to Husabah, *Chakravarty, Rawi, Khatib & Alizzi* 31740!; Rutba, *Rawi* 15621!; 3 km SW of Rutba, *Rawi* 21001!; 4 km SW of Rutba, *Rawi* 21009!; 135 km NW of Ramadi towards Rutba, *Rawi* 21281!; 100 km from Rutba to Ramadi, *Botany Staff* 41824!; H1, 160 km E of Rutba, *Chakravarty, Rawi, Khatib & Alizzi* 31587!; Wadi Al-Qaiser, *Omar, Qaisi, Hamad & Hamid* 44362!; 15 km W of Falluja, *Rawi* 30304!; bottom of Wadi Hawran, 12 km SE of H1, *Chakravarty, Rawi, Khatib & Alizzi* 31678!; 17 km W of Rutba, *K. Hamad* 48865! **DSD**: 50 m N by W of Aidhaha, *Guest, Rawi & Rechinger* 19185!; 30 km from Waksa to Aidaha, *Karim, Sharif, Hamad & Hamid* 47954!; 8 km N by W of Aidaha, *Guest, Rawi & Rechinger* 19159!; nr Aidaha, *Guest, Rawi & Rechinger* 19147! Zarqa, 12 km SE of Samawa, *Alkas* 17687!; 5 km SE by E of Zubair, *Guest, Rawi & Rechinger* 16805!; Zubair, *Rawi & Gillett* 6068!; 44 km S of Zubair, *Thamer* 47415!; Faidhat Um al Minaikher nr Salman, *Karim, Nasri, Hamid & Kadhim* 40208!; Wadi al-Khirr, 90 km NW by W of Shabicha, *Guest, Rawi & Rechinger* 19413!; 23 km SE by S of Zubair, *Guest, Rawi & Rechinger* 16855!; Safwan, *Botany Staff* 42437!; 6 km W of Safwan, *Guest, Rawi & Rechinger* 16942!; nr Jabal Sanam, ± 30 km W by S of Safwan, *Guest, Rawi & Rechinger* 17015!; around Jabal Sanam, *Qaisi, Hamad & Hamid* 48561! 48630!; Umm Qasr, *Khatib & Alizzi* 33446! *Alizzi & Omar* 35016!; nr Safai al-Maghif, NE of Ghazlani, *Guest, Rawi & Rechinger* 17255!; Al-Batin nr Jarishan, ± 65 km SW of Basra, *Guest, Rawi & Rechinger* 17103! 17075!; 33 km WNW of Ansab, by Saudi border, *Guest, Rawi & Rechinger* 19035!; 17 km from Shabicha to Salman, *Karim, Noori, Hamid & Kadhim* 40068!; 30 km NW of Ansab, *Qaisi, Hamad & Hamid* 48163!; Al-Samah, *Alizzi & Omar* 35728!; Jelaida well near Samah, *Karim, Hazim & Hamid* 38908! **LCA**: Falluja district, *Haines* W. 166!; between Baghdad and Falluja, *Omar & Widad* 47517!; Iskanderiya, *Agnew & Shawqi* 188!; 85 km W of Shaikh Sa'ad, *Rawi & Haddad* 25530! **LEA**: 45 km E of Tib, *Thamer* 47706! **LSM**: Fakha, 70 km E of Amara, *Rawi* 25796!

N Africa, Arabia, Palestine, Syria, S Iran, Afghanistan, Pakistan.

No vernacular names have been recorded on Iraqi specimen labels, but this species is known in Kuwait as 'AIN AL BUQR (camel's eye) according to Shuaib (1995).

TRIBE **CICHORIEAE** LAM. & DC.

31. **SCOLYMUS** L.

Sp. Pl.: 813 (1753); Gen. Pl. ed. 5: 355 (1754)
Lack in Kubitzki, Fam. Gen. Vasc. Pl. 8: 182 [2006] (2007).

Frances G. Davies

Annual to perennial spiny herbs. Leaves rigid, thistle-like, ± decurrent to form wings along stem. Capitula sessile, large, enclosed in and concealed by upper leaves; phyllaries spiny, rigid; receptacle paleaceous. Florets ligulate, exceeding phyllaries, corollas yellow, bearing one or more kinds of hairs. Achenes smooth, elliptical, 3–4-ribbed, chestnut-brown, separate from or intimately enclosed in paleae; pappus absent or of bristles and scales.

3 species in Europe, Macaronesia, Asia and N Africa; 1 species in Iraq.

Scolymus (from Gr. σκολημος, *skolimos*, a kind of thistle; the name *cactus* is originally derived from the Gr. for the artichoke, true cactuses being unknown in the ancient world despite featuring in 'classical' scenes by Lear, Alma-Tadema and other artists).

Ref. Vasquez, F. M. 2000. The genus *Scolymus* Tourn. ex L. (Asteraceae): taxonomy and distribution. Ann. Jard. Bot. Madrid 58(1): 83–100.

1. **Scolymus maculatus** *L.*, Sp. Pl.: 813 (1753); Boissier, Fl. Orient. 3: 713 (1875); Nábělek in Publ. Fac. Sci. Univ. Masaryk 52: 47 (1925); Dinsmore in Post, Fl. Syria, Palest. & Sinai ed. 2, 2: 123 (1933); Guest in Dep. Agr. Iraq Bull. 27: 89 (1933); Anthony in Notes Roy. Bot. Gard. Edinburgh 18: 292 (1935); Blakelock in Kew Bull. 4: 62 (1949); Zohary in Dep. Agr. Iraq Bull. 31: 157 (1954); Rawi, ibid. Tech. Bull. 14: 123 (1964); Matthews in Fl. Turkey 5: 624 (1975); P.D. Sell in Fl. Europ. 4: 304 (1976); Rechinger, Fl. Iranica 122: 5 (1977); Feinbrun-Dothan, Fl. Palaest. 3: 407 (1978); Boulos & Hind in Fl. Egypt 3: 269 (2002); Taifour & El-Oqlah, Pl. Jordan Ann. Checkl.: 47 (2017).

Annual herb to 50 cm tall. Stem scarcely branched, white, continuously winged except towards base. Leaves sessile, decurrent as broad leafy wings below free part oblanceolate to oblong, 4–20 × 3–7 cm, lobed and dentate, very rigidly spiny with all lobe and teeth tips hardened on lamina and on stem wings, lamina sparsely and shortly pubescent, lobes and wings with white marginal veins. Capitula borne towards apex of stem, solitary or in clusters of 2–3, surrounded by an involucre-like ring of regularly 7-pectinate spiny leaves. Paleae 3 × 2 mm, oblong, transparent, completely enclosing achene. Florets 15–17 mm long, both tube and ray with short white and black hairs[11]. Pappus absent. Fig. 31: 1–5.

HAB. Waste ground, sandy soils, fallow fields; 90–1000; fl. Apr.-Sep.
DISTRIB. Occasional in the mountains and foothills of Iraq. **MAM**: Sarsang, Anders 15. **MSU**: Susa, *F. Karim* 39280!; sine loc. in Rawi l.c. **FNI**: Rewens village between Zakho and Jasira ibn Umar (Cizre), *Nábělek* 3981; **FKI/FPF**: Jabal Hamrin, in Anthony, l.c. **LCA**: Daltawa (Khalis), Blakelock in Kew Bull. 4: 62 ex *Guest* 2478!; Hilla, *Graham* 55. **LEA**: Muqdadiya, *Rechinger* 9708.

Guest (1933) explained that the young plant is eaten by people and also grazed by animals; the plant is usually called KASSŪB (a general name for various thistly plants) or UMM AL-HALAIB "mother (i.e. provider) of milk" (a general name for plants which yield milky latex).

Mediterranean Europe (E to Crimea), Cyprus, Syria, Palestine, Turkey, Iran, Macaronesia, N Africa.

32. **CICHORIUM** L.

Sp. Pl.: 813 (1753); Gen. Pl. ed. 5: 825 (1754)
Lack in Kubitzki, Fam. Gen. Vasc. Pl. 8: 182 [2006] (2007)

C. C. Townsend

Annual, biennial or perennial herbs with widely divaricate branches and pedicels. Leaves rosulate and cauline, simple and denticulate to runcinatte-pinnatifid and dentate, glabrous to setose or multicellular, and sometimes glandular-hairy. Capitula homogamous. Involucre cylindrical, phyllaries biseriate, outer series somewhat to considerably broader and ± indurate at base in fruit, variable in length in relation to the inner; receptacle flat, naked. Corollas bright sky-blue, rarely white. Achenes glabrous, obovoid to obpyramidal, ± 5–angled, truncate at apex, pappus of 1–2 rows of short obtuse scales, rarely bristly.

The basally indurate phyllaries apparently serve to retain the achenes, clasping them until softened by rain.

Six species in Europe, the Mediterranean region, SW Asia and NE tropical Africa; 2 species in Iraq.

[11] The black hairs are unique among the Compositae of Iraq, being composed of biseriate, longitudinally multicellular structures with oblique walls.

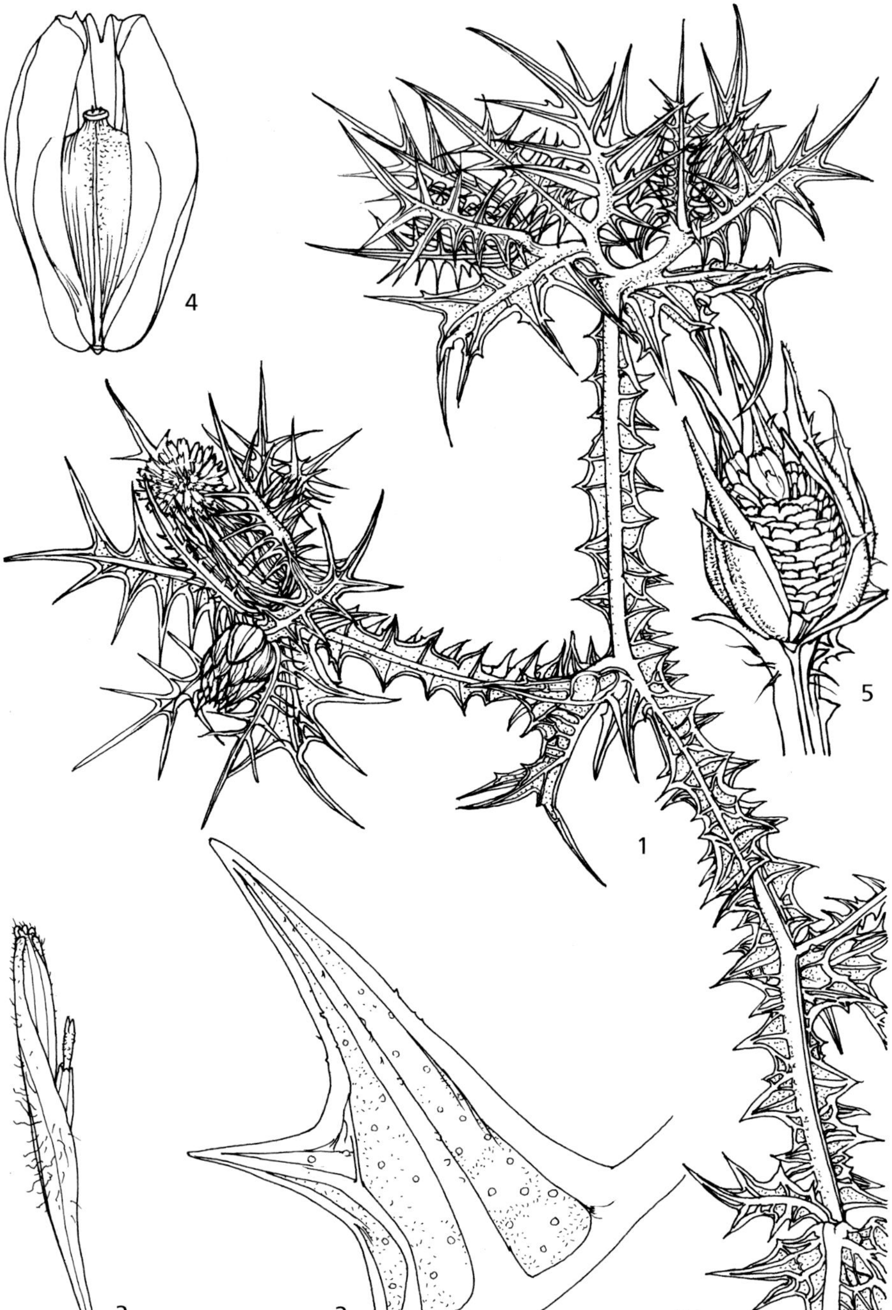

Fig. 31. **Scolymus maculatus**. 1, habit × 2/3; 2, section of leaf × 6; 3, floret × 6; achene × 12; 5, seed head × 3. Drawn by M. Millar Watt.

Cichorium (from a latinized version of an old Ar. name for the plant, possibly cognate with the Kurdish CHAQCHAQA (q.v.), now known in Arabic as هندباء HINDIBA).

A.M. Kiers. (2000). Endive, Chicory, and their wild relatives. A systematic and phylogenetic study of *Cichorium* (Asteraceae). Gorteria, Suppl. 5: 1–77.

Outer phyllaries considerably shorter than inner, forming a distinct calyculus; plants perennial . . . . .1. *C. intybus*
Outer phyllaries subequalling inner, not forming a distinct calyculus; plant annual . . . . 2. *C. pumilum*

1. **Cichorium intybus** *L.*, Sp. Pl.: 813 (1753); Handel-Mazzetti in Ann. Naturh. Mus. Wien 27: 432 (1913); Dinsmore in Post, Fl. Syria, ed. 2, 2 : 124 (1933); Guest in Dep. Agr. Iraq Bull. 27: 22 (1933); Bornmüller in Beih. Bot. Centralbl. 60B: 216 (1940); Rawi in Dep. Agr. Iraq Tech. Bull. 14: 111 (1964); Rawi & Chakraverty, ibid. 15: 26 (1964); Rechinger, Fl. Lowland Iraq: 663 (1964); Tsvelev in Fl. S.S.S.R. 29: 16 (1964); Matthews in Fl. Turkey 5: 627 (1975); Lack in Fl. Iranica 122: 6 (1977); Collenette, Wild Flow. Saudi Arabia: 180 (1999); Kilian & Ghazanfar in Fl. Oman 3: 220 (2015); Abdulridha, Taha & Wedad, Ecology and Fl. Basra: 203 (2016); Taifour & El-Oqlah, Pl. Jordan Ann. Checkl.: 39 (2017).

Perennial herb with a stout, tough taproot, (20–)30–120 cm tall, branched from near base upwards, branches widely divaricate. Basal leaves rosulate, usually ± withered at flowering time, oblanceolate in outline, commonly runcinate-pinnatifid with wide rounded sinuses and very acute lobes and apex, sometimes entire and denticulate, mostly 10–30 × 2.5–5 cm; cauline leaves rapidly reducing and becoming less divided, uppermost small and bract-like, amplexicaul; all leaves rather roughly whitish-hairy, rarely subglabrous, indumentum of stem and branches similar but often sparser. Capitula 2.5–4 cm in diameter, sessile, in small axillary or lateral clusters or solitary on tumid pedicels 2–10 cm. Outer phyllaries ovate to oblong-ovate, often ± spreading, 5–7 mm; inner phyllaries lanceolate, about twice as long as outer, glabrous or pilose above with bristly hairs, or with soft hairs about apex. Corollas bright sky-blue, rarely white, withering very rapidly on picking. Achenes 2–3 mm, frequently dark-mottled; pappus only about a tenth to an eighth as long.

HAB. Disturbed areas, along roadsides, dry slopes and pastures; alt. 1100–1700(–2000) m; fl. & fr. (May-) Jun.-Aug. (-Oct.).
DISTRIB. Quite common in the mountains, occasional in the steppe region, also common in the irrigated desert region. **MJS/FUJ**: between Tal Afar and Balad Sinjar, *Field & Lazar* 517! **MAM**: Zawita, *E. Chapman* 26433!, *Guest* 3718!; Sharanish, *Rawi, Noori & Tikriti* 29070!; nr Dinarta, *E. Chapman* 26142! **MRO**: Gali Ali Beg (Rowanduz gorge), *Gillett* 9422!; by Bekhal falls, nr Rowanduz, *Guest & Husham (Alizzi)* 15913!; Chiya-i Mandan, *Guest* 2746!; Qura Shina on the lower slopes of Qandil range, *Rawi & Serhang* 23822!; Kudu, nr Haji Umran, NE of Rania, *Rawi* 9173! **MSU**: Jarmo, *Helbaek* 1716!, 1762!; 13 km SW of Sulaimaniya, *Rawi* 9173!; Hauraman (Avroman) mts., *Rawi, Chakraverty, Alizzi & Noori* 19816! **FNI**: by r. Zab, nr Eski Kellek, *Gillett* 8202!; Mindan, *Rawi* 11353!, *E. Chapman* 26237! **FKI**: 10 km N of Kirkuk, *Juma' Brahim* 6131! **FPF**: E of Badra, *Haddad* 9965! **LCA**: Baghdad, *Janan al-Mukhtar* 34472!; nr Baghdad, *Omar & Sahira* 34845!; Abu Ghraib, *Rawi* 30736!, *Alizzi* 33183!; Rustamiya, *Lazar* 1171!; Suwaira, *Muhammad Shabul* 35774!; 10 km N of Diwaniya, *Bharucha, Rawi & Tikriti* 29325! **LEA**: 5 km NW of Ba'quba, *Alkas* 18552!; Ba'quba, *Omar, Janan & Abdulla* 37601!; Jabal Hamrin, *Sutherland* 88. **LSM**: 30 km E of Amara, Bharucha, *Rawi & Tikriti* 29303!

Most of Europe, Turkey, Lebanon, Saudi Arabia, Iran, Caucasia, Afghanistan, Pakistan, Siberia, C Asia (Turkmenia to Sinkiang), N Africa (Morocco, Algeria, Libya). Naturalised in many other regions of the world.

2. **Cichorium pumilum** *Jacq.*, Obs. Bot. 4: 3, t 80 (1771); Boissier, Fl. Orient. 3: 216 (1875); Nábělek in Publ. Fac. Sci. Univ. Masaryk 35: 217 (1928); Bornmüller in Beih. Bot. Centralbl. 60B: 216 (1940); Blakelock in Kew Bull. 4: 50 (1949); Zohary in Dep. Agr. Iraq Bull. 31: 157 (1950); Rawi, ibid. Tech. Bull. 14: 111 (1964); Rechinger, Fl. Lowland Iraq: 663 (1964); Tsvelev in Fl. S.S.S.R. 29: 18 (1964); Matthews in Fl. Turkey 5: 627 (1975); Lack in Fl. Iranica 122: 8 (1977); Feinbrun-Dothan, Fl. Palaest. 3: 408 (1978).

C. divaricatum Schousb., Vext. Marok.: 197 (1800).
[*C. glandulosum* (non Boiss. & Huet) Hand.-Mazz. in Ann. Naturh. Mus. Wien 27: 452 (1913); Guest in Dep. Agr. Iraq Bull. 27: 22 (1933); Zohary, ibid. 31: 175 (1950); Rawi, ibid. Tech. Bull. 14: 111 (1964)].

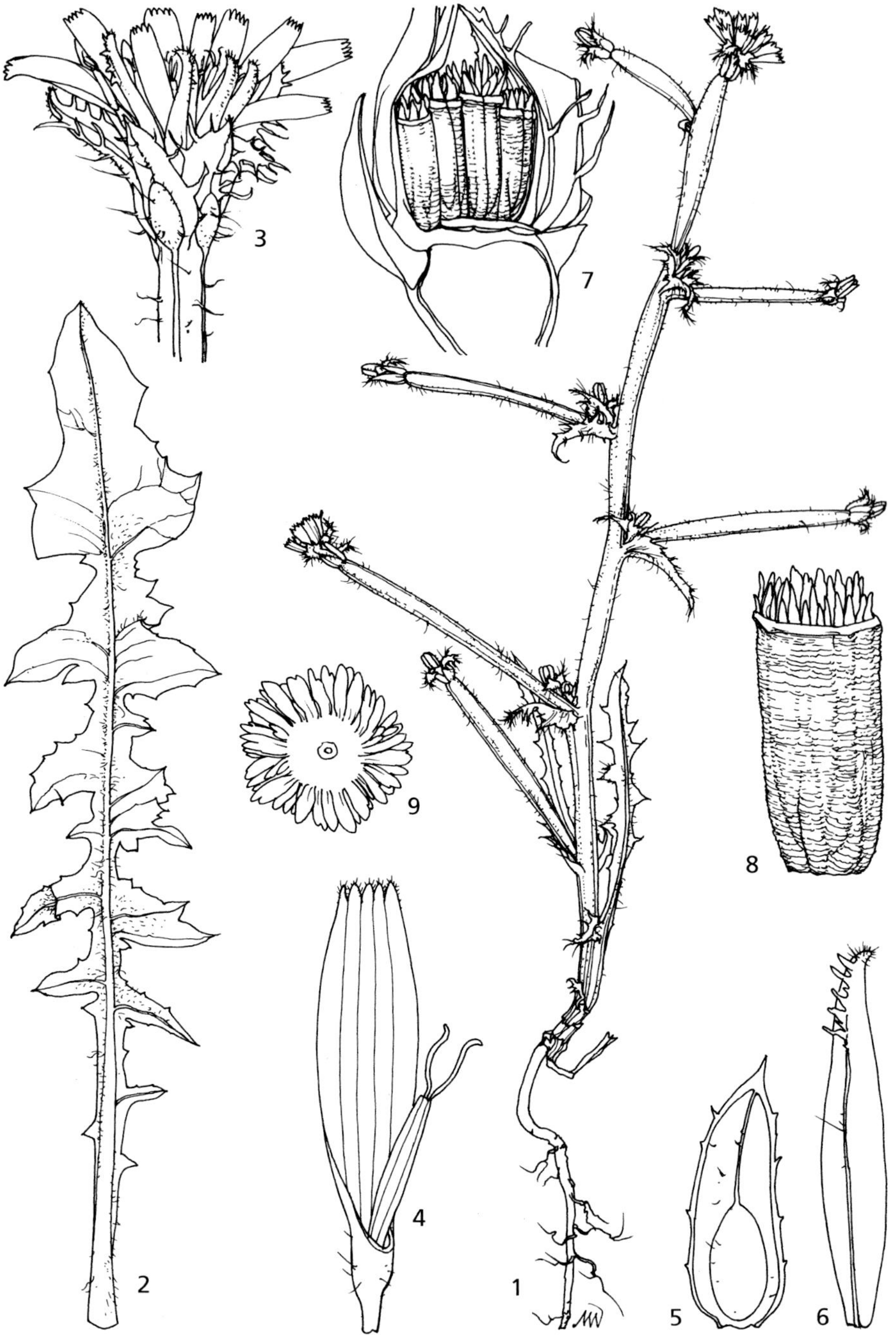

Fig. 32. **Cichorium pumilum**. 1, habit × 1; 2, leaf × 1; 3, capitulum × 3; 4, floret × 4; 5, phyllary × 6; 6, phyllary × 6; 7, capitulum with mature achenes × 6; 8, achene × 18; 9, pappus × 18. All from *Sufer* 5530. Drawn by M. Millar Watt.

Annual herb, 5–55 cm, with a slender wiry taproot, simple with divaricate pedicels only or branched from base upwards, branches also divaricate. Basal leaves rosulate, present at flowering or not, oblanceolate in outline, runcinate-pinnatifid with wide sinuses to entire and denticulate, mostly 5–14 × 1.5–3 cm; cauline leaves rapidly reducing, narrower and less divided, uppermost small and bract-like, all amplexicaul; all leaves subglabrous to rather roughly whitish-hairy, margins often stiffly shortly ciliate, indumentum of stem and branches similar or sparser. Capitula 2–3 cm in diameter, in axillary or lateral clusters or solitary on tumid pedicels 2–12 cm long; outer phyllaries lanceolate, 7–14 mm, subequalling inner, glabrous to pilose with rather stiff hairs, inner sometimes softly hairy at tip. Corollas bright sky-blue, soon withering. Achenes densely congested, frequently dark-mottled, obpyramidal, 2–3 mm; pappus ¼ as long or in inner achenes sometimes longer and bristle-like. Fig. 32: 1–9.

HAB. In fields and meadows in the hills and on the plains; alt. up to ± 1000 m; fl & fr. May-Sep.
DISTRIB. Quite common in the lower mountains, upper plains and foothills of Iraq, occasional on the northern alluvial plain. **MJS**: Asi, *Rawi* 8462! **MAM**: 5 km S of Zakho, *Rawi* 23107!; Ba'dhara, nr Shaikhan, *Salim* 2599!; Sersang, *E. Chapman* 26440!; 10 km S of Aqra on the road to Mosul, *id.* 26165!; **MRO**: Khalana, *Rawi* 13771!; Koi Sanjaq, *Rawi, Noori & Kass* 28116! **MSU**: 20–25 km E of Qaranjir, *Rawi* 21666!; Jarmo, *Helbaek* 2006!; Saruchan (Saraw Khan?), on road to Halabja, *anon.* 4348!; Avroman, *Gillett* 11760! **FUJ**: Telegraph pole M. 90 on Mosul to Baiji road, *Field & Lazar* 397!; Shargil, *Handel-Mazzettii* 1056! **FNI**: Tal Kaif, *Mudir of Tal Kaif* 3175!; Mindan bridge, *E. Chapman* 26221! **FKI**: 10 km N of Kirkuk, *Rawi* 21575!; 31 km N of Kirkuk on road to Koi Sanjak, *Rawi, Noori & Kass* 28010!; Jabal Naft, nr Kirkuk, *Nábělek* 3752!; Kirkuk province, *Ali al-Haidari* 3945!; **FPF**: Mandali, *Guest* 783!; Badra, *N Sufer* 5530! **DLJ**: 20 km N of lake Thirthar, *Barclay & Agnew* 6016B! **LCA**: Khalis (Daltawa), *Guest* 2464!, *Darwish al-Haidari* 2464A!; Baghdad, *Colvill* s.n.; Rustamiya, *Guest* 268A!; 20 km W of Shaikh Sa'ad, *Rawi & Haddad* 25517!; among ruins of Babylon, *Reuter* in *Nábělek* 3753! **LEA**: 20 km E of Baghdad, *Alizzi & Omar* 34882!; W10 (area near Ba'quba), *Hunting Technical Services* 6!; Aziziya, *Guest* 0257!; on right bank of r. Tigris nr Kut, *Morck* 28.

Southern Europe from Portugal to European Turkey, Syria, Lebanon, Palestine, Turkey, Iran, Transcaucasia, N Africa (Libya to Morocco), Macaronesia.
TALA SHÎR, Shaikhan, (Kurd.), *Salim* 2599!; CHAQCHAQA, Kirkuk (Kurd.), *Ali Haidari* 3945: "a useful grazing plant which is also mixed with an unguent and applied to skin diseases".

See Kiers (2000, p. 66) for a more extensive synonymy. She has suggested that *C. pumilum* is 'representative of the ancestor of *C. endivia*'. *C. endivia* L. is the cultivated Endive, which closely resembles *C. intybus* but has the leaves glabrous and dentate to pinnatifid rather than runcinate-lobed; the upper bract-like leaves are commonly larger and exceed the capitula. Some recent authors have regarded *C. pumilum* as a subspecies of *C. endivia*. This is a reasonable proposition, but the relationship between these two and *C. intybus* needs thorough biological investigation.

C. glandulosum Boiss. & Huet was described from two syntypes, one of which was Kotschy's "Pl. Mesopot., Kurdistan & Mossul" no. 138. This was, however, gathered at Diyarbakır in Turkey and not in Iraq. As indicated above, Handel-Mazzetti's record of this species in Iraq is based on misidentification of material of *C. pumilum*. *C. glandulosum* is distinguished by a rather dense indumentum of soft, multicellular, gland-tipped hairs on pedicels, phyllaries and often also branches and leaves. Such hairs do not occur on the pedicels and phyllaries of *C. pumilum*.

33. **CATANANCHE** L.

Sp. Pl.: 812 (1753); Gen. Pl.: 354 (1754)
Lack in Kubitzki, Fam. Gen. Vasc. Pl. 8: 183 [2006] (2007).

Patricia Halliday

Erect annual or perennial herbs. Leaves mostly basal. Capitula homogamous, ligulate, many-flowered, 1–5 per inflorescence but usually solitary, terminal. Involucre cylindrical-campanulate; phyllaries multiseriate, imbricate, scarious; receptacle flat, pitted, bristly. Rays blue or yellow. Achenes turbinate, 5-angled, 5–10-ribbed; pappus persistent, of 5–7, long-aristate scales.

A genus of 5 species, mainly Mediterranean in distribution. One species is cultivated in Europe and America, and perhaps occasionally in Iraq.

Catananche (from Gr. καταναγκε, *catananke*, the name used by Dioscorides for a species of that genus).

1. **Catananche caerulea** *L.*, Sp. Pl.: 812 (1753); Ba'li in Gard. & Fls. 1: 399 (1946); Sell in Fl. Europ. 4: 305 (1976); Husain & Kasim, Cult. Pl. Iraq: 110 (1975).

Erect, simple or branched, appressed-hairy perennial herb, 22–76 cm tall; stems striate. Leaves mostly basal, linear-lanceolate, (2–)5–20(–32) × 0.2–1.5(–3.5) cm, entire or remotely toothed, teeth up to 2.5 × 0.2–0.5 cm; base attenuate, petiolate; apex acute; main veins (1–)3; cauline leaves few, sessile, diminishing in size upwards. Inflorescence simple or more usually branched. Capitula 4–5 cm in diameter, solitary at branch ends, on ± bracteate pedicels up to 30(–52) cm long; bracts lanceolate, aristate, scarious with dark midrib and arista. Phyllaries silvery, shining, with dark midrib excurrent in a cusp or arista; outer ovate, 6 × 4 mm, with thickened midrib and base; inner oblong-obovate, 1.2–1.5 × 1–2 mm at half-way point, with scarious apical 'wing' and thickened base. Florets hermaphrodite; rays blue, often purple at base, strap-shaped, densely long-hairy dorsally, 18 × 4–7 mm; tube 3 mm long; apex truncate with 3–4(–5) teeth; anthers 4.5 mm long, sagittate with ovate-semicircular apical appendage; style shortly bifid, column slender, pubescent. Achenes 4 × 1–2 mm, densely stiffly long-hairy; pappus of 5–6(–7) aristate scales, 8–9 × 1 mm at base; aristae barbellate.

HAB. Said to be occasionally cultivated in gardens in Iraq by Ba'li (1946) and Husain & Kasim (1975).
DISTRIB. SW Europe (Spain, Balearic Is., France, Italy).

Said to be known as Cupidone or Cupid's Dart because of its alleged aphrodisiac properties. This plant is mentioned by Bailey (1939) who says it was used in love philtres or potions and was often preserved as an "everlasting flower"; it is also mentioned by Chittenden (1951) and Marshall Cavendish (1968) as a garden plant.

A different species, *C. lutea* L., occurs in Spain, Italy, Greece, N Africa, Palestine, Lebanon and Turkey, but not in Iraq. Dinsmore (1933) states that it is known in Palestine as DUWANAT AL-JIDI, straw-flower.

34. **CHONDRILLA** L.

Sp. Pl.: 796 (1753); Gen. Pl. ed. 5: 815 (1754)
Lack in Kubitzki, Fam. Gen. Vasc. Pl. 8: 184 [2006] (2007).

C. C. Townsend

Biennial or perennial herbs with branched, green, twiggy stems which are usually leafless by fruiting stage. Leaves various, coriaceous. Capitula cylindrical, few-flowered, inner phyllaries uniseriate and outer phyllaries in several imbricate, much shorter series. Corollas yellow. Achenes homomorphic, terete, body is sharply distinct from very long beak, ribbed, muricate above, uppermost processes usually scaly; pappus of simple, white, very minutely scabrous bristles, detaching with beak from achene body

Chondrilla (from the Gr. χονδρος, chondros, cartilage (Jackson has "grain"); *-illa* is a diminutive suffix.

1. **Chondrilla juncea** *L.*, Sp. Pl.: 796 (1753); Rawi in Dep. Agr. Iraq Tech. Bull. 14: 110 (1964); Mathews in Fl. Turkey 5: 812 (1975); Feinbrun-Dothan, Fl. Palaest. 3: 428 (1978); Mouterde, Nouv. Fl. Liban et Syrie 3: 527 (1983); Taifour & El-Oqlah, Pl. Jordan Ann. Checkl.: 39 (2017).

A tap-rooted perennial herb about 1 m tall with many rush-like branches. Stems glaucous, with a waxy bloom, finely striate, often with rigid downward-pointing brown bristles near base. Basal leaves in a short-lived rosette, oblanceolate, 70 × 13 mm, runcinate-pinnatisect. Cauline leaves sessile, narrowly elliptic to linear, entire or denticulate. Capitula numerous, in groups of 1–3(–5) on short naked stalks in leaf axils, ± 10-flowered. Involucre 8–12 mm long, cylindrical, spreading distally; outer phyllaries numerous, minute; inner phyllaries 8,

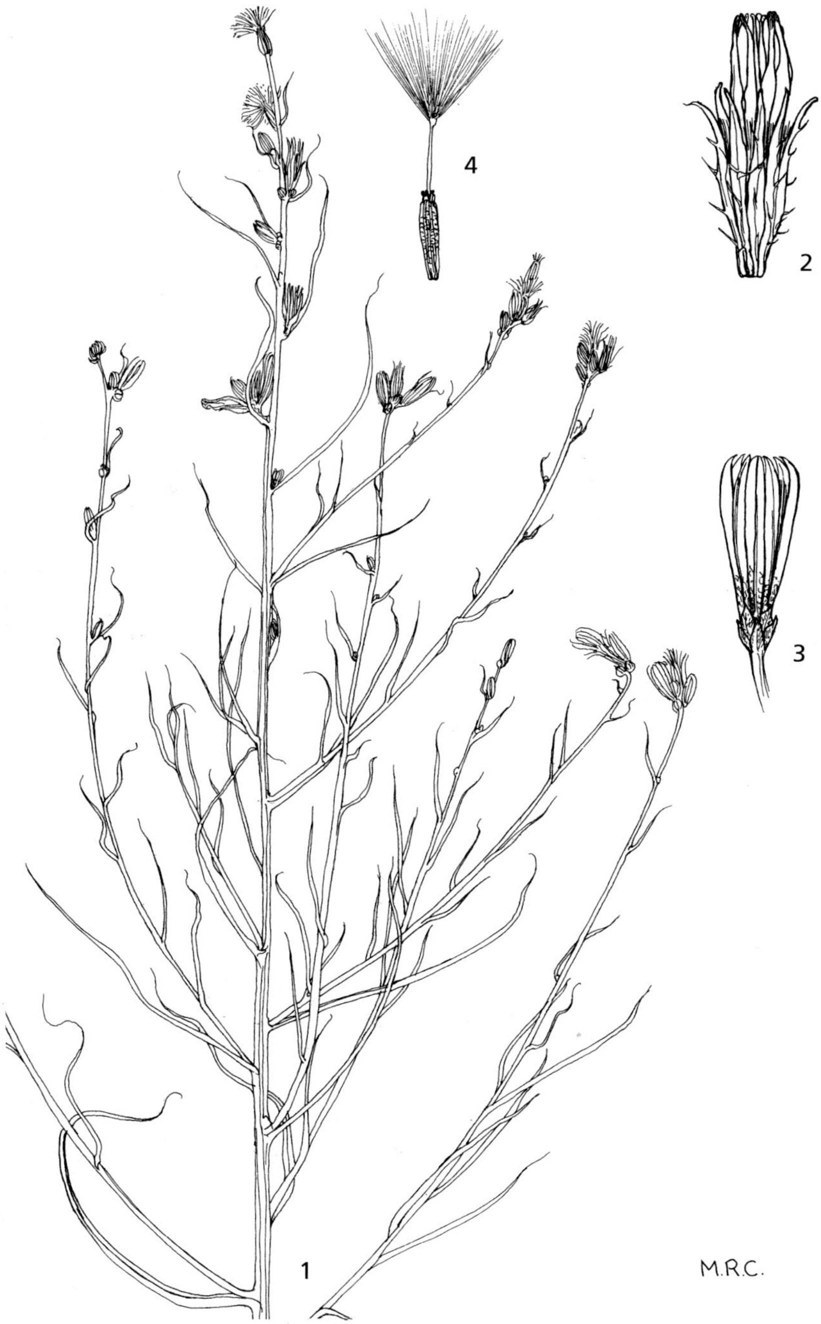

Fig. 33. **Chondrilla juncea**. 1, habit × ½; 2, capitulum × 4; 3, outer & inner phyllaries × 6; 4, achene with pappus × 4. 1, 3 from *Furse* 3115 (Iran material); 2, 4 from *Botany Staff* 43855. Drawn by Moira Rae Carter.

linear, tapering from 1 mm wide at base to a fine point, farinose-pubescent on each side of prominent midrib; phyllaries spreading untidily after achene dispersal; receptacle naked. Achenes strongly muricate and scaly at apex, body 3–4 mm, beak 5–8 mm; pappus ± 7 mm, silky, whitish. Fig. 33: 1–4.

HAB. On rocky mountainsides and stony river beds; alt. (600–)1100–1800 (–2000) m; fl. & fr. Jul.-Sep.
DISTRIB. Scattered but rather local in the mountain region of Iraq. **MAM**: Sharanish, 25 km NE of Zakho, *Rawi, Tikriti & Noori* 28987!, 29037!; Chiya-i Gara (Gara Dagh), *Kotschy* 384!; Atrush, *Guest* 3661!; Sersang, *Haines* W. 414! **MRO**: Algurd Dagh, *Gillett* 12951!; Rowanduz gorge nr Khalirfan, *Guest* 15937!; Rayat, *id.* 13067!; Haji Umran, *Guest & Husham (Alizzi)* 15895!; between Haji Umran and Rayat, *Gillett* 9474!; between Pushtashan and Surade, *Rawi & Serhang* 26521! **MSU**: S of Shu'aiba, *Qaisi & Khayat* 51136!; Mt. Hauraman (Avroman), *Rawi, Chakravarty, Noori & Alizzi* 19813!

C & S Europe (N France to SC Russia, Mediterranean region from Portugal to Turkey), Cyprus, Palestine, Syria, Lebanon, Iran, Turkmenia, Afghanistan, N Africa (Morocco, Algeria, Tunisia).

35. **CREPIS L.**

Sp. Pl.: 805 (1753); Gen. Pl. ed. 5: 350 (1754)
Lack in Kubitzki, Fam. Gen. Vasc. Pl. 8: 136 [2006] (2007)

Frances G. Davies & C. C. Townsend

Annual, biennial or perennial herbs, scapose or with leafy stems, erect or ascending, glabrous or with a variable indumentum on vegetative parts and phyllaries ranging from canescent or tomentose to setose, or of simple or glandular hairs. Leaves simple and toothed to pinnately, sometimes runcinately, lobed. Capitula solitary to many; phyllaries bi- or multiseriate, outer shorter than inner and sometimes distinct in texture and attitude, inner indurate in fruit in some species, developing corky thickenings of dorsal surface particularly along midrib; receptacle naked or short-ciliate or with long setae or paleae, these remaining after achenes have fallen. Florets few to numerous, tube hairy, ray glabrous; styles yellow or greenish-black; corollas yellow (in Iraq). Achenes all uniform or dimorphic, fusiform, beaked or attenuate or truncate, longitudinally ribbed, outer occasionally much modified and enclosed by inner phyllaries; pappus normally white and very fine, setae minutely barbellate, coherent at base, exserted from involucre at maturity or included, persistent or caducous.

A large genus with over 200 species worldwide of which 11 occur in Iraq.

Crepis (from Gr. κρεπις, *krepis*, a boot, foundation or pedestal; used by Theophrastus for a plant which is believed to be *Picris echioides*).

Babcock, E. B. 1947. The genus *Crepis*. Univ. Calif. Publ. Bot. 21-22: 1–197 (Part 1 – The taxonomy, phylogeny, distribution and evolution of *Crepis*); 199–1030 (Part 2 – Systematic treatment).
- (1949). Supplementary Notes on Crepis, 1. Mem. Soc. Hist. Nat. Afrique Nord 20 : 9–19.
- (1949). Evolution 3: 374–376.
- (1950). Evolution. 4: 358–359.
- (1951). Univ. Calif. Publ. Bot. 23: 383–404 (1951).
- (1951). Kew Bull. 6: 411–417.
- (1952). Univ. Calif. Publ. Bot. 26 : 213–222.

1. Receptacle with long setae or paleae between the florets, exceeding fertile part of achene 2
   Receptacle glabrous or short-ciliate 5
2. Achenes strongly dimorphic, marginal tapering above with no distinct beak, inner with a distinct, long, slender beak; receptacle paleae narrow but flattened, simple or forked above . . . . 6. *C. foetida* subsp. *commutata*
   Achenes not dimorphic, inner with no long, slender beak; paleae long, not or scarcely broader than pappus, rarely forked above 3
3. Plant annual with a slender taproot, crown not clothed with scaly remains of previous years' leaf-bases; plant of lower ground and foothills . . . . 9. *C. sancta*
   Plant perennial, tough rootstock crowned with scaly remains of previous years' leaf-bases; high alpines, in Iraq not recorded below 2900 m 4

4. Leaves glabrous or with sparse, white, simple hairs, eglandular; plant 20–50 cm; achenes 6–7 mm. . . . . 2. *C. sahendi*
Leaves with a dense indumentum of simple and (in Iraq) glandular hairs; plant 6–25 cm; achenes 5–5.5 mm . . . . 3. *C. elbrusensis*
5. Outer phyllaries clearly demarcated from inner, broadly ovate or elliptic, membranous, finally ± reflexed, longest mostly c. 9–12 mm. . . . . 5. *C. alpina*
Outer phyllaries not thus clearly demarcated; if broad then much shorter 6
6. Achenes strongly dimorphic, marginal not beaked, inner with a long slender beak clearly marked off from fertile part; pappus ± wholly exserted from involucre at maturity 7
Achenes not or weakly dimorphic, inner lacking long, slender beaks; pappus not or slightly exserted at maturity 9
7. Short outer phyllaries ovate, membranous with a brown midrib; stem clothed with long, rigid, divergent hairs . . . . 11. *C. aspera*
Outer phyllaries linear to lanceolate, not membranous with a brown midrib; hairs, if present, finer and less rigid 8
8. Receptacle ciliate; florets considerably exceeding the involucre; outer phyllaries not deciduous . . . . .6. *C. foetida* subsp. *foetida*
Receptacle glabrous; florets not considerably exceeding the involucre; outer phyllaries usually deciduous . . . . .7. *C. kotschyana*
9. Involucre at maturity 3.5–6 mm long; achenes 1.5–2 mm . . . . 10. *C. micrantha*
Involucre at least 7 mm long; achenes 3–6 mm. 10
10. Inner phyllaries densely white-tomentose and shortly glandular-pilose, never setose . . . . 1. *C. kurdica*
Inner phyllaries ± white-tomentose and long-setose, never glandular. . . 8. *C. quercifolia*

1. **Crepis kurdica** *Rech.f.* in Anz. Öst. Akad. Wiss., Math.-Nat. Kl. 109: 174 (1972); Rechinger, Fl. Iranica 122: 305 (1977).– Type. Iraq, Kurdistan, Mt Qandil, supre Goam-e-Kirmosoran, 3200 m, *Rechinger* 11128 (W, holo.).

Perennial herb 15–40 cm tall from a somewhat woody rootstock, stem base sparsely clothed with brownish, membranous remains of previous years' leaf-bases. Stems green, little-branched, glabrous, striate, branches slender, divaricate. Basal leaves in a rosette, shortly petiolate below, narrowly oblanceolate, 5–16 × 1–2.5 cm, pinnately lobed with runcinate or divergent lobes, lobes lanceolate to oblong, terminal largest, acute or obtuse, glabrous, midrib prominently white. Cauline leaves similar, smaller, acutely hastate, lower with greater dissection of lamina at least towards base; upper leaves smaller, ± entire. Capitula few, many-flowered, pedicels not incrassate but shortly glandular and tomentose at top. Involucre 11–13 × 9–10 mm, campanulate, rounded at base; outer phyllaries dark green, 2.5–5 mm, linear; inner phyllaries lanceolate, narrowly white-margined; all phyllaries with dense shortly stalked glands and white-tomentose hairs, tips bearing tufts of white hairs. Receptacle shortly ciliate. Corollas yellow, 13–20 mm; styles yellow. Mature achenes not seen, young achenes elliptic-oblong, 5 mm, glabrous, all similar; pappus 6–7 mm.

HAB. Rocky limestone mountain tops; alt. 1800–3200 m; fl. Jul.-Aug.
DISTRIB. Rare in the mountain region of Iraq. **MRO**: Germasur lake, *Rechinger* 11128!, *Rawi & Serhang* 24123! **MSU**: E of Qala Diza, *Thesiger* 1153!

Endemic.

2. **Crepis sahendii** *Boiss. & Buhse* in Aufz. Reise Transkauk. Pers.: 141 (1860); Boissier, Fl. Orient. 3: 838 (1875); Babcock in Univ. Calif. Publ. Bot. 22: 620 (1947); Rechinger, Fl. Iranica 122: 313 (1977); Lamond in Fl. Turkey 5: 826 (1977).

C. huetii Boissier, Fl. Orient. 3: 838 (1875).
Lagoseris sahendii (Boiss. & Buhse) Cherepanov in Fl. S.S.S.R. 29: 707 (1964).

Perennial herb 20–50 cm tall with a woody rootstock crowned with the scaly remains of previous years' leaf-bases and with 1-several slender, glabrous or very shortly and sparingly white-pilose stems with a few long and slender branches. Leaves in a rosette, petiolate at base, oblanceolate, 3–12(–17) × 0.8–3.5 cm, narrowly toothed to deeply runcinate-pinnatifid, acute, glabrous or with a few white simple hairs, midrib prominently white; cauline leaves bract-like, subtending the branches. Pedicels slender, not thickened beneath many-flowered capitula. Involucre cylindrical-campanulate, 10–12 × 5–6 mm; outer phyllaries ovate, 1–4 mm long, dark green or black, pale-margined or not; inner phyllaries lanceolate, acuminate, blackish-green centrally, white-margined, glabrous or with spreading white hairs, hardly altering in form but somewhat spreading in fruit; receptacle paleaceous, paleae setiform, exceeding achenes. Corollas 10–17 mm, yellow; styles yellow. Achenes 6–7 mm, all similar, oblong-fusiform, slightly narrowed towards both ends, brown or stramineous, feebly striate; pappus white, partly exserted from involucre, 4.5–6 mm.

HAB. Stony mountain slopes in the *Astragalus* thorn-cushion zone; alt. 2900–3400 m; fl. & fr. Aug.-Sep.
DISTRIB. Local in the high mountain region of Iraq. **MRO**: Algurd (Helgord) Dagh, *Gillett* 12377!; ibid., nr Bermasand (Germasur?) lake, *Rawi & Serhang* 24800!; top of Qandil range between top of mt. Perrish and Bardanas, on the Iranian frontier, *id.* 24571!; NE of Qandil, *id.* 24374!; Kermasur lake-Qandil range, *Rawi* 24122!; Qandil range, *Rawi & Serhang* 24476!

Turkey (E Anatolia), NW Iran, Transcaucasia (Nakhichevan).

3. **Crepis elbrusensis** *Boiss.*, Diagn. Pl. Orient. ser. 1, 11: 58 (1849); Boissier, Fl. Orient. 3: 842 (1875); Babcock in Univ. Calif. Publ. Bot. 22: 626 (1947); Rechinger, Fl. Iranica 122: 314 (1977).

Caespitose perennial herb 6–25 cm with numerous sterile and fertile shoots arising from top of tough rootstock, crown with crowded remains of previous years' leaf-bases. Stems slender, 6–25 cm, scapose or branches very few, long and subtended by very reduced bract-like leaves, sparingly setose, often glandular below capitula. Basal leaves in a rosette, lyrate-pinnatipartite, oblanceolate in outline, 2.2–4(–8) × 0.5–1.5 cm, lateral segments linear-oblong to deltoid, terminal ovate to lanceolate, dentate, obtuse to subacute, ± densely clothed with short or long eglandular and sometime short glandular hairs, midrib not prominently white. Capitula solitary or few, many-flowered, stem or pedicels not incrassate above. Involucre 7–12 × 4–5 mm, cylindrical-campanulate; outer phyllaries short, broadly pale-margined; inner phyllaries narrowly lanceolate, pale-margined, ± white-tomentose and setose, occasionally glandular; receptacle with long setiform paleae ± equal the florets. Corollas 10–13 mm, yellow, sometimes drying reddish; styles yellow. Achenes ± monomorphic, glabrous, brownish, many-ribbed, equally narrowed to the base and apex, 5–5.5 mm, marginal 5.5 mm and curved; pappus 5–6 mm, scarcely exserted.

HAB. Stony mountain summits; alt. 3300–3700 m; fl. Jul.-Aug.
DISTRIB. Very rare in the mountain region of Iraq. **MRO**: Qandil mts., *Rechinger* 11796!; Qandil mts. NE of Rania, *Rawi & Serhang* 18280!

Iran.

Rechinger in *Fl. Iranica* changed the spelling of the specific epithet of this plant to "elbursensis". However Boissier used the spelling "elbrusensis" not only in the original description in the *Diagnoses*, but also in *Flora Orientalis*, so there seems no doubt that this was deliberate and should be maintained.

The indumentum of this plant is more variable than the original description or those in *Fl. Iranica* and Babcock's monograph allow. Boissier described it as being white-hairy on the leaves, making no mention of glands except on the phyllaries; Babcock described the leaves as "densely pubescent with white glandless setules", while Rechinger also described them as "eglandular-pubescent". However, Iraqi material, like *Bornm*üller 7536b from Iran's Alborz mts., has the leaves for the most part with a shorter eglandular indumentum than that of the type and also with numerous shortly stipitate glands. Glands may be present or absent on the phyllaries and the pedicels immediately below the capitula.

4. **Crepis pulchra** *L.*, Sp. Pl.: 806 (1753)
var. **pulchra**; Boissier, Fl. Orient. 3: 846 (1875); Dinsmore in Post, Fl. Syria, Palest. & Sinai ed. 2, 2: 154 (1933); Babcock in Univ. Calif. Publ. Bot. 22: 661 (1947); Zohary in Dep. Agr. Iraq Bull. 31: 161 (1950); Rawi, ibid. Tech. Bull. 14: 113 (1964); Cherepanov in Fl. S.S.S.R. 29: 674 (1964); Rechinger, Fl. Lowland Iraq: 683 (1964) & Fl. Iranica 122: 315 (1977); Lamond in Fl. Turkey 5: 827 (1977); Meikle, Fl. Cyprus 2: 1006 (1985); Ghafoor & Abedin in Fl. Pakistan 222: 47 (2017).

C. pulchra subsp. *turkestanica* Babcock in Univ. Calif. Publ. Bot. 19: 402 (1941); Rechinger, Fl. Iranica

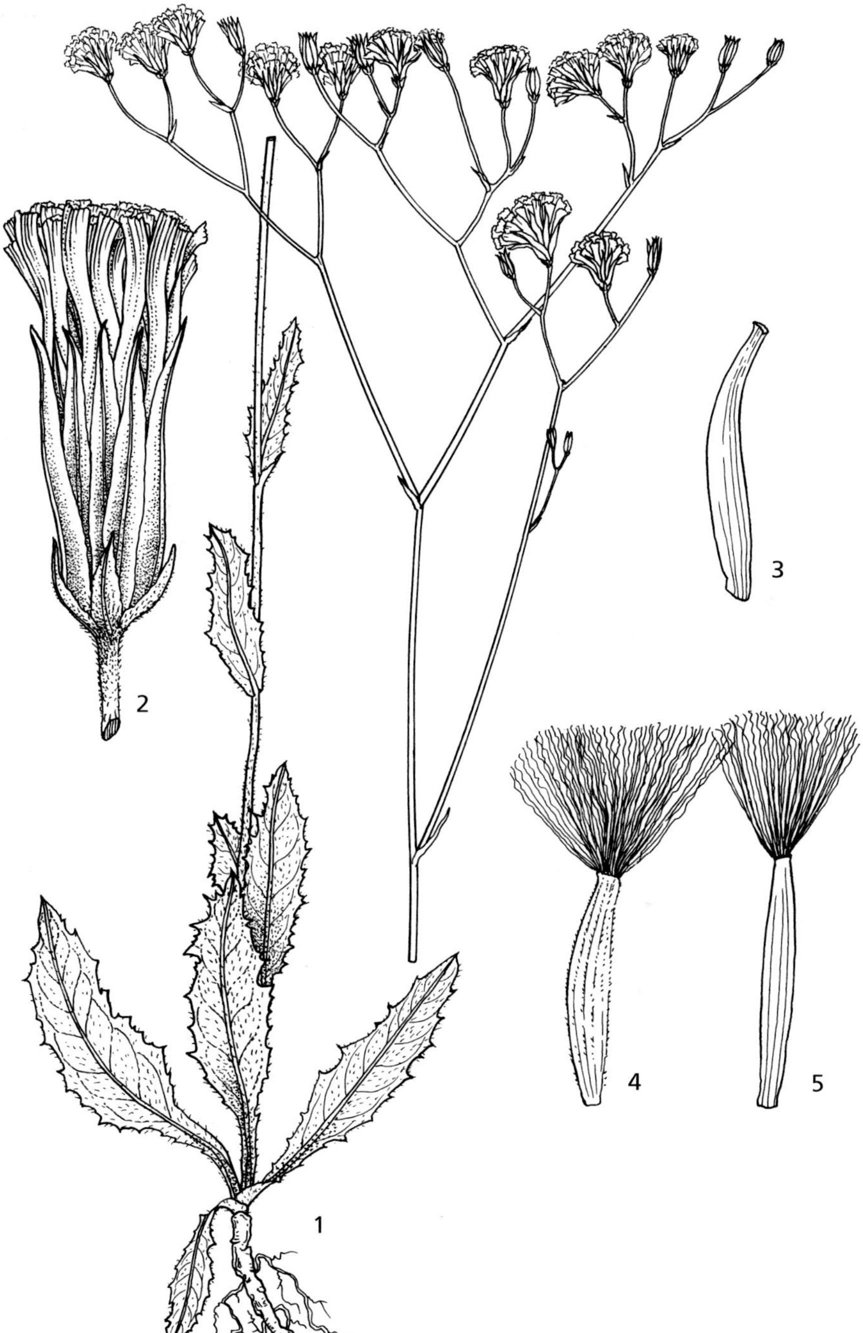

Fig. 34. **Crepis pulchra** var. **pulchra** 1, habit × 1; 2, capitulum × 6; 3, outer achene × 9; 4, middle achene × 9; 5, inner achene × 9. 1 from *Archibald* 1682; 2 from *Gillett* 11005; 3–5 from *Rawi* 23500. Drawn by Sue Wickison.

122: 316 (1977).

Annual herb 30–80(–100) cm tall. Stems 1-many, slender, ± glabrous above, shortly white-pubescent or glandular-pubescent below, with numerous slender branches from base or above. Basal leaves petiolate, obovate to oblanceolate, 3.5–15(–20) × 0.6–4.5(–6.5) cm, usually withered by flowering time, entire or commonly toothed, rarely pinnatifid, ± pubescent, hairs either simple or glandular or often both. Cauline leaves sessile, hastate to subsagittate, simple and toothed, rarely pinnatifid. Inflorescence a wide, often much-branched, cymose panicle; pedicels slender, to about 6 cm, glabrous or rarely glandular-pilose, not thickened above. Capitula rather few- to many-flowered; involucre 7–11 × 3–4 mm, cylindrical at anthesis, pale green; outer phyllaries very short, ovate or lanceolate, pale-margined; inner phyllaries linear-oblong, central part green with a strong vein, glabrous or (in generally pubescent specimens) glandular-hairy but not setose, margins membranous, brownish or hyaline; fruiting inner phyllaries developing a pale or brown spongy keel over midrib, spreading after fruit-fall. Receptacle naked. Corollas yellow, (6–)8–13 mm; styles greenish. Outer achenes in typical form 4.5–5.5 mm, enclosed in phyllaries, dimorphic, some greenish-striate dorsally and smoother and pale ventrally, glabrous, without pappus and others scabrid, with pappus (one or both forms of outer achenes may be absent). Inner achenes obscurely ribbed, brown, with a very persistent pappus of soft, pure white hairs or with about 6 broad-based stiff, scabrid, brownish setae, 3–5 mm. Fig. 34: 1–5.

HAB. On limestone mountain slopes and plains among *Quercus*, *Crataegus* and *Ficus* gardens, on clay mountains, limestone ledges in gorge, stream banks; alt. 450–1900 m; fl. & fr. Apr.-Jul.
DISTRIB. Widespread in the mountain regions of Iraq, rare in the foothills. **MJS**: Jabal Sinjar, *Haussknecht* (1867) s.n!, *Gillett* 11005!; *Sharif & Hamad* 50189! **MAM**: Baidaho village 15 km W of Kani Masi, *Omar* 45750! **MRO**: Shakh-i Harir, *Emberger, Guest et al.* 15452!; Haji Umran, *Chapman* 11948!; Handrian, nr Diana Rowanduz, *Field & Lazar* 870!; Shaqlawa, *Haines* 691!, *Karim* 37521!; 31 km N of Kirkuk towards Koi Sanjaq, nr Schwan village, *Rawi, Noori & Kass* 27976!; Chinaruk, *id.* 28312!; Sewaka (Sewok), *Kass & Noori* 27551!; Gali Ali Beg, *Omar, Qaisi & Wedad* 49547!; **MSU**: Qara Dagh village, *Gillett* 7969!; 11 km N of Penjwin, *Rawi* 22905!; Qopi Qaradagh, *Haines* 1551! 38 km S of Sulaimaniya, *Barkley & Haddad* 5857!; *Azmir, Omar & Karim* 38059!; Darband-i Khan, *Haines* 1870!; Ani, *E. D. & F. A. Barkley* 133!; Kamarspa, *Rawi* 22235! **FUJ**: Jabal Khatchra, *Field & Lazar* 639!

S Europe from France and Portugal to Romania and the Crimea, Cyprus, Turkey, Syria, Iran, Caucasia, Afghanistan, C Asia (Turkmenia to Tian Shan), N Africa (Algeria, Morocco).

Babcock, op. cit. 19: 402 (1941) distinguished subsp. *turkestanica* by its larger flower-capitula, which remain open all day, all Iraqi gatherings being cited under this subspecies by Rechinger in *Fl. Iranica*. Capitulum size, however, appears to overlap; the other character is of course impossible to determine in the herbarium and might well vary with the weather. *Haines* 1870, *E. D. & F. A. Barkley* 133 and *Rawi* 22235 have pappus of few (*c.* 6) broad-based stiff, scabrid, brownish setae. The achenes are all slightly scabrid, resembling the second form of outer achene in *pulchra*.

5. **Crepis alpina** *L.*, Sp. Pl.: 806 (1753); Boissier, Fl. Orient. 3: 854 (1875); Nábělek in Publ. Fac. Sci. Univ. Masaryk 52: 55 (1925); Dinsmore in Post, Fl. Syria, Palest. & Sinai ed. 2, 2: 156 (1933); Babcock in Univ. Calif. Publ. Bot. 22: 678 (1947); Zohary in Dep. Agr. Iraq Bull. 31: 161 (1950); Rawi, ibid. Tech. Bull. 14: 113 (1964); Cherepanov in Fl. S.S.S.R. 29: 688 (1964); Rechinger, Fl. Iranica 122: 318 (1977); Lamond in Fl. Turkey 5: 829 (1977).

Robust annual herb 0.2–0.9(–12) m tall. Stems sulcate, often reddish below, scabridulous and arachnoid especially towards top, simple or usually with a few long branches above. Basal leaves oblanceolate to narrowly oblong-oblanceolate, gradually expanding from a petioloid base to the broadest part above middle, up to c. 15 × 4 cm, finely denticulate, mostly withered at flowering time, glabrous or puberulent; middle cauline leaves as large as or slightly larger than basal, lanceolate to oblong, base clasping or auriculate, otherwise similar, or more coarsely toothed, especially towards base; uppermost leaves much reduced, bract-like. Capitula solitary or few, many-flowered; involucres campanulate, 12–22 × 8–14 mm at flowering; phyllaries of two kinds, outer broadly ovate or elliptic, up to 12 × 4 mm, papery, light brown, glabrous or slightly arachnoid, loosely erect at first, spreading later, inner linear, pale-margined, gradually narrowed to a blunt or truncate tip, densely hairy with three kinds of pubescence: yellowish eglandular setae, more shortly stalked glandular hairs and a dense, short, white tomentum; inner phyllaries in the fruiting capitulum hardened below, increasing to as much as 25 mm long, remaining erect; receptacle shortly ciliate. Corollas 12–16 mm long, yellow; styles greenish. Achenes 14–20 mm, dimorphic, outer partly enclosed in inner phyllaries, compressed and somewhat angular, densely shortly white pilose, gradually attenuate above; inner achenes scabrid, golden brown, finely ribbed and scabrid, pale slender beak 7–11 mm long; pappus 6–9 mm, exserted from involucre.

HAB. *Quercus* forests in valley and on limestone ridge, orchard/pasture and edge of vine field, rocky clay hill slopes; alt. 850–1700 m; fl. Jul., fr. Aug.-Sep.
DISTRIB. Widespread, locally common, in the mountain regions of Iraq: **MAM**: Dohuk, *Haines* 686A!; Amadiya, *Botany staff* 43409!, *Rechinger* 11950; Gara Dagh, *Dabbagh & Jasim* 46943!, *Hadač* 6272; Gali Mazurka, *Dabbagh & Jasim* 46877! **MRO**: N slopes of Handren above Jindian, *Gillett* 8284!; Shaqlawa, *Haines* 686!; Jabal Baradost, *Thesiger* 897!; Chinaruk, *Rawi, Noori & Kass* 28273!; Berd Agha Gin, Rawi, *Noori & Kass* 28713!; Gali Warta, *Rawi, Noori & Kass* 28813! **MSU**: 8 km N of Kani Spi, *Rawi* 22398!; ridge NE of Darband-i Khan, *Poore* 538!; Qaradagh, *Karim, Hamid & Jasim* 40568A!; mt. Goizha (Gweija), *Omar & Karim* 37941!; Spring bet. Erbit & Saeid Sadiq, *Botany Staff* 43180 (BAG!)

Crimea, Syria, Turkey, W Iran, Caucasus.

6. **Crepis foetida** *L.*, Sp. Pl.: 807 (1753); Boissier, Fl. Orient. 3: 851 (1875); Dinsmore in Post, Fl. Syria, Palest. & Sinai ed. 2, 2: 155 (1933); Babcock in Univ. Calif. Publ. Bot. 22: 687 (1947); Rawi in Dep. Agr. Iraq Tech. Bull. 14: 113 (1964); Cherepanov in Fl. S.S.S.R. 29: 689 (1964); Rechinger, Fl. Iranica 122: 319 (1977); Lamond in Fl. Turkey 5: 831 (1977); Feinbrun-Dothan, Fl. Palaest. 3: 443 (1978); Meikle, Fl. Cyprus 2: 1008 (1985); Wild Flow. Saudi Arabia: 184 (1999); Jeffery & Beentje in Fl. Trop. E. Africa, Comp. part 1, 72 (2000).

Annual herbs with or without a leaf rosette present at flowering time, (5–)10–50(–70) cm tall. Stem simple or branched, erect to prostrate, glabrescent with coarse eglandular setae towards base, pedicels sometimes swollen and shortly pilose and/or setose under capitula. Basal leaves petiolate, very variable in size and form, 3–20(–30) × 0.4–4(–6) cm, narrowly oblanceolate to elliptic-obovate, simple to pinnatisect, with terminal lobe usually larger, acute or obtuse, sparsely to rather densely pubescent or setose with glandular and eglandular hairs; median cauline leaves sessile, auriculate, acute, commonly more sharply toothed or dissected than lower leaves; upper leaves much reduced. Capitula solitary or numerous in few-headed corymbs on branches, many-flowered; involucre campanulate, 8–12(–15) × 7–10(–13) mm, outer phyllaries linear to lanceolate, multiseriate and up to ⅔ as long as inner, ± spreading in fruit; inner phyllaries linear, acute, finally incrassate and navicular, ± shortly arachnoid-tomentose and bearing spreading eglandular setae and/or glandular hairs, ± erect in fruit; receptacle paleaceous or ciliate. Corollas yellow, mostly 7–12 mm, outer rays often purple dorsally; styles yellow or green. Achenes dimorphic, outer 5–8 mm long, shortly beaked and 2-celled as in *C. kotschyana*, sometimes lacking, inner 10–15 mm, with a slender 6–10 mm beak; pappus 3–7 mm, ± wholly exserted.

DISTRIB. (of species). Throughout most of Europe except the north, Syria, Turkey, Iran, Caucasus, C Asia (Turkmenia), N Africa (Morocco, Algeria). Introduced in Macaronesia (Madeira, Canary Is.).

2 subspecies in Iraq

Receptacle shortly pilose around the achene-pits, epaleaceous . . . . subsp. **foetida**
Receptacle with long, narrow, entire or forked paleae almost as long as
the achenes, and much exceeding their fertile part . . . . subsp. **commutata**

subsp. **foetida**

Barkhausia foetida (L.) F.W. Schmidt, Samml. phys.-ökon. Aufs. 1: 283 (1795).
Crepis rhoeadifolia M. Bieb., Fl. Taur.-Cauc. 3: 538 (1819); Nábělek in Publ. Fac. Sci. Univ. Masaryk 52: 55 (1925); Zohary in Dep. Agr. Iraq Bull. 30: 161 (1950); Rawi, ibid. Tech. Bull. 14: 113 (1964); Cherepanov in Fl. S.S.S.R. 29: 690 (1964).
C. foetida var. *vulgaris* Bisch., Beitr. Fl. Deutsch.: 252 (1851).
C. foetida var. *glandulosa* (Guss.) Bisch., ibid. (1851); Dinsmore in Post, ed. 2, 2: 155 (1933); Bornmüller, in Beih. Bot. Centralbl. 60B: 227 (1939); Babcock in Univ. Calif. Publ. Bot. 22: 695 (1947); Rechinger, Fl. Iranica 122: 321 (1977); Lamond in Fl. Turkey 5: 832 (1977).
C. foetida var. *rhoeadifolia* (M.Bieb.) Boiss., Fl. Orient. 3: 851 (1875); Dinsmore in Post, Fl. Syria, Palest. & Sinai ed. 2, 2: 155 (1933).
C. foetida subsp. *vulgaris* (Bisch.) Babc. in Journ. Bot. 76: 205 (1938).

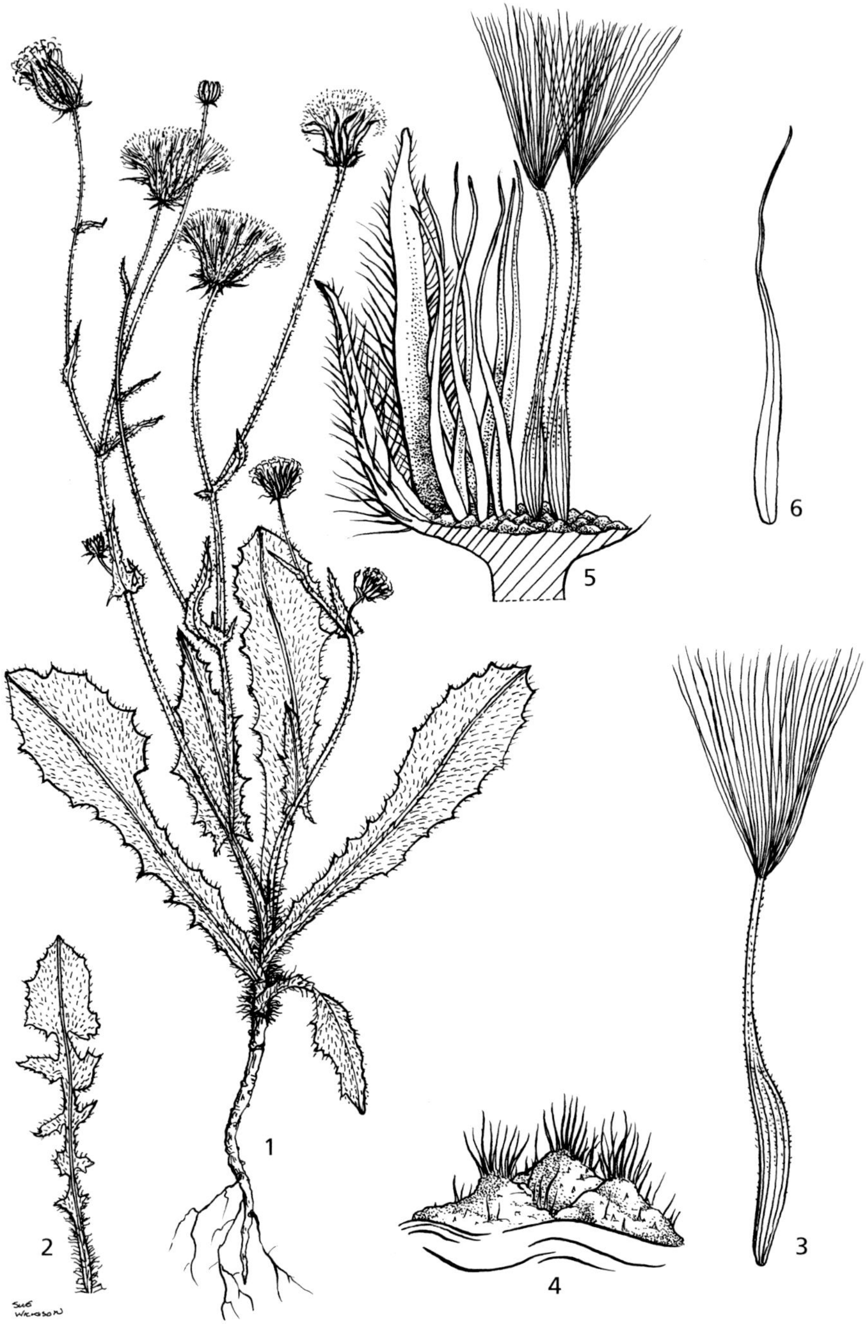

Fig 35. **Crepis foetida** subsp. **foetida**. 1, habit × ½; 2, basal leaf × 1; 3, achene with pappus × 9; 4, receptacle surface (enlarged). **Crepis foetida** subsp. **commutata**. 5, receptacle with achenes and paleae × 9; 6, paleae × 9. 1 from *Chapman* 641 (Cyprus); 2 from *Kennedy* 1464 (Cyprus); 3–4 from *Rawi* 22534; 5–6 from *Haines* W690. Drawn by Sue Wickison.

[*C. strigosa* (non Hall. f.) Anth. in Notes Roy. Bot. Gard. Edinburgh 18: 292 (1935); Rawi in Dep. Agr. Iraq Tech. Bull. 14: 113 (1964)].

Receptacle shortly pilose around achene-pits, epaleaceous. Fig. 35: 1–4.

HAB. Open *Quercus* or *Rhus/Quercus* forests, riverside scrub between cultivation, streambanks, cultivated ground, hillsides, on limestone, sandstone, rocky clay or sandy/gravelly soil; alt. 150–1300 m; fl. & fr. Apr.-Sep.

DISTRIB. Widespread, locally frequent, in the mountains and rather less so in the foothill regions of Iraq; rare at the edge of the Lower Mesopotamian Plain. **MAM**: 84 km N of Mosul, *Rawi* 22960!; Jabal Bekher, *id.* 22998! & 23154!; 5 km S of Zakho, *id.* 23112!; Kani Mazi (Ain Nuni), *Hamad & Fadhil* 45593!; Gara Dagh, *Dabbagh & Jasim* 46939! **MRO**: 31 km N of Kirkuk towards Koi Sanjaq, *Rawi, Noori & Kass* 27976A!; Koi Sanjaq, *id.* 28101!; Shaqlawa, *Haines* 690! **MSU**: Jarmo, *Helbaek* 588!, 1273!; Penjwin, *Rawi* 22534!; Darband-i Khan, *Haddad & Barkley* 7348! **FNI**: Zab R. nr Eski Kellek, *Gillett* 8204! **FPF**: Qaraghan (Jalaula), *Rogers* 0222!; Chilat Police Station, *Rawi & Haddad* 25725!; Buskaya, *id.* 25573! **LSM/FPF**: Shatt at-Tib, 70 km N of Amara, *Rechinger* 14142.

Throughout the range of the species.

subsp. **commutata** (*Spreng.*) *Babc.* in Journ. Bot. 76: 207 (1938); Rawi in Dep. Agr. Iraq Tech. Bull. 14: 113 (1964); Rechinger, Fl. Iranica 122: 321 (1977); Lamond in Fl. Turkey 5: 833 (1977); Meikle, Fl. Cyprus 2: 1009 (1985).

Rodigia commutata Spreng., Neue Entdeck. 1: 275 (1820); Boissier, Fl. Orient. 3: 880 (1875); Dinsmore in Post, Fl. Syria, Palest. & Sinai ed. 2, 2: 160 (1933); Anthony in Notes Roy. Bot. Gard. Edinburgh 18: 292 (1935); Rawi in Dep. Agr. Iraq Tech. Bull. 14: 123 (1964).
Crepis brachycarpa Bornm. in Beih. Bot. Centralbl. 32 (2): 418 (1914).

Receptacle with long, narrow, entire or forked paleae almost as long as achenes, and much exceeding their fertile part. Fig. 35: 5–6.

HAB. Rocky and stony hill slopes, flat ground near water, grassy slopes amid *Quercus*, in valley vegetation and as a weed of cultivation, on clay, limestone, sand/conglomerate or alluvium; alt. 390–1360 m; fl. & fr. Apr.-Jun.

DISTRIB. Widespread and locally frequent in the mountain and foothill regions of Iraq. **MAM**: Dohuk, *Makki Beg* 3272!; Shaikhan, *Salim Effendi* 2547! **MRO**: Plingan, *Rawi, Noori & Kass* 28650!; Shaqlawa, *Haines* 690!; Khulifan, *Omar, Qaisi & Wedad* 49579!; below Salahuddin, *Barkley* 5636!; between Rowanduz and Akoyan, *Kass & Noori* 27257!; 2 km N of Darband, *Botany Staff* 43208! **MSU**: Avroman (Hawraman) mts., Susakhan nr Tawila, *Rechinger* 10177; Tawila, *Rawi* 21894!; Qaradagh to Ja'afaran, *Gillett* 7876!; 19 km S of Darband-i Khan, *Barkley & Haddad* 7392!; nr Sulaimaniya, *Barkley* 8483! **FUJ**: Wadi Thirthar nr Hadhr, *Rawi & Gillett* 7257!; Tal Kochak, *Gillett* 10811! **FNI**: Tal Kaif, *Mudir of Tal Kaif* 3198! **FKI**: Kani Dolman hills, *Guest* 4342! **FPF**: Khanaqin, *Haines* 930!

SE Europe (Greece, Bulgaria), Crete, Rhodes, Syria, Turkey, W Iran.

HŌDHĀN (*Mudir of Tal Kaif* 3198, *Gillett* 100811), KALÎLK ZAR (Kurd., *Makki Beg* 3272), GULLIKAH ZAR (Kurd., *Salim Effendi* 2547). There are somewhat contradictory notes on the effect of this plant on livestock. Notes only occur on sheets of the subsp. *commutata*. *Mudir of Tal Kaif* 3198 notes that it is "bad for sheep and causes their death"; *Makki Beg* 3272 observes "when green, gives sheep a fever from which they may eventually die if they continue eating it". *Gillett* 10811 bears a note "shepherd boys said this is not a poisonous plant". Older Bedouin said that it is, and that it does much damage. They say it is particularly common in years with late rain in May – while *Salim Effendi* 2547 states that it is a "fodder plant". Perhaps it is a plant which is poisonous when fresh but wholesome dry; however, the matter clearly needs further investigation.

7. **Crepis kotschyana** (*Boiss.*) *Boiss.*, Fl. Orient. 3: 852 (1875); Babcock in Univ. Calif. Publ. Bot. 22: 707 (1947); Zohary in Dep. Agr. Iraq Bull. 30: 161 (1950); Rawi in Dep. Agr. Iraq Tech. Bull. 14: 113 (1964); Cherepanov in Fl. S.S.S.R. 29: 693 (1964); Rechinger, Fl. Iranica 122: 325 (1977); Lamond in Fl. Turkey 5: 834 (1977); Feinbrun-Dothan, Fl. Palaest. 3: 443 (1978); Wild Flow. Saudi Arabia: 184 (1999); Breckle, Hedge, Podlech & Rafiqpoor, Vasc. Pl. Afghan. checklist: 169 (2013); Kilian & Ghazanfar in Fl. Oman 3: 220 (2015); Taifour & El-Oqlah, Pl. Jordan Ann. Checkl.: 40 (2017); Ghafoor & Abedin in Fl. Pakistan 222: 49 (2017).

Barkhausia kotschyana Boiss., Diagn. Pl. Orient. ser. 1, 7: 13 (1846).
Crepis bureniana Boissier, Fl. Orient. 3: 852 (1875); Hand.-Mazz. in Ann. Naturh. Mus. Wien 27: 69

(1913); Dinsmore in Post, Fl. Syria, Palest. & Sinai ed. 2, 2: 155 (1933); Zohary in Dep. Agr. Iraq Bull. 30: 161 (1950); Rawi in Dep. Agr. Iraq Tech. Bull. 14: 113 (1964).

Annual herb with an ascending or erect simple or branched stem 10–30 cm tall, ± canescent-tomentose and with longer glandular and eglandular hairs. Basal leaves usually rather small, attenuate to base, sinuate-dentate or rarely pinnatipartite, mostly withered at time of flowering; cauline leaves sessile, narrowly elliptic or oblong, lower 2–8 × 0.5–2 cm, dentate or pinnately and sometimes runcinately lobed, cartilaginous-denticulate, upper becoming simple and smaller, apex obtuse or acute with a pronounced tooth-like point, glabrous or with a variable indumentum from sparingly canescent and shortly glandular to similar to that of stem. Capitula many-flowered, solitary to numerous, pedicels only slightly thickened upwards; involucre campanulate, 8–13 × 7–10 mm; outer phyllaries short, linear, soon caducous, inner phyllaries subequal, linear to lanceolate, becoming hardened, curved-navicular and separating from each other laterally in fruit, with a dense indumentum of long eglandular setae, usually shorter glandular hairs and white arachnoid tomentum, remaining erect after fruit dispersal; receptacle naked. Corollas 5–7 mm, rays yellow but often mauve in dried material; styles yellow. Achenes dimorphic: outer 5–6 mm long, enclosed by inner phyllaries, curved, unbeaked, laterally compressed, ribbed, spinulose; inner achenes 8–12 mm long, body 5–7 mm, fusiform, golden-brown, ribbed and spinulose, beak slender; pappus white, 4–6 mm long, ± wholly exserted from involucre.

HAB. On rocky and sandy hillsides and disturbed places, depressions near water, along stream beds and on gentle sandy slopes, also in recently cultivated fields; alt. 50–1400 m; fl. & fr. Apr.-Jun.
DISTRIB. Widespread and locally abundant in the mountain, foothill and desert plateau regions of Iraq, apparently rarer in the desert. **MRO**: Haibat Sultan Dagh, *Rawi, Noori & Kass* 28181! **MSU**: ± 10 km W of Tawila, *Rawi* 22143!; Jarmo, *Helbaek* 1219!; Qopi Qaradagh, *Haines* 1185!; 5 km W of Chemchemal, *Rawi* 21682!; Darband-i Bazian, *Rawi* 22808! **FUJ**: Jazira S of Sinjar nr Ba'aj, *Guest* 13391!; Jabal Khatchra, *Field & Lazar* 645!; 23 km N of Hadhr, *Bharucha* 8674!; Syrian desert, *Hausskn.* (1867) s.n.! **FNI**: Zab R. nr Eski Kellek, *Gillett* 8204A! **FKI**: Jabal Hamrin, Injana, *Omar & Karim* 37825!; 10–15 km N of Kirkuk, *Rawi* 21571!; nr Tauq bridge, *Guest* 15539! **FPF**: Fakka-Laqlaq valley, *Rawi* 25836!; Khanaqin, *Haines* 932!; Koma Sank police station nr Mandali, *Rawi* 20615! **DGA**: Asib valley, 50 km E of Samarra, *Rawi* 20450A! **DWD**: Wadi al-Ajrumiya, *id.* 30505!

Syria, Turkey (extreme S Anatolia), Iran, Afghanistan, Pakistan, C Asia (Turkmenia to Tajikistan).

8. **Crepis quercifolia** *Bornm. & Gauba* in Repert. Spec. Nov. Regni Veg. 41: 319 (1937); Rechinger, Fl. Iranica 122: 322 (1977).

Annual or biennial herb 25–60 cm tall, with long divergent branches from base or in upper half only, clothed especially (and often only) below with fine, white, subtomentose hairs and divergent setae. Radical leaves (not seen from Iraq) obovate to oblong, attenuate but scarcely petiolate below, widely sinuate or dentate, 12–16 × 5–6 cm; cauline leaves 4–10 × 2–7 cm, ± lyrate-pinnatisect with narrow, acute, acutely dentate lateral lobes and a large, acute, sharply dentate terminal lobe, acutely hastate at base, finely setose-hairy; upper leaves rapidly diminishing, uppermost bract-like. Capitula numerous, many-flowered, pedicels not incrassate above, ± glabrous; involucre campanulate, 8–12 × 7–9 mm; outer phyllaries mostly over half length of inner, 4–8 mm, linear-lanceolate to linear, appressed, feebly floccose with an occasional seta; inner phyllaries linear, acute, pale-margined, ± floccose-tomentose and long-setose; receptacle glabrous. Corollas 11–15 mm, yellow; styles yellow. Achenes homomorphic, 13–15 mm, body finely ribbed and scabrid, ½ to ⅔ length of slender beak; pappus 3–5 mm.

HAB. Wooded hillside and slopes in the hills; alt. (only one recorded) up to 1515 m; fl. Apr.-May.
DISTRIB. Rare in the mountain and foothill regions of Iraq. **MAM**: Jabal Khair above Mirsidah village, 40 km W of Aqra, *Thesiger* 584! **FNI**: Shaikh Adi, 10 km N.W. of Ain Sifni, *Thesiger* 701!

W, C & S Iran.

9. **Crepis sancta** (*L.*) *Babc.* in Univ. Calif. Publ. Bot. 19: 403 (1941) & 22: 730 (1947); Rawi in Dep. Agr. Iraq Tech. Bull. 14: 113 (1964); Rechinger, Fl. Lowland Iraq: 684 (1964); Rechinger, Fl. Iranica 122: 328 (1977); Lamond in Fl. Turkey 5: 834 (1977); Feinbrun-Dothan, Fl. Palaest. 3: 444 (1978); Meikle, Fl. Cyprus 2: 1009 (1985); Wild Flow. Saudi

Arabia: 185 (1999); Boulos & Hind in Fl. Egypt 3: 302 (2002); Taifour & El-Oqlah, Pl. Jordan Ann. Checkl.: 40 (2017); Ghafoor & Abedin in Fl. Pakistan 222: 54 (2017).

Hieracium sanctum L., Cent. Pl. 2: 30 (1756).
Crepis nemausensis Gouan, Illustr.: 60 (1773).
Trichocrepis bifida Vis., Stirp. Dalm.: 19, t. 7 (1826).
Lagoseris bifida (Vis.) W.D.J. Koch, Syn. Fl. Germ. Helv. ed. 1: 435 (1837).
Pterotheca bifida (Vis.) Fisch. & C.A.Mey., Ind. Sem. Horti Petrop. 4: 43 (1838).
P. obovata Boiss. et Noë in Boissier, Diagn. Pl. Orient ser. 2, 3: 98 (1856).
Lagoseris orientalis Boissier, Fl. Orient. 3: 882 (1875).
L. obovata (Boiss. et Noë) Bornm. in Beih. Bot. Centralbl. 20 (2): 176 (1906); Dinsmore in Post, Fl. Syria, Palest. & Sinai ed. 2, 2: 160 (1933); Zohary in Dep. Agr. Iraq Bull. 30: 161 (1950); Cherepanov in Fl. S.S.S.R. 29: 712 (1964).
L. sancta (L.) K. Maly in Glasn. mus. Bosn. Herc. 20: 5565, 562 (1908); Dinsmore in Post, Fl. Syria, Palest. & Sinai ed. 2, 2: 160 (1933); Zohary in Dep. Agr. Iraq Bull. 30: 161 (1950); Cherepanov in Fl. S.S.S.R. 29: 708 (1964).
L. marschalliana (Reichb.) Thell. in Mem. Soc. Nat. Sc. Cherbourg ser. 4, 38: 576 (1911–12); Handel-Mazzetti in Ann. Naturh. Mus. Wien 27: 66 (1913).
C. sancta subsp. *bifida* (Vis.) Babc. in Univ. Calif. Publ. Bot. 22: 736 (1947); Rawi in Dep. Agr. Iraq Tech. Bull. 14: 113 (1964).
C. sancta subsp. *obovata* (Boiss. et Noë) Babc. in Univ. Calif. Publ. Bot. 22: 741 (1947); Rawi in Dep. Agr. Iraq Tech. Bull. 14: 113 (1964); Rechinger, Fl. Iranica 122: 333 (1977); Feinbrun-Dothan, Fl. Palaest. 3: 444 (1978).
C. sancta subsp. *azerbaijanica* Rech. f. in Fl. Iranica 122: 334 (1977).

Scapose annual herb with several to many stems 5–60 cm tall. Leaves of basal rosette persistent, subpetiolate, 1–18 × 0.5–4 cm, obovate, oblong or oblanceolate, simple and denticulate to lobed or lyrate-pinnatifid, acute or obtuse, lamina pubescent especially on margins and major veins with eglandular spreading setae, sometimes ± glabrous. Stems glabrous to glandular and setose-hairy. Capitula rather few- to many-flowered, borne singly or in cymose groups; involucre (4–)6–11 × (3–)4–7 mm, at first cylindrical, finally campanulate to turbinate, then reflexed after fruit fall. Outer phyllaries ovate to linear, 1.5–2 mm long, broadly white-margined, with spreading glandular and eglandular white or black setae and with or without thin, floccose white hairs; receptacle long-setose, setae persisting after fruit dispersal, very rarely absent. Corollas yellow, 8–12 mm, rays sometimes reddish-tinged dorsally; styles green or yellow. Achenes 3–5(–7) mm long, outer whitish, compressed, angled or winged, or absent, inner ribbed, rugose or innermost smooth, medium-brown; all achenes truncate or narrowed but not distinctly beaked; pappus not exserted from fruiting involucre, 2.5–4 mm.

HAB. Open steppe with *Poa bulbosa, Stipa tortilis* and *Artemisia herba-alba,* waste ground by roadsides, river banks and damp depressions, hill slopes, on clay, alluvium or sandy conglomerate, also on stony hills and denuded limestone slopes with scattered *Quercus*; alt. 90–1360 m; fl. & fr. Mar.-Jun.
DISTRIB. Widespread, though only once noted as abundant, in the mountains and foothills of Iraq, rarely collected in the desert. **MAM**: Amadiya, *Guest* 1251! **MRO**: Tawila, *Rawi* 21941!; Jarmo, *Helbaek* 344!; Jasina, *Omar* 42719!; 19 km S of Darband-i Khan, *Barkley & Haddad* 7391! **MJS**: Karsi, *Gillett* 10872! **FUJ**: Wadi Thirthar E of Hadhr, *Rawi & Gillett* 7258!; Mosul, *Rathi* 5547!; Jazira S of Sinjar nr Ba'aj, *Guest* 13394!; Ain Husan, *Guest* 4212!; Balad Sinjar to Tal Afar, *Guest* 4186! **FNI**: Jabal Maklub, *Haines* 1413!; Eski Kellek, *Gillett* 10381!, *Shahwani* 25157! **FKI**: Sar Qal'a, *Poore* 389! **FPF**: Khanaqin, *Haines* 929!; Kani Kirmaj, *Poore* 388!, 470!; Pilkhana, *Rawi* 12780!, 12788!; 22 km S of Badra, *id.* 18428!; Jabal Muwaila, *Guest, Rawi & Rechinger* 17588!, 17601!; Qaraghan (Jalaula), *Guest* 1880! **DWD**: 70 km W of Nukhaib, *Rawi* 31385A!; Tal al-Nisir, 48 km NE of Rutba, *id.* 31197! 20 km to *Ana, Omar, Abass, Kaisi & Hamad* 55825 (BAG!).

SE Europe from Balkans to Crimea (widely naturalised also in western Mediterranean Europe), Syria, Lebanon, Palestine, Cyprus, Aegean Is., Turkey, Iran, Caucasus, Afghanistan, Pakistan, C Asia (Turkmenia, Kazakhstan).

The subspecies and other infraspecific taxa of this protean species seem to run into one another so much as to make their recognition impractical with any degree of certainty – and practicality must surely be the aim of a regional Flora.

10. **Crepis micrantha** *Czerep.* in Fl. S.S.S.R. 29: 684 (1964); Rechinger, Fl. Iranica 122: 336 (1977); Lamond in Fl. Turkey 5: 837 (1977); Feinbrun-Dothan, Fl. Palaest. 3: 445 (1978); Meikle, Fl. Cyprus 2: 1011 (1985); Boulos, Fl. Egypt 3: 300 (2002).

C. parviflora Desf. ex Pers., Syn. 2: 376 (1807) nom. illeg., non Moench (1794) [= *Crepis capillaris* (L.) Wallr.]; Boissier, Fl. Orient. 3: 849 (1875); Handel-Mazetti in Ann. Naturh. Mus. Wien 27: 69 (1913); Dinsmore in Post, Fl. Syria, Palest. & Sinai ed. 2, 2: 154 (1933); Anthony in Notes Roy. Bot. Gard. Edinburgh. 18: 292 (1935); Babcock in Univ. Calif. Publ. Bot. 22: 775 (1947); Zohary in Dep. Agr. Iraq Bull. 30: 161 (1950); Rawi, ibid. Tech. Bull. 14: 113 (1964); Rechinger, Fl. Lowland Iraq: 685 (1964); Täckholm, Stud. Fl. Egypt ed. 2: 606 (1974); Taifour & El-Oqlah, Pl. Jordan Ann. Checkl.: 40 (2017).

Annual herb (10–)20–75(–110) cm tall. Stem slender or more robust, sparingly to considerably branched, leafy, clothed with strong or weak setose hairs, most frequently ± glabrous above and pilose below. Basal leaves elliptic, tapering to a petioloid base, variable in form from denticulate to deeply runcinate-pinnatisect, 1.5–17 × 0.6–4 cm, acute, ± pubescent, mostly ± withered at flowering time; cauline leaves auriculate or sagittate, usually less dissected or toothed than the lower; uppermost leaves much reduced. Inflorescences usually broad with many capitula; pedicels slender, to 3 cm. Capitula many-flowered; involucre cylindrical-campanulate, 3.5–6 × 2.5–4 cm. Outer phyllaries 1–2 mm long, glabrous or setose; inner phyllaries lanceolate, dark green centrally, with setae or short tomentose hairs, slightly corky-indurate in fruit, finally spreading after fruit-fall, margins pale; receptacle naked. Corollas yellow, 5–8.5 mm; styles green or yellow. Achenes beakless, all alike, pale or golden-brown, 1.5–2 mm long, truncate-fusiform, 10-ribbed; pappus 3–4 mm long, not exserted from involucre, deciduous, white.

HAB. Weed of roadsides and cultivation, often near irrigation ditches or streams, occasionally on hill or mountain slopes, and recorded in destroyed *Quercus* forest, but probably on disturbed ground, on clay, limestone and sandy gravelly soil, apparently saline-tolerant in the south.; alt. 5–1060 m; fl. & fr. Apr.-Jul. DISTRIB. Apparently scattered in the mountain region of Iraq, only once recorded in the foothills, local in the Lower Mesopotamian region. **MAM**: Zawita, *Chapman* 26287!; Sersang, *Omar* 37744!, *Haines* 583! **MRO**: Shaqlawa, *Gillett & Rawi* 11579!, *Karim* 37521!, *Haines* 692! **MSU**: Talan region nr Dukan, *Karim* 39289!; Tainan, *Rawi & Gillett* 11647!; 6 km W of Sulaimaniya, *Rechinger* 10066!; 13 km W of Sulaimaniya, *Rawi* 21708!; Qaranjir, *Rawi & Gillett* 7508! **FNI**: Nineveh, *E. Chapman* 25374! **LEA**: 5 km NW of Ba'quba, *Alkas* 18556!; Zafariniya, *Hamid* 42683! **LCA**: Beni Saad, 2 km N of Baghdad, *Anderson* 2004!; Daltawa, *Guest* 2346!, *Darwish Haidari* 2340!; Tarmiya, *Haines* 117! **LBA**; 40 km S of Basra on road to Fao, *Alizzi & Omar* 35042!; Abu'l-Khasib, *Khatib & Alizzi* 32655!, *Thamer & Abdel Jabar* 50088!; Siba, *Hamad, Hamid & Kadhim* 46360!

Greece, Crete, Crimea, Turkey, Aegean Is., Cyprus, Syria, Egypt, Transcaucasia, Iran, C Asia (Turkmenia).

MARAIR (*Guest* 2462). WARD-AL-FUTÎR . *Darwish Haidari* 2340 bears the note "stems eaten by sheep, but the leaves are bitter".

11. **Crepis aspera** *L.*, Sp. Pl. ed. 2: 1132 (1763); Boissier, Fl. Orient. 3: 857 (1875); Dinsmore in Post, Fl. Syria, Palest. & Sinai ed. 2, 2: 157 (1933); Babcock in Univ. Calif. Publ. Bot. 22: 878 (1947); Zohary in Dep. Agr. Iraq Bull. 30: 161 (1950); Rawi, ibid. Tech. Bull. 14: 113 (1964); Rechinger, Fl. Lowland Iraq: 685 (1964); Täckholm, Stud. Fl. Egypt ed. 2: 606 (1974); Rechinger, Fl. Iranica 122: 337 (1977); Lamond in Fl. Turkey 5: 840 (1975); Feinbrun-Dothan, Fl. Palaest. 3: 447 (1978); Meikle, Fl. Cyprus 2: 1012 (1985); Collenette, Wild Flow. Saudi Arabia: 184 (1999); Boulos & Hind in Fl. Egypt. 3: 305 (2002); Taifour & El-Oqlah, Pl. Jordan Ann. Checkl.: 40 (2017).

Annual herb 8–55 cm tall with a leaf rosette and one to numerous usually branched leafy stems which are pubescent throughout with a mixed indumentum of appressed white hairs (sometimes visible only beneath the capitula) and stout spreading brown or white setae up to 7 mm. Rosette leaves oblanceolate to narrowly oblong-obovate, 2–15(–20) × 0.5–4 cm, coarsely dentate to lobed or pinnatisect, lobes oblong or lanceolate, sharply acute, base petioloid, glabrous or with setose hairs on midrib and some secondary veins. Cauline leaves sessile, more deeply divided than rosette leaves, auriculate, upper much reduced. Pedicels not thickened beneath capitula, up to *c.* 4 cm but primary capitula short-stalked to subsessile. Involucre campanulate to turbinate, 6–9 × 3.5–6 mm; outer phyllaries membranous, short, with a brown midrib, glabrous or sparsely pubescent; inner phyllaries lanceolate with spreading dark triangular tips, midrib area dark green or blackish, with spreading setae and appressed-tomentose hairs, yellowish towards margin with a narrow hyaline border; receptacle ciliate. Corollas yellow, 6–9 mm; styles yellow. Outer achenes (sometimes absent) enfolded within phyllaries, pale, 4–6 mm long, curved, laterally compressed with a pale

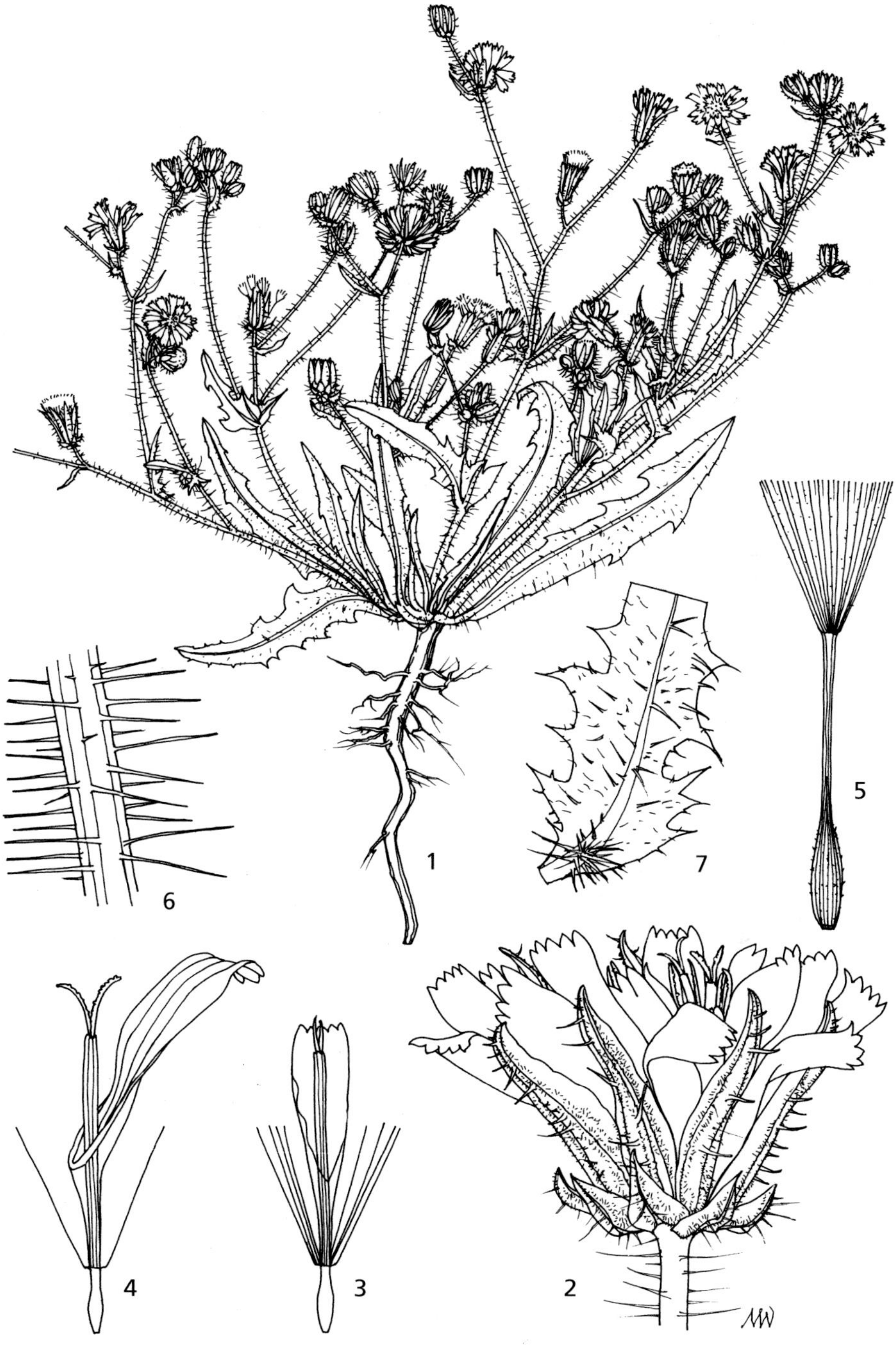

Fig. 36. **Crepis aspera**. 1, habit × 2/3; 2, capitulum × 6; 3, floret (inner) × 9; 4, floret (outer) × 9; 5, inner achene with pappus × 9; 6, part of stem × 6; 7, section of leaf × 3. All from *Guest, Rawi & Long* 14123. Drawn by M. Millar Watt.

keel on inner surface, attenuate in upper half; inner achenes brown, 4–7 mm long, body 10–12-ribbed, scabrid, subequalling slender beak; pappus partly exserted from involucre, ± 4 mm. Fig. 36: 1–7.

HAB. Silty and clay depressions and flats in the desert, low hills, fallow, roadside waste and between irrigated cultivation; alt. 25–340 m; fl. & fr. Mar.-Jun.

DISTRIB. A predominantly southern species in Iraq, widespread in the Southern Desert region and no doubt extending to the adjacent Western Desert, also in the Lower Mesopotamian Plain and occasionally in the foothills. **FUJ**: Shergat, *Graham* (1920), s.n.! **FPF**: 10 km N of Jalaula, *Hadač* 2984; 43 km W of Khanaqin, *Cowan & Darlington* 1811!; 15 km N of Zurbatiya, *Kaisi & Khayat* 50629!; N of Badra 3.5 km S of Al-Fateh Police Station, *Kaisi, Haloob, Adnan & Riyadh* 58849 (BAG!) **DSD/DWD**: Wadi al-Khirr 15 km N of Lussuf, *Guest, Rawi & Rechinger* 19455! **DSD**: 30 km W by N of Busaiya, *Guest, Rawi & Long* 14169!; Shabicha, Karim, *Nuri, Hamid & Kadhim* 40002!; 52 km E by S of Salman, *Guest, Rawi & Long* 14123!; 70 km W of Nukhaib, *Rawi* 31387!; c. 20 km WNW of Ansab, on Saudi frontier, *Guest, Rawi & Rechinger* 18999! **LEA**: Fakka, *Rawi* 14991A!; Fakka-Laqlaq valley, *id.* 25836B!; Shahraban, *Haines* 853!; 10 km SW of Ba'quba, *Anderson* 1968! **LCA**: 23 km W of Shaikh Sa'ad, *Rawi & Haddad* 25514!; Daltawa, *Guest* 2483!; Baghdad, *Colvill* s.n. **LSM**: 70 km E of Amara nr Rshaida (Musaida?) well, *Ani* 39754! **LBA**: Maqil, *Haines* s.n.

Cyprus, Turkey (SE Anatolia), Syria, Palestine, Jordan, Sinai, Egypt, NW Saudi Arabia.

Crepis connexa Babc. was listed for Iraq by Zohary in Dep. Agr. Iraq Bull. 31: 161 (1950) and Rawi in Dep. Agr. Iraq Tech. Bull. 14: 119 (1964), under the name *Lagoseris hieracioides* Boiss. & Hausskn. Although some specimens of the type gathering (the only one known) are labelled as "Mt. Avroman et Schahu" and might thus have come from the Iraqi side of the Hawraman mts., these labels are general printed ones used either by the collector or subsequent curators. The "Pir Omar Gudrun" label which appears on one isotype in herb. Boissier (G) is of the same order. The plant has only been collected from "in rup. calc. m. Schahu 6000', Jul. 1867, *Haussknecht* 621" according to the fuller details written on both sheets in G; it is not an Iraqi species.

36. **HETERODERIS** (Bunge) Boiss.

Fl. Orient. [Boissier] 3: 793 (1875)
Barkhausia sect. *Heteroderis* Bunge in Mem. Acad. Sci. Petersb. 7: 384 (1854)

Frances G. Davies

Annual herb with a variable indumentum of pubescent or glandular and coarse setose hairs, leaves with cartilaginous teeth. Capitula nodding or inclined, many-flowered; phyllaries biseriate, outer ones very reduced; inner phyllaries lengthening in fruit, spreading after fruit-fall; receptacle naked. Achenes strongly dimorphic, outer attenuate and ± recessed into inner phyllaries, inner with a long slender beak and muricate especially towards apex; pappus of simple hairs.

A monotypic genus.

Heteroderis (from Gr. έτερος, *eteros*, another and δερης, deris, for δερριστ, derris (a member of the Asteraceae).

Leonard, J. (1983). Contribution a la connaiscance de la flore de l'Iran. Compositae: *Heteroderis, Scorzonera, Taraxacum.* Bull. Jard. Bot. Nat. Belg. 53(3–4): 437–440.

1. **Heteroderis pusilla** (*Boiss.*) *Boiss.*, Fl. Orient. 3: 794 (1875); Rawi in Dep. Agr. Iraq Tech. Bull. 14: 117 (1964); Rechinger, Fl. Lowland Iraq: 673 (1964); Rechinger, Fl. Iranica 122: 290 (1977); Mandaville, Fl. E Saudi Arabia: 320 (1990); Boulos & Hind in Boulos, Fl. Egypt 3: 306 (2002); Breckle, Hedge & Rafiqpoor, Vasc. Pl. Afghan: 174 (2013); Ghafoor & al. in Fl. Pakistan 222: 60 (2017).

Chondrilla pusilla Boiss., Diagn. Pl. Orient. ser. 1, 7: 11 (1846).
Barkhausia leucocephala Bunge, Mém. Acad. Imp. Sci. St.-Pétersbourg Divers Savans 7: 385 (1851); Beitr. Fl. Russl.: 209 (1852).
Heteroderis stocksiana Boiss., Fl. Orient. 3: 794 (1875).
H. aegyptiaca Schweinf. in Aschers. & Schweinf., Mém. Inst. Egypt, 2, Suppl.: 766 (1889).
H. leucocephala (Bunge) Leonova, Fl. URSS 29: 590 (1964).

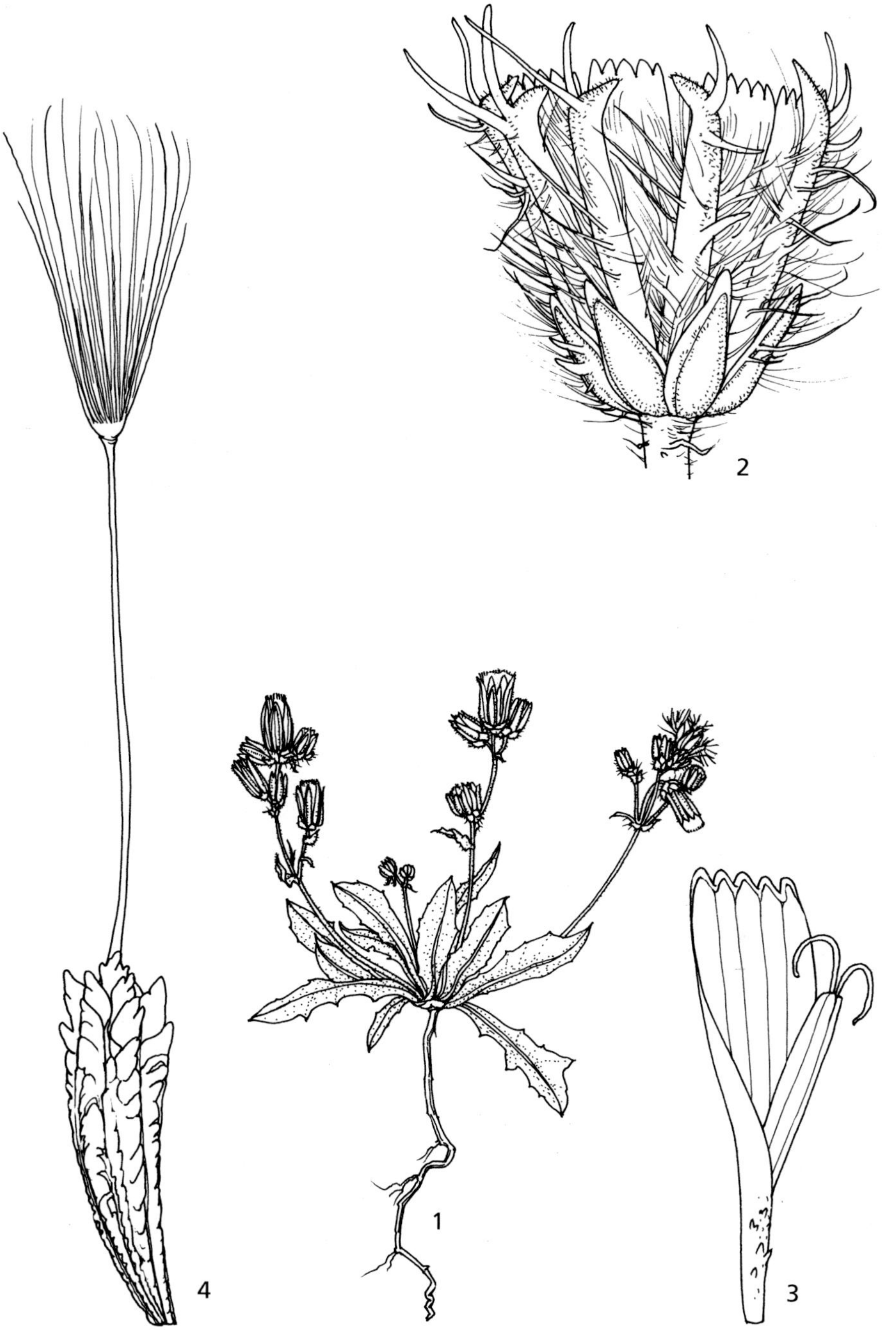

Fig. 37. **Heteroderis pusilla**. 1, habit × 1/2; 2, capitulum × 12; 3, floret × 18; 4, achene and pappus × 18. Artist and specimen details unknown.

Crepis aegyptiaca (Schweinf.) Täckh. & Boulos, Publ. Cairo Univ. Herb. 5: 31 (1974).
Heteroderis pusilla var. *gymnocephala* Rech.f. in Fl. Iranica 122: 292 (1977).

Annual herb with basal leaves in a rosette and a few cauline leaves. Stems few, weak, decumbent to ascending or more rarely erect, (4–)10–25(–35) cm tall, subglabrous to puberulent and/or setose. Basal leaves narrowly obovate to obovate-elliptic, 2–9(–12) × 0.5–3 cm, contracted gradually to a petioled base, obtuse, sinuate-pinnatifid to dentate, teeth white-cartilaginous, lamina glabrous or almost so. Stem leaves smaller, auriculate to amplexicaul, often lacking or bract-like. Capitula in terminal clusters of 1–5, subsessile, pendent or horizontal; involucre oblong, 6–8 × 2.5–4.5 mm at flowering; outer phyllaries 2–3 mm long, broadly ovate, green with a hyaline margin; inner phyllaries lanceolate, green with narrow pale margins, glabrous or bearing coarse setae at least along midrib, sometimes floccose-puberulent, lengthening in fruit and lower part of midrib becoming corky and prominent. Outer achenes 4 mm long, slightly curved, with 5 broad, pale ribs; inner achenes grey, body 2.5–3 mm long, coarsely ribbed, muricate especially above, expanding to a small disc at apex of filiform, 4–5.5 mm beak; pappus white, *c.* 4 mm, partially exserted from involucre, sometimes absent from outer achenes. Fig. 37: 1–4.

HAB. In the desert; alt. 100–200 m; fl. & fr. Apr.–Jun.
DISTRIB. Rather rare in the desert region of Iraq. **DLJ**: 3 km S of Lake Thirthar, *Chakraverty & Rawi* 30456! **DSD**: 30 km E of Salman, *Rechinger* 13905; Jarishan, *id.* 14462.

Saudi Arabia, Iran, Afghanistan, Pakistan, C Asia (Turkmenia to Tajikistan).

37. **LAPSANA** L.

Sp. Pl.: 811 (1753); Gen. Pl. ed. 5: 353 (1754)
Lack in Kubitzki, Fam. Gen. Vasc. Pl. 8: 186 [2006] (2007).

C. C. Townsend

Annual, biennial or perennial herbs with alternate leaves, glabrous or clothed with whitish eglandular hairs and/or glandular hairs. Leaves simple to lyrate-pinnatisect, ± toothed or upper entire. Inflorescence of small capitula in panicles or corymbs, capitula rather few-flowered, homogamous, ligulate; involucre obconical to cylindrical-campanulate; phyllaries biseriate, inner long, outer small and short; receptacle naked. Rays yellow. Achenes ± compressed to subterete from outer to inner, finely ribbed, soon falling; pappus absent.

Monotypic; N temperate region of the Old World.

Lapsana (from the Gr. λαψανα, *lapsana,* otherwise λαμψανα, *lampsana,* a name given to an edible herb in Dioscorides).

J.-H. Pak, J.-K. & K. Bremer, K. (1995). Phylogeny and reclassification of the genus *Lapsana* (Asteraceae: Lactuceae). Taxon 44 (1): 13–21.

1. **Lapsana communis** *L.*, Sp. Pl.: 811 (1753); Boissier, Fl. Orient. 3: 720 (1875); Dinsmore in Post, Fl. Syria, Palest. & Sinai ed. 2, 2: 126 (1933); Rawi in Dep. Agr. Iraq Tech. Bull. 14: 199 (1964); Vassilchenko in Fl. S.S.S.R. 29: 401 (1964); Sell in Fl. Turkey 5: 785 (1975); Rechinger, Fl. Iranica 122: 221 (1977); Ghafoor in Fl. Pakistan 222: 64 (2017).

Lapsana sonchifolia Gilib., Fl. Lit. Inch. 2: 243 (1782), nom. inval., opera utique oppressa.
L. crispa Willd., Sp. Pl., ed. 4, 3(3): 1624 (1803).
L. grandiflora M.Bieb., Fl. Taur.-Caucas. 2: 261 (1808).
L. lyrata Willd., Enum. Pl. [Willdenow] 2: 832 (1809).
L. pubescens Hornem., Hort. Reg. Bot. Hafn. Suppl.: 91 (1819).
L. intermedia M. Bieb., Fl. Taur.-Caucas. 3: 540 (Dec. 1819 or early 1820).
L. [as *Lampsana*] *glandulifera* Cass., Dict. Sci. Nat., ed. 2, 25: 211 (1822), nom. illeg. pro *Lapsana* [as *Lampsana*] *lyrata* Willd.
L. sylvatica Wallr., Erst. Beitr. Fl. Hercyn. 2: 125 (1840).
L. adenophora Boiss., Diagn. Pl. Orient., ser. 1, 4: 23 (1844).
L. pisidica Boiss. & Heldr., Diagn. Pl. Orient., ser. 1, 11: 33 (1849).
L. cassia Boiss., Diagn. Pl. Orient., ser. 1, 11: 33 (1849).
L. ramosissima Boiss., Diagn. Pl. Orient., ser. 1, 11: 34 (1849), nom. illeg. later homonym, non Crantz (1766).

L. alpina Boiss. & Balansa, Diagn. Pl. Orient., ser. 2, 6: 117 (1859).
L. macrocarpa Coss., Bull. Soc. Bot. France 9: 173 (1862).
L. peduncularis Boissier, Fl. Orient. 3: 720 (1875), nom. illeg. superfl. pro parte, pro *L. pisidica* & *L. cassia*, pro parte *L. ramosissima.*
L. communis L. subsp. [as *] *macrocarpa* (Coss.) Nyman, Consp. Fl. Europ. pt. 4: 474 (Oct. 1879).
L. communis L. ?subsp./?var. β *pubescens* (Hornem.) Arcang., Comp. Fl. Ital.: 411 (1882).
L. intermedia M.Bieb. subsp. *glandulosa* Freyn & Sint., Oesterr. Bot. Zeit. 42(8): 266 (1892). Note: Originally published as '*Lapsana glandulosa* Freyn et Sint. – Subspecies *Lapsanae intermedia* MB., ...' but alterable under Art. 24.4.
L. cancellata Borbás, Oesterr. Bot. Zeit. 43: 69 (1893), nom. nud. pro syn.
L. olympica P. Candargy, Bull. Soc. Bot. France 44(6): 147 (1897).
L. grandiflora M. Bieb. f. *pinnatisecta* Sommier & Levier, Trudy Imp. S.-Peterburgsk. Bot. Sada 16: 289 (1900).
L. communis L. subsp. *intermedia* (M.Bieb.) Hayek, Prodr. Fl. Pen. Balc. 2: 803 (1931).
L. communis L. f. *adenoclados* (Borbás) Hayek, Prodr. Fl. Balc. 2: (1931).
L. communis L. subsp. *adenophora* (Boiss.) Rech.f., Fl. Aeg.: 673 (1943).
L. communis L. subsp. *pisidica* (Boiss. & Heldr.) Rech.f., Fl. Aeg.: 674 (1943).
L. communis L. subsp. *ramosissima* Rech.f., Fl. Aeg.: 674 (1943).
L. cappadocica Bornm., Repert. Spec. Nov. Regni Veg. Beih. 89: 392 (1944).
L. pinnatisecta (Sommier & Levier) Ter-Chatsch., Zametki Sist. Geogr. Rast.. 15: 77 (1949).
L. aipetriensis Vassilcz., Fl. U. R. S. S. 29: 404, 727 (1964).
L. communis L. subsp. *alpina* (Boiss. & Balansa) P.D.Sell, Notes Roy. Bot. Gard. Edinburgh 33(3): 432 (1975).
L. communis L. subsp. *grandiflora* (M.Bieb.) P.D.Sell, Notes Roy. Bot. Gard. Edinburgh 33(3): 432 (1975).
L. communis L. subsp. *pinnatisecta* (Sommier & Levier) Greuter, Willdenowia 36(2): 713 (2006).

Annual, biennial or sometimes perennial herb, (10–)30–90(–125) cm tall, with slender branches from below or more commonly above middle, stem and branches glabrous or glandular, or densely white-pilose below. Basal leaves petiolate, ovate and dentate or more frequently lyrate-pinnatifid to lyrate-pinnate, very variable in size, both the large deltoid-ovate to ovate ± acute terminal lobe and the much smaller mostly 1–3 pairs of lateral lobes dentate, as is the wing along the rhachis in pinnatipartite leaves; lower and median cauline leaves similar and somewhat reduced and sessile, upper similar or lanceolate and dentate or entire; inflorescence leaves rapidly reducing and entire, above small and bract-like; all leaves thinly to more densely hairy. Upper branches and pedicels slender, glabrous or glandular-hairy. Capitula usually numerous, in a corymbose panicle; involucre c. 3–10 mm wide; phyllaries glabrous or glandular, pale green, outermost few very short, ovate-lanceolate, inner linear-oblong, contracted to a rather blunt, sometimes shortly pilose apex, 5–10 mm long, erect after fruit-fall. Corollas c. 6–18 mm, rays 4–14 mm, tube pubescent; styles blackish-green to greenish-yellow. Achenes finely ribbed, usually slightly compressed, outer c. 3.5–4.5 mm, inner shorter.

HAB. Rocky mountain slopes, valleys in the mountains, amongst *Quercus, Juglans* and *Salix* trees, usually near water; 700–2400+ m; fl. & fr. May-Sep.

DISTRIB. Frequent, locally abundant, in the mountain region of Iraq. **MAM**: Gara Dagh, *Dabbagh & Jasim* 46952!; Sarsang, *Haines* 1724!, *Rechinger* 11686. **MRO**: Kani Khanjar Khan, *Rawi & Serhang* 24710!; Algurd Dagh, valley above Nawanda, *Rechinger* 11894; Pushtashan, *Rawi & Serhang* 26557!, *Rechinger* 11061; Hisar-i Sakran, *Hadač et al.* 5632!; Sakran mt., 15 km SE of Chuman, *Dabbagh & Hamad* 46252!, 46224!; Haji Umran, *Qaisi, Karim & Hamad* 49771!; Magar Range, Haj Omran, *Rawi & Sarhang* 24311!; Haj Omran, *Al-Dabbagh & K. Hamid* 46264! **MSU**: Qopi Qaradagh, *Haines* 1162!; Tawila, *Rechinger* 13402! **FPF**: Biggatoda village 6 km S Kani Masi, *S. Omar & Al-Dabbagh* **45653!**

Almost throughout Europe; Turkey, Lebanon, Caucasia, Iran, Pakistan, C Asia (Tian Shan), E & W Siberia, Macaronesia (Canary Is., Madeira), N Africa (Libya to Morocco). Introduced in the Azores, Australasia, tropical Africa, N America, W Indies and probably elsewhere.

Rechinger in Fl. Iranica 122: 221–113 (1977) recognized three species of *Lapsana* in Iraq: *L. communis, L. intermedia* M. Bieb. and *L. grandiflora* M. Bieb. Sell in Tutin et al;. Fl. Europaea 4: 344 (1976) and Davis, Fl. Turkey 5: 785–787 (1975) regarded all of these a subspecies of *L. communis,* and "rather poorly defined" subspecies at that. I have been unable confidently to assign the Iraqi herbarium material (much of which is past flowering) to these taxa even at subspecific rank, and indeed Rechinger observed of the material which he assigned to *L. grandiflora* that it agrees with Caucasian material of this species well on involucre length but agrees better with *L. communis* in aspect and colour. Although extremes of variation can look very different (for example, "L. intermedia" is introduced in Great Britain and looks different from local *L. communis*), throughout the range variation appears clinal and impossible to classify. The genus only occurs in the limited mountain region in Iraq, yet specimens referred to the two extremes – *L. communis* and *L. grandiflora* – by Rechinger occur in a single locality, Sarsang. I have thus taken a broad and pragmatic view of the species here. I am aware, however, that this may merely be a confession of failure which future field studies by Iraqi botanists will rectify. Bornmüller also records *L. intermedia* for Iraq in Beih. Bot. Centralbl. 60B: 216 (1939), and Sell also notes it for our country in Fl. Turkey.

Nipplewort (Eng.), so called as the plant was formerly used for treating ulcerated or sore breasts. It was also used as a salad herb.

38. **TARAXACUM** H.G. Weber ex Wiggers

Prim. Fl. Holsat.: 56 (1780)
Lack in Kubitzki, Fam. Gen. Vasc. Pl. 8: 186 [2006] (2007)

R. Doll

Perennial herbs with a milky latex, often arachnoid-pilose. Leaves numerous, congested into a rosette. Scape almost invariably with a single head. Phyllaries biseriate, outer unequal, spirally arranged, some or all callose, corniculate or with a larger horn-like process, sometimes flat, inner equal in length and with broadly membranous margins; receptacle naked. Florets hermaphrodite, almost always ligulate; style bifid. Achenes most frequently with a spinulose apex which passes into a conical or cylindrical pyramid often with a colourless beak, rarely erostrate; pappus of very many numerous delicate barbellate persistent setae, white or somewhat coloured.

Worldwide in distribution, with the exception of the tropical and subtropical regions; there are over 2000 species and microspecies.

Taraxacum (possibly derived from the Pers. *talkh chakok,* "bitter herb", or from the Gr. ταραξος, *taraxos,* or ταραξια, *taraxia,* 'heckler' or 'agitator' in modern Gr. but more likely a reference to a medical disorder, as the plant provided a traditional cure for eye problems.)

So far 28 species are known from Iraq but this list is certainly incomplete, and a number of species must await discovery, especially in the north of the country. Unfortunately, only a small quantity of material has been available for study. Further gatherings of *Taraxacum* from Iraq are much to be desired. I have therefore included in the following account eight species not hitherto found in Iraq but which are very likely to occur there. These are included in the key (without numbers) and are described at the end.

Handel-Mazzetti, H.F. (1907). Monographie der Gattung *Taraxacum* & Österr. Bot. Zeitschr. 72: 254–275.
- (1923). Nachträge zur Monographie der Gattung *Taraxacum.*
Dahlstedt. H. (1926). Über einige orientalische *Taraxacum*-Arten. Acta Horti Bergiani 9: 1–36.
Doll, R. (1973). Revision der Sect. *Erythrosperma* Dahlst. emend. Lindb. f. der Gattung *Taraxacum* Zinn. Teil 1, in Fedde Rep. 83: 673–740.
- (1976); Revision der Sect. *Erythrosperma* Dahlst. emend. Lindb. f. der Gattung *Taraxacum* Zinn. Teil 2, in Feddes Repert. 84: 553–585.

The following should be noted in using the account:
Pyramid: conical or cylindrical apical part of the achene between the achene body and the beak.
Interlobes: the attenuate parts of the lamina of the leaf occurring between the lobes.
Outer phyllaries: are said to be "narrowly margined" if the margins are 0.05–0.25 mm broad, "broadly margined" if the margins are broader than 0.3 mm.
Measurements: *capitulum* width includes the ligules; *achene length* is measured excluding the pyramid unless indication is given to the contrary.

For certain identification both flowers and mature achenes are necessary, so collectors should take care to collect such material. Pappus colour must always be observed on mature achenes.

1. Achenes with either pyramid or beak lacking 2
   Achenes with both pyramid and beak 3
2. Pappus white . . . . 3. *T. kurdicum*
   Pappus reddish-brown . . . . 1. *T. assemanii*
3. Pappus ± coloured, tinged reddish or pale brown, not white 4
   Pappus white 8
4. Outer phyllaries ± ovate, broader than inner, broadly white-margined, ± appressed . . . . T. oliganthum
   Outer phyllaries distinctly narrower 5
5. Leaves fleshy; plant halophilous 6
   Leaves leathery or herbaceous, not fleshy; plants not halophilous 7
6. Achenes 6–7 mm long including pyramid, stramineous or orange, gradually passing into pyramid . . . . 7. *T. anatolicum*
   Achenes 4–5 mm long including pyramid, stramineous to pale brown, pyramid abruptly demarcated . . . . T. bessarabicum
7. Beak 3–5 mm long; mountain species . . . . 2. *T. microcephaloides*
   Beak 6–9 mm long; steppe and ruderal species . . . . T. serotinum
8. Beak 1–4(–5) mm long 9
   Beak longer than above 11
9. Leaves undivided, margins entire . . . . 4. *T. crepidiforme*
   Leaves with margins not entire, divided 10
10. Style arms yellowish-green, pyramid indistinct, beak 1–3 mm long . . . . . 5. *T. brevirostre*
    Style arms ± dark, pyramid 1 mm long, beak ± 4 mm . . . . 6. *T. iranicum*
11. Leaves leathery, closely felted; outer phyllaries pruinose, mostly distinctly imbricate and broad; scape often with 2–3 capitula and bracts; achenes to 10 mm 12
    Leaves rarely leathery, usually herbaceous, glabrous or slightly hairy; outer phyllaries not or only slightly pruinose, not distinctly imbricate; scape always with only a single capitulum and without bracts; achenes to 5 mm . . 16
12. Corollas pale rose-pink, rarely white . . . . 8. *T. roseum*
    Corollas yellow 13
13. Achenes 3.5–4 mm long . . . . T. farinosum
    Achenes distinctly longer 14
14. Style and style arms purple; achenes fulvous . . . . 9. T. *neospurium*
    Style arms yellow; achenes light brown 15
15. Scape almost always with a single head; outer phyllaries lanceolate; leaves hairy, pyramid 1.2–1.5 mm, beak ± 10 mm . . . . 10. *T. syriacum*
    Scape very frequently divided; outer phyllaries ovate; leaves almost glabrous, pyramid 1.5–2.5 mm long, beak 13–18 mm long . . . . 11. *T. montanum*
16. Pappus snowy-white; phyllaries mostly with a large horn-like process; achenes with pyramid usually only to 3.5 mm long 17
    Pappus white to dirty white; phyllaries exappendiculate to thickened; achenes often larger 21
17. Style arms always yellow; outer phyllaries ovate, with a large horn-like process; leaves mostly only toothed or with very small lobes (to ± 10 mm) . . . . 12. *T. monochlamydeum*
    Style arms yellow to yellow-green or blackish; outer phyllaries ± lanceolate, smooth to horned; leaves always lobed, lateral lobes longer 18
18. Style arms pale yellow . . . . 13. *T. wallichii*
    Style arms yellow-green to blackish 19
19. Pyramid ± 0.7 mm long . . . . 14. *T. nevskii*
    Pyramid 1–1.3 mm long 20
20. Involucre blackish-green; phyllaries with black horns; style arms dark brownish . . . . 15. *T. stanjukoviczii*
    Involucre green; phyllaries with a green swelling; style arms yellowish-green . . . . 16. *T. afghanicum*
21. Achenes stramineous, with a short conical pyramid to 0.5 mm long; outer phyllaries exappendiculate, ± patent to squarrose, without or with only a narrow border . . . . 18. *T. kurdiciforme*

Achenes stramineous or red-brown, with a cylindrical pyramid usually longer than 0.5 mm long; outer phyllaries often appresssed, with or without horn-like processes and often with a distinct to broad border 22
22. Margins of outer phyllaries distinct to broad, often to 1.5 mm broad............. 23
Margins of outer phyllaries narrow, at most to 0.5 mm broad 26
23. Achenes stramineous; plants of damp places .... 19. *T. scaturiginosum*
Achenes brownish to red-brown; plants of dry places.......................... 24
24. Pyramid 1.2–1.5 mm long .... 17. *T. delphicum*
Pyramid only 0.8–1 mm long 25
25. Lobes of leaves blunt; achenes ± 3.5 mm long ....T. minimum
Lobes of leaf pointed, frequently directed backwards; achenes 3.5–4 mm long ....T. hellenicum
26. Plants cushion-forming, with numerous scapes, each surrounded at base by withered remains of stalks and thus strongly thickened........T. dissectum
Plants not cushion-forming, with few scapes and these not so strongly surrounded by withered remains of stalks at base 27
27. Achenes mostly about 4 mm long; plants usually large; fibrous remains of previous years' leaves lacking or weakly developed at base; border of outer phyllaries mostly more than 0.25 mm wide 28
Achenes 2.3–3.5 mm long; plants ± small and short; base usually strongly thickened with remains of previous years' leaves; margins of outer phyllaries at most to. 0.25 mm wide (compare T. dissectum)............. 34
28. Achenes stramineous or brown 29
Achenes reddish-brown or orange 31
29. Outer phyllaries ± patent .... 24. *T. pseudodissimile*
Outer phylleries appressed.. 30
30. Achenes *c.* 4.5 mm long.................................. T. pseudonigricans
Achenes *c.* 3 mm long .... 25. *T. purpureopetiolatum*
31. Plants closely hairy throughout ....20. *T. aellenii*
Plants almost glabrous or only leaves hairy 32
32. Achenes light orange-brown to yellowish-brown....................22. *T. kabulense*
Achenes brownish-red to brick-red.. 33
33. Terminal lobes of leaf relatively small, pyramid 1 mm long...23. *T. pseudocalocephalum*
Terminal lobes large, pyramid 1.5 mm long ....21. *T. calocephalum*
34. Marginal florets tubular; achenes 3–3.5 mm long, stramineous to olivaceous-brown .... 26. *T. laxum*
Marginal florets not tubular; achenes 2.5–2.8 mm long, reddish-yellow to brick-red. 35
35. Achenes reddish-yellow; outer phyllaries narrowly green-bordered; pollen present; involucre (aggregate of inner phyllaries) light green27. *T. hepaticolor*
Achenes brick-red; outer phyllaries narrowly white- or purple-bordered; anthers empty; involucre olive-green, pruinose......................28. *T. trottii*

SECTION **RHODOTRICHA** HAND.-MAZZ. Outer phyllaries mostly reddish, narrowly bordered, narrower than inner. Achenes short-beaked (absent in *T. assemanii*), grey or brownish; pappus brownish-red.

1. **Taraxacum assemanii** *Blanche* ex *Boiss.*, Fl. Orient. 3: 791 (1875); van Soest in Fl. Turkey 5: 794 (1975).

T. primigenium Hand.-Mazz., Monogr. Tarax.: 17, t. 1 f. 2, s. str.; t. 4 f. 2, p.p. (1907) & Ann. Naturh. Mus. Wien 27: 456 (1913); Bornmüller in Beih. Bot. Centralbl. 60B: 221 (1939); Blakelock in Kew Bull. 4: 64 (1949); Rawi in Dep. Agr. Iraq Tech. Bull. 14: 126 (1964); van Soest in Fl. Iranica 122: 230 (1977).
T. primigeniforme van Soest in Proc. Kon. Nederl. Akad. Wetensch. ser. C, 69: 364 (1966).

Plants small, 2–6 cm tall. Leaves glabrous, scarcely toothed, with a pale petiole. Scape slender, pilose. Involucre small; outer phyllaries appressed, broadly lanceolate, broadly white-bordered. Florets ± 30 per capitulum; corollas deep yellow; pollen present; style arms yellowish-green. Achenes smooth, about 5 mm long, beakless, pale stramineous; pappus 3–5 mm, brownish-red to dirty white

HAB. Damp places on high mountains; alt. 2800–3500 m; fl. & fr. Jul.-Aug.

DISTRIB. Occasional in montane Iraq. **MRO**: Algurd Dagh (Helgord), *Guest* 2884!, *Gillett* 9603!; Qandil range, *Rawi & Serhang* 24408!, 24477!, 26766!; Qandil range, E side of slope, *Rawi & Serhang* 24500!

Bulgaria, Syria, Lebanon, Turkey, Iran.

2. **Taraxacum microcephaloides** *van Soest* in Notes Roy. Bot. Gard. Edinburgh 33: 263 (1974); van Soest in Fl. Turkey 5: 794 (1975) & Fl. Iranica 122: 229 (1977); Breckle, Hedge & Rafiqpoor, Vasc. Pl. Afghan. Checkl.: 192 (2013).

T. officinale var. *microcephalum* Boiss., Fl. Orient. 3: 788 (1875);
[*T. microcephalum* (non Pomel) Hand.-Mazz., Monogr. Tarax.: 18 (1907), p.p. incl. t. 1 f. 3 a-b].
T. fulvipile Hand.-Mazz. in Österr. Bot. Zeitschr. 72: 258 (1923), non Harv. (1865).

Plants 3–9 cm tall, hairy. Leaves ± erect, glabrous or hairy, linear, shortly lobed. Scapes slender, numerous, closely hairy to glabrescent. Involucre small, often strongly hairy; outer phyllaries ± appressed, ± ovate, distinctly white-bordered, frequently thickened. Corollas yellow; pollen present; style arms yellowish-green. Achenes *c.* 5 mm, sparingly spinulose, stramineous, pyramid indistinctly demarcated, beak thick, *c.* 3 mm; pappus white.

HAB. On the high mountains; alt. 1700–2400 m; fl. & fr. Aug.-Sep.
DISTRIB. Rare. **MRO**: Hisar-i Rost, *Guest & Husham (Alizzi)* 15811!; Haji Umran, *Rawi & Serhang* 24304!, 24319!

Turkey, Syria, Lebanon, Iran, Afghanistan, Transcaucasia, N Africa (Algeria, Morocco).

The material from Haji Umran referred here has a dirty white pappus and deserves further study.

SECTION **ORIENTALIA** HAND.-MAZZ. Leaves not or little toothed, lobed or entire. Outer phyllaries ± appressed, mostly broadly white-margined. Corollas yellow to orange, rarely white or rose-lilac. Achenes large, smooth to sparingly spinulose, light brown or stramineous to blackish-grey, beaked (except for *T. kurdicum*), beak mostly shorter than fruit, rarely longer; pappus ± white.

Subsection ORIENTALIA. Plants larger, glabrous or thinly hairy. Outer phyllaries narrowly or broadly white-bordered, ± appressed, ± ovate to broadly lanceolate. Achenes 3–5 mm long, beak absent, longer or shorter than fruit; pappus white.

3. **Taraxacum kurdicum** *Hand.-Mazz. ex Nábělek* in Publ. Fac. Sci. Univ. Masaryk 52: 52 (1925), non Haglund (1939); Rawi in Dep. Agr. Iraq Tech. Bull. 14: 125 (1964).

T. paradoxum Hand.-Mazz. in Ann. Naturh. Mus. Wien 27: 457, f. 5, 1 (1913).
T. crepidiforme DC. subsp. *kurdicum* (Hand.-Mazz. ex Nábělek) van Soest in Act. Bot. Neerl. 17: 487 (1968), Fl. Turkey 5: 803 (1975) & Fl. Iranica 122: 246 (1977).

Leaves ± without teeth, entire-margined, with a pale petiole. Outer phyllaries ± appressed, exappendiculate, often broadly white-margined. Pollen present; style arms dark. Achenes stramineous, erostrate; pappus white.

HAB. Damp places in the alpine zone of Iraq; alt. 3500 m; fl. & fr. Aug.
DISTRIB. Only hitherto found on the highest mountain in Iraq. **MRO**: Algurd Dagh (Helgord), *Guest* 3070!, *Rechinger* 11457!.

Turkey, Iran.

4. **Taraxacum crepidiforme** *DC.*, Prodr. 7 : 149 (1838); Boissier, Fl. Orient. 3: 789 (1875); Schischkin in Fl. S.S.S.R. 29: 543, t. 25 f. 2 (1964); van Soest in Fl. Turkey 5 : 802 (1975) & in Fl. Iranica 122 : 246 (1977); Breckle, Hedge & Rafiqpoor, Vasc. Pl. Afghan. Checkl.: 191 (2013).

T. psychrophilum Boiss., Diagn. Pl. Orient., ser. 1, 11: 48 (1849).
T. integrifolium C. Koch in Linnaea 23 : 676 (1850).
T. stevenii Hand.-Mazz., Monogr. Tarax. : 52, t. 2 f. 4, t. 4 f. 9 (1907), s. str.
T. litvinovii Shishk. in Grossh., Fl. Kavk. 4 : 250 (1934); Shishkin in Fl. S.S.S.R. 29 : 555, 752 (1964).

Plants ± glabrous. Leaves ± without teeth, entire-margined, with a pale, often winged petiole. Involucre dark green; outer phyllaries ± appressed, exappendiculate, often broadly white-margined, sometimes also green with a black central area. Corollas purple on outer surface; pollen present; style arms dark. Achenes *c.* 4 mm, stramineous, beak 3–6 mm; pappus white.

HAB. On damp high alpine meadow at or above ± 3000 m.
DISTRIB. Only collected once in Iraq. **MRO**: Algurd Dagh (Helgord), Shirzawa (?Shahrzur), *Hadač et al.* 596887!

Greece, Turkey, Caucasia, Iran, Afghanistan.

5. **Taraxacum brevirostre** *Hand.-Mazz.*, Monogr. Tarax.: 46, t. 1 f. 18 (1907); van Soest in Fl. Turkey 5: 796 (1975) & in Fl. Iranica 122: 234 (1977); Breckle, Hedge & Rafiqpoor, Vasc. Pl. Afghan. Checkl.: 191 (2013); Sultan Abedin in Fl. Pakistan 222: 135 (2017).

Plants 3–9 cm tall, hairy. Leaves ± erect, glabrous or hairy, linear, shortly lobed. Scapes slender, numerous, closely hairy to glabrescent. Involucre small, often strongly hairy; outer phyllaries ± appressed, ± ovate, distinctly white-bordered, frequently thickened. Corollas yellow; pollen present; style arms yellowish-green. Achenes *c.* 5 mm, sparingly spinulose, stramineous, pyramid indistinctly demarcated, beak thick, *c.* 3 mm; pappus white. Fig. 38: 1–2.

HAB. On or near the summit of a high mountain; alt. 3000–3800 m; fl. & fr. ?Aug.
DISTRIB. Very rare in Iraq; only found once in the central region of the alpine zone. **MRO**: Algurd Dagh (Helgord), *Rechinger* 11879!

Turkey, Iran, Afghanistan, N Pakistan, Kashmir, C Asia, Tibet.

T. brevirostre can be expected to occur more widely in Iraq. Abedin in Fl. Pakistan p. 136, recognizes var. lanatum (Soest) Abedin, of plants hairy with involucre densely hairy.

Subsection OLIGANTHA (van Soest) Doll [Sect. *Oligantha* van Soest]. Outer phyllaries broadly white-bordered, appressed, ± ovate. Plants small and often strongly hairy. Achenes 4–5 mm long, beak shorter; pappus brownish or white.

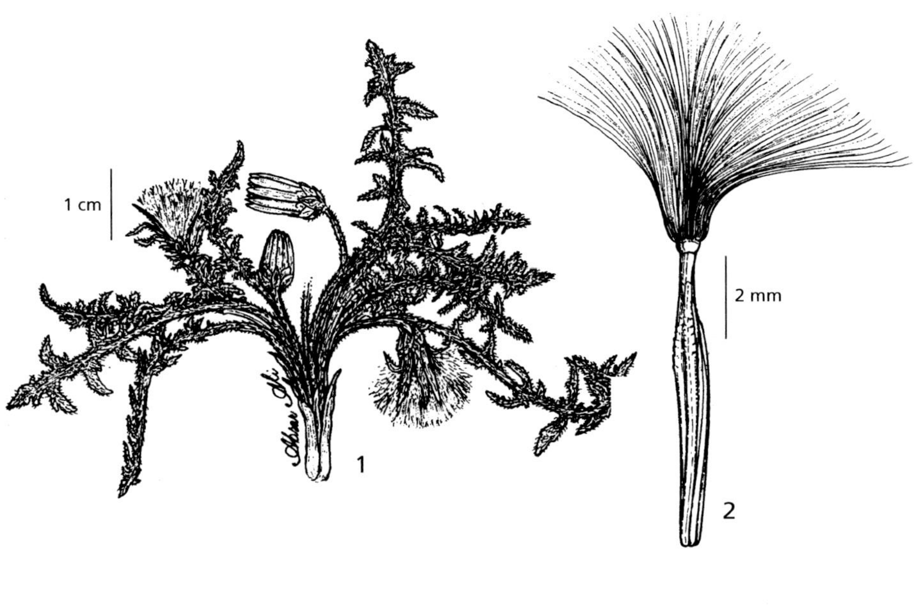

Fig. 38. **Taraxacum brevirostre**. 1, habit; 2, achene. Drawn by Abrar Ali. Reproduced with permission from Flora of Pakistan 222, fig. 4 D–E.

6. **Taraxacum iranicum** *van Soest* in Act. Bot. Neerl. 9: 302, f. 1 (1960) & Fl. Iranica 122: 235 (1977); Breckle, Hedge & Rafiqpoor, Vasc. Pl. Afghan. Checkl.: 192 (2013).

Plants ± 4 cm tall, densely hairy. Leaves lobed, with a pink petiole; lobes with coarse teeth. Involucre large; outer phyllaries narrowly ovate, broadly bordered, gibbous. Corollas yellow; pollen present; style arms ± dark. Achenes olivaceous-brown to stramineous, 5.5 mm including pyramid, beak *c.* 4 mm; pappus white.

HAB. Roadsides on the upper plain of Mosul; alt. 250–300 m; period of fl. & fr. not recorded.
DISTRIB. Rare in Iraq, and only recorded twice. **FUJ**: 20 km and 65 km E of Balad Sinjar, *Hadač* 596875! & 596874!

Iran, Afghanistan.

SECTION **LEPTOCEPHALA** VAN SOEST. [*Taraxaca rhodotricha* Hand.-Mazz., p.p.]. Plants glabrous, short, halophilous. Involucre often reddish. Corollas deep yellow. Achenes beaked; pappus reddish.

7. **Taraxacum anatolicum** *van Soest* in Act. Bot. Neerl. 17: 483, f. 1 (1968), Fl. Turkey 5: 793 (1975) & Fl. Iranica 122: 231 (1977).

Plants hairy below, *c.* 10 cm tall. Leaves grass-green, leathery, terminal lobe very large. Involucre red; outer phyllaries loosely appressed, lanceolate. Corollas pale yellow; pollen present; style arms pale yellow. Achenes to 7 mm, orange or stramineous, beak 3–4 mm; pappus reddish-brown.

HAB. &. Very rare in Iraq; only found once in the alpine region on a mountain slope near the frontier with Iran; alt. 2800–3000 m; fl. Aug.
DISTRIB. **MRO**: Qandil range, *Rawi & Serhang* 26767!

Turkey, Iran.

SECTION **SPURIA** DC. Achenes beaked, very large, mostly stramineous, ± strongly spinulose; pappus white. Scape hairy, often branched. Involucre pruinose. Leaves leathery, densely hairy, sinuately dentate.

8. **Taraxacum roseum** *Bornm. ex Hand.-Mazz.*, Monogr. Tarax.: 125, t. 3 f. 16 (1907); van Soest in Fl. Iranica 122: 281 (1977); Breckle, Hedge & Rafiqpoor, Vasc. Pl. Afghan. Checkl.: 194 (2013).

Plants only 5–8 cm tall, hairy towards base. Leaves leathery, grey-green, almost glabrous, terminal lobe large. Involucre grey-green; outer phyllaries appressed, narrowly ovate, distinctly bordered, thickened. Corollas rose-pink; pollen present; style arms yellow. Achenes light brown, 6 mm, pyramid *c.* 1 mm long, beak *c.* 7 mm; pappus white, *c.* 10 mm.

HAB. Dry hill slopes in the alpine region; alt. ± 2000 m; fl. & fr. Oct.
DISTRIB. Only collected twice in one restricted area. **MSU**: E slope of Pira Magrun, *Hadač* 596883!; slope above the cemetery, Zewiya, *Hadač* 596864!

Iran, Afghanistan.

9. **Taraxacum neospurium** *van Soest* in Act. Bot. Neerl. 9: 314, f. 11 (1960). van Soest in Fl. Iranica 122: 281 (1977).

Plants ± 8 cm tall, hairy. Leaves ± dark green, hairy, toothed, lobed, lateral lobes acute. Scape densely hairy, sometimes forked, shorter than leaves. Involucre to 25 mm long; outer phyllaries loosely appressed, imbricate, ovate-lanceolate, with bright pale green borders, thickened; anthers empty; style arms ± purple. Achenes fulvous, *c.* 9 mm including pyramid, pyramid conical, beak *c.* 9 mm; pappus white, *c.* 10 mm.

HAB. Only found once in Iraq, in a dry spot in the alpine region; alt. 2600–3000 m; fl. ?
DISTRIB. **MRO**: Algurd Dagh (Helgord), in the valley above Nawanda, *Rechinger* 11891!

Iran.

10. **Taraxacum syriacum** *Boiss.*, Diagn. Pl. Orient., ser. 2, 3 : 96 (1856); Hand.-Mazz., Monogr. Tarax. 125, t. 3 f. 17, t. 5 f. 9 (1907); van Soest in Fl. Turkey 5: 797 (1975) & Fl. Iranica 122 : 282 (1977); Breckle, Hedge & Rafiqpoor, Vasc. Pl. Afghan. Checkl.: 194 (2013).

T. denudatum Boiss. & Noë in Boiss., Diagn. Pl. Orient., ser. 2, 3: 95 (1856).
T. montanum (C.A. Mey.) DC. var *denudatum* (Boiss. & Noë) Boiss., Fl. Orient. [Boissier] 3: 786 (1875).

Plants 8–25 cm tall, grey. Leaves leathery, usually densely hairy, grey-green, with few teeth. Scape densely hairy. Involucre blue-green, pruinose, to 4 cm long; outer phyllaries imbricate, ± appressed, lanceolate, bordered. Corollas pale yellow; pollen present; style arms yellow. Achenes *c.* 8 mm including pyramid, pyramid cylindrical, 1.2–1.5 mm, beak 9–12 mm; pappus white, 8–12 mm.

HAB. Dry places in the mountains; alt. 1060–3400 m; fl. & fr. Jun.-Aug.
DISTRIB. Apparently widespread in the mountains of Iraq. **MAM**: 30 km NE of Zakho, *Rawi* 23516!; Sersang, *Haines* 579! **MRO**: NE of Rania, *Rawi & Serhang* 26822!; Algurd Dagh (Helgord), *Gillett* 12367!, *Rechinger* 11856!; Qandil range above Pushtashan, *Rechinger* 11806! **MSU**: Mt. Avroman (Hawraman), *Rawi, Chakravarty & Noori* 19787!

Syria, Lebanon, Turkey, Transcaucasia, Iran, Afghanistan, C Asia, Kashmir (Karakoram).

11. **Taraxacum montanum** (*C.A.Mey.*) *DC.*, Prodr. 7: 154 (1838); Hand.-Mazz., Monogr. Tarax.: 127, t. 3 f. 18 (1907); Shishkin in Fl. S.S.S.R. 29: 430, t. 25 f. 3 (1964); Rawi in Dep. Agr. Iraq Tech. Bull. 14: 126 (1964); van Soest in Fl. Turkey 5: 797 (1975) & Fl. Iranica 122: 280 (1977); Gabrielian & Fragman-Sapir, Flow. Transcaucasus & Adj. Areas: 90 (2008).

Leontodon montanus C.A.Mey., Verz. Pfl. Cauc.: 58 (1831).
Lasiopus sonchoides D.Don in Sweet, Br. Fl. Gard. 7: t. 346 (1836).
Dubyaea sonchoides (D.Don) DC., Prodr. 7: 247 (1838).
T. sonchoideum Sch. Bip., Del. Sem. Hort. Bot. Heidelberg: 4 (1850), sec. Pollichia: 318 (1866).

Plants very large, to 50 cm tall. Leaves large, broad, grey-green, lobed or undivided. Scape usually densely hairy, often forked. Involucre large, pruinose; outer phyllaries imbricate, ± appressed to patent-erect, ovate, obscurely bordered; pollen present; style arms yellow. Achenes 7–8.5 mm, strongly spinulose, pyramid 1.5–2.5 mm, beak (10–)1–18 mm; pappus white, 8–12 mm.

HAB. On dry rocky slopes and near streams in the mountains; alt. 1100–2200 m; fl. & fr. Jul.-Aug.
DISTRIB. Apparently quite frequent in northern Iraq. **MAM**: Betanora vallage, 7 km E Kani-Masi, *Botany staff* 43855 (BAG!) **MRO**: Seri Hasan Beg, *Guest & Ludlow-Hewitt* 2914!; Baradost, *Thesiger* s.n.!; Zirva, N of Rowanduz, *Gillett* 9683!; Kani Rash in Qurnago valley, *Rawi & Serhang* 26673!; NE of Rania, on lower slopes of Qandil, *id.* 18303!; Haji Umran, *AL-Kaisi, Karim & Hamad* 49784 (BAG!) **MSU**: 9 km from Chuwarta, *Noori & Hamid* 41259!

Palestine, Lebanon, Syria, Turkey, Transcaucasia, Iran, Afghanistan, C Asia.

SECTION **MONGOLICA** DAHLST. Achenes stramineous, rarely brownish or yellowish-brown, 2–4.5 mm long, ± strongly spinulose, beak longer than body of fruit; pappus white, pyramid shortly conical to cylindrical. Outer phyllaries appressed to patent, narrowly ovate to lanceolate, gibbous to furnished with a horn-like process, with a network of veins.
Subsection MACROCORNUTA (van Soest) Shishkin [Sect. Macrocornuta van Soest]. Achenes small, to 3.5 mm long including pyramid, stramineous to brownish, beak long; pappus snow-white. Involucre in young state often almost coroniform; phyllaries with large horn-like processes. Capitulas small; Ray florets yellow.

12. **Taraxacum monochlamydeum** *Hand.-Mazz.*, Monogr. Tarax.: 43, t. 16 (1907), emend. Haglund in Persson, Bot. Not. 1938: 309 (1938); Rawi in Dep. Agr. Iraq Tech. Bull. 14: 126 (1964); Rechinger, Fl. Lowland Iraq: 674 (1964); van Soest in Fl. Iranica 122: 253 (1977); Breckle, Hedge & Rafiqpoor, Vasc. Pl. Afghan. Checkl.: 192 (2013); Sultan Abedin in Fl. Pakistan 222: 106 (2017).

T. aschabadense Shishkin in Fl. Turkmen.: 318, 382 (1960).

Leaves pale green or yellow-green, glabrous, dentate or with very small lobes, petioles pale or pink. Involucre pale green; outer phyllaries appressed, broadly ovate, grey-violet at tip, with a horn-like process. Corollas pale yellow; pollen present; style arms pale yellow. Achenes 2.2–2.4 mm, pyramid 0.8–1 mm, beak 5–8 mm; pappus snow-white.

HAB. In the lower plains and foothills and the dry steppe and desert region of Iraq; alt. 50–200 m; fl. & fr. Mar.-Jun.

DISTRIB. Occasional and scattered. **FKI/FPF**: Jabal Hamrin, *Bornmüller* 1516! **FPF**: Mandali, *Rawi* 12589! **LCA**: Baghdad, *C. Regel* s.n.!; Babylon, *Handel-Mazzetti* 876!, *Bornmüller* 1516!

Iran, Afghanistan, Pakistan, Turkmenia.

13. **Taraxacum wallichii** *DC.*, Prodr. 7 (1): 147 (1838); Handel-Mazzetti, Monogr. Tarax.: 42, t. 1 f. 15 (1907); van Soest in Fl. Iranica 122: 258 (1977); Breckle, Hedge & Rafiqpoor, Vasc. Pl. Afghan. Checkl.: 194 (2013). Sultan Abedin in Fl. Pakistan 222: 110 (2017).

Plant short, ± glabrous. Leaves dark green, lobed, mid-vein violet, petiole pale to pink, lateral lobes numerous and narrow. Involucre small; outer phyllaries lanceolate, small, bordered, with horn-like processes. Corollas pale yellow; pollen present; styles pale yellow. Achenes stramineous, 2.4 mm, pyramid 0.6 mm, beak 7–9 mm; pappus snow-white, 4 mm.

HAB. Soil containing humus, usually on limestone; alt. in Iraq low, elsewhere to 3500 m; fl. Mar.
DISTRIB. Apparently very rare in Iraq, only found once in the subdesert region. **DSD**: nr Karbala, *Hadač* 596876!

Iran, Afghanistan, Pakistan, Kashmir, India (Himachal Pradesh).

14. **Taraxacum nevskii** *Yuzep.* in Act. Inst. Bot. Acad. Sci. U.R.S.S. ser. 1, 4: 293 (1937); van Soest in Fl. Iranica 122: 255 (1977).

Leaves glabrous, pale green, shortly lobed, petioles pale. Involucre olive-green; outer phyllaries spreading, gibbous to horned, broadly lanceolate, bordered. Corollas pale yellow; pollen present; style arms brownish. Achenes stramineous, 3.5–4 mm, pyramid c. 0.7 mm, beak 8 mm; pappus 6 mm, snow-white.

HAB. In Iraq (when details given) as a weed of cultivation; alt. from near s.l. to the mountains; fl. & fr. Feb.-Apr.
DISTRIB. Occasional in the forest zone, rare in the steppe region, but more common near cities in the desert zone of Iraq. **MRO**: Shaqlawa, *Hadač* 1437!; Sakri-Sakran, *Hadač & Kader* 596873!, *Hadač et al.* 1551! **FPF**: nr Mandali, *Rawi* 12589! **LCA**: "Diyala", *Rawi & Gillett* 10220!; Daltawa (Khalis), *Polunin* 5176!, *Rawi & Gillett* 10220!; Baghdad, *Hadač & Haines* 596878!, 596885!; *Hadač & Agnew* 596879!; Karrada Mariam, *Ahmed* 9826! **LBA**: Basra, *Rogers* 0464!

Iran, Afghanistan, C Asia (Turkmenia to Tadjikistan).

T. nevskii is one of the most frequent species in Iraq.

The name HINDIBA recorded for a specimen of this plant (*Rawi & Gillett* 10220) is almost certainly incorrect – no doubt due to an error on the part of their informant – since HINDIBA is the well-known common name of chicory, *Cichorium intybus* L., in Lower Iraq.

15. **Taraxacum stanjukoviczii** *Shishkin* in Fl. S.S.S.R. 29: 508, 737 (1964); van Soest in Fl. Iranica 122: 258 (1977) Breckle, Hedge & Rafiqpoor, Vasc. Pl. Afghan. Checkl.: 194 (2013).

Plants glabrous. Leaves shortly and bluntly lobed, coarsely toothed, terminal lobe larger. Scape hairy below the capitulum. Involucre blackish-green; outer phyllaries ± ovate, bordered, ± horned. Corollas yellow, margins with dark lines. Achenes stramineous-brownish, 3.5–4 mm long, pyramid 1 mm, beak 7–10 mm; pappus 7–8 mm, snow-white.

HAB. Very rare in Iraq (habitat not recorded); only one record from the central section of the mountains; alt. ± 900 m; fl. (?).
DISTRIB. **MRO**: Shaqlawa, *Hadač* 1437 p.p.!

Iran, Afghanistan, C Asia (Tadjikistan).

The identification of the Iraqi material remains uncertain. I adopt the determination of van Soest only with much hesitation.

16. **Taraxacum afghanicum** *van Soest*, Acta Bot. Neerl. 9: 305 (1960). van Soest in Fl. Iranica 122: 251 (1977); Breckle, Hedge & Rafiqpoor, Vasc. Pl. Afghan. Checkl.: 190 (2013); Sultan Abedin in Fl. Pakistan 222: 104 (2017).

Taraxacum lipskyi Shishkin, Fl. Turkmen. 7: 318, 383 (1960).

Plants 10–25 cm tall. Leaves lobed, toothed, recurved. Involucre green; outer phyllaries broadly lanceolate, spreading, bordered, gibbous to horned. Corollas pale yellow. Achenes stramineous-brownish, 2.2–2.8 mm, pyramid 1–1.4 mm, beak 8–11 mm; pappus 5–6 mm, white.

HAB. Very rare; only found once on the alluvial plain in the sub-desert region; alt. fl. & fr. Mar.
DISTRIB. **LCA**: Daltawa (now known as Khalis), *Hadač et al.* 596880!

N Iran, Afghanistan, C Asia (Turkmenia, Kara Kum).

SECTION **SCARIOSA** Hand.-Mazz. emend. Dahlst. ['*Taraxaca scariosa*' Hand.-Mazz.]. Outer phyllaries broadly ovate to ovate-lanceolate, ± appressed, distinctly and usually broadly white-margined, non-thickened to horned. Corollas yellow; pollen usually present. Achenes red-brown, yellowish-red or stramineous, pyramid (0.5–)1–2 mm, beak 4–9 mm; pappus white, rarely brownish.

17. **Taraxacum delphicum** *Dahlst.* in Acta Hort. Bergiani 9: 9, f. 4 t. 1 (1926); Doll in Fedde Rep. 87: 572, f. 4 t. 25b (1976).

Plants c. 15 cm tall. Leaves deep green, sparingly hairy, lobed; lateral lobes broadly triangular with straight patent teeth, frequently blunt; petiole violet. Outer phyllaries ovate to ovate-lanceolate, loosely appressed, ± broadly white- or violet-margined, thickened or not. Pollen present; style arms yellowish-green. Achenes pale red-brown, 3.5–4 mm, pyramid 1.2–1.5 mm, beak 7–8 mm; pappus 5–6 mm, white.

HAB. & DISTRIB. Only found once in Iraq, presumably as a ruderal, over a century ago. **LCA**: Baghdad, *W.H. Colvill* s.n.!

Greece, Bulgaria, Turkey.

SECTION **TARAXACUM** [Sect. *Vulgaria* Dahlst.; Ser. b. *Vulgaria* Hand.-Mazz.].
Plants mostly hairy and robust, base not thickened with remains of previous years' leaves. Outer phyllaries mostly squarrosse or spreading, unbordered or narrowly bordered, not thickened. Achenes stramineous to light brownish-stramineous, small, with a ± conical pyramid, beak long; pappus white.

This section is widespread in Europe, and coming from the North at the time of the Ice Age, became disseminated. In Iraq the species of this section are only synanthropic. Many of the Asiatic species that have been placed here do not belong to this section. Similarly, *T. kurdiciforme* has yet to be critically reviewed from a taxonomic point of view. In the future further European species will certainly be introduced into Iraq by human agency.

18. **Taraxacum kurdiciforme** *Haglund* in Bot. Not. 1939: 536, f. 1 & 2 (1939); van Soest in Fl. Turkey 5: 804 (1975) & Fl. Iranica 122: 277 (1977).

[*T. kurdicum* (non Hand.-Mazz. ex Nábělek) Haglund in Bot. Not. 1939: 537 (1939).]

Leaves yellow-green, hairy; petiole violet. Involucre ± 15 mm long; outer phyllaries lanceolate, without border, ± squarrose. Corollas deep yellow; pollen absent; style arms green. Achenes *c.* 3.3 mm, stramineous, pyramid 0.7 mm, beak 7–9 mm; pappus white.

HAB. Near a mountain summit; alt. ± 3300 m; fl. & fr. Jul.
DISTRIB. Very rare in Iraq; only found once. **MRO**: Algurd Dagh (Helgord), *Rechinger* 11457!

Turkey, Iran.

SECTION. **PALUSTRIA** DAHLST. Leaves undivided, denticulate or dentate, more rarely lobed or linear. Outer phyllaries broadly ovate, always ± appressed, broadly white- or pink-margined, smooth. Corollas pale- to deep-yellow. Achenes 3–4.5 mm, frequently not spinulose, stramineous, pyramid 1–2.3 mm, ± cylindrical, beak 5–10 mm.

19. **Taraxacum scaturiginosum** *Haglund* in Ark. Bot. 26A (5): 26 (19340; van Soest in Fl. Turkey 5: 804 (1975).

T. murbeckianum Haglund in Bot. Not. 1939: 538, f. 3 (1939).
T. albanicum van Soest in Act. Bot. Neerl. 14: 13, f. 1 (1965).

Plants ± glabrous. Leaves narrow, shortly lobed, terminal lobe elongate; petiole red. Involucre dark green; outer phyllaries ± ovate, loosely appressed, broadly white- or greenish-margined. Corollas yellow; pollen present; style arms dark grey to blackish. Achenes 2.8–3.8 mm long, stramineous, pyramid 1–1.7 mm, beak *c.* 7–8 mm; pappus pure white, c. 6 mm.

HAB. Damp places in the mountains; alt. to ± 2500 m; fl. & fr. Mar.-Jun.
DISTRIB. Rare in Iraq; only found once in the central sector of the montane-forest zone. **MRO**: nr Rowanduz, in the valley Hassar-i Sakran, *Hadač et al.* 596872!

C & S Europe (Switzerland, Poland, Italy, Serbia, Albania, Greece), Turkey.

SECTION **ERYTHROCARPA** (HAND.-MAZZ.) DAHLST. emend. VAN SOEST. Plants ± large and robust. Base not or only scarcely thickened with remains of previous years' leaves. Outer phyllaries mostly broadly lanceolate, ± appressed, more rarely spreading, broadly or narrowly margined, not thickened or gibbous. Corollas light to deep yellow. Achenes mostly about 4 mm long, deep red, more rarely light brown or stramineous, pyramid cylindrical, (0.3–)0.6–1.2 mm, beak *c.* 5–8 (–10) mm; pappus usually white, but not snow-white.

20. **Taraxacum aellenii** *van Soest* in Acta Bot. Neerl. 9: 307, f. 4 (1960). van Soest in Fl. Iranica 122: 261 (1977); Breckle, Hedge & Rafiqpoor, Vasc. Pl. Afghan. Checkl.: 190 (2013).

Plants small, densely clothed at base with pale brown hairs. Leaves hairy; petioles winged, violet; lateral lobes short. Scape hairy. Involucre dark green; outer phyllaries spreading-erect, ovate-lanceolate, ± broadly margined, in part thickened. Pollen present; style arms dark. Achenes red-brown, 3.2 mm, pyramid 0.8 mm, cylindrical, beak 8 mm.

HAB. &. Rare in Iraq; only found once in the mountain region; alt. fl. Mar.
DISTRIB. **MRO**: Jindian, nr Rowanduz, *Guest* 733!

Iran, Afghanistan.

21. **Taraxacum calocephalum** *Hand.-Mazz.*, Monogr. Tarax.: 106, t. 3 f. 8 (1907), emend. Dahlst. in Acta Hort. Bergiani 9 (1): 24, f. 10 & 11 (1926); van Soest in Fl. Iranica 122: 263 (1977).

Plants almost glabrous. Leaves olive-green, lobed; petiole purple, narrow, lateral lobes acute. Involucre c. 12 mm long; outer phyllaries ovate, broadly bordered, ± appressed to somewhat spreading, thickened to horned. Pollen present; style arms green, drying black. Achenes red-brown, rarely stramineous, 4 mm, pyramid cylindrical, 1.5 mm, beak 8–9 mm.

HAB. Very rare in Iraq; only collected twice in the central montane forest region; alt. 1000–2000 m; fl. & fr. May.
DISTRIB. **MRO**: NW of Haji Umran, *Alkas* 18578!; 35 km NW of Arbil, *Barkley* 33. Ir. 4594!

Italy, Greece, Bulgaria, Cyprus, Lebanon, Iran.

22. **Taraxacum kabulense** *van Soest* in Acta Bot. Neerl. 19: 25, f. 3 (1970). van Soest in Fl. Iranica 122: 264 (1977); Breckle, Hedge & Rafiqpoor, Vasc. Pl. Afghan. Checkl.: 192 (2013).

Plants neat, ± hairy towards base. Leaves glabrous; petiole narrow, reddish-violet; lateral lobes narrowly triangular, toothed, terminal lobe elongate. Involucre dark bluish-green; outer phyllaries erect-spreading, ovate, distinctly white-margined, ± thickened. Pollen present; stigmas blackish. Achenes pale orange-brown, *c.* 3–3.5 mm, densely spinulose, pyramid 1–1.5 mm, beak 12 mm.

HAB. Very rare in Iraq; only collected once, presumably as a ruderal; alt. not given; fl. Feb.
DISTRIB. **LCA**: Safiya, nr Baghdad, at the R. Tigris, *Hadač & Agnew* 596877!

Iran, Afghanistan.

The above specimen is very close to this species but may not be identical with it. Further material is needed.

23. **Taraxacum pseudocalocephalum** *van Soest* in Acta Bot. Neerl. 9: 308, f. 5 (1960); Fl. Iranica 122: 266 (1977); Breckle, Hedge & Rafiqpoor, Vasc. Pl. Afghan. Checkl.: 193 (2013).

Plants of moderate size, hairy. Leaves ± glabrous; petiole purple; lateral lobes triangular. Involucre c. 13 mm long; outer phyllaries lanceolate, very distinctly bordered, in part thickened, ± appressed to erect-spreading. Pollen present; stigmas greenish-yellow. Achenes 3 mm long, red-brown, densely spinulose, pyramid 1 mm long, cylindrical, beak 8–9 mm.

HAB. Very rare in Iraq; only collected once in the central sector of the montane forest zone; alt. ± 2200 m; fl. Jun.
DISTRIB. **MRO:** S slope of Karoukh, *Kass, Noori & Serhang* 27466!

Iran, Afghanistan, C Asia (Turkmenia).

24. **Taraxacum pseudodissimile** *van Soest* in Acta Bot. Neerl. 9: 309, f. 6 (1960); Fl. Iranica 122: 267 (1977); Breckle, Hedge & Rafiqpoor, Vasc. Pl. Afghan. Checkl.: 193 (2013); Sultan Abedin in Fl. Pakistan 222: 82 (2017).

Plants of moderate size, ± glabrous. Leaves lobed, lateral lobes triangular to falcate, acute, toothed. Involucre ± pale green; outer phyllaries lanceolate, loosely appressed to erect-spreading, pink with a broad white margin or entirely pink, ± smooth. Pollen present; style arms greenish-yellowish, drying dark. Achenes brownish-stramineous, 3 mm, pyramid cylindrical, 2 mm; pappus *c.* 5 mm, dirty white.

HAB. On damp ground near streams and on slopes in the mountains; alt. 2190–2610 m; fl. & fr. Jun.-Aug.
DISTRIB. Probably quite common in the central sector of the mountain-forest region. **MRO**: Algurd (Helgord) Dagh, at "Sirzawa" (?Shahrazur), *Hadač* 2457 p.p.!; Sarcal (Tenji Saipayan), at foot of Helgord, *Hadač* 59688!; Lake Kermasur, on Qandil Dagh, *Rawi & Serhang* 24141!; Qandil Dagh, *Rechinger* 11827 p.p.!

Iran, Afghanistan, C Asia (Turkmenia, Tajikistan).

25. **Taraxacum purpureipetiolatum** *van Soest* in Act. Bot. Neerl. 9: 310, f. 7 (1960); Fl. Turkey 5: 808 (1975) & Fl. Iranica 122: 269 (1977); Breckle, Hedge & Rafiqpoor, Vasc. Pl. Afghan. Checkl.: 194 (2013). Sultan Abedin in Fl. Pakistan 222: 83 (2017).

Plants relatively small and neat, sparingly hairy. Leaves glabrous, lobed; lateral lobes shortly triangular, acute; petiole narrow, both it and the mid-vein purplish. Scape glabrous. Involucre to 12 mm long; outer phyllaries appressed, ovate, narrowly white-margined, in part thickened. Pollen present; style arms blackish when dry. Achenes stramineous, 3 mm, pyramid cylindrical, 1 mm, beak *c.* 7 mm; pappus white.

HAB. On the lower mountain slopes; alt. 900 m; fl. Apr.
DISTRIB. Very rare; only found once in Iraq. **MRO**: Shaqlawa, *Hadač* 1437 p.p.!

Turkey, Iran, Afghanistan.

SECTION **ERYTHROSPERMA** Dahlst. emend. Lindb. f. [Sect. *Erythrocarpa* Hand.-Mazz., p.p.; Grupp *Erythrosperma* Dahlst.; Subsect. *Erythrosperma* (Dahlst.) Shishkin].]. Plants ± small and short, base mostly strongly thickened with remains of previous years' leaves. Outer phyllaries ovate or lanceolate, rarely linear, appressed, patent or squarrose, usually narrowly white-bordered, thickened to horned. Corollas mostly pale yellow. Achenes c. 2.3–3.5 mm, red-brown, brown-red, red-violet, grey-red, brown, yellowish-red, pinkish-brown or stramineous, pyramid usually cylindrical, c. (0.5–)0.8–1.7 mm, beak c. (5–)7–12 mm; pappus white.

There are few members of this section yet to be discovered in Iraq, as it lies at the easternmost limit of the area of distribution of Sect. *Erythrosperma*. This is not the case with Sect. *Erythrocarpa*; this is of Central Asiatic origin, and I expect this will prove to be represented by more than 7 species in Iraq.

26. **Taraxacum laxum** *Haglund* in Ark. Bot. 5 (1): 470, t. 38 (1959).

Plant 10–15 cm tall. Leaves green; petiole narrow, hairy, shining red-violet; lateral lobes acute, often backwards-directed, toothed, terminal lobe generally large. Scape sparingly hairy, red-violet. Involucre small, dark green; outer phyllaries loosely appressed to spreading, ovate-lanceolate to lanceolate, pale green, narrowly white-margined, thickened. Corollas deep yellow, marginal florets tubular; pollen present; style arms yellow (-greenish). Achenes stramineous (to olive-brown), 3–3.5 mm, pyramid cylindrical, 0.6–0.8 mm, beak *c.* 8 mm; pappus white.

HAB. Very rare in Iraq; only found once in the central sector of the forest zone; alt. ± 1000 m; fl. May.
DISTRIB. **MRO**: Warta, ± 30 km NW of Rania, *Rawi, Noori & Kass* 28794!

Turkey, Lebanon.

Closely related to this species, but perhaps not identical and needs further study. The Warta gathering has no flowers, and the plant should be re-collected.

27. **Taraxacum hepaticolor** *van Soest* in Proc. Kon. Nederl. Akad. Wetensch. Ser. C, 69: 380, f. 11 (1966); Fl. Turkey 5: 809 (1975) & Fl. Iranica 122: 272 (1977).

Plants small, hairy towards base. Leaves shortly lobed, mid-vein ± densely hairy; petiole pink. Involucre pale green; outer phyllaries ovate-lanceolate, narrowly green-bordered, almost appressed, tips recurved, all thickened. Corollas pale yellow; pollen present; style arms dark green. Achenes 2.5 mm, yellowish-red, pyramid cylindrical, 1 mm, beak 8 mm; pappus white.

HAB. Very rare in Iraq; only found once in the lower thorn-cushion zone; alt. ± 2100 m; fl. Jun.
DISTRIB. **MRO**: in the valley of Hassar-i Sakran (Sakri Sakran), *Hadač & Kader* 596872!

Turkey, Iran.

28. **Taraxacum trottii** *van Soest* in Act. Bot. Neerl. 19: 24, f. 1 (1970); Fl. Iranica 122: 273 (1977); Breckle, Hedge & Rafiqpoor, Vasc. Pl. Afghan. Checkl.: 194 (2013).

Plants *c.* 10 cm tall. Leaves grey-green, hairy throughout; petiole narrow, red-violet; terminal lobe sagittate. Involucre olive-green, pruinose; outer phyllaries ovate or broadly lanceolate, squarrose, narrowly white- or purple-bordered, in part thickened or gibbous, inner horned. Corollas pale yellow; anthers empty; style arms black when dry. Achenes 2.8 mm, brick-red, densely spinulose, pyramid cylindrical, *c.* 0.8 mm, beak *c.* 9 mm; pappus white, 6–7 mm.

HAB. Very rare in Iraq; only found once in the central sector of the forest zone in an irrigated garden; alt. ± 950 m; fl. Mar.
DISTRIB. **MRO**: Shaqlawa, *Rawi* 10456!

Iran, Afghanistan.

SPECIES NOT YET RECORDED BUT EXPECTED TO OCCUR IN IRAQ

Taraxacum bessarabicum (*Hornem.*) *Hand.-Mazz.* in Monogr. Tarax.: 26, t. 1 f. 7 (1907); van Soest in Fl. Turkey 5: 792 (1975) & Fl. Iranica 122: 231 (1977); Sultan Abedin in Fl. Pakistan 222: 72 (2017).

Leontodon bessarabicus Hornem., Suppl. Hort. Hafn.: 88 (1819).
Taraxacum leptocephalum Reichb., Fl. Germ. Excurs.: 270 (1831).
T. officinale var. *leptocephalum* (Reichb.) W.D.J. Koch, Syn. Fl. Germ. Helv. ed. 2: 492 (1844); Ic. Fl. Germ. Helv. 19: 25, t. 55 (1406) f. 4 (1860).
T. fulvipile Harv., Fl. Cap. 3: 527 (1859) pro parte.
T. stenocephalum Boiss. & Kotschy in Boissier, Fl. Orient. 3: 790 (1875).
T. serotinum subsp. *bessarabicum* (Hornem.) Hegi, Illustr. Fl. Mitteleur. 6 (2): 1081, f. 764, b & e (1928).

Plants 5–20 cm tall, hairy below but otherwise glabrous. Leaves succulent, grey-green, glabrous, toothed or lobed, lobes triangular. Involucre small; outer phyllaries narrowly lanceolate, appressed or patent, pink- or green-bordered, smooth or thickened. Corollas light yellow, *c.* 40 per capitulum; pollen present; style arms yellow (-green). Achenes 4–5 mm long including pyramid, stramineous to brownish, pyramid abruptly demarcated, subcylindrical; pappus brownish-red, beak *c.* 4 mm.

HAB. In saline places, in steppe or meadows, or by lakes; fl. summer-autumn.
DISTRIB. This species, which has a wide distribution from France to Siberia, China and Mongolia, and also S. Africa (Cape), has not yet been recorded from Iraq; it is very probable, however, that it occurs.

Taraxacum dissectum (*Ledeb.*) *Ledeb.*, Fl. Ross. 2: 814 (1846); van Soest in Fl. Iranica 122: 239 (1977).

Leontodon dissectus Ledeb. in Mem. Acad. Sci. Petersb. 5: 555 (1812).
Taraxacum zermattense Dahlst. in Acta Hort. Bergiani 4: 19, t. 2 f. 46-51 (1907).

Plants small, cushion-forming, neck of root thick and arising from remains of numerous old leaves; petioles pale green; leaf-lobes numerous, coarsely toothed. Involucre small, pale green; outer phyllaries appressed, thickened, bordered. Corollas pale yellow; pollen absent. Achenes pale brown, small, 3–3.5 mm, gradually attenuate into pyramid, beak *c.* 8 mm.

HAB. Subalpine to alpine meadows over 1000 m; fl. summer to autumn.
DISTRIB. Not yet recorded, but to be expected in N. Iraq.

In various forms, difficult of separation, from the Altai through the high Asiatic mountains to the Alps, Pyrenees and Sierra Nevada.

Taraxacum farinosum *Hausskn. & Bornm. ex Penth. & Zederb.* in Ann. Naturh. Mus. Wien 20: 427 (1906); Handel-Mazzetti, Monogr. Tarax.: 124, t. 5 f. 8 (1907); van Soest in Fl. Turkey 5: 797 (1977).

Plants succulent, 5–15 cm tall. Leaves rather thick, grey-green, hairy, broadly lanceolate to ovate, not lobed or toothed. Scape densely arachnoid. Outer phyllaries light reddish, pruinose, with a broad pale margin, appressed, imbricate. Florets *c.* 60 per capitulum; corollas sulphur-yellow; pollen present; style arms yellow. Achenes grey (-brown), 3.5–4 mm long, pyramid to 0.6 mm, beak thick, *c.* 4–5 mm long; pappus white.

HAB. Damp saline meadows and steppes, 800–1200 m; fl. Jun.-Sep.
DISTRIB. Not yet discovered in Iraq, but to be expected.

Turkey.

Taraxacum hellenicum *Dahlst.*, Acta Hort. Bergiani 9: 11-14, f. 5 t. 1 (1926); Doll in Feddes Repert. 87: 572, t. 26a (1976).

Plants 12–15 mm tall. Leaves light green, ± lobed, lobes acute, often backwardly directed; petiole violet. Involucre olive-green; outer phyllaries short, appressed, broadly white-bordered, ovate, tips violet, gibbous to horned. Corollas pale yellow; pollen present; stigmas greenish. Achenes stramineous-brownish, 3.5–4 mm, pyramid *c.*1 mm, beak *c.* 5 mm; pappus white.

HAB. Waysides, dry slopes and hills, saline meadows, waste places; fl. spring to autumn.
DISTRIB. Not yet found in Iraq, but very probably occurs.

Spain, Greece, Serbia, Cyprus, Turkey, Syria, Palestine, Lebanon, N Africa (Algeria).

Taraxacum minimum (*Briganti ex Guss.*) *Terrac.* in Atti Ist. Sc. Nat. Nap. ser. 2, 6: 352 (1869); Dahlstedt in Acta Hort. Bergiani 9: 16, f. 7 t. 1 (1926); van Soest, Fl. Turkey 5: 801 (1975); Doll in Feddes Repert. 87: 571, f. 3a (1976) Boulos & Hind in Fl. Egypt 3: 316 (2002).

Leontodon minimum Briganti ex Guss., Fl. Sic. Syn. 2, 1: 397 (1843).

Plants c. 10 cm tall. Leaves ± glabrous, dentate, lobes blunt, petiole pale. Involucre olive-green, pruinose; outer phyllaries ± appressed, ± ovate, broadly white-bordered, slightly thickened to gibbous. Capitula small. Pollen present; style arms yellowish-green. Achenes yellowish-red to red-brown, *c.* 3.5 mm, pyramid 0.8–1 mm long, beak 5–6 mm.

HAB. Waysides and roadsides, pastures and other dry places, reaching the subalpine zone (± 2000 m); fl. spring to autumn.
DISTRIB. Not yet recorded in Iraq, but to be expected.

S Europe (Spain, Portugal, France, Italy, Greece), Cyprus, Lebanon, Syria, W Turkey.

Taraxacum oliganthum *Schott & Kotschy ex Hand.-Mazz.*, Monogr. Tarax.: 32, t. 1 f. 10 (1907); van Soest in Fl. Turkey 5: 796 (1975) & in Fl. Iranica 122: 236 (1977).

Plants 4–6 cm tall, hairy. Leaves leathery, densely hairy, with few teeth, lobes narrowly triangular. Involucre light green; outer phyllaries appressed, ± ovate, broadly white-bordered, often hairy and thickened. Florets 15–30 in each capitulum; corollas yellow; pollen present; style arms yellow. Achenes stramineous, *c.* 5 mm, beak 3–5 mm; pappus ± brownish-red.

HAB. Dry localities in the alpine region; fl. summer.
DISTRIB. Not yet recorded from Iraq, but to be expected.

Syria, Turkey, Iran.

Taraxacum pseudonigricans *Hand.-Mazz.* in Ann. Naturh. Mus. Wien 20: 427 (1907) & Monogr. Tarax.: 51, 5. 2 f. 3 & t. 4 f. 8 (1907); van Soest in Fl. Turkey 5: 807 (1975).

Plants *c.* 10 cm tall. Leaves deeply lobed, closely toothed (as are interlobes); petiole coloured. Involucre c. 12–15 mm; outer phyllaries appressed, ovate, broadly bordered, central area coloured with a dark mid-vein, indistinctly thickened. Scape glabrous. Pollen present; style arms green. Achenes stramineous, *c.* 4–5 mm, pyramid 1–1.5 mm, attenuate-demarcated, beak *c.* 5 mm.

HAB. Dry places and cliffs; alt. 1500–3000 m.
DISTRIB. Not yet discovered in Iraq, but very probably occurs.

Turkey, recorded as endemic, but no doubt more widely distributed.

Taraxacum serotinum (*Waldst. & Kit.*) *Poir.* in Lam., Encycl. Suppl. 4: 420 (1816); van Soest in Fl. Turkey 5: 795 (1975) & Fl. Iranica 122: 233 (1977).

Leontodon serotinum Waldst. ex Kit., Descr. Ic. Pl. Rar. Hung. 2: 119, t. 114 (1805).
T. libanoticum DC., Prodr. 7: 149 (1838).

Plants to 20 cm tall, densely hairy. Leaves leathery, grey-green, shortly lobed, few-toothed. Scape densely hairy throughout. Involucre 15–20 mm long; outer phyllaries brownish-reddish, lanceolate, patent to erect, scarcely bordered. Pollen present; style arms clear yellow. Achenes *c.* 5 mm, stramineous-brownish, pyramid *c.* 1.2 mm long, beak 6–9 mm long; pappus 6–9 mm, light brownish-red.

HAB. Dry places in steppe, also as a ruderal; alt. 400–2400 m.
DISTRIB. Not yet recorded from Iraq, but should certainly occur.

C & SE Europe (Slovakia, Austria, Hungary, Romania, Bulgaria, Serbia, Ukraine), Crimea, Turkey, Caucasus, Lebanon, W Siberia.

39. **CICERBITA** Wallr.

Sched. Crit.: 433 (1822)
Lack in Kubitzki, Fam. Gen. Vasc. Pl. 8: 187 [2006] (2007)
Lactuca sect. *Cicerbita* (Wallr.) Benth. in Benth. & Hook. f., Gen. Pl. 2: 525 (1873), p.p.
Lactuca sect. *Scariola* Hoffm. in Pflanzenr. IV. 5: 371 (1891), p. min. p.

C. C. Townsend

Perennial herbs, with a taproot or rhizomatous, with a setose or glandular-setose or glandular-pubescent indumentum, sometimes glabrous. Leaves simple to runcinate-pinnatifid or lyrate-pinnatisect, at least the cauline usually distinctly auriculate, glabrous, setulose, glandular-setulose or glandular-pubescent. Involucre ± biseriate, inner phyllaries subequal and longer, outer phyllaries much shorter and of varying length, all glabrous to glandular-setose, glandular-pubescent or crisped-pubescent. Florets homogamous, ligulate; receptacle naked; corolla commonly blue, rarely yellow (not in Iraq). Achenes all similar, elliptic or elliptic-oblong to linear-oblong, ± compressed, ribbed, with a stout beak; pappus biseriate, scabrid or barbellate, inner series long, outer series very short.

About 15 species in temperate Eurasia; one species in Iraq.

Cicerbita (from medieval Latin *cicirbita*, origin unknown).

Beauverd, G. (1910). Revision systematique des sections, especes et varietes du genre *Cicerbita*. Bull. Soc. Bot. Gen. ser. 2, 2: 115–144.
Tuisl, G. (1968). Der Verwandtschaftkreis der Gattung *Lactuca* L. im iranischen Hochland. Ann. Naturh. Mus. Wien 72: 612–638.

1. **Cicerbita adenophora** (*Boiss. et Kotschy*) *Beauv.* in Bull. Soc. Bot. Gen. sér. 2, 2: 125 (1910); Jeffrey in Fl. Turkey 5: 768 (1975).

Lactuca adenophora Boiss. & Kotschy in Boissier, Fl. Orient. 3: 814 (1875).
L. aff. *racemosa* auctt. non Willd.: Blakelock in Kew Bull. 4: 57 (1949); Rawi in Dep. Agr. Iraq Tech. Bull. 14: 118 (1964).

Perennial, probably rhizomatous, 0.5–1 m tall. Stem fistular, finely striate, glabrous and usually streaked or suffused with purple below, becoming ± stipitate-glandular upwards and in the inflorescence. Basal leaves simple to runcinate-pinnatifid, 15–30 × 3.5–10 cm or probably larger, with broad-based patent teeth, broadly to narrowly elliptic, attenuate and subpetiolar below with a sheathing base, broadly acute at apex; lower cauline leaves similar but with a shorter, auriculate base; upper cauline leaves rapidly narrowing and acuminate, much reduced and bract-like in the corymbiform inflorescence, ± glandular-pilose especially on lower surface or rarely glabrous. Capitula numerous (up to 60). Involucre 10–13 mm long, glandular-setose or rarely glabrous. Ligules pale to pale purple-blue, 2–2.4 cm, with scattered fine hairs dorsally at least above. Achenes 6–6.5 mm, mid-brown, abruptly beaked, beak 1 mm, concolorous, elliptical or obovate, compressed with marginal and 1–2 facial ribs, glabrous; pappus very fine, white, inner 7–8 mm, outer very short. Fig. 39: 1–4.

HAB. & DISTRIB. Very rare in Iraq, only found once, well above the tree line, over 80 years ago, in the central sector of the subalpine region of Iraq. **MRO**: Algurd (Helgord) Dagh, *Guest* 2965!

Turkey (SE Anatolia).

40. **LACTUCA** L.

Sp. Pl.: 79 (1753); Gen. Pl. ed. 5: 348 (1754)
Lack in Kubitzki, Fam. Gen. Vasc. Pl. 8: 188 [2006] (2007)
Scariola F.W. Schmidt, Samml. Phys.-Oekon. Ausatze 1: 270 (1795)

Frances G. Davies

Annual, biennial or (not in Iraq) perennial herbs. Stems scapose or leafy, glabrous or with spreading or deflexed bristles. Leaves simple or lobed or runcinate-pinnatisect, base often sagittate and occasionally decurrent. Inflorescence a spike or panicle of erect capitula. Involucre cylindrical; phyllaries multiseriate, gradate, imbricate; receptacle naked. Ray florets yellow or blue, c. 2× as long as tube. Achenes ± compessed, beak shorter to much longer than body, terminating in a disc which bears pappus and floral remains; pappus setae uniseriate, smooth or barbellate, white, wholly deciduous from disc.

About 60 species, in dry habitats and those with a moderate water supply in the northern hemisphere and Africa; 10 species in Iraq.

Lactuca (from Gr. γαλακτικό, *galactico*, having milk).

Feráková, V. (1975). The genus *Lactuca* L. in Europe. Bull. Jard. Bot. Belg. 52: 367–382.
Tuisl, G. (1968). Der Verwandschaftskreis der Gattung *Lactuca* L. im iranischen Hochland und seinen Randgebieten. Ann. Natur. Mus. Wien 72: 587–638.

1. Pappus uniseriate 3
   Pappus biseriate, outer setae small, forming a corona 2
2. Outer phyllaries triangular to lanceolate; achenes 5–7 mm . . . . 9. *L. rechingeriana*
   Outer phyllaries narrowly lanceolate; achenes 8–9 mm . . . . 10. *L. microcephala*
3. Stems woody, longitudinally green and white-banded with decurrent leaf bases; corollas yellow 4
   Stems not woody; corollas yellow or blue 5
4. Biennial up to 1 m tall; mature achenes reddish-black, 10–15 mm . . . . 7. *L. viminea*
   Perennial up to 0.5 m tall; mature achenes light brown, 8–9 mm . . . . 8. *L. orientalis*
5. Lower leaves over 5 cm wide, succulent; cultivated plant . . . . 4. *L. sativa*
   Lower leaves narrower, less than 5 cm wide, not succulent 6
6. Midribs of lower leaves glabrous or weakly hairy; corollas blue when fresh . . . . 6. *L. undulata*
   Midribs of lower leaves frequently bearing stiff bristles on lower surface; corollas yellow when fresh 7

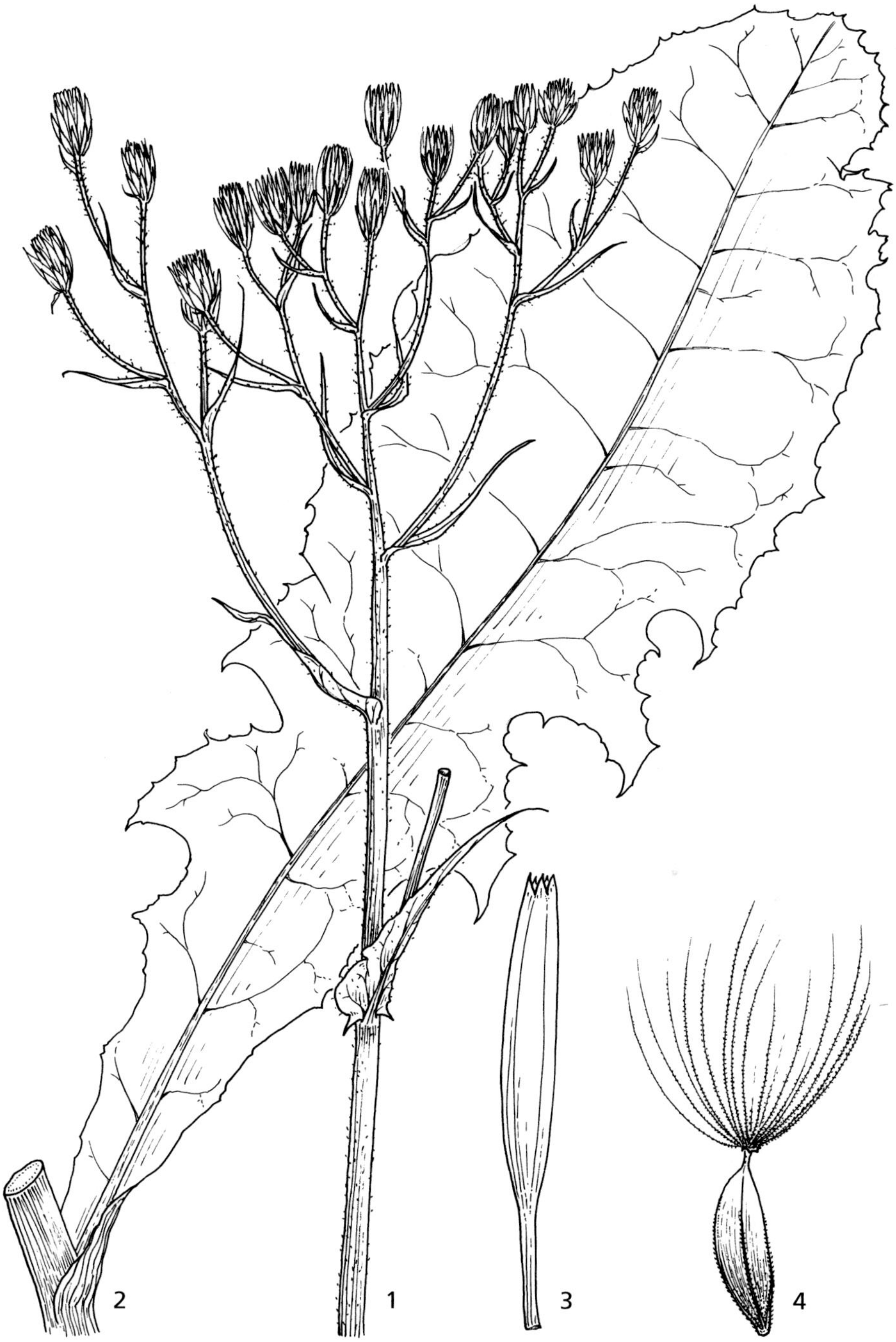

Fig. 39. **Cicerbita adenophora**. 1, habit × 1/2; 2, basal leaf × 1; 3, floret × 6; 4, achene with pappus × 6. 1–3 from Davis 45043; 4 from Davis 24239 (Turkish material). Drawn by Christine M. Grey-Wilson.

7. Stems densely prickly throughout .... 1. *L. aculeata*
   Stems lacking bristles or with a few basal prickly ones 8
8. Achene beak less than half body length; ray limb 12–15 mm. .... 2. *L. scarioloides*
   Achene beak equalling body length; ray limb 6–10 mm. 9
9. Capitula ± sessile; stem leaves usually linear, simple, ± entire .... 5. *L. saligna*
   Capitula pedunculate; stem leaves broader, usually lobed to half width of lamina and sinuate-denticulate. .... 3. *L. serriola*

1. **Lactuca aculeata** *Boiss. & Kotschy ex Boiss.*, Fl. Orient. 3: 809 (1875); Handel-Mazzetti in Ann. Naturh. Mus. Wien 3: 809 (1875); Rawi in Dep. Agr. Iraq Tech. Bull. 14: 118 (1964); Jeffrey in Fl. Turkey 5: 779 (1975); Rechinger, Fl. Iranica 122: 193 (1977).

A robust, erect, prickly biennial herb to 1 m tall. Stem sparingly and widely branched, white, shining, prickly throughout with horizontal (to deflexed near base of plant) rigid setae, shortly glandular above. Leaves sessile, clasping stem and shortly decurrent on it, auriculate, lower leaves obovate, upper leaves oblong, to 7 × 3 cm (in Iraq), coriaceous, pale green, obtuse, margins compoundly prickly-dentate, midrib strongly setose below and lesser veins more weakly setose. Inflorescence with spreading branches and remote, solitary capitula. Involucre about 7 mm long at flowering, increasing to 11 mm long in fruit; phyllaries oblong, greenish-brown, glandular, darker-tipped. Corolla yellow, ligule length not determined. Achenes compressed, body 3–3.5 mm long, fusiform, 10–15-ribbed, brown, with white spinules in upper part, beak 4.5 mm, pale; pappus 3–5 mm, white.

HAB. not recorded; alt. 1100–1500 m; fl. & fr. Aug.-Sep.
DISTRIB. Rare in Iraq; only found from a few localities in the mountain region. **MAM**: Dohuk, *Haines* W.1214!; above Sawara Tuka, *Rechinger* 11604. **MRO**: Safarin (Jabal Sefin), nr Salah ad-Din, *Barkley* 3370!; Haji Umran, *Rechinger* 11304. **MSU**: Dukan to Mirza Rustam, *Rechinger* 10990.

Syria, Turkey (C Anatolia), W Iran.

2. **Lactuca scarioloides** *Boiss.*, Diagn. Pl. Orient. ser. 1, 2: 49 (1847); Boissier, Fl. Orient. 3: 810 (1875); Guest in Dep. Agr. Iraq Bull. 27: 54 (1933); Rawi in Dep. Agr. Iraq Tech. Bull. 14: 118 (1964); Jeffrey in Fl. Turkey 5: 778 (1975); Rechinger, Fl. Iranica 122: 196 (1977); Breckle, Hedge & Rafiqpoor, Vasc. Pl. Afghan. Checkl: 177 (2013).

Biennial erect herb, *c.* 1 m tall. Stem unbranched below, pale, shining, glabrous. Basal leaves oblanceolate, entire or sinuately lobed to runcinate, clasping stem at base. Median leaves runcinate, auriculate, up to c. 16 × 5 mm, obtuse, margin denticulate, midrib white, setulose below, lamina thin, glabrous, bright green or glaucous. Inflorescence with diverging branches bearing capitula on short pedicels furnished with clasping bracts. Capitula oblong to infundibuliform; involucre 11–15 mm long at flowering; outer phyllaries 1–2-seriate, inner phyllaries 8, uniseriate, green, glandular. Corollas yellow; ligules 12–15 mm. Achene body 6 mm, fusiform, black, many-ribbed, apiculate, beak 1.5–2 mm, colourless; pappus 5.5 mm.

HAB. not noted; alt. (1700–)2000–3200 m; fl. & fr. Jun.-Aug.
DISTRIB. Occasional in the mountains. **MRO**: Chiya-i Mandau, nr Walza, *Guest* 2800!; NE of Qandil, *Rawi & Serhang* 24388!; by Kani Mazu Shirin nr Zeyta at foot of mt. Botin, *Hadač et al.* 5979; mts. N of Rania, *Thesiger* 1199; mt. Qandil above Gom-i Kirmosoran lake, *Rechinger* 11127; above Pushtashan, *Rechinger* 11093; Algurd Dagh above Nowanda, *Rechinger* 11409!; Rowandaz, *S. Omar, Sahira, F. Karim & H.* Hamid 38363! **MSU**: Hauraman (Avroman), *Rawi, Noori & Husham (Alizzi)* 29362!; Malakawa, Rawi, *Husham & Noori* 29342!

Turkey (SE Anatolia), Iran, Afghanistan.

3. **Lactuca serriola** *L.*, Cent. Pl. 2: 29 (1756); Rechinger, Fl. Lowland Iraq: 660 (1964); Rawi in Dep. Agr. Iraq Tech. Bull. 14: 118 (1964); Jeffrey in Fl. Turkey 5: 778 (1975); Rechinger, Fl. Iranica 122: 194 (1977); Feinbrun-Dothan, Fl. Palaest. 3: 436 (1978); Boulos & Hadidi, Weed Fl. Egypt: 37 (1984); Boulos & Hind in Fl. Egypt 3: 315 (2002); Roohi & Abid in Fl. Pakistan 222 (Asteraceae IV): 218 (2017).

L. scariola L., Sp. Pl. ed. 2: 1119 (1763), nom. illegit.; Boissier, Fl. Orient. 3: 809(1875), excl. var. β sativa; Dinsmore in Post, Fl. Syria, Palest. & Sinai ed. 2, 2: 145 (1933); Anthony in Notes Roy.

Bot. Gard. Edinburgh 18: 292 (1935); Blakelock in Kew Bull. 4: 59 (1949); Zohary in Dep. Agr. Iraq Bull. 31: 160; Rawi & Chakravarty in Agr. Tech. Bull. 15: 57 (1964); Husain & Kasim, Cult. Pl. Iraq: 116 (1975); Feráková in Fl. Europ. 4: 330 (1976).

Annual or biennial herb, 50–150(–200) cm tall. Stem whitish, shining, entirely glabrous or prickly-setose towards base. Leaves in a short-lived basal rosette and throughout stem, clasping and hastate-auriculate at base, simple or runcinate with 5–11 lobes, 7–10 × 3–5 cm, midrib coarsely setose below, veins with shorter scabrid hairs, lamina ± glabrous, glaucous, rather thick-textured. Upper leaves transitional to simple bracts with decurrent rounded auricles. Inflorescence a panicle with widely-diverging branches. Involucre 11–15 × 4–5 mm at flowering, elongating to 20 mm; phyllaries purple-spotted, margins pale, abruptly contracted above level of achene bodies. Florets 17–25 per capitulum, corollas yellow, tube 3 mm, ligule 6 mm. Achene body 3–4 mm, abruptly narrowed above, strongly compressed, tending to develop lateral wings, many-ribbed, mottled greenish- or yellowish-brown, with brittle, fine spinules on uppermost part, beak equalling body; pappus caducous, 3–5 mm.

HAB. On mountains, in waste land, and as a weed in cultivated areas in southern Iraq; common; alt. 1300–2800 m in the north and 10–360 m in the south; fl. & fr. (May-)Jul.-Aug.(-Sep.).
DISTRIB. Quite common almost all over Iraq, particularly in the northern mountains. **MAM**: Merga Qus, 36 km NE of Zakho, *Rawi* 23528!; Sarsang, *Haines* W.1247! **MRO**: Rowanduz gorge, *Gali Ali Beg, Rawi & Serhang* 23972!; Shaikh wa Shaikhan valley, *id.* 24877!; Galala, nr Rayat, *Guest & Husham (Alizzi)* 15933!; Kermasur lake, Qandil range, *Rawi & Serhang* 24134! **MSU**: Qala Diza, *Rawi* 9366!;Jarmo, *Helbaek* 1998!; Penjwin, *Al-Kaisi & Al-Khayat* 51162! **FUJ**: in an ochard nr Sinjar, *Alizzi & Omar* 35317! **FUJ/FNI**: in an ahrash plantation at Nineveh, *E. Chapman* 25368! **FNI**: Eski Kellek, *Gillett* 8365! **LCA**: Baghdad, *Rogers* 0337!, *Guest* 3229!, *Al-Kaisi* 59284!; Tarmiya, NW of Baghdad, *Haines* W. 18!; Abu Ghraib, *Gill:tt* 10123! **LBA**: Basra, *Memerian* 5144!; Abul-Khasib, *Omar* 34496!

Europe (from Scandinavia southwards to the Mediterranean), Syria, Palestine, Turkey, Caucasia, Iran, Afghanistan, Pakistan, India, China, C Asia (widespread), Siberia, N Africa (Egypt to Morocco). Naturalized in N America and elsewhere.

4. **Lactuca sativa** *L.*, Sp. Pl. ed. 1: 795 (1753); Dinsmore in Post, Fl. Syria, Palest. & Sinai ed. 2, 2: 146 (1933); Guest in Dep. Agr. Iraq Bull. 27: 54 (1933); Anthony in Notes Roy. Bot. Gard. Edinburgh 18: 292 (1935; Rawi in Dep. Agr. Iraq Tech. Bull. 14: 118 (1964); Rawi & Chakravarty, ibid. 15: 57 (1964); Feráková in Fl. Europaea 4: 330 (1976). Zhu Shi & Norbert Kilian in Fl. China 20–21: (2011).

L. scariola L. var. β *sativa* Boissier, Fl. Orient. 3: 809 (1875).

The cultivated lettuce may be recognized by the thin-textured, broad, glabrous rosette leaves. Its fertile parts resemble *L. serriola,* from which it is believed to have originated, except that the achenes vary with the varieties cultivated and may be dark grey or white and often lack the spinules so characteristic of *L. serriola.*

HAB. & DISTRIB. Cultivated in many parts of Iraq and eaten as a vegetable.
LCA: Karrada, S of Baghdad, *Paranjpye* (comm. Graham) s.n.; Za'faraniya, 15 km SW by S of Baghdad, formerly the Horticultural Experiment Station, *Sahira* C. 927!; Hilla, *Paranjpye* (comm. Graham) s.n.; Baghdad, (Karradah) *Paranjpye* 51! **LEA**: 1 km NE of Kut Al-Imara, *F. & E. Barkley* 6402! **LSM**: Amara, *Khatib & Alizzi* 32616!

Native distribution unknown; exists only in cultivation and as an escape.

Lettuce, Garden Lettuce, KHASS, KAHU (Kurd.). "Wild Lettuce", as opposed to cultivated, are covered by such common names as MURRAIR "bitter weed", UMM AL-HALÎB, "milky weed" (literally "mother of milk") etc.

5. **Lactuca saligna** *L.*, Sp. Pl.: 796 (1753); Boissier, Fl. Orient. 3: 810 1(875); Nábělek in Publ. Fac. Sci. Univ. Masaryk 5: 54 (1925); Dinsmore in Post, Fl. Syria, Palest. & Sinai ed. 2, 2: 146 (1933); Anthony in Notes Roy. Bot. Gard. Edinburgh 18: 292 (1935; Bornmüller in Beih. Bot. Centralbl. 60B: 224 (1939); Zohary in Dep. Agr. Iraq Bull. 31: 160 (1950); Rawi in Dep. Agr. Iraq Tech. Bull. 4: 118 (1964); Jeffrey in Fl. Turkey 5: 777 (1975); Ferakova in Fl. Europ. 4: 330 (1976); Rechinger, Fl. Iranica 122: 192 (1977); Feinbrun-Dothan, Fl. Palaest.

3: 437 (1978); Boulos & Hind in Fl. Egypt 3: 313 (2002); Taifour & El-Oqlah, Pl. Jordan Ann. Checkl.: 43 (2017).

Annual or rarely biennial with woody base, with many slender ascending whitish stems, to 0.5 m tall, smooth or prickly in lower part. Leaves sessile, oblong-ovate (basal) to linear (cauline), 3–20 × 0.4–0.5 cm, sagittate-auriculate at base, basal runcinate-pinnatifid, cauline leaves usually simple, all dark green, underside of midrib sometimes prickly-setose. Inflorescence spicate, with sessile groups of (1–)2–4 capitula, about 11-flowered; involucres cylindrical; phyllaries glabrous, purple-tinged around midrib; inner phyllaries 4–6, lanceolate, increasing in length at base in fruit to become linear, *c.* 14 mm long and spreading, fruiting involucre 1–2.5 mm wide, not abruptly contracted above achene bodies. Corollas yellow, ligules 10 mm. Achenes obovoid, body 3.5–4.5 mm, tapering into very slender beak, compressed, many-ribbed, dull brown; beak equaling body.

HAB. Occasional by rivers or in other damp situations; alt. up to 250 m; fl. & fr. not known.
DISTRIB. Occasional, mainly in irrigated areas. **MSU**: Chemchemal, *Hadač* 2777. **FNI**: Nineveh, *Bornmüller* 1519!; Nineveh, *Chapman* 25383! **FKI/FPF**: Jabal Hamrin, *Sutherland* 68! **FPF**: Kani Masi, *Al-Kaisi et al.* 43900! **LCA**: R. Tigris river plain, *Low* 380!; Babylon, *Reuter* 3842 (in Nábělek). **LSM**: Amara, *Anthony* loc. cit. **LBA**: Doorah village between Basra and Fao, *Alizzi & Omar* 35867!

Mediterranean Europe, Arabia, Palestine, Lebanon, Syria, Turkey, Caucasia, Iran, N Africa (Morocco, Algeria, Tunisia, Egypt); naturalized in N America.

6. **Lactuca undulata** *Ledeb.*, Ic. Fl. Ross. 2: 12, t. 129 (1830); Boissier, Fl. Orient. 3: 813 (1875); Handel-Mazzetti in Ann. Naturh. Mus. Wien 27: 458 (1913); Dinsmore in Post, Fl. Syria, Palest. & Sinai ed. 2, 2: 146 (1933); Rawi in Dep. Agr. Iraq Tech. Bull. 4: 119 (1964); Rechinger, Fl. Lowland Iraq: 681 (1964); Jeffrey in Fl. Turkey 5: 780; Rechinger, Fl. Iranica 122: 187 (1977); Feinbrun-Dothan, Fl. Palaest. 3: 437 (1978). Boulos & Hind in Fl. Egypt 3: 313 (2002); Zhu Shi & Norbert Kilian in Fl. China 20–21: (2011); Breckle, Hedge & Rafiqpoor, Vasc. Pl. Afghan. Checkl: 177 (2013); Taifour & El-Oqlah, Pl. Jordan Ann. Checkl.: 43 (2017); Roohi & Abid in Fl. Pakistan 222: 205 (2017).

Small annual herb, 4–15(–35) cm tall. Stem simple or branched, slightly fleshy, leafy, pubescent throughout with white, spreading or deflexed, glandular hairs. Rosette leaves numerous, sessile, broadly clasping stem, oblanceolate, 10–30 × 3–6 mm, margin sinuate-dentate to pinnatisect with 3–9 dentate lobes, apex acute or obtuse, mucronate. Cauline leaves smaller, purplish. Inflorescence a cymose panicle, capitula long-fusiform, borne on pedicels about same length as involucre. Outer phyllaries ovate, inner lanceolate, 20 mm long, glabrous, light green, purple-tinged, dark-spotted at apex. Corolla blue. Achene body 3.5 × 1 mm, 4-ribbed, lateral ribs tending towards wings, transversely rugose, beak white, dilated at its junction with body, 10 mm; pappus 3.5 mm. Fig. 40: 1–3.

HAB. On light silty disturbed soil on the lower mountains and foothills and by a river in the desert; alt. (50–)200–300(–500) m; fl. & fr. Mar.-Apr.
DISTRIB. Occasional in the dry steppe zone. **MAM**: Umm al-Matin, 74 km NW of Baiji-Haditha road, *Chakravarty, Rawi, Khatib & Alizzi* 32027! **MSU**: Jarmo, *Helbaek* 704! **FKI**: Kirkuk, *Bornmüller* 1523! **FPF**: Sa'diya, *Haines* W1905!; Pilkhana, *Rawi* 12781, 12795!, 12860!; Mandali, *Noë* s.n. **DLJ**: Hit, *Handel-Mazzetti* 829. **DWD**: between Ana and Al-Qaim, *H. Hamid* **39166!**

Sinai, Palestine, Turkey, Caucasia, Iran, Pakistan, Afghanistan, C Asia; widespread.

7. **Lactuca viminea** (*L.*) *J. & C. Presl*, Fl. Čechica: 160 (1819); Boissier, Fl. Orient. 3: 818 (1875); Dinsmore in Post, Fl. Syria, Palest. & Sinai ed. 2, 2: 147 (1933); Bornmüller in Beih. Bot. Centralbl. 60B; 224 (1939); V. Ferakova in Fl. Europ. 4: 329 (1976); Feinbrun-Dothan, Fl. Palaest. 3: 438 (1978); Taifour & El-Oqlah, Pl. Jordan Ann. Checkl. 43 (2017).

Prenanthes viminea L., Sp. Pl.: 797 (1753).
Scariola viminea (L.) F.W.Schmidt in Samml. phys.-ökonom. Aufs. 1: 270 (1795); Kirpichnikov in Fl. S.S.S.R. 29: 318 (1964); Jeffrey in Fl. Turkey 5: 783 (1975); Rechinger, Fl. Iranica 122: 205 (1977).

Biennial herb 0.5–1 m tall. Stems many, erect or ascending, virgate, branching, straw-coloured to whitish, shining, glabrous. Basal and median cauline leaves sessile, 5–8 × 1.2–3 cm, elliptic, runcinate-pinnatisect, decurrent through a whole internode, lamina

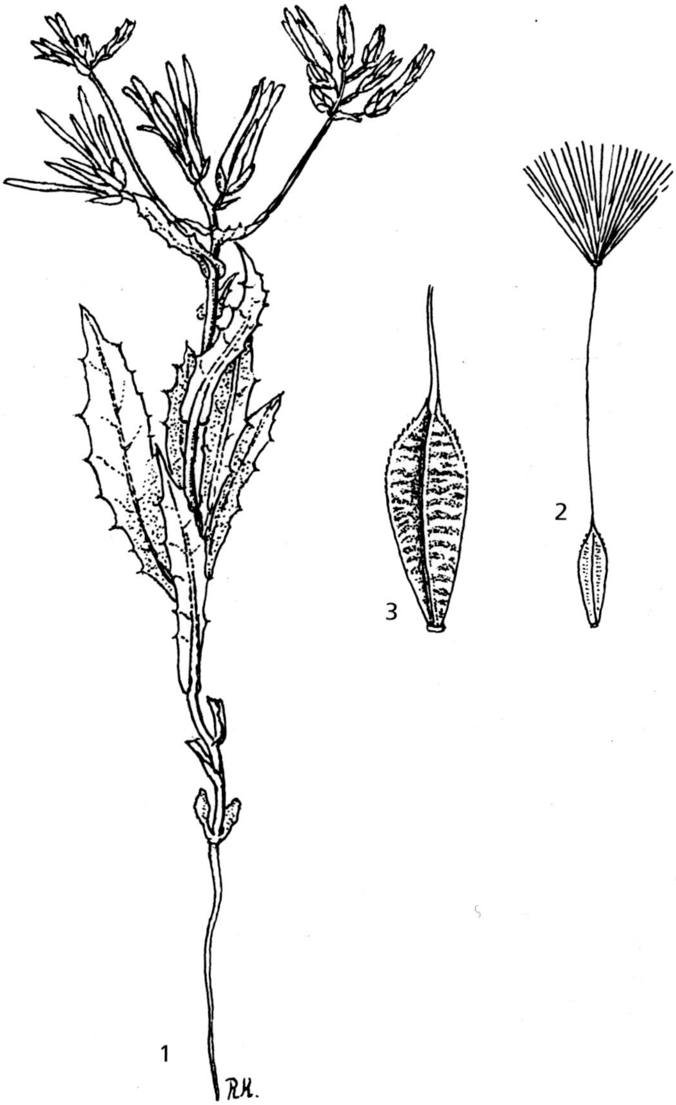

Fig. 40. **Lactuca undulata**. 1, habit × 1; 2, achene with pappus × 4; 3, achene × 10. Drawn by Ruth Koppel. From Fl. Palaestina 3: plate 741. From Naomi Feinbrun-Dothan, *Flora Palaestina,* Part three: Ericacaceae to Compositae, Plates, Jerusalem: The Israel Academy of Sciences & Humanities, 1977, pl. 741. © The Israel Academy of Sciences and Humanities. Reproduced with permission.

deep green, glabrous or with crispate hairs on midrib and sometimes on the lamina itself. Upper leaves smaller, lamina reduced. Inflorescence branched, branches spiciform with many capitula solitary or in small sessile groups; involucre linear, *c.* (10–)13–23 × 2–3 mm; phyllaries scarious-margined, glabrous or floccose at tip; mostly 4 long inner phyllaries and *c.* 5–7 shorter and broader, pale green outer ones; inner phyllaries lengthening in fruit. Achenes 10–15 mm, dark reddish-black, finely ribbed, minutely scabrid, beak making up at least ½ of the length; pappus 6–7 mm, white, readily caducous as single hairs. Fig. 41. 1–4.

Fig. 41. **Lactuca viminea**. 1, habit × 1; 2, flowering branch × 1; 3, capitulum × 2½; 4, achene with pappus × 4. Drawn by Ruth Koppel. From Naomi Feinbrun-Dothan, *Flora Palaestina*, Part three: Ericacaceae to Compositae, Plates, Jerusalem: The Israel Academy of Sciences & Humanities, 1977, pl. 742. © The Israel Academy of Sciences and Humanities. Reproduced with permission.

HAB. Weed in vineyard on a steep stony slope, 1000–1800 m; fl. Jul.-Oct.
DISTRIB. Rare in Iraq; only found in two or three places in the mountain region. **MAM**: Jabal Khantur, 20–25 km E of Zakho (nr Turkish frontier), *Rawi, Tikriti & Noori* 29018!; Sarsang, *Haines* 1230. **MRO**: Kuh-i Sefin, nr Shaqlawa, *Bornmüller* 1504. **MSU**: Pira Magrun, nr Zewiya, *Haines, Hadač & Walid* W. 1864! Qaradagh, *Haines et al.* s.n.

S & C Europe (Portugal to European Turkey, N to C France & SE Russia), Mediterranean Is., Palestine, Turkey, Iran, Caucasia, Turkmenia, N Africa (Morocco, Algeria, Tunisia).

8. **Lactuca orientalis** (*Boiss.*) *Boiss.*, Fl. Orient. 3: 819 (1875); Dinsmore in Post, Fl. Syria, Palest. & Sinai ed. 2, 2: 147 (1933); Zoh. in Dep. Agr. Iraq Bull. 31: 160 (1950); Rawi in Dep. Agr. Iraq Tech. Bull. 14: 118 (1964); Rechinger, Fl. Lowland Iraq: 681 (1964); Feinbrun-Dothan, Fl. Palaest. 3: 438 (1978). Zhu Shi & Norbert Kilian in Fl. China 20—21: (2011); Kilian & Ghazanfar in Fl. Oman 3: 223 (2015); Taifour & El-Oqlah, Pl. Jordan Ann. Checkl.: 43 (2017); Roohi & Abid in Fl. Pakistan 222: 207 (2017).

Phaenopus orientalis Boiss., Voy. Bot. Espagne 2: 390 (1839).
Scariola orientalis (Boiss.) Soják in Nov. Bot. Delect. Seminum Horti Bot. Univ. Carol. Prag. 1962: 46 (1962); Kirpichnikov in Fl. S.S.S.R. 29: 320 (1964); Jeffrey in Fl. Turkey 5: 783 (1975); Rechinger, Fl. Iranica 122: 202 (1977); Boulos & Hind in Fl. Egypt 3: 315 (2002).

Perennial, stiff, divaricately much-branched herb up to 0.5 m tall; stem white, shining, glabrous or thinly white-floccose when young. Leaves ovate to runcinate-pinnatisect, sessile, long-decurrent, lamina 1–8 × 0.3–3 cm, apex acute, glabrous or white-floccose when young. Upper leaves with a reduced lamina. Inflorescence branches with subsessile narrowly cylindrical capitula 6–16 mm long, which lengthen in fruit. Phyllaries green to glaucous, sometimes purplish towards apex, glabrous or somewhat floccose especially about tip, narrowly scarious-margined, long inner phyllaries usually 4, outer 4–7. Achenes 8–9 mm, light brown, beak concolorous, 2–3 mm, rather thick; pappus 6–9 mm, white, readily caducous as single hairs. Fig. 42. 1–2.

HAB. In the mountains on rocky places, alt. 1000–2600 m; also in the Western Desert, alt. 500 m upwards; fl. & fr. May-Aug.
DISTRIB. Occasional in two disjunct areas in Iraq, occurring in the forest/moist steppe zone and also in the Western Desert. **MJS**:L Jedala on Jabal Sinjar, *Handel-Mazzetti* 1548; Jabal Sinjar, *Alizzi & Omar* 35320! **MAM**: Sarsang, *Haines* W1230! **MRO**: Lolan, NE of Algurd, *Gillett* 12450!; Algurd (Helgord) Dagh, *Rechinger* 11381; Haji Umran, *Rawi* 24286!; between Haji Umran and Rayat, *Gillett* 9470!, *Guest* 13056!, 13058A!; Magar range (SE of Haji Umran), *Rawi* 24317!; Qandil range, NE of Rania, *Rawi & Serhang* 26827!; between Pushtashan and Surade, *id.* 26536! **MSU**: Pira Magrun, *Haines* W.2045!, *Agnew, Haines & Walid* W.1852; Penjwin, *Guest* 12952! **DWD**: Ubaila, 7 km N of Rutba, *Dabbagh, Tahir & Taufiq* 47884!; 5 km S of H.3, *Rawi* 21204!; 250 km W of Ramadi, *Rechinger* 9853!; 160–190 km NW of Ramadi, on road to Rutba, *Rawi* 31353!; from Haditha to Ana, *S. Oman, Adel & Al-Kaisi* 55960!

Egypt, Sinai, Palestine, Lebanon, Syria, Turkey, Transcaucasia, Afghanistan, Pakistan, C Asia (Turkmenia to W Siberia), Tibet.

9. **Lactuca rechingeriana** (*Tuisl*) *N.Kilian & Greuter*, Willdenowia 36(2): 713 (2006). Type: Iraq, Kurdistan, Mt. Avroman (Hawraman), near Tawila, *Rechinger* 10386a (W, holo.).

Cephalorrhynchus rechingerianus Tuisl, Ann. Naturhist. Mus. Wien 72: 614 (1968); Rechinger, Fl. Iranica 122: 210 (1977).

Perennial herb 40–90 cm tall, paniculate-branched above; stem erect, hollow, leafy, the lowest leaves shriveled at flowering time, petiolate; lower cauline leaves 10–25 × 5–13 cm, oblong-elliptic to obovate, broadly auriculate and amplexicaul at base, margins irregularly divided into oblong or polygonal segments, obtuse or acute; upper cauline leaves smaller, mostly pinnately lobed or ± entire. Capitula 9–11 mm, narrowly cylindrical, densely to sparsely glandular-hairy, outer 4–6 phyllaries triangular to lanceolate, inner 3–4 × longer than outer, linear. Corolla yellow, outer parts sometimes tinged with purple. Achenes 5–7 mm, dark brown, with 15 ribs; upper part clothed with antrorse setae, tapering into a 2–3 mm yellowish beak; pappus 3–3.5 mm, white. Fig. 43: 5.

HAB. On mountains in rocky places; alt. 1200–2000 m; fl. Jun.
DISTRIB. Occasional in the upper forest zone of NE Iraq. **MAM**: Sersang, *Haines* 1818, *Rechinger* 11916. **MRO**: Shaqlawa, *Haines* 699; Shaykhan to Sakri Sakran, *Hadač* et al. 5419; Kani Mazu Shirin nr Shirwan

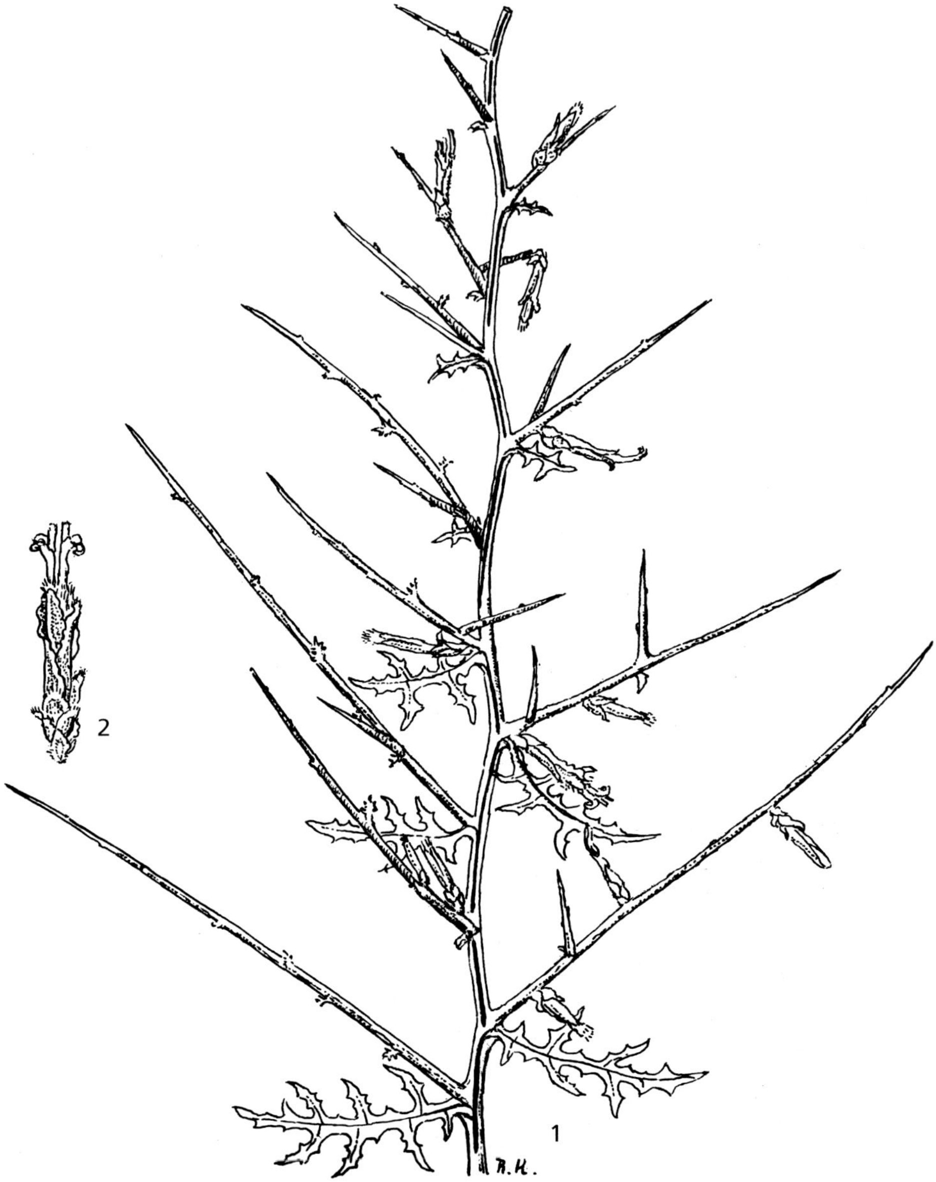

Fig. 42. **Lactuca orientalis**. 1, habit × 1; 2, capitulum × 2. Drawn by Ruth Koppel. From Naomi Feinbrun-Dothan, *Flora Palaestina*, Part three: Ericacaceae to Compositae, Plates, Jerusalem: The Israel Academy of Sciences & Humanities, 1977, pl. 744. © The Israel Academy of Sciences and Humanities. Reproduced with permission.

Mazin, *Haines et al.* s.n.; Sakri Sakran, *Hadač et al.* 5541; Mt. Botin, *Hadač et al.* 6057; Rowanduz gorge, *Hadač* 2603, 2608; *Rechinger* 11832; Khani Jindi, *Low* 251. MSU: Pira Magrun, *Prosser* in Hadač 2008; Penjwin, *Rechinger* 12263; Benawah Sutah nr Penjwin, *Hadač* 4953; Qopi Qaradagh, *Haines* 1081; Mt. Avroman (Hawraman) nr Tawila, *Rechinger* 10386a (type); Qarachitan, *Gillett* 7790!; Kamar spa, on rd. between Halabja & Tawela, *Rawi* 22174!

W & S Iran.

10. **Lactuca microcephala** *DC.*, Prodr. 7(1): 134 (1838).

Cephalorrhynchus microcephalus (DC.) Schchian, Fl. Gruz. 8: 693 (1952) et Zametki Sist. Geogr. Rast. 23: 101 (1963); Rechinger, Fl. Iranica 122: 211 (1977).

Perennial herb 25–80 cm tall, paniculate-branched above; stem erect, hollow, leafy; leaves glabrous, lowest ones petiolate, 6–30 × 1.5–10 cm, lyrate-pinnatisect, often shriveled at flowering time; lower cauline leaves smaller, auriculate-amplexicaul at base, oblong, sparsely lobed to pinnatisect; uppermost acute, subentire. Capitula 10–18 mm, narrowly cylindrical, usually glabrous; phyllaries acute, 2–3-seriate, outer 4–5 narrowly lanceolate, inner 7–11, 3–5 × longer than outer, linear. Corolla yellow, drying purplish, margins often tinged with purple. Achenes 8–9 mm, pale brown, with 5 ribs; upper part tapering to a 4–5 mm whitish or yellowish beak; pappus 5–5.5 mm, white. Fig. 43: 1–4.

HAB. On mountains in rocky places; alt. 900–2050 m.; fl. May–Jul.
DISTRIB. Occasional in the upper forest zone of NE Iraq. **MAM**: Mt. Khantur, *Rechinger* 10382. **MRO**: Mt. Karokh nr Handren, *Kass & Noori* 27354; Mt. Handren, *Thesiger* 1070; Rowanduz gorge, Emberger, *Guest et al.* 15480; Mt. Hawara Blinda NE of Hajji Umran, *Kass & Noori* 27817; Serkabkhan 7 km NW of Rania, *Rawi et al.* 28549; Mt. Sefin nr Shaqlawa, *Bornmüller* 1518; Sefin Dagh, *Gillett* 8158!; Mt. Baradost, *Thesiger* 891; Gara Dagh, *Al-Dabbagh & M. Jasim* 46990! **MSU**: Nr Tawila, *Rechinger* 10285 & 10386; Tawana, *Rawi, Noori & Husham* 29564!; Kamar spa, on rd. between Halabja & Tawela, *Rawi* 22180!; Pira Magrun, NW of Sulaimaniya, *Haussknecht* s.n. **MJS/FUJ:** Gali Ali Beg, *F. Karim, H. Hamid & M. Jasim* 40925!

C, W & S Iran.

41. **STEPTORHAMPHUS** Bunge

Beitr. Fl. Russ.: 205 (1852) & Mém. Sav. Etr. Petersb. 7: 381 (1854)
Lack in Kubitzki, Fam. Gen. Vasc. Pl. 8: 189 [2006] (2007)

Lactuca sect. *Cicerbita* Benth. & Hook. f., Gen. Pl. 2: 526 (1873), p.p.
Lactuca sect. *Tuberosae* Boissier, Fl. Orient. [Boissier] 3: 804 (1875), p.p.
Cicerbita sect. *Steptorhamphus* (Bunge) Beauv. in Bull. Soc. Bot. France 2, 2: 117 (1910).

Frances G. Davies

Perennial herbs, root tuberous or rhizomatous. Stem erect, hollow, leafy. Leaves fleshy or not, glabrous, pilose or glandular-pilose, simple and dentate to pinnatifid or pinnatisect. Inflorescence a cylindrical paniculate raceme. Capitula homogamous, large, many-flowered; phyllaries 3–5-seriate, broad, imbricate, inner series clearly longer and elongating in fruit; receptacle naked. Ray florets yellow or purple. Achenes homomorphic, brown to blackish, very compressed, margins broad and flat, beak half to twice as long as body, pale; pappus uniseriate, hairs white, almost smooth to minutely scabrid, deciduous by basal ring.

6 species in W & C Asia; one species in Iraq.

(*Steptorhamphus*, from Gr. στέφω, stepho, to crown, ράμφος, ramphos, a beak).

1. **Steptorhamthus tuberosus** (*Jacq.*) *Grossh.*, Fl. Kavk. 4: 258 (1934); Kirpichnikov in Fl. S.S.S.R. 29: 330 (1964); Jeffrey in Fl. Turkey 5: 775 (1975); Rechinger, Fl. Iranica 122: 199 (1977); Taifour & El-Oqlah, Pl. Jordan Annot. Checkl.: 48 (2017)

Lactuca tuberosa Jacq., Hort. Vindob. 1: 18 (1770); Boissier, Fl. Orient. 3: 806 (1875); Nábělek in Publ. Fac. Sci. Univ. Masaryk 52: 54 (1925); Dinsmore in Post, Fl. Syria, Palest. & Sinai ed. 2, 2: 145 (1933); Bornmüller, in Beih. Bot. Centralbl. 60B: 223 (1939); Zohary in Dep. Agr. Iraq Bull. 31: 50 (1950); Rawi in Dep. Agr. Iraq Tech. Bull. 14: 119 (1964); Feinbrun-Dothan, Fl. Palaest. 3: 436 (1978).
Lactuca petraea Fisch. & C.A.Mey. ex DC., Prodr. 7: 134 (1838).

Perennial herb 25–80 cm tall, from a tuberous rhizome. Stem glabrous, light brown, leafy. Basal and lower cauline leaves obovate or oblanceolate, 12–25 × 4–10 cm, lobed-dentate, lyrate or pinnatisect, 5–17-lobed, acute, margins denticulate and ciliate, lamina green, glabrous or sparsely pilose especially on undersides of veins, lower clasping stem by a petiolar base, cauline leaves rapidly reducing, upper auriculate at base. Inflorescence a compound

Fig. 43. **Lactuca microcephala**. 1, habit× 1; 2, leaf × 1; 3, capitulum × 3; 4, achene with pappus × 10. **Lactuca rechingeriana**. 5, achene with pappus × 10. 1–4 from *Kass & Noori* 27354; 5 from *Haines* 699. Drawn by Moira Rae Carter.

raceme with 6–40 capitula, capitula sessile or on pedicels shorter than fruiting capitulum. Involucre oblong-cylindrical or finally urceolate, 18–25 mm; phyllaries regularly imbricate, membranous-margined, elongating to 25–30 mm or more in fruit, outer phyllaries ovate, inner lanceolate, all acute, green, purple-tinged. Ray florets 25–30 mm, lilac or yellow with a lilac suffusion dorsally. Achenes very compressed, body 7–9 × 3–4 mm, brown to blackish, margins inflated into wings, thick-margined, each face single-ribbed, rugulose, beak slender, up to ± 15 mm long, colourless; pappus 10–12 mm, white.

HAB. In the mountains, on rocky slopes, in *Quercus* and *Pinus* forest, sometimes by a stream, under *Juglans* trees; alt. 800–1700 m; fl. & fr. May-Jul.
DISTRIB. Frequent in the lower forest zone of the mountain region of Iraq. **MAM**: Jabal Bekher nr Zakho, *Field & Lazar* 827!; Mar Yaqub, Nábělek 3740!; Ain Nuni, *id.* 3739!; Ghara (Qara?) Dagh, *Dabbagh & Jasim* 46944!; Aqra, *Rawi* 11450A; Kani Mazu Shirin, *Dabbagh & K. Hamad* 45542!; nr Zeyta at foot of Mt. Botin, *Hadač et al.* 5979; Bekma gorge, *Gillett* 8237!; Ser Amadiya, *Agnew & Haines* W.7064, *Haines* 2064; Zawita, *Omar, Qaisi & Widad* 49683!; Dohuk, *Rawi* 8702. **MRO**: Hab as-Sultan Dagh, *Omar, Qaisi & Widad* 49487!; mt. Qandil above lake Goam-e Kirmosoran (Kermasur), *Rechinger* 11127; NE of Qandil, *Rawi* 24388; mountains N of Rania, *Thesiger* 1199; Algurd (Helgord) Dagh above Nowanda, *Rechinger* 11409; Kuh-i Sefin, *Bornmüller,* 1517!; Hawara Blinda, *Rawi, Noori & Kass* 27787!; above Pushtashan, *Rechinger* 11093; **MSU**: mt Avroman (Hauraman), *Rawi* s.n.; Pira Magrun, *id.* 12038!; Tawila, *id.* 21890!; Dara Tri, between Halabja and Tawila, *id.* 22037!; Balkha, 7 km W of Tawila, *id.* 21890!; Amoret nr Qaradagh, *Haines* W.1130! **MJS**: Jabal Sinjar, *Gillett* 11081! **MJS/FUJ**: in the Sinjar desert, *Haussknecht* s.n.!

Bulgaria, Greece, Crete, Crimea, Palestine, Lebanon, Turkey, Transcaucasia, Iran, C Asia.

42. **LAUNAEA** Cass.

Dict. Sci. Nat., ed. 2, 25 : 321 (1822)
Lack in Kubitzki, Fam. Gen. Vasc. Pl. 8: 190 [2006] (2007)

Microrhynchus Less., Syn. Gen. Compos.: 139 (1832)
Zollikoferia DC. non Nees, Prodr. 7: 183 (1838)
Paramicrorhynchus Kirp., Fl. URSS 29: 725 (1964)
Hexinia H. L. Yang, Fl. Desert. Reipubl. Popul. Sin. 3: 472 (1992)

Frances G. Davies

Annual or perennial, erect or procumbent herbs, rosette-forming or stems leafy; some species fleshy. Leaves lobed or deeply divided, usually with a white cartilaginous margin bearing white teeth. Capitula remote or in sessile imbricate, clusters, few- to many-flowered; phyllaries biseriate, outer shorter and generally broad, inner lanceolate, both ± glabrous; receptacle naked. Corollas yellow. Achenes all similar or di- or trimorphic, ornamentation decreasing towards center of capitulum, truncate; pappus homo- or heteromorphic, white, persistent or caducous as a ring.

About 54 species in S Europe, the Mediterranean, Asia and Africa, one species in Australia; 5 species in Iraq.

Launaea (named in honour of the librarian at the Museum d'Histoire Naturelle in Paris, Jean Claude Michel Mordant de Launay (c. 1750–1816).).

Kilian, N. (1997). Revision of *Launaea* Cass. (Compositae, Lactuceae, Sonchinae). Englera 17: 1–478.

1. Outer phyllaries without broad scarious margins, the outermost cordate-auriculate at base; leaves irregularly 1–3 pinnatisect or pinnatilobed 2
   Outer phyllaries with a broad scarious margins, the outermost not cordate-auriculate at base; leaves ± regularly lobed or somewhat runcinately divided 3
2. Involucre about as broad as long; achenes all similar with 4 papery wings; achene plus pappus less than 10 mm long ..... 1. *L. angustifolia*
   Involucre longer than broad; achenes trimorphic, unwinged; achene plus pappus over 10 mm long ..... 2. *L. mucronata*
3. Capitula at flowering time less than 2 × as long as wide, almost sessile, majority clustered ..... 3. *L. capitata*

Capitula at flowering time 2 × as long as wide, pedicels 2–20 mm; capitula mostly borne singly 4

4. Pappus setae thin flexuous and rigid, persistent; achenes oblong, truncate, cream, outer prickly-rugose; plant ± erect . . . . 4. *L. nudicaulis*

Pappus setae uniform, rigid, caducous; achenes tapered upwards, brown, finely prickly; plant rooting at nodes . . . . 5. *L. procumbens*

1. **Launaea angustifolia** (*Desf.*) *O.Kuntze*, Revis. Gen. Pl. 1: 350 (1891); Dinsmore in Post, Fl. Syria, Palest. & Sinai ed. 2, 2: 149 (1933): Rawi in Dep. Agr. Iraq Tech. Bull. 14: 119 (1964); Rechinger, Fl. Lowland Iraq: 675 (1964); Feinbrun-Dothan, Fl. Palaest. 3: 431 (1978); Kilian in Englera 17: 361 (1997); Collenette, Wild Flow. Saudi Arabia: 200 (1999); Boulos & Hind in Fl. Egypt 3: 295 (2002); Abdulridha, Taha & Wedad, Ecol. & Fl. Basra: 215 (2016); Taifour & El-Oqlah, Pl. Jordan Ann. Checkl.: 44 (2017).

Sonchus angustifolia Desf., Fl. Atlant. 2 : 225 (1799).
Zollikoferia arabica Boiss., Diagn. Pl. Orient., ser. 1, 7 : 12 (1846) & Fl. Orient. [Boissier] 3: 823 (1875).
Z. angustifolia (Desf.) Coss. & Durieu, Bull. Soc. Bot. France 2: 254 (1855).
Atalanthus angustifolius (Desf.) Pomel, Nouv. Mat. Fl. Atl., pt. 1: 8 (1874).
Rhabdotheca angustifolia (Desf.) Pomel, Nouv. Mat. Fl. Atl., pt. 2: 263 (1875).
R. squarrosa Pomel, [Bull. Soc. Climatol. Alger: 13] Nouv. Mat. Fl. Atl., pt. 2: 263 (1875).
Zollikoferia angustifolia (Desf.) Coss. & Durieu [unranked] *squarrosa* (Pomel) Batt., Fl. Algérie 1: 558 (1889).
Z. foxii Post, Fl. Syria, Pal., Sinai: 19 (1896).
Z. arabica Boiss. [var.] β *squarrosa* (Pomel) Batt., Fl. Algérie Tunisie: 216 (1904).
Launaea foxii (Post) Eig, Repert. Spec. Nov. Regni Veg. Beih. 63(1): 48 (1931).
L. arabica (Boiss.) H. Lindb., Acta Sci. Fenn., ser. B, Opera Biol. 1(2): 163 (1932).
L. angustifolia (Desf.) Kuntze subsp. *arabica* (Boiss.) N.Kilian, Willdenowia 25: 274 (1995).

A rather fleshy tap-rooted herb, 5–30 cm tall, with stout leafy branches. Lowermost leaves with a broad petiole-like base, middle and upper leaves sessile, clasping stem and broadly auricled, elliptic, 2–10 × 0.5–2 cm, margins variously crenate to lobed to irregularly 1- or 2-pinnatisect, bearing white cartilaginous teeth, glabrous. Capitula borne singly on pedicels with minute bracts; involucre ± globose; phyllaries multiseriate, imbricate, ovate, outermost deeply cordate-auriculate at base and transitional to peduncle bracts, middle and inner 6–11 × 3–6 mm, green or purplish, without white margins, sometimes with a white apex. Florets c. 40, exceeding involucre by about 1 cm. Achenes 3.5 mm long, brown, with 4 grey, papery wings and 4–8 subsidiary wings between these; a few achenes abortive; pappus persistent, 3–4 mm long, thin bristles flexible, fluffy, greatly outnumbering rigid ones. Fig. 44: 1–2.

HAB. Stony desert with sandy soil, on a limestone plateau with calcareous sandy patches; alt. up to 350 m; fl. & fr. (Feb.)Mar.-Apr.
DISTRIB. Occasional in the Southern Desert of Iraq. **DSD**: 13 km W by N of Shabicha, *Guest & Rawi* 19289B!; *Rawi & Gillett* 6315!; 10–15 and 20–30 km SW of Shabicha, *Alizzi & Omar* 35454!; 5 km from Salman on route to Samarra, *Al Kaisi et al.* 48134!; 12 km ESE of Salman, *Guest & Rawi* 18837!; 6 & 25 km SW of Salman, *Al Kaisi et al* 42145!, 42190!; 30 km E of Golaib, *Guest, Rawi et al.* 18741!; nr Ansab, *id.* 18929!; Jabal Sanam, *Watson* s.n. (comm. Sharples), *Rawi & Gillett* 6148!, *Guest, Rawi & Schwan* 14373!

Following Kilian's revision (1997), Iraq plants would be *L. angustifolia* subsp. *arabica*, the more easterly and widespread of the two subspecies.

MURRAIR ("bitter"), *Guest & Rawi* 18837.

Palestine, Jordan, Sinai, Egypt, Arabia.

2. **Launaea mucronata** (*Forssk.*) *Muschl.*, Man. Fl. Egypt 2: 1057 (1912); Handel-Mazzetti in Ann. Naturh. Mus. Wien 27: 458 (1913); Nábělek in Publ. Fac. Sci. Univ. Masaryk 52: 54 (1925); Dinsmore in Post, Fl. Syria, Palest. & Sinai ed. 2, 2: 148 (1933); Dickson, Wild Fls. Kuwait: 58 (1955); Rawi in Dep. Agr. Iraq Tech. Bull. 14: 119 (1964); Rechinger, Fl. Lowland Iraq: 675 (1964); Rechinger, Fl. Iranica 122: 159 (1977); Feinbrun-Dothan, Fl. Palaest. 3: 431 (1978); Kilian in Englera 17: 399, f. 106–107 (1997); Collenette, Wild Flow. Saudi Arabia: 201 (1999); Boulos & Hind in Fl. Egypt 3: 297 (2002); Kilian & Ghazanfar in Fl. Oman 3: 229 (2015); Abdulridha, Taha & Wedad, Ecol. & Fl. Basra: 217 (2016); Haloob & Al-Kaisi, Illustr. Fl. Lowland Iraq 1(1): 47 (2016); Ghafoor in Fl. Pakistan 222: 175 (2017); Taifour & El-Oqlah, Pl. Jordan Ann. Checkl.: 44 (2017).

Fig. 44. **Launaea angustifolia**. 1, habit × 1; 2, achene with pappus × 5. Drawn by Ruth Koppel. From Naomi Feinbrun-Dothan, *Flora Palaestina*, Part three: Ericacaceae to Compositae, Plates, Jerusalem: The Israel Academy of Sciences & Humanities, 1977, pl. 729. © The Israel Academy of Sciences and Humanities. Reproduced with permission.

Fig. 45. **Launaea mucronata**. 1, habit ×1; 2, flowering stem × 1; 3, capitulum × 3. Drawn by Ruth Koppel. From Naomi Feinbrun-Dothan, *Flora Palaestina,* Part three: Ericacaceae to Compositae, Plates, Jerusalem: The Israel Academy of Sciences & Humanities, 1977, pl. 728. © The Israel Academy of Sciences and Humanities. Reproduced with permission.

Leontodon mucronatum Forssk., Fl. Aegypt.-Arab. : 144 (1755).
Zollikoferia mucronata (Forssk.) Boiss., Diagn. Pl. Orient. ser. 1, 7 : 12 (1846) & Fl. Orient. [Boissier] 3: 822 (1875).
[*L. chondrilloides* (non (Desf.) DC.) Anthony in Notes Roy. Bot. Gard. Edinburgh 18 : 292 (1935).]
Launaea resedifolia (L.) Druce ssp. *mucronata* (Forssk.) Maire, Bull. Soc. Hist. Nat. Afr. Nord 28: 366 (1937).

subsp. **mucronata**

Rhabdotheca tenuiloba Pomel, [Bull. Soc. Climatol. Alger. : 13] Nouv. Mat. Fl. Atl. 2: 263 (1875).
Zollikoferia tenuiloba (Pomel) Batt., Fl. Algérie 1: 557 (1889), non Boiss. (1849).
Launaea resedifolia (L.) Druce var. *pulchella* Pamp., Agric. Colon. 22: 366 (1928).

Perennial, tap-rooted, woody at base, erect, much-branched, stems leafy, 30–50 cm tall. Leaves both in a rosette and cauline, elliptic or oblanceolate, clasping stems at base, up to 15 × 6 cm, strongly irregularly dissected, toothed, glabrous, green or greyish. Capitula about 45-flowered, terminal on ultimate branches, usually singly, fusiform to oblong; phyllaries multiseriate, imbricate, ovate, acuminate, not white-margined, with or without white apices, lengthening in fruit. Florets exceeding involucre by about 0.5 cm. Achenes cylindrical, 4–5 × 0.3 mm, 4-partite, outer grey, papillose, middle brown, rugulose, inner empty and abortive; pappus dimorphic, 6–10 mm, persistent. Fig. 45: 1–3.

HAB. In the foothills on stony ground, sandstone hills, on sandy gravel in the desert, sometimes with calcareous substrate; alt. up to ± 400 m; fl. & fr. (Feb.-)Mar.-Apr.(-May).
DISTRIB. Occasional in the steppe region, very common in the desert region and on irrigated alluvial plains. **FUJ**: by telegraph pole 90 on road from Baiji to Mosul, *Field & Lazar* 391!; Sharqat, *Guest* 13452! **FKI**: 7 km N.E. of Kirkuk, *Rawi, Kass & Noori* 27904!; Jabal Hamrin, nr Injana, *Guest* 1551! **FKI/FPF**: Jabal Hamrin NE of Baghdad, *Evans* M/294! **FPF**: Koma Sang frontier post, nr Mandali, *Rawi* 20625!; Lajama post, 25 km SE of Mandali, *Rawi* 20712!; Chilat post, *Rawi & Haddad* 25726!; 25 km E of Badra, *Rawi* 20770!; nr Wadi al-Tib post, ± 70 km N of Amara, *Guest & Rawi* 17528! **DLJ**: nr Samarra, on Wadi Thirthar canal, *C. & E. Guest & Sharif Muhammad Amin Haider* 18320!; on S shore of Thirthar lake, *Chakravarty & Rawi* 30437!; 85 km NNW of Falluja, *Rawi* 20308!; 34 km NW of Falluja, *Rawi & Shahwani* 25474! **DWD**: 50 km NE of Amara, *id.* 25491!; nr Rutba, *Alizzi & Husain* 34094!; Massad depression, *Alizzi & Omar* 35371!; Rutba, *Rawi* 31316!; 60 km E of Rutba, *Barkley & Jum'a Brahim* 4925!; H.1, *Chakravarty, Rawi, Khatib & Alizzi* 31576!; Habbaniya lake, *Rawi & Alizzi* 34454!; 17 km W by S of Ramadi, *Guest & Rawi* 15951!; 20 km E of Falluja, *Guest, Rawi & Abdul-Karim Noori* 13509! **DSD**: 12 km ESE of Salman, *Guest & Rawi* 18844!; Waksa well, 85 km W by N of Salman, *Guest & Rawi* 19218A!; 35 km SE of Salman, *id.* 18860!; 15 km N by W of Aidaha, ± 90 km SW of Salman, *id.* 19178!; 25 km from Aidaha on road to Ansab, *Fauzi, Hazim & Hamid* 38821!; 69 km from Ansab, on way to Samah, *id.* 38876!; nr Chilawa, *Guest & Rawi* 17200!; 75 km ESE of Busaiya, *Guest & Mahalhal* 15233B!; 25 km from Busaiya, on way to Khadr al Mai, *Qaisi, Hamad & Hamid* 48516!; 60 km SW by W of Zubair, *Guest & Rawi* 14304A!; Samah, *Alizzi* 34402! **LEA**: W of Mandali, *Haddad* 10013!; Shihabi post, 45 km NNE of Shaikh Sa'd, *Rawi & Haddad* 25610!; 45 km E of Tib, *Thamir* 47705!; Iskandariya, *Haines* W.705!; Fakka, 70 km E of Amara, *id.* 25767! **LCA**: Garma, on Falluja-Baghdad road, *Al-Kaisi et al.* 47201! **LSM**: Musaida, *Rawi* 14947! 8 km N of Qurna, *Thamir* 46778!

MURARA (MURRĀR, MURRAIR), *Rawi* 26825!; 'ADHÎD (Southern Desert, *Guest & Rawi* 18860!) – a name also recorded by Carter from Kuwait.

Palestine, Jordan, Egypt, Arabia (Saudi Arabia, Kuwait, Yemen, Oman, Bahrain), N Africa.

3. **Launaea capitata** (*Spreng.*) *Dandy* in Andrews, Fl. Pls. Sudan 3: 40 (1956); Rechinger, Fl. Lowland Iraq: 677 (1964); Täckholm, Stud. Fl. Egypt: ed.2, 602 (1974); Feinbrun-Dothan, Fl. Palaest. 3: 432 (1978); Rechinger, Fl. Iranica 122: 146 (1977); Mandaville, Fl. East. Saudi Arabia: 321 (1990); Kilian in Englera 17: 296, 4f, 9e–g (1997); Collenette, Wild Flow. Saudi Arabia: 200 (1999); Boulos & Hind in Fl. Egypt 3: 294 (2002); Kilian & Ghazanfar in Fl. Oman 3: 226 (2015); Breckle, Hedge & Rafiqpoor, Vasc. Pl. Afghan. Checkl.: 177 (2013); Abdulridha, Taha & Wedad, Ecol. & Fl. Basra: 217 (2016); Ghafoor in Fl. Pakistan 222: 170 (2017); Taifour & El-Oqlah, Pl. Jordan Ann. Checkl. 44 (2017).

Sonchus capitatus Spreng., Syst. Veg. ed. 16, 3 : 650 (1826).
S. mareoticus Del., Descr. Egypte, Hist. Nat., f. 64: 12 (1826), nom. nud.
Lomatolepis glomerata Cass., Dict. Sci. Nat., ed. 2, 48 : 423 (1827), nom. illeg.
Chondrilla capitata (Spreng.) Sieb. ex DC., Prodr. 7: 180 (1838), nom. inval., comb. illeg. pro syn.
Microrhynchus glomeratus (Cass.) Jaub. & Spach, Ill. Pl. Orient. 3 : 105 (1848), comb. illeg.
Zollikoferia glomerata (Cass.) Boiss., Fl. Orient. [Boissier] 3: 826 (1875), comb. illeg.

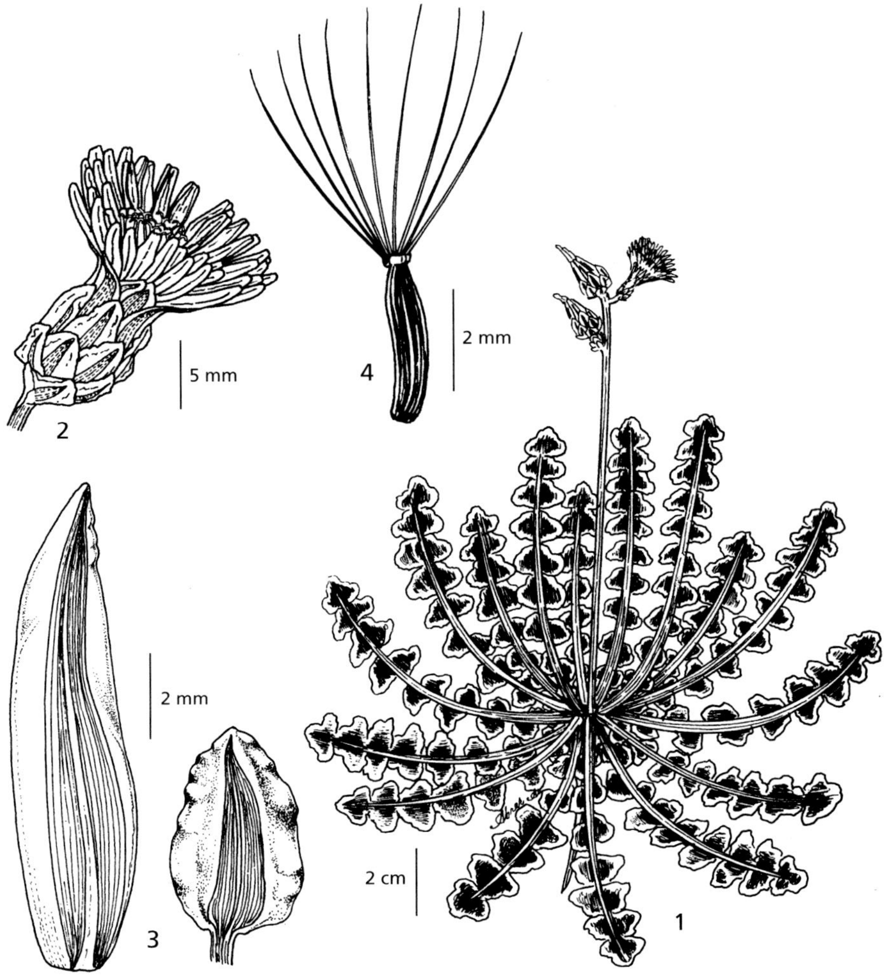

Fig. 46. **Launaea capitata**. 1, habit; 2, capitulum, 3, phyllaries, 4, achene with pappus. Reproduced with permission from Fl. Pakistan 222: fig. 10 D–G. Drawn by Abrar Ali.

Launaea glomerata (Cass.) Hook.f., Fl. Brit. India 3: 417 (1881), comb. illeg.; Dinsmore in Post, Fl. Syria, Palest. & Sinai ed. 2, 2: 150 (1933); Anthony in Notes Roy. Bot. Gard. Edinburgh. 18: 293 (1935); Zohary in Dep. Agr. Iraq Bull. 31: 160 (1950).

A dwarf rosette herb, tap-rooted, with one or several erect stems or erect stems lacking and capitula at ground level, stems c. 15 (–40) cm. long. Rosete leaves 2–16 × 0.5–5.5 cm, oblong to spatulate, base narrowed, margins white cartilaginous, sharp-toothed, apices obtuse, glabrous, flowering stem leaves reduced to inconspicuous bracts. Inflorescence bearing groups of 3–12 ± sessile capitula. Capitula c. 45-flowered, terminal on ultimate branches, campanulate; involucre 8–12 × 7–10 mm; outer phyllaries 13, ovate, 3–8 mm long, central ⅓ brown and outer ⅔ papery-white, inner 8, lanceolate, broadly white-margined, all phyllaries lengthening in fruit and midribs become hardened. Corollas yellow. Achenes all similar, whitish, with a corky texture, 4-winged, 4 mm; pappus 6 mm, homomorphic, caducous. Fig. 46: 1–4.

HAB. In the desert, on sand-hills and dunes, and on sandy gravel soils; alt up to 400 m; fl. & fr. (Feb.-) Mar.-Apr.

DISTRIB. Occasional in the Southern Desert of Iraq. **DSD**: 6 km S of Salman, *Rawi, Khatib & Tikriti*

29213!; nr Jumaima, ±140 km SW of Salman, *Guest & Rawi* 19111!; 15 km WSW of Samawa, *id.* 18701!; Zarqa, *Alkas* 17684!; Ichrishi, 35 km E by N of Busaiya, *Guest & Rawi* 14181!; nr Ur, 30 km SSW of Tal Lahm, *Guest & Mahalhal* 15210C!; 20 km from Busaiya, *Fauzi, Muhammad, Hamid & Kadhim* 40363!; Batin, nr (WNW of) Jarishan, ± 85 km SW of Basra, *Guest & Rawi* 17105B!; nr (W of) Jarishan, ± 90 km SW of Basra, *id.* 17145!; Umm Qasr, *H. Hamid* 38558! Ukhaidhir, *Kaisi, Khayat, Jenan, Sahira & Wedad* 51204!; Jumaima Police Station, *Kaisi, Hamad & Hamid* 48272! **DSD/LBA**: "Mesopotamia", *Watson* s.n. **LCA**: "Falluja desert" (i.e. E of Falluja on road to Baghdad), *Haines* W.206! (BAG!); **LEA/FPF**: Bet. Ali-AL-Gharbi & Teb *Kaisi & Khayat* 50545 (BAG!).

Palestine, Jordan, Egypt, Arabia (Kuwait, Bahrain, Qatar, Saudi Arabia, Abu Dhabi, Yemen), Iran, Pakistan.

4. **Launaea nudicaulis** (*L.*) *Hook.f.*, Fl. Br. Ind. 3: 416 (1881); Carter in Rec. Bot. Surv. India 6, 6: 205 (1917); Dinsmore in Post, Fl. Syria, Palest. & Sinai ed. 2, 2: 149 (1933); Anthony in Notes Roy. Bot. Gard. Edinburgh 18: 293 (1935); Blakelock in Kew Bull. 4: 58 (1948); Zohary in Dep. Agr. Iraq Bull. 31: 160 (1950); Dickson in Wild Fls. Kuwait: 59 (1955); Rawi, ibid. Tech. Bull. 14: 119 (1964); Rechinger, Fl. Lowland Iraq: 677 (1964); Täckholm, Stud. Fl. Egypt ed. 2: 602 (1974); Boulos in Fl. Europ. 4: 326 (1976); Rechinger, Fl. Iranica 122: 156 (1977); Feinbrun-Dothan, Fl. Palaest. 3: 431 (1978); Kilian in Englera 17: 217, f. 7g–i (1997); Collenette, Wild Flow. Saudi Arabia: 202 (1999); Boulos & Hind in Fl. Egypt 3: 291 (2002); Breckle, Hedge & Rafiqpoor, Vasc. Pl. Afghan. Checkl.: 177 (2013); Abdulridha, Taha & Wedad, Ecol. & Fl. Basra: 217 (2016); Taifour & El-Oqlah, Pl. Jordan Ann. Checkl.: 44 (2017).

Chondrilla nudicaulis L., Mant. Alt. : 278 (1771).
Lactuca nudicaulis (L.) Murray, Commentat. Soc. Regiae Sci. Gott. 3: 74, t. 4 (1780).
Lomatolepis nudicaulis (L.) Cass., Dict. Sci. Nat., ed. 2, 48: 424 (1827).
Microrhynchus nudicaulis (L.) Less., Syn. Gen. Compos.: 139 (1832).
Sonchus nudicaulis (L.) Sch.Bip. in Webb & Berth., Phyt. Canar. 2(2): 427 (1849-50).
Atalanthus nudicaulis (L.) Pomel, Nouv. Mat. Fl. Atl.: 7 (1874).
Rhabdotheca nudicaulis (L.) Pomel, Nouv. Mat. Fl. Atl.: 262 (1875).
Zollikoferia nudicaulis (L.) Boiss., Fl. Orient. [Boissier] 3 : 824 (1875).

A tap-rooted rosette herb, with or without stem leaves. Leaves 6–20 × 0.5–4 cm, oblanceolate to linear, runcinately c. 8-lobed, the largest lobes, nearer the apex occasionally bearing secondary lobes, margins closely and sharply toothed, teeth white-tipped, lamina glabrous. Stems 15–20 cm tall, weakly ascending; capitula borne singly on slender pedicels 0.2–1 cm long in upper parts of stems. Capitula cylindrical, 16–30-flowered; involucre 11–16 mm long × 4 mm diam.; phyllaries strongly segregated into short outer and long inner groups, outer c. 12, erect to patent, 1–5 mm long, ovate, inner ⅓ green, dull, outer ⅔ hyaline, inner phyllaries c. 8, 12 mm long, dull green centrally with a yellow line between this and outer hyaline parts. Corollas yellow, exceeding involucre by c. 6 mm. Achenes 2.6–5.5 mm long, cream to brown, 4-partite, heteromorphic, outer prickly-rugulose, 5-ribbed, inner faintly rugulose, some of innermost abortive and cylindrical, 4-ribbed; pappus 8 mm, 4-partite, dimorphic, but not strikingly so, inner setaceous, persistent, white or cream to brown.

HAB. On sandy desert soils; alt. up to 650 m; fl. & fr. (Feb.-)Mar.-Apr.(-Jun.).
DISTRIB. Rare in the steppe region, common in the desert region. **FPF**: Koma Sang, on the frontier, nr Mandali, *Rawi* 20627!; 3 km E of Badra, *Haddad* 9964!; 1 km S of Badra, *Rawi* 18148!; 16 km SE of Badra, *Rechinger* 13949. **DLJ**: nr Hit, *Hamid* 39041!; 13 km E of K.3, *Chakravarty, Rawi, Khatib & Alizzi* 32946! **DGA**: Samarra, *Haines* W.1485! **DWD**: 10 km E of Rutba, *Barkley* 33218!; 160–190 km NW of Ramadi on way to Rutba, *Rawi* 31338!; 16 km E of Rutba, *Barkley* 33!; 35 km W of Ramadi on way to Rutba, *Rawi & Hamada* 33387!; 30 km W of Ramadi on road to Rutba, *Guest & Rawi* 13953!; 15 km W of Ramadi on road to Rutba, *Rawi & Gillett* 6780!; Nukhaib, *Guest, Rawi & Long* 14034!; Wadi Ghadaf (Nakhla district), *Rawi* 14849!; opposite Rawa, *Rawi & Gillett* 7061!; 12 km W of Ukhaidhir, *Rawi* 30859!; 24 km SW of Ukhaidhir, *Rawi & Hamada* 33497!; 50 km WSW of Karbala, *Rawi* 30819!; 20 km from Ruhaba, *Karim, Nuri, Hamid & Kadhim* 39965! **DSD**: 80 km W of Shabicha, *Chakravarty, Rawi, Khatib & Alizzi* 30033!; 13 km W by N of Shabicha, Guest, *Rawi et al.* 19276!; Shabicha, *Rawi & Gillett* 6304!; 6 km W of Shabicha, *Rawi* 13846!; 17 km ESE of Shabicha, on way to Salman, *Karim, Nuri, Hamid & Kadhim* 40048!; 28 km ESE of Shabicha, *id.* 40087!; 50 km N by W of Aidaha, 80 km SSW of Salman, *Guest & Rawi* 19183!; nr Jumaima, 140 km SW of Salman, *id.* 19112!; Waksa well, 85 km W by N of Salman, *id.* 19218!; 'Ain al-Halwa, Salman, *Karim, Noori et al.* 40132!; 25 km SE of Salman, *Rawi* 14884!; Saddat, *Guest & Rawi* 18858!; Jalib Bakhur well, *id.* 18899!; 12 km WNW of Ansab, *id.* 18994!; nr Ansab, *id.* 18972!; 4 km SW of Umaiyid, *id.* 18717!; 80 km W by N of Busaiya, *Guest, Rawi & Long* 14131!; 67 km WNW of Busaiya, *Chakravarty, Rawi, Khatib & Tikriti* 29979!; 3 km NW of Zubair, *Barkley & Hikmat Abbas (al-Ani)* 6493!; Zubair, *Haines* W.1675!; 5 km SE by E of Zubair, *id.* 16789!; between Zubair and Safwan, *Alizzi*

34351!; Chubaishiya (al Chabaysh), *Haddad* 9515!; Shu'aiba, *id.* 9507!; 23 km SE by S of Zubair, *Guest & Rawi* 16850!; Jabal Sanam, *Guest, Rawi & Schwann* 14385!; 12 km NE of Safwan, *Rawi* 25983! **LCA**: "Falluja desert" (E of Falluja, on road to Baghdad, *Haines* W. 210! **LBA**: Tanuma, *Thamir & Jabar* 50076.

S Spain, Syria, Palestine, Jordan, Sinai, Egypt, Arabia (Bahrain, Kuwait, Saudi Arabia, Qatar, Yemen), Eritrea, Iran, India, N Africa (Sahara, Sudan), Macaronesia (Canary Is.).

Carter (1917) says that this plant, known as Eve's Dandelion, is very common in Kuwait where it is eaten by townspeople and Badawin alike, while all grazing animals freely devour it; it is known as HUWĀ which Dickson confirms, noting the name as HOW, for HUWĀ AL-GHAZĀL.

5. **Launaea procumbens** (*Roxb.*) *Rajagopal & Ramayya* in Kew Bull. 23: 465 (1969); Täckholm, Stud. Fl. Egypt: ed.2, 602 (1974); Rechinger, Fl. Iranica 122: 155 (1977); Kilian in Englera 17: 278, f. 9h–k, 17b–c (1997); Collenette, Wild Flow. Saudi Arabia: 202 (1999); Boulos & Hind in Boulos, Fl. Egypt 3: 291 (2002); C. Shih & Kilian in Fl. China 20-21: 239, f. 239.1 & f. 240, 1–4 (2011); Breckle, Hedge & Rafiqpoor, Vasc. Pl. Afghan. Checkl.: 178 (2013); Kilian & Ghazanfar in Fl. Oman 3: 231 (2015); Haloob & Al- Kaisi, Illustr. Fl. Lowland Iraq 1(1): 48 (2016); Taifour & El-Oqlah, Pl. Jordan Ann. Checkl.: 44 (2017); Ghafoor in Fl. Pakistan 222: 165 (2017).

Prenanthes procumbens Roxb., Fl. Ind. ed. 2 (Carey), 3 : 404 (1832).
Brachyrhamphus obtusus DC., Prodr. 7: 177 (1838).
Microrhynchus patens DC., Prodr. 7: 181 (1838).
Youngia procumbens (Roxb.) DC., Prodr. 7: 193 (1838).
Lactuca obtusa (DC.) Sch.Bip., Flora 25: 423 (1842).
Ammoseris patens (DC.) D. Dietr., Syn. Pl. 4: 1319 (1847).
Microrhynchus fallax Jaub. & Spach, Ill. Pl. Orient. 3 : 106 (1848).
Microrhynchus arabicus Jaub. & Spach, Ill Pl. Orient. 3: 107, fig. 277 (1848).
Zollikoferia fallax (Jaub. & Spach) Boiss., Fl. Orient. [Boissier] 3 : 824 (1875).
Lactuca fallax (Jaub. & Spach) Kuntze, Revis. Gen. Pl. 1 : 350 (1891); Muschl., Man. Fl. Egypt : 1060 (1912) – comb. superfl.; Handel-Mazzetti in Ann. Naturh. Mus. Wien 27 : 458 (1913); Nábělek in Publ. Fac. Sci. Univ. Masaryk 52 : 54 (1925).
Sonchus mairei H. Lév., Bull. Acad. Int. Géogr. Bot. 25: 16 (1915), nom. illeg., non H. Lév. (1913).
Launaea procumbens (Roxb.) Amin in Täckholm, Stud. Fl. Egypt: 84 (1956).
Paramicrorhynchus procumbens (Roxb.) Kirp. in Siskin & Bobrov, Fl. SSSR 29: 237 (1964).
Sonchus lakouensis S. Y. Hu, Quart. J. Taiwan Mus. 21: 164 (1968), nom. nov. pro S mairei H .Lév.
Launaea procumbens (Roxb.) Lack in Rech.f., Fl. Iranica 122: 155 (1977), comb. superfl.
L. obtusa (DC.) Mundhe & Almeida, J. Bombay Nat. Hist. Soc. 85: 523 (1989).

A prostrate or procumbent herb, producing rosettes of leaves at the rooting nodes, with stems 15–60 cm; in dried specimens, plant is commonly brownish or reddish in colour. Leaves oblong, oblanceolate or linear, or basal ones spatulate, 4–8 × 0.8–2 cm, with 16–18 angular lobes, minutely white-toothed. Capitula borne singly or in groups, on a short peduncle. Involucre cylindrical, 10–12 × 3–5 mm; phyllaries differentiated into a short outer and longer inner series; outer 1–8 mm long, erect, tips curved outwards, inner 10–12 mm long, slightly spreading, all phyllaries white-margined and reddish tinged. Achenes homomorphic, 2.5 mm long, brown, longitudinally ribbed, transversely rugulose to give a prickly outline, tapered towards apex; pappus homomorphic, 7 mm long, caducous, white, shining.

HAB. On conglomerate hills, sandy gravel, gypsiferous soil and sometimes on clay; alt. up to ± 300 m; fl. & fr. (Jan.-)Mar.-Apr.(-Jun.).
DISTRIB. Occasional in the foothills and desert and on the irrigated alluvial plains. **DWD**: 10 km W of Kubaisa, *Omar, Qaisi, Hamad & Hamid* 44537!; S of Shithatha, *Rawi & Chakravarty* 32569! **DWD/LCA**: between Musaiyib and Karbala, *Nábělek* 3934. **DSD**: Wishah, 103 km SE of Ansab, *Al-Kaisi et al.* 42294!; Jabal Sanam, *Omar, Qaisi, Hamid & Hamad* 43977! **LEA**: *Shihabi, Rawi & Haddad* 25617! **LSM**: 5 km from Qurna, *Bharucha, Rawi & Tikriti* 29317! **LBA**: Basra, *Bornmüller* 427, *Guest* 289!; Ma'qil, *Guest & Rawi* 16731!; 80 km ESE of Basra on way to Fao, *Rawi* 25905!; 17 km NW of Fao, *Khatib & Alizzi* 33411!

Jordan, Egypt, Sinai, Arabia (Kuwait, Saudi Arabia, Oman), Bahrain, Iran, Pakistan, Afghanistan, Turkmenistan, NW India, China, S Africa.

MURRAIR, Basra (*Guest & Rawi* 16731); HUWWA, Muscat.

Of the common names recorded in the Arab countries for various species of *Launaea* the most common is MURRAIR, this being the name used for many different plants which have a bitter taste, derived from the verb MURR (to be bitter). Another name 'ADHÎD has been recorded in Kuwait for *L. mucronata,* both by Carter (1917) and by Dickson (1955) – and in the southern desert of Iraq for *L. capitata* by Guest & Rawi.

A third name HAUWA (in various forms and phonetic spellings, e.g. HAUWA, HUWWA, HUWWAH, HUWEI and even HOW – neither the letter E nor O exist in written Arabic) has also been noted by various collectors, including the above, mainly for *L. nudicaulis* but also for *L. procumbens.* H.W. Hume, in his "The topography and geology of the Peninsula of Sinai", published in Cairo by the Survey Department of Egypt in 1906, records the name HALAWĀT AL-GHIZLAIN (sweetmeat of the gazelles) which might also be translated as "gazelles' delight" and indicates that it is a plant on which gazelles browse avidly. The term HAUWA is not recorded in the Arabic-English dictionaries, however Professor al-Rawi who was then working in Kuwait ascertained that the name HAW is a colloquial contraction of HALAWĀT (noted by Evan Guest).

43. **REICHARDIA** Roth

Bot. Abhandl., 35 (1785)
Lack in Kubitzki, Fam. Gen. Vasc. Pl. 8: 190 [2006] (2007)
Picridium Desf., Fl. Atlant. 2: 220 (1799); Benth. & Hook.f., Gen. Plant. 2: 527 (1873)

Patricia Halliday

Erect annual or perennial herbs. Stems 1–many-branched. Leaves radical or alternate, entire to deeply pinnatisect, lobes ciliate-denticulate, often with crisped-undulate margins; cauline leaves usually amplexicaul. Capitula few to many, homogamous, ligulate, pedunculate, peduncle often swollen just beneath capitulum; phyllaries multiseriate, imbricate, outer shorter and broader with scarious margins, often grading into bracts of peduncle, inner lanceolate, subequal, unaltered after anthesis; receptacle flat, without scales. Rays yellow, often with a purplish tinge on outer face, sometimes purplish at base, apex truncate, usually 5–toothed; anthers with sagittate hairs; style-arms slender. Achenes of 2 types: outer oblong or subcylindrical, 4–5–angled and transversely rugose-verrucose, glabrous with a slightly contracted base, apex slightly constricted beneath projecting rim; inner paler, usually smooth; pappus setae white, copious, multiseriate, soft, simple, connate at base into a ring, caducous.

Some 6 species, mostly found in S Europe, SW Asia and N Africa, but also extending to Macaronesia (Canary Is.), Afghanistan, Pakistan, India, Australia, Polynesia, Arabia, Soqotra and NE Tropical Africa; 1 species in Iraq.

Reichardia (named in honour of the German botanist and physician Johann Jacob Reichard (1743–82).

Galego, M.J., Talavera, S. & Silvestre, S. (1980). Revisión del género *Reichardia* Roth. (Compostiae). Lagascalia 9: 159–217.

1. **Reichardia tingitana** (*L.*) *Roth,* Bot. Abhandl., 35 (1787); Dinsmore in Post, Fl. Syria, Palest. & Sinai ed. 2, 2: 151 (1933); Zohary in Dep. Agr. Iraq Bull. 31: 160 (1950); Dickson, Wild Flowers Kuwait: 77 (1955); Rawi in Dep. Agr. Iraq Tech. Bull. 14: 123 (1964); Täckholm, Stud. Fl. Egypt ed. 2: 604 (1974); Matthews in Fl. Turkey 5: 695 (1975); Sell in Fl. Europ. 4: 325 (1976); Feinbrun-Dothan, Fl. Palaest. 3: 439 (1978); Chaudhary, Fl. Kingd. Saudi Arabia 2(3): 247 (2000); Boulos & Hind in Fl. Egypt 3: 299 (2002); Breckle, Hedge & Rafiqpoor, Vasc. Pl. Afghan. Checkl.: 183 (2013); Kilian & Ghazanfar in Fl. Oman 3: 231 (2015); Haloob & Al-Kaisi, Illustr. Fl. Lowland Iraq 1(1): 53 (2016); Abdulridha, Taha & Wedad, Ecol. & Fl. Basra: 223 (2016); Taifour & El-Oqlah, Pl. Jordan Ann. Checkl.: 46 (2017); Sultan Abedin in Fl. Pakistan 222: 231 (2017).

Scorzonera tingitana L., Sp. Pl.: 791 (1753).
S. orientalis L., Sp. Pl. ed. 2, 2: 113 (1763).
Picridium orientale (L.) DC., Fl. France 4: 96 (1805).
Sonchus tingitanus (L.) Lam., Encycl. 3: 397 (1791); Smith, Fl. Graec. Prodr. 2: 126 (1813); Fl. Graec. 8: 67, t.792 (1833).
Picridium tingitanum Desf., Fl. Atl. 2: 220 (1799); Boissier, Fl. Orient. 3: 827 (1875).

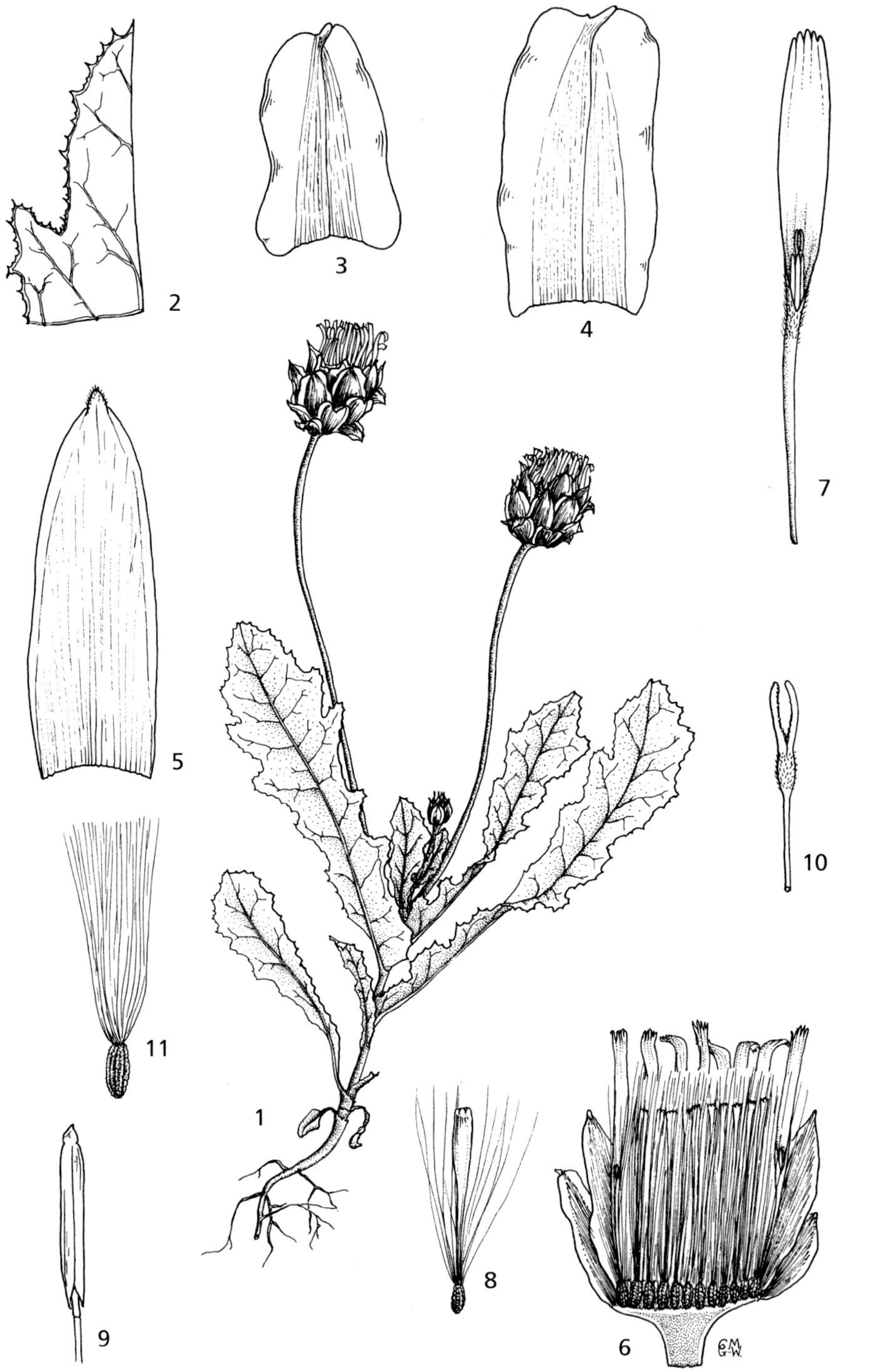

Fig. 47. **Reichardia tingitana**. 1, habit × 1; 2, leaf margin × 3; 3, outer phyllary × 6; 4, median phyllary × 6; 5, inner phyllary × 6; 6, capitulum (dissected) × 6; 7, ray floret × 6; 8, disc floret × 6; 9, anther × 18; 10, style × 12; 11, outer achene and pappus × 6. 1–2 from *Watson* 1919; 3–11 from *Guest & Rawi* 13699A. Drawn by Christine M. Grey-Wilson.

Reichardia orientalis (L.) Hochr. in Ann. Cons. Jard. Bot., Genève, 7–8: 238 (1904); Rechinger, Fl. Iranica 122: 169 (1977).

Erect annual, 1.8–53 cm tall, glaucous (var. *subintegra*) or not, with or without basal rosettes. Stems stout, sometimes unbranched, glabrous with leafy lateral shoots or plant almost stemless. Seed leaves, when present, obovate, entire with cartilaginous border. Leaves green above and beneath or somewhat glaucous, obovate, oblong, elliptic or linear-lanceolate, occasionally ± spatulate, 1–21 × 0.3–6 cm; blade gradually tapering from about half way or abruptly contracted into winged petiole 2–3.5 cm wide, apex obtuse to acute, spinulose; margins entire to shallowly dentate, deeply lobed or pinnatifid, lobes toothed and spinulose, surrounded by a narrow, usually pale, spinulose cartilaginous border. Cauline leaves sessile, often large, elliptic to obovate-oblong, 5–14.2 × 2.2–6.5 cm, upper smaller and narrowly ovate to lanceolate, often with large flaring auricles, apex obtuse to acute, and spinulose, often acuminate, margins entire and smooth to deeply toothed and spinulose; uppermost cauline leaves reduced to lanceolate-acuminate bracts up to 1 cm. Peduncles 0.7–4.5(–12) × 1–6 mm in diameter just beneath capitulum, usually ebracteate. Capitula solitary, up to 4 cm diam., terminal, crowded at centre of plant on very short pedicels; involucre narrowly cylindrical, c. 1 cm tall; phyllaries 3(–5)-seriate, margins scarious, up to 2 mm wide, glabrous, minutely hairy inside at base of mucro, outer bracts broader, ovate, inner ones lanceolate; receptacle enlarging somewhat in fruit. Ray florets hermaphrodite, fertile; corollas yellow to orange, linear to elliptic, 8.5-19 mm. Achenes tetragonal, 4-sulcate, tuberculate, base and apex truncate, outer achenes 1.5–2.2, body greyish-brown, inner achenes 2.5–3 mm, white; pappus 8–9 mm, setae deciduous, white. Fig. 47. 1–11.

HAB. Degraded conglomerate hills, sandy gravel, gypsiferous soil with small gravel, sandy clay, sometimes on silty soil; alt. to 500 m; fl. & fr. (Feb.-)Mar.-Apr.

DISTRIB. Occasional in the dry steppe zone of Iraq, common in the western desert and very common in the southern desert region. **FKI**: Kifri, *Rawi & Gillett* 7390! **FPF**: Jabal Muwaila, 70 km N of Amara, *Guest & Rawi* 17537!, 17580A!; Chilat, on frontier between Mandali and Badra, *Rawi & Haddad* 25694! **DWD**: Qa'ra, 75 km N of Rutba, *Rawi & Khatib* 32301!; 95 km NE of Rutba, *id.* 32320!, 32345!; 20 km E of Ana, on road to Qaim, *Omar, Qaisi, Hamad & Hamid* 44421!; Wadi Hauran 12 km SE of H.1, *Chakravarty, Rawi, Khatib & Alizzi* 31638!; 15 km W of Ramadi, *Guest & Rawi* 13699A!; 20 km W of Nukhaib, *Rawi* 30987!; 175 km NW of Ramadi, *Rawi* 31374!; 12 km N of Ukhaidhir, *Rawi* 30881!; E of Shithatha, *Rawi & Gillett* 6437!; 5 km W of Karbala, *id.* 6371! **DSD**: Ruhaba, between Najaf and Shabicha, *Alizzi & Omar* 35399!; 80 km W of Shabicha, *Chakravarty et al.* 30057!; ±15 km SE of Ashuriya, *Guest et al.* 19371!; 22 km E of Shabicha, *Chakravarty et al.* 30118!; 6 km SE of Shabicha, *Guest & Rawi* 19239!; 3 km N of Salman, *Rawi et al.* 29239!; 10 km E of Salman, *Guest & Rawi* 18788!; 60 km S of Salman, *anon.* 42197!; Jal al-Rawaq, *Guest & Rawi* 18914!; 25 km N of Buraiya on road to Salman, *Qaisi, Hamad & Hamid* 48539!; 75 km W of Khadr al-Ma'i, *Rawi, Khatib & Tikriti* 29139!; between Zubair and Safwan, *Rawi* 25976!; nr Baniya, 65 km WSW of Basra, *Guest et al.* 17304!; S of Zubair, *Rawi & Gillett* 6086!; Jabal Sanam, *Guest et al.* 17036!; Ansab, *id.* 18928!; 50 km S of Ansab, *Botany Staff* 42245! **LCA/FPF**: Jabal Hamrin, nr new dam, *Qaisi* 48679! **LEA**: Fakka, 70 km E of Amara, *Rawi & Haddad* 25764! **LBA**: 70 km E of Zubair, *Thamir* 47439!

Mediterranean Europe (Spain, Portugal, Balearic Is., Greece), Cyprus, Egypt, Arabia, Palestine, Jordan, Syria, S Turkey, S Iran, Afghanistan, Pakistan, Macaronesia (Canary Is.), N Africa.

var. SUBINTEGRA (Boiss.) Pamp. apparently does not occur in Iraq, although one sheet at K (*Watson* s.n. from Jabal Sanam, 23 Aug. 1919) could possibly be this variety. The specimen has some leaves toothed, but is somewhat glaucous, is taller than the Iraqi plant and has the general appearance of var. SUBINTEGRA which differs from var. *tingitana* in its greater height (up to 53 cm), usually glabrous and somewhat glaucous leaves and stems, subentire leaves and longer rays. The cauline leaves of var. SUBINTEGRA are almost always elliptic and larger than those of var. *tingitana.*

44. **SONCHUS** L.

Sp. Pl.: 793 (1753); Gen. Pl. ed. 5: 347 (1754)
Lack in Kubitzki, Fam. Gen. Vasc. Pl. 8: 190 [2006] (2007)

C.C.Townsend

Annual, biennial or perennial herbs, often ± thistle-like, with copious milky latex and hollow stems. Leaves simple or pinnatifid to pinnatisect, often runcinately lobed, lowermost petiolate, cauline leaves sessile and auriculate. Inflorescence a panicle containing few to many capitula. Involucre campanulate in flower, conical in fruit, often with white or tawny woolly pubescence at least at base; phyllaries multiseriate, scarcely elongating in fruit. Florets many, corollas yellow. Achenes compressed, ± narrowed at each end, erostrate, with 1–4(-5) ribs on each face, yellow or brown, glabrous, often scabridulous; pappus white or off-white, dimorphic, with a few strong early-caducous hairs and many weak, flexuose, cottony, persistent ones.

Sonchus (from the Gr. name for the sow-thistle, σόγχος (*sogchos*). Sow-thistle (Eng.).

About 54 species distributed widely in Eurasia and Africa and introduced to N America and elsewhere; 4 species are known from Iraq.

Boulos, L. Revision systematique du genre *Sonchus* L. s.l., 1–6. Bot. Not. 125: 287–319 (1972); 126: 155–196 (1973); 127: 7–37 & 402–451 (1974).

1. Perennial with a creeping rhizome; leaves narrow, linear to lanceolate, simple, entire to denticulate or rarely shallowly sinuate-lobate . . . . . . 4. *S. maritimus*
   Annual or biennial, rarely perennating; leaves variously pinnately lobed or divided; if simple then broader 2
2. Achenes strongly compressed and ± winged, surface smooth between ribs; cauline leaves with broad, rounded auricles . . . . 2. *S. asper*
   Achenes neither strongly compressed nor winged, surface rugose or tuberculate between ribs; cauline leaves with acute auricles 3
3. Terminal lobe of leaf not much larger than laterals; lateral lobes narrowly linear; if somewhat broader then distinctly narrowed at base 1. *S. tenerrimus*
   Terminal lobe of leaf mostly much larger than laterals; lateral lobes broader, not narrowed to base . . . . 3. *S. oleraceus*

1. **Sonchus tenerrimus** *L.*, Sp. Pl.: 764 (1753); Boissier, Fl. Orient. 3: 797 (1875); Nábělek in Publ. Fac. Sci. Univ. Masaryk 52: 53 (1925); Dinsmore in Post, Fl. Syria, Palest. & Sinai ed. 2, 2: 144 (1933); Anthony in Notes Roy. Bot. Gard. Edinburgh 18: 292 (1935); Bornmüller in Beih. Bot. Centralbl. 60B: 223 (1939); Zohary in Dep. Agr. Iraq Bull. 31: 159 (1950); Rawi in Dep. Agr. Iraq Tech. Bull. 14: 125 (1964); Rechinger, Fl. Lowland Iraq: 679 (1964); Boulos in Bot. Not. 126: 158 (1973); Matthews in Fl. Turkey 5: 692 (1975); Rechinger, Fl. Iranica 122: 163 (1977); Feinbrun-Dothan, Fl. Palaest. 3: 434 (1978); Meikle, Fl. Cyprus 2: 1038 (1985); Chaudhary, Fl. Kingd. Saudi Arabia 2(3): 242 (2000); Boulos & Hind in Fl. Egypt 3: 308 (2002); Haloob & Al- Kaisi, Illustr. Fl. Lowland Iraq 1(1): 58 (2016); Abdulridha, Taha & Wedad, Ecol. & Fl. Basra: 227 (2016); Taifour & El-Oqlah, Pl. Jordan Ann. Checkl.: 48 (2017); Ghafoor in Fl. Pakistan 222: 237, f. 21A–D (2017).

Annual, biennial or perennial herb 0.2–1 m tall with a glabrous stem. Basal leaves lyrate with few lobes, petiolate; cauline leaves clasping stem, 4–20 × 2–7 cm, elliptical in outline, pinnatisect, lobes oblong or linear, entire or weakly toothed to dentate, sometimes unilaterally hastate at base, terminal and lateral lobes not strongly differentiated; auricles ending in a sharp tip; lamina glabrous, green above, glaucous below. Inflorescence irregular, subcorymbose with groups of 1–7 capitula; pedicels 0.5–2 cm long, white-woolly towards top, sometimes glandular (not seen in Iraq). Involucre 10–12 mm long; inner phyllaries linear, blunt and ciliate at apex. Achenes obovate, ± 3 mm long, muricate, yellow-green or brownish, tapering below; pappus hairs soft, persistent, 5–8 mm.

HAB. Damp places, in a date garden, on a semi-disturbed silt island in the Shatt al-Arab; alt. up to c. 10 m; fl. & fr. Mar.-Apr.
DISTRIB. Rather rare, mainly in southernmost Iraq. **LCA**: Babylon, *Bornmüller* 437; Deltawa (Khalis), *Haines* 1669! **LSM**: Musaida, near Amara, *Field & Lazar* 53!; nr. Qalat Salih, *Guest, Rawi & Rechinger* 17411! **LBA**, Basra, *Bornmüller* 438; Basra, on a silt island opposite Ma'qil, *Guest, Rawi & Rechinger* 16330!; 10 km S of Basra, *Rawi* 25913!

Mediterranean Europe (from Portugal and Spain to Italy, Sicily & the Balkans), Cyprus, Syria, Lebanon, Palestine, Jordan, Egypt, Turkey, Iran, Pakistan, Macaronesia.

HARFASH (Abdulridha et.al. l.c.)

Clammy Sowthistle (Eng., Dinsmore).

2. **Sonchus asper** (*L.*) *Hill*, Herb. Brit. 1: 47 (1769); Boissier, Fl. Orient. 3: 796 (1875); Dinsmore in Post, Fl. Syria, Palest. & Sinai ed. 2, 2: 144 (1933); Anthony in Notes Roy. Bot. Gard. Edinburgh 18: 292 (1935); Zohary in Dep. Agr. Iraq Bull. 31: 159 (1950); Rawi in Dep. Agr. Iraq Tech. Bull. 14: 125 (1964); Rechinger, Fl. Lowland Iraq: 678 (1964); Kirpichnikov in Fl. S.S.S.R. 29: 256 (1964); Boulos in Bot. Not. 126: 164 (1973); Matthews in Fl. Turkey 5: 691 (1975); Rechinger, Fl. Iranica 122: 164 (1977); Meikle, Fl. Cyprus 2: 1037 (1985); Chaudhary, Fl. Kingd. Saudi Arabia 2(3): 243 (2000); Boulos & Hind in Fl. Egypt 3: 311 (2002); Ghafoor in Fl. Pakistan 222: 237, f. 21A–D (2017).

S. oleraceus var. *asper* L., Sp. Pl.: 794 (1753).
(For additional synonymy see Boulos (1973)).

Annual or biennial, commonly spinous herb, 10–120 cm tall. Basal leaves simple to pinnatipartite with a petioloid winged base; cauline leaves sessile with rounded auricles, 6–26 × 2–8 cm, elliptical or obovate in outline, usually more deeply divided but otherwise similar to basal, lobes triangular in outline, spines stiff to rather soft. Peduncles 0.7–4.5 cm, glandular-pilose or glabrous. Involucre 9–12 mm long; inner phyllaries linear, somewhat lanose-pubescent at base, outer often glandular. Achenes 2–3 mm long, strongly compressed, winged, faces 3-ribbed, glabrous and smooth between ribs, margins and ribs ciliate or not; pappus hairs fine, 7–10 mm. Fig. 50: 2–3.

The species is widespread as a weed in temperate Eurasia, N Africa and Macaronesia, and has been widely introduced in many other regions of the world.

subsp. ASPER, an annual with the margins and ribs of the achenes smooth or almost so, occurs mostly in the western and northern parts of the range of the species. In Iraq, as in SW Asia generally, the species is represented by:

subsp. **glaucescens** (*Jord.*) *Ball* in Linn. Soc. Bot. 16: 548 (1878); Boulos in Bot. Not. 126: 165 (1973); Matthews in Fl. Turkey 5: 691 (1975); Rechinger, in Fl. Iranica 122: 164 (1977); Meikle, Fl. Cyprus 2: 1037 (1985).

S. nymanii Tineo & Guss. in Guss., Fl. Sic. Syn. 2: 860 (1844); Feinbrun-Dothan, Fl. Palaest. 3: 435 (1978).
S. glaucescens Jord., Obs. Pl. Crit. 5: 75, t. 5 (1847); Boissier, Fl. Orient. 3: 796 (18875); Zoh. in Dep. Agr. Iraq Bull. 31: 159 (1950); Rawi in Dep. Agr. Iraq. Tech. Bull. 14: 125 (1964); Rechinger, Fl. Lowland Iraq: 678 (1964); Taifour & El-Oqlah, Pl. Jordan Ann. Checkl.: 47 (2017).
S. asper subsp. *nymanii* (Tin. & Guss.) Hegi, Illustr. Fl. Mitteleur. 6: 1110 (1929).

Biennial. Margins and often also ribs of achenes ± densely retrorsely scabrid.

HAB. Most frequent as a weed of irrigated cultivation, also on a rocky hill, on mountain slope among *Quercus*, on an alluvial flat by a lake, and by seepage in a wadi, on clay loam, silt, and alluvium, sometimes on slightly saline sand; alt. 5–1250 m; fl. & fr. Dec.-Jul.
DISTRIB. Occasional in the lower mountains, rare in the steppe and desert regions, quite frequent on the irrigated alluvial plain. **MAM**: Baidahu, 15 km W of Kani Mazi, *Omar* 45751! **MRO**: Shaqlawa, *Haines* s.n.! **MSU**: Qara Dagh, *Gillett* 7863!; Jarmo, *Helbaek* 1015! **DSD**: Jabal Sanam, *Khatib & Alizzi* 33471! **FPF**: 1 km S of Badra, *Nat. Herb. Iraq Rawi* 18126!; 5 km to Fakka, *Qaisi, Thamer & Salah* 51300! **LCA**: Abu Ghraib, *Alizzi, Thaib & Janan* 32588!; Dukhaila, 30 km N of Baghdad, *Anderson* 1984! **LEA**: 64 km NE of Baghdad, *Hunting Technical Services* 25!; 3 km NE of Baghdad, *Gillett & Rawi* 5918!; 86 km N of Kut, *Rawi* 18456!; 37 km S of Ali Gharbi on road to Amara, *Rawi & Haddad* 25744! **LSM**: nr Azair, between the Hor and R. Tigris, *Guest, Rawi & Rechinger* 17401!; 15 km E of Amara, *Chakravarty, Rawi, Tikriti & Khatib* 29843! **LBA**: Basra, *Rogers* 70457!; Abu Al-Khasib, 20 km E of Basra, *Rawi* 25929!

S & C Europe, Cyprus, Lebanon, Syria, Egypt, Turkey, Iran, Afghanistan, Pakistan, India, N Africa (Libya to Morocco), Macaronesia.

3. **Sonchus oleraceus** *L.*, Sp. Pl.: 794 (1753) p.p. – excluding var. *asper*; Blatter in Journ. Ind. Bot. Soc. 9, 1: 38 (1932); Dinsmore in Post, Fl. Syria, Palest. & Sinai ed. 2, 2: 144 (1933); Guest in Dep. Agr. Iraq Bull. 27: 95 (1933); Anthony in Notes Roy. Bot. Gard. Edinburgh 18: 292 (1935); Blakelock in Kew Bull. 4: 64 (1949); Zohary in Dep. Agr. Iraq Bull. 31: 159 (1950); Rawi in Dep. Agr. Iraq Tech. Bull. 14: 125 (1964); Rechinger, Fl. Lowland Iraq: 679 (1964); Rawi & Chakravarty in Dep. Agr. Iraq Tech. Bull. 15: 90 (1964); Kirpichnikov in Fl. S.S.S.R. 29: 254 (1964); Boulos in Bot. Not. 126: 155 (1973); Matthews in Fl. Turkey 5: 691 (1975); Rechinger, Fl. Iranica 122: 162 (1977); Feinbrun-Dothan, Fl. Palaest. 3: 434 (1978);

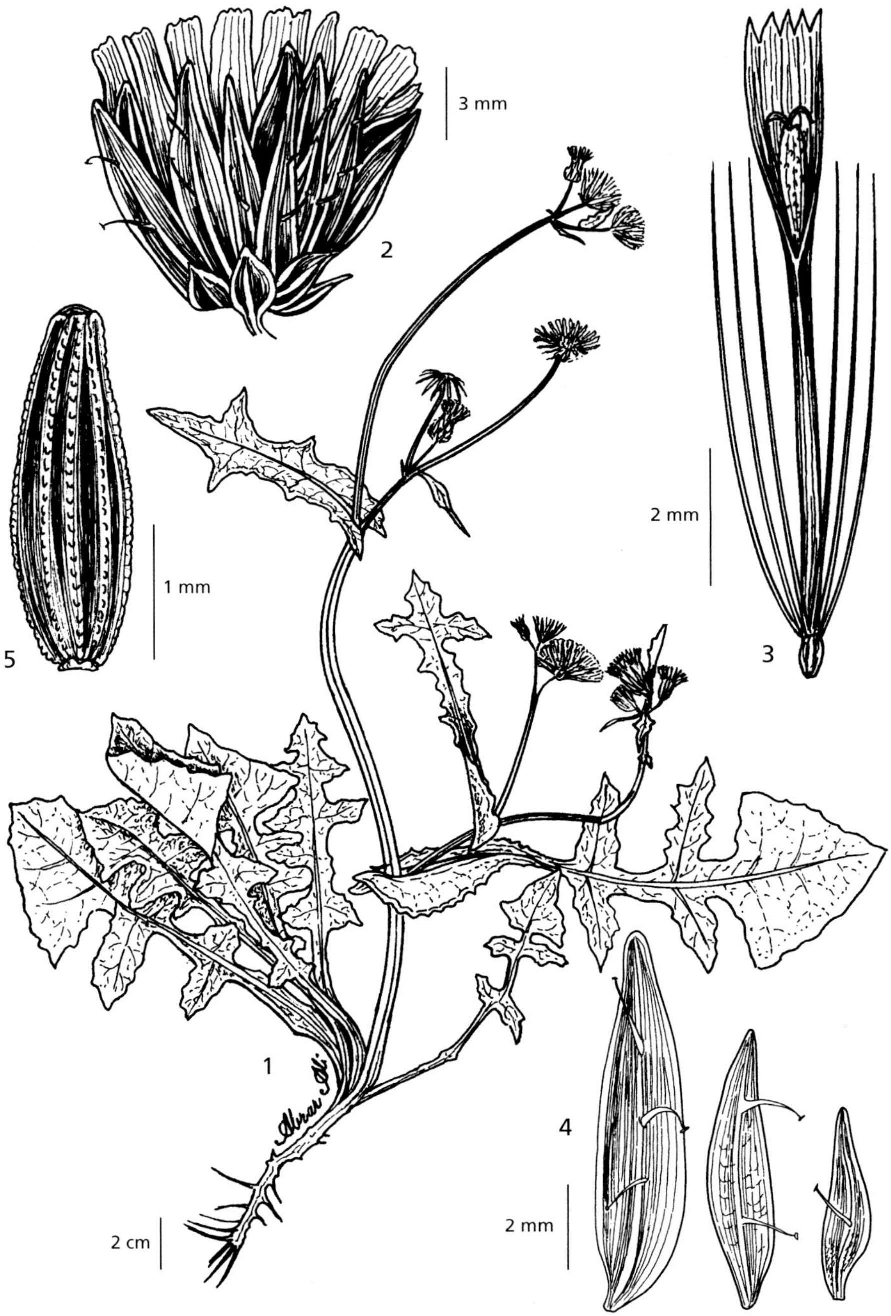

Fig. 48. **Sonchus oleraceus**. 1, habit; 2, capitulum; 3, floret; 4, phyllaries; 5, achene with pappus removed. Drawn by Abrar Ali. Reproduced with permission Fl. Pakistan 222, fig. 20, A–E.

Meikle, Fl. Cyprus 2: 1035 (1985); Chaudhary, Fl. Kingd. Saudi Arabia 2(3): 242 (2000); Boulos & Hind in Fl. Egypt 3: 309 (2002); Tadese in Fl. Ethiop. & Eriter. 4(2): 69 (2004); Zhu Shih & Kilian in Fl. China 20–21: 242 (2011); Breckle, Hedge & Rafiqpoor, Vasc. Pl. Afghan. Checkl.: 188 (2013); Haloob & Al- Kaisi, Illustr. Fl. Lowland Iraq 1(1): 57 (2016); Abdulridha, Taha & Wedad, Ecol. & Fl. Basra: 227 (2016); Taifour & El-Oqlah, Pl. Jordan Ann. Checkl.: 48 (2017); Ghafoor in Fl. Pakistan 222: 237, f. 21A–D (2017).

S. laevis (L.) Gars., Fig. Pl. Anim. Med. 4: t. 564 (1764); Handel-Mazzetti in Ann. Naturh. Mus. Wien 27: 458 (1913). [A considerable synonymy has been omitted but may be found in Boulos (1973).]

Annual or biennial herb, 10–150 cm tall. Basal leaves simple to deeply pinnatipartite with a petioloid winged base. Cauline leaves sessile with acuminate auricles, 6–30 × 3–12 cm, elliptical or obovate in outline, usually more deeply divided but otherwise similar to basal, lobes ovate or deltoid to oblong, entire or dentate, teeth softly spinous or not. Inflorescence irregular, subcorymbose, with groups of 1–6 capitula; peduncles 0.5–3.5(–7) cm, glabrous or glandular-pilose. Involucre 8–12 mm, inner phyllaries linear, obtuse, glabrous or glandular-pilose. Achenes yellowish to brown, obovate or oblanceolate, compressed but not winged, faces 3-ribbed, scabrid, 2.5–3.5 mm; pappus hairs fine, persistent, 6–8 mm. Fig. 48: 1–5.

HAB. In fields and date gardens, river banks, by roadsides on waste ground etc.; alt. up to 350(–450) m; fl. & fr. (Sept.-)Mar.-Apr.(-May).
DISTRIB. Occasional in the steppe and lower mountain region, quite common on the irrigated alluvial plain in the desert region of Iraq. **MRO**: Qala No, nr Koi Sanjaq, *Rawi, Noori & Kass* 28047! **MSU**: Sayyid Sadiq, *Omar, Karim et al.* 37432! **FUJ**: Mosul, *A. Low* 168! **FUJ/FNI**: Mosul, *Calder* 2050. **FKI**: Kirkuk, *P. Grigg* 1!, *Barkley & Jum'a Brahim* 1472! **FPF**: Sudur, *Botany staff* 47158!; Mandali, *Rawi* 12669!; Makatu nr Mandali, *Rawi* 12676!; Mandali, *Kaisi & Yahya* 45200 (BAG!); **DWD**: Qaim, Sharif, Hamad & Hamid 47320! 17 km SE of K3 (Albu Hayat) *Kaisi & Hamed* 53139! **DSD**: sine loc., *Watson* s.n.! **LCA**: Falluja, *Guest* 2427!; Abu Ghraib, *Samir Munim* 1649!, *Alizzi & Janan* 32806!; Baghdad, *Paranjpye* s.n. (comm. Graham); between Ba'quba and Rustamiya, *Rogers* 071!, *Guest* 3511!; Suwaira, *Alizzi & Omar* 34655!; Kut, on bank of Gharaf river, *Karim & Hamid* 37803!; 46 km S of Musayib, *Thamir & Widad* 47260!; between Ali al-Gharbi and Amara, *Rawi & Haddad* 25744! **LSM/LBA**: 50 km from Basra on road to Nasiriya, *Husham, Alizzi & Muhammad* 29625! **LSM**: Amara, *Alizzi & Omar* 34925!; Qalat Salih, *Botany staff* 42476! **LBA**: 10 km S of Basra, *Rawi* 25914!

Almost throughout Europe, Aegean Is., Cyprus, Syria, Lebanon, Palestine, Jordan, Sinai, Egypt, N Africa, Arabia, Turkey, Caucasia, Iran, Afghanistan, Pakistan, India, C Asia (widespread), Siberia, China, Japan. Introduced world-wide as a weed in cultivated fields.

According to Lindley & Moore (1870) this species was at one time eaten as a salad plant, though it has long since given place to more palatable herbs. Guest has recorded the local common names for the plant as MURRAIR ("bitter plant") and also as UMM AL-HALÎB (lit. "mother of milk", a generic name for plants with white milky juice); HARFASH, MERRER (Abdulridha et.al., l.c.). Rawi & Chakravarty (1964) state that the brownish gum left after the evaporation of the juice of the plant is said to be a powerful hydragogue cathartic and has been used as a "so-called cure" of opium addiction.

4. **Sonchus maritimus** *L.*, Syst. Nat. ed. 10, 2: 1192 (1759); Handel-Mazzetti in Ann. Naturh. Mus. Wien 27: 458 (1913); Nábělek in Publ. Fac. Sci. Univ. Masaryk 52: 53 (1925); Dinsmore in Post, Fl. Syria, Palest. & Sinai ed. 2, 2: 144 (1933); Zohary in Dep. Agr. Iraq Bull. 31: 159 (1950); Rawi in Dep. Agr. Iraq Tech. Bull. 14: 124 (1964); Rechinger, Fl. Lowland Iraq: 679 (1964); Boulos in Bot. Not. 126: 174 (1973); Matthews in Fl. Turkey 5: 693(1975); Rechinger, Fl. Iranica 122: 165 (1977); Feinbrun-Dothan, Fl. Palaest. 3: 435 (1978); Chaudhary, Fl. Kingd. Saudi Arabia 2(3): 243 (2000); Boulos & Hind in Fl. Egypt 3: 309 (2002); Zhu Shih & Kilian in Fl. China 20–21: 242 (2011); Breckle, Hedge & Rafiqpoor, Vasc. Pl. Afghan. Checkl.: 188 (2013); Haloob & Al- Kaisi, Illustr. Fl. Lowland Iraq 1(1): 57 (2016); Abdulridha, Taha & Wedad, Ecol. & Fl. Basra: 227 (2016); Taifour & El-Oqlah, Pl. Jordan Ann. Checkl.: 47 (2017); Ghafoor in Fl. Pakistan 222: 242 (2017).

Creeping, rhizomatous perennial herb (15–)25–110 cm tall. Basal leaves broadly to narrowly linear, entire, denticulate or rarely shallowly sinuate-lobate, sheathing. Cauline leaves similar, becoming narrower upwards, rounded-auriculate at base, 4–26 × 0.4–2.5 cm. Inflorescence few- (up to 8-)headed; pedicels 0.5–5 cm, tomentose below capitulum. Involucre 10–12 mm, inner phyllaries linear, glabrous or ciliate at apex. Achenes yellowish to brown, 2.5–3 mm, with fine ribs on each face and broader, incrassate margins, smooth or almost so; pappus hairs fine, persistent, 6–9 mm.

HAB. In wet places in the mountains, in ditches below a perennial spring, in a wet wadi, on irrigation channels in orchards and date gardens in the desert region and similar habitats on the irrigated alluvial plain; alt. up to ± 700 m; fl. & fr. (Oct.-)Mar.-Apr.(-Jul.).
DISTRIB. Rare in the lower forest zone, occasional in the desert region, more common on the irrigated alluvial plain. **MAM**: Dohuk, *Haines* W.458! **MSU**: Jarmo, *Helbaek* 599! **DLJ**: 17 km E of Haditha, *Qaisi* 44237! **DWD**: Hit, *Rawi & Gillett* 6844!; Shithatha, *Rawi & Chakravarty* 32567! **LSM**: Hor Hawiza, 16 km E of Qalat Salih, *Thamir* 46693!; Musharrah (Halfaya), *id.* 47111!; Abu Chisaf, 25 km S of Chahala, *id.* 46659!; Hor ash-Kaba'ish, *Qaisi* 44034! **LBA**: 50 km SE of Basra, on road to Fao, *Alizzi & Omar* 35052!; Daura village, between Basra and Fao, *id.* 35863!; 17 km NW of Fao, *Khatib & Alizzi* 33406!

Mediterranean Europe (Portugal, Spain, W & S France, Italy), Cyprus, Syria, Palestine, Jordan, Egypt, Saudi Arabia, Iran, Afghanistan, Pakistan, N Africa (Libya to Morocco), S Africa, Australia.

HARFASH (Abdulridha et.al., l.c.); Sea Sow-thistle (Eng. Dinsmore, l.c.).

45. **PILOSELLA** Hill

Brit. Herb. 441 (1756)
Lack in Kubitzki, Fam. Gen. Vasc. Pl. 8: 195 [2006] (2007).

C. C. Townsend[12]

Perennial rhizomatous herbs, often (though not usually in Iraq) stoloniferous. Plant ± hairy, indumentum of simple, glandular or stellate hairs. Stems usually few to several, more rarely solitary, each with 1-numerous capitula. Leaves entire to slightly denticulate, often (not in Iraq) all basal; margins smooth, tapering gradually into the petiole. Capitula homogamous, ligulate; involucre ± campanulate; phyllaries numerous, multiseriate, imbricate; receptacle epaleaceous, flat, pits short-dentate. Corollas yellow, sometimes red-streaked on outer surface. Achenes homomorphic, 10-ribbed; pappus hairs uniseriate with some shorter ones 1.5–2.5 mm long, scabrid to barbellate, dirty white to brownish.

About 20 species world-wide, native to most north-temperate regions; two species in Iraq.

Named from the Lat. *pilosus*, hairy and *-ella*, diminutive suffix.

Bräutigam, S. & Greuter, W. (2007). A new treatment of *Pilosella* for the Euro-Mediterranean flora. Willdenowia 37: 123–138.
Gaskin, J.F. & Wilson, L.M. (2007). Phylogenetic Relationships among native and naturalized *Hieracium* (Asteraceae) in Canada and the United States Based on Plastid DNA Sequences. Syst. Bot. 32 (2): 478–485.

1. Plant lacking stolons; pedicels and involucre densely glandular-hairy but with few or no simple eglandular hairs .... 1. *P. verruculata*
   Stoloniferous; pedicels and involucre with numerous simple eglandular hairs 2
2. Involucre with few or no glandular hairs............................2. *P. echioides*
   Involucre with numerous glandular hairs ....3. *P.* × *maschukensis*

1. **Pilosella verruculata** (*Link*) *Soják*, Folia Geobot. Phytotax. 6 (2): 218 (1971).

Hieracium incanum M.Bieb. in Fl. Taur.-Cauc. 2: 253 (1808), non L. (1753).
H. verruculatum Link, Enum. Hort. Berol. 2, 2: 287 (1822); Lack & Rechinger in Fl. Iranica 122: 176 (1977).
[*H. bauhinii* auctt. non Bess., Agnew et al. in Bull. Iraq Nat. Hist. Inst. 2 (2): 6 (1962); Lack & Rechinger in Fl. Iranica 122: 173 (1977).]

Perennial, lacking stolons. Rosette leaves usually withered by flowering time, 3–20 × 0.5–4 cm, varying from elliptic to linear-lanceolate, acute to subacute, entire, gradually narrowed

---

[12] The determinations of most of the specimens were made by P. D. Sell, and the account is based on his treatment in Fl. Turkey.

beneath, clothed with numerous stellate hairs, few to numerous simple eglandular hairs and scattered glandular hairs. Flowering stems, 50–80(–110) cm tall, with numerous stellate and a variable number of glandular hairs, but with simple hairs only near base. Cauline leaves few to numerous (up to ± 12), narrowly to broadly elliptic to oblanceolate, lower attenuate below, upper increasingly distinctly amplexicaul, very variable in size. Inflorescence laxly paniculate with longer lower branches to shorter and sub-umbellate, with up to 60 capitula. Involucre 8–11 mm, campanulate; phyllaries linear-lanceolate, obtuse (not seen in Iraq) to acute or acuminate, with dense stellate and glandular hairs but without or with very few simple hairs. Corollas yellow. Achenes narrowly oblong, ± 2 mm, dark chestnut-brown; pappus dirty white.

HAB. Clearings in *Quercus* woodland, dried-up gravel outwash; alt. 9500–2100 m.
DISTRIB. Rare; known from four records from mountainous areas of Iraq. **MAM**: Sarsang, *E. Chapman* 26430! **MRO**: Rowanduz, above Zeita, *Hadač et al.* 5953. **MSU**: nr Penjwin, *Rechinger* 10519; Qopi Qaradagh, *Haines* W.1552!

Turkey (E Anatolia), Caucasia, NW Iran, Russia (SE European part).

The *Bornmüller* specimen cited by Lack & Rechinger (loc. cit.) under this species is now placed under the hybrid taxon.

2. **Pilosella echioides** (*Lumn.*) *C.H.& Sch.Bip.*, Flora 45: 431 (1862); Sell & West in Fl. Turkey 5: 756 (1975); C. Shih & Gottschlich in Fl. China 20–21: 353 (2011); Sultan Abedin in Fl. Pakistan 222: 259 (2017).

Hieracium echioides Lumn., Fl. Poson.: 348 (1791); Boissier, Fl. Orient 3: 864 (1875); Juxip & Bobr., Fl. U.S.S.R. 30: 418 (1960).
H. echioides subsp. *asiaticum* Nägeli & Peter, Hierac. Mitt.-Eur. 1: 486 (1885).
H. asiaticum Nägeli & Peter, Uksip in Fl. U.S.S.R. 30: 419 (1960).
Pilosella asiatica (Nägeli & Peter) Schljakov, Fl. Evropeiskoi Chasti SSSR 8: 329 (1989).

Perennial, compactly stoloniferous, stolons slender. Stems tall, slender but rigid, ± setose, leafless or with 1–3 leaves; inflorescence a lax corymb. Leaves oblong-lanceolate, acute, glaucous, ± setose on margins and lower surface, this surface also finely stellate-hairy. Peduncles sparsely stellate-hairy and glandular. Involucre 7–8 mm, cylindrical, blackish and weakly clothed with long eglandular hairs. Corollas yellow, with reddish veins on abaxial surface. Fig. 49: 1–3.

HAB. Not noted; alt. 1350–2200 m; fl. Jun.
DISTRIB. Very rare; known from two records in Iraq. **MRO**: Tenji Saipaiyan (Sarchal) in upper part of valley near Nawanda, *Hadač* 2306; in valley between the villages Hasitha and Amadia, *Nábělek* #3966 (SAV – 0001222).

C & E Europe, Turkey, Kazakhtan, Kyrgistan, N Caucasus, China, India.

3. **Pilosella** × **maschukensis** (*Litw. & Zahn*) *Soják* in Folia Geobot. Phytotax. 6: 218 (1971); Sell in Fl. Turkey 5: 759 (1975).

Hieracium maschukense Litw. & Zahn in Feddes Repert. 4: 182 (1907).
H. incanum var. *carduchorum* Hausskn. & Bornm. nomen in sched. *Bornmüller* 1470.

This plant is a hybrid between *P. verruculata* and *P. echioides* (Lumn.) C.H. & F.W. Schulz, a species occurring from S Russia, Turkey and the Lebanon to Afghanistan and C Asia but not yet found in Iraq. It may not be present, because many *Pilosella* hybrids spread by seed or stolons – the latter, however, being absent in *P. verruculata* and infrequent in *P. echioides.* It should, however, be looked for. The two species and the hybrid may be separated as shown in the key.

The hybrid appears to be as common in Iraq as its other parent, as may be seen from the specimens cited.

HAB. On rocky mountain slopes, sometimes in woodland, or cultivated ground on the lower slopes of a hill; alt. 1250–2100 m.
DISTRIB. Locally common in the mountain region of Iraq. **MAM**: Haisi village 10 km S of Kani Masi, *Dabbagh & Hamad* 45550A!; Baidaho, 15 km W of Kani Masi, *Omar* 45755! **MRO**: Sakri-Sakran, *Bornmüller*

Fig. 49. **Pilosella echioides**. 1, habit lower portion of plant × 1; 2 & 3, flowering stems × 1. From Flora of China, Asteraceae 20-21, fig. 330. Reproduced with permission from Missouri Botanical Garden Press, St. Louis, and Science Press, Beijing.

1470! **MSU**: Penjwin, *Rawi* 22632!; Tawila, 1867, *Haussknecht* s.n.!; Avroman & Shahu, 1867, *Haussknecht* s.n.!

Turkey, Caucasia, Iran.

Hieracium incanum var. *carduchorum* Hausskn. & Bornm. nomen in sched. *Bornmuller* 1470 Two other specimens from Iraq, *Omar* 37724 from a mountain side at Sulaf and *Haines* 1335 from Sarsang, are not determinable, the *Haines* sheet being quite sterile. Efforts should be made to re-collect both of these plants. Collections of *Pilosella* should always be carefully collected to show the basal parts, especially as to whether or not stolons are present.

46. **GARHADIOLUS** Jaub. & Spach

Illustr. Pl. Orient. 3: 119, t. 284, 285 (1849)
Lack in Kubitzki, Fam. Gen. Vasc. Pl. 8: 196 [2006] (2007)

Patricia Halliday
Revised by D. J. Nicholas Hind

Annual herbs. Leaves usually rosulate, entire, dentate-pinnatifid or lobed. Inflorescences of stout cymes. Capitula ligulate; phyllaries biseriate, outer minute forming a calyculus-like structure, inner accrescent in fruit; receptacle naked. Ligules yellow; style arms long, with very short hairs. Achenes dimorphic, outer straight or incurved, terete, subulate, persistent and partly enclosed by persistent phyllaries, pappus of minute hairs, central acenes incurved, beaked, caducous, pappus of a crown-like structure or of bristles.

A genus of two/three species from the Caucasus, Near East, central Asia and eastern Asia; 3 species in Iraq.

1. Pappus coroniform; marginal achenes 12–18, shorter than inner achenes, strongly incurved . . . . 1. *G. papposus*
   Pappus ciliate; marginal achenes 6–8, longer than inner achenes, straight, spreading, or incurved 2
2. Marginal achenes c. 8 mm long, erect, straight; leaves subentire to pinnatifid; ligules 4 mm long . . . . 2. *G. angulosus*
   Marginal achenes 10–11 mm long, spreading, incurved; leaves dentate; ligules 6–10 mm long . . . . 3. *G. hamosus*

1. **Garhadiolus papposus** *Boiss. & Buhse*, Aufz. Transk. Pers.: 135, t. 7 (1860); Boissier, Fl. Orient. 3: 723 (1875); Rawi in Dep. Agr. Iraq Tech. Bull. 14: 116 (1964); Rechinger, Fl. Iranica 122: 142 (1977); Zhu Shi & Kilian in Fl. China 20–21: 270, f. 261.1 (2011); Breckle, Hedge & Rafiqpoor, Vasc. Pl. Afghan. Checkl.: 173 (2013); Sultan Abedin & Ghafoor in Fl. Pakistan 222: 263 (2017).

Rhagadiolus papposus (Boiss. & Buhse) O.Kuntze in Act. Hort. Petrop. 10: 205 (1837); Kupicha in Fl. Turkey 5: 689 (1975).
Koelpinia scaberrima Franch., Ann. Sci. Nat. ser. 6, 16 (2): 331 (1883).
Garhadiolus scaberrimus Franch. ex O.Fedtsch. & B.Fedtsch., Consp. Fl. Turkestanicae [O.A. Fedchenko & B.A. Fedchenko] 4: 306 (1912).

Erect, sometimes decumbent, annual, (14–)20–38(–45) cm tall, with inconspicuously ribbed hispid stems; inflorescence spreading, much-branched. Basal leaves lanceolate, oblanceolate or obovate, 5–12 × 1.6–5 cm, runcinate with large terminal lobe and smaller lateral lobes, obtuse to acuminate, minutely mucronate; base clasping, with often large auricles, margins subentire to deeply lobed or toothed, terminal lobe crenate to shallowly toothed; midrib hispid below, otherwise glabrous. Cauline leaves similar or narrower, linear-lanceolate, 3.5–12 × 0.8–5.2 cm, usually acute or acuminate, subentire (or entire and linear). Capitula cylindrical, sessile or very shortly pedunculate, 5–6 mm long, 2–3 mm diam., solitary or in congested axillary clusters, somewhat arachnoid at base; involucre cylindrical to urceolate, biseriate, outer phyllaries lanceolate to linear-lanceolate, erect, never spreading, 2–3 × 0.5 mm, glabrous; inner phyllaries lanceolate, incurved, 3–6 × 1.5 mm, with scarious margins, incrassate, becoming indurated in fruit, boat shaped, variably setose, tips reflexed. Receptacle naked, flat. Corollas yellow, oblong, strap-shaped, truncate, c. 4

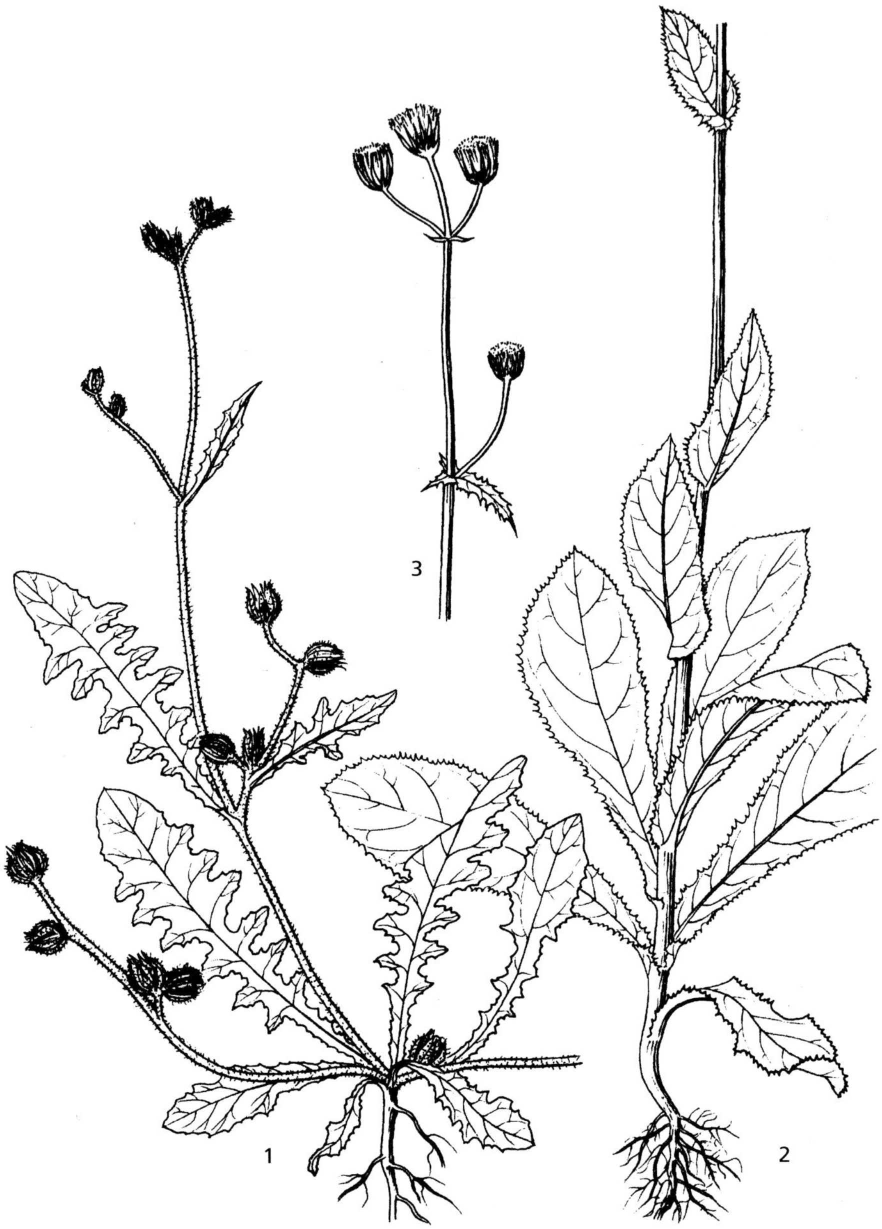

Fig. 50. **Garhadiolus papposus**. 1, habit × 1. **Sonchus asper**. 2, lower portion of stem × 1; 3, flowering stem × 1. Drawn by Zhang Taih. From Flora of China, Asteraceae 20-21, fig. 264. Reproduced with permission from Missouri Botanical Garden Press, St. Louis, and Science Press, Beijing.

× 1–1.5 mm, tubular for c. 1 mm, with 4–5 teeth and 4 veins; anthers cylindrical, 1.25 mm long; appendages oblong, rounded; style dark (in dried material), 3 mm long, forked for c. 0.5 mm, papillose for 1 mm at apex, style arms obtuse. Achenes (8–)12–18 mm, incurved, of two types: marginal enclosed by phyllaries, smooth, equalling them in length, 6–7 mm long, glabrous except for short strigose beak; inner achenes naked, exceeding phyllaries by *c.* 3 mm, slender, attenuate, ribbed, scabrid, with a longer beak 7–10 mm. Outer achenes ± epappose; inner achenes with bowl-shaped, coroniform, persistent pappus of a few patent scabrid hairs c. 1 mm. Fig. 50: 1

HAB. Muddy river bank; alt. ± 40 m; fl. & fr. Mar.
DISTRIB. Very rare in Iraq, one confirmed record on the irrigated alluvial plain in the NE corner of the desert region and two others cited from the literature. **MSU**: Sulaimaniya, fide Rawi (op. cit., 1964). **LCA**: Dawudiya nr Khan Bani Sa'd, *Haines* W.1901! **LEA**: Shahraban (Muqdadiya), *Haines* s.n. (fide Rech. f., op. cit. 142, 1977).

Xinjiang (China), C Asia (Kazakhstan, Kyrgyzstan, Tajikistan, Turkmenistan, Uzbekistan), Dagestan, Armenia, Azerbaijan, Turkey, Iran, Afghanistan, Pakistan, Syria.

2. **Garhadiolus angulosus** *Jaub. & Spach*, Illustr. Pl. Orient. 3: 122, t. 285 (1850); Feinbrun-Dothan, Fl. Palaest. 3: 414 (1978); Boulos & Hind in Fl. Egypt 3: 273 (2002).

Rhagadiolus hedypnois Fisch. & C.A.Mey., Ind. Sem. Horti Petrop. 4: 46 (1838), nom. illeg., non All. (1785).
Garhadiolus hedypnois (Fisch. & C.A.Mey.) Jaub. & Spach, Illustr. Pl. Orient. 3: 120, t. 284 (1850), comb. illeg.; Dinsmore in Post, Fl. Syria, ed. 2, 2: 128 (1933); Bornmüller in Beih. Bot. Centralbl. 60B: 217 (1939); Zohary in Dep. Agr. Iraq Bull. 31: 158 (1950); Rechinger, Fl. Lowland Iraq: 666 (1964); Rawi in Dep. Agr. Iraq Tech. Bull. 14: 116 (1964); Täckholm, Stud. Fl. Egypt, ed. 2: 590 (1974).
Hedypnois minutissima Bunge, Beitr. Kenntn. Fl. Russl.: 372 (1852).
Garhadiolus minutissimus (Bunge) Kitam. in Acta Phytotax. Geobot. Kyoto 17: 36 (1957).
Rhagadiolus angulosus (Jaub. & Spach) Kupicha in Notes Roy. Bot. Gard. Edinburgh 33: 433 (1975); Kupicha in Davis, Fl. Turkey 5: 688 (1975).

Erect herb, occasionally ± prostrate with spreading branches, 2–23 cm tall (young plants sometimes with a single stem and terminal capitulum), sparsely to densely hispid, especially in lower part, hairs to 2 mm long, or occasionally glabrous; rarely flowers and fruits ± stemless and clustered at centre of rosette. Basal leaves oblanceolate, 1.8–1.9 × 0.5–3.6 cm, subentire to dentate, lobed or pinnatifid, lobes often toothed, apex usually obtuse; petiole (0–)3–24 mm. Cauline leaves 1.5–6 × 0.5–3.4 cm, oblanceolate, ovate or elliptic, dentate, usually sessile, upper leaves sometimes linear. Leaf margins often fringed with hairs. Peduncles usually swollen, to 4 mm diam., glabrous. Capitula axillary and terminal, subcylindrical, often sessile, (5–)6(–7) mm long, 4–5.5 mm diam.; outer phylaries 1–1.5 mm, lanceolate, inner phyllaries erect or somewhat spreading, keeled and shortly beaked, 3.5–6 mm long, enclosing base of marginal achenes, glabrous or hispidulous. Ligules lemon-yellow, elliptic or oblong, 3.5–4 × 1.25 mm, truncate with (3–)5 apical teeth and 5–6 veins; anther cylinder 1.5 mm, anther bases sagittate with long slender acute tails, apical appendages indistinct, oblong, ± as long as tails, obtuse to rounded; style forked, dark and papillose for upper 0.5–1 mm. Marginal achenes 6–8(–10) mm long, stout, straight, smooth, arching out of enclosing bract and exceeding it, glabrous or hispidulous, at least above. Inner achenes usually not more than 3, hispidulous dorsally, 3–7 × <1 mm; pappus minute, ciliate, less than 1 mm.

HAB. In the mountains among coppiced *Quercus* on limestone, by stream in a *Juglans* grove, on dry sandstone and conglomerate hills, sometimes gypsiferous, weed among irrigated crops and vegetable gardens including an *Allium cepa* plot, in a clay depression; alt. 50–1500(–1850) m; fl. & fr. Mar.-May(-Jun.) on the plains, Apr.-Jun. in the hills.
DISTRIB. Quite common in the mountain region only. **MRO**: Rawanduz Gorge, *Guest* 2105!; Hamam Alil, *Rawi* 5685!; 33 km S of Samarra, *Rawi & Gillett* 6511!; 17 km NW of Rania, *Rawi, Nuri & Kass* 28653! **MSU**: Kamarspa (on road bet. Halabja &Tawela) *Rawi* 22266! MAM: Shkho to Sharanish, *Karim, Hamid & Jassim* 21129! **MJS**: 40 km S of Sinjar, *Chakravarty, Rawi, Khalil & Alizzi* 32113!

Palestine, Turkey, Caucasia, Iran, Afghanistan, C. Asia.

? PECHWALAKA (Kurd.-Sul., Rawi 12033; ? KHAMISHA (Kurd.-Arbil, Radhi 3868 – "grazed by animals").

3. **Garhadiolus hamosus** *Boiss. & Hausskn.* in Boissier, Fl. Orient. 3: 723 (1875); Dinsmore in Post, Fl. Syria, ed. 2, 2: 128 (1933).

Rhagadiolus hamosus (Boiss. & Hausskn.) Kupicha in Notes Roy. Bot. Gard. Edinburgh 33: 433 (1975); Kupicha in Fl. Turkey 5: 689 (1975).

Erect, branched, often rosette-forming, 13.5–15 cm tall, with strongly ribbed stems, sparsely to densely strigose, especially in lower part. Basal leaves oblanceolate, obtuse, mucronate, 4–8 × 0.8–1.6 cm, with ± distinct winged petiole, dentate, glabrous above, hispidulous below, mainly on veins. Cauline leaves oblong, obtuse, dentate, often exceeding basal leaves in size,

3–4.6 × 1.4–1.9 cm. Inflorescence a spreading panicle. Capitula usually narrowly cylindrical, 8–10 mm long; outer phyllaries 1.5 mm long, ovate, acute, ± erose; inner phyllaries stellately spreading as achenes develop, oblong, incurved, enclosing base of achenes only, 6(–7) mm long, keeled, with a crest of flattened setae along spine. Ligules 6–10 × 2 mm, exceeding involucre by c. 4 mm, broadly strap-shaped, truncate with 4–5 apical teeth and 6–7 indistinct veins; anthers 3 mm long, apical appendages oblong, obtuse; style arms papillose and dark in upper half. Marginal achenes usually 8, stellate-patent, incurving and thus curving out of enclosing phyllary, 10–11 × 1 mm, attenuate above. Inner achenes c. 4, strongly arcuate, hispidulous, 5–7 × 0.5 mm; pappus minute, ciliate, less than 1 mm.

HAB. On a lower mountain slope and plain, on clay soil; alt. 600 m; fl. & fr. Apr.-May.
DISTRIB. Rare in Iraq, only found twice in the lower forest zone. **MAM**: 8 km S of Dohuk, *Botany staff* 43365! **MRO**: 6 km SW of Rania, *Rawi, Noori & Kass* 28428! Zaniter mountain bet. Rania & Shaqlawa, *Karim, Hamid & Jassim* 40832! **MJS**: Sinjar N plain of the mountain opposite to Andalus village, *Widad & AL-Khayat* 53465A!

SW Turkey, NE Syria, Lebanon.

47. **HEDYPNOIS** Mill.

Gard. Dict. ed. 4, 2 (1754)
Lack in Kubitzki, Fam. Gen. Vasc. Pl. 8: 196 [2006] (2007)

Patricia Halliday

Erect or straggling, variably hispid, often rosulate annual herb. Leaves alternate. Capitula solitary, homogamous, ligulate, many-flowered, terminal and axillary, pedunculate, pedicels commonly somewhat swollen beneath capitulum; involucre subcylindrical or campanulate; phyllaries biseriate, outer small, inner much larger, boat-shaped and becoming thickened in fruit, enclosing marginal achenes; receptacle flat, naked. Ligules yellow. Achenes cylindrical, ribbed, strigose along ribs, incurved; pappus (of marginal achenes) coroniform or (of inner achenes) of 5 free scales.

A monotypic genus.

Hedypnois (from the Gr. Ηδυπνοής, ēdupnoēs, sweet exhalation, referring to the scent of the plant. According to Miller's *Gardener's Dictionary*, Hudson preferred to adopt this name for the genus rather than Schreber's *Apargia.*).

Nordenstam, B. 1971. Cytogeography of the genus *Hedypnois* (Compositae). Bot. Not. 124(4): 483–489.

1. **Hedypnois rhagadioloides** (*L.*) *F.W. Schmidt*, Samml. Phys. Aufs. 1: 279 (1795); Dinsmore in Post, Fl. Syria, Palest. & Sinai ed. 2, 2: 125 (1933); Nordenstam in Fl. Iranica 122: 138 (1977); Feinbrun-Dothan Fl. Palaest. 3: 4113 (1978); Meikle, Fl. Cyprus 2: 997 (1985); Boulos & Hind in Fl. Egypt 3: 272 (2002) & Checklist: 246 (2009); Haloob & Al-Kaisi, Illustr. Fl. Lowland Iraq 1(1): 45 (2016); Taifour & El-Oqlah, Pl. Jordan Ann. Checkl.: 42 (2017).

Hyoseris rhagadioloides L., Sp. Pl.: 809 (1753).
H. hedypnois L., Sp. Pl.: 809 (1753).
H. cretica L., Sp. Pl.: 810 (1753).
Hedypnois cretica (L.) Dum.-Cours., Bot. Cult. 2: 339 (1802); Boissier, Fl. Orient. 3: 719 (1875); Handel-Mazzetti in Ann. Naturh. Mus. Wien 27: 62 (1913) ; Nábělek in Publ. Fac. Sci. Univ. Masaryk 52: 47 (1925); Dinsmore in Post, Fl. Syria, Palest. & Sinai ed. 2, 2: 126 (1933); Anthony in Notes Roy. Bot. Gard. Edinburgh 18: 292 (1935); Blakelock in Kew Bull. 4: 56 (1949); Zohary in Dep. Agr. Iraq Bull. 31: 157 (1950); Rawi, ibid. Tech. Bull. 14: 116 (1964); Vassilchenko in Fl. S.S.S.R. 29: 230 (1964); Rechinger, Fl. Lowland Iraq: 665 (1964); Täckholm, Stud. Fl. Egypt, ed. 2: 587 (1974); Kupicha in Fl. Turkey 5: 686 (1975).
H. monspeliensis Willd., Sp. Pl., ed. 4, 3(3): 1616 (1803).
H. cretica (L.) Willd., Sp. Pl., ed. 4, 3(3): 1617 (1803), comb. superfl.
H. pendula Willd., Sp. Pl., ed. 4, 3(3): 1618 (1803).
H. coronopifolia Ten., Cat. Hort. Neapol. App. 1813 (ed. alt.): 60 (1819).
H. tubiformis Ten., Fl. Napol. 2: 173 (1820).
H. polymorpha DC., Prodr. 7: 81 (1838), nom. illeg.
H. sabulorum Pomel, Nouv. Mat. Fl. Atlant.: 17 (1874).
H. cretica var. *gracilior* Boiss., Fl. Orient. [Boissier] 3: 719 (1875).

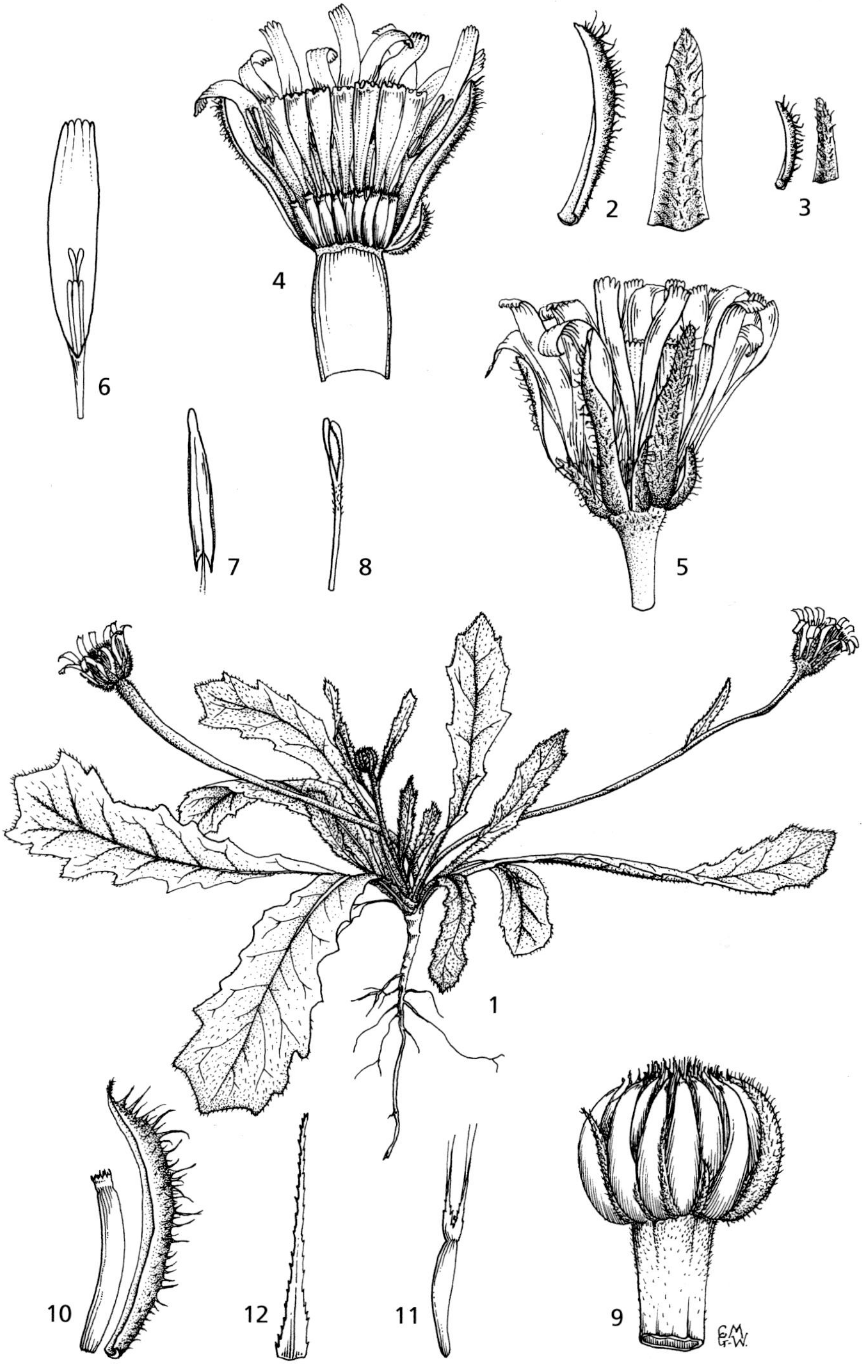

Fig. 51. **Hedypnois rhagadioloides**. 1, habit × ½; 2, inner phyllaries × 4; 3, outer phyllaries × 4; 4, capitulum (scale swollen) × 4; 5, capitulum (scape not swollen) × 4; 6, ray floret × 7; 7, anther × 16; 8, style × 7; 9, fruiting capitulum × 3; 10, marginal achene and phyllary × 6; 11, inner achene × 4; pappus × 6. 1–8 from *Lazar* 1574; 9–12 from 1 from *Chakravarty et al.* 31678; 2–9 from *Rogers* 0392. Drawn by Christine M. Grey-Wilson.

H. persica Fisch., Catal. Hort. Gorenk.: 34 (1808/1812); M.Bieb., Fl. Taur.-Caucas. 3: 539 (1819-20); Vassilchenko in Fl. S.S.S.R. 29: 230 (1964).
H. cretica (L.) Dum. Cours. forma *persica* (Fisher) Cout., Fl. Port. 664 (1913).
H. rhagadioloides (L.) F.W.Schmidt subsp. *cretica* (L.) Hayek, Prodr. Fl. Penin. Balc. 2 (1929).
H. cretica (L.) Dum. Cours. subsp. *rhagadioloides* (L.) Täckholm & Boulos, Publ. Cairo Univ. Herb. 5: 29 (1972) [1974].

Erect or decumbent annual, (2–)4–14(–41) cm tall with smooth or sparsely strigose branching stems. Leaves greyish-green, green-purple or purple; basal leaves ± narrowly obovate, (2.5–)5–7(–14) × (0.6–)1.2–2 cm sessile or petiolate with ±winged petiole, base attenuate, decurrent, apex obtuse to acute, margins subentire to dentate (often with 1 apical tooth between 2 subapical teeth), teeth diminishing in size downwards, both leaf surfaces hispid, upper part of upper leaf surface often densely so, hairs confined to veins and margins of lower surface, hairs longish and stiff and often with a 2–3-furcate tip. Cauline leaves similar to basal or linear-lanceolate to lanceolate or oblong, decreasing in size upwards, usually sessile but occasionally lowermost petiolate, base broad and clasping stem. Inflorescence branched or simple; pedicels usually swollen and hollow just beneath capitulum (to 7 mm diam.), striate and or linear-lanceolate to lanceolate, subcylindrical in flower, 8–14 mm long, (6–)8–9(–13) mm diam., suborbicular or hemispherical in fruit, 1–1.5(1.7) cm diam.; outer phyllaries few, linear-lanceolate to awl-shaped, 2–5 mm long, with a tuft of minute hairs at apex and sometimes with long white bristles on surface and/or margins; inner phyllaries narrower, 7(–9) × 1–1.5 mm, glabrous or hispid, acuminate, often dark-coloured and with minute hairs at tip. Corollas yellow, outer surface of ligules purple (in dried material), elliptic-oblong, 5–7 mm long, tube ⅓ length of floret, ± glabrous. Fruiting capitulum resembling a hand of bananas, but spherical, achenes erect and incurved, not spreading. Marginal achenes 5–7 × 1 mm wide a base, each enclosed by an acrescent, indurate, boat-shaped phyllary 11 × 1.5 mm and with bowl-shaped coroniform erose pappus c. 1 mm long; inner achenes less curved, naked, undeveloped, 5–7 mm long, with 5 erosulose pappus scales produced into well-developed scabrid awns, 4–6 mm, usually with many minute teeth at base of each scale. Fig. 51: 1–12.

HAB. On mountain slopes including *Quercus* forest, stony hillsides, dry steppe, in a wadi, along rivers and in depressions, on sand dunes, and as a weed of pasture, fallow and cultivation and along irrigation channels; alt. 10–1400 m; fl. Mar.–May, fr. Mar.–Jun.

DISTRIB. Widespread on lower mountain slopes and in foothills, extending to the Mesopotamian Plain, rare in the desert region. **MJS**: Jabal Sinjar, *Qaisi & Wedad* 52073!, *Gillett* 10933!; Digila, nr Mosul, *F. & E. Barkley* 55141!; Zakho to Sharanish, *Karim, Hamid & Jasim* 41086!; Aqra, *Qaimaqam of Aqra* 3105! **MRO**: Koi Sanjaq, *Rawi, Noori & Kass* 28107!; Khanzad, *Omar, Qaisi & Wedad* 49616! **MSU**: Jarmo, *Helbaek* 586!, 600!, 1226!; ± 10 km W of Tawila, *Rawi* 23131! **FUJ**: Shargat, *Thomas* s.n. comm. Graham!; Qaiyara, *Bayliss* 108! **FNI**: Eski Kellek, *Herb. Univ. Mosul* 836!; Nimrud, *Helbaek* 893! **FKI**: nr Daroznah, N of Kifri, *Poore* 657!; 87 km NE of Kirkuk, Rawi, *Kass & Noori* 27921!, 27938!; Injana, *Omar & Karim* 37828! **FPF**: Chlat Police Station, *Rawi & Haddad* 25665!; Sa'adiya, *Qaisi* 46712!; Kurdarra, *id.* 48841! **DLJ**: 10 km S of Baiji, *Rawi & Gillett* 7267! **DGA**: 33 km S of Samarra, *Rawi & Gillett* 6521! **LEA**: 21 km NE of Ba'quba, *Cowan & Darlington* 192!; 20 km N of Amara, *Rawi* 15015!; nr Shahraban, *Hunting Technical Services* 18!; Baghdad, *Rogers* 0392!; Karrada, *Paranjpye* s.n. comm. Graham. **LSM**: Musaida, *Rawi* 14943!

S Europe from Portugal to Crimea including the islands, Cyprus, Turkey, Syria, Lebanon, Palestine, Egypt, Caucasia, Iran, N Africa (Libya to Morocco), Macaronesia (Azores, Canaries, Madeira).

Täckholm & Boulos (1972) and Sell (1976) both used *Hedypnois cretica* for this entity, even though Schmidt's combination was significantly earlier than that of Dumont de Courset. An early use of the name *H. rhagadioloides* was in Webb & Berthelot's Phytogr. Canar. 3(2): 397 (1848), which included *H. cretica* as a synonym; Boulos & Hind (2002) used *H. rhagadioloides.*

48. HELMINTHOTHECA Zinn.

Cat. Plant. Hort. Acad. Agric. Goetting.: 430 (1757)
Helminthia Juss., Gen. Pl.: 190 (1789); *Deckera* Sch.Bip., Flora 17: 478 (1834); *Vigineixia* Pomel, Nouv. Mat. Fl. Abl. 12 (1874); Lack in Kubitzki, Fam. Gen. Vasc. Pl. 8: 196 [2006] (2007).

H. W. Lack

Taprooted annual or perennial herbs; indumentum of rigid hairs with hooked branches, simple rigid hairs, and very rarely also a few forked or stellate hairs intermixed. Stem solitary to many, branched, leafy. Leaves basal and cauline, oblanceolate to obovate, very variable, margins entire to dentate or simple, cauline usually well-developed, sessile, similar to basal. Capitula terminating stems and branches, numerous, homogamous, ligulate, many-flowered, involucre ovoid, urceolate; phyllaries biseriate, 3–5 outer foliaceous, broadly ovate to cordiform, acuminate, inner ones lanceolate; margins of phyllaries entire; phyllary length variable, but outer ones always well developed and overtopping inner; receptacle naked. Ligules yellow, marginal florets often with reddish stripes on outer surface. Achenes heteromorphic: outer fusiform, straight to slightly incurved, hairy with persistent pappus of many very short setose rays; inner cylindrical to obovoid, straight, ribbed with a few transverse ridges, abruptly narrowed at apex into a thin, smooth, fragile, threadlike beak; pappus of many persistent long whitish plumose hairs.

Helminthotheca (from Gr. ἔλμινς, *helmins*, worm and θήκη, thēkē, case or capsule.

1. **Helminthotheca echioides** (*L.*) *Holub*, Folia Geobot. Phytotax. 8: 176 (1973); Lack in Dissertationem Univ. Wien 116 (1975) & Fl. Turkey 5: 684 (1975) & in Fl. Iranica 122: 136 (1977); Feinbrun-Dothan, Fl. Palaest. 3: 421, t. 711 (1978); Boulos & Hind in Fl. Egypt 3: 279, pl. 67 (2002); Taifour & El-Oqlah, Pl. Jordan Ann. Checkl.: 42 (2017).

Picris echioides L., Sp. Pl.: 792 (1753); Juss., Gen. Pl.: 190 (1789); Sell in Fl. Europ. 4: 316 (1976).
Crepis echioides (L.) All., Fl. Pedem.: 222 (1783).
Helminthia echioides (L.) Gaertn., Fruct. Sem. Pl. 2: 368, t. 159 (1791); Boissier, Fl. Orient. 3: 742 (1875); Dinsmore in Post, Fl. Syria, Palest. & Sinai ed. 2, 2: 133 (1933); Blakelock in Kew Bull. 4: 57 (1949); Zohary in Dep. Agr. Iraq Bull. 31: 158 (1950); Rawi, ibid. Tech. Bull. 14: 117 (1964); Rechinger, Fl. Lowland Iraq: 668 (1964); Vasiliev in Fl. S.S.S.R. 29: 208 (1964); Täckholm, Stud. Fl. Egypt ed. 2: 596 (1974) sub *Helmintia*, sphalm.
Helminthia tuberculata Moench, Methodus (Moench): 540 (1794), nom. illeg.
Helminthia spinosa DC. in Lam. & DC., Fl. Franç., ed. 3, 4: 58 (1805).
Picris humifusa Willd., Mag. Neuesten Entdeck. Gesammten Naturk. Ges. Naturf. Freunde Berlin 1: 137 (1807).
Picris spinosa (DC.) Poir. in Lam., Encycl. Suppl. 4: 408 (1816).
Helminthia humifusa (Wild.) Trevir., Nova Acta Phys.-Med. Acad. Caes. Leop.-Carol. Nat. Cur. 13: 195 (1826).
Helminthia pratensis Chevall., Fl. Gén. Env. Paris 2: 552 (1828).

Plants 10–80 cm tall. Hairs predominantly 2- and 4- hooked, often tubercle-based. Peduncles slender; involucre *c.* 1 cm in diameter in flower, *c.* 2 cm in diameter in fruit. Inner phyllaries with broad hyaline margins. Outer achenes 3–3.5 mm, pale yellowish, gradually attenuate into a beak 2–3 mm, persistent, enfolded within inner phyllaries; pappus ± 2 mm. Inner achenes *c.* 3 mm long, brownish, beak *c.* 4 mm long; pappus deciduous, *c.* 6 mm. Fig. 52: 1–2.

HAB. Weed in fields, gardens and ditches, on alluvial soil; alt. up to 50 m; fl. & fr. May.
DISTRIB. Rare in Iraq; only found a few times on the irrigated alluvial plain in the desert region. **FKI**: sine loc., *Zohary* s.n. **LCA**: Daltawa (Khalis), *Guest* 2440!; Za'faraniya, *Janan & Sahira* 47568! **LEA**: Ba'quba, *Haines* W. 1304!; "Babylonia", *Noë* s.n.

S Europe (Spain to Crimea, widely naturalized also in N Europe), Syria, Lebanon, Palestine, Egypt, Turkey, Caucasia, Iran, N Africa, Macaronesia (Canary Is., Madeira). Introduced as a weed into N America, Australia, New Zealand & S. Africa.

LIZAIJ ("sticky", "cling weed"), *Guest* 2440.

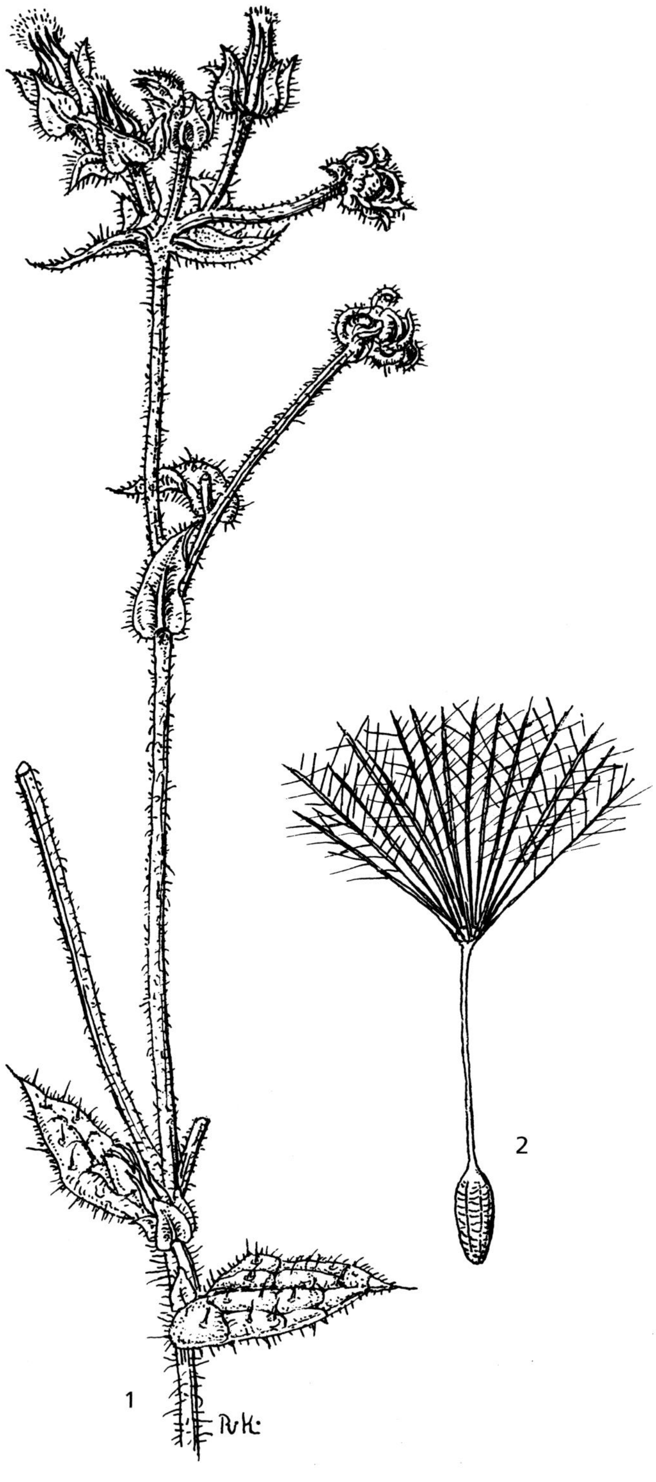

Fig. 52. **Helminthotheca echioides**. 1, habit × 1; 2, achene with pappus × 6. Drawn by Ruth Koppel. From Naomi Feinbrun-Dothan, *Flora Palaestina*, Part three: Ericacaceae to Compositae, Plates, Jerusalem: The Israel Academy of Sciences & Humanities, 1977, pl. 711. © The Israel Academy of Sciences and Humanities. Reproduced with permission.

49. **LEONTODON** L.

Sp. Pl.: 798 (1753); Gen. Pl. ed. 5: 349 (1754)
Lack in Kubitzki's Fam. Gen. Vasc. Pl. 8: 196 (2007)

Scorzoneroides Moench, Methodus (Moench): 549 (1794); *Thrincia* Roth, Arch. Bot. [Leipzig] 1(1): 36 (1796); *Asterothrix* Cass., Dict. Sci. Nat., ed. 2, 48: 434 (1827); *Microderis* DC., Prodr. 7: 127 (1838); *Kalbfussia* Sch.Bip., Flora 16(2): 723 (1833)

H. W. Lack

Rosulate annual or perennial herbs, tap-rooted or rhizomatous. Indumentum of simple, forked or stellate hairs (but never of hairs with hooked branches), sometimes glabrate. Stem solitary to many, mostly simple to few-branched, usually leafless or bearing very small filiform, linear triangular to subulate bracts, in particular beneath capitula. Leaves all basal, oblanceolate to obovate, ± entire to deeply pinnatipartite. Capitula terminating stems and branches, solitary to numerous, homogamous, ligulate, many-flowered; involucre hemispherical, ovoid or turbinate; phyllaries multiseriate, rarely biseriate, outer considerably shorter than inner, lanceolate, margins entire, imbricate; receptacle naked. Corollas yellow, rarely pink, dull red to purple, marginal florets with greyish to reddish stripes on outer surface. Achenes narrowly cylindrical to fusiform, straight, longitudinally striate or ribbed, with numerous transverse ridges, beaked at apex, homomorphic; pappus persistent, whitish, of 10 to many long plumose setae.

A genus of c. 40 spp.; 3 species occur in Iraq.

Recent molecular studies have suggested that each of the two subgenera, subgen. *Oporinia* and *Leontodon* are discrete. According to Greuter et al. (2006) *Leontodon* subgen. *Oporinia* should be bestowed the 'correct' generic name of *Scorzoneroides* Vaill. Since Vaillant's names were originally provided in a pre-Linnean work, and their post-Linnaean 'validation' was only through a German translation of the French original, none of these names should be accepted; '*Scorzoneroides* Vaill.' is not recognized here. Moench's *Scorzoneroides* is equivalent to Greuter's concept, but is still not recognized here.

Leontodon (derived from Gr. λεων, *leon,* lion and οδούς, *odous,* tooth, because of its dentate leaves.) Hawkbit (Eng.), one of the composites often called WARD HODHĀN and variants of this name.

Greuter, W., Gutermann, W. & Talavera, S. (2006). A preliminary conspectus of *Scorzoneroides* (Compositae, Cichorieae) with validation of the required new names. Willdenowia 36(2): 689–692.

Samuel, R., Gutermann, W., Stuessy, T. F., Ruas, C.F., Lack, H.W., Tremetsberger, K., Talavera, S., Hemanowski, B., & Ehrendorfer, F. (2006). Molecular phylogenetics reveals *Leontodon* (Asteraceae, Cichorieae) to be diphyletic. Amer. J. Bot. 93: 1193–1205.

Smalla, M. (2000). Studies in the Compositae of the Arabian Peninsula and Socotra – 6. The Hypochaeridinae (Lactuceae) in the Arabian Peninsula. Willdenowia 30(2): 315–337.

Widder, F. J. (1931). Beiträge zur Kenntnis der Gattung *Leontodon.* II. Die "nickenden Knospen" einiger *Leontodon*-Arten in ihrer Bedeutung für das System der Gattung. Österr. Bot. Zeit. 80: 136–148.

Widder, F.J. (1975). Die Gliederung der Gattung *Leontodon.* Phyton (Austria) 17(1–2): 23–29.

1. Annual, rosette leaves deeply pinnatipartite with narrow to filiform lobes; pappus of ±10 setae, conspicuously broadened at base .... 1. *L. laciniatus*
   Perennial, rosette leaves subentire to dentate, never deeply pinnatipartite with very narrow to filiform lobes; pappus of many setae, not conspicuously broadened at base.................................. 2
2. Dense indumentum of stellate hairs on leaves and stems; margins of outer phyllaries densely covered with numerous, very short, stout, stellate hairs; involucre turbinate, 15–25 mm long at flowering stage 2. *L. asperrimus*
   Sparse indumentum, mainly of forked hairs, on leaves and stems; margins of phyllaries usually naked; involucre hemispherical, 10–15 mm long at flowering stage.......................................3. *L. hispidus*

1. **Leontodon laciniatus** (*Bertol.*) *Widder ex Bornm.*, Beih. Bot. Centralbl. 60B: 217 (1939); Rechinger, Fl. Lowland Iraq: 667 (1964); Täckholm, Stud. Fl. Egypt: ed. 2, 596 (1974); Kupicha in Fl. Turkey 5: 672 (1975); H. Pittoni in Rechinger, Fl. Iranica 122: 123 (1977);

Fig. 53. **Leontodon laciniatus**. 1, habit × 1; 2, achene × 12; 3, pappus × 5. Drawn by Ruth Koppel. From Naomi Feinbrun-Dothan, *Flora Palaestina*, Part three: Ericacaceae to Compositae, Plates, Jerusalem: The Israel Academy of Sciences & Humanities, 1977, pl. 703. © The Israel Academy of Sciences and Humanities. Reproduced with permission.

Feinbrun-Dothan, Fl. Palaest. 3: 416 (1978); Boulos & Hind in Fl. Egypt 3: 278 (2002); Abdulridha, Taha & Wedad, Ecol. & Fl. Basra: 219 (2016); Haloob & Al- Kaisi, Illustr. Fl. Lowland Iraq 1(1): 49 (2016); Taifour & El-Oqlah, Pl. Jordan Ann. Checkl.: 44 (2017).

Oporinia laciniata Bertol., Novi Comment. Acad. Sci. Inst. Nononiensis 6: 235 (1843); Misc. Bot. (Bertol.) 2 : 21 (1843).
Millinia persica Boiss., Diagn. Pl. Orient., ser. 1, 7 : 3 (1843).
Millinia arabica Boiss., Diagn. Pl. Orient., ser. 1, 11: 38 (1849).

Leontodon hispidulus (Del.) Boiss. var. *tenuiloba* Boiss., Diagn. Pl. Orient., ser. 1, 11 : 38 (1849). Boissier, Fl. Orient. 3: 728 (1875); Handel-Mazzetti in Ann. Naturh. Mus. Wien 27 : 453 (1913); Nábělek in Publ. Fac. Sci. Univ. Masaryk 52 : 48 (1925); Dinsmore in Post, Fl. Syria, ed. 2, 2 : 129 (1933): Guest in Dep. Agr. Iraq Bull. 27 : 56 (1933); Anthony in Notes Roy. Bot. Gard. Edinburgh 18 : 292 (1935); Zohary in Dep. Agr. Iraq Bull. 31 : 158 (1950); Dickson, Wild Fls. Kuwait : 59 (1955); Rawi in Dep. Agr. Iraq Tech. Bull. 14 : 119 (1964); Rechinger, Fl. Iranica 122: 123 (1977); Feinbrun-Dothan, Fl. Palaest. 3: 416 (1978).
L. arabicus (Boiss.) Boiss., Fl. Orient. [Boissier] 3: 728 (1875).
Scorzoneroides laciniata (Bertol.) Greuter, Willdenowia 36(2): 691 (2006).

Annual, tap-rooted, glabrous, (2–)5–20(–50) cm tall. All leaves in a rather small basal rosette, variable, 2–4(–12) × 5–10(–25) mm broad, very deeply pinnatipartite to pinnatifid, lobes narrowly linear-lanceolate, filiform, straight, curved towards apex or base, often sickle-shaped, only very rarely with secondary filiform lobes; scapes decumbent, ascending or erect, rarely single, mostly few to many, mostly simple, in vigorous specimens apically branched or repeatedly branched. Very young pedicels, young capitula, bases of leaves (especially below) and of stems densely to sparsely villous with short simple flexuose hairs or glabrous, all other parts always glabrous. Capitula few to many (rarely one); involucre ± hemispherical; phyllaries (7–)10–12 mm long, glabrous above, below glabrous or with a few flexuose, thin, villous hairs. Ligules yellow, 11–15 mm. Achenes brown, fusiform, 5–6(–8) mm long, 0.3–0.4 mm diam., tapering gradually into a long thin beak, with 5 deep longitudinal furrows and very many transverse ridges; pappus of c. 10 setae, 5–6 mm, conspicuously broadened at base. Fig. 53: 1–3.

HAB. Steppic hills and desert plains, often in or near slight hollows where rain water collects, on sandy gravel soils, sometimes between calcareous rocks, gypsiferous, silty or clay soils; alt. up to 650 m; fl. & fr. (Feb.–)Mar.–Apr.(–Jun.).
DISTRIB. Quite common in the steppe region, abundant all over the desert region. **FUJ**: Tal Julaimid, *Guest* 13431!; Hadhr, Dabbagh, *M. Taufiq & A. Lafta* 44162!; between Shargat and Baiji, *Barkley & Haddad* 6774!; nr Baka, *Alizzi & Husain* 33842! **FUJ/FNI**: nr Mosul, *Handel-Mazzetti* 1278. **FKI**: SE of Tuz, *Rawi & Gillett* 10694!; 61 km S of Kirkuk, *F. & E. Barkley* 4659! **FKI/FPF**: Fatha, *Calder* 2080; **FPF**, Pilkhana, 12 km E of Khanaqin, *Rawi* 12784!; Qaraghan (Jalaula), *Guest* 1874!; Jabal Hamrin, *Bornmüller* 1499; nr Chafta wadi, 20 km SE of Khazinah, *Rawi & Haddad* 25649!; 16 km S of Badra, *Rawi* 18210!; **DWD**: 35 & 20 km WNW of Rutba, *Rawi* 21148!, 21109!; Amrah, 120 km NW of Rutba, *Omar* 34229!; 20 km N of Rah, *Rawi* 31293! Masad, 15 km S of Rutba, *Chakravarty Rawi, Khatib & Alizzi* 32882!; between Rutba and H.2, *Omar & Hamid* 36723!; 16 km S of Ana, *id.* 45037!; 35 km SW of K.3, *Omar, Qaisi, Hamad & Hamid* 44691!; 16 km S of Ana, *id.* 45037!; 18 km NW of T.1 on road to Qaim, *id.* 45086!; 4 km NW of K.3, *Alizzi & Husain* 34024!; between Salahiya (Kifri) and Haditha, *Gillett & Rawi* 6872!; 108 km W of Nukhaib, *Rawi* 31408A!; Nukhaib, *Rawi* 15737!; 180 km W of Ramadi, *Rawi & Noori* 27031!; 15 km W of Ramadi, *Guest & Rawi* 13951!; 18 km W of Karbala, *Rawi* 30918! **DLJ**: Wadi Thirthar, E of Hadhr, *Rawi & Gillett* 7256!; W of railway between 'Ain Dibs and Baiji, *id.* 7167!; 70 km W of Tikrit, *Alizzi & Husain* 33726!; nr Tikrit, *Omar & Hamid* 36297!; 155 km NE of K.3, nr Sha'bani, *Chakravarty, Rawi, Khatib & Alizzi* 32987!; 12 km W of Rawa, *Omar, Qaisi, Hamid & Hamad* 44478!; nr R. Euphrates, 15 km E of Ana, *Barkley & Hikmat 'Abbas (al-Ani)* 565A!; 23 km E of Haditha, *K. Hamad* 49988!; Jazira desert, *Edmonds* 3798!; 65 km NW of Falluja, *Chakravarty & Rawi* 30376! **DGA**: Musaiyif al Ghizlan, Janan, Sahira, *Widad & Khalaf* 49918!; 13 km E of Samarra, *Rawi* 20368!; **DSD**: 95 km W of Shabicha, *Khatib & Hazim* 30537!; 30–40 km from Najf on road to Shabicha, *Alizzi & Omar* 35423!; Salman, *Guest, Rawi & Long* 14082!; Takhadid, ± 60 km S of Salman, *Qaisi, Hamad & Hamid* 48076!; nr Jalib Bakur (well), *Guest et al.* 18884!; 5 km from Jumaima, *Qaisi, Hamid & Hamad* 48187!; nr Ansab, *Guest et al.* 18953!; 50 km S of Ansab, *Al-Kaisi et al.* 42243!; 67 km WNW of Salman, *Chakravarty, Rawi, Khatib & Tikriti* 29978!; Busaiya, *Haddad & Danail Araha* 9583!; 20 km from Busaiya on road to Samah, *Karim, Sharif, Hamad & Hamid* 48005!; 20 km SW of Safwan, *Guest & Rawi* 8744!; Baniya, *id.* 17285!; Batin, 65 km SW of Basra, *id.* 17050!; Zubair, *Rawi & Walter* 16219!; between Zubair and Jaliba, *Guest et al.* 5057! **LCA**: Baghdad, *Colvill* s.n., *Paranjpye* s.n. (comm. Graham); Rustamiya, *Guest* 1104!; nr Babylon, *Bornmüller* 421! **LEA**: nr Sharaban (Muqdadiya), *Hunting Technical Services* 12!; Baquba, *Paranjpye* s.n.! (comm. Graham); 70 km SSE of Baghdad, nr Aziziya, *Khatib & Alizzi* 32608!; 55 km N of Amara, *Guest et al.* 17447! **LSM**: 70 km E of Amara, nr Rashida well, *Husain al-'Ani* 39724!

Syria, Palestine, Jordan, Egypt, Kuwait, Saudi Arabia, Iran, Caucasus.

WARD HODHĀN (Rawi & Gillett 7256); MURRĀN (*Guest, Rawi & Long* 14138); MURRĀR (Kuwait, *Dickson*); MURAIR (*Guest & Rawi* 17097); UMRĀR (*Guest & Rawi* 18884). The first name is commonly used for many composites; the last three all have the connotation of a "bitter" taste.

2. **Leontodon asperrimus** (*Willd.*) *Endl.*, Cat. Horti Vindob. 1: 415 (1842).

Scorzonera asperrima Willd., Sp. Pl. 3 : 501 (1803).
Asterothrix asperrima (Willd.) Cass., Dict. Sci. Nat., ed. 2, 48: 434 (1827).
Leontodon asperrimus (Willd.) Boiss. ex Ball, Ann. Mag. Nat. Hist. ser. 2, 6 : 15 (1850), comb. superfl.; Fl. Orient. [Boissier] 3 : 723 (1875); Nábělek in Publ. Fac. Sci. Univ. Masaryk 52 : 48 (1925); Dinsmore in Post, Fl. Syria, ed. 2, 2 : 130 (1933); Zohary in Dep. Agr. Iraq Bull. 31 : 158 (1950); Rawi in Dep. Agr. Iraq Tech. Bull. 14 : 119 (1964); Vasiliev in Fl. S.S.S.R. 29 : 216 (1964); Kupicha in Davis, Fl. Turkey 5: 674 (1975); Pittoni in Fl. Iranica 122 : 126 (1977).
L. masmenaeus Boiss. & Balansa in Boissier, Diag. Pl. Orient., ser. 2, 5: 117 (1856).
L. crispus Vill. subsp. *asperrimus* (Willd.) Finch & P.D.Sell, Bot. J. Linn. Soc. 71(4): 246 (1976).

Perennial, rhizomatous, densely hispid, 15–30 cm tall. Indumentum consisting of rigid, shortly branched, 2–5-stellate hairs. All leaves in a basal rosette, oblanceolate, remotely dentate, densely hispid on both surfaces. Scapes erect, rarely ascending, stout, single or few, simple to sparsely branched at base, densely hispid. Peduncles slightly inflated at flowering and fruiting stages, densely hispid. Capitula 1 to very few; involucre narrowly turbinate, at flowering stage 15–25 mm long; phyllaries multiseriate, glabrous above, densely hispid below, margins of outer phyllaries very densely covered with very short, stout, rigid stellate hairs. Ligules yellowish, pink, dull red to purple. Achenes brown, fusiform, 15–18 mm long, c. 1 mm diam., gradually attenuate towards apex, with many longitudinal furrows and, in particular in upper part, transverse ridges; pappus of many setae, up to 15 mm, not broadened at base.

HAB. On mountain slopes, in coppiced *Quercus* forest, on limestone; alt. 1100–2050 m; fl. May-Jul.
DISTRIB. Occasional in the upper forest zone of Iraq, very rare in the lower thorn-cushion zone. **MJS**: Tschil Miran, Jabal Sinjar, *Handel-Mazzetti* 1522; Jabal Sinjar, *Gillett* 23627! **MAM**: Gali Zawita, NE of Zakho on Turkish frontier, *Rawi* 23629!; Ain Nuni, *Nábělek* 3940; Aradin, *Nábělek* 3993; Kani Masi, *K. Hamid & Fadhil* 45611! **MRO**: Zaytah, nr Shirwan Mazin, *Haines et al.* s.n.; Rayat, *Thesiger* 1045.

Europe, Balkans, Syria, Lebanon, Turkey, Caucasia, Iran, Turkmenia, Kazakhstan.

3. **Leontodon hispidus** *L.*, Sp. Pl.: 799 (1753); Boissier, Fl. Orient. 3: 370 (1875); Kupicha in Fl. Turkey 5: 674 (1975); Finch & Sell in Fl. Europ. 4: 313 (1976); Pittoni in Rechinger, Fl. Iranica 122: 124 (1977).

L. hastilis L., Sp. Pl. ed. 2: 1123 (1762).

Perennial, rhizomatous, glabrate to sparsely hispid, 15–40 cm tall. Indumentum consisting of few, rigid, long-stalked, shortly branched, 2–4-forked to stellate hairs. All leaves in a basal rosette, 5–20 × 1–2 cm, oblanceolate, tapering at base and obtuse to acute at apex, remotely dentate to runcinate dentate, glabrous above, sparsely hispid beneath. Scapes erect to ascending, single or few, unbranched, glabrous. Peduncles not inflated. Capitula 1-very few; involucre ± hemispherical, at flowering stage 10–15 mm; phyllaries few-seriate, glabrous above, glabrous to hispid below, margins glabrous. Ligules yellow. Achenes brown, fusiform, 5–8 mm long, c. 0.5 mm diam., gradually attenuate at apex, with many longitudinal furrows and inconspicuously transverse-rugulose; pappus of many setae, 8–10 mm, not broadened at base.

HAB. In the mountains; alt. 1450 m; fl. Aug.
DISTRIB. Very rare, not far from the Iranian frontier, found only once in Iraq. **MRO**: between Darband and Haj Umran, *Hadač* 2554.

Most of Europe to C & S. Russia, Turkey, Caucasia, Iran, N Africa.

L. crispus Vill. was recorded from Jabal Sanam by Anthony in Notes Roy. Bot. Gard. Edinburgh 18: 292 (1935) under the name "L. asperus", i.e. *L. asper* (Waldst. & Kit.) Poir. No specimen from this locality of this or any other species of the genus is to be found in BM, E or K herbaria, in which institutions material seen by Anthony may be expected.

50. **PICRIS** L.

Sp. Pl.: 792 (1753)
Lack in Kubitzki, Fam. Gen. Vasc. Pl. 8: 197 [2006] (2007)

H. W. Lack

Annual to perennial herbs, taprooted. Indumentum of long rigid hairs with hooked (anchor-shaped) branches, simple rigid hairs and very rarely also a few forked hairs intermixed. Stem solitary to many, erect to decumbent, branched, mostly leafy. Leaves basal and cauline, oblanceolate to obovate, spatulate, very variable in size, form and margins, cauline leaves usually well developed, sessile, similar to basal but smaller, reduced or missing. Capitula terminating stems and branches, homogamous, ligulate, many-flowered, numerous; involucre ovoid, urceolate or campanulate; phyllaries multiseriate, all linear-lanceolate, imbricate; inner phyllaries always longer than outer, ± subequal, after achene loss (in heterocarpic species central achenes) phyllaries usually star-like, reflexed; receptacle naked. Corollas yellow, marginal floretss often dark on lower surface. Achenes homomorphic or heteromorphic; inner 5-ribbed, straight to incurved, fusiform or long attenuate at apex, transversely rugose, never narrowed into a slim, smooth thread-like beak, in heterocarpic species outer achenes pubescent, broad based, firmly coalesced with receptacle, enfolded by naviculiform inner phyllaries, persistent on dead plant; pappus of numerous long plumose setae, about as long as achene, often easily deciduous, in heterocarpic species pappus of outer achenes consisting of very short setose setae.

Heath, P. V. (1994). Commentary on the proposal to conserve *Picris* Linne. Calyx 4(3): 117.
Holzapfel, S. (1994). A revision of the genus *Picris* (Asteraceae: Lactuceae) s.l. in Australia. Willdenowia 24(1–2): 97–218.
Lack, H.W. (1975). Die Gattung *Picris* L. sensu lato im ostmediterranwestasiatischen Raum. Wien: VWGO. Diss. Univ. Wien 116: 1–184.
Lack, H.W. (1979). The genus *Picris* (Asteraceae, Lactuceae) in Tropical Africa. Pl. Syst. Evol. 131 (1–2): 35–52.

Picris (named from Gr. πικρός, *pikros*, bitter, because the plant has a bitter taste).

1. Robust biennial or perennial plants . . . . . 1. *P. strigosa*
   Annual herbs 2
2. Indumentum of pedicels and phyllaries consisting of 2-, 3- and 4-hooked hairs . . . . . 2. *P. kotschyi*
   Indumentum of pedicels and phyllaries consisting of 2-hooked hairs only 3
3. In fruiting stage pedicels always inflated, constricted below base of capitulum; outer achenes strongly incurved . . . . . 3. *P. pauciflora*
   In fruiting stage pedicels not inflated, not constricted below base of capitulum; outer achenes straight or slightly incurved 4
4. Ripe achenes 4–5 mm long, lower part 0.5–0.7 mm in diameter, undulate, ± abruptly tapering into a short beak; central strip of phyllaries greyish-whitish arachnoid . . . . . 4. *P. babylonica*
   Ripe achenes (5–)6–7 mm long, 0.3–0.5 mm in diameter, transversely rugose, gradually tapering into a long beak; central strip of phyllaries greenish, rarely arachnoid, midrib often prominent . . . . . 5. *P. longirostris*

1. **Picris strigosa** *M. Bieb.*, Fl. Taur.-Cauc. 2: 250 (1808); Boissier, Fl. Orient. 3: 736 (1875); Dinsmore in Post, Fl. Syria, Palest. & Sinai ed. 2, 2: 130 (1933); Anthony in Notes Roy. Bot. Gard. Edinburgh 18: 292 (1935); Zohary in Dep. Agr. Iraq Bull. 31: 158 (1950); Agnew in Bull. Coll. Sci. Baghdad 6: 68 (1961); Täckholm, Stud. Fl. Egypt ed. 2: 597 (1974); Lack in Fl. Turkey 5: 680 (1975) & in Fl. Iranica 122: 131 (1977), (incl. vars. *kurdica* and *matrotricha*); Taifour & El-Oqlah, Pl. Jordan Ann. Checkl.: 46 (2017).

P. glaucescens DC., Prodr. 7: 130 (1838)
P. glomerata C.Koch, Linnaea 23: 666 (1851)
P. persica Gaud., Bull. Soc. Bot. France 65: 53 (1918)
P. turcomanica Gaud., Bull. Soc. Bot. France 65: 53 (1918)
P. subinflata (Bornm. & Gauba) Rech.f., Österr. Bot. Z. 97: 261 (1950)

Stout erect biennial to perennial; massive woody rootstocks with dried bases of stems and leaves of previous seasons; indumentum of rigid 2- and 4-hooked hairs, less frequently 3-hooked hairs. Stems (10–)30–75 cm long, markedly angular, stiff, strictly branched. Foliage often reduced. Basal leaves withered at anthesis. Cauline leaves oblanceolate, 3–10 cm long, margins entire to distantly and shallowly dentate, at anthesis sometimes already fallen. Peduncles not thickened in fruit. Involucres obconical, turbinate at base, 1–1.5 × 1–1.5 cm; phyllaries with rigid hooked hairs and few to many stiff simple to inconspicuously fork-tipped elongate hairs in central strip, mostly densely whitish arachnoid. Achenes homomorphic, fusiform, attenuate at base and apex, 4–6.5 mm long, 0.5–1 mm in diameter, transversely rugose, brown, glabrous.

HAB. On dry mountain slopes in *Quercus* forest and on stony hillsides; alt. 900–1700 m; fl. & fr. Jun.-Aug. (-Sep.).
DISTRIB. Occasional in the forest zone of Iraq. **MAM**: Gara Dagh, *Kotschy* 392!; Sarsang, *Haines* W.1213!; Mar Ya'qub, *Handel-Mazzetti* s.n. **MRO**: Lolan, *Gillett* 12462!; between Rayat and Haji Umran, *Gillett* 9471!; Haji Umran, *Guest & Husham (Alizzi)* 15893! 15903!; NE of Rania, N of Pushtashan, *id.* 26512!; Kew-a Rash, *id.* 23777! **MSU**: Zewiya, *Hadač* 2830; Qaradagh, *Haines* E.1514!; Balkha, *Rawi* 22332!; Tawila, *Rawi* 21982!; Zalam, *Rawi, Husham (Alizzi) & Noori* 29403!; Avroman range, *Rawi, Chakravarty, Noori & Alizzi* 19792! **FUJ**: Ashur, in *Anthony* l.c. **FUJ/FKI**: Fatha, in *Anthony* l.c.

Syria, Lebanon, Turkey, Iran, Caucasia, Turkmenia.

No attempt is made here to present an infraspecific classification for the Iraqi material of this species; there is large variation in the indumentum – in particular in the number, length, distribution and density of the stiff, simple to inconspicuously fork-tipped hairs of the involucre – which however does not seem to merit formal taxonomic recognition.

2. **Picris kotschyi** *Boiss.*, Fl. Orient. 3: 738 (1875); Bouloumoy, Fl. Liban Syrie 208 (1930); Dinsmore in Post, Fl. Syria, Palest. & Sinai ed. 2,2: 131 (1933); Thiébaut, Fl. Libano-Syrienne 2: 287 (1940); Agnew in Bull. Coll. Sci. Baghdad 6: 67 (1961); Lack in Fl. Turkey 5: 378 (1975); Lack, Diss. Univ. Wien 116, t. 22 & in Fl. Iranica 122, t. 96 (1977).

Erect to ascending annual, 5–40(–80) cm; indumentum on stems and leaves consisting of rigid 2-hooked hairs, of pedicels and phyllaries consisting of rigid 2- and 4-hooked hairs, less frequently 3-hooked hairs. Basal leaves mostly withered at anthesis, cauline leaves linear-lanceolate or missing. Peduncles thickened at fruiting time, but constricted at base of involucre. Fruiting capitula 10–16 mm. Achenes heteromorphic, brown-blackish, fusiform, attenuate at apex; outer 5.5–7 mm long, pubescent, broad-based, firmly coalesced with receptacle, enfolded by naviculiform inner phyllaries, persistent on the dead plant; inner 6–8 mm long, transversely rugose, attenuate at base, easily detached from receptacle.

HAB. In the foothills and upper steppic plains; alt. 250–1200 m; fl. & fr. May-Jul.
DISTRIB. Rare in Iraq, only found a few times in the mountain and steppe regions. **MAM**: Bekher, *Rawi* 23070! **FNI**: Hinnis, *Thesiger* 625! **FUJ**: S of Mosul, *Anders* 2541A; S of Badush, *Anders* 2074.

Syria, Turkey.

3. **Picris pauciflora** *Willd.*, Sp. Pl., ed. 4, 3(3): 1557 (1803); Boissier, Fl. Orient. 3: 337 (1875); Dinsmore in Post, Fl. Syria, Palest. & Sinai ed. 2,2: 130 (1933); Anthony in Notes Roy. Bot. Gard. Edinburgh 18: 292 (1935); Agnew in Bull. Coll. Sci. Baghdad 6: 61 (1961); Rawi in Dep. Agr. Iraq Tech. Bull. 14: 121 (1964); Lack in Fl. Turkey 5: 681 (1975) & in Fl. Iranica 122: 317 (1976).

Erect, rarely ascending annual, 10–50 cm; indumentum consisting of rigid 2-hooked hairs. Basal leaves mostly withered at anthesis, cauline leaves linear-lanceolate or absent. Peduncles conspicuously elongated, markedly thickened in fruiting stage, but constricted at base of involucre. Fruiting capitula 10–15 mm long, urceolate, markedly constricted above. Achenes homomorphic, 4–6 mm, dark brown, transversely rugose, attenuate at base and apex, outer achenes conspicuously incurved.

HAB. On mountain sides, by river in foothills; alt. unrecorded; fl. & fr. Apr.-May.
DISTRIB. Occasional in the mountain region of Iraq. **MJS**: crest of Jabal Sinjar, *Anders* 2033; **MAM**: Bekher, nr Zakho, *Rawi* 23070! **MRO**: Rowanduz gorge, *Agnew, Hadač & Haines* 2606. **MSU**: Tawila, *Rawi* 22295!; Avroman, *Rechinger* 10253. **FPF**: Qaraghan (Jalaula), *Rogers* 0222!

S Europe (SE France, Greece, Crete, Balkans), Crimea, Syria, Turkey, Caucasia, Iran, Arabia.

4. **Picris babylonica** *Hand.-Mazz.* in Ann. Naturh. Mus. Wien 27: 453, fig. 5/2 (1913); Nábělek in Publ. Fac. Sci. Univ. Masaryk 52: 49 (1925); Blakelock in Kew Bull. 4: 60 (1949); Zohary in Dep. Agr. Iraq Bull. 31: 58 (1950); Agnew in Bull. Coll. Sci. Baghdad 6: 62 (1961); Rechinger, Fl. Lowland Iraq: 670 (1964); Rawi in Dep. Agr. Iraq Tech. Bull. 14: 121 (1964); Lack in Diss. Univ. Wien 116: t. 13 (1975) & in Fl. Iranica 122: 135 (1977); Limpscombe

Vincett, Wild Fls. of C Saudi Arabia: 100 (1979); Collenette, Wild Flow. Saudi Arabia: 207 (1999); Abdulridha, Taha & Wedad, Ecology & Fl. Basra: 219 (2016); Taifour & El-Oqlah, Pl. Jordan Ann. Checkl.: 46 (2017)

Decumbent, ascending to erect annual, 4–40 cm; indumentum consisting of rigid 2-hooked hairs. Basal leaves in a rosette, 2–10(–15) × 5–20 mm, oblong-lanceolate to spatulate, margins rarely entire, mostly sinuate, lobate, erose to irregularly laciniate, cauline leaves linear-lanceolate, entire to sinuate, rarely deeply and irregularly laciniate, lobes rounded, often reduced or missing. Peduncles long, slender, often elongated and creeping. Capitula homocarpic to heterocarpic, fruiting capitula 9–12(–14) mm long; phyllaries with central strip of densely covered with stout 2-hooked hairs and conspicuously whitish to greyish arachnoid. Ligules yellow (according to a few scattered field notes later becoming brown), in dry state sulphur yellow, dull brown to brown-blackish. Ripe central achenes 4–5 mm long, blue-blackish to brown-blackish, in lower part 0.5–0.7 mm diam., transversely rugose and ± abruptly attenuate into a short beak, firmly coalesced with receptacle, enfolded by naviculiform inner phyllaries, persistent on dead plants.

HAB. Usually on sandy desert soil, but also on gypsiferous, stony, rocky limestone gravel, etc.; up to alt. ± 350 m; fl. & fr. (Jan.-)Mar.-May(-Jun.).

DISTRIB. Very common in the desert region of Iraq. **FPF**: between Khanaqin and Baghdad, *Nábělek* 3951. **FPF/FKI**: Jabal Hamrin, *Rawi & Gillett* 7350! **DLJ**: near the right hand bank of R. Tigris, between Balad and Samarra, *Handel-Mazzetti* 986!; 3 km S of Hit, *Rawi & Gillett* 6833!; Wadi Thirthar canal, *C & ER Guest & Sharif Muhamad Amin Haidar* 18319!; on S shore of Wadi Thirthar, *Rawi & Chakravarty* 30453!; 30 km NW of Falluja, *Rawi* 20221! **DGA**: 26 km E of Samarra, *Rawi* 20373!; Asila, 50 km E of Samarra, *id.* 20450! **DWD**: Ajrumiya, *Alizzi & Husain* 34045!; 80 km NW of Nukhaib, *Aftan Rawi* 23710!; 5 km S of Shanana, *Rawi* 14811!; Habbaniya, *Rawi & Alizzi* 34456!; 140 km W of Karbala, *Chakravarty & Rawi* 30970!; 12 km W of Karbala, *Rawi* 30869!; between Karbala and Najaf, *id.* 20169. **DSD**: Wadi al-Khirr, 90 km NW of Shabicha, *Guest et al.* 19419!; 40 km WNW of Shabicha, *id.* 19281!; 6 km E of Shabicha, *Rawi* 13875!; 35 km S of Samawa, *Rawi & Gillett* 6519!; Abu Ghar, 60 km S of Ur, *Guest & Mahalhal* 15297B!; 12 km ESE of Salman, *Guest et al.* 18835!; between Aidaha and Ansab, *Alizzi & Omar* 35639!; 12 km SSW of Busaiya, *Guest & Mahalhal* 15221B!; ?50 km E of Busaiya, *Karim et al.* 48037!; 85 km ESE of Busaiya, *Guest & Rawi* 14286!; Jaliba, nr Suq ash-Shiyukh, *id.* 5072!; 15 km S by W of Zubair, *Guest, Rawi & Schwan* 14355!; nr Safai al-Maghif, NE of Ghazlani, *Guest et al.* 17254!; nr Jarishan, *id.* 17123!; Jabal Sanam, *id.* 17007!; Umm Qasr, *Khatib & Alizzi* 32672!; 20 km E of Salman, on road, to Takhadid, *Omar, Kaisi et al.* 44662!; 30 km NW of Ansab, *Qaisi, Hamad & Hamid* 48162!; Rikhamiya, 41 km N of Samah, *Botany staff* 42330!; nr Samah, *Alizzi* & *Khatif* 34389! **LBA**: 80 km SE of Basra, *Rawi, Tikriti, Khatib & Chakravarty* 29934! **LBA**: 80 km NW of Zubair, *Thamir* 47388!

UMRĀR ("very bitter"), nr Busaiya, *Guest & Mahalhal* 15221B; MURRAIR ("bitter"), Batin, nr Jarishan, *Guest & Rawi* 17098. The Kurdish name recorded on the Khanaqin road GULA ZERDE (*Rawi & Gillett* 7350) simply means "yellow flower" in Kurdish.

Saudi Arabia, Kuwait, Iran.

5. **Picris longirostris** *Sch.Bip.*, Mus. Senckenb. 3: 60 (1839); Lack in Fl. Iranica 122: 134 (1977); Boulos & Hind in Boulos, Fl. Egypt 3: 282 (2002).

P. damascena Boiss. & Gaill. ex Boiss., Fl. Orient. 3: 740 (1875); Dinsmore in Post, Fl. Syria, Palest. & Sinai ed. 2,2: 132 (1933); Zohary in Dep. Agr. Iraq Bull. 31: 158 (1950); Agnew in Bull. Coll. Sci. Baghdad 6: 66 (1961); Rechinger, Fl. Lowland Iraq: 669 (1964); Rawi in Dep. Agr. Iraq Tech. Bull. 14: 21 (1964); Feinbrun-Dothan, Fl. Palaest. 3: 419 (1978); Collenette, Wild Flow. Saudi Arabia: 207 (1999); Haloob & Al-Kaisi, Illustr. Fl. Lowland Iraq 52 (2016); Abdulridha, Taha & Wedad, Ecology & Fl. Basra: 219 (2016).

P. blancheana Boiss. in Fl. Orient. 3: 741 (1875); Handel-Mazzetti in Ann. Naturh. Mus. Wien 27: 454 (1913); Dinsmore in Post, Fl. Syria, Palest. & Sinai ed. 2, 2: 132 (1933); Zohary in Dep. Agr. Iraq Bull. 31: 158 (1950); Rechinger, Fl Lowland Iraq: 676 (1964).

P. desertorum Nábělek in Publ. Fac. Sci. Univ. Masaryk 52: 49 (1925); Dinsmore in Post, Fl. Syria, Palest. & Sinai ed. 2,2: 132 (1933).

P. damascena Boiss. & Gaill. ex Boiss. var. *persistens* Eig, Palestine J. Bot., Jerusalum Ser. 1: 72 (1938).

P. damascena Boiss. & Gaill. ex Boiss. subsp. *blancheana* (Boiss.) Eig in Palestine J. Bot., Jerusalem Ser. 1 : 73 (1938); Agnew in Bull. Coll. Sci. Baghdad 6 : 66 (1961); Rawi in Dep. Agr. Iraq Tech. Bull. 14 : 121 (1964).

P. damascena Boiss. & Gaill. ex Boiss. var. *diffusa* Eig in Palestine J. Bot., Jerusalem Ser. 1: 73 (1938).

P. damascena Boiss. & Gaill. ex Boiss. var. *homocapra* Eig in Palestine J. Bot., Jerusalem Ser. 1: 73 (1938).

P. damascena Boiss. & Gaill. ex Boiss. subsp. *desertorum* (Nábělek) Agnew in Bull. Coll. Sci. Baghdad 6 : 66 (1961); Rawi in Dep. Agr. Iraq Tech. Bull. 14 : 121 (1964).

Fig. 54. **Picris longirostris**. 1, habit × 1; 2, marginal achene × 4; 3, inner achene with deciduous pappus × 4. Drawn by Ruth Koppel. From Naomi Feinbrun-Dothan, *Flora Palaestina*, Part three: Ericacaceae to Compositae, Plates, Jerusalem: The Israel Academy of Sciences & Humanities, 1977, pl. 708. © The Israel Academy of Sciences and Humanities. Reproduced with permission.

Decumbent, ascending to erect annual, 10–40 cm; indumentum consisting of 2-hooked hairs. Basal leaves rosettiform, 2–10 × 5–20 mm, obovate to spatulate, margins entire to sinuate, erose, sometimes withered at anthesis, cauline leaves reduced to absent. Peduncles long, slender, sometimes elongated, creeping. Capitula usually heterocarpous, rather slender, fruiting capitula 8–12 mm long; phyllaries with central strip of densely covered with rigid 2-hooked hairs, but scarcely arachnoid, greenish-looking, often with prominent midrib. Ligules yellow, in dry state pale yellow to ochre, not darkening. Ripe central achenes (5–)6–7 mm long, dark brown to blackish, 0.3–05 mm diam., transversely rugose, attenuated at base, easily dehiscent from the receptacle, outer achenes hairy, broad-based, firmly coalesced with receptacle, enfolded by naviculiform inner phyllaries, persistent on dead plants. Fig. 54:

HAB. On steppic hills and dry stony slopes, steep rolling gravelly desert hills of degraded conglomerate, sandstone and conglomerate, sandy gravel plains and silty depressions; alt. 50–650 m; fl. & fr. (Jan.-) Mar.-Jun.

DISTRIB. Quite common in the steppe region of Iraq, very common in the desert region. **FUJ**: Qaiyara, *Bayliss* 102! **FUJ/FAR/FKI**: Makhlat, *Guest* 4245! **FKI**: Kani Domlan hills, *id.* 4335!, 4367!; on bank of R. Zab, *Grigg* 6!; *Hadač* 1282. **FKI/FPF**: Jabal Hamrin, *Haines* W.214! **FPF**: Dar-i Diwan, *Poore* 503!; on left bank of R. Zab, between Sangur and Tanura, *id.* 491!; near Khanaqin, *Cowan & Darlington* 231!, 246!; Jabal Darawishka, nr Khanaqin, *Guest* 13974!; Pilkhana, *Rawi* 12790!; 2–16 km S of Badra, *id.* 18160!, 18187!, 18189!; Jabal Muwaila, nr village of Kuwait, 70 km N of Amara, *Guest, Rawi & Rechinger* 17541!, 17650! **DLJ**: Hadhr, *Chakravarty, Rawi, Khatib & Alizzi* 33025!; 15 km E of Rawa, *Omar, Qaisi et al.* 45135!; midway between Rawa and Haditha, *id.* 45163! **DWD**: Tall al-Nisr, 48 km NE of Rutba, *Rawi* 31217!; 20 km W of Rutba, *id.* 14696!; 10 km SW of Rutba, *Rawi* 21080!; 260 km NW of Ramadi, *id.* 200949!; Nahiya, *Handel-Mazzetti* 717; Ana, *id.* 744; Wadi Fuhaimi, between Haditha and Ana, *Rawi & Gillett* 6933!; 140 km W of Karbala, *Chakravarty & Rawi* 30966!; 20 km from Mugrianam, *Rawi & Hamada* 33544!; Wadi Khirr, 15 km N of Lusuf, ± 100 km NNW of Shabicha, *Guest et al.* 19456!; 13 km W by N of Shabicha, *id.* 19289!; 40 km WNW of Shabicha, *id.* 19298!; 20 km WNW of Ansab, 140 km S of Salman, *id.* 18998! **LCA/LEA**: Sudur (to) Mansuriya, *F. & E. Barkley & Brahim al-Jabri* 6807! **LEA**: Garmashaya, 77 km NE of Kut, *Rawi* 18046!

Syria, Palestine, Jordan, Sinai, Egypt, Arabia, SW Iran.

HODHĀN (Ir., Eg. etc.), a common name for many Compositae in Iraq (and also elsewhere, e.g. Egypt) as explained by Täckholm in Stud. Fl. Egypt ed. 2: 81 (1974) who says that it is also used there for such species as *Launaea* and *Leontodon* as well as *Picris*. In Iraq this name has been recorded at Makhlat for *Senecio*, which also has yellow flowers. Dickson (1955) has noted the same name as HŌWĪTHAN, "grazed by all animals" for *Picris saharae* Batt. in Kuwait.

51. **RHAGADIOLUS** Juss.

Gen. Pl., 168 (1789), *nom. cons.*
Lack in Kubitzki, Fam. Gen. Vasc. Pl. 8: 197 [2006] (2007)

Patricia Halliday
Revised by D. J. Nicholas Hind

Annual erect herbs, usually rosette-forming. Leaves variously toothed or lobed or pinnatisect, occasionally subentire. Capitula sessile or pedunculate, homogamous, ligulate with a ± cylindrical involucre of 2 series of phyllaries, outer *c.* 5, minute, inner much longer, boat-shaped, accrescent, variably thickened in fruit, enclosing marginal achenes; receptacle flat, naked. Florets hermaphrodite; ligules yellow. Achenes of 2 types: marginal subulate or narrowly fusiform, smooth or ribbed, straight or curved, persistent, erect or stellately patent; inner strongly arcuate, naked, often hispidulous, caducous; pappus minute, persistent, ciliolate or of scabrous hairs or absent.

A genus of about 2 species centred around the Mediterranean; 2 species occur in Iraq.

Brummitt, R. K. (1982). Report of the Committee for Spermatophyta: 24. Proposal 516. 9566 *Rhagadiolus* Juss. (1789) (Compositae) vs. *Rhagadiolus* Zinn (1757). Taxon 31(3): 544.
Meikle, R.D. (1979). *Rhagadiolus*: a misunderstood genus. Taxon 28(1–3): 133–141.
Meikle, R.D. (1980). Proposal to conserve 9566 *Rhagadiolus* Juss. over *Rhagadiolus* Zinn (Compositae). Taxon 21(1): 159–160 (1980).

Outer phyllaries glabrous; marginal achenes 5–6, 10–13 mm long, straight, fusiform; leaves with large, remote, suborbicular terminal lobe . 1. *R. edulis*
Outer phyllaries hispid or crested in upper half; marginal achenes 7–8, about 16 mm long, subulate; leaves repand-dentate, terminal lobe acute . . . . .2. *R. stellatus*

1. **Rhagadiolus edulis** *Gaertn.*, Fruct. Sem. Pl. 2, 3: 354 (1791); Willd., Sp. Pl. ed. 4, 3(3): 1625 (1804).

Lapsana rhagadiolus L., Sp. Pl.: 812 (1753); Smith, Fl. Graec. Prodr. 2, 1: 144 (1813); Lindl., Fl. Graec. 9: 12, t. 818 (1837).

Lapsana lapsanifolius (as lapsanaefolius) Mill., Gard. Dict., ed. 8, no. 3 (1768), nom. illeg.
Rhagadiolus lapsanoides Desf., Fl. Atlant. 2: 234 (1799).
R. lampsanifolius (as lampsanaefolius) (Mill.) Mirbel, Hist. Nat. Pl. 11: 328 (1805).
R. stellatus (L.) Gaertn. var. *edulis* (Gaertn.) DC., Prodr. 7, 1: 77 (1838); Rawi, Dep. Agr. Iraq Tech. Bull. 14: 123 (1964).
R. rigidus Pomel, Nouv. Mat. Fl. Atlant.: 17 (1874).
R. stellatus (L.) Gaertn. subsp. *edulis* (Gaertn.) Holmboe in Bergens Mus. Skr. N. R. 1(2): 191 (1914).
R. stellatus (L.) Gaertn. var. *hebelaenus* auct. non DC., Prodr. 7, 1: 78 (1838); Boissier, Fl. Orient. 3: 722 (1875) pro parte.

Erect or spreading, much-branched, 12–60 cm tall with ribbed stems, variably hispid especially in lower part. Basal leaves usually forming a rosette, oblanceolate in outline, 2.6–11 cm long, green, lyrate, pinnatisect, often divided almost to midrib, terminal lobe much the largest, 1–3 cm wide, suborbicular (occasionally reniform), obtuse, minutely mucronate, sinuate or bluntly angled with a truncate to cordate base, lateral lobes few, remote, oblong and rounded or bluntly triangular, sparsely strigose on both surfaces, usually more densely so on veins beneath, but indumentum very variable. Cauline leaves obovate, as basal leaves or with suborbicular to cordate terminal lobe narrowing into sinuate winged petiole, often clasping and auriculate, entire to shallowly toothed, reducing in size upwards, finally becoming lanceolate to linear, entire, obtuse and mucronate, 2.7–7(–10.5) × 0.3–3.8 cm. Petiole distinct, slender, hispidulous, 0.3–2.3 cm. Inflorescence spreading, pedicels c. 0.8 cm or to 7.5 cm when terminal. Capitula cylindrical, c. 6 mm; outer phyllaries ovate, acute or acuminate, c. 1 mm long, with erose, scarious margins, glabrous; inner phyllaries oblong, (3–)5–7 × 1 mm, obtuse but with short incurved beak or spur at tip and scarious margins, glabrous, enclosing achene and thickening in fruit, especially at base. Ligule strap-shaped, 6–8 × 1 mm, truncate with 4–5 apical teeth, floret tube more than 1 mm. Anthers cylindrical, appendages minute, oblong-ovate. Marginal achenes 5–6(–8), fusiform, straight, glabrous, equalling enclosing phyllary in length, 10–14 × 1–1.75 mm; inner achenes 1–3, occasionally 0, arcuate, hispidulous, 4–10 × 0.5–1 mm. Fig. 55: 8–10.

HAB. Grassy patches among *Quercus* scrub, irrigated grassy pasture in mountain orchard, on moist black loam among rocks; alt. (± 50–)600–1000 m; fl. & fr. Apr.-May.
DISTRIB. Occasional in the lower forest zone of Iraq. **MAM**: Zakho valley, *Guest* 2254!; Bakirma (Bakurman), *Rawi* 8565A! **MRO**: Shaqlawa, *Haines* W.697!; Rowanduz gorge, *Guest* 2030A! **MSU**: NE of Darband-i Khan, *Poore* 548!; just E of Darband-i Khan, *Barkley & Haddad* 7753!; Qarachitan, *Gillett*, 7713! **MJS**: Jabal Sinjar, N slope oppost to Andalus village, *Widad & Al-Khayat* 53451!; Jabal Sinjar, W. slop, *Al-Kaisi & Wedad* 52223! **LEA**: Ba'quba, *Paranjpye* s.n.! (comm. Graham); "Mesopotamia", *Aucher-Eloy* 3449! **FPF**: Mandali, *Kaisi, Thamir & Salah* 51487!

Europe, Cyprus, Syria, Palestine, Turkey, Caucasia, Iran, N Africa (Morocco).

2. **Rhagadiolus stellatus** (*L.*) *Gaertn.*, Fruct. Sem. Pl. 2, 3: 354 (1791); Willd., Sp. Sp., ed. 4, 3(3): 1625 (1804); Nábělek in Publ. Fac. Sci. Univ. Masaryk 52: 48 (1925); Dinsmore in Post, Fl. Syria, Palest. & Sinai ed. 2, 2: 127 (1933); Zohary in Dep. Agr. Iraq Bull. 31: 157 (1950); Rechinger, Fl. Lowland Iraq: 665 (1964); Rawi, Dep. Agr. Iraq Tech. Bull. 14: 123 (1964); Täckholm, Stud. Fl. Egypt, ed. 2: 590 (1974); Kupicha in Fl. Turkey 5: 687 (1975); Rechinger, Fl. Iranica 122: 143 (1977); Feinbrun-Dothan, Fl. Palaest. 3: 412 (1978); Boulos & Hind in Fl. Egypt 3: 275 (2002).

Lapsana stellata L., Sp. Pl. ed. 1, 2: 811 (1753); Smith, Fl. Graec. Prodr. 2, 1: 144 (1813); Lindl., Fl. Graec. 9: 12, t. 817 (1837).
Rhagadiolus intermedius Ten., Curso Bot. Lez. 4 (1 sect. 2): 25 (1822).
R. lyratus Lam., Tabl. Encycl. 3: t. 655, fig. 2 (1823).
R. stellatus (L.) Gaertn. var. *leiocarpus* DC., Prodr. 7: 77 (1838).
R. stellatus (L.) Gaertn. var. *hebelaenus* DC., Prodr. 7: 78 (1838); Blakelock in Kew Bull. 4: 61 (1949); Rawi in Dep. Agr. Iraq Tech. Bull. 14: 116 (1964).
R. stellatus (L.) Gaertn. var *indivisus* Vis., Fl. Dalmat. 96 (1847).
R. stellatus (L.) Gaertn. var. *leiolaenus* Boissier, Fl. Orient. [Boissier] 3: 722 (1875).
R. stellatus (L.) Gaertn. forma *hebelaenus* (DC.) Murbeck, Acta Reg. Soc. Phsiogr. Lund 8: 110 (1897).
R. stellatus (L.) Gaertn. forma *intermedius* (Ten.) Hayek, Prodr. Fl. Pen. Balc. 2(6): 806 (1931).
R. stellatus (L.) Gaertn. var. *eriocarpus* Faure & Maire in Jahand. & Maire, Cat. Pl. Maroc. 3: 830 (1934).

Erect, much-branched, 12.5–34(–60) cm, with ribbed stems, variably hairy, but usually densely so towards base. Basal leaves obovate or oblanceolate, deeply irregularly toothed,

Fig. 55. **Rhagadiolus stellatus**. 1, habit × $^{2}/_{3}$; 2, 3, basal leaves outlines × $^{2}/_{3}$; 4, capitulum × 4; 5, fruiting branch × $^{2}/_{3}$; 6, achene × 3; 7, phyllary × 2. **Rhagadiolus edulis**. 8, 9, basal leaf outlines; 10, capitulum × 5. 1, 4 from *Hennipman* et al. 273 (Turkish material); 2 from s.n. 1892 (French material); 3, 6 from *Lindberg* 2739 (Moroccan material); 5 from *Boissier* 1869 (Spanish material); 7, *Hepper* 3434 (Palestine material); 8, 9 from *Feinbrun & Grizi* 705 (Palestine material); 10 from *Alston & Sandwith* 1428 (S Albanian material). Drawn by A. Langhorn (Fl. Cyprus) and Pat Halliday.

lobed or pinnatisect, (3.7–)6–18 × (0.6–)1.4–6.4 cm wide, lateral lobes or teeth gradually reducing in size towards (usually) indistinctly winged petiole, terminal lobe acute and shallowly toothed, occasionally suborbicular with a continuously toothed or lobed decurrent lamina. Lowermost cauline leaves as basal, uppermost ovate, lanceolate to linear-lanceolate and often entire, reducing in size upwards and becoming bractiform, acute to acuminate, mucronate, shallowly to deeply toothed, 3.3–13.4 × (0.2–)1–4.4 cm. Inflorescence spreading, paniculate. Capitula with erect or occasionally spreading florets 6–9 mm; outer phyllaries minute, 1.5 × 1 mm, ovate-cordate, acute to acuminate with scarious, erose margins, hispidulous or glabrous; inner phyllaries accrescent, at first erect, becoming patent as achenes develop, eventually stellate, linear-oblong, 4–6(–10) × 0.5–1 mm, with short inturned beak or spur at tip and scarious margins, becoming thickened in fruit, keeled, with crest of ± flattened setae along spine at least in upper half, or whole phyllary setose-hispidulous. Ligule yellow-orange, ± equalling involucre, (5–)7–8 × c. 2 mm, strap-shaped, truncate with 3–5 apical teeth and 3–4 veins, glabrous; anthers to 2 mm long, sagittate, appendages oblong, rounded; style dark (when dried) and papillose for at least 0.5 mm (⅓ of style-length). Marginal achenes 7–8, slender, incurved, 2–4 mm shorter than enclosing phyllary, attenuate above with a long narrow neck, subulate, smooth, (12–)14–16(–18) × 1–1.5 mm, inner achenes 3–4 or fewer, strongly arcuate, 6–9 × 0.5–1 mm, enclosing phyllary without an apical spur or beak, dorsally hispidulous (glabrous in var. *leiocarpos*, which is unknown from Iraq). Fig. 55: 1–7.

HAB. On the lower mountain slopes and valleys, limestone ledges on a cliff, in open *Quercus* woods, weed in a grassy orchard, mountain pasture and hill fields; alt. (50–)500–800(–1350) m; fl. & fr. Apr.-May (-Jul.).

DISTRIB. Quite common in the lower forest zone of Iraq, rare on the plain. **MJS**/**FUJ**: Sinjar, *Omar & Hamid* 36507! **MAM**: Bikhair (Bekher) mountain, nr Zakho, *Rawi* 22992!; Dohuk gorge, *Emberger et al.* 15412!; Sawara Tuka, *O. Polunin* 5147!; Olaka, *Rawi* 8439A!; Aqra, *Qaimaqam of Aqra* 3103! **MRO**: Salah ad-Din, *G. Chapman* 11984!, *Helbaek* 738! Shaqlawa, *Haines* W.696!; Diyana, *Emberger et al.* 155281!; Chinaruk, *Rawi, Noori & Kass* 28292!; Gali Warta, 30 km NW by N of Rania, *id.* 28831! **MSU**: Bakrajo, *Rawi & Gillett* 7672!; sine loc., *Rawi* 8897!; Tasluja, *Rawi* 21712!; between Sesar and Dukan, *Botany staff* 43054! **FPF**: Jasina (?Jassan), *Omar* 42805! **FUJ**: Baidha, 78 km S of Sinjar, *Chakravarty, Rawi, Khatib & Alizzi* 32101! **DWD**: nr Ana, *Barkley, Andrew, Agnew & Elmer Palmatier* 529B! **LCA**: Daltawa (Khalis), *Darwish al-Haidari* 2635!; Baghdad, *Colvill* s.n.

Mediterranean Europe, Cyprus, Syria, Lebanon, Palestine, Jordan, Egypt, Turkey, Caucasia, Iran, Macaronesia (Canary Is., Madeira), N. Africa (Morocco, Algeria, Tunisia, Libya).

52. **UROSPERMUM** Scop., *nom. cons.*

Introd. Hist. Nat.: 122 (1777)
Lack in Kubitzki, Fam. Gen. Vasc. Pl. 8: 197 [2006] (2007)

Tragopogonoides Vaill., Königl. Akad. Wiss. Paris Phys. Abh. 5: 729 (1754), nom. rej. prop.
Arnopogon Willd., Sp. Pl. ed. 4 (Willdenow), 3(3): 1496 (1803)
Tragopogonodes Kuntze, Revis. Gen. Pl. 1: 370 (1891)
Daumailia Arènes, Bull. Soc. Bot. France 95: 277 (post 22 Oct. 1948)

Frances G. Davies

Annual or perennial, leafy erect herbs. Leaves large, dentate. Capitula homogamous, ligulate, solitary and rather showy; involucre campanulate to urceolate; phyllaries uniseriate, subequal; receptacle convex, ciliate. Ligules yellow. Achenes bipartite, lower fertile cell rugose to verrucose, upper part a sterile, basally inflated beak; pappus hairs uniseriate, plumose, connate at base and deciduous in a ring.

Two species in Macaronesia, the Mediterranean region and W Asia; one species in Iraq.

Urospermum (from the Gr. ουρά, oura, a tail and σπέρμα, sperma, a seed – referring to the long beak of the achene).

1. **Urospermum picroides** (*L.*) *Scop. ex F.W.Schmidt*, Samml. Phys. Aufn. Naturk. 1: 276 (1795); Boissier, Fl. Orient. 3: 743 (1875); Dinsmore in Post, Fl. Syria, Palest. & Sinai, ed.

Fig. 56. **Urospermum picroides**. 1, habit × 1; 2, bud × 1; 3, capitulum × 1; 4, mature capitulum × 1; 5, floret × 9; 6, achene × 9; 7, pappus × 9. 1–2 from Helback 1008; 3, 6 from *Rawi, Noori & Kass* 28859; 4, 6–7 from *Kass & Noori* 27225. Drawn by M. Millar Watt.

2, 2: 133 (1933): Anthony in Notes Roy. Bot. Gard. Edinburgh 18: 293 (1935); Bornmüller in Beih. Bot. Centralbl. 60B: 217 (1940); Zohary in Dep. Agr. Iraq Bull. 31: 158 (1950); Rawi, ibid. Tech. Bull. 14: 126 (1964); Rechinger, Fl. Lowland Iraq: 666 (1964); Vasil'ev in Fl. S.S.S.R. 29: 229 (1964); Täckholm, Stud. Fl. Egypt, ed. 2: 590 (1974); Fl. Iranica 122: 137 (1977); Chamberlain in Fl. Turkey 5: 685 (1977); Feinbrun-Dothan, Fl. Palaest. 3: 416 (1978); Boulos & Hadidi, Weed Fl. Egypt: 44 (1984); Meikle, Fl. Cyprus 2: 1043 (1985); Mandaville, Fl. Eastern Saudi Arabia: 316 (1990); Boulos & Hind in Fl. Egypt 3: 275–276 (2002); Abdulridha, Taha & Wedad, Ecol. & Fl. Basra: 229 (2016); Haloob & Al- Kaisi, Illustr. Fl. Lowland Iraq 1(1): 60 (2016); Ghafoor in Fl. Pakistan 222: 284 (2017); Taifour & El-Oqlah, Pl. Jordan Ann. Checkl.: 48 (2017).

Tragopogon picroides L., Sp. Pl.: 790 (1753).
T. asperum L., Sp. Pl.: 790 (1753).
T. capensis Jacq., Icon. Pl. Rar. 3(3): pl. 577 (March 1789); Collect. 2: 820 (April 1789).
Arnopogon picroides (L.) Willd., Sp. Pl., ed. 4 [Willdenow] 3(3): 1496 (1803).
A. asper Willd., Sp. Pl. ed. 4 [Willdenow] 3(3): 1497 (1803).
A. capensis (Jacq.) Willd., Sp. Pl. ed. 4 [Willdenow] 3(3): 1497 (1803).
Urospermum capense (Jacq.) Spreng., Syst. Veg., ed. 16 [Sprengel] 3: 662 (1826).
Daumailia spinulosa Arènes, Bull. Soc. Bot. France 95: 277 (post 22 Oct. 1948).

Erect annual herb 15–50 cm tall. Stem leafy throughout, thinly to moderately hispid, sulcate, simple or branched. Leaves thin in texture, setose along lower surface of midrib, hispid on both surfaces but especially beneath, margins densely hispidulous, lower leaves oblong-obovate, 3–20 × 0.8–5(–8) cm, subentire to coarsely dentate or lyrate-pinnatisect, tapering and subpetiolate below, median and upper cauline leaves sessile, with large acute or more rarely ± obtuse auricles, reducing upwards. Capitula solitary, on 2–14 cm pedicels slightly swollen at tip; involucre ± urceolate, sparingly to moderately but conspicuously long-hispid, broad, 15–20 × 10–16 mm; phyllaries ± 7, equal, deltoid-lanceolate, acute, reflexed after fruit-fall, outer series commonly purplish-, inner broadly membranous-margined. Ligule 16–20 mm, exceeding phyllaries. Achenes 11–15 mm, ± curved below, fertile cell 3–5 mm, brown, strongly rugose, articulated to the swollen, less rugose base of beak; pappus very delicate, white, 8–10 mm. Fig. 56: 1–7.

HAB. Rocky hill slopes, degraded *Quercus* forest, dry steppe, orchards, and as a weed of roadsides and cultivation, on sandy gravel or rocky clay; alt. 20–950 m; fl & fr. Mar.-May(-Jun.).
DISTRIB. Quite common in the foothills of Iraq and also in the Lower Mesopotamian Plain. **MAM**: W of Aqrah, *Anders* 1226; Jabal Khaira, *Thesiger* 593; **MRO**: Haibat Sultan Dagh, *Rawi, Noori & Kass* 28287!, *Omar, Karim, Hamza & Hamid* 37191!; 17 km N of Rania, *Rawi, Noori & Kass* 28616!; Gali Ali Beg (Rowanduz Gorge), *Omar, Qaisi & Widad* 49543!; between Akoyan and Rowanduz, *Kass & Noori* 27225!; Gali Warta, *Rawi, Noori & Kass* 28859!; Kuh Sefin, *Bornm.* 1500. **MSU**, Jarmo, *Helbaek* 587!; Qara Chitan, *Gillett* 7713!; mt. Avroman (Hauraman), *Rechinger* 12438! **MJS**: Karsi, *Gillett* 10877!; Jabal Sinjar, *Qaisi & Widad* 52065! **FUJ**: Jabal Makhul, nr Ain Dibis, *Rawi & Gillett* 7215! **FKI**: Kifri, *id.* 7419!; 25 km SW of Badra, *Qaisi, Thamer & Salah* 51363! **FPF**: Badra, *Rechinger* 9189; **DWD**: Hit, *Gillett & Rawi* 6848! **LCA**: Baghdad, *Colvill* s.n.!, *Paranjpye* s.n.!; Abu Ghraib, *Alizzi & Janan* 32810!; Za'faraniya, *Omar & Sahira* 37816!; Hilla, *Paranjpye* s.n.!, *Bornmüller* 445! **LEA**: near Ba'quba, *Hunting Technical Services* 28!; Kut al-Imara, *Gillett* 6586! **LSM**, Amara, *G.A.Watson* s.n.!

S Europe from Portugal to Greece, Mediterranean Is., Egypt, Arabia, Palestine, Syria, Turkey, Caucasia, Iran, N Africa (Libya to Morocco), Macaronesia (Canary Is.). An introduced weed in Australia, S Africa and temperate S America.

53. **EPILASIA** Benth. & Hook.f.

Gen. Pl. [Bentham & Hooker f.] 2(1): 532 (1873)
Lack in Kubitzki, Fam. Gen. Vasc. Pl. 8: 198 [2006] (2007)
Scorzonera sect. *Epilasia* Bunge, Beitr. Kenntn. Fl. Russ. : 200 (1851)

Frances G. Davies

Floccose annual herbs similar in appearance to small *Scorzonera* species. Leaves simple, linear or lanceolate, ± parallel-veined. Capitula homogamous, hemispherial; phyllaries *c.* 2-seriate. Florets exceeding involucre or not, corollas yellow or pink or purple; receptacle alveolate, naked. Achenes columnar, ribbed, glabrous in lower part, shaggy above from long plumose hairs set on a circumferential ring-like callus; pappus arising from below apex,

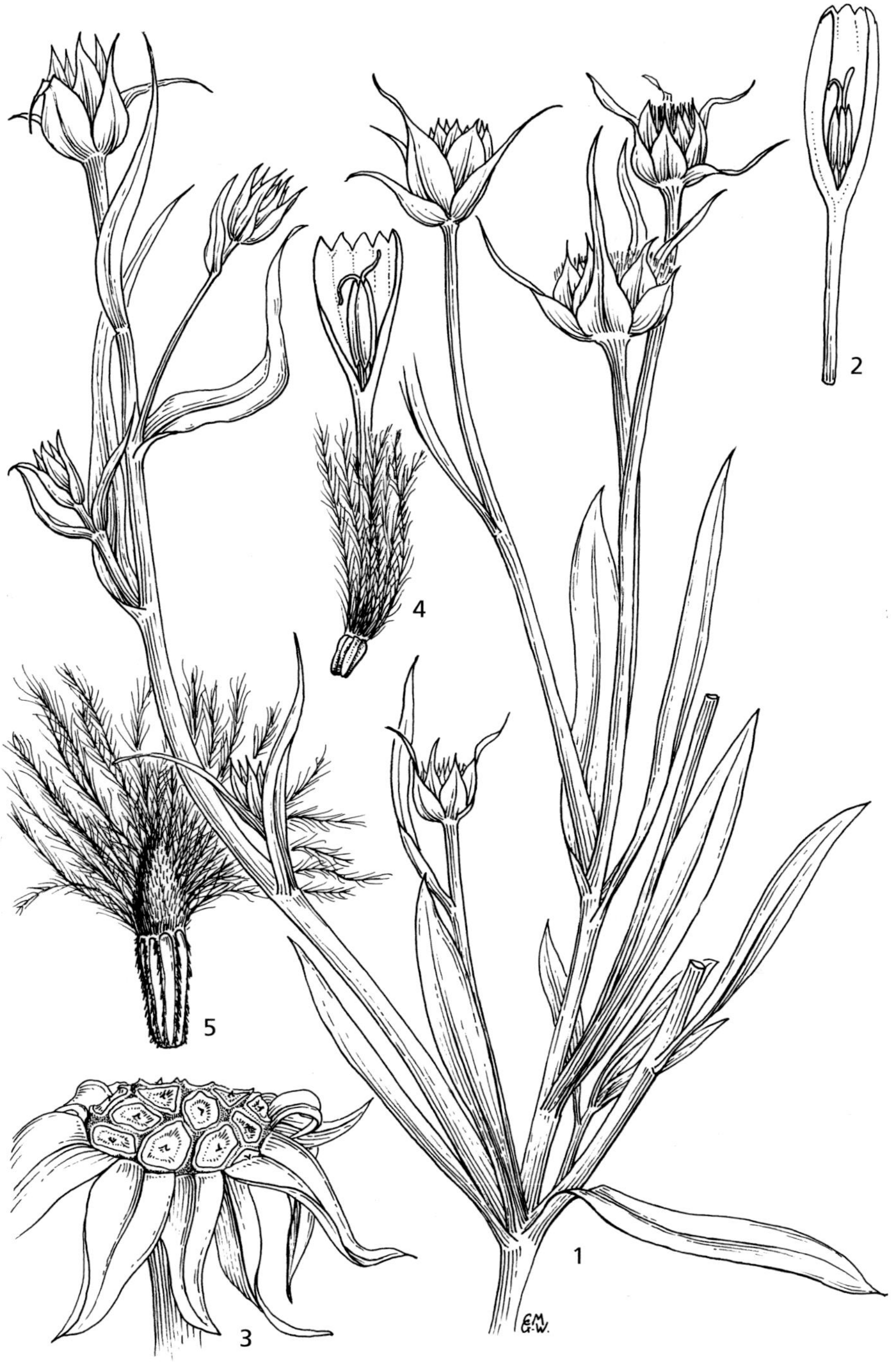

Fig. 57. **Epilasia hemilasia**. 1, habit × 1; 2, floret × 6; 3, capitulum (after flowering) × 6; 4, floret with achene × 6; 5, achene with pappus × 6. 1–2 from *Omar et al.* 44462; 3, 5 from *Davis & Bokhari* (Turkish material) 56024; 4 from *Hewer* 3569. Drawn by Christine M. Grey-Wilson.

outer hairs plumose throughout, inner scabrid above. Note: This 'pappus' may be derived from achene trichomes rather than homologous with a true pappus.

Three species in W & C Asia.

Epilasia (from Gr. επί, epi, over or above and λάσιος, lasios, woolly or shaggy, from the woolly upper part of the achenes.)

1. **Epilasia hemilasia** (*Bunge*) *C.B.Clarke*, Compos. Ind.: 279 (1876); Lipshitz in Fl. S.S.S.R. 29: 114 (1964); Rechinger, Fl. Iranica 122: 81 (1977); Breckle, Hedge & Rafiqpoor, Vasc. Pl. Afghan. Checkl.: 172 (2013); Ghafoor & Sultan Abedin in Fl. Pakistan 222: 281 (2017).

Scorzonera hemilasia Bunge, Beitr. Kenntn. Fl. Russ. : 201 (1851).
S. cenopleura Bunge, Beitr. Kenntn. Fl. Russ. : 202 (1851).
S. nana Boiss. & Buhse in Nouv. Mém. Soc. Nat. Imp. Mosc. 12 : 138 (1860).
Epilasia nana (Boiss. & Buhse) Grossh. in Bot. Sborn. Azerb. G.N.I.I. 1 : 56 (1932).

Herb with hollow stems 4–40(–50) cm tall, simple or branched from base or above, stem and branches with a single terminal capitulum. Leaves simple, with a broad petioloid base, (1.5–)5–10 × (0.2–)1–1.5 cm, linear to elliptic, acute to acuminate, entire or sub-serrate with cartilaginous teeth, sometimes undulate, lateral veins parallel to midrib, lamina initially floccose on both surfaces. Involucre 10–18 × 6–20 mm, campanulate; outer phyllaries leaf-like, lanceolate, up to 30 × 4 mm, sometimes only equalling lanceolate, narrowly white-margined, glabrous to sparsely floccose inner phyllaries. Rays somewhat exceeding inner phyllaries, said to be yellow or whitish when living, sometimes pink when dried. Achenes 7 mm long, upper part obscured by indumentum; pappus grey, c. 6–8 mm. Fig. 56: 1–5.

HAB. Sandy soil in a valley, clay soil in a wheat field, locally common; alt. (noted on only one gathering) 170 m; fl. Apr.–May, fr. May–Jun.
DISTRIB. Only found twice in Iraq. **FPF/LEA**: 25 km E of Badra, *Rawi* 29785K! **DLJ**: 21 km W of Rania, *Omar, Qaisi, Hamad & Hamid* 44462!

Iran, Transcaucasia, Afghanistan, Pakistan, C Asia.

54. **GEROPOGON** L.

Sp. Pl. ed. 2: 1109 (1763)
Lack in Kubitzki's Fam. Gen. Vasc. Pl. 8: 198 [2006] (2007)

C. C. Townsend

Annual herb with a leafy stem, leaves grass-like and parallel-veined, subamplexicaul at base, plants very similar in appearance to *Tragopogon*. Phyllaries long, narrow, green; receptacle rather sparingly furnished with long setaceous paleae ± attaining base of beak of mature achenes. Ligules mauve to purplish or pink, much shorter than phyllaries. Achenes dimorphic, all beaked, outer achenes with beak much longer and bearing a pappus of (3–)5(–7) long, rigid, compressed, scabridulous awns, inner achenes with a shorter but still slender beak and a longer pappus of plumose hairs.
A monotypic genus.

Geropogon (from Gr. γερο, *gero*, old and πωγόν, pōgon, a beard – a reference to its "old man's beard" indumentum).

Diaz de la Guardia, C. & Blanca, G. (1997). Lectotypification of names of Linnaean species of *Scorzonera and Geropogon*. Taxon 46(4): 759–762.
– (1988). La posicion sistematica de *Geropogon* L. (Compositae) en la subtribu Scorzonerinae Dumort: 1. Lagascalia 15 (Suppl.): 361–367.
– (1986). El genero *Geropogon* L. (Compositae: Lactuceae)". Lazaroa 9: 31–44.
Wilson, F. D. (1982). A cytological basis for the separation of *Geropogon* from *Tragopogon* (Compositae: Lactuceae)." Brittonia 34(3): 290–293 (1982).

1. **Geropogon hybridus** (*L.*) *Sch.Bip.* in Webb & Berth., Phytogr. Canar. 2: 472 (1850); Borisova in Fl. S.S.S.R. 29: 197 (1964); Matthews Fl. Turkey 5: 668 (1975); Rechinger, Fl. Iranica 122: 121 (1977); Feinbrun-Dothan, Fl. Palaest. 3: 423 (1978); Meikle, Fl. Cyprus 2: 1041 (1985); Boulos & Hind in Fl. Egypt 3: 286 (2002).

Fig. 58. **Geropogon hybridus**. 1, habit × 1; 2, 3, fruiting head × 1½. 1 from *Guest & al-Haidari* 2437; 2–3 from *Haines* W.1398. Drawn by Christine M. Grey-Wilson.

Tragopogon hybridum L., Sp. Pl.: 1: 789 (1753); Richardson in Fl. Europ. 4: 325 (1976).
Geropogon hirsutum L., Sp. Pl. ed. 2: 1109 (1753).
G. glabrum L., Sp. Pl. ed. 2: 1109 (1753); Boissier, Fl. Orient. 3: 744 (1875); Dinsmore in Post, Fl. Syria, Palest. & Sinai ed. 2, 2: 134 (1933); Zohary in Dep. Agr. Iraq Bull. 31: 158 (1950); Rawi in Dep. Agr. Iraq Tech. Bull. 14: 116 (1964); Rechinger, Fl. Lowland Iraq: 671 (1964); ; Täckholm, Stud. Fl. Egypt, ed. 2: 599 (1974).

Tragopogon glaber (L.) O.Hoffm. in Nat. Pflanzenfam. 4(5): 365 (1893).
T. geropogon Rouy, Fl. France 10: 8 (1908), nom. illeg.
T. longirostris auct. non Sch.Bip.: Guest in Dep. Agr. Iraq Bull. 27: 100 (1933).

Erect annual herb, 12–50 cm, simple or branched up to about middle, glabrous or more rarely sparingly pilose, stems striate. Leaves grass-like, median 8–18 × (1–)2–6(–7) mm, acuminate, ± amplexicaul-vaginate at base. Peduncles 2–7(–14) cm, ± thickened in fruit but constricted below capitulum. Phyllaries 7–8, glabrous, ± equal,2–4 cm, linear-subulate. Corollas mauve to purplish or pink, 12–18 mm, rays 8–15 mm; anthers blackish-purple. Achenes very narrowly fusiform, with numerous fine, scabrid ribs, gradually tapering into a beak; outer achenes 2.5–4 cm including the beak, exceeding inner by (0.6–)1(–1.5) cm, rigid pappus-awns very unequal, 2–10 mm; inner achenes 1.5–3 cm including the beak; pappus-hairs (12–)15–21 mm, softly plumose. Fig. 58: 1–3.

HAB. Most commonly on clay or limestone on hill slopes in degraded *Quercus* forest or on grassy or stony ground, also frequently as a weed of cereal crops; alt. 50–1080 m; fl. & fr. Apr.-May.
DISTRIB. Abundant in an area on the Lower Mesopotamian Plain centered on Khalis and Ba'quba, widespread in the mountain region and occasional in the foothills.
MAM: Dohuk, *Makki Beg* 3249!; Mahad, nr Shaikhan, *Salim Effendi* 2618! **MRO**: N slopes of Handren, *Gillett* 8303!; 2 km N of Chinaruk, *Rawi, Noori & Kass* 28372!; Salah ed Din, *Helbaek* 734!, *Rawi* 36216!, *Karim* 37537! **MSU**: Khormal, *Rawi* 8963!; Pira Magrun, *Gillett* 7747!; Bakrajo, *anon.* (?Mooney) 4336!; Darband-i Khan, *Poore* 536!; Jasina, *Omar* 42726!; Tasluja, *Qaisi* 49432! **LEA**: Ba'quba, *Schläfli* 91!, *Haines* W.1398! **LCA**: Daltawah (Khalis), *Guest & Darwish al-Haidari* 2437!; between Ba'quba and Baghdad, *Rogers* 0148!

S Europe from Portugal to Turkey, Mediterranean Is., Cyprus, Syria, Palestine, Egypt, Turkey, Transcaucasia, W Iran, N Africa (Libya to Morocco), Macaronesia (Canary Is., Madeira).

DHANAB AL-FARAS, horse-tail (Ir., *Guest & Haidari* 2437!), ZABÎB AL-KHAIL, horse's raisins (Ir., *id.*), ASPINK, (Kurd., *Makki Beg* 3249), SHADDANAK (Kurd., *Salim Effendi* 2618).

There are contradictory reports on the wholesomeness of this plant. *Salim Effendi* 2618 bears a note stating "harmful to sheep and causes death in spring (when green)" while *Guest & Haidari* 2437 bears the observation "green immature flowering shoots are eaten by horses and animals – also by mankind".

55. **KOELPINIA** Pall.

Reise 3, 2: 755 (1776)
Lack in Kubitzki's Fam. Gen. Vasc. Pl. 8: 198 (2007)

Patricia Halliday

Erect or sprawling branched annual herbs with linear, entire, grass-like leaves; flowering stems bearing several shortly pedunculate capitula. Capitula few- to many-flowered, ligulate; involucre narrowly cylindrical; phyllaries biseriate, outer small, inner much longer, equalling florets, enlarging slightly as achenes develop, but not enclosing them; receptacle flat, naked. Corollas yellow. Achenes much exceeding involucre, spreading, deciduous, slightly incurved to strongly arcuate, cylindrical or attenuate, ribbed and clothed with few to many hooked spines; pappus of several minute, stout, blunt bristles, evident on immature achenes and developing into a crown of radiating hooked spines as fruit matures.

Koelpinia (named in honour of Alexander Bernhard Koelpin (1737/9–1801), botanist and friend of Peter Simon Pallas, author of a Flora of Griefswald and translator of Peter Löfling's *Iter Hispanicum* from Swedish to German.

1. **Koelpinia linearis** *Pall.*, Reise 3, 2: 755 (1776); Boissier, Fl. Orient. 3: 721 (1875); Handel-Mazzetti in Ann. Naturh. Mus. Wien 27: 62 (1912); Carter in Rec. Bot. Surv. Ind. 6, 6: 204 (1917); Nábělek in Publ. Fac. Sci. Univ. Masaryk 52: 47 (1925); Dinsmore in Post,

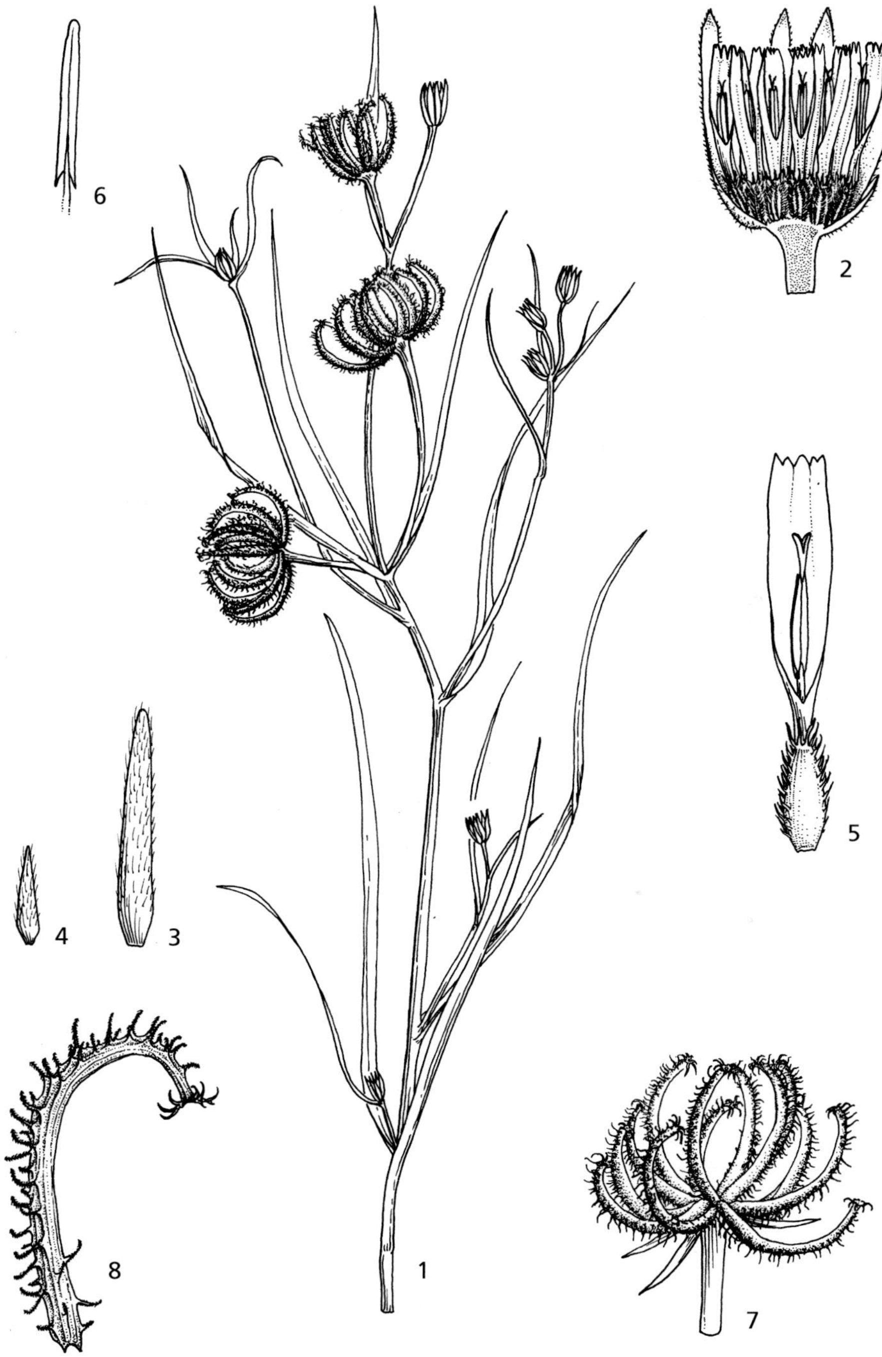

Fig. 59. **Koelpinia linearis**. 1, habit × 1; 2, capitulum (dissected) × 6; 3, outer phyllary × 6; 4, inner phyllary × 6; 5, floret × 12; 6, anther × 18; 7, fruiting capitulum × 3; 8, achene × 6. 1–4 from *Botany Staff* 42362; 5–6 from *Rawi* 29916; 7–8 from *Guest, Rawi & Rechinger* 16953. Drawn by Christine M. Grey-Wilson.

Fl. Syria, Palest. & Sinai ed. 2, 2: 127 (1933); Guest in Dep. Agr. Iraq Bull. 27: 53 (1933); Anthony in Notes Roy. Bot. Gard. Edinburgh 18: 292 (1935); Bornmüller in Beih. Bot. Centralbl. 60: 219 (1939); Zohary in Dep. Agr. Iraq Bull. 31: 167 (1950); Dickson, Wild Fls. Kuwait: 58 (1955); Rechinger, FI. Lowland Iraq: 664 (1964); Rawi in Dep. Agr. Iraq Tech. Bull. 14: 144 (1964); Sell in Fl. Europ. 4: 387 (1976); Kupicha in FI. Turkey 5: 631 (1975); Rechinger, FI. Iranica 122: 11 (1977); Feinbrun-Dothan, Fl. Palaest. 3: 410 (1978); Boulos & Hind in Fl. Egypt 3: 273 (2002); Kilian & Ghazanfar in Fl. Oman 3: 222 (2015); Haloob & Al- Kaisi, Illustr. Fl. Lowland Iraq 1(1): 46 (2016); Taifour & El-Oqlah, Pl. Jordan Ann. Ann. Checkl.: 43 (2017). Ghafoor & Abedin in Fl. Pakistan 222: 273 (2017).

Lapsana koelpinia L.f., Suppl. Pl. Syst. Veg. 348 (1781).
Rhagadiolus koelpinia Willd., Sp. Pl. ed. 4, 3: 1626 (1804).
Koelpinia latifolia C. Winkl. in Acta Horti Petrop. 11, 5: 284 (1890).
K. linearis Pall. var. *filiformis* Eichen. ex Anth. in Notes Roy. Bot. Gard. Edinburgh 18: 292 (1933).
K. chrysoglochis Rech. f., Fl. Iranica 122: 11 (1977).

Erect or sprawling, branched, sparsely pubescent annual herb 5-33(-37) cm tall, with linear or lanceolate, acute, entire, 3-veined, grass-like leaves with clasping base, lanceolate leaves abruptly narrowing to a slightly winged petiole. Basal leaves 3–13.5 × 1–2(–6) mm; cauline leaves up to 15 × 2(–6) mm. Capitula homogamous, 5–9 mm long, 6–8 mm diam.; involucre cylindrical; phyllaries biseriate, outer 1(–2) × 0.5 mm, narrowly lanceolate or subulate, inner phyllaries 5–7(–9) × c. 1 mm in fruit, equaling florets in length, ± narrowly lanceolate; receptacle flat, naked. Corollas yellow, glabrous, ligule oblong, tapering slightly to base, 6 × 1.5–2 mm, apex truncate, veins 4–6; anthers cylindrical, 1.5 mm long, with sagittate base and long, slender, appressed tails; style bifid, style arm apices rounded. Fruits resembling birds' claws, a cluster of 11–18 achenes, at first straight and erect, becoming stellate-patent and incurved to arcuate as they develop, eventually much exceeding involucre, 7–19 × 1–2 mm at base, strongly ribbed, minutely pubescent, with few to many hooked spines to 2 mm long arising from ribs, and bearing a crown of radiating similar spines of varying lengths at apex; pappus of several minute, stout, blunt bristles. Fig. 59: 1–3.

HAB. On sandy plains, sometimes gypseferous, occasionally in fields on silt, clay, sandstone or limestone; alt. (4–)20–700 m; fl. (Feb.-)Mar.-May(-Jun.).

DISTRIB. Rare in the lower mountain region, quite common in the steppe and very common in the desert region of Iraq: **MJS**: Jabal Qilat betwen Balad Sinjar and Ain Tellawi, *Field & Lazar* 416!; **MSU**: Jarmo, *Helbaek* 5321!; **FUJ**: between Balad Sinjar and Tal Afar, *Guest* 4187!; Tal Julaimid, *Alizzi & Husain* 33945!; Sharzat, *Y.R. Rao* s.n.!; Qaiyara, *Bayliss* 56! **FKI**: Kani Domlan, *Guest* 4307!; 14 km E of Kirkuk, *Rawi & Gillett* 10664!; Tuz, *Guest* 1408!; 7 km NE of Kirkuk, *Rawi, Alkas & Noori* 27883!; Jabal Hamrin nr Kifri on Khanaqin road, *Rawi & Gillett* 7347! **FPF**: Pilkhana, *Rawi* 12776!; Mandali, *Guest* 780!; Qaraghan (Jalaula), *Rogers* 6138!; Zurbatiya, *Gillett* 6723! **DLJ**: 12 km SE of Rawa, *Omar, Qaisi, Hamid & Hamad* 45157!; between Rawa and Haditha, *id.* 44471!; 70 km NW of Falluja, nr Thirthar, *Chakraverty & Rawi* 30408!; nr Tikrit, *Omar & Hamid* 36286!; nr Baiji, *Rawi & Hamada* 33594!; 12 km W of Rawa, *S. Omar, Al-Kaisi, K. Hamid & H. Hamid* 44471! **DGA**: Thirthar dam, Samarra, *Ani & Hadid* 9670!; 4 & 20 km E of Samarra, *Rawi* 20337! & 20402!; between Samara and Balad, *Handel-Mazzetti* s.n.. **DWD**: between Jordanian frontier and Rutba, *Ali Askari & Bharucha* s.n.!; Rutba, *Springfield & Jaf* 21298!; H.3, *Janan al-Mukhtar* 33378!; 40 km SE of Rutba, *Omar* 34205!; 180 km SW of Rutba, *Rawi & Hamada* 33533!; Nuqaib, *Rawi* 13860!; 15 km W of Ramadi, *Guest & Rawi* 13689!; Wadi Hauran, 12 km SE of H.2, *Chakravarty, Rawi, Khatib & Alizzi* 31676!; 20 km S of Nukhaib, *Alizzi & Omar* 35511!; 5 km W of Karbala, *Rawi & Gillett* 6373!; 112 km W and 50 km SW of Karbala, *Rawi* 30931! & 30781! **DSD**: Wadi Khirr, 90 km N.W. of Shabicha, *Guest et al.* 19430!; 40 km WNW of Shabicha, *id.* 19320!; Shabicha, *Fauzi, Hazim & Hamid* 38730!; 25 km SW of Shabicha on road to Salman, *Rawi* 14911!; 50 km N by W of Aidaha, *Guest, Rawi & Rechinger* 19184!; nr Golaib, *Guest et al.* 18754!; 25 km N of Busaiya, *Qaisi, Hamad & Hamid* 48527!; nr Chilawa, *Guest et al.* 17209!; nr Safai al-Maghif, *id.* 17262!; nr Zubair, *Rawi & Gillett* 6079!; Jabal Sanam, *Guest, Rawi & Rechinger* 17010!; Khadhar al-Ma'i, *Qaisi, Hamad & Hamid* 48493!; between Ansab and Samah, *Alizzi & Omar* 35709!; Um Qasr, *Guest et al.* 16891! **LCA**: Sudur Mansuriya, *Majid Brahim & Barkley* 6797!; between Falluja and Baghdad, *Widad* 47555!; Rustamiya (Baghdad), *Lazar* 38961!; W of Shaikh Sa'd, *Rawi & Haddad* 25533!; Babylon, *Handel-Mazzetti* s.n., *Nábělek* 3853. **LEA**: 45 km E. of Tib, *Thamir* 47696!; Garmashaya (Karmashiya), NE of Kut, *Rawi* 18016!; 60 km N of Tib, *Guest et al.* 17470! **LSM**: Musaida, *Rawi* 14941!; Amara in Anth., l.c. **LBA**: 65 km W of Basra, *Al-Kaisi* 42404!

Syria, Palestine, Arabia, Turkey, Iran, Turkmenia, Afghanistan.

LAHYA AT-TAIS ("goat's beard"), Carter, Dickson; DH'ILŪQ Carter "grows chiefly on sandy plains; Badawin are fond of the plant, eating all but the root, and all grazing animals feed on it".

The achenes of this species are sometimes galled.

56. **SCORZONERA** L.

Sp. Pl.: 790 (1753); Gen. Pl. ed. 5: 346 (1754)
Lack in Kubitzki, Fam. Gen. Vasc. Pl. 8: 198 [2006] (2007)
Gelasia Cass., Bull. Sci. Soc. Philom. Paris 181: 33 (1818); *Lasiospora* Cass., Dict. Sci. Nat., ed. 2, 25: 306 (1822); *Achyroseris* Sch.Bip., Nov. Actorum Acad. Caes. Leop.-Carol. Nat. Cur. 21(1): 165 (1845); *Avellara* Blanca & C. Diáz, Candollea 40(2): 453 (1985); *Takhtajaniantha* Nazarova, Biol. Zhurn. Armenii 43(3): 179 (1990).

C. C. Townsend

Biennial or perennial herbs, more rarely annual or suffruticose, scapose to tall and leafy, rootstock stout and sometimes tuberous, basal part of stem sometimes with a scaly or fibrous collar formed of remains of previous years' leaf-sheaths. Leaves variable in form from linear and entire to broadly elliptic or ovate and toothed or deeply pinnatipartite. Capitula homogamous, ligulate; involucre narrowly cylindrical to broadly obconical; phyllaries multiseriate or occasionally sub-biseriate, imbricate; receptacle flat or slightly convex, epaleaceous. Ligules yellow to pink or purplish, ± pilose at junction of tube and ray. Achenes glabrous to densely hairy, ribbed, smooth or ± rugose or transversely lamellate, or outer rugose and inner smooth, all ± cylindrical, with or without a hollow inflated base; pappus setae multiseriate, setae entirely plumose, plumose below and scabrid above or more rarely entirely scabrid.

Perhaps about 150 species in Europe and temperate Asia, a few species extending to tropical SW Asia; 15 species in Iraq.

Scorzonera (from the It. for *S. hispanica* L., the Black Salsify, derived from the It. *scorza*, skin or peel and *nero*, black, referring to the colour of the root of that species).

Lipschitz, S. Yu. (1935). Fragmenty k monografii roda Scorzonera" I. in Trans. Rubber & Gutta Percha Inst., Moscow, s.n. : 1–164 (1935).
- (1939). II. Societas Naturae Curiosorum Mosquensis: 1–167.
Mavrodiev E. V. et al. (2004). Phylogenetic relationships in subtribe Scorzonerinae (Asteraceae: Cichorioideae: Cichorieae) based on ITS sequence data. Taxon 53(3): 699–712.

An extremely difficult genus for the taxonomist because of the limited number of characters of proven systematic significance. As Chamberlain observed in Fl. Turkey, there are a number of difficult complexes. In Iraq the tuberous species other than *S. papposa* present considerable difficulties not only among themselves but with regard to their relationship to closely allied species in adjacent countries. For example, *S. phaeopappa* and *S. semicana* as here understood are scarcely separable except on flower colour, the taxonomic value of which is only upheld since it appears to be constant in other species of the genus. The present account should thus be regarded only as provisional pending a genuine critical revision – which Lipschitz' work did not claim to be since he had not even seen material of a significant proportion of the species which he enumerated.

In a detailed study of the subtribe Scorzonerinae Mavrodiev et al. (2004) indicated much of the generic synonymy, including recognizing a separate *Geropogon* and *Podospermum*; Lack (2006), however, synonymised *Podospermum* with *Scorzonera*. This synonymy is not accepted here, although Mavrodiev et al. indicated that they were not prepared to provide any formal taxonomic revision before carrying out additional research.

In the descriptions which follow, "outer" phyllaries are the outermost one or two, and "inner" phyllaries similarly the innermost; intermediate phyllaries are transitional.

1. At least some of outer or median phyllaries with a tooth-like appendage on outer upper surface; leaves obviously pinnatipartite almost to midrib, of if linear and entire then plant a dwarf, caespitose high alpine; staminal and corolla tube yellow . . . . .1. *S. cana*
   All phyllaries lacking tooth-like dorsal appendages; leaves entire (though often undulate or crispate), if occasionally pinnately divided then only to around halfway and staminal and corolla tube purple (*S. incisa*) 2

2. Corolla limb pink or lilac on both surfaces 3
Corolla limb yellow at least on upper surface 4
3. Phyllaries glabrous; pappus pure white; lower leaves ± abruptly contracted at base .... 7. *S. papposa*
Phyllaries ± tomentose at least in flower; main pappus hairs purplish-brown at least towards the base; lower leaves gradually narrowed above .... 3. *S. phaeopappa*
4. At least some leaves sharply toothed, or pinnatipartite to a little more than half way .... 8. *S. incisa*
Leaves entire, margins plane to erose-crispate, never toothed or pinnatipartite 5
5. Ovaries and achenes ± densely lanate or sericeous 6
Ovaries and achenes glabrous 9
6. Plant with a scapose or subscapose stem; root a large, globose tuber ..... 15. *S. lanata*
Plant with stem, even if short, clearly branched or leafy; root not tuberous 7
7. Basal leaves widest, linear and grass-like, usually 3–veined, to 4 mm wide; caespitose mountain plants with many wiry stems .... 9. *S. ramosissima*
Widest leaves not linear and grass-like, narrowly to broadly elliptic, 5–20–veined, > 4 mm wide; not caespitose plants of montane habit 8
8. Plant with a dense whitish-pannose indumentum; lower leaves 12–20–veined .... 13. *S. veratrifolia*
Plant with a short greyish tomentum to ± glabrous; lower leaves 5–10–veined .... 14. *S. latifolia*
9. Plant an intricately branched desert chasmophyte with a bushy habit 10. *S. tortuosissima*
Plant not an intricately branched, bushy desert chasmophyte 10
10. Achenes without a short, inflated, hollow basal "beak"; rootstock not tuberous; leaves glabrous or shortly and densely tomentose 11
Achenes with a short, inflated hollow basal "beak"; rootstock tuberous; leaves with an arachnoid, stellate indumentum at least when young 12
11. Leaves glabrous, with a conspicuous yellowish midrib and 1–2 much finer veins on each side .... 11. *S. davisii*
Leaves densely shortly tomentose with several equally conspicuous veins. . 12. *S. cinerea*
12. Stem taller, leafy; cauline leaves broader, to 2.8 cm wide .... 2. *S. semicana*
Plant subscapose; leaves narrowly linear, not more than 8 mm wide.............. 13
13. Plant with a long basal collar formed of scaly remains of previous years' leaf-sheaths; leaves strongly crispate-margined .... 4. *S. mucida*
Plant with leaves feebly crispate-margined, or if more so then with no long basal collar of scaly leaf-remains 14
14. Pappus white or dirty white when mature; achenes finely rugose when ripe; plant of foothills .... 5. *S. mollis*
Pappus ± purplish-brown when mature; achenes more coarsely transversely lamellate-rugose when ripe; desert plant........... 6. *S. schweinfurthii*

1. **Scorzonera cana** (*C.A.Mey.*) *O.Hoffm.* in Pflanzenfam. 4, 5: 365 (1893): Lipshits in Fl. S.S.S.R. 29: 44 (1964); Chamberlain in Fl. Turkey 5: 635 (1975); Rechinger, Fl. Iranica 122: 27 (1977).

Podospermum canum C.A.Mey, Verz. Pflanz. Cauc.: 62 (1831).

Perennial herb 5–50(–60) cm. tall with a stout cylindrical taproot with 1-several crowns, clothed about base of stem with brown scaly remains of previous years' leaves but no distinct collar, vegetative parts white-tomentose or arachnoid to glabrescent; stems solitary or several, subscapose to distinctly foliose, simple or with a few long, ascending branches. Leaves very variable in size and shape, from small and simple in high mountain forms to up to 35 × 7 mm in outline with lamina deeply and narrowly pinnatipartite almost to midrib into almost filiform to broadly linear, narrowly acute segments; radical and lower stem leaves petiolate with pale expanded sheath, upper cauline leaves sessile, they and uppermost cauline leaves rapidly reduced and bract-like. Involucres cylindrlcal in flower, soon campanulate, solitary on branches; outer phyllaries 4–6 mm, lanceolate to deltoid-ovate, at least some (and some of the median phyllaries) with subapical corniculate appendages; inner phyllaries lanceolate, 10–18 mm in flower, 14–28 mm in fruit and deflexed after fruit-fall; all phyllaries ± arachnoid to glabrous, commonly with a white tuft of tomentum at apex initially. Florets 12–22 mm, at least 1.5 × involucre; ligules 9–17 mm, yellow, those of marginal florets commonly reddish or purplish dorsally. Achenes stramineous, cylindrical, 8–12 mm, strongly ribbed, glabrous, ribs smooth; pappus c. 10–18 mm, setae barbellate above, plumose below, pure white.

DISTRIB. (of species). E, C & SE Europe from Austria and Italy to S Russia including the Crimea, Turkey, Syria, Caucasia, Iran.

var. **cana**

Stems rarely exceeding 20 cm in height, leafy, usually numerous and branched from the base and/or decumbent.

HAB. & DISTRIB. Only one specimen of this plant has been seen, *Chesney* 2021 of his Expedition to the Euphrates; this was probably of Syrian rather than Iraqi origin, and no habitat information appears on the label.

Turkey, Caucasus, Iran.

var. **jacquiniana** (*W.D.J. Koch*) *Chamberlain* in Notes Roy. Bot. Gard. Edinburgh 33: 433 (1975) & in Fl. Turkey 5: 636 (1975).

Podospermum jacquinianum W.D.J. Koch, Syn. Fl. Germ. Helv. ed. 1: 425 (1837).
Scorzonera jacquiniana (W.D.J. Koch) Čelak., Prodr. Fl. Böhm. 1: 218 (1867); Dinsmore in Post, Fl. Syria, Palest. & Sinai ed. 2, 2: 136 (1933); Meikle, Fl. Cyprus 2: 1044 (1985).
S. laciniata auct. non L.: Nábělek in Publ. Fac. Sci. Univ. Masaryk 52: 50 (1926)
S. luristanica Rech.f. in Anz. Math.-Nat. Ki. Öst. Akad. Wiss. 88: 267 (1951) & Rechinger, Fl. Iranica 122: 24 (1977).

Stems 20–60 cm in height, rarely less, erect, leafy, simple or branched above base with long, ascending branches.

HAB. Rocky mountain slopes. sometimes in *Quercus* forest, once near water; alt. 1100–1650 m.; fl. May-Jun., fr. Jun.-Jul.
DISTRIB. Scattered, probably not uncommon, in the forest zone of the mountain regions of Iraq. **MAM**: Ser Amadiya (Gali Mazurka), *Dabbagh, Qaisi & Hamid* 45989! **MRO**: Gali Warta, *Rawi, Noori & Kass* 28737!; between the slope of Karoukh Mt. and Dargala village. *Noori & Kass* 27746!; between Kujar and Kani Grawi region, E of Karoukh, *Kass, Noori & Serhang* 27636!; between Siewaka and Dargala villages, E of Karoukh, *Kass, Noori & Serhang* 27577!; Handren Dagh above Jindian, *Gillett* 8282!; Haji Umran, *Chapman* 11940!, *Rawi, Noori & Kass* 27773!; *Kotschy*, Pl. Mesopot., Kurdistan & Mossul 268!

Throughout the range of the species.

I have not been able to find any means of separating *S. luristanica* from this variety. Though I have not seen the type, I have seen several other specimens cited by Rechinger in *Flora Iranica*. All seem to me to be perennial rather than biennial – indeed one, *Furse* 2040, bears a note to that effect – and I find the broad outer phyllaries and very narrow leaf segments said to characterize *S. luristanica* readily matched in Oriental *S. cana* var. *jacquiniana*.

The *Nábělek* specimen recorded from Sapna (MAM) as *S. laciniata* has not been seen, but is presumably referable to *S. cana* var. *jacquiniana*.

var. **radicosa** (*Boiss.*) *Chamberlain* in Notes Roy. Bot. Gard. Edinburgh 33: 433 (1975) & in Fl. Turkey 5: 637 (1975).

S. radicosa Boiss., Diagn. Pl. Orient., ser. 1, 11: 43 (1849); Rechinger, Fl. Iranica 122: 29 (1977).

Densely caespitose alpine plant mostly 5–10 cm tall, with shortly pinnatisect and/or simple leaves, ± canescent. Stems scapose or subscapose.

HAB. High slopes and summits of mountains; alt. 3000–3700 m; fl. Jul.-Aug., fr. Aug.-Sep.
DISTRIB. Only in the alpine zone of the highest mountains along the Iranian frontier of Iraq, where it is locally widespread. **MRO**: Qandil Range, *Rawi & Serhang* 18282!, 18282A!, 24112!, 24478!, 24495!, *Rechinger* 11797!

Turkey, Iran.

Meikle [Fl. Cyprus 2: 1044 (1985)] regarded *Podospermum canum* C.A.Mey. as a nomen superfluum illegitimum, presumably on the grounds that Meyer cited "Scorzonera laciniata M.B." following his own new binomial. However, referring to Bieberstein's two references to *S. laciniata* [Fl. Taurico-Cauc. 2: 237 (1808) & 3: 524 (1819)] convinces me that he had no intention of publishing a "Scorzonera laciniata M.B." but was merely discussing Treviranus' notes on the two plants which had been confused under the name *S. laciniata*. These were the true *S. laciniata* L. and "Sc. laciniata agri Vindobonensis atque Jacquini", to which latter Bieberstein referred all the Tauro-Caucasian plants without giving them a new specific name. It appears to me that Meyer was formally describing the latter as *Podospermum canum*, and the reference to "Scorzonera laciniata M.B." was merely a citation such as we would today give more precisely – viz., "Scorzonera laciniata sensu M.B. quoad pl. taur.-cauc.". I thus continue to regard *Podospermum canum* C.A. Mey. as a legitimate name.

2. **Scorzonera semicana** *DC.*, Prodr. 7: 119 (1838); Boissier, Fl. Orient. 3: 784 (1875); Chamberlain in Fl. Turkey 5: 642 (1975); Taifour & El-Oqlah, Pl. Jordan Ann. Checkl.: 47 (2017).

S. syriaca Boiss. & C.I.Blanche in Boissier, Diagn. Pl. Orient. ser. 2, 3: 93 (1856); Rechinger, Fl. Iranica 122: 32 (1977); Feinbrun-Dothan, Fl. Palaest. 3: 425 (1978).

S. mollis non M.B.: Guest in Dep. Agr. Iraq Bull. 27: 90 (1933); Zohary, ibid. 31: 158 (1950)

S. mollis var. *longifolia* Boiss., Fl. Orient. 3: 762 (1875); Dinsmore in Post, Fl. Syria, Palest. & Sinai ed. 2, 2: 136 (1933); Bornmüller in Beih. Bot. Centralbl. 60B: 219 (1939).

S. mollis var. *platyphylla* Bornm. in Beih. Bot. Centralbl. 28(2): 264 (1911) & 60B: 220 (1939); Blakelock in Kew Bull. 4: 62 (1949); Rawi in Dep. Agr. Iraq Tech. Bull. 14: 124 (1964).

S. abdulazizii auctt. non Eig: Zohary in Dep. Agr. Iraq Bull. 31: 159 (1950); Rawi, in Dep. Agr. Iraq. Tech. Bull. 14: 123 (1964), *nom. nud.*

Erect, ± caulescent, simple or branched perennial herb (12–)18–35(–45) cm tall, stem sulcate-striate, rootstock a narrowly cylindrical to oblong-ovoid tuber 4–10 × 1–3 cm., base of plant with scaly remains of previous years' leaves only occasionally forming a collar up to 4 cm. long; vegetative parts sparsely arachnoid-lanate (rarely moderately so) when young, soon glabrescent. Basal leaves narrowly linear, grass-like, long-attenuate below and caudate-attenuate above, 15–30 × 3–6 mm (sometimes broader in very robust specimens), margins plane to very crispate; cauline leaves broader, up to 2.8 cm wide, midrib prominent with up to 4 veins on each side, margins plane to wavy, occasionally crispate. Involucre narrowly ovoid-cylindrical; outer phyllaries ovate-lanceolate to ovate, acute to acuminate, (8–)10–15 mm; inner phyllaries narrowly oblong-lanceolate, ± acute, (15–)18–24 mm in flower, 30–50 mm in fruit; all phyllaries finely stellate arachnoid, somewhat glabrescent in fruit. Florets 18–24 mm; ligules 13–17 mm, corollas yellow, concolorous or the marginal with orange stripes dorsally. Achenes 17–28 mm, glabrous, cylindrical, ribbed, the outer rather coarsely transversely lamellate-rugose, inner smooth; pappus 18–25 mm, hairs barbellate above, plumose below, pale-stramineous or whitish.

HAB. Rocky mountains, earthy/stony hill slopes, frequently in degraded or coppiced *Quercus* or *Rhus/Quercus* forest, extending to the *Daphne/Astragalus* community; alt. 454–1840 m; fl. Apr.-May, fr. Apr.-Jun.

DISTRIB. Frequent in the forest zone, just extending to the thorn-cushion zone of the mountain region of Iraq, occasionally somewhat lower. **MJS**: Jabal Sinjar, *Qaisi & Hamad* 49140! **MRO**: Sefin Dagh, *Bornmüller* 1509!, *Polunin* 5023!; Rowanduz Gorge, *Guest* 2140!; 31 km N of Kirkuk on way to Kol Sanjaq, *Rawi, Noori & Kass* 27970!; Hab-as Sultan Dagh, *Rawi, Noori & Kass* 28160!, 28250B!; Gali Dargala, *Rawi, Noori & Kass* 28904!; 5 km N of Darband, *Botany Staff* 43196!; Chinaruk, *Omar, Karim, Hamza & Hamid* 37238!; Baradost Dagh, *Field* 34b. **MSU**, between Chemchemal and Darband-i Basian, *Rawi & Gillett* 7606!; Qarachitan, *Gillett* 7727!; Pir Omar Gudrun, *Gillett* 7820!; Kamarspa, *Rawi* 22187!; Avroman Mt., *Rawi* 22090!; Mela Kowa, *Rawi* 22490!; Qopi Qaradagh, *Barkley & Haddad* 8537!; Bakrajo, *Omar & Karim* 38001!; Jasnia, *Omar* 42725!

Turkey, Syria, Iran.

SHIN (Kurd., *Gillett* 7727!), HALALOK (Kurd., *Gillett* 71120!). According to a note on both of these sheets the plant is "edible", though whether the tuber (as often in this genus) or the foliage is eaten is not stated.

Differs from *S. phaeopappa* in its yellow flowers and lighter-coloured pappus. The taxonomic value of both of these characters needs testing by field and breeding studies.

3. **Scorzonera phaeopappa** (*Boiss.*) *Boiss.*, Fl. Orient. 3: 764 (1875); Dinsmore in Post, Fl. Syria, Palest. & Sinai ed. 2, 2: 137 (1933); Chamberlain in Fl. Turkey 5: 641 (1975); Rechinger, Fl. Iranica 122: 37 (1977).

Podospermum phaeopappum Boiss., Diagn. Pl. Orient. ser. 1, 7: 5 (1846).

Subscapose or shortly caulescent perennial herb 12–40 cm tall, the rootstock thick and cylindrical or an oblong tuber, *c.* 8–12 × 1–2.5 cm, base of plant with scaly remains of a few previous years' leaves but not often forming a distinct neck; vegetative parts rather sparingly arachnoid or glabrescent. Leaves mostly in basal part of stem, narrowly to broadly linear, lower *c.* 9–27 × 0.2–1.5 cm, flexuose-undulate and margins ± crispate, with 1–2 slender veins on each side of midrib, sheathing base broad and subamplexicaul especially in broader-leaved forms; upper cauline leaves reducing, all leaves longly and finely narrowed. Stem and long pedicels sulcate-striate, usually stout. Involucre cylindrical; outer phyllaries few, broadly lanceolate or lanceolate-ovate, acute, 7–10 mm; inner phyllaries narrowly oblong-lanceolate, acute, 15–22 mm in flower, 30–60 mm in fruit; all phyllaries arachnoid, usually glabrescent. Florets 22–27 mm; ligules 15–18 mm, marginal ligules lilac with purple stripes dorsally. Achenes c. 20–28 mm, outer black with coarse transverse flap-like rugosities, inner stramineous and less rugose or generally smooth and longitudinally ribbed; pappus c. 20–28 mm, barbellate above, plumose below, main hairs purplish-brown towards base or throughout.

HAB. Mountain slopes, among rocks on red-brown or black loamy soil, on calcareous sandstone spur or in *Quercus* scrub; alt. 950–1060 m; fl. & fr. Apr.-May.
DISTRIB. Apparently rather rare in the forest region of Iraq. **MRO**: Shaqlawa, *Haines* 695!; Haibat Sultan Dagh, *Rawi, Noori & Kass* 28250!,28250C! **MSU**: mountainside just E of Darband-i Khan, *Barkley & Shehbaz* 7744!; Palegawra, *Helbaek* 655!

Turkey, Lebanon, Iran.

4. **Scorzonera mucida** *Rech.f., Aellen & Esfand.* in Österr. Bot. Z. 97: 264 (1950); Rechinger, Fl. Iranica 122: 33 (1977).

Scapose or subscapose perennial herb, rootstock a large subglobose to subcylindrical tuber 2.5–7 × 1.8–3 cm, base of plant with a thick neck up to 4.5 cm long formed of scaly remains of previous years' leaves; vegetative parts ± arachnoid-lanate throughout. Leaves linear, 7–16 × 0.3–0.8 cm, frequently flexuose with a ± uncinate tip, undulate and margins ± densely erose-crispate, with 1–2 slender veins on each side of conspicuous yellow midrib, glabrous, striate, sheathing base narrow. Stem 3.5–12 cm tall, scapose or with up to about 5 stem leaves in lower half, shorter than to somewhat exceeding leaves when in flower. Involucre shortly cylindrical, inner phyllaries ± spreading above; outer phyllaries few, broadly lanceolate, acuminate, 7–12(–15) mm; inner phyllaries lanceolate-oblong, 18–24 mm in flower, 27–33 mm in fruit. Florets 20–22 mm; ligules 13–15 mm, bright yellow, marginal with darker lines dorsally. Achenes 22 mm, glabrous, black, outer coarse with transverse flap-like rugosities, inner smooth; pappus 20 mm, hairs scabrid above, plumose below, pale-stramineous.

HAB. In steppic grassland, in a wadi, on silty and clay slopes and once in *Rhus-Quercus* forest on hillside; alt. 335–1320 m; fl. Mar.-May, fr. May-Jul.
DISTRIB. Scattered in the upper steppe and lower forest zones of Iraq. **MSU**: Mela Kowa, *Rawi* 22486!; Jarmo, *Helbaek* 506! 632! 691!; Dukhan, *Meade* 118! **FKI**: near Bawa Nur, *Poore* 397A! **FPF**: Kani Kirmaj, *Poore* 318!

Iran.

Helbaek 506 was cited in *Flora Iranica* as *S. phaeopappa.* However, the collector clearly gives the flower colour as "yellow", and the specimen is a fine match for western Iranian material of *S. mucida* such as *Archibald* 1392. I am rather doubtful if this species will survive the test of time: it is very close to *S. mollis,* but the mature outer achenes seen have the coarse flap-like lamellate rugosities of *S. phaeopappa* rather than the much finer ones of *S. mollis,* which rarely develops the basal "collar" of dried leaf remains, though I have seen this in some Crimean specimens of *S. mollis.*

5. **Scorzonera mollis** *M.Bieb.*, FI. Taur.-Cauc. 3: 522 (1819); Boissier, FI. Orient. 3: 761 (1875); Dinsmore in Post, Fl. Syria, Palest. & Sinai ed. 2, 2: 136 (1933); Blakelock in Kew

Bull. 4: 62 (1949); Rawi in Dep. Agr. Iraq Tech. Bull. 14: 124 (1964); Lipshits in Fl. S.S.S.R. 29: 49 (1964); Chamberlain in Fl. Turkey 5: 640 (1975); Taifour & El-Oqlah, Pl Jordan Ann. Checkl.: 47 (2017).

Subscapose or rather shortly caulescent simple or somewhat branched perennial herb (4–) 10–25(–30) cm tall, including the terminal peduncle, rootstock a narrowly cylindrical to oblong-ovoid tuber c. 3–10 × 1–2 cm tall, base of plant with a few scaly remains of previous years' leaves but in Iraq no distinct collar; vegetative parts ± arachnoid-lanate or occasionally ± glabrescent. Leaves narrowly linear, grass-like, usually in basal part of stem or sometimes plant with a distinct leafy stem; largest leaves c. 4.5–20 × 0.15–0.5 cm, upper not conspicuously reducing, all caudate-attenuate, slightly expanded at base, margins smooth and entire to strongly crispate. Involucre cylindrical; outer phyllaries lanceolate, usually rather narrow, acute, c. 7–11 mm; inner phyllaries narrowly oblong-lanceolate, ± acute, 15–22 mm in flower, c. 25–45 mm in fruit; all phyllaries ± arachnoid, often glabrescent. Florets 18–22 mm, ligules 10–15 mm, yellow, marginal darker- or bluish-veined dorsally, frequently with a bluish dorsal vitta. Achenes c. 14–18(–20) mm, glabrous, cylindrical, ribbed, outer finely transversely lamellate-rugose, inner rugose or smooth; pappus 17–25 mm, hairs all barbellate above, plumose below, white or pale stramineous.

HAB. Not recorded in Iraq; elsewhere on dry limestone hills, stony slopes, scree, on broken steppe etc.; alt. (recorded once in Iraq) 1700 m.; fl. & fr. May.
DISTRIB. Very rare in the extreme northern mountain region of Iraq. **MAM**: Jabal Rus & Jabal Kashan (nr Gilli), Zakho, not common, *Mudir of Gilli* 3149!; Kantur, *Rawi* 8783!

Balkans, Romania, S Russia (including Crimea), Turkey, Syria, Transcaucasia, probably also Iran.

KÎRISCHIK (Kurd., *Mudir of Gilli* 3149). This sheet bears a note to the effect that this plant is eaten by sheep and also by shepherds as a vegetable.

6. **Scorzonera schweinfurthii** *Boiss.*, Fl. Orient. Suppl.: 320 (1888); Dinsmore in Post, Fl. Syria, Palest. & Sinai ed. 2, 2: 131 (1933); Feinbrun-Dothan, Fl. Palaest. 3: 425 (1978); Taifour & El-Oqlah, Pl. Jordan Ann. Checkl.: 47 (2017).

S. rawii Rech.f. & Guest in Bull. Coll. Sci. Baghdad 2: 10 (1957); Rech. f. in Ber. Deutsch. Bot. Ges. 72: 284 (1959); Rechinger, Fl. Lowland Iraq: 673 (1964): Rawi in Dep. Agr. Iraq Tech. Bull. 14: 124 (1964).
S. deserti-syriaci auctt. non Eig: Zohary in Dep. Agr. Iraq Bull. 31: 159 (1950); Rawi in Dep. Agr. Iraq. Tech. Bull. 14: 123 (1964), *nom. nud.*

Scapose or subscapose perennial herb, simple or branched about base, rootstock a large ovoid or oblong-ovoid tuber c. 5–9 × 2.5–3.5 cm, base of plant with a few scaly remains of previous years' leaf-sheaths but no distinct collar; vegetative parts ±densely arachnoid with close, fine stellate hairs, glabrescent with age. Leaves mostly from base, linear, c. 8–30 × (1–)1.75–5 mm (lowest leaves rarely wider), flexuose and apex sometimes hamate, margins entire to undulate, with a conspicuous midrib and 1–2 obscure veins on each side, upper leaves not conspicuously reduced. Involucre narrowly cylindrical or ovoid-cylindrical; outer phyllaries ovate, 6–8(–10) mm, inner phyllaries oblong-lanceolate, (14–)17–20 mm in flower, (26–)30–34 mm in fruit, all phyllaries moderately to densely stellate-arachnoid especially along margins and apex and obtuse or bluntly pointed in flower, somewhat glabrescent and sometimes more acute in fruit. Florets 17–22(–27) mm; ligules 12–15(–17) mm, yellow, marginal with reddish stripes and often ± reddish-suffused dorsally. Achenes 13–20 mm, glabrous, cylindrical, ribbed, outer sublamellate-rugose, inner rugose or smooth; pappus 20–26 mm, hairs barbellate above, plumose below, deep-stramineous to dark purplish-brown. Fig. 60: 1–4.

HAB. On rocky low hills, rocky gullies from the desert plateau, in silty or sandy gravelly depressions, in wadi, on clay and gravelly degraded steppe and by dry roadsides; alt. 130–710 m; fl. & fr. Mar.-Jun.
DISTRIB. Common in the Western Desert region of Iraq. **DWD**: between Jabal Tanf (Syria) and the Euphrates, *Edmonds* 3791!; 5 km S of H.3, *Rawi* 21213!; 30 km N of Rutba, *Chakravarty, Rawi, Alizzi & Khatib* 31500A!; 20 km W of Rutba, *Rawi* 15636!; Tall an-Nisr (Tall Al-Nasir), *Rawi* 31194!, 31218!; Rutba, *Rawi* 14650!; 40 km ESE of Rutba on way to Ramadi, *Rawi* 31314!; Faggart Al-Ghadaf, *Rawi* 31147!, 31150!; 35 km S of Rutba, *Chakravarty, Rawi, Khatib & Alizzi* 3142B!; 20 km from Mugriannam, *Rawi & Hamada* 33542!; Wadi al-Tibal, *Rawi* 31104!; 24 km SW of Ukhaidhir, *Rawi & Hamada* 33502!;

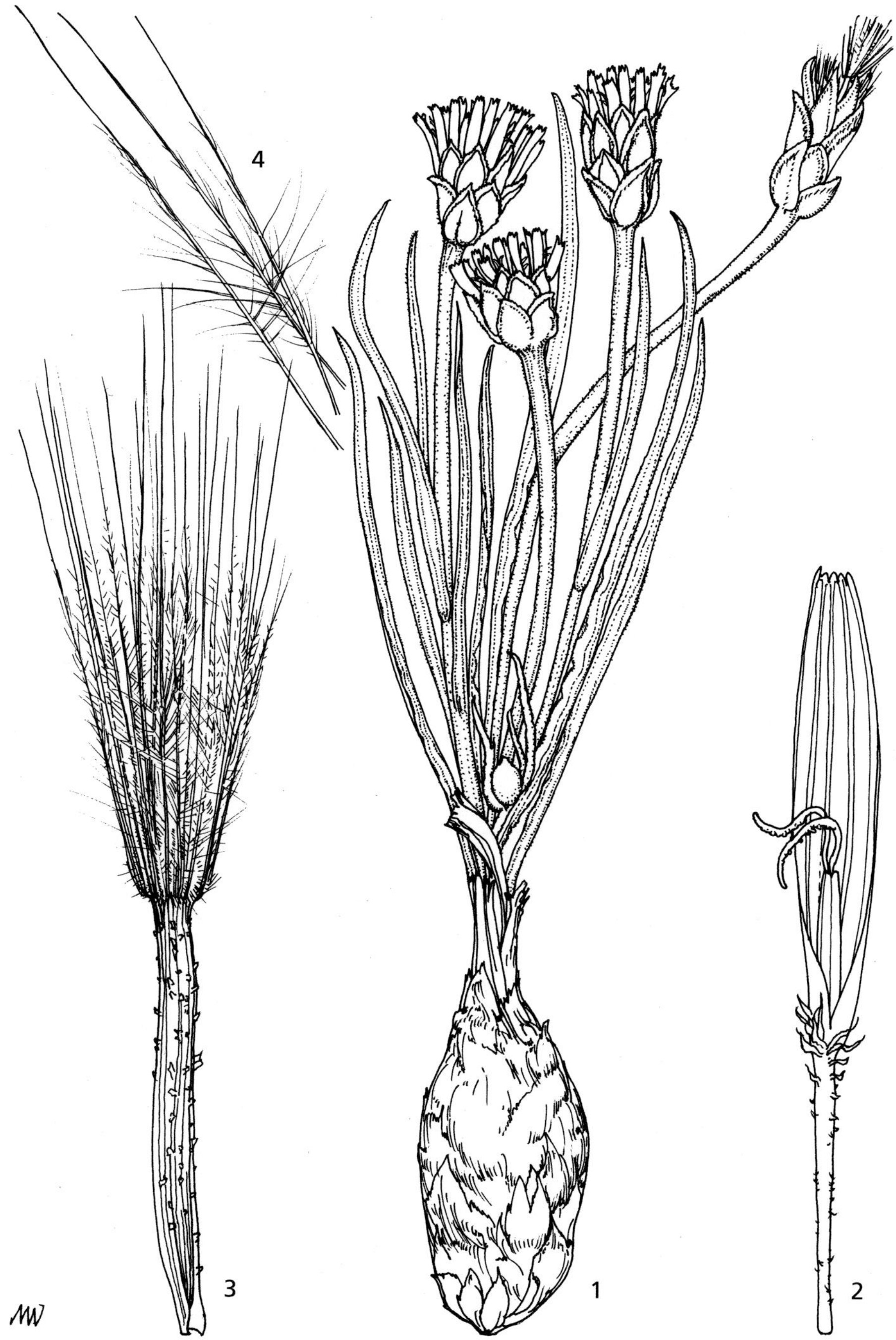

Fig. 60. **Scorzonera schweinfurthii.** 1, habit × 1; 2, floret × 6; achene and pappus × 6; 4, pappus hairs × 6. Drawn by M. Millar Watt.

12 km N of Nukhaib, *Rawi* 31116!; Nukhaib, *Agnew, Hadač & Haines* 2174!; 120 km W of Karbala, *Chakravarty & Rawi* 30959!

Jordan, Saudi Arabia.

Very close to *S. mollis*, differing in its blunter phyllaries, more coarsely rugose achenes (though less so than *S. phaeopappa*), lamellae seldom broader than long) and darker pappus. These and its restricted desert habitat seem to render it worthy of recognition. Subspecific rank might be more suitable, but would add nothing to our knowledge of the plant, and its final destiny must be left for a monographer.

In his supplementary description of *S. rawii*, Rech.f. (Ber. Deutsch. Bot. Ges. 72: 284, 1959) describes the achenes of this plant as "glabra et laevia", but this is certainly not true of the type specimen. It may be that the rugose marginal achenes are sometimes absent; the occasional absence of characteristic marginal achenes has been noted in numerous other Compositae.

The type specimen of *S. schweinfurthii* is very fragmentary but having seen a specimen collected recently in the type locality, Galala, Egypt (*Hassan* 2495, CAI) by courtesy of the collector, I have no doubt that *S. rawii* must fall into the synonymy of this species.

7. **Scorzonera papposa** *DC.*, Prodr. 7: 119 (1838); Boissier, Fl. Orient. 3: 765 (1875); Handel-Mazzetti in Ann. Naturh. Mus. Wien 27: 65 (1912); Nábělek in Publ. Fac. Sci. Univ. Masaryk 52: 51 (1925); Dinsmore in Post, Fl. Syria, Palest. & Sinai ed. 2, 2: 137 (1933); Guest in Dep. Agr. Iraq Bull. 27: 89 (1935); Blakelock in Kew Bull. 4: 62 (1949); Zohary in Dep. Agr. Iraq Bull. 31: 159 (1950); Rechinger, Fl. Lowland Iraq: 672 (1964); Rawi in Dep. Agr. Iraq Tech. Bull. 14: 124 (1964); Lipshits in FI. S.S.S.R. 29: 96 (1964); Chamberlain in Fl. Turkey 5: 642 (1975); Rechinger, Fl. Iranica 122: 61 (1977); Feinbrun-Dothan, Fl. Palaest. 3: 426 (1978); Gabrielian & Fragman-Sapir, Flow. Transcaucasus & Adj. Areas: 86 (2008); Haloob & Al-Kaisi, Illustr. Fl. Lowland Iraq 1(1): 54 (2016); Abdulridha, Taha & Wedad, Ecology and Fl. Basra: 225 (2016); Taifour & El-Oqlah, Pl. Jordan Ann. Checkl.: 47 (2017);

S. kurdica Boiss. & Noë in Boissier, Diagn. Pl. Orient. ser. 2, 3: 93 (1856).

Perennial herb 10–45 cm tall, branched from near base or at about the middle with short to slender and ascending branches; rootstock forming an ellipsoid to cylindrical rarely ± globose tuber immediately below ground, abruptly narrowed to a slender root below this; stem, branches and leaves glabrous or initially sparsely floccose and then glabrescent. Basal leaves variable in size and shape, with a distinct petiole up to 7 cm long, lamina ovate to narrowly elliptic, rarely linear, 4–14(–35) × (0.15–)2–6 cm, ± long-acuminate, margins minutely crispate; lower cauline leaves similar, rapidly reducing to lanceolate and finally linear, finely pointed uppermost cauline and branch leaves. Capitula solitary or more usually 2–8; outer phyllaries broadly ovate, 5–10(–15) mm; inner phyllaries lanceolate-oblong, ± obtuse, 15–30 mm; all narrowly hyaline-margined, glabrous or rarely outermost faintly ciliate-margined. Florets 22–30 mm; ligules 11–25 mm, pink to lilac with dark purplish veins, teeth glabrous or minutely papillate-hairy. Achenes 7–8 mm, ± tetragonous with outer face of outer achenes more convex, strongly muricate on angles, less so on faces, frequently smooth towards apex, pale-stramineous, pappus white, longer, scabrid straight central setae 15–24 mm, outer setae entirely softly long-plumose, often ± reflexed when in young fruit.

HAB. On rocky lower slopes of mountains, rocky valleys, in coppiced *Quercus* forest; on gravelly clay hills, loamy plains in sandy wadis, sandy patches in gravelly plains, in cereal fields, roadsides etc.; alt. 150–1700 m.; fl. & fr. Mar.-May.

DISTRIB. Widespread and locally abundant in many parts of Iraq but particularly in the steppe and desert regions. **MJS**: Jabal Sinjar, *Qaisi & Hamad* 49121!, *Thesiger* 502! (BM). **MAM**: between Faida and Al Baqaq, *Karim, Dabbagh & Hamid* 44947!; Gali Mazurka, *Dabbagh & Jasim* 46853! **MRO**: 5 km N of Darband, *Botany Staff* 43195!; Shaqlawa, *Gillett* 11560! **MSU**: 20 km NW of Sulaimaniya, *Rawi* 21743!; Tasluja, *Omar & Karim* 47071!; Jarmo, *Helbaek* 692! **FUJ**: 6 km S of Qal'a Sharqat, *Rawi & Hamada* 336031; Tal Afar, *Qaisi & Hamad* 49235! **FKI**: 6 km E of Kirkuk on way to Sulaimaniya, *Botany Staff* 42992!; Bawanur, *Poore* 397!; near Kirkuk on the road to Baba Gurgur, *Guest* 1347!; **FPF**: Pilkhana, *Rawi* 12846!; 30 km NE of Mandali, *Qaisi, Thamer & Salah* 51436! **DLJ**: near Tharthar, *Chakravarty & Rawi* 30390!; 13 km N of Baka Police Station, *Alizzi & Husain* 33851!; in Jazira midway between Rawa and Haditha, *Omar, Qaisi, Hamad & Hamid* 33851! **DWD**: 20 km W of Nukhaib, *Rawi* 30996!; 13 km E of K.3.

Chakravarty, Rawi, Khatib & Alizzi 32957!; Faggart al-Ghadaf, *Rawi* 31152!; 8 km E of Shithatha, *Rawi & Gillett* 6473!; mountainside 12 km S of Rutba, *Barkley* 33 IR 5314!; 18 km S of Rutba, *Rawi* 146081. **DSD**: Shabicha, *Rawi & Gillett* 6293!; Wadi al-Khirr, *Guest, Rawi & Rechinger* 19407!; ± 15 km SE of Ashuriya, *id.* 193621; around Jabal Sanam, *Qaisi, Hamad & Hamid* 48560! **LSM**: nr. Shatt at-Tib 70 km N of Amara, *Rechinger* 14206.

Turkey, Syria, Lebanon. Palestine, Jordan, Saudi Arabia, Kuwait, Iran, Transcaucasia.

The vernacular names HALAKOK (*Botany Staff* 43195) and RUBA HELA or ERBAHLE (*Rawi & Gillett* 6293) have been recorded. According to Guest (1933) this plant is much relished by stock, and the root is eaten by the people as a vegetable.

Only a selection of the specimens examined have been cited. One Iraqi specimen, *Botany Staff* 42077 from Al-Rubaitha, 20 km N of Salman (DSD), has a scatter of floccose indumentum on the dorsal surfaces of the phyllaries, but this is the only such example which I have seen.

8. **Scorzonera incisa** *DC.*, Prodr. 7: 119 (1838); Boissier, Fl. Orient. 3: 766 (1875); Chamberlain in Fl. Turkey 5: 644 (1975).

S. bulbipes Boiss. & Hausskn. in Boissier, Fl. Orient. 3: 759 (1875); Zohary in Dep. Agr. Iraq Bull. 31: 158 (1950); Rawi in Dep. Agr. Iraq Tech. Bull 14: 123 (1964); Rechinger, Fl. Iranica 122: 31 (1977).

S. calyculata Boiss., Diagn. Pl. Orient. ser. 1, 11: 42 (1849) & Fl. Orient. 3: 766 (1875); Rechinger, Fl. Iranica 122: 41 (1977).

Perennial herb, 25–60 cm. tall, rootstock non-tuberous, slender but tough, base or stem with a wide, dense collar formed of remains of previous years' leaves. Stems 1-few, angular, striate, simple or with a few slender, erect branches, these and upper part of stem naked or with very few bract-like leaves. Basal leaves broadly oblong-elliptic to more narrowly elliptic, sharply and narrowly toothed to pinnatisect with rather distinct, narrow lobes, abruptly or usually gradually tapering to a distinctly petiolate base, lamina outline 7.5–26 × 2.4–9 cm, lobes in pinnatisect forms up to 3.5 × 0.4 cm.; lowest cauline leaves similar, above gradually to very abruptly reduced, in larger plants ± amplexicaul; all vegetative parts at first thinly to moderately floccose, finally ± glabrescent. Capitula solitary on stems and branches; involucre ovate-cylindrical; outer phyllaries broadly ovate, 5–10 mm; inner phyllaries 18–26 mm, to ± 35 mm in fruit; indumentum rather thinly floccose dorsally and along margins, denser at apex. Florets 26–34 mm; ligules 16–23 mm, dark blackish-purple at base, yellow within, marginal blackish or reddish-striped or much reddish-suffused dorsally, staminal tube dark blackish-purple at least in outer florets. Achenes 12–18 mm, pale, glabrous, broadly ribbed, cylindrical, ribs bluntly muriculate at least in outer achenes, pappus ± equalling achenes or a few inner hairs longer, hairs scabrid above and plumose below, brownish.

HAB. Usually in destroyed or coppiced *Quercus* forest, on sandstone or limestone, also on a clay hillside but perhaps here also where once forested; alt. 750–1700 m.; fl. Apr.-May; fr. May-Jun.
DISTRIB. Rather rare in the forest zone of the lower mountain region of Iraq. **MRO**: Haji Omran, *Wiltshire* s.n.; 5 km before Chemchemal on the road from Kirkuk, *Omar & Karim* 37879! **MSU**: between Chemchemal and Darband-i Basian, *Rawi & Gillett* 7603!; Darband-i Basian, *Rawi & Gillett* 7621!; Qopi Qaradagh, *Poore* 586!; Sarchinar, *Gillett* 7688A!; Qaradagh, *Gillett* 7907!

Turkey, Transcaucasia, Iran.

9. **Scorzonera ramosissima** *DC.*, Prodr. 7: 125 (1838); Boissier, Fl. Orient. 3: 781 (1875); Blakelock in Kew Bull. 4: 63 (1949); Rawi in Dep. Agr. Iraq Tech. Bull. 14: 124 (1964); Lipshits in FI. S.S.S.R. 29: 104 (1964); Rechinger, Fl. Iranica 122: 559 (1977).

S. arlgirdagi Eig sensu Zohary in Dep. Agr. Iraq Bull. 31: 159 (1950): Rawi in Dep. Agr. Iraq Tech. Bull. 14: 123 (1964), *nom. nud.*

Caespitose perennial 9–45 cm tall with tough rootstock, base densely clothed with brown or blackish remains of previous years' leaves. Stems numerous. slender, striate, simple or furcate with long, slender, ascending branches, ± crisped-pubescent or glabrescent. Basal leaves narrowly linear and grass-like, 4–15 × 2–4 mm, usually 3-veined; cauline leaves rapidly reducing to about middle, upper parts of stem and branches naked or with very few subulate bract-like leaves; all leaves ± crisped-pubescent. Capitula solitary and terminal on stem and branches; involucre obconical to cylindrical in flower and young fruit; outer phyllaries lanceolate to ovate, acuminate, 1.5–5 mm; inner phyllaries narrowly lanceolate, 10–16 mm (up to 20 mm in fruit); all narrowly hyaline-margined, ± appressed-sericeous. Florets 12–14 mm; limb 9–11 mm, yellow, marginal with slightly darker veins, teeth papillose. Achenes 5–8 mm, narrowly cylindrical, densely sericeous; pappus 5–9 mm, hairs scabrid above and silky-plumose below, dirty white.

HAB. On rocky mountain slopes; alt. 2200–2730 m.; fl. & fr. Jun.-Aug.
DISTRIB. Rare on the highest mountain areas of Iraq. **MRO**: Chia-i Mandali, *Guest* 2690!; Algurd Dagh, *Guest* 2955!; Kudu, *Gillett* 9229A! **MSU**: Beribadan to Margamir, *Ludlow-Hewitt* 1526!

W & C Iran, Transcaucasia (Nakhichevan).

10. **Scorzonera tortuosissima** *Boiss.*, Fl. Orient. 3: 775 (1875); Rawi in Dep. Agr. Iraq Tech. Bull. 14: 124 (1964); Rechinger, FI. Iranica 122: 49 (1977); Feinbrun-Dothan, Fl. Palaest. 3: 427 (1978); Breckle, Hedge, Podlech & Rafiqpoor, Vasc. Pl. Afghan. checklist: 186 (2013); Taifour & El-Oqlah, Pl. Jordan Ann. Checkl.: 47 (2017); Abid, Mehdi & Ghafoor in Fl. Pakistan 222: 307, f. 29F–K (2017).

S. musilii auct. non Vel.: Rech.f., Fl. Lowland Iraq: 672 (1964)

Chamaephytic perennial 15–60 (–75) cm tall, much-branched from base upwards, when well-grown forming a ± rounded bush with branches intricately interwoven, furnished throughout its vegetative parts with appressed ± crispate hairs. Stem and branches slender but wiry, angular, striate. Leaves few, basal withered by flowering time; cauline leaves linear-filiform, sometimes uncinate-tipped, lower 3–10 cm × 0.5–2.5 mm, upper rapidly diminishing, those of upper stem and branches very small and bract-like. Capitula terminal and solitary on stem and branches. Involucre obconical or cylindrical; outer phyllaries 2–5 mm, lanceolate, acuminate; inner phyllaries narrowly lanceolate, 10–16 mm; all narrowly pale-margined, appressed-sericeous. Florets (9–)12–14 mm; limb (7–)9–10 mm, yellow, marginal with darker, sometimes purplish stripes dorsally, occasionally purplish-suffused at least when dry. Achenes 9–12 mm, narrowly cylindrical, pale, strongly ribbed, glabrous; pappus (10–)13–19 mm, hairs scabrid above, silky-plumose below, dirty white to brownish.

HAB. In sandy wadis and on silty flats, also on gravelly and rocky clay soil and on hillocks in the desert; alt. 150–500 m; fl. & fr. Mar.-Apr.
DISTRIB. Scattered, locally common, in the desert regions of Iraq. **DWD**: Jadida, W of Nukhaib, *Rawi* 31043!; 20 km W of Nukhaib, *Rawi* 30495!; bottom of Hawran valley, 12 km SE of H.1, *Chakravarty, Rawi, Khatib & Alizzi* 31697!; Wadi al-Ajrumiya, *Rawi & Noori* 27167! **DSD**: 62 km WNW of Ansab, on Saudi border, *Guest, Rawi & Rechinger* 19062!; 80 km W of Shabicha, *Chakravarty, Rawi, Khatib & Tikriti* 30077!; 9 km W of Shabicha, *Khatib & Hazim* 30544!; 99 km E of Shabicha towards Najaf, *Rawi, Khatib &Tikriti* 29277!; Rutba, *Agnew, Hadač & Haines* 21671; Jumaima police station, *Qaisi, Hamad & Hamid* 48262!

Jordan, Iran, Afghanistan.

The Arabian *S. musilli* appears to be a more slender plant than *S. tortuosissima* in all its parts, and the Iraqi material is certainly similar to the latter. These two taxa need critical appraisal.

11. **Scorzonera davisii** *Lipschitz* in Not. Syst. Herb. Inst. Bot. Acad. Sc. U.R.S.S. 22: 291 (1963); Chamberlain in Fl. Turkey 5: 645 (1975).

[*S. elata* auctt. non Boiss.: Blakelock in Kew Bull. 4: 62 (1949); Rawi in Dep. Agr. Iraq Bull. 14: 123 (1964)].
S. tenax Rech.f., Fl. Iranica 122: 59 (1977).

Erect perennial herb (15–)20–70 cm tall, root thick and cylindrical, not at all tuberous. Stems foliose, mostly 2–4 from same stock, branched from base upwards or above only, glabrous, striate, subterete, clothed at extreme base with scaly remains of leaf-sheaths but no distinct collar; branches slender, erect or ascending. Leaves slender, narrowly linear and grass-like, caudate-attenuate to apex, with a prominent yellow midrib and 1–2 finer veins on each side, plane or with margins ± involute. Margins smooth or (especially in basal leaves) minutely scabrid; basal leaves 10–30 cm × 1.5–6 mm, attenuate below and finally expanded to a pale, striate sheath; cauline leaves similar but progressively shorter and narrower, lower with a subamplexicaul narrowly expanded base. Involucres cylindrical, solitary on branches; outer phyllaries deltoid-ovate, broad, 3–8 mm, mucronate, acute; inner phyllaries oblanceolate, 24–30 mm in flower, 32–40 mm in fruit; all phyllaries coriaceous, greenish centrally, glabrous or inner minutely marginally ciliate and sometimes ± lanate around apex. Florets 22–32 mm; limb 14–19 mm, yellow, outer with orange or deep yellow veins. Achenes stramineous, glabrous, 15–20 mm., linear, ± tetragonous with thick ribs on angles and finer ones on faces, outer with minute sharp and upwardly directed or blunter denticulations along the angle or all ribs, inner smooth; pappus fragile, long, barbellate-tipped central hairs 19–25 mm, shorter outer hairs 13–16 mm and plumose, pure white or creamy.

HAB. On rocky mountain slopes, frequently in *Quercus*, *Quercus/Juniperus* or *Pinus* forest on red marl or clay; alt. 750–1500 m.; fl. May-Jul., fr. Jun.-Jul.
DISTRIB. Widespread and locally common in the forest zone of the extreme northern mountain region of Iraq. **MAM**: Atrush, *Guest* 4395!: Zawita valley, *Guest* 4433!; Zawita, *Gillett* 8349!, *Chapman* 26293!, *Omar, Dabbagh & Qaisi* 45313!, *Omar, Qaisi & Wedad* 49685!; Kani Masi, *Hamad & Fadhil* 45625!; Ravena village, 15 km before Kani Masi, *Botany Staff* 43780!; 23 km NE of Zakho, *Rawi* 231871; pass between Dagh al Radzji and Sharanish, *Rechinger* 10807!; above Suwara Tuka, *id.* 11572!; Sarsang, *Haines* 1241!, 1715!, *Dabbagh* 46174!; Sandur, *Qaisi & Khalid* 46103!; Badi, *Omar, Qaisi & Wedad* 49638!; "Mesopot., Kurdistan & Mossul", *Kotschy* 235!

Turkey.

Differs from *S. elata* Boiss. in its larger involucre and non-tuberous root in addition to its larger achenes. The larger involucre was commented upon by Blakelock, l.c., in recording *Guest* 4395 & 4433 as *S. elata*. From the holotype of *S. davisii* Lipschitz was only able to say that the root was probably non-tuberous, but the Iraqi material abundantly confirms this. None of the Iraqi material is as broad-leaved as the holotype, which attains 6 mm in width while the Iraqi maximum is 4 mm.; this, however, is the only detectable difference. The Kani Masi locality is only about 50 km from the type locality of *S. davisii* (in the R. Zab gorge ± 46 km S of Başkale in the Turkish province of Hakkâri).

12. **Scorzonera cinerea** *Boiss.*, Diagn. Pl. Orient. ser. 1, 11: 44 (1849); Boissier, Fl. Orient. 3: 771 (1875); Chamberlain in Fl. Turkey 5: 646 (1975); Rechinger, Fl. Iranica 122: 66 (1977).

Perennial herb 20–55 cm tall, with a stout rootstock; stems leafy, several to very many from a branched base, usually clad at base with the dark remains of previous years' leaves; stems, branches and leaves densely clothed with short, greyish tomentum; branches erect or ascending, short to long and slender, each bearing a single capitulum. Basal leaves linear and attenuate below to narrowly obovate and petiolate, 6–18 × 0.4–1.5 cm, acute, entire, margins plane or (in broader-leaved forms) crispate-undulate, longitudinally finely 3–9-veined. Cauline leaves similar but sessile, reducing upwards, finally small, very narrow and bract-like on branches; outer phyllaries lanceolate, acuminate, 4–13 mm; inner phyllaries linear-lanceolate, 10–18 mm; all phyllaries ± tomentose. Florets 14–16 mm; limb 9–13 mm, yellow, marginal with darker veins, teeth hirtellous. Achenes 6–12 mm, glabrous, with broad smooth ribs; pappus (8–)10–16 mm, hairs barbellate above, plumose below, dingy white becoming brownish.

HAB. Only found once on a rocky mountain along the Iranian frontier, gathered in June with no note of altitude.
DISTRIB. **MSU**: Avroman Mt., *Rawi, Chakravarty, Noori & Alizzi* 19742!; *id.* 19723!

Iran.

Although the single Iraqi specimen is very poor, with short, obviously malformed branches, the narrow, multinervose leaves, glabrous achenes and pappus with the hairs scabrid above and plumose below leave no doubt as to its identity.

13. **Scorzonera veratrifolia** *Fenzl* in Flora 26: 399 (1843); Boissier, FI. Orient. 3: 778 (1875); Zohary in Dep. Agr. Iraq Bull. 31: 159 (1950); Rawi in Dep. Agr. Iraq Tech. Bull. 14: 124 (1964); Chamberlain in Fl. Turkey 5: 654 (1975); Rechinger, Fl. Iranica 122: 68 (1977).

Perennial herb 0.25–1.1 m, with a stout annulate rootstock; stems 1–4 from incrassate base, procumbent to ascending, slender to stout, leafy, whitish-pannose or brownish towards base, paniculate-racemose above. Basal leaves narrowly to broadly elliptic, lamina 7–15 × 1.3–5.5 cm, narrowed below, distinctly petiolate with a petiole up to 5 cm, acute to rather blunt, entire, plane or undulate, longitudinally finely but conspicuously 12–20-veined. Cauline leaves similar but sessile, lowest attenuate below, median subamplexicaul; upper stem and branch leaves rapidly reducing and more acute, upper stem leaves subtending branches, sometimes reflexed; all leaves with an indumentum similar to that of stem. Branches few to numerous, lowest widely divergent, up to 5 cm long, bearing 1–5 capitula which are ± sessile or on pedicels up to 1.5 cm. Phyllaries densely pannose or inner less so, outer lanceolate, 5–7 mm, inner linear-lanceolate, 13–16 mm. Florets 13–17 mm.; limb 10–13 mm, yellow, marginal with darker veins. Achenes 6–9 mm, densely lanate; pappus 8–11 mm, hairs barbellate above, plumose below; yellowish.

HAB. Rocky mountain slopes and *Quercus* forest; alt. 1200–2060 m.; fl. & fr. Jun.-Jul.
DISTRIB. Apparently not rare in the forest region of the extreme northern sector of the mountain region of Iraq, collected once also in the eastern sector. **MAM**: Gara Dagh, *Kotschy* 4181; Sarsang, *Haines* s.n.!; Shiwarana, between Sharanish and Marsis, *Rechinger* 10873!; Shivara Hisar (nr Marsis?) 25 km NE of Zakho, *Rawi* 23452!; Haisy village, 10 km S of Kani Mazi (Ain Nuri), *Dabbagh & Hamad* 45541! **MSU**: Pira Magrun, *Haussknecht* s.n.!; Zawita,NE of Zakho, *Rawi* 123570!

E Turkey, W Iran.

14. **Scorzonera latifolia** (*Fisch. & C.A.Mey.*) *DC.*, Prodr. 7, 1: 124 (1838); Lipshits in Fl. S.S.S.R. 29: 99 (1964); Chamberlain in FI. Turkey 5: 653 (1975); Rechinger, Fl. Iranica 122: 69 (1977).

Lasiospora latifolia Fisch. & C.A.Mey., Ind. Sem. Hort. Petrop. 1: 30 (1835).
Scorzonera paniculata Boiss., Diagn. Pl. Orient. ser. 1, 7: 8 (1846).
S. lasiospora Boiss. & Balansa in Boissier, Diagn. Pl. Orient. ser. 2, 6: 118 (1859).
Lasiospora nervosa Sch.Bip. in Nov. Act. Acad. Caes. Leop. Carol. 21: 166 (1845).
Scorzonera nervosa Trevir., Del. Sem. Hort. Bonn.: 125 (1833); Boissier, Fl. Orient. 3: 779 (1875); Bornmüller in Beih. Bot. Centralbl. 60B: 221 (1940), non Lam. (1779).

Perennial herb (0.2–)0.3–1 m with a stout rootstock; stems solitary or few from incrassate base, erect, slender to stout, leafy, ± glabrous throughout to densely shortly greyish-tomentose below and less so above. Inflorescence a simple or branched rather lax raceme with suberect to somewhat divergent slender branches and pedicels. Basal leaves narrowly to broadly elliptic, mostly withered at flowering time, lamina c. 5–15 × 0.6–4.5 cm, narrowed below or distinctly petiolate with a petiole up to 4 cm, acute to rather blunt, entire or margins minutely crispate-denticulate, longitudinally finely c. 5–10-veined. Cauline leaves similar but sessile, lower attenuate below, median subamplexicaul, upper stem and branch leaves rapidly reducing, those of branches usually linear and bract-like, all leaves green, shortly tomentose to ± glabrescent. Branches few to numerous, usually shorter and ascending in smaller forms, longer and spreading in larger plants, lowest up to c. 16 cm long, bearing 1–5 capitula on slender, 1.4–5 cm pedicels; outer phyllaries lanceolate, c. 3–5(–7) mm, inner phyllaries linear-lanceolate, c. 12–15(–17) mm, all phyllaries densely tomentose or inner ± glabrescent, or tomentose only along median line. Florets c. 15–18 mm; ligules c. 11–14 mm, yellow, marginal with darker veins. Achenes c. 6–9 mm, densely lanate; pappus 9–19 mm, hairs barbellate above, plumose below, yellowish-brown.

HAB. Rocky mountain slopes, sometimes in *Quercus* forest; alt. 1200–1700 m; fI. May-Jul., fr. Jun.-Aug.
DISTRIB. Apparently rather rare in the forest zone of the mountain region of Iraq. **MAM**: Sarsang, *Haines* 1204!, *Dabbagh* 46179! **MRO**: Handren, *Bornmüller* 1514; **MSU**: Molla Khort (Malakawa) Mt., *Rawi, Husham & Noori* 29373!; Avroman Mt., *Rawi, Chakravarty, Noori & Alizzi* 197801 19783!

E Turkey, NW Iran, Transcaucasia.

15. **Scorzonera lanata** *Hoffm.* in Comm. Soc. Phys. Mosc. 1: 9 (1808); Fl. Taur.-Cauc. 2: 237 (1808) & 3: 524 (1819); Boissier, Fl. Orient. 3: 776 (1875); Dinsmore in Post, Fl. Syria, Palest. & Sinai ed. 2, 2: 139 (1933); Guest in Dep. Agr. Iraq Bull. 27: 90 (1933); Rawi in Dep. Agr. Tech. Bull. 14: 124 (1964); Taifour & El-Oqlah, Pl. Jordan Ann. Checkl.: 47 (2017).

Leontodon lanatus L., Cent. Pl. 1: 26 (1755).
Tragopogon lanatum (L.) L., Sp. Pl. ed. 2: 1111 (1763).
Scorzonera judaica Eig in Eig, Zohary & Feinbrun, Fl. Pal.: 399 (1931), in clav.; Feinbrun-Dothan, Fl. Palaest. 3: 427 (1978).
S. pseudolanata Grossh. in Trudy Bot. Inst. Baku 2: 258 (1936): Lipshits in FI. S.S.S.R. 29: 105 (1964); Täckholm, Stud. FI. Egypt ed. 2: 599 (1974); Chamberlain in Fl. Turkey 5: 652 (1975); Rechinger, Fl. Iranica 122: 72 (1977); Breckle, Hedge, Podlech & Rafiqpoor, Vasc. Pl. Afghan. checklist: 146 (2013).
S. lanata var. *latifolia* Bornm. in Beih. Bot. Centralbl. 60B: 220 (1939).
S. mesopotamica Eig sensu Zohary in Dep. Agr. Iraq Bull. 31: 159 (1950); Rawi in Dep. Agr. Iraq Tech. Bull. 14: 124 (1964), *nomen nudum.*

Scapose perennial herb, rootstock a globose tuber 1.5–2.5 cm in diameter, with a slender 1–4 cm neck between tuber and basal leaves, ± densely furnished throughout its vegetative parts with long, white, sericeous-lanate hairs, only dark-striate leaf-sheaths glabrous. Leaves all basal, linear to narrowly lanceolate, 3.5–20 × 0.1–1.5 cm, 3–5-veined, attenuate above to a long, narrow, frequently circinnate tip, margins ± strongly undulate or crispate. Scapes 1-several, 2.5–16 cm, often deflexed in fruit with capitulum inclined. Capitula solitary on scapes; involucre broadly obconical; outer phyllaries few, narrowly lanceolate, 2–6 mm, often ± concealed in lanate indumentum; inner phyllaries 9–15 mm in flower, 15–22 mm in fruit. Florets 8–12 mm; limb 7–11 mm, yellow, marginal reddish-striped dorsally. Achenes 5–6 mm, completely concealed by long, dense hairs; pappus 9–12 mm, hairs scabrid only about apex, densely plumose and merging with achene-hairs below, brownish.

HAB. *Poa*-steppe, rocky hillsides on limestone, stony clay slopes, grassy places in oak scrub, extending to the upper limits of the *Quercus* zone; alt. 260 (or lower)–1500 m.; fl. Mar.-Apr., fr. Apr.-May.
DISTRIB. Widespread in the lower parts of the mountain region of Iraq and the steppe region, only one record (perhaps from seeds brought down by river action) in the plain. **MAM**: Sawara Tuka, *Guest* 2212!; Sarsang, *Botany Staff* 43453! **MRO**: Shaqlawa, *Guest* 559!, *Haines* 694!: Pirmum Pass, *Guest* 2049!; Algurd Dagh, *Thesiger* 991; Salah ad-Din, *Rawi & Gillett* 10426! **MSU**: Pir Omar Gudrun, *Gillett* 7809!; 16 km W of Chemchemal, *Rawi & Gillett* 10600!; Jarmo, *Helbaek* 367!; Azmir, *Qaisi & Hamad* 49317! **FUJ/MJS**: S Sinjar mountain, *Rawi* 41369! **FKI**: 31 km N of Kirkuk on way to Koi Sanjaq, *Rawi, Noori & Kass* 27958! **FKI/MSU**: 14 km E of Kirkuk, *Rawi & Gillett* 10654! **DLJ**: 40 km NW of Shbaichan, *Chakravarty, Rawi, Khatib & Alizzi* 32004! **DLJ**: Shbaichan, *id.* 32011! **LCA**: nr Khalis, *Rogers* 0278!

Turkey, Syria, Jordan, Palestine, Transcaucasia, Iran, Afghanistan.

57. **TRAGOPOGON** L.

Sp. Pl.: 789 (1753); Gen. Pl. ed. 5: 346 (1754)
Lack in Kubitzki, Fam. Gen. Vasc. Pl. 8: 199 [2006] (2007)

C. C. Townsend

Erect annual, biennial or perennial herbs. Leaves simple, entire, linear or lanceolate to elliptic-oblong, parallel-veined, sessile, usually ± expanded and subamplexicaul at tip. Capitula homogamous, ligulate, solitary on ± elongate pedicels which may be ± incrassate below capitula; phyllaries biseriate, all ± equal, lengthening in fruit; receptacle naked. Ligules purplish to yellowish or more rarely white, with a ring of hairs where ray joins tube, outer corollas, if yellow, frequently with dark lines dorsally or ± purplish-suffused. Achenes narrowly fusiform, glabrous, (5–)10-ribbed, outer achenes with ribs ± scaly or muricate and 5 sometimes expanded into keels or wings, inner achenes less muricate or smooth, beak long and slender to very short, when long commonly ± expanded above, with an annulus of lanate hairs below pappus; pappus homomorphic, setae plumose, pinnnulae intertwined, dingy.

About 110 species in Europe, temperate Asia and N Africa, some widely introduced elsewhere in the World; 8 species are known from Iraq,

Tragopogon (from the Gr. τσαγος, tragos, a goat, πογον, pogon, a beard – no doubt referring to the conspicuous pappus). Goat's Beard (Engl.), DHANAB AL-FARAS ("Mare's Tail", Ir., Guest, 1933).

A taxonomically difficult genus. To be of real use, specimens must include the rootstock, notes on flower colour and mature fruit. Species are often in flower and fruit together, and to obtain both is frequently easy.

1. Corollas purple or lilac 2
   Corollas yellow (outer sometimes purplish-lined or suffused dorsally) 3
2. Mature achenes with alternate ribs expanded into 5 conspicuous obliquely veined wings, veins terminating in scale-like teeth; outer florets more than 2 cm long .... 8. *T. pterocarpus*
   Mature achenes muriculate with ribs all ± equal; outer florets less than 2 cm long .... 1. *T. coelesyriacus*
3. Fruiting peduncle distinctly incrassate at apex 4
   Fruiting peduncle not or scarcely incrassate at apex 5
4. Stem leaves only shortly expanded at base, semi-amplexicaul not sheathing stem; achene 22–37 mm (including 12–20 mm beak) .... 2. *T. dubius*
   Stem leaves ± abruptly expanded to a long widened base which fits close to stem and ± sheathing it; achene 15–20 mm (including 8–10 mm beak) .... 3. *T. vaginatus*
5. Corollas white; mature achenes broadly 5-winged, wings abruptly truncate or even emarginate into base of beak .... 7. *T. bornmuelleri*
   Corollas yellow, or if white then mature achenes unwinged or wings narrow and gradually tapering into beak 6
6. Corollas 11–14 mm, ligules 8–10 mm. .... 6. *T. rechingeri*
   Corollas 16–35 mm, ligules 14–26 mm. 7
7. Neck at base of stem below basal leaves almost always with ± reticulate fibres of previous years' leaves; mature achene ribbed but not with 5 keels or narrow wings. .... 4. *T. reticulatus*
   Neck at base of stem below basal leaves with scaly remains of previous years' leaves; mature achene with 5 toothed keels or narrow wings. .... 5. *T. buphthalmoides*

1. **Tragopogon coelesyriacus** *Boiss.*, Diagn. Pl. Orient. ser. 1, 11: 47 (1849).

T. longirostris Bisch. ex Sch.Bip. in Webb & Berth., Phyt. Canar. 2: 469 (1850). Boissier, Fl. Orient. 3: 745 (1875); Nábělek in Pub. Fac. Sci. Univ. Masaryk 52: 50 (1925); Guest in Dep. Agr. Iraq Bull. 27: 100 (1933); Dinsmore in Post, Fl. Syria, ed. 2, 2: 134 (1933); Anthony in Notes Roy. Bot. Gard. Edinbburgh 18: 293 (1935); Bornmüller in Beih. Bot. Centralbl. 60B: 217 (1939); Zohary in Dep. Agr. Iraq Bull. 13: 158 (1950); Rawi in Dep. Agr. Iraq Tech. Bull. 14: 126 (1964); Täckholm, Stud. Fl. Egypt, ed. 2: 599 (1974); Matthews in Fl. Turkey 5: 659 (1975); Fl. Iranica 122: 94 (1977); Feinbrun-Dothan, Fl. Palaest. 3: 422 (1978); Breckle, Hedge & Rafiqpoor, Vasc. Pl. Afghan. Checkl.: 195 (2013); Taifour & El-Oqlah, Pl. Jordan Ann. Checkl.: 48 (2017).
[*T. pterocarpus* (non DC.) Rech. f. in Fl. Iranica 122: 110 (1977), quoad spec. iraq.].

Biennial herb, (10–)25–80 cm, glabrous or sparingly arachnoid when young, glabrescent, rootstock rather slender and tapering, base of stem sparsely clothed with scaly remains of previous years' leaves. Stem terete, striate-subsulcate, simple or with a few long, slender branches, rather sparingly leafy. Basal leaves few, withered at flowering; stem leaves similar, grass-like, ± gradually expanded below into a subamplexicaul base, up to 18 × 13 cm, finely tapering above. Peduncle strongly (occasionally less so) incrassate beneath fruiting head, 3.5–17 cm. Phyllaries 6–8, gradually attenuate, 4–9 cm in fruit, usually subequalling to exceeding pappus. Corollas 12–15 mm, ray 6–9 mm, pink to dark purple. Outer achenes including beak 35–50 mm, beak 18–34 mm, slender, abruptly expanded at tip, achene-body terete or somewhat angled, muriculate with rows of acute or subspatulate white scales; inner achenes becoming less roughened and smooth; pappus stramineous, 21–35 mm.

HAB.: On rocky slopes of hills, in open *Quercus* forests and a *Juglans* grove, in a ravine, in fields and on banks, always on limestone and frequently near streams; alt. 750–1800+ m; fl. & fr. Apr.-Jun.
DISTRIB.: Widespread in the mountainous regions of Iraq, apparently rare in the foothills. **MAM**: nr Zakho, *Rechinger* 10724!, 1208!, 12151!; Bakirma, *Rawi* 8792! **MRO**: Saran village, *Noori & Kass* 27323!; nr Chinaruk, *Rawi, Noori & Kass* 28354!, *Omar et al.* 37237!; Gali Warta, *Rawi, Noori & Kass* 28802!; Gali Ali Beg, *Kass, Noori & Serhang* 27207!; mts. S of Rayat, *Thesiger* 1058 p.p. **MSU**: Qara Dagh, *Gillett* 7906!; Jarmo, *Helbaek* 1004!; Avroman, Sosakon (Shushami?) nr Tawila, *Rechinger* 10160!, 10267!; Tawila, *Rawi* 21933!; nr Surdash, *Rechinger* 12456!; Cham-i Gogosur, 18 km W of Penjwin, *id.* 10405!; Bakrajo, *Omar & Karim* 38002! **MJS**: Jabal Sinjar, *Qaisi & Hamad* 49168!; Kursi, *Gillett* 11000! **FNI**: around Sheikh Adi, 9 km NW of Ain Sifni, *Thesiger* s.n.!

Cyprus, Sinai, Palestine, Lebanon, Syria, Jordan, Turkey, Iran, Transcaucasia, Afghanistan, C Asia.

It is unfortunate that the type of *T. coelesyriacus* Boiss. is so stunted and atypical, but there seems no doubt that it is the same in essential characters as *T. longirostris,* as opined by Feinbrun and Fertig on the attached label.

SHIN (Kurd. – Qara Dagh, *Gillett* 7906), SPING (Kurd. – Kursi, *Gillett* 11000). According to Guest (1933), this plant has been reported from Shaikhan as harmful to sheep, even causing death in its green state; it is said to be less dangerous later in the season when it has begun to dry.

2. **Tragopogon dubius** *Scop.*, Fl. Carn. ed. 2, 2: 95 (1772); Borisova in Fl. S.S.S.R. 29: 131 (1964); Matthews in Fl. Turkey 5: 662 (1975); Chaudhary, Fl. Kingd. Saudia Arabia 2(3): 228 (2001); Zhu Shi, Sukhorukov & Evgeny V. Mavrodiev in Fl. China 20–21: 209 (2011); Ghafoor & Sultan Abedein in Fl. Pakistan 222: 287 (2017).

T. majus Jacq., Fl. Austr. Ic. 1: 19, t. 29 (1773); Boissier, Fl. Orient. 3: 747 (1875); Rawi in Dep. Agr. Iraq Tech. Bull. 14: 126 (1964), sphalm. "major".

Biennial (sometimes annual?) herbs, 30–85 cm, glabrous or sparingly floccose when young, glabrescent, rootstock stout, widely fistular, terete, base of stem with few or no remains of previous years' leaves. Stem stout, widely fistular, terete, striate, simple or with a rather few long, ascending branches from near base or only above, moderately to rather densely leafy throughout. Basal leaves few, generally only partly or not withered at flowering, narrow and long, grass-like, stem leaves similar, up to 40 × 7 cm, ± gradually expanded below into a subamplexicaul but not ± appressed-vaginate base. Peduncle in well-grown plants very strongly incrassate beneath fruiting head, 3–18 cm. Phyllaries 8–10 (in Iraq, more recorded elsewhere), gradually attenuate (sometimes more abruptly so above base), 3.5–8 cm in fruit, shorter than to subequalling pappus. Corollas 21–34 mm, ligules 14–25 mm, yellow with dark dorsal stripes. Outer achenes including beak 22–37 mm, 12–20 mm beak gradually though shortly clavate-expanded at tip, achene-body terete or commonly angled, muriculate with rows of acute or subspatulate white scales, inner achenes becoming less muriculate and smooth; pappus 20–32 mm, dirty white to stramineous.

HAB. A weed of *ahrash* (forestry) plantation, palm groves, fields and river banks; alt. 40–250 m; fl. & fr. Apr.-Jun.
DISTRIB. Local in the Lower Mesopotamian Plain, occasional in the foothills of Iraq. **FUJ**: Nenawa (Nineveh) Farm, *Ani* 9773! **FNI**: between Nineveh and Mosul, *E. Chapman* 25376! **LEA**: Ba'quba, *Paranjpye* s.n.! **LCA**: Abu Ghraib, *Rawi* 10752!, *Khanim* 33186!; 20 km NNE of Baghdad, *E. Chapman* 24929!

C & S Europe from Portugal and N France to C Russia and Crimea, Saudi Arabia, Turkey, Afghanistan, Pakistan, Kashmir, Tibet, C Asia (Turkmenia to W Siberia). Introduced in S Africa and N America.

A few other specimens from Iraq may belong here, but in the absence of either notes on flower colour or good mature fruit are impossible of certain determination.

This species may well be adventive in Iraq.

3. **Tragopogon vaginatus** *Ownbey & Rech.f.* in Fl. Iranica 122: 89, t. 52, 204 f. 4 (1977). – Type. Kurdistan, Sulimaniyah distr., inter Kirkuk et Sulaimaniya. 34 km a Sulaymaniyah occidentem versus, ad versura, 13 June 1957, *Rechinger* 10571 (B, E, K!, W).

Biennial herbs, (15–)30–70 cm, glabrous, rootstock stout and tapering, base of stem with some scaly remains of previous year's leaves. Stem terete, striate-subsulcate, simple or a few long ascending branches arising below, ± densely foliose below, leaves reducing and fewer above. Basal leaves rather few, mostly withered at flowering time, narrow, ± abruptly broadly expanded at base, stem leaves similar, grass-like, up to 20 × 3(–4) mm, expanded below into a long (in median leaves) 3–4 cm, ± 5-veined base which is appressed and sheaths the stem, rapidly narrowed to finely tapering blade. Peduncle very long, 11–30 cm, incrassate below fruiting head. Phyllaries 8–10, gradually attenuate above, 4.5–6 cm in fruit, subequalling pappus. Corollas 22–30 mm, ligules 15–22 mm, yellow (fading whitish on drying), dorsally

Fig. 61. **Tragopogon vaginatus**. 1, fruiting capitulum × 1½; 2, achene and pappus × 3. **Tragopogon buphthalmoides** var. **buphthalmoides**. 3, lower portion of stem × 1; 4, flowering stem × 1. **Tragopogon bornmuelleri**. 5, achene and pappus × 3. **Tragopogon pterocarpus**. 6, achene and pappus × 3. 1–2 from *Haines* W.241; 3–4 from *Gillett* 11018; 6 from *Samuelson* 5622 (Turkish material). Drawn by Christine M. Grey-Wilson.

purple-striped. Outer achenes including beak 15–20 mm, beak 8–11 mm slender, only slightly and gradually incrassate at tip; achene-body terete or only slightly angled, muriculate with rows of dense white scales; inner achenes becoming less muriculate, sometimes almost smooth; pappus 18–25 mm, dirty white to stramineous. Fig. 61: 1–2.

HAB. Streamside between high wheat-fields, in marshy ground, on roadside and on hill slopes; alt. 700–900 m; fl. & fr. Apr.-Jun.
DISTRIB. Apparently scattered in the eastern mountain region of Iraq. **MRO**: Khalakan village, *Omar, Karim, Hamza & Hamid* 37327!; 2 km from Haji Umran towards the Iranian frontier, *Alkas* 18593!; Sefin Kuh, nr Shaqlawa, *Bornmüller* 1505. **MSU**: Jarmo, *Haines* 241!; 19–34 km W of Sulaimaniya, *Rechinger* 10571 (type)!; 10 km E of Darband-i Basian, *Rawi & Gillett* 7660!

Turkey, W Iran.

One specimen from Iraq which Rechinger cited under this species in Fl. Iranica is *Haddad & Barkley* 8552 from Qopi Qaradagh. This is in flower only and the leaf-bases are shortly expanded and non-sheathing; I believe it to be a tall form of *T. buphthalmoides*.

The above description has been drawn up using only wild material. Specimens cultivated from *Rechinger* 10571 by Ownbey and preserved in herb. W have the peduncles more widely incrassate at the apex and slightly longer achenes.

4. **Tragopogon reticulatus** *Boiss. & Huet* in Boissier, Diagn. Pl. Orient., ser. 2, 3: 90 (1856); Boissier, Fl. Orient. 3: 752 (1875); Borisova in Fl. S.S.S.R. 29: 176 (1964); Matthews in Fl. Turkey 5: 665 (1975); Rechinger, Fl. Iranica 122: 106 (1977).

T. kurdicus Blakelock in Kew Bull. 4: 64 (1949); Rawi in Dep. Agr. Iraq Tech. Bull. 14: 126 (1964).
[*T. pusillum* auct. non M.Bieb.: Rawi in Dep. Agr. Iraq Tech. Bull 14: 126 (1964)].

Perennial herbs, 9–60 cm, glabrous or frequently ± arachnoid and glabrescent, rootstock slender to stout, base of stem sparingly or commonly considerably fibrous-reticulate with remains of previous years' leaf-sheaths. Stem erect, slender, commonly simple and bearing a single capitulum even when tall, sometimes with numerous long, slender, erect branches from near base upwards, leafy chiefly in lower part, upper leaves rapidly reducing. Basal leaves narrow, long, few to numerous, mostly withered at flowering; stem leaves similar or wider, grass-like, ± gradually expanded below into a subamplexicaul base, up to 21 × 7 mm (usually narrower), finely tapering above. Peduncle slender, not at all incrassate below fruiting head, 1.5–12(–16) cm according to size of plant. Phyllaries mostly 6–9, acute to attenuate, 2–3.3 cm in fruit, slightly to considerably shorter than pappus. Corollas 16–35 mm, ligule 14–26 mm, yellow, purplish-streaked dorsally, sometimes turning bluish when dry. Outer achenes including beak 10–20 mm, beak ill-defined, almost obsolete to *c.* 7 mm long, achene-body ± terete or angular, ± striate, strongly muriculate to almost smooth; inner achenes usually smooth or almost so; pappus 12–23 mm, pale or dark stramineous.

HAB. On rocky mountainsides and ridges, grassy slopes and summits, chiefly in the *Astragalus* zone on limestone, clay, slate, metamorphic rock (incl. serpentinite); alt. 1600–3100+ m; fl. Jun.-Sep., fr. Jul.-Sep.
DISTRIB. Common chiefly in the high alpine zone of the eastern mountain region of Iraq. **MRO**: Chiya-i Mandau, *Guest* 2704!, 2719!; E. of Qala Diza, Qandil Mt., *Thesiger* s.n.!; Qandil range, NE of Rania, *Rawi* 26817!; NE of Qandil, *Rawi & Serhang* 24401!, 24436!; Algurd Dagh, *Guest & Ludlow-Hewitt* 2858!, *Rawi & Serhang* 24925!, *Rechinger* 11871!; Hisar-i Rost, *Guest & Husham* (*Alizzi*) 15818! **MSU**: bare hills near Penjwin, *Rechinger* 12276!; Avroman mts. near Tawila, *Rechinger* 10357!

Turkey, NW Iran, Caucasia.

A very variable species in the size of its parts, and without fruit difficult to separate from small high-altitude forms of *T. buphthalmoides* unless rootstock and the base of the stem are present. Few of the other specimens are in fruit.

5. **Tragopogon buphthalmoides** (*DC.*) *Boiss.*, Fl. Orient. 3: 750 (1875); Dinsmore in Post, Fl. Syria, Palest. & Sinai ed. 2, 2: 135 (1933); Zohary in Dep. Agr. Iraq Bull. 31: 158 (1950); Rawi in Dep. Agr. Tech. Bull. 14: 126 (1964); Borisova in Fl. S.S.S.R. 29: 156 (1964); Matthews in Fl. Turkey 5: 664 (1975); Rechinger, Fl. Iranica 122: 99 (1977); Feinbrun-Dothan, Fl. Palaest. 3: 423 (1978); Taifour & El-Oqlah, Pl. Jordan Ann. Checkl.: 48 (2017).

Perennial herbs, (5–)10–70(–120) cm, simple or branched with short to long, slender, ascending branches, rootstock slender, tapering, base of stem with scaly (not fibrous) remains of previous years' leaves. Stem terete or somewhat angled, slender to stout, commonly arachnoid-floccose at first, ± glabrescent. Basal leaves withered at flowering time, stem leaves very variable in size and shape, grass-like to elliptic-oblong, acuminate to obtuse-subapiculate, 3–30 cm × 0.15–2.2 cm, very shortly widened and base not or scarcely subamplexicaul. Peduncles rather slender, 3.5–7(–11) cm, only very slightly incrassate below involucre. Phyllaries mostly ± 8, rarely more, attenuate above, 18–42 mm in fruit, glabrescent or ± persistently floccose, usually distinctly shorter than pappus. Corollas 22–30 mm, ligule 16–20 mm, yellow, marginal with dark purplish stripes or reddish or bluish-suffused, occasionally marginal whitish. Outer achenes including beak 15–35 mm, beak 4–20 mm slender to rather stout, usually shortly and abruptly expanded at apex, achene-body narrowly 5-winged (often little more than carinate) with denticulate wings, muriculate between wings or almost smooth, inner achenes becoming less denticulate along angles, and finally ± smooth throughout; pappus 20–30 mm, stramineous, sometimes purplish-tinged.

DISTRIB. (of species): Syria, Lebanon, Palestine, Turkey, Iran, Transcaucasia.

var. **buphthalmoides**

T. persicus Boiss., Diagn. Pl. Orient. ser. 1, 7: 4 (1846).
T. buphthalmoides var. *stenophyllus* Boiss., Fl. Orient. 3: 750 (1875).
T. buphthalmoides var. *persicus* (Boiss.) Dinsm., Fl. Syria, Palest. & Sinai ed. 2, 2: 135 (1933).

Leaves narrow, not more than 1 cm wide. Fig. 61: 3–4.

HAB. (of var.): Rocky mountain slopes (including in fragments of *Quercus* forest and among *Crataegus*), clay hillside, valleys and streamside; alt. 320–2000 m; fl. & fr. Jun.-Sep.
DISTRIB. (of var.): Scattered, probably frequent, in the mountain region of Iraq. **MRO**: Rayat mt., S of Rayat, *Thesiger* 1058C!; Shaikh wa Shaikhan, *Rawi & Serhang* 24884!; Sie-waka, nr S foot of Karoukh, *Alkas, Noori & Serhang* 27579!; Karoukh, *Kass & Noori* 27359!; Hawara Blinda, *Kass & Noori* 27841!; Algurd Dagh, *Gillett* 9526! **MSU**: Pir Omar Gudrun, *Rawi* 12119!, *Haussknecht* s.n.; Qopi Qaradagh, *Haines* 1310!; Azmir, *Omar & Karim* 38063! **MJS**: Kursi, *Gillett* 11018!, 11019!; Jabal Sinjar, *Thesiger* s.n.!; Top of Sinjar, *S. Omar, Al-Khayat & Al- Kaisi* 52523!

Throughout the range of the species.

var. **latifolius** *Boiss.*, Fl. Orient. 3: 751 (1875); Dinsmore in Post, Fl. Syria, Palest. & Sinai ed. 2, 2: 135 (1933); Bornmüller in Beih. Bot. Centralbl. 60B: 218 (1939); Matthews in Fl. Turkey 5: 664 (1975); Rechinger, Fl. Iranica 122: 100 (1977).

T. plantagineum Boiss. & Huet in Boissier, Diagn. Pl. Orient. ser. 2, 3: 91 (1856); Borisova in Fl. S.S.S.R. 29: 155 (1964).
[*T. albinervis* (non Freyn & Sint.) Bornm. in Beih. Bot. Centralbl. 60B: 218 (1939)]

Leaves broad, more than 1 cm wide.

HAB. (of var.): Rocky mountain slopes in degraded *Quercus* forest; alt. 1360–1820+ m; fl. May-Jun., fr. Jun.-Jul.
DISTRIB. Scattered in the mountain region of Iraq. **MAM**: Khantur, *Rawi* 23496! **MRO**: Rayat mt., S of Rayat, *Thesiger* 1058A!; **MSU**: Pir Omar Gudrun, *Rawi* 12119!; Qopi Qaradagh, *Haines* 1309!

Probably throughout the range of the species. The above are weak varieties, and do little more than emphasise the great variety of leaf breadth in this species.

One specimen cited by Rechinger in *Fl. Iranica*, *Thesiger* 1149 from Qala Diza, would push the altitude range above that given; but I agree with Ownbey's annotation on this specimen that it is too dwarfed for determination, especially as Thesiger collected *T. reticulatus* in the same place on the same day.

6. **Tragopogon rechingeri** *Ownbey ex Rech.f.*, Fl. Iranica 122: 102, t. 63 (1977). – Type. Haji Umran, *Rechinger* 11844 (W).

Very similar in general facies to shorter forms of *T. bornmuelleri*, with similar floret and ray ligule length, differing in its fruit. Outer achenes including beak 15–18 mm long, beak rather stout or more slender, gradually slightly thickened above, 3–5 mm; achene-body 5-angled, extremely narrowly winged (scarcely more than carinate), angles minutely toothed, surface muriculate; inner achenes rapidly less muriculate and finally smooth; pappus 18 mm, dirty white to stramineous.

HAB. On stony mountain slope; alt. ± 1700 m; fl. & fr. Jul.-Aug.
DISTRIB. **MRO**: Haji Umran, *Rechinger* 11844 (type)!

W Iran.

A perplexing and unsatisfactory species. According to Rechinger the corollas are yellow, whilst those of *T. bornmuelleri* are white. This information is presumably gleaned from Ownbey's cultivated material and a Wendelbo gathering from Iran, since no specimen I have seen from Iraq bears colour notes and in the dried state both species look identical in this respect. For non-fruiting material Rechinger relied heavily on the abbreviated lateral branches of *T. rechingeri*, producing many broad-based, abruptly narrowed leaves. However, two of the three stems of the type have the tips chopped or grazed off, as have certain specimens of similar appearance from herb. K and herb. BAG (e.g. *Rawi* 24269 from Haji Umran) as well as *Rechinger* 12003 (from mt. Zawita nr Sharanish N. of Zakho, 1800–2000 m). Abbreviated secondary growth branchlets of this kind are frequent in other *Tragopogon* species such as *T. pratensis* after lopping and in fact many other plants, and Rechinger's statement that in cultivation Ownbey found that *T. rechingeri* produced numerous long branches indicates that this character is probably worthless. Thus the form of fruit, seen apparently only on the type and one W. Iranian specimen, *Pabot* 1562 (which has long slender branches and is thus of very different habit from the type) seems to be the sole significant character; floret colour may vary from whitish, at least in the marginal florets, to yellow in the closely related *T. buphthalmoides*. The fruit of the type of *T. rechingeri* is certainly different from the type of *T. bornmuelleri* in its narrow wings tapering into the beak, and in fact very similar to that of *T. buphthalmoides*. There are only two fruiting capitula on the type of *T. rechingeri*, and one of these bears a number of abortive achenes. It is thus tempting to postulate that *T. rechingeri* is a hybrid between *T. buphthalmoides* and *T. bornmuelleri*, but this reduced fertility may also be due to the lopping of this particular stem. The whole matter needs accurate field study, and new collections of Iraqi *Tragopogon* with flower colour notes (very few collections give this) and both flower and fruit are needed.

Of the Iraqi specimens cited under this species by Rechinger, I have cited only the type above. *Rechinger* 12003 from Zawita is in flower only and I cannot allocat it to *T. rechingeri* or *T. bornmuelleri* on this basis, while *Rechinger* 11760 from E slopes of mt. Qandil above Pushtashan has young fruit which nevertheless appear to me to show oblique-ribbed wings developing and I would refer it to *T. bornmuelleri*. *T. bornmuelleri* also occurs at the latter locality. The description is also prepared from the type, whose achenes were found to be smaller than in Rechinger's description.

7. **Tragopogon bornmuelleri** *Ownbey & Rech.f.*, Fl. Iranica 122: 102, t. 64 (1977). –Type. Kurdistania, Riwandous ad fin. Persiae, in monte Sakri-Sakran, 2100 m, 23 June 1893, *Bornmüller* 1507 (B, holo.; JE, K!, iso.).

Perennial herbs, 28–70 cm, glabrous or thinly arachnoid when young and ± glabrescent rootstock slender to rather stout, base of stem with ± fibrous and scaly remains of previous years' leaves. Stem terete, striate, slender, simple or usually with a few long, ascending branches, occasionally with abbreviated branches. Basal leaves withered at flowering time, narrow, stem leaves similar, grass-like, up to 22 cm × 6 mm, only uppermost ± distinctly expanded at base. Peduncle slender and up to 10 cm long, or on some short lateral branches sometimes ± obsolete, only very slightly incrassate above. Phyllaries 7–10, gradually attenuate above, 2–3.8 mm in fruit, glabrous or ± persistently floccose, shorter than to subequalling pappus. Corollas 11–16 mm, ligule 8–12 mm, white (sometimes creamy). Outer achenes including beak 20–35 mm, beak 0–17 mm, slender, abruptly expanded at apex; achene-body broadly 5-winged, wings prominently obliquely ribbed with ribs ending in short teeth, abruptly narrowed to base of beak or even deeply emarginate, muriculate with a longitudinal row of scales between wings; inner achenes becoming less muriculate and denticulate-margined to ± smooth; pappus dirty 17–25(–30) mm, white to stramineous. Fig. 61: 5.

HAB. On mountain slopes, on and about summits and in a ravine, on limestone and rocky basic clay; alt. 1480–2100+ m; fl. & fr. May-Aug.

DISTRIB. Scattered in the mountain region of Iraq. **MAM**: Jabal Khantur, nr Sharanish, *Rechinger* 12028a!; Gali Zawita, *Rawi* 23628! **MSU**: Avroman mt., *Rawi, Husham & Noori* 29353! **MRO**: Sakri-Sakran, *Bornmüller* 1507 (type)!; Zernakau mt. N of Rania, *Thesiger* s.n.!; Algurd (Helgord) Dagh, valley above Nawanda, *Rechinger* 11353b!; Qandil mts. above Pushtashan, *Rechinger* 11190!; Pushtashan, *Rawi & Serhang* 24266!

Turkey, W Iran.

Two specimens cited by Rechinger in *Fl. Iranica* 122: 103 (1977) as belonging to *T. buphthalmoides* var. *latifolius* Boiss. have the leaf form of that plant but the broadly winged outer achenes, short florets and short pappus (18–19 mm) of the present species. Either broad- or narrow-leaved forms of *T. bornmuelleri* occur as they do in *T. buphthalmoides*, or these two specimens are of hybrid origin. Ownbey (in Rechinger, l.c.) suggests that *T. buphthalmoides* is an amphidiploid species based on *T. rechingeri* and *T. bornmuelleri*. The specimens in question are *Rechinger* 11126 from above the Gom-i Kirmosaran lake on the Qandil range and *Rechinger* 11353a from the Nawanda valley on Algurd Dagh. Since typical *T. bornmulleri* occurs at the latter locality (see above) I suspect this is merely a question of variation in leaf shape, the leaves of *Rechinger* 11126 attaining 2.3 cm in width.

8. **Tragopogon pterocarpus** *DC.*, Prodr. 7: 112 (1838); Boissier, Fl. Orient. 3: 751 (1875); Dinsmore in Post, Fl. Syria, Palest. & Sinai ed. 2, 2: 135 (1933); Matthews in Fl. Turkey 5: 660 (1975); Rechinger, Fl. Iranica 122: 110 (1977).

T. straussii Bornm., Beih. Bot. Centralbl. 20, 2: 172 (1906).

Biennial herbs, (10–)15–30 cm, glabrous or glabrescent, rootstock stout to slender, base of stem ± clothed with scaly remains of first year's leaves; stem terete, striate, simple or with a few ascending branches, stout, densely leafy, pale or purplish. Basal leaves numerous, narrowly linear and grass-like, withered by flowering time, stem leaves similar, with a broad, expanded, semi-amplexicaul sheath, up to 25 cm × 2 mm, shorter than to overlapping fruiting capitula. Peduncle usually ± strongly incrassate upwards in fruit, 2–6 cm. Phyllaries 6–8, gradually attenuate, 3.5–8(–10) cm long in fruit, subequalling to exceeding pappus. Corollas *c.* 22–25 mm, ligules 16–18 mm, lilac to purple. Outer achenes including beak 20–40 mm, beak *c.* 7–23 mm, scarcely or shortly and abruptly widened at apex, achene-body distinctly 5-winged, wings obliquely ribbed with ribs terminating in pale teeth, also muriculate between wings; inner achenes angled and white-muriculate but not winged; pappus 22–40 mm, stramineous to purplish. Fig. 61: 6.

HAB. Only found once on rocky clay in the mountain region of Iraq, where however it was said to be common.
DISTRIB. **MJS**: Jabal Sinjar, *Qaisi* 49744!

sub

Turkey, Syria, W & C Iran.

The specimen cited for Iraq in *Fl. Iranica* as belonging to this species, *Thesiger* 1058 p.p. from S of Rayat, 1800–2400 m, has only very immature achenes in which I see no trace of wings. It is moreover a tall, slender plant and is I believe probably *T. longirostris*. However, the above specimen has mature achenes identical with those of *T. pterocarpus* and the squat habit of the latter with disproportionately large capitula. There are unfortunately, as so often, no notes on flower colour and both pedicels and fruit beaks are slender for *T. pterocarpus*. Such variation occurs, however, in other species of the genus.

TRIBE **GUNDELIEAE** DC. EX LECOQ & JUILLET

58. **GUNDELIA** L.

Sp. Pl. 814 (1753); Gen. Pl. ed. 5: 356 (1754)
Jeffrey in Kubitzki's Fam. Gen. Vasc. Pl. 8: 199 [2006] (2007)

Patricia Halliday
Revised by D. J. Nicholas Hind

Erect, perennial herbs with a stout woody rootstock and yellowish or milky latex. Leaves alternate, pinnatifid, toothed and spiny, decreasing in size upwards, uppermost involucrate. Flowering structure a spherical or subspherical or ovoid tertiary-level involucrate synflorescence (or syncephalium), composed of secondary level capitula of one central fertile single-flowered capitulum subtended by 5–7, single-flowered, primary capitula forming a discrete disseminule; primary capitula homogamous and discoid and each subtended by a single spiny bract and 2 subsidiary bracts. Corollas white, yellow, pink, purple, maroon or brown outside, sometimes yellow within. Subsidiary single-flowered capitula functionally male; anthers linear, brown or yellow, appendage ovate-lanceolate, base shortly sagittate, tails appressed, obtuse; style bifid, thickened and papillose above, arms reflexed, somewhat compressed, slender, linear or somewhat widened apically, acute. Fruit complex or (compound) disseminule a turbinate cupule, somewhat compressed and ± semicircular in cross-section, tapering to base, smooth, glabrous; pappus setae uniseriate, persistent, bowl-shaped, margins usually minutely toothed, ± thickened at centre.

A genus of perhaps 3 or 4 species; 2 species are recogized in Iraq. The recognition of 2 more new species from Turkey may justifiably add to the total allthough the authors limited their commentary and key to a small region of Turkey (Vitek et al., 2014).

There is wide variation in accounts as to the exact number of species in the genus. Some discussion is provided following the account of *G. tournefortii* below. In recognizing the more distinctive colour extremes two species are considered to be present in Iraq, although some workers may prefer to see these as subspecies, or simply one variable species. Determining colour from herbarium material is very difficult, and may prove impossible based on label data, hence accurate species determination may well prove nigh on impossible in many instances when dealing with limited herbarium material.

Gundelia (named in honour of the German botanist Andreas von Gundelsheimer, 1668–1715, a contemporary of Tournefort who travelled with him on his expedition through Turkey to Armenia).

Al-Taey R. A. & Hossain, M. (1984). Studies in *Gundelia* 1. a new species from Iraq. Notes Roy. Bot. Gard. Edinburgh 41(1): 39–44.
Claßen-Bockhoff, R., Froebe, H. A. & Langerbeins, D. (1989). Die Infloreszenzstrucktur von *Gundelia tournefortii* L. (Asteraceae). Flora 182(5/6): 463–479.
Claßen-Bockhoff, R. (1996). Functional units beyond the level of the capitulum and cypsela in Compositae. In: P. D. S. Caligari & D. J. N. Hind (eds), Compositae: Biology & Utilization. Proc. Internat. Compositae Conf., Kew, 1994; 2: 129–160.
Karis, P. O., Eldenäs, P. & Källersjö, M. (2001). New evidence for the systematic position of *Gundelia* L. with notes on the delimitation of Arctoteae (Asteraceae). Taxon 50(1): 105–114.
Kilian, N., Gemeinholzer, B. & Lack, H.W. (2009). Cichorieae. In: V.A. Funk et al., Syst. Evol. Biogeog. Compositae: 343–383.
Lev-Yadum, S. & Abbo S. (1999). Traditional use of A'kub (*Gundelia tournefortii*, Asteraceae), in Israel and the Palestinian Authority Area. Econ. Bot. 53(2): 217–219.
Vitek, E. & Jarvis, C.E. (2007). The typification of *Gundelia tournefortii* L. (Compositae). Ann. Naturhist. Mus.Wien, B, 108: 267–272.
Vitek, E. & Noroozi, J. (2017). *Gundelia rosea* (Compositae), a new record from Iran. Ann. Naturhist. Mus. Wien, B 119 249–256
Hind, D.J.N. (2013). Plate 763. *Gundelia tournefortii*. Curtis's Bot. Mag. 30(2): 114–138.

Corollas yellow, cream, brown or maroon; disseminule, obconical, 10–15 mm . . . . *G. tournefortii*
Corollas pink to deep purple; disseminule obovoid, 10–20 mm . . . . *G. rosea*

Fig. 62. **Gundelia tournefortii**. 1, leaf and juvenile flowering stem with two synflorescences × 1 1/2; 2, secondary capitulum and associated bract × 1; 3, secondary capitulum showing central female capitulum floret × 3; 4, floret × 3; 5, anther (enlarged); 6, style (enlarged); 7, fruiting central capitulum × 2. Drawn from an herbarium specimen at (K) by Christine M. Grey-Wilson. Reproduced with permission from Curtis's Botanical Magazine 30(2): 130 (2013).

1. **Gundelia tournefortii** *L.*, Sp. Pl.: 814 (1753); Boissier, Fl. Orient. 3: 421 (1875); Nábělek in Publ. Fac. Sci. Univ. Masaryk 52: 27 (1925); Dinsmore in Post, Fl. Syria, Palest. & Sinai ed. 2, 2: 74 (1933); Guest in Dep. Agr. Iraq Bull. 27: 40 (1933); Anthony in Notes Roy. Bot. Gard. Edinburgh 18: 291 (1935); Blakelock in Kew Bull. 4: 55 (1949); Zohary in Dep. Agr. Iraq Bull. 31: 151 (1950); Vassilchenko in Fl. S.S.S.R. 26: 862 (1961); Rawi in Dep. Agr. Iraq Tech. Bull. 14: 116 (1964); Rechinger, Fl. Lowland Iraq: 635 (1964); Kupicha in Fl. Turkey 5: 325 (1975); Chakravarty, Pl. Wealth Iraq: 268 (1976); Feinbrun-Dothan, Fl. Palaest. 3: 360 (1978); Meikle, Fl. Cyprus 2: 937 (1985); Boulos & Hind in Fl. Egypt 3: 143 (2002) & Boulos, Checklist: 236 (2009); Breckle, Hedge & Rafiqpoor, Vasc. Pl. Afghan: 173 (2013); Taifour & El-Oqlah, Pl. Jordan Ann. Checkl.: 42 (2017).

Gundelia glabra Mill., Gard. Dict., ed. 8, n. 2 (1768).
G. tournefortii L. var. *glabra* (Mill.) DC., Prodr. 5: 88 (1836).
G. tournefortii L. var. *araneosae* DC., Prodr. 5: 88 (1836), nom. illegit. effectively based on *G. tournefortii* L.
G. tournefortii L. [unranked/var.] β *tenuisecta* Boiss., Fl. Orient. [Boissier] 3: 421 (1875).
G. tournefortii L. var. *asperrima* Trautv., Acta Horti Petrop. 4(1): 146 (1876).
G. tournefortii L. var. *armata* Freyn & Sint., Oesterr. Bot. Zeit. 42(5): 168 (1892).
G. tenuisecta Freyn & Sint., Oesterr. Bot. Zeit. 42(5): 168 (1892).

Leafy, thistle-like herb to 40(–100) cm, rootstock clothed with remains of old leaves and membranous scales. Stems erect, branched, stout, smooth, glabrous or with arachnoid multicellular hairs. Leaves ± lanceolate in outline, pinnatifid to pinnatisect, decurrent, uppermost often somewhat keeled, basal petiolate, up to 42 × 20 cm, cauline sessile, to 36 × 19 cm, acute, base auricled and/or occasionally winging stem for a short distance, auricles and wings spiny-margined, veins conspicuous, raised on both surfaces, white, excurrent in rigid, yellow spines, midrib sometimes purple; margins toothed, spiny, ± densely multicellular-hairy between spines, surfaces of leaf glabrous or ± multicellular-hairy. Inflorescence of several capitula clustered together forming a dense compound head, these clusters terminal, solitary, subglobose to ovoid, up to 4.5 cm in diameter, 5–7 cm long, each secondary capitulum up to 2 cm in diameter, subtending bracts scarcely exceeding involucre or with a long apical spine up to 5 cm long, and then much exceeding involucre; phyllaries multiseriate, fused except for tips which are free and spinescent. Central florets hermaphrodite, 7–10(–13) mm long, top of tube minutely pubescent externally, lobes narrow, linear, acute, 3–5 × *c.* 1 mm. Fruiting body (= diseminule) turbinate and fluted to ovoid and smooth, 1.2–1.5 cm long, 7–9 mm in diameter, apical openings horizontal or oblique, fringed with spines of varying length, subsidiary capitula *c* . 7, becoming woody as achene develops, achene of central primary capitulum surmounted by a tuft of ± fused apical spines, 0.5–7 mm long, or ± unarmed. Achene somewhat compressed, subtetragonal, 6 × 3 mm; pappus up to 3.5 mm. Fig. 62: 1–7.

HAB. Grassy and stony mountain slopes in the moist steppe region, often associated with coppiced *Quercus*, also in the dry steppe and subdesert on stony plains and hills, desert flats, along roadsides and as a weed of arable land, on limestone and grey sandstone, clay, sandy loam and gravel; alt. 80–1050(–2500) m; fl. Mar.-Jun., fr. Apr.-Jul.
DISTRIB. Widespread and locally abundant in the mountain and foothill regions of Iraq, also in the western desert, rare in the Lower Mesopotamian Plain. **MJS**: Jabal Sinjar, above the town, *Gillett* 11150! **MAM**: 5 km SE of Zakho, *Rawi* 23100!; Sarsang, *Alkas* 18685!; **MRO**: Sefin Dagh, *Bornmüller* 1407!, *Karim* 39404!; Rowanduz gorge, *Guest* 2036!; Khanzad valley, *id.* 2011! **MSU**: Chemchemal, *Rawi & Gillett* 7571!; Jarmo, *Helbaek* 631!; Qara Dagh, Timana village, *Karim, Hamid & Jasim* 40542! **FUJ**: Balad Sinjar, *Guest* 4124!; Mosul, *id.* 1487!; 50 km N of Mosul towards Tel Afar, *Ani* 9779! **FNI**: 12 km N of Mindan bridge, *Chapman* 26212! **FAR**: Arbil, *Guest* 2126! **FKI**: Kani Domlan hills, Kirkuk, *id.* 4291!; Kirkuk, on the road to Baba Gurgur oil wells, *id.* 1374!; 13 km N of Kirkuk, *Gillett* 10577!; 10 km N of Tuz Khurmatu, *Ani & Barkley* 7016! **FKI/FPF**: Jabal Hamrin, *Haines* 304! **FPF**: Jabal Qizil Robat, *Guest* 1785!; Mafraz (Tarsaq) nr Mandali, *Rawi* 12595!; Khanaqin, *Fawzi & Noori* 39582! **DLJ**: nr Tikrit, *Omar & Hamid* 36284! **DWD**: ± 65 km W of Ramadi, *Guest & Rawi* 13959!; 12 km W of Ukhaidhir, *Rawi* 30494!; H.1, 160 km E of Rutba, *Chakravarty, Rawi, Khatib & Alizzi* 31575! **LCA**: nr Daltawa (Khalis), *Rogers* 0287!

Cyprus, Syria, Lebanon, Palestine, Egypt (rare), Turkey, Iran, Transcaucasia, Afghanistan, Turkmenia (very rare); adventive in Algeria.

In most accounts this species is generally recognized as showing considerable variation, especially in corolla colour, largely unrelated to gross morphology. In lectotypifying the name Vitek & Jarvis (2007) anchored the name to just one part of the corolla colour variation (cream to yellowish corollas), and later Vitek et al. (2010) proposed that another part of the colour variation was accounted for in the redder entities. However, the brownish colour variant, at one time thought more restricted in its distribution is clearly present in Turkey many hundreds of kilometers away from its expected distribution. For this reason I have chosen to once again accept a broader view of *G. tournefortii*, but also recognize the presence of other recognizable entities easily separated (at least when fresh) based on corolla colour, although clearly other characters can be used (see Vitek et al. 2010, 2014).

Until the detailed work of Vitek (2007, 2010, pers. comm.), *G. tournefortii* has been considered to be a variable species, both in habit, disseminule shape and size, and most certainly in corolla colour. Having seen good evidence (courtesy of Vitek in 2015) of the link between habit, inflorescence and corolla colour one of us (DJNH) has recognized two species in this account. With several new species having been described it remains to be seen if more than these are present in Iraq.

Guest (1933) noted that the fleshy rootstock of this plant, given the Kurdish names KAŪB and KANGAR is sold in the northern markets as a vegetable, while a note on *Gillett* 10577 states that the young shoots are an excellent vegetable like *Asparagus*. *Rawi & Gillett* 7571 also notes that the disseminules are also eaten as nuts after the spines have been singed off, and both these and the young shoots are sometimes sold in Baghdad markets. Meikle (Fl. Cyprus 2: 938, 1985) quotes P.H. Davis's note that in Cyprus the plant is called SILIFA, and the capitula are eaten like globe artichokes; Kupicha (*Fl. Turkey*, loc. cit.) observes that on the Anatolian plateau the species is collected and dried in summer, and stacked for winter fodder, while a chewing gum is prepared from the latex and the seeds are used (after drying) for making a drink.

The fruiting inflorescences of this species behave as a "tumbleweed", each of the secondary capitula finally dispersing as a one-seeded disseminule. *Rawi & Gillett* 10577 notes that these inflorescences are dispersed by whirlwinds, and have been seen by Gillett carried high into the air.

Fruits (disseminules) have been used in the recent past as a source of oil. 'Charred fruits have been found at Neolithic sites in Iraq and Turkey, and are evidence that oil extraction dates back at least 10,000 years' (Prance & Nesbit, 2005).

2. **Gundelia rosea** *Houssain & Al-Taey* in Notes Roy. Bot. Gard. Edinburgh 42: 41 (1984). –Type. Iraq, N of Atrush, ± 60 km NE of Mosul, ± 900 m, 10.v.1978, *Mosharraf Hossain s.n.* (Biol. Dept. Herb. Univ. Mosul, holo; BAG, E, K!, iso.).

G. tournefortii L. var. β L., Sp. Pl.: 814 (1753).
G. tournefortii L. f. *purpurascens* Bornm. in Beih. Bot. Centralbl. 60B: 197 (1939).
G. purpurascens Bornm., Beih. Bot. Centralbl. lx. R. 197, in obs., pro syn. 197, 1939, nom. inval.; Rawi, in Dep. Agr. Tech. Bull. 14: 116 (1964); Blakelock in Kew Bull. 4: 1 (1949).

Differs from *G. tournefortii* in possessing dark pink to purplish phyllaries, corollas that are bright to light pink inside and purple outside, florets 5–8(–11) per secondary capitulum, disseminules obovoid, 10–20 mm long.

HAB. Grassy mountain slopes and pine woodland; alt. 800–900 m; fl. Apr.-May.
DISTRIB. Restricted to *Pinus* woodlands. **MAM**: N of Atrush, ± 60 km NE of Mosul, *M. Hossain* s.n. (Biol. Dept. Herb. Univ. Mosul, holo; BAG, E, K!, iso.); ± 2 km N of Zawita, *M. Hossain* s.n. (Biol. Dept. Herb. Univ. Mosul). **MRO**: nr. Koi-Sanjak, E of Erbil, *M. Hossain* s.n. (Biol. Dept. Herb. Univ. Mosul).

The species appears to be restricted to pine woodland, compared with *G. tournefortii* which occurs in more open, ruderal habitats, often appearing weedy, at somewhat lower to mid-range altitudes (± 800–900 m) than much of the distribution of *G. tournefortii* (0–2500 m).

Vitek & Jarvis (2007) originally considered that *G. rosea* was 'possibly restricted to the area near the Iranian and Armenian borders, and most probably illustrated the species (Vitek et al. 2010: 89 considering it a new species).

Iran?

TRIBE **ARCTOTIDEAE** CASS.

59. **ARCTOTIS** L.

Sp. Pl.: 911 (1753); Gen. Pl. 5: 394 [2006] (1754)
Karis in Kubitzki, Fam. Gen. Vasc. Pl. 8: 203 [2006] (2007)
Venidium Less., Linnaea 6: 91 (1931)

C. C. Townsend

Annual or perennial herbs or subshrubs, ± arachnoid, with alternate, subentire or lobed to pinnatifid leaves. Capitula terminal on stem and branches, radiate, heterogamous; involucre hemispherical or campanulate; phyllaries multiseriate; receptacle alveolate, fimbrillate. Ray floret limbs white, yellow or bluish, pistillate and fertile or very rarely neuter and sterile; disc florets hermaphrodite, fertile, or central often sterile. Achenes glabrous or pubescent, 5-ribbed; pappus biseriate, delicate, paleaceous, inner row about 8, sometimes very small or absent, pappiferous achenes with a copious tuft of hairs at base.

About 55 species in S Africa and a few in Australia.

Arctotis (from the Gr. ἄρκτος, arktos, bear, αυτί, auti, ear – from the form of the pappus-scales).

1. **Arctotis venusta** *Norl.* in Bot. Not. 118: 403–411 (1965).

Arctotis stoechadifolia auct. non Berg.: Lewin in Feddes Repert. Beih. 11: 69 (1922); Husain & Kasim, Cult. Pl. Iraq: 108 (1975).

Erect or ascending annual herb, 10–60 cm tall, usually branched from below middle, rarely simple, vegetative parts ± arachnoid-tomentose or glabrescent with age. Lowest leaves petiolate, coarsely sinuate-dentate to pinnatilobed, up to 20 × 4.5 cm, but usually smaller; median cauline leaves similar, elliptic to obovate in outline, sessile and ± amplexicaul; upper leaves reducing. Capitula rather few; involucre hemispherical, 2–3 cm in diameter, outer and median phyllaries ± arachnoid, herbaceous with a linear apical appendage; inner phyllaries oblong-obovate with a broad hyaline border. Ray limbs 1.5–2 cm, white above with a yellow base, lower surface reddish or purplish. Disc florets pinkish or purplish-lobed.

HAB. & DISTRIB. Cultivated at a Horticultural Research Station: **LCA**: Za'faraniya, *Omar & Sahira* 34808!

A native of S Africa occasionally cultivated, but mostly in botanic gardens. African Daisy (Eng.). According to Husain & Kasim this plant is "commonly cultivated" in Iraq, but only the specimen above has been seen.

60. **GAZANIA** Gaertn. *nom. cons.*

Fruct. Sem. Pl. 2: 451 (1791)

C. C. Townsend

Rhizomatous perennial herbs, more rarely annuals or subshrubs, with ± rosulate or alternate, oblanceolate to elliptic, entire to pinnately lobed leaves. Capitula solitary, long- or short-pedunculate, radiate, heterogamous; involucre campanulate; phyllaries 2–4-seriate, connate for at least half their length into a basal cup; receptacle convex or conical, alveolate, naked. Ray florets' rays neuter and sterile, usually yellow, orange or red, often dark-spotted at base; disc florets hermaphrodite, fertile. Achenes turbinate or clavate, ± densely sericeous; pappus scales biseriate, delicate, equal or outer series somewhat shorter.

17 species in southern Africa, extending to Angola and Tanzania.

Gazania (said to be named after Theodore of Gaza, 1398–1478, who translated Aristotle and Theophrastus).

Roessler, H. (1959). Revision der Arctotideae – Gorteriinae (Compositae). Mitt. Bot. Staatssamml. Munchen 3: 71–500.
– (1973). Mitt. Bot. Staatssamml. Munchen 11 : 91–99 (1973).

1. **Gazania rigens** (*L.*) *Gaertn.*, Fruct. 2: 451, t. 173 (1791); Roessler, op. cit. 370 (1959); Husain & Kasim, Cult. Pl. Iraq: 114 (1975).

Othonna rigens L., Pl. Rar. Afr.: 24 (1760).

Creeping perennial herb with a decumbent, simple or branched, glabrescent to ± tomentose, slender stem. Leaves congested and subrosulate below pedunculcs to scattered along stem, entire and oblanceolate to linear or pinnatisect with 1–2 pairs of linear to broadly oblanceolate leaflets, very variable in size, margins ± revolute, densely white-tomentose to glabrous above; pedicels with or without bract-like leaves. Involucre campanulate, 14–25 × 5–10 mm at top of fused portion, glabrous or ± tomentose, phyllary tips lanceolate-acuminate. Rays (1.5–)2–3 mm, yellow or usually orange, with a white-centre and black spot at base or rarely absent (not in cultivated plants).

HAB. & DISTRIB. Cultivated at Horticultural Research Stations and in gardens.
LCA: Abu Ghraib, *Janan Mokhtar* 357171!, *Sahira* C. 838!; Za'faraniya, *Omar & Sahira* C. 387!, C 388!; public garden, Baghdad, *Thamer & Wedad* C. 875!

A native of southern Africa widely cultivated.

Said by Husain & Kasim to be cultivated in the middle and south of Iraq, but only the above specimens, all from the Baghdad region, have been seen.

TRIBE **SENECIONEAE** CASS.

61. DORONICUM L.

Sp. Pl.: 885 (1753); Gen. Pl. ed. 5: 377 (1754)
Nordenstam in Kubitzki, Fam. Gen. Vasc. Pl. 8: 215 [2006] (2007)

J. R. Edmondson

Stout perennial herbs with a short horizontal rhizome, often clothed in a mixture of glandular and eglandular hairs. Basal leaves ovate-cordate or oblong, petiolate; cauline leaves shortly petiolate and reducing upwards, becoming sessile, amplexicaul. Capitula heterogamous, radiate, solitary at ends of branches or (in Iraq) several on each branch of stem; involucre depressed-hemispherical; phyllaries 2–3–seriate, herbaceous; receptacle convex, pilose. Ray florets ligulate, yellow; disc florets yellow, tubular. Achenes cylindrical to obovate-cylindrical with 10 longitudinal ribs, bearing a pappus of white, minutely scabrid fine bristles; pappus sometimes absent from ray florets.

Some 30+ species extending from S & C Europe through S.W. Asia to the Himalayas and China.

Doronicum (said to be from the Ar., though no equivalent can now be traced; the oldest scientific epithet within the genus, *pardalianches* (Lat.) literally meaning "leopard-killer", is possibly derived from the Ar. *fahad*, leopard.).

Cavillier, F. (1911). Nouvelles études sur le genre *Doronicum*. Annu. Cons. Jard. Bot. Genève 13: 195–367.
Edmondson, J. (1978). The genus *Doronicum* in Iran. Notes Roy. Bot. Gard. Edinburgh 37: 67–73.
Fernandes, I.Á. (2003). Systematics of Eurasian and North African *Doronicum* (Asteraceae): Senecioneae. Ann. Missouri Bot. Gard. 90 (3): 319–389.

1. **Doronicum bracteatum** *Edmondson*, Notes Roy. Bot. Gard. Edinburgh 32: 257 (1973). – Type. Iraq, Mt. Helgord (Algurd), nr Rost, 25 km ENE of Rowanduz, *Guest & Ludlow-Hewitt* 2928 (K!, holo. & iso.). Rechinger, Fl. Iranica 164: 47 (1989).

Stem 65–70 cm, glabrous. Basal leaves 6.5–8 × 6.5–9 cm, suborbicular to reniform, rounded at apex, margins weakly dentate, upper surface ± glabrous, lower with sparse short eglandular hairs especially on veins; median cauline leaves *c.* 12 × 10 cm, broadly ovate, margins most distinctly sinuate-dentate, petiole 17–20 cm; upper cauline leaves somewhat smaller, ovate, sessile, amplexicaul; branch leaves large, elliptical, sessile, grouped around the pedicels; lateral capitula long-pedunculate, pedicels 10–35 cm, primary capitula subsessile. Capitula 5–9, 30–45 cm in diameter; phyllaries lanceolate, acute, margins clothed with long-stalked glandular hairs, surface covered in short-stalked glandular hairs. Fig. 63: 1–9.

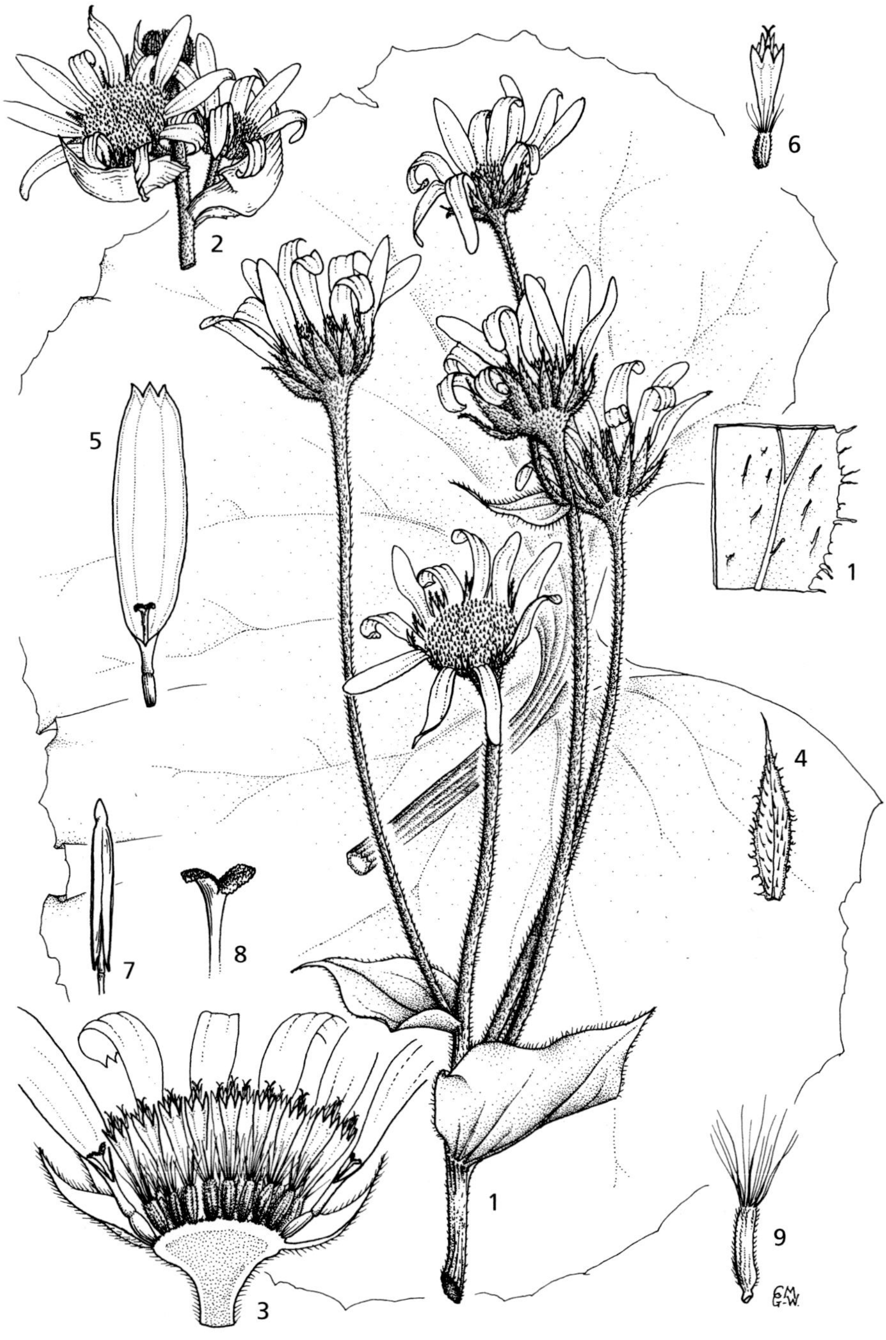

Fig. 63. **Doronicum bracteatum**. 1, habit with leaf × 1 & portion of leaf enlarged; 2, inflorescence × 1; 3, capitulum × 3; 4, phyllary × 3; 5, ray floret × 3; 6, disc floret × 3; 7, anther × 9; 8, style × 9; 9, achene × 9. 1, 3–9 from *Rawi & Serhang* 24553; 2 from *Guest* et al. 2928. Drawn by Christine M. Grey-Wilson.

HAB. By torrents in the alpine zone; alt. (1800–)2300–3000 m; fl. Aug.
DSTRIB. Rare, on high mountains in the NE sector of the mountain zone of Iraq. **MRO**: Mt. Helgord (Algurd), nr Rost, 25 km ENE of Rowanduz, *Guest & Ludlow-Hewitt* 2928 (type)!; Mt. Qandil E of Qala Diza on the Iranian frontier, *Thesiger* 1154.

W Iran. A late-gathered specimen from **MRO**: Mt. Helgord, 3000–3300 m, 3 ix 1957, *Rawi & Serhang* (K!) may also belong to this species; in Fl. Turkey I tentatively assigned it to *D. hakkaricum* J.R.Edmondson, which is not otherwise recorded from Iraq.

62. **PERICALLIS** (D.Don) DC.

in Sweet, Brit. Fl. Gard. 6 (ser. 2, 3): t. 228 (1834)
Nordenstam in Kubitzki, Fam. Gen. Vasc. Pl. 8: 229 [2006] (2007)

C. Jeffery

Perennial herbs or shrubs with palmately veined leaves. Rays purple or blue, 4–7-veined. Involucres not calyculate. Achene-cells not papillose; pappus caducous, often absent in the ray-florets.

15 species native to the Canary Islands, Madeira and the Azores, but often found elsewhere in cultivation.

Pericallis (perhaps from Gr. περι, *peri*, around and Lat. *kallis*, rough, perhaps a reference to the involucre).

1. **Pericallis hybrida** (*Regel*) *B.Nord.*, Opera Bot. 44: 21 (1978).

Cineraria hybrida Willd., Enum. hort. bot. reg. Berol.: 893 (1809), nom. illegit.; Husain & Kasim, Cult. Pl. Iraq: 112 (1975).
[*Senecio cruentus* (non (L'Hérit.) DC.) auctt. hort., incl. Husain & Kasim, Cult. Pl. Iraq: 119 (1975)].
[*S.* × *hybridus* (Willd.) Bailey, Hortus Third: 1036 (1976)[13]]

Perennial with a short stem. Lower leaves petiolate (base of lamina confluent with winged petiole), floccose on lower surface, glabrous above, cordate-triangular, dentate, variable in size; uppermost leaves lanceolate, sessile, amplexicaul, auriculate, entire, indumentum similar. Inflorescence cymose. Pedicels thickened near base of involucre. Capitula radiate, 1.6–3.5 cm in diameter, involucre campanulate. Ray limbs purple to purple-pink, 8–18 mm.; disk florets tubular, corollas purple; phyllaries linear, acuminate, 4–6 mm, glabrous to pilose; calyculus absent.

HAB. & DISTRIB. Said to be in cultivation at horticultural stations: **LCA**: Baghdad (cult.), *Omar* C. 395!

This plant is a hybrid derivative of *P. cruenta* (L'Hérit.) Bolle and one or more other wild species from the Canary Islands, best known as a pot plant and known as "Cineraria" or "florist's Cineraria" (Eng., Am.).

Also closely related to *Senecio* is the genus *Emilia* Cass., which differs most obviously in having neither calyculus nor peduncular bracts; it also has central tufts of papillae at the ends of the style-branches which are lacking in *Senecio*.

63. **SENECIO** L.

Sp. Pl.: 866 (1753); Gen Pl. ed. 5: 373 (1754)
Nordenstam in Kubitzki, Fam. Gen. Vasc. Pl. 8: 230 [2006] (2007)

S. T. Al-Kaisi

Herbs or subshrubs with alternate, entire, dentate or 1–2-pinnatipartite or -pinnatisect,

---

[13] The above is almost certainly not the earliest valid publication of this binomial. *Index Kewensis* cites Regel in Gartenflora 12: 113 (1863), but here Regel describes a new var. *holtzeri* of *Senecio hybridus* "hort." Since *S. hybridus* was apparently not at that time validly published, this name is illegitimate under Article 43 of the Shenzhen Code (2018). The same applies to the earlier publication of the name in Gartenfl. 8: 310 (1859). Bailey attributed the above combination to Regel, but Regel nowhere mentions *Cineraria hybrida* Willd.

glabrous to densely hairy leaves. Inflorescences usually corymbose, dense or lax, capitula rarely solitary. Capitula heterogamous and radiate or homogamous and discoid; involucre campanulate, obconical or cylindrical; phyllaries uniseriate, with or without an outer calycullus, which when present, is usually shorter than phyllaries; receptacle flat or convex, naked. Ray florets (when present) female, rays yellow, orange, pink, red or purplish; disc florets tubular, hermaphrodite, 4- or 5-lobed, corolla usually yellow. Achenes ribbed, subcylindrical, usually beakless, glabrous or setuliferous; pappus usually present, multiseriate, barbellate, persistent or caducous, usually white.

Once one of the largest genera of flowering plants, of about 3000 species distributed throughout the world, but now considered to contain about 1250 species, *Iranecio* and *Pericallis* being among those now recognized as separate genera; 10 native species in Iraq, 2 other taxa cultivated.

Senecio (a name in Pliny, probably for *S. vulgaris*; from Lat. *senex*, an old man, referring to the conspicuous white pappus of these plants); Groundsel, also Ragwort or sometimes Ragweed (more generally applied to species of Ambrosia). Guest (1933) noted the vernacular names HODHĀN (or WARD HODHĀN) and MURRAIR for certain species of *Senecio* in N Iraq, also known in the Kurdish foothills as GUZÎLKA ZERD ("little yellow flower", Kurd.) which are palatable by livestock only when dry but injurious, even at times causing death.

Alexander, J.C.M. (1979). The Mediterranean species of *Senecio* sections *Senecio* and *Delphinifolius*. Notes Roy. Bot. Gard. Edinburgh 37: 387–428.

1. Rays absent . . . . 10. *S. vulgaris*
   Rays present 2
2. Rays purple or purplish-pink, rarely white, disk-florets yellow; cultivated plants. . . . . 7. *S. arenarius*
   Rays and disc florets both yellow 3
3. Perennials with lower and middle leaves simple, entire, denticulate or dentate (sometimes ± coarsely so at base, at least 1 cm wide, green and glabrous or almost so 4
   Annuals or perennials with lower and middle leaves distinctly lobed, pinnatifid, pinnatipartite or pinnatisect (rarely linear and toothed in annual *S. glaucus* and then less than 1 cm. wide), if glabrous then leaves pinnatisect or plant annual 6
4. Capitula 1.5–2 cm in diameter, phyllaries 4.5–6 mm long; rays 7–11 mm . . . . 3. *S. doriiformis*
   Capitula (2–)2.2–4 cm in diameter, phyllaries (6–)7–12 mm long; rays 9–15(–18) mm 5
5. Inflorescence a lax terminal corymb; leaves entire or regularly denticulate or dentate . . . . 1. *S. pseudo-orientalis*
   Inflorescence a compound raceme; leaves denticulate or dentate with teeth coarser towards base in at least some leaves . . . . 2. *S. racemosus*
6. Involucre, upper part of stem and lower surface of leaves densely white-tomentose; achenes glabrous; cultivated plant . . . . 6. *S. cineraria*
   Involucre, upper part of stem and lower surface of leaves not white-tomentose, if arachnoid then leaves only deeply toothed or pinnatifid; achenes ± hairy; wild plants 7
7. Perennials; leaves only broadly toothed to ± pinnatifid or lyrately lobed, teeth or lobes rarely with a few marginal denticulations. 8
   Annuals; leaves pinnatisect, pinnatipartite or pinnatifid, the segments either very narrow or with marginal denticulations 9
8. Phyllaries 7–10 mm long; achenes 4.5(–6) mm; disk florets 4-lobed; basal petiolate leaves up to 20 cm long . . . . 5. *S. eriospermus*
   Phyllaries 5–6(–7) mm long; achenes 1.7–2(–2.5) mm; disc florets 5-lobed; basal petiolate leaves up to 10 cm long . . . . 4. *S. mollis*
9. Leaf-lobes wide, < 4 × longer than broad, margins distinctly sharply denticulate . . . . 9. *S. leucanthemifolius*
   Leaf-lobes linear to narrowly linear, at least 4 × as long as broad, margins entire or with a few very remote denticulations . . . . 8. *S. glaucus*

In the descriptions that follow, measurements of the petiolate leaves include petioles, and the diameter of the capitulum of flowers includes rays at anthesis.

SECTION **CROCISERIS** (RCHB.) HALL & WOHLF.

1. **Senecio pseudo-orientalis** *Shishkin* in Fl. S.S.S.R. 26: 727 (1961); Matthews in Fl. Turkey 5: 159 (1975); B. Nordenstam in Fl. Iranica 164: 63 (1989); Qaiser & Abid in Fl. Pakistan 218: 34 (2011).

S. orientalis Willd., Sp. Pl. ed. 4, 3(3): 2006 (1804); Boissier, Fl. Orient. 3: 406 (1875), non Miller (1768).

Erect glabrescent perennial 65–75(–100) cm tall. Stem ± densely floccose, slightly swollen at base, with many fibrous roots. Lower and basal leaves petiolate, linear-oblanceolate to narrowly elliptic-oblong, minutely denticulate or entire, 13–25 × 1–3.5 cm; median cauline leaves similar but sessile, 8–16 × 0.5–3 cm; upper leaves smaller and narrower, entire to denticulate with teeth more acute. Inflorescence a lax terminal corymb and sometimes also solitary branches. Capitula 2.2–3 cm in diameter; involucre campanulate; phyllaries ± glabrescent, 7–10 mm, linear, acuminate, apices dark and hairy; calycular bracts linear, 3–7 mm, darker and hairy at tip. Rays yellow, 9–15(–18) mm. Achenes glabrous, 3–5 mm; pappus 7–8 mm, white.

HAB. & DISTRIB. The claim of this plant to be an Iraqi species rests entirely on one gathering: Pl. Mesopot., Kurdistan & Mossul, *Kotschy* 269! However, Kotschy certainly gathered it around Diyarbakir in SE Turkey, and this may have been the source of this gathering also.

Turkey, Iran, Pakistan, C Asia.

2. **Senecio racemosus** (*M.Bieb.*) *DC.*, Prodr. 6: 358 (1838); Boissier, Fl. Orient. 3: 402 (1875); Shishkin in Fl. S.S.S.R. 26: 731 (1961); Rechinger, Fl. Lowland Iraq: 633 (1964); Matthews in Fl. Turkey 5: 157 (1975); B. Nordenstam in Fl. Iranica 164: 66 (1989).

Cineraria racemosa M.Bieb., Fl. Taur.-Cauc. 2: 314 (1808).
Senecio racemosus var. *latronum* Boiss., Fl. Orient. 3: 402 (1875); Handel-Mazzetti. in Ann. Naturh. Mus. Wien 27: 437 (1913).

Erect perennial 25–75(–100) cm with a thick rootstock and numerous fibrous roots. Stem rigid, stout, glabrescent. Lower cauline and basal leaves obovate-oblong to narrowly elliptic, 8–18 × 2.5–5 cm, ± acute, remotely sinuate-dentate or dentate (teeth increasing in size towards base), sparsely hairy on lower surface; median cauline leaves oblong-lanceolate, 7–14 × 1–3.5 cm, sessile, amplexicaul with rather wide, obtuse basal auricles; upper leaves similar, reducing. Inflorescence a compound raceme. Capitula (2–)3–4 cm in diameter; involucre hemispherical-ovoid or campanulate; phyllaries ± glabrescent, narrowly lanceolate to lanceolate-oblong, 6–12 mm, acuminate, darker and hairy at tip; calycular bracts linear, 2.5–7 mm, darker and floccose at tip. Rays yellow, 10–15 mm. Achenes glabrous or sparsely hairy, ± 4 mm; pappus 6–8 mm, white.

HAB. River island of sandy silt and riverine plain; alt. 240 m (in the one precise locality); fl. Nov.
DISTRIB. Only rarely recorded along the R. Tigris. **FNI**: island in R. Tigris opposite Mosul, *Handel-Mazzetti* 1259; Mosul, *Thompson* s.n. **DLJ/DGA**: Tigris plain, *Mrs A. Low* 409!

Turkey, W Iran, Caucasus, Transcaucasia.

Iraqi material is referable to var. *latronum* Boiss., with the lower leaves obovate, inflorescence a widely corymbose panicle, leaf margins remotely sinuate-dentate and hemispherical-ovoid capitula ± 2 cm wide, included here in synonymy.

3. **Senecio doriiformis** *DC.*, Prodr. 6: 352 (1838); Boissier, Fl. Orient. 3: 407 (1875); Dinsmore in Post, Fl. Syria, Palest. & Sinai ed. 2, 2: 71 (1933); Matthews in Fl. Turkey 5: 160 (1975); Nordenstam in Fl. Iranica 164: 64 (1989).

S. doriiformis var. *megalophron* Boiss., Fl. Orient. 3: 407 (1875); Zohary in Dep. Agr. Iraq Bull. 31: 151 (1950); Rawi in Dep. Agr. Iraq Tech. Bull. 14: 124 (1964).

Erect glabrous perennial 65–75(–100) cm tall, stem angular-striate. Basal and lower cauline leaves ovate to elliptic- or lanceolate-oblong, petiolate, 8–18 × 0.8–4 cm, entire or denticulate; median cauline leaves similar or somewhat smaller; upper leaves sessile, ovate to elliptic or lanceolate, entire. Inflorescence of a lax terminal corymb and sometimes ± remote solitary branches, with small reduced leaves. Capitula 1.5–2 cm in diameter, involucre campanulate; phyllaries linear, ± acuminate, 4.5–6 mm, apex darker and hairy. Calycular bracts several, linear, 3–5 mm. Rays yellow, 7–11 mm. Achenes glabrous, 2.5–4 mm; pappus 7–9 mm, white.

HAB. Shaded rocky places on mountains; alt. 1500–2120 (–2700) m.; fl. Jun.-Sep., fr. Jul.-Oct.
DISTRIB. Only found once or possibly twice in Iraq. **MSU**: Pira Magrun, *Agnew, Hadač, Haines & Waleed* 1847.

Lebanon, Turkey, Iran.

Nordenstam (l.c.) recognized 2 subspecies: subsp. *doriiformis* and subsp. *orientalis* (Fenzl) Matthews (Notes R.B.G. Edinburgh 33: 434 1975), citing *Agnew et al.* 1847 for the former subspecies from Iraq, and *Haussknecht* s.n.! Iraq or Iran, Mts. Avroman et Schahu for the latter subspecies. Subsp. *orientalis* is separated on its upper cauline leaves being lanceolate with apex acute to acuminate, as opposed to elliptic-ovate leaves with apex obtuse in subsp. *doriiformis*.

SECTION **JACOBAEA** (MILL.) DUMORT.

4. **Senecio mollis** *Willd.*, Sp. Pl. ed. 4, 3(3): 2011 (1804); Boissier, Fl. Orient. 3: 393 (1875); Dinsmore in Post, Fl. Syria, Palest. & Sinai ed. 2, 2: 70 (1933); Matthews in Fl. Turkey 5: 150 (1975); Nordenstam in Fl. Iranica 164: 75 (1989).

Erect perennial with a subterete-striate, ± floccose stem 0.6–1 m tall, usually with long, slender, divergent branches. Leaves ± floccose on both surfaces, basal and lower cauline ± petiolate, 7–10 × (1.5–)2.5–4.5 cm, elliptic to elliptic-obovate in outline, dentate to ± lyrately lobed, lobes with a few teeth; median leaves similar but sessile, 4–6 × 2–3 cm, amplexicaul with dentate auricles; upper leaves also auriculate, denticulate to subentire. Inflorescence a wide, lax corymb. Capitula 1.5–3 cm in diameter, involucre campanulate; phyllaries narrowly oblong or oblanceolate, 5–7 mm, glabrescent to ± floccose; calycular bracts numerous, linear, mostly 4–5 mm, glabrescent to floccose. Rays yellow, 7.5–11 mm, 4-veined. Achenes hairy, 1.7–2(–2.5) mm; pappus 4–6 mm, dirty white.

HAB. In the mountains on metamorphic rock, in damp shady places, near springs and in vineyards; alt. 1240–1500 m.; fl. & fr. Jul.-Aug.
DISTRIB. Occasional in the middle forest zone of Iraq. **MRO**: Gali Ali Beg, *Rawi* 24264!; Darband, *Gillett* 9458! **MSU**: Penjwin, *Guest* 12996!, *Haines* 2035!, *Rechinger* 12250. *Al- Kaisi & Al- Khayat* 51172 (BAG!)

Turkey, N & W Iran.

5. **Senecio eriospermus** *DC.*, Prodr. 6: 358 (1838); Boissier, Fl. Orient. 3: 395 (1875); Matthews in Fl. Turkey 5: 150 (1975); Nordenstam in Fl. Iranica 164: 74 (1989).

Erect perennial with a woody rootstock; stem 30–70 cm., subterete or sulcate-striate, ± floccose. Basal and lower cauline leaves broadly ovate to elliptic, congested above stock, petiolate, 12–26 × 3–7.5 cm, deeply toothed or pinnatifid, acute, subcordate to attenuate at base, sparsely floccose on upper surface, densely so below; median cauline leaves similar but smaller, becoming sessile and amplexicaul; upper leaves becoming linear and bract-like. Inflorescence a lax corymb with slender branches. Capitula 2.5–4.5 cm in diameter, involucre hemispherical; phyllaries linear-lanceolate, 7–10 mm, floccose; calycular bracts few, linear, 3–6 mm. Rays yellow, 10–18 mm, 4-veined. Achenes hairy, 4.5–6 mm; pappus ± 9 mm, white.

HAB. On a rocky slope above the tree line; alt. 1800–2100 m.; fl. & fr. Jun.
DISTRIB. Very rare in Iraq; found once in the central sector of the mountain-forest region. **MRO**: Mt. Botin, nr Shirwan Mazna (Mazin), *Agnew, Hadač & Haines* 2098! *Hadač* 6114.

Turkey (E Anatolia).

8. **Senecio cineraria** *DC.*, Prodr. 6: 355 (1838); Boissier, Fl. Orient. 3: 395 (1875); Husain & Kasim, Cult. Pl. Iraq: 112 (1975).

Cineraria maritima L., Sp. Pl. ed. 2: 1244 (1763); Husain & Kasim, Cult. Pl. Iraq: 112 (1975).

Senecio bicolor (Willd.) Tod. subsp. *cineraria* (DC.) Chater in Bot. J. Linn. Soc. 68: 273 (1974); Matthews in Fl. Turkey 5: 151 (1975).

Erect bushy perennial up to 70 cm. tall. Stems much-branched from base and often also above, stout, densely white-tomentose to glabrescent, non-flowering stems bearing rosettes of leaves. Main cauline leaves ovate in outline, 4–13 × 2.5–7 cm, pinnatisect with pinnae frequently pinnatifid or pinnatipartite with blunt segments to sinuate-dentate or lyrate, densely white-tomentose on lower surface, green and arachnoid or glabrescent above, distinctly petiolate; upper leaves reducing, sessile. Inflorescence a terminal compound corymb, many-headed, usually dense, more rarely lax with long pedicels. Capitula 1–1.5 cm in diameter, involucre campanulate; phyllaries narrowly oblong, 5–7 mm, densely white-tomentose to very sparsely so or glabrous; calycular bracts 1–2 (–4) mm, indumentum similar to that of phyllaries. Rays yellow, 4–6 mm, 4-veined. Achenes glabrous, 1.5–2.5 mm; pappus 4–5 mm, dirty white.

HAB. & DISTRIB. Sometimes cultivated on the alluvial plain in Iraq at Experimental Stations and perhaps also occasionally in gardens; alt. ± 35 m.; fl. & fr. Mar.-May.
LCA: Baghdad, Abu Ghraib, *Janan & Sahira* C. 265!, *Omar* 278!, *Fawzi* 391!, *Janan* 485!; Za'faraniya, *Shakir Sabir as-Sabbagh* 20184!

Native of the Mediterranean region from Spain to Turkey, locally naturalised, especially on sea cliffs, from Britain to the Crimea.

Cineraria (Engl.); Silver Ragwort (Engl.); Dusty Miller (Engl., Am.). The plant cultivated in Iraq and elsewhere is subsp. *cineraria*, with pinnatisect leaves, a densely white-tomentose involucre and short pedicels.

According to Husain & Kasim (1975) this plant (as *Cineraria maritima*) "cultivated throughout the country"; but since they give for it the English narne "Florist's Cineraria", which is *S.* × *hybridus*, and state that "in the north protection is necessary", there seems to be some confusion here.

For the nomenclature of this species see Mabberley in Watsonia 14: 279–280 (1983), where it is shown that *S. bicolor* (Willd.) Tod. is a homonym of *S. bicolor* Viv.

SECTION **SENECIO**

9. **Senecio arenarius** *Thunb.*, Prodr. Fl. Cap.: 158 (1800); Bailey, Hortus Third: 1034 (1976).

Annual, viscid-pubescent, 15–40 cm., simple or branched from a little above base upwards. Stem and leaves somewhat rough with rather coarse glandular hairs. Basal leaves petiolate, usually pinnatisect or pinnatifid with lobes dentate or pinnatifid, rarely acutely lobed; median stem leaves sessile, mostly pinnatifid, auriculate, amplexicaul, 3.5–7 × 0.8–4 cm; upper leaves similar but reducing, pinnatifid to linear and dentate. Inflorescence a lax corymb with several showy capitula. Capitula 1.75–3.5 cm in diameter; involucre broadly obconical to shortly cylindrical; phyllaries narrowly lanceolate, subacute, glandular-hairy, 7 mm, apices often purplish; calycular bracts few, 4–6, linear, 1.5–2.5 mm, also often purplish at tip. Rays pinkish-purple to purple, rarely white, (5–)7–12(–15) mm, disk florets yellow. Achenes hairy, 3–4 mm; pappus dimorphic, 6 mm, white.

HAB. & DISTRIB. Said to be sometimes cultivated on the irrigated alluvial plain; alt. ± 35 m; fl. Mar.
LCA: Za'faraniya Horticultural Experiment Station, Baghdad, *Rawi* (X253!).

A South African species which has been reported widely cultivated in gardens in Iraq; no other specimens have been seen.

10. **Senecio glaucus** *L.*, Sp. Pl.: 868 (1753) subsp. **coronopifolius** (*Maire*) *C.Alexander*, Notes Roy. Bot. Gard. Edinburgh 37: 412 (1979); Boulos & Hadidi, Weed Fl. Egypt: 40 (1984); Boulos & Hind in Fl. Egypt 3: 263 (2002); Kilian & Ghazanfar in Fl. Oman 3: 240 (2015); Abdulridha, Taha & Wedad, Ecology and Fl. Basra: 225 (2016); Haloob & Al-Kaisi, Illustr. Fl. Lowland Iraq 1(1): 55 (2016); Taifour & El-Oqlah, Pl. Jordan Ann. Checkl.: 47 (2017).

S. coronopifolius Desf., Fl. Atlant. 2: 273 (1799), non Burm.f. (1786); Boissier, Fl. Orient. 3: 390 (1875); Handel-Mazzetti in Ann. Naturh. Mus. Wien 27: 437 (1913); Carter in Rec. Bot. Surv. India 6:

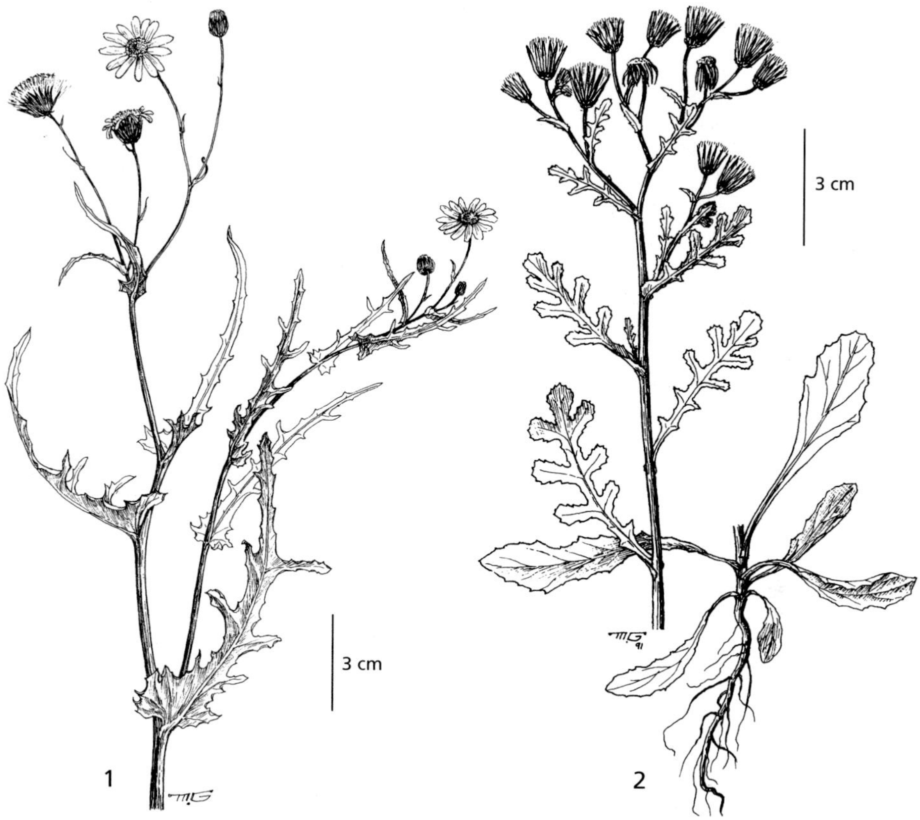

Fig. 64. **Senecio glaucus** subsp. **coronopifolius**. 1, habit. **Senecio vulgaris**. 2, habit. Drawn by Magdy Al Gohary. Reproduced with permission from Boulos & Hadidi, Weed Flora Egypt: 1994, p. 83 & 85.

203 (1917); Nábělek in Publ. Fac. Sci. Univ. Masaryk 52: 25 (1925); Guest in Dep. Agr. Iraq Bull. 27: 90 (1933): Blakelock in Kew. Bull. 4: 63 (1949); Zohary in Dep. Agr. Iraq Bull. 31: 151 (1950); Dickson, Wild Fls. Kuwait: 86 (1955); Rawi in Dep. Agr. Iraq Tech. Bull. 14: 124 (1964).

S. desfontainei Druce, Br. Pl. List ed. 2: 61 (1928); Dinsmore in Post, Fl. Syria, Palest. & Sinai ed. 2, 2: 69 (1933); Anthony in Notes Roy. Bot. Gard. Edinburgh 18: 291 (1935); Rawi in Dep. Agr. Iraq Tech. Bull. 14: 124 (1964); Rechinger, Fl. Lowland Iraq: 632 (1964); Täckholm, Stud. Fl. Egypt ed. 2: 581 (1974).

S. gallicus subsp. *coronopifolius* Maire in Jahandiez et Maire, Cat. Pl. Maroc 3: 784 (1934).

Erect or decumbent-erect annual, ± fleshy, (6–)10–50(–60) cm tall, glabrous to sparingly arachnoid, simple or branched mostly from about base. Leaves mostly arachnoid when young but soon glabrescent; basal leaves petiolate, obovate to narrowly lanceolate or linear in outline, pinnatifid with dentate lobes to simple and entire or dentate; middle leaves usually pinnatipartite rarely attaining 6 cm in length, lobes entire or with small remote teeth, apical tooth acute or rounded; sessile, amplexicaul, auriculate; upper leaves more narrowly lobed with lobes ± acute. Inflorescence a lax corymb. Capitula 1.5–2(–2.5) cm in diameter, not indurated below in fruit, involucre campanulate. Phyllaries 4–6 mm, glabrous or almost so, black-tipped or usually concolorous; calycular bracts up to 12 in number, sometimes black-tipped, 1–4 mm. Ligules yellow, 4–8(–12) mm. Achenes appressed-pilose, 1.75–3.3 mm; pappus dimorphic, white, 3.5–7 mm. Fig. 64: 1.

HAB. Mainly on sandy soils in stony and rocky places, on gravelly hillsides, near wadis and river banks and under trees of *Zizyphus nummularia*; alt. 50–650 m.; fl. & fr. Feb.-Jun.

DISTRIB. Very common in the Desert Region of Iraq, occasional in the Steppe Region. **FUJ/MJS**: Jabal Qilat, *Field & Lazar* 408l. **FUJ**: nr. Ba'aj, *Guest* 13392!; 40 km from Sha'bani along the road to Sinjar, *Qaisi & Hamid* 48986!; between Shargat and Tal Julaimid, *Alizzi & Husain* 33937!; between Shargat and Hadhr, *F. & E. Barkley* 6606!, *Rawi & Hamada* 33637!; Hadhr, *Ani* 9864!, *Chakravarty, Rawi, Khatib & Alizzi* 33027!. **FUJ/DLJ**: 25 km N of Baiji along road to Mosul, *Ani* 9701!; **FKI**: bank of R. Zab at Kirkuk, *Mrs P. Grigg* 14!. **FPF**: Qizil Robat (Sa'diya), *Guest* 1771!, *Qaisi* 42515!, 44195!; Khanaqin, *Ibrahim al-Jabri* 348!; Jabal Darawishka, *Poore* 457!; Jabal Hamrin, nr Table Mt., *Guest* 1969!; Jabal Hamrin between Baghdad and Khanaqin, *Nábělek* 3526; between Naft Khana and Kani Masi, nr the frontier NE of Baghdad, *Morck* 12 in Hand.-Mazz., l.c.; Mandali, *Haines* 542!; Zurbatiya, *Gillett* 6704!; Baqsaya, *Rawi & Haddad* 35595!; SE of Badra, *Rawi* 18418!; Jabal Muwaila, *Guest, Rawi & Rechinger* 17640B!; Wadi at-Tib police post, 70 km N of Amara, *id.* 17503! **DLJ**: 5 and 50 km S of Hadhr, *Alizzi & Husain* 33869!, *Chakravarty, Rawi, Khatib & Alizzi* 33079!; Rawa, *Gillett & Rawi* 6957!; between Baiji and Haditha, *Omar, Nuri, Hamid, Muhammad & Jasim* 41856!; Hit, in Ant., l.c.; Khamran, 30 km S of Tikrit, *Alizzi & Husain* 33755!; Tikrit, *Guest* 393!; 6 km E of Wadi Thirthar, *Rawi & Gillett* 7159! **DGA**: 13 km and 50 km E of Samarra, *Rawi* 20369!, 20435! **DLJ/DWD**: 35 km N of Rutba, *Rawi* 21160! **DWD**: 15–17 km W of Rutba, *Hamad* 50443!, *Rawi* 27096!; 20 km N of Rutba, *Rawi* 21102, *Rawi & Khatib* 32206!; Wadi al-Ajramiya, 73 km N of Rutba, *Rawi* 27164!; Masad, 15 km S of Rutba, *Chakravarty, Rawi, Khatib & Alizzi* 32876!; between Rutba and H.2, *Omar & Hamid* 36710!; 50 km E of Rutba, *Springfield & Akram Jaf* 21317!; Wadi al-Batikha, 18 km N of T.1, *Chakravarty, Rawi, Khatib & Alizzi* 31790!; T.1, *Omar, Qaisi, Hamid & Hamad* 45065!; Ana, *Omar & Hamid* 36608!; Wadi Fuhaimi, S of Ana, *Omar, Qaisi et al.* 44388!; 145 km NE of K.3, *Chakravarty, Rawi et al.* 33019!; 180 km E of Rutba, *Rawi & Noori* 27041!; 100 km W of Ramadi, *Omar & Hamid* 36737!; 15 km W of Ramadi, *Guest & Rawi* 13942!; 120 km from Nukhaib, *Rawi & Hamada* 33517!; nr Shithatha, *Rawi & Gillett* 6453!; Ukhaidhir, *Hamid Ali & Barkley* 4348!; 20 km from Mugrianam, *Rawi & Hamada* 335540! **DSD**: 80 km W of Shabicha, *Chakravarty, Khatib, Rawi & Tikriti* 30056!; 40 km WNW of Shabicha, *Guest, Rawi & Rechinger* 19300!; Chibritiya, 25 km S of Maniya, *Botany Staff* 42015!; Shabicha, *Rawi & Gillett* 6268!; 20–30 km W of Shabicha on road to Salman, *Alizzi & Omar* 35581!; 30 km E of Salman, nr Golaib, *Guest, Rawi & Rechinger* 18745!; 25 km N of Busaiya, *Qaisi, Hamad & Hamid* 42376!; 50 km E of Busaiya, *Chakravarty, Khatib, Rawi & Tikriti* 29982!; 50 km E of Busaiya on road to Khadr al-Mai, *Karim, Sharif & Hamid* 48040!; Khadr al-Mai, *Rawi, Tikriti & Khatib* 29099!; 5 km E of Jumaima, *Qaisi et al.* 48192!; Jabal Sanam, *Watson* s.n. (comm. Sharples); nr Ansab, *Guest, Rawi & Rechinger* 18942!; Samah, *Alizzi* 34401! **LCA**: Babylon, *Reuter* 3525 in Nábělek, l.c. **LEA**: 60 km N of Amara, nr Wadi Tib, *Guest, Rawi & Rechinger* 17468! **LBA**: 15 km NE of Ma'qil, Basra, *id.* 16764!; **LBA/DSD**: Shuwaiba, *Rogers* 01!

Sicily, SW Asia from Syria and Palestine to Afghanistan and Pakistan, N Africa from Egypt to Morocco, Canary Is.

HODHĀN, Ba'aj, *Guest* 13392 (or WARD HODHĀN), a common name in Iraq for various Compositae with yellow flowers, particularly *Senecio*, according to Guest (1933), also in Abdulridha et. al. l.c.; RIJL (or RIJLA) AL-GHURAB "crow's foot" (lr. – from Rutba, Wadi Thirthar and Shabicha, *Rawi* 7159, 14664 & 6258 respectively). H. E. Schwan, who collected this species from near Riyadh in Saudi Arabia, where he noted that it was called 'ALOGA, states that the plant "has a sharp peppery taste and the Badawin say it is never grazed. It is one of the species which now populate the Arabian desert landscape because it is not eaten" – a comment which would seem equally appropriate in our territory. Carter (1917) recorded the local name ZIMLUG in Kuwait – a name confirmed by Dickson (1955).

Subsp. GLAUCUS is a fleshy maritime plant with phyllaries 6–8 mm long, capitula often indurated at base in fruit, middle cauline leaves usually more than 6 cm long and lobes of leaves dentate or bi- or tri-furcate at tip. It is confined to Egypt and Palestine.

11. **Senecio leucanthemifolius** *Poir.*, Voy. Barb. 2: 238 (1789), non Phil. (1873) var. **vernalis** (*Waldst. & Kit.*) *C.Alexander* in Notes Roy. Bot. Gard. Edinburgh 37: 403 (1979); Meikle, Fl. Cyprus 2: 932 (1985); Taifour & El-Oqlah, Pl. Jordan Ann. Checkl.: 47 (2017).

S. vernalis Waldst. & Kit., Pl. Rar. Hung. 1: 23. t. 24 (1800); Boissier, Fl. Orient. 3: 389 (1875); Nábělek in Publ. Fac. Sci. Univ. Masaryk 52: 25 (1925); Dinsmore in Post, Fl. Syria, Palest. & Sinai ed. 2, 2: 69 (1933); Guest in Dep. Agr. Iraq Bull. 27: 90 (1933); Anthony in Notes Roy. Bot. Gard. Edinburgh 18: 291 (1935); Blakelock in Kew Bull. 4: 63 (1949); Zohary in Dep. Agr. Iraq Bull. 31: 151 (1950); Schischkin in Fl. S.S.S.R. 26: 783 (1961); Rawi in Dep. Agr. Iraq Tech. Bull. 14: 124 (1964); Täckholm, Stud. Fl. Egypt ed. 2: 583 (1974); Matthews in Fl. Turkey 5: 167 (1975); Feinbrun-Dothan, Fl. Palaest. 3: 355 (1978).

S. leucanthemifolius Poir subsp. *vernalis* (Waldst. & Kit.) Greuter in Willdenowia 33(2): 247 (2003).

Erect or decumbent-erect annual, (20–)6–45(–60) cm, simple (sometimes subacaulescent) or slightly to considerably branched, stem and branches arachnoid to sparsely hairy to subglabrous. Leaves mostly arachnoid when young, glabrous when mature; basal leaves spatulate-obovate to oblong-elliptic, pinnatifid with deltate, deltoid or oblong lobes, or dentate or denticulate, petiolate, 0.8–7.5 × 0.3–5 cm; middle and upper cauline leaves oblong in outline, shallowly lobed or usually pinnatifid to pinnatipartite with rounded to deltoid, usually ± denticulate lobes, (1–)3–6(–10) × 1–4 cm, sessile, amplexicaul and auriculate. Inflorescence a lax or dense corymb. Capitula 1.5–2.5 cm in diameter; involucre campanulate; phyllaries glabrous to slightly pilose or floccose, 4–6.5(–8) mm, black-tipped or concolorous; calycular bracts 6–22, 0·5–3 mm, frequently black-tipped. Rays yellow, 2–15 mm, rarely absent. Achenes sparsely to densely appressed-hairy, (1.5–)2–3 mm; pappus 2.5–6 mm, white.

HAB. Grassy places among coppiced *Quercus* on limestone, overgrazed mountain slopes, in fields and vineyards in the mountains, steppic grassland on conglomerate hills; alt. (200–)450–1800 m; fl. & fr. (Mar.-)Apr.-May (-Jun.)

DISTRIB. Common in the forest zone of Iraq, occasional in the steppe region. **MAM**: Jabal Bekher, nr Zakho, *Field & Lazar* 832!; 20 km from Zakho, *Qaisi & Hamad* 49294!; Amadiya, *Guest* 1248!; Sharifa, *O. Polunin* 5120!; Dohuk, *Guest* 2156! Kani Mazi, *Al-Khayat, Jawhar & Abdulla* 57965 (BAG!) **MRO**: Salah ad-Din, *E. Chapman* 11988!, *Barkley & (Hikmat Abbas al-) 'Ani* 7127!; Shaqlawa, *Gillett* 10467!, *Haines* W.687!; between Harir and Kani Watman, *Nábělek* 3527; Gali Ali Beg, *Omar, Widad & Kaisi* 49558!; Diyana, *Guest* 713!; Rowanduz, *id.* 2045!; **MRO/FAR**: 4 km S of Taktak bridge on road to Koi Sanjaq, *Rawi, Noori & Kass* 28086!; Koi Sanjaq, *id.* 28120!; Plingan, NW of Rania, *id.* 2862!, *F.K. Jackson* 15075!; 'Ain Shaikh Barakat, *Alkas s.n.*!; Haji Umran, *Chapman* 12301! **MSU**: Jarmo, *Helbaek* 342!, 402!, 1740!; 16 km W of Chemchamal, *Rawi & Gillett* 7582!,10594!; Bakrajo, *Omar & Karim* 37410!; Sulaimaniya, *Graham* 725!, 751!; Gilazarda, *Karim, Hamid & Jasim* 40672!; above Dukan, *Barkley & Haddad* 75931; Kani Watman, 20 km SW of Dukan, *Qaisi & Hamad* 49329!; between Sulaimaniya and Arbat, *Omar & Karim* 38105!; Darband-i Khan, *Qaisi & Hamad* 49347!; 1 km S of Darband-i Khan, *Barkley & Shehbaz* 7706!; 2 km E of Halabja, *Rawi* 21822!; Avroman, N of Biyara, *Gillett* 11761!; 10 km W of Tawila, *Rawi* 22117!; 110 km N of Jalaula, *F. & E. Barkley* 4426!; Shewa Sur (Wadi Ahmar, i.e. red river), *Omar, Karim, Hamza & Hamid* 37107!; Jasina, *Omar* 42086! **MJS**: Jabal Sinjar, *Gillett* 11118!; **FUJ**, between Nisilrin (on Syrian frontier) and Mosul, *Loftus* 3! **FUJ/FNI**, Mosul, *Shahwani* 25172, 5174! **FNI**, Jabal Maklub, *Haines* 1414!; Ain Sifni, *Salim* 2572! **FKI**, 14 km E of Kirkuk, *Rawi & Gillett* 10671!; 21 km N of K.3 on road to Koi Sanjak, *Rawi, Noori & Kass* 27975! **FPF**, Kani Kirmaj, *Poore* 319! **LCA**, Baghdad, *Rogers* 0391!; Rustamiya, *Guest* 1105!

Native in C & E Europe, spreading westwards as a naturalized species to Cyprus, Syria, Lebanon, Palestine, Jordan, Egypt, Turkey, Caucasia, Iran, C Asia.

? HALULA (Ar.), *Loftus*; ?GULÎLKA HASPI ("little white (!) flower", Kurd., Ain Sifni), *Salim* 2572; GULÎLKA ZARDA ("little yellow flower"), Kurd., Haji Umran), *F. K. Jackson* 15075, perhaps a name used for this plant.

12. **Senecio vulgaris** *L.*, Sp. Pl.: 867 (1753); Boissier, Fl. Orient. 3: 386 (1875); Dinsmore in Post, Fl. Syria, Palest. & Sinai ed. 2, 2: 68 (1933); Anthony in Notes Roy. Bot. Gard. Edinburgh 18: 291 (1935); Blakelock in Kew Bull. 4: 63 (1949); Zohary in Dep. Agr. Iraq Bull. 31: 151 (1950); Schischkin in Fl. S.S.S.R. 26: 780 (1961); Rawi in Dep. Agr. Iraq Tech. Bull. 14: 124 (1964); Rawi & Chakravarty, ibid. 15: 83 (1964); Rechinger, Fl. Lowland Iraq: 632 (1964); Täckholm, Stud. Fl. Egypt ed. 2: 582 (1974); Matthews in Fl. Turkey 5: 165 (1975); Feinbrun-Dothan, Fl. Palaest. 3: 355 (1978); Alexander, op. cit. 408 (1979); Boulos & Hadidi, Weed Fl. Egypt: 84 (1984); C. Jeffery & Y.L. Chen in Kew Bull. 39(2): 351 (1984); Nordenstam in Fl. Iranica 164: 83 (1989); Meikle, Fl. Cyprus 2: 930 (1985); Boulos & Hind in Fl. Egypt 3: 263 (2002); Qaiser & Abid in Fl. Pakistan 218: 32 (2011); Taifour & El-Oqlah, Pl. Jordan Ann. Checkl.: 47 (2017).

Erect annual 6–40(–50) cm tall; stem erect, branched or simple, ± glabrous to sparsely arachnoid (as are the leaves). Basal and lower cauline leaves obovate to oblong in outline, 1.2–9.5 × 0.3–1.5 cm, pinnatifid or dentate-lobate, shortly petiolate; median cauline leaves similar but sessile, amplexicaul, auriculate, reducing upwards; upper leaves ± similar, frequently not much reduced. Inflorescence corymbose. Capitula discoid, rarely (not in Iraq) shortly ligulate; involucre cylindrical, 0.5 cm in diameter; phyllaries glabrous, linear, acute, (5–)6–8 mm, often black-tipped; calycular bracts numerous, (1–)1.5–3 mm, usually black-tipped or occasionally entirely black. Achenes 1.5–2 mm, densely appressed-pilose; pappus ± 6 mm, white. Fig. 64: 2.

HAB. A weed of cultivation, in gardens, orchards and in cereal fields; alt. 35–300 m.; fl. & fr. Jan.-May, probably in gardens throughout the year.

DISTRIB. Occasional in the dry steppe zone and in the desert region of Iraq, on the irrigated alluvial plain. **FKI**: Kirkuk, *Meade* 46. **FPF**: Khanaqin, *Rechinger* 14124, 14129; Mandali, *Rawi* 12673!; Mandali, *Al-Kaisi, Thamer & Salah* 51483!; 2 km NE of Mandali id. 51497! **LCA**: Sudur, *Rawi* 20017! **LEA**: Baghdad, *Colvill* s.n.!, *Paranjpye* s.n. (comm. Graham), *Rogers* 013!, *Guest* 1107!; between Baghdad and Shahraban (Muqdadiya), *Rechinger* 15822.

Throughout Europe, Aegean Is. to Cyprus, Syria, Lebanon, Palestine, Egypt, Arabia, Turkey, Caucasia, Iran, Afghanistan, E & W Siberia, N Africa (Morocco to Libya), Macaronesia (Madeira, Canary Is.). Introduced in E Asia, N America and elsewhere.

Common Groundsel, a frequent weed of cultivated ground in Europe and elsewhere and, according to Johns (1908), a favourite food of many small birds.

? SHAIKH AR-RABI (which in this case might be loosely translated as "Harbinger of Spring". Wren (1956) states that this plant is diuretic and diaphoretic; it is said to relieve bilious pains and, in strong infusions, acts as a purgative and emetic.

64. **IRANECIO** B.Nord.

Fl. Iranica 164: 53 (1989)
Jeffrey in Kew Bull. 47(1): 102–103 (1992); Nordenstam in Kubitzki, Fam. Gen. Vasc. Pl. 8: 241 [2006] (2007)

S. T. Al-Kaisi & J. R. Edmondson

Rhizomatous perennial herbs with stout stems. Leaves alternate, petiolate, glabrous on upper surface, shortly hairy on lower, pinnatisect with lanceolate to ovate segments, margins dentate. Capitula homogamous or heterogamous, ligulate or discoid, numerous in a much-branched corymbose inflorescence; involucre obconical to campanulate; phyllaries ± uniseriate; receptacle flat, glabrous, pitted. Marginal florets few, female, corolla yellow; disc florets hermaphrodite; anthers distinctly caudate at base; style shortly branched towards stigma. Achenes oblong, shallowly grooved; pappus with numerous barbellate setae.

Around 18 species distributed in SE Europe, SW & C Asia; two species in Iraq.

Iranecio (from Iran and *-ecio*, part of the Lat. name *Senecio, q.v.*)

Rays present; achenes sparsely hairy . . . . .1. *I. othonnae*
Rays absent; achenes glabrous . . . . 2. *I. paucilobus*

1. **Iranecio othonnae** (*M.Bieb.*) *B.Nord.*, Fl. Iranica 164: 58 (1989).

Senecio othonnae M. Bieb., Fl. Taur.-Cauc. 2: 308 (1808); Boissier, Fl. Orient. 3: 410 (1875); Dinsmore in Post, Fl. Syria, Palest. & Sinai ed. 2, 2: 71 (1933); Shishkin in Fl. S.S.S.R. 26: 741 (1961); Rawi in Dep. Agr. Iraq Tech. Bull. 14: 124 (1964); Matthews in Fl. Turkey 5: 163 (1975).

Erect perennial with a stout, glabrous, sulcate-striate stem 0.7–2 m tall, branched above to form a broad compoundly corymbose inflorescence. Leaves glabrous on upper surface, shortly hairy on lower, pinnatisect with lanceolate to ovate segments; lower cauline leaves 35–70 cm long, pinnatisect with segments sharply and irregularly dentate, up to 14 × 5 cm wide; median cauline leaves mostly 15–30 cm long, similar in form; upper leaves smaller, more narrowly dissected, sessile, bract-like and entire or almost so in the inflorescence. Inflorescence a compound corymb with many capitula. Involucre campanulate, 6–10 mm in diameter, phyllaries narrowly oblong or oblong-obovate, ± 6 mm, hairy at tip and sometimes below. Calycular bracts few, linear, 1–3 mm. Rays few (4–6), yellow, ± 8 mm. Achenes sparsely hairy, ± 3 mm; pappus ± 6 mm, dirty white.

HAB. Very rare in Iraq; only found once near a spring on metamorphic rocks; alt. 1840 m.; fl. & fr. Aug.
DISTRIB. **MRO**: S of Ser Kurawa, between Zirva and Ari, *Gillett* 9794!

SE Europe (southern Balkan Peninsula), Turkey, Caucasus, Iran.

2. **Iranecio paucilobus** (*DC.*) *B.Nord.*, Fl. Iranica 164: 56 (1989).

Senecio paucilobus DC., Prodr. 1: 352 (1838); Boissier, Fl. Orient. 3: 410 (1875); Blakelock in Kew

Bull. 4: 63 (1949); Zohary in Dep. Agr. Iraq Bull. 31: 151 (1950); Shishkin in Fl. S.S.S.R. 26: 741 (1961); Rawi in Dep. Agr. Iraq Tech. Bull. 14: 124 (1964); Matthews in Fl. Turkey 5: 163 (1975). *Pojarkovia pauciloba* (DC.) Askerova in Nov. Sist. Vyssh. Rast. 21: 187 (1984).

Perennial with a woody rootstock; stem (15–)30–55 cm tall, angular-striate, subglabrous or ± floccose. Basal leaves petiolate, 12–22 (–32) × 2.5–7 cm, congested and clasped about base of stem above stock, usually pinnatisect with dentate lobes, terminal lobe the largest and with a cordate base, lateral lobes with broad, decurrent bases, in 1–2 pairs or rarely absent, sparsely floccose on upper surface, more densely so beneath; median cauline leaves similar but smaller and more shortly petiolate; upper leaves small, lanceolate to obovate, entire or dentate. Inflorescence a compound corymb with many capitula. Capitula discoid, with a narrowly obconical-cylindrical involucre 3–5 mm. wide; phyllaries narrowly oblong, 5–7(–9) mm, acuminate to ± obtuse, glabrous or usually ± hairy at tip. Calycular bracts linear, 1.5–2(–4) mm. Achenes glabrous, (2–)3–4 mm; pappus 6–7 mm, white.

HAB. On mountainside, on rocks and cliffs in the *Astragalus* zone, and among metamorphic rocks; alt. 2100–3200 m; fl. & fr. Jun.-Aug.
DISTRIB. Occasional in the thorn-cushion zone of Iraq. **MAM**: Gara Dagh, *Kotschy* 414! **MRO**: Ser Kurawa, *Gillett* 9787!; Chiya-i Mandau, *Guest* 2681!; Qandil Range NE of Rania, *Rawi & Serhang* 18263!, 26715!, 26740!; mt. Qandil above Goam-e Kirmosoran lake, *Rechinger* 11132; Warshanka-Magar range, *Rawi & Serhang* 24322!; Kermasur lake, *Rawi & Serhang* 18295! **MSU**: Hauraman (Avroman), *Chakravarty, Rawi, Noori & Alizzi* 19733!; Mt. Avroman & Shahu, *Haussknecht* s.n.(1867).

Turkey, Iran, Caucasus.

TRIBE **CALENDULEAE** CASS.

65. **CALENDULA L.**

Sp. Pl.: 921 (1753); Gen. Pl. ed. 5: 885 (1754)
Nordenstam in Kubitzki, Fam. Gen. Vasc. Pl. 8: 243 [2006] (2007)

Patricia Halliday

Annual or perennial, aromatic, often straggling herbs, sometimes woody at base, usually glandular–multicellular–hairy, sometimes also arachnoid. Leaves entire or shallowly toothed, sessile, often ± amplexicaul. Inflorescences many-headed. Capitula heterogamous, radiate, solitary, terminal and axillary. Involucre hemispherical; phyllaries biseriate, imbricate, subequal, linear–lanceolate, acuminate, margins narrow, scarious. Receptacle flat, naked. Ray florets female, fertile, corolla limbs, yellow or orange; disc florets functionally male, tubular with 5 apical teeth, yellow, orange, brown or purple. Achenes 2–3-seriate, heteromorphic, usually incurved; pappus absent.

A widely cultivated genus of some 9 species; 3 native species in Iraq.

Calendula (from the Lat. *calendae* and *–ula*, the diminutive suffix. The Calends are the first days of each month in the Medieval Roman calendar; it is not easy to see why such a name should have been chosen for this genus, though Lindley & Moore (1870) suggested that the connection perhaps originated because species of the common plant may be found in flower on the calends of every month in the year.

Ohle, H. (1974). Beiträge zur Taxonomie der Gattung *Calendula*, II: Taxonomische Revision der südeuropäischen perennierenden Calendula–Sippen. Feddes Repert. 85: 245–283.
– (1975). Beiträge zue Taxonomie und Evolution der Gattung *Calendula* L., III: Revision der marokkanischen perennierenden Sippen unter Berücksichtigung einiger marokkanischer Annuellen. Feddes Repert. 86: 1–17.

1. Annual or short–lived perennial, with stems often woody at base; capitula large, 3–5 cm in diameter; cultivated .... 3. *C. officinalis*
   Capitula smaller, to 2 cm in diameter; plant annual 2
2. Leaves toothed, glandular–hairy; ray florets with limb 5–7.5 mm, equalling or only slightly exceeding phyllaries; marginal achenes broadly 3–winged; no beaked achenes present .... 1. *C. tripterocarpa*
   Leaves entire or subentire, glandular–hairy and often arachnoid; ray

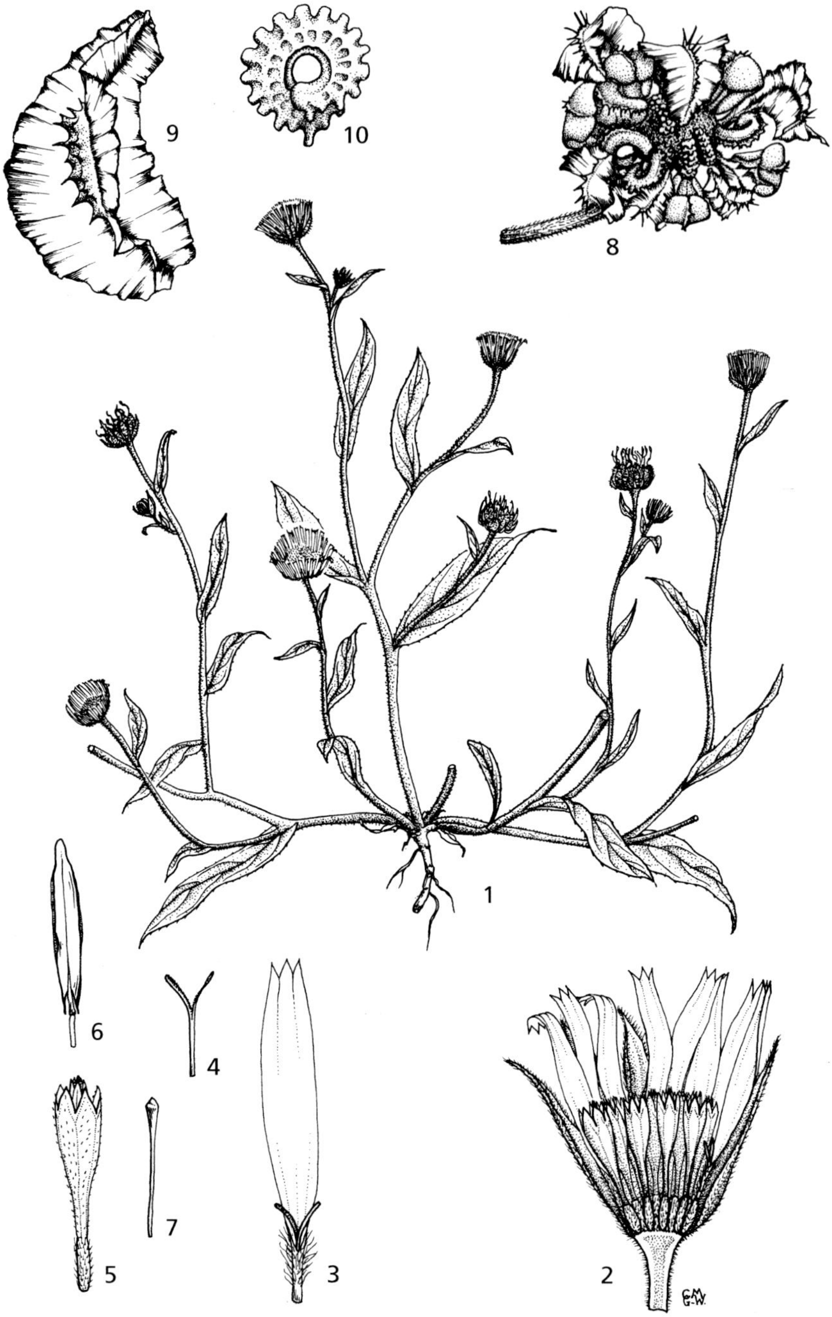

Fig. 65. **Calendula tripterocarpa**. 1, habit × 1; 2, capitulum (dissected) × 6; 3, ray floret × 6; 4, style (ray floret) × 9; 5, disc floret × 9; 6, anther (disc floret) × 18; 7, style (disc floret) × 9; 8, capitulum with mature achenes × 3; 9, marginal achene × 9; 10, inner achene × 9. 1–7 from *Botany Staff* 42010; 8–10 from *Barkley et al.* 5039. Drawn by Christine M. Grey-Wilson.

limbs 8–12 mm, up to 2 × length of phyllaries; marginal achenes beaked and cymbiform or cymbiform, not 3–winged .... 2. *C. arvensis*

1. **Calendula tripterocarpa** *Rupr.*, Bull. Phys.-Math. Acad. Pétersb. 14: 231 (1856); Meikle in Fl. Europ. 4: 207 (1976); Feinbrun–Dothan, Fl. Palaest. 3: 358 (1978); Collenette, Wild Flow. Saudi Arabia: 175 (1999); Boulos & Hind in Fl. Egypt 3: 267 (2002); Abdulridha, Taha & Wedad, Ecol. & Fl. Basra: 199 (2016).

Calendula aegyptiaca Desf. subsp. *tripterocarpa* (Rupr.) Lanza, Monogr. 110 (1919).

Much-branched, spreading, often decumbent, sparsely glandular–hairy annual herb to 10(–25) cm tall, resembling *C. arvensis* but usually smaller. Stems very slender, ribbed. Leaves basal and cauline, linear-oblong, 19–75(–105) × 2–9(–10) mm, sessile, cuneate, often attenuate, acute, usually repand-dentate, rarely subentire, reducing in size upwards. Capitula 6–10(17) mm in diameter, including rays; fruiting capitulum 7–21 mm in diameter. Phyllaries distinctly biseriate, outer 5–8 × 1–2 mm, margins narrow, scarious, inner 4–7 × 1 mm, the scarious margins more pronounced. Receptacle 3 mm in diameter. Ray florets uniseriate, only slightly exceeding involucre, (4–)5–75 × 1–1.5 mm, rays ± spreading, oblanceolate or strap–shaped, with (2–)3 apical teeth and (3–)4 veins, base of florets long-multicellular-hairy outside; disc florets 2.5–5 × 1 mm at apex, papillose outside; styles of ray florets bifid, arms compressed, acute, papillose, styles of disc florets imperfect, at most slightly divided, club-shaped, papillose. Marginal achenes 8–9 × 6–10 mm, sometimes stained purple, unbeaked with 3 broad, flaring, variably lacerate wings and a somewhat rugose spine, glabrous or minutely pubescent; inner achenes ± cylindrical, all annulate, transversely rugose-tuberculate dorsally. Fig. 65. 1–10.

HAB. On limestone desert plateau in calcareous sandy patches between rocky ridges, sandy washes and silty depressions, gypsum desert with gravel surface etc.; alt. to 650 m; fl. & fr. (Feb.-) Mar.-Apr.(-Jun.). DISTRIB. Quite common in the SW sector of the desert region of Iraq, occasional as a weed in cultivation on the irrigated alluvial plains. **DWD**: 35 km WNW of Rutba, *Rawi* 21155!; 25 km W of Rutba, *Rawi* 14698!; 15 km W of Rutba, *Rawi & Noori* 27083!; Masud, 15 km S of Rutba, *Chakraverty, Rawi, Khatib & Alizzi* 32868!; between Rutba and H.2, *Omar & Hamid* 36721!; between H.3 and Wadi Hauran, *Rawi & Gillett* 6889!; 30 km N of Ramadi, *Guest & Rawi* 13952!; 22 km W of Ramadi, *F. & E. Barkley & Fuad Safwat* 5039!; Wadi al-Khirr, 15 km N of Lusuf, *Guest, Rawi & Rechinger* 19454! **DSD**: Chibritiya, 25 km S of Maniya, *Botany Staff* 42010!; 40 km WNW of Shabicha, *Guest, Rawi & Rechinger* 19315!; 6 km SSW of Shabicha, *Guest, Rawi & Rechinger* 19237!; 50 km N by W of Aidaha, c. 80 km SSW of Salman, *Guest, Rawi & Rechinger* 19196!; Khadhr al-Ma', *Rawi, Khatib & Tikriti* 29106!; 65 km W of Basra, *Guest, Rawi & Rechinger* 42412!; ubair, *Rawi & Gillett* 6042!; 5 km SE by E of Zubair, *Guest, Rawi & Rechinger* 16782! **LCA**: Sudur, *Yalda Matti* 297!; Baghdad, *Haddad* 9448!; **LSM**: between Qa'lat Salih and Qurna, *Chakravarty, Khatib, Rawi & Tikriti* 29862! **LBA**: 80 km SE of Basra, *Chakravarty et al.* 29929!

W Mediterranean Europe (Spain, Balearic Is., France, Corsica, Italy), Macaronesia, N Africa (Morocco), Jordan, Egypt, Arabia.

2. **Calendula arvensis** *L.*, Sp. Pl. ed. 2: 1303 (1763); Boissier, Fl. Orient. 3: 418 (1875); Carter in Bot. Surv. India 6 (6): 204 (1917); Dinsmore in Post, Fl. Syria, ed. 2, 2: 72 (1933); Blakelock in Kew Bull. 4: 44 (1949); Rawi in Dep. Agr. Iraq Tech. Bull. 14: 107 (1964); Tackholm, Stud. Fl. Egypt: 585 (1974); Meikle in Fl. Turkey 5: 171 (1975) & Fl. Europ. 4: 207 (1976); Feinbrun-Dothan, Fl. Palaest. 3: 359 (1978); Boulos & Hadidi, Weed Fl. Egypt: 25 (1984); Collenette, Wild Flow. Saudi Arabia: 175 (1999); Fragman, Levy-Yamamori & Christodoulou, Fl. Eastern Med.: 400 (2001); Boulos & Hind in Fl. Egypt 3: 266 (2002); Kilian & Ghazanfar in Fl. Oman 3: 241 (2015); Abdulridha, Taha & Wedad, Ecol. & Fl. Basra: 199 (2016).

Calendula sancta L., Sp. Pl. ed. 2: 1304 (1763).

C. aegyptiaca Desf., Tabl. Hort., Paris ed. 1: 100 (1804); Boissier, Fl. Orient 3: 419 (1975); Dinsmore in Post, Fl. Syria, ed. 2, 2: 72 (1933); Bornmüller in Beih. Bot. Centralbl. 60B: 197 (1940); Blakelock in Kew Bull. 3: 44 (1949); Zohary in Dep. Agr. Iraq 31: 151 (1950); Dickson, Wild Fls. Kuwait: 29 (1955); Rechinger, Fl. Lowland Iraq: 634 (1964); Rawi in Dep. Agr. Iraq Tech. Bull. 14: 107 (1964); Täckholm, Stud. Fl. Egypt ed. 2: 583 (1974).

C. persica C.A.Mey., Verz. Pfl. Cauc. 72 (18310; Boissier, Fl. Orient. 3: 418 (1875); Dinsmore in Post, Fl. Syria, ed. 2, 2: 72 (1933); Guest in Dep. Agr. Iraq Bull. 27: 17 (1933); Bornmüller in Beih. Bot. Centralbl. 60B: 196 (1940); Blakelock in Kew Bull. 4: 45 (1949); Zohary in Dep. Agr. Iraq Bull. 31: 151 (1950); Dickson, Wild Fls. Kuwait: 29 (1955); Rechinger, Fl. Lowland Iraq: 634 (1964).

C. arvensis L. var. *bicolor* DC., Prodr. 6: 452 (1838); Anthony in Notes Roy. Bot. Gard. Edinburgh 18: 291 (1935); Rawi in Dep. Agr. Iraq Tech. Bull. 14: 207 (1953).
C. micrantha Boiss., Diagn. Pl. Orient. ser. 2 (6): 111 (1859).
C. persica C.A. Mey. var. *gracilis* (DC.) Boissier, Fl. Orient. 3: 418 (1875); Nábělek in Publ. Fac. Sci. Univ. Masaryk 52: 26 (1925); Dinsmore in Post, Fl. Syria, ed. 2, 2: 72 (1933); Bornmüller in Beih. Bot. Centralbl. 60B: 197 (1939).
C. persica C.A. Mey. var. *microcephala* Boiss., Fl. Orient. 3: 419 (1875).
C. aegyptiaca Desf. var. *microcephala* auct. non Boiss.: Rawi in Dep. Agr. Iraq Tech. Bull. 14: 107 (1964).
C. palaestina auct. non Boiss.: Rawi, in Dep. Agr. Iraq Tech. Bull. 14:107 (1964).

Much-branched, aromatic, annual herb to 18(–40) cm tall, ± erect and spreading or decumbent, sparsely glandular-multicellular-hairy, sometimes also arachnoid. Leaves basal and cauline, usually oblong, occasionally narrowly oblanceolate or linear-lanceolate (1–)1.7–7(–14.5) × 0.3–1.2(–2.5) cm, sessile, attenuate (especially basal leaves), clasping, subentire to shallowly or sinuate-dentate. Capitula 0.9–2 cm in diameter, much smaller than those of *C. officinalis*; fruiting capitulum 1–2 cm in diameter. Involucral bracts 5–7.5 × 1–2 mm, ciliate, often ± erose. Receptacle up to 4 mm in diameter. Ray florets usually < 2 × involucre, yellow to orange, oblanceolate or ± strap–shaped, (5–)8–12 × 1–2.5 mm wide, with (2–)3 apical teeth and (3–)4 veins, base of floret long glandular-multicellular-hairy. Disc florets yellow or purple, (2–)2.5–5 × 1–1.5 mm, glandular-hairy externally. All achenes glabrous to minutely pubescent, all or inner achenes annulate, or with marginal row of cymbiform and beaked achenes, or beaked achenes absent; annulate achenes 3–5 mm in diameter, rugose along spine; cymbiform achenes 3–10 mm long, inflated, sometimes narrowly winged; beaked achenes incurved or sometimes erect, 1–2(–2.2) cm long, echinate on spine.

HAB. In the foothills and upper plains on calcareous soil, degraded conglomerate and other stony soils, also along the banks of streams and ditches, on silty plains by roadsides and on old dry fallow, in the desert on sandy, silty and gypsiferous soils, a weed in cultivation on the alluvial plains; alt. up to 650 m; fl. & fr. (Jan.-) Feb.-May (-Jul.)
DISTRIB. Uncommon in the lower forest zone of Iraq, common in the Steppe and Desert Regions including the lower Mesopotamian irrigated plains. **MAM**: Dohuk, *Guest* 1303!; **MRO**: Qara No, on road to Koi Sanjaq, *Rawi, Noori & Kass* 28050! 2 km N of Darband, *Botany Staff* 43209!; **MSU**: Jarmo, *Helbaek* 693!; between Sesar and Darband, *Botany Staff* 43046!; valley near R. Silwan, 7 km S of Darband-i Khan, *A. Mughair, Ani & Barkley* 118! **FUJ/MJS**: 1 km N of Balad Sinjar, *Barkley & Haddad* 6720! **FUJ**: by R. Tigris, S of Mosul, *Barkley & Majid Brahim* 6772!; Qaiyara, *Bayliss* 12!; Qal'a Sharqat, by mounds at Ashur, *Guest* 394!; 6 km S of Sharqat, *Rawi & Hamada* 33604!; **FUJ/DLJ**: among ruins at Hadhr, *Alizzi & Husain* 33894! **FNI**: Nineveh, *Stocks* 4!; **FAR**: nr Eski Kellek, c. 30 km W of Arbil, *Shahwani* 25159!; Arbil, *Guest* 1459!; **FKI**: Jabal Hamrin, nr Kirkuk, *Bornmüller* 1400; bank of R. Zab, nr Kirkuk, *P. Grigg* 12!; Tuz, *Rawi & Gillett* 10299! **FKI/LCA**: 25 km S of Tuz on road to Khalis, *Ani* 9814!; Jabal Hamrin nr Table Mt., *Ludlow-Hewitt* 1074! **FPF**: 20 km 'from' Khanaqin, *Fauzi & Noori* 39597!; Mandali, *Rawi* 12668!;, 1 km S of Badra, *anon.* 18131!; Buskaya (Bagsaya), *Rawi & Haddad* 25593A!; Wadi Dawarly, *Rawi* 14954! **DLJ**: 30 km NW of Falluja, *Rawi* 20220! **DGA**: Samarra, *Jack Isaac (Ishaq)* 314!; 10 km S of Samarra, *Fauzi & Noori* 39670!; **DWD**: Ajramiya 65 km N of Rutba, *Rawi & Khatib* 32247!; 'Ud an-Nisr, 43 km N of Rutba, *Rawi & Khatib* 32241!; 20 km N of Rutba, *Rawi & Khatib* 32205!; Qara, N of Rutba, *Alizzi & Husain* 34071!; 10 km SW of Rutba, *Rawi* 21081!; Masad S of Rutba, *Alizzi* 35194!; Hazimiya, *Omar* 34216!; 70 km W of Nukhaib, *Rawi* 31393!; 30 km W of Rahhaliya, *Chakravarty & Rawi* 29816!; Ukhaidhir, *Barkley & Jum'a Brahim* 361!; **DSD**: S of Samawa on Salman-Shabicha track, *Rawi & Gillett* 6159!; Rubaitha, 20km N. of Salman, *Botany Staff* 42091!; 10 km E of Salman, *Guest, Rawi & Rechinger* 18783!; Salman, *Guest, Rawi & Long* 14079!; Ma'niya, *Alizzi & Omar* 35539!; E of Shabicha, *Rawi & Gillett* 6435!; Shabicha, *Alizzi & Omar* 35439!; 87 km from Busaiya along road to Salman, *Chakravarty, Khatib, Rawi & Tikriti* 29992!; N of Ansab, *Botany Staff* 42236!; Zubair, *Rawi & Gillett* 6042!; between Zubair and Imam Ali, *Rawi & Walter* 16212!; Shaiba, *Rogers* 02!; Baniya, 65 km WSW of Basra, *Guest, Rawi & Rechinger* 17296!; 75 km S of Khadhr al-Ma, *Alizzi* 34427!; "Mesopotamia" (probably nr Jabal Sanam), *Watson* s.n.. **LCA/DWD**: between Baghdad and Syrian desert, *F.K. Jackson* H. 240! **LCA**: Sudur, *Omar, Fauzi, Qaisi & Noori* 44293!; Khan Bani Sa'd, *Alizzi & Hazim* 33683!; Daltawa (Khalis), *Guest* 1707!; Baghdad, *Colvill* 7!; *Paranjpye* s.n. (comm. Graham), *Rogers* 0719!; Babylon, *Nábělek* 3790; Hilla, *Bornmüller* 783; nr Garmashaya, 70 km NE of Kut, *Rawi* 18031! **LEA**: nr Wadi-Tib, *Guest, Rawi & Rechinger* 17466!; Fakka, N of Amara, *Rawi* 14914!; **LSM**: Majarr, *Alizzi* 34431!; 70 km E of Amara, nr Rushida well, *Husain al-Ani* 39758!; **LBA**: between Basra and Zubair, *Botany Staff* 33278!

S & SC Europe (naturalized further N), Cyprus, Syria, Lebanon, Palestine, Jordan, Sinai, Egypt, Arabia (Kuwait, Bahrain, Qatar, Oman, Yemen), Turkey, Caucasia, Iran, Afghanistan, N Africa (Morocco, Algeria, Tunisia, Libya).

KHUZNA (Turk., Tuz, *Rawi & Gillett* 10299. HANWA (Kuw., *Carter*, "camels feed on it but the badawin hold it is fatal if they eat too much; a local name confirmed by Dickson as HANUWA). According to Sharaf (1928) the name KAHLĀ is recorded by Ibn al-Baitar (c. 1240).

3. **Calendula officinalis** *L.*, Sp. Pl.: 921 (1753); Curtis's, Bot. Mag. 59: t. 3204 (1832); Guest in Dep. Agr. Iraq Bull. 27: 17 (1933); Dinsmore in Post, Fl. Syria, ed. 2, 2 : 72 (1933); Rawi & Chakravarty, Dep. Agr. Iraq Bull. 14: 21 (1964); Meikle in Fl. Turkey 5: 171 (1975) & Fl. Europ. 4: 207 (1976); Husain & Kasim, Cult. Pl. Iraq: 109 (1975); Chakravarty, Plant Wealth Iraq 1: 93 (1976); Zhu Shi et al. in Fl. China. 20-21: 819 (2011).

C. aurantiaca Kotschy ex Boiss. in Kotschy, Pl. Pers. Austr. no. 295 (1845).
C. sinuata Boiss. & Gaill. in Boissier, Diagn. Pl. Orient. ser. 2, 6: 109 (1859).
C. sinuata Boiss. & Gaill. var. *aurantiaca* Kotschy ex Boissier, Fl. Orient. 3: 416 (1875).

Erect, spreading annual or short-lived perennial herb to 32 cm tall, herbaceous or woody at base only, variably glandular-multicellular-hairy. Leaves all cauline, generally oblanceolate or spatulate, 4.5–7(–13) × 1.1–1.9(–4) cm, sessile, ± amplexicaul, acute to obtuse, apiculate, lower leaves attenuate, uppermost sometimes oblong-linear, ± truncate at base, entire or subentire with minute callouses at intervals along margin. Capitula radiate, solitary, terminal at the end of branches, 3–5(–6.4) cm in diameter; outer phyllaries 1.4 × (1.5–)2–3 mm, inner 1–1.2 × 1–1.2 mm. Ray floret limbs usually oblanceolate, corollas 11–24 × 1–10 mm, with 2–3 apical teeth and usually 4 veined; base of floret densely glandular-multicellular-hairy; disc florets 4–6 mm long, deeply 5-dentate, exterior variably papillose or glandular-multicellular-hairy all over or at base only. Achenes trimorphic, marginals beaked and cymbiform or all cymbiform, beaked achenes 1.7 cm long, (to 2 cm in cultivated material), ± cylindrical, strongly incurved, stout, spine muticous, sometimes ± echinate with or without narrow wings, beak usually short; cymbiform achenes 1.2 cm long, inflated with a somewhat rugose spine; inner achenes annulate, 4 mm in diameter, dorsally strongly rugose to tuberculate.

HAB. Widely cultivated as a garden ornamental; alt. up to 35 m; fl. & fr. Mar.-Apr.(-Jul.).
DISTRIB. Known only in cultivation from Lower Iraq. **LCA**: Baghdad (cult.) *Janan* C. 410!; Zafariniya, *Omar & Sahira* 34811! **LEA/FPF**: 45 km E of Tib, *Thamer* 47709. Zafariniya, Rawi 12876!, *Guest, Rawi & K.H. Kreeb* 14555!, *Shakir Sabir* 20187!; Abu-Ghraib, *Ahmed* 9861!, *AiIzzi* 34852!

Widely cultivated throughout S Europe; also said to be cultivated in Syria, Lebanon, Palestine, Egypt, Turkey, Iran, N Asia, China, Japan and elsewhere. The origin of this species is unknown; it is often recorded as a garden escape; a proliferous form is sometimes found. There is a doubtfully localised specimen at Kew from 'Mosul, *Stocks* 32'. J.E. Stocks is not known to have collected in Iraq, but a Stocks specimen is cited under *C. arvensis.*

Marigold (Eng.), sometimes Pot or Garden Marigold, a plant long cultivated for its medicinal properties. According to Wren (1956) these are stimulant and diaphoretic. Campbell Thompson (1949) included it in his catalogue of Ancient Assyrian drugs; Lindley & Moore (1870) referred to its reputation as a domestic remedy. Stearn (1971) mentioned that the plant was used in English country cooking to flavour thick soups since a single head thrown into the pan imparted a pleasing taste.

Parsa (1946) gives the Persian name of this species as GUL HAMÎSHA BAHAR (Pers.) which is similar to the Kurdish name recorded for Marigold, HEMÎSHE BEHĀR (Kurd.) by Wahby & Edmonds (1966).

According to Husain & Kasim (1975) this species is "cultivated all over the country".

66. **DIMORPHOTHECA** Moench

Meth. Pl.: 585 (1794)
Nordenstam in Kubitzki, Fam. Gen. Vasc. Pl. 8: 245 [2006] (2007)

C. C. Townsend

Annual (or perennial, suffruticose or shrubby, not in Iraq) plants with alternate, entire to pinnatipartite leaves. Capitula terminal, heterogamous, radiate. Involucre broadly campanuate or hemispherical; phyllaries ± uniseriate, herbaceous with a scarious border; receptacle flat or almost so, naked. Marginal florets radiate, pistillate, fertile; rays yellow, or white above and ± bluish-suffused beneath. Disc florets hermaphrodite, fertile or innermost sterile, corolla lobes yellow or tipped with purplish-black. Achenes dimorphic, those of ray-florets terete to 3-winged, smooth or usually ± rugose or tuberculate, those of disc florets strongly compressed with a thickened border; pappus absent.

Seven species native to southern Africa; one species in Iraq.

Dimorphotheca (from the Gr. δυμορφή, dimorfi, two-formed; θήκη, thēkē, a case, alluding to the dimorphic achenes). The plants of this genus are known in cultivation as Cape Marigold (Eng.). Cultivated in Iraq.

Norlindh, T. (1943). Studies in the Calenduleae. 1. A monograph of the genera *Dimorphotheca, Castalis, Osteospermum, Gibbaria* and *Chrysanthemoides*. Lund.

1. **Dimorphotheca sinuata** *DC.*, Prodr. 6: 72 (1838); Husain & Kasim, Cult. Pl. Iraq: 113 (1975).

Annual herb, 10–30 cm, branched from base or up to about middle, stems erect or ascending, subglabrous to ± puberulous and minutey glandular. Lower and medium cauline leaves oblong-oblanceolate, 1.5–8(–10) × 0.4–2(–3) cm, sinuate-dentate, with 2–5 remote, coarse, obtuse teeth along margin, indumentum similar to that of stem. Capitula long-pedunculate, terminal on stem and branches; involucre hemispherical, 15–25 cm in diameter, phyllaries subglabrous to ± pubescent and glandular. Rays 16–30 mm, yellow above, whitish and ± suffused with purple beneath especially towards apex. Disc florets golden-yellow above with lobes tipped with purplish-black. Marginal achenes very rugose; inner achenes very compressed, smooth with an incrassate margin.

HAB. & DISTRIB. Cultivated, presumably as an ornamental. **DWD**: Kerbela, *Rawi & Gillett* 6337!

A native of S & SW Africa N to Angola.

According to Husain & Kasim this species is cultivated all over Iraq, but this is the only specimen seen.

TRIBE **GNAPHALIEAE** (CASS.) LECOQ & JULIET

67. **BOMBYCILAENA** (DC.) Smolj.

Not. Syst. (Leningrad) 17: 447 (1955)
Bayer et al. in Kubitzki, Fam. Gen. Vasc. Pl. 8: 245 [2006] (2007)
Micropus L. sect. *Bombycilaena* DC., Prodr. 5: 460 (1836)

Frances G. Davies

Lanate annual herbs. Leaves simple. Capitula heterogamous, disciformis, in clusters at stem apices and surrounded by a whorl of leaves; phyllaries hyaline, woolly, inner phyllaries prominent, woolly, completely enclosing female florets and deciduous with them; receptacle peg-like, epaleaceous. Tubular florets few at apex of naked receptacle, functionally male. Achenes compressed, ovoid, greenish brown, glabrous; pappus absent.

Three species in the Mediterranean region, Asia and N America; one species in Iraq.

Bombycilaena (from Lat. *bombycinus*, silky, derived from *Bombyx*, the name of the silk-worm).

1. **Bombycilaena discolor** (*Pers.*) *Lainz* in Bol. Inst. Estud. Astur. Supl. Cienc. 16: 194 (1973); Holub in Fl. Turkey 5: 114 (1975) & Fl. Europ. 4: 125 (1976); Wagenitz in Fl. Iranica 145: 11 (1980).

Micropus discolor Pers., Synops. Pl. 2: 423 (1807).
M. bombycinus Lag., Gen. Sp. Nov. 32 (1816); Dinsmore in Post, Fl. Syria, Palest. & Sinai ed. 2, 3: 37 (1933); Anthony in Notes Roy. Bot. Gard. Edinburgh 18: 290 (1935); Rawi in Dep. Agr. Iraq Tech. Bull. 14: 120 (1964).
M. erectus L. subsp. *bombycinus* (Lag.) Rouy, Fl. France [Rouy & Foucaud] 8: 170 (1903).
Bombycilnaena bombycina (Lag.) Soják in Novit. Bot. Del. Sem. Horti Bot. Univ. Carol. Prag. 1962: 50 (1962).

Branched herb 10–40 cm tall. Leaves sessile, oblong to oblanceolate, 1–2 × 1.5–3 mm, acute, entire. Capitula ± spherical, 2–4 mm across; outer phyllaries 5, hyaline, 2–3 mm long, spread in a star-shaped pattern after fruiting, inner phyllaries 3–3.5 mm long, ovate,

Fig. 66. **Bombycilaena discolor**. 1, habit × 2; part of stem × 12; 3, inner phyllary × 18; 4, hermaphrodite floret × 36; female floret × 36; 6, achene × 6. All from *Chapman* 349. Drawn by Christine M. Grey-Wilson.

concave, densely lanate, enfolding female floret so that only style is visible. Achenes 1 mm. 'Hermaphrodite' florets 2 mm long, without achenes. Fig. 66: 1–5.

HAB. Among cut *Quercus* scrub above a small mountain town; alt. c. 1500+ m; fl. & fr. Jul.
DISTRIB. Rare in Iraq – only found once in the upper Forest Zone. **MSU**: Penjwin, *Haines* W. 2083A!

Distinguished from the rather similar *Cymbolaena* by the individual capitula being visible and becoming separate at maturity.

S Europe, Cyprus, Syria, Lebanon, Turkey, Iran, N Africa (Morocco, Algeria, Libya).

68. **CYMBOLAENA** Smoljan

Not. Syst. (Leningrad) 17: 452 (1955)
Bayer et al. in Kubitzki, Fam. Gen. Vasc. Pl. 8: 262 [2006] (2007)

C. C. Townsend

Small grey-woolly annual herbs. Capitula aggregated into clusters each subtended by a whorl of leaves; receptacle very narrow and elongated, woolly below. Female florets in several undefined series, each enclosed except for the style within an inner phyllary resembling a receptacular scale; apex of receptacle naked, bearing several functionally male tubular 4-lobed florets, without achenes; pappus of female florets consisting of capillary babellate setae with patent cilia; male florets lacking a pappus.

A monotypic genus of SW & C Asia.

Wagenitz, G. (1971). Zur Taxonomie und Nomenclatur von *Micropus* (Compositae-Inuleae). Öst. Bot. Zeitschr., 119: 399–403.

1. **Cymbolaena griffithii** (*A.Gray*) *Wagenitz* in Österr. Bot. Zeitschr. 119: 402 (1971); Holub in Fl. Turkey 5: 115 (1975); Wagenitz in Fl. Iranica 145: 13 (1980); Collenette, Wild Flow. Saudi Arabia: 186 (1999); Breckle, Hedge, Podlech & Rafiqpoor, Vasc. Pl. Afghan. checklist: 170 (2013); Kilian & Ghazanfar in Fl. Oman 3: 278 (2015); Taifour & El-Oqlah, Pl. Jordan Ann. Checkl.: 41 (2017).

Stylocline griffithii A.Gray in Proc. Amer. Acad. Art. Sci. 8: 652 (1873).
Micropus longifolius Boiss. & Reut in Boissier, Fl. Orient. 3: 242 (1875); Handel-Mazzetti in Ann. Naturh. Mus. Wien 27: 431 (1913); Nábělek in Publ. Fac. Sci. Univ. Masaryk 52: 11 (1925); Dinsmore in Post, Fl. Syria, Palest. & Sinai ed. 2, 2: 37 (1933); Guest in Dep. Agr. Iraq Bull. 27: 63 (1933); Bornmüller in Beih. Bot. Centralbl. 60B: 189 (1939); Blakelock in Kew Bull. 4: 59 (1949); Zohary in Dep. Agr. Iraq Bull. 31: 146 (1950); Rechinger, Fl. Lowland Iraq: 598 (1964); Rawi in Dep. Agr. Iraq Tech. Bull. 14: 120 (1964); Holub in Fl. Turkey 5: 115 (1975).
Cymbolaena longifolia (Boiss. & Reut.) Smoljan, Not. Syst. (Leningrad) 17: 452 (1955).

Small ephemeral herb 1–15 cm tall, adpressed white-woolly; larger specimens branching at base and with overtopping lateral inflorescence branches, reduced specimens acaulescent and unbranched. Stem floccose to glabrescent, pale, not altogether concealed by leaves. Leaves 6–20 × 1–2.5 mm, simple, oblong, oblanceolate or linear, entire, acute, grey-lanate on both surfaces. Capitula disciform, 2.5 mm, pedunculate, aggregated into dense terminal or axillary sensely woolly glomerules, 10–20 mm across, and surrounded by leaves. Involucre *c.* 4 mm; phyllaries densely white-woolly, acute; outer phyllaries minute; median 8–10 phyllaries in 2 series, with a scarious glabrous appendage, enveloping each a marginal flower; the inner 5 phyllaries surrounding the 4 central flowers, slightly plicate, linear-lanceolate with a dorsally prominent, yellowish, basally protruding midrib. Flowers purple, marginal female, filiform; central functionally male, tubular. Achenes of marginal flowers 1–1.5 mm long, obovoid, glabrous; of central flowers rudimentary, sparsely hairy. Fig. 67: 1–4.

HAB. Stony mountain slopes, hillsides and valleys, dry stony stream beds, roadsides and along edges of fields; alt. up to 1100(–1200) m; fl. & fr. (Mar.-)Apr.-May(-Jun.).
DISTRIB. Very common in the Steppe and Desert Regions of Iraq, quite common in the Lower Forest Zone. **MJS**: Jabal Golar (Qilat), *Field & Lazar* 483! **MAM**: 27 km NNE of Zakho, nr Sharanish, *Rawi* 23600A!; Oqur, *Alkas* 18668!; nr Suwara Tuka, *E. Chapman* 26349! **MRO**: Salah ad-Din, *Gillett* 11266!; Shaqlawa, *Haines* W. 702!; Kuh-i Sefin, above Shaqlawa, *Bornmüller* 1387!; Koi Sanjak, *Rawi, Noori & Kass* 28135!; between Rowanduz and Akoyan, *Kass & Noori* 27255!; Bird Agha Gin, W of Rania, *Rawi et al.* 28694! **MSU**: Jarmo, *Helbaek* 1346!; Pira Magrun, *Rawi* 12129! **FUJ**: 'Ain al-Husan, *Guest* 4188!; nr Balad Sinjar, *id.* 4103!; Qaiyara, *Bayliss* 69!; Shargat, *Y.R. Rao* s.n. (comm. Graham)! **FNI**: Tal Kaif, *Mudir of Tal Kaif* 3210! **FAR**: Arbil, *Shams ad-Din* 5601!; 7 km W of Altun Kupri, *Botany staff* 43251! **FKI**: 13 km N of Kirkuk, *Rawi & Gillett* 10576!; Kirkuk, *Bornmüller* 1385; 10 km N of Tuz, *Botany staff* 42938!; Tuz, *Guest* 1418!, 1980!; Sulaiman Beg, *Guest, Eig & Zohary* 5089!; 25 km S of Tuz on road to Kifri, *Ani* 9824!; Salahiya (Kifri), *Nábělek* 3883; Jabal Hamrin, N of Samarra, *Barkley & Juma Brahim* 9824!; NE of

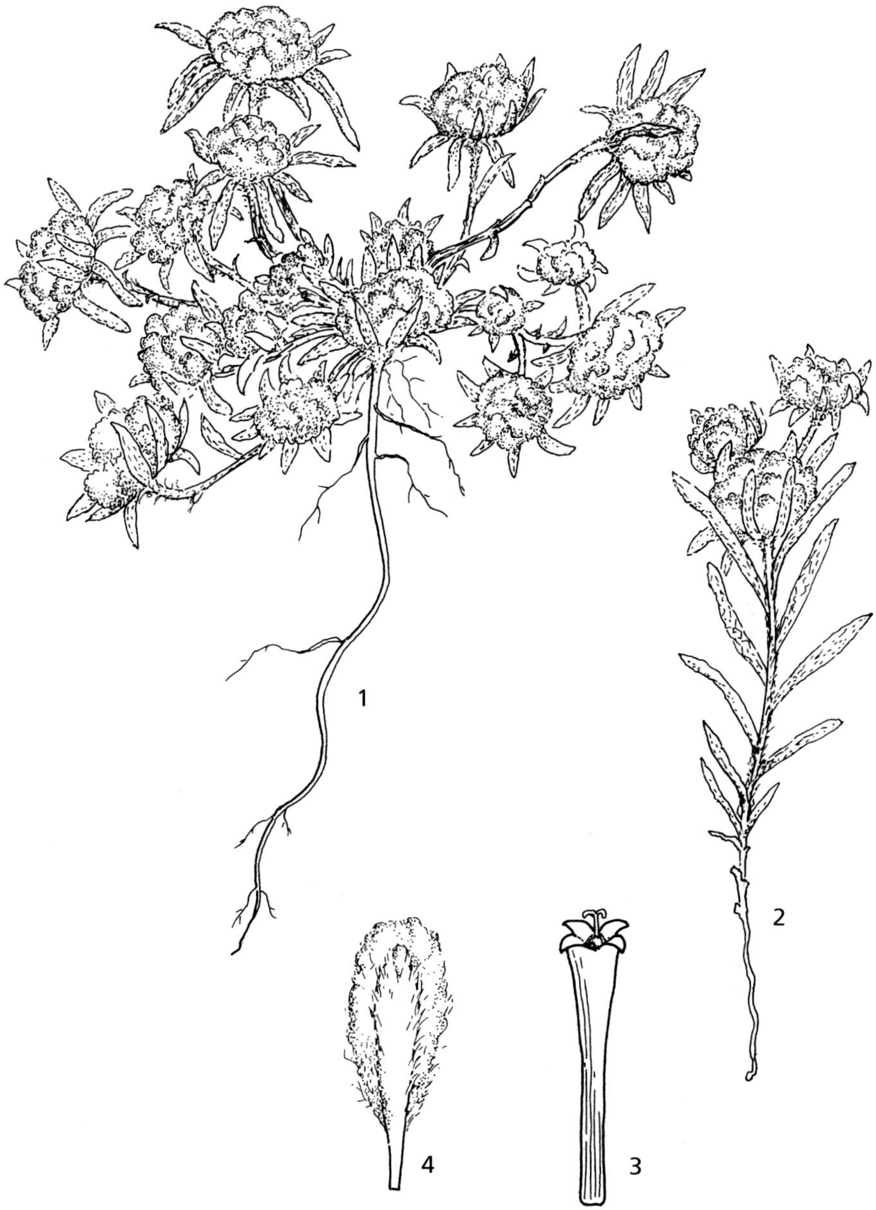

Fig. 67. **Cymbolaena griffithii.** 1, habit (prostrate form) × 1; 2, habit (erect form) × 1; 3 floret (erect form) × 21; 4, bract (erect form) × 6. 1 from *Kaisi, Thamer & Salah* 51360; 2–4 from *Kaisi* 50535.

Sar Qal'a in valley of Narin Chai, *Poore* 525! FPF: 4 km N of Sa'diya, *Kaisi* 42580!; Khanaqin, *Cowan & Darlington* 209!; Pilkhana, 10 km E of Khanaqin, *Rawi* 12783!; 16 km SE of Badra, *anon.* 18215!; Jabal Muwaila, 70 km N of Amara, *Guest, Rawi & Rechinger* 17573! FPF/LEA: nr Wadi at-Tib, *id.* 17525! DLJ/FUJ: between Ain Dibs and Baiji, *Rawi & Gillett* 7177! DLJ: 40 km SW of Ilaimid (Tal Julaimid), *Alizzi & Husain* 33967!; 80 km N of Rawa on road to Shbaichan, *Chakravarty, Rawi, Khatib & Alizzi* 31968!; Jazira, *Edmonds* 3819! DGA: 4 km E of Samara, *Rawi* 20319! DWD: Tal al-Nisr, NE of Rutba, *id.* 31199!; nr Ana, *Hamid* 39074!; 20 km W of Ramadi, *Omar* 34306!; Nukhaib, *Rawi* 14791! LCA: 15 km W of Falluja, *Chakravarty & Rawi* 30325A!; Baghdad, *Bornm.* 1386; Sudur Mansuriya, *Barkley, Haddad et al.* 6816! LEA: Ba'quba, *Rogers* 044! LBA/DSD: "Mesopotamia" (probably near Basra and/or Shuwaiba), *Watson* s.n.!

Syria, Lebanon, Jordan, Arabia, Turkey, Iran, Turkmenia, Afghanistan, Pakistan.

QUTAINA ("cottony", diminutive of QUTN, cotton, Ar. – sometimes further diminished to UQTAINA Ir., or even debased to UKTAINA). It appears to be prized as quite a useful forage plant, "useful for sheep grazing", *Mudir of Tal Kaif* 3210, "grazed by livestock", *Radhi* 3870 from Arbil; 'ALAICH Al-GHAZĀL ("gazelle fodder", Ir., Falluja, *Guest* 945). The nomadic tribesmen collect the dry tow (fibre) from this plant and use it as tinder to make fire for smoking or other purposes, igniting it by sparks from two stones struct together – see also note under *Filago pyramidata.*

69. FILAGO L. *nom. cons.*

Sp. Pl.: 927 (1753); Gen. Pl. ed. 5: 397 (1754)
Bayer et al. in Kubitzki, Fam. Gen. Vasc. Pl. 8: 264 [2006] (2007)
Logfia Cass., Bull. Soc. Philom. Paris 1819: 142 (1819), p.p.; *Oglifa* Cass., loc. cit. 143; *Gifola* Cass., loc. cit. 143

C. C. Townsend

Greyish annual herbs, erect or prostrate. Leaves simple, arachnoid or felted-hairy, in rosettes or alternate. Capitula aggregated into ovate or globose terminal or axillary clusters sometimes overtopped by subtending leaves; phyllaries few, minute, scarious. Receptacles often united into compound receptacles; paleae each subtending a female filiform epappose floret; middle florets female, filiform, enclosed within innermost scales, with a pappus; and a small number of central tubular hermaphrodite (fertile or functionally male) florets with a pappus. Achenes cylindrical to obovoid, papillose; pappus (when present) of bristles coherent at base, barbellate or plumose, or epappose.

Around 50 species in Europe, Asia, N Africa and N America; 7 species in Iraq.

Filago, from the lat. *filum*, a thread, with reference to its cottony indumentum.

Chrtek J. & Holub, J. (1963). Poznámky k taxonomii a nomenlaktuře rodů Evax Gaertn. a Filago L. Preslia 35: 1–17.
Wagenitz, G. (1969). Abgrenzung und Gliederung der Gattung Filago L. s.l. (Compositae – Inuleae). Willdenowia 5: 395–444.
Wagenitz, G. (1970). A key to the species of Filago L. sensu lato (Compositae) in Palestine. Israel Journ. Bot. 19: 260–265.
Fayed A. & Zareh, M. (1988). Systematic revision of Compositae in Egypt. 3. Inuleae: *Filago* and *Ifloga*. Willdenowia 17: 115–123.

1. Usually acaulescent, forming a rosette around the inflorescence; achene epappose. . 2
   Plant caulescent, usually erect; some achenes (at least those of hermaphrodite and inner female flowers) with a pappus. 4
2. Upper leaves linear, < 3 mm wide; groups of capitula ± spherical . . . . . . 1. *F. palaestina*
   Upper leaves linear-lanceolate to elliptic or spatulate, longest 3 mm or more wide; groups of capitula depressed. 3
3. Leaves erect, spatulate, narrowed and subpetiolate below, expanded at the base to a short sheath . . . . 2. *F. contracta*
   Leaves ± spreading, linear-lanceolate to elliptic, gradually narrowed at each end, neither subpetiolate nor sheathing below . . . . 3. *F. anatolica*
4. Aristate tips of phyllaries very short, inconspicuous . . . . 7. *F. arvensis*
   Aristate tips of phyllaries conspicuous, at least ⅓ of total scale length 5
5. Innermost phyllaries brownish or purplish on glabrous inner surface, lanate on outer surface and margins (not separating easily) . . . . 4. *F. desertorum*
   Innermost phyllaries glabrous 6
6. Phyllaries ± 5 mm, ovate, concave, abruptly contracted to a 1.5–2 mm long hair-tip, glabrous . . . . .6. *F. pyramidata*
   Phyllaries ovate to lanceolate, ± 3 × 1 mm, keeled at apex, shortly aristate, scarious and glossy . . . . .5. *F. eriocephala*

1. **Filago palaestina** (*Boiss.*) *Chrtek & Holub* in Preslia 35: 3 (1963); Wagenitz in Willdenowia 5: 418 (1969); Holub in Davis, Fl. Turkey 5: 112 (1975).

Fig. 68. **Filago palaestina**. 1, habit × 3; 2, leaf × 3; 3, central involucre bract × 15. 1 from *Anders* 2278; 2–3 from *Guest* 13379.

Anthony in Notes Roy. Bot. Gard. Edinburgh 18: 290 (1935); Rechinger, Fl. Lowland Iraq: 599 (1964); Rawi in Dep. Agr. Iraq Tech. Bull. 14 : 115 (1964)]; Feinbrun-Dothan, Fl. Palaest. 3: 307 (1978); Wagenitz in Fl. Iranica 145: 21 (1980); Taifour & El-Oqlah, Pl. Jordan Ann. Checkl.: 42 (2017).

Evax palaestina Boiss., Diagn. Pl. Orient. ser. 1, 11: 2 (1849); Boissier, Fl. Orient. 3: 244 (1875);

Dwarf tomentose annual, 2–8 cm, acaulescent or rarely with stems up to 4 cm, with prostrate lateral branches radiating from beneath a central capitulum, or simple. Leaves linear, 6–24 × 1–2.5 mm, floral up to twice as long as clusters of capitula. Capitula in groups of 10–20, ± spherical, 6–8 mm wide; phyllaries concave, carinate, 4–5 mm, dorsally lanate and ciliate, abruptly narrowed to long, setaceous points which protrude from the woolly clusters. Female florets 2.5 mm long; central hermaphrodite florets 2 mm long, bearing achenes. Achenes 1 mm, oblong-obovoid, papillose. Fig. 68: 1–3.

HAB. Apparently very local in Iraq, found in an open hillside and an open *Poa* steppe in the Upper Jazira; alt. 250–320 m; fl. Apr.
DISTRIB. **FUJ**: between Balad Sinjar and Tal Afar, *Guest* 4188A!; Ain-al-Husan, 30 km E of Balad Sinjar, *Guest* 4207!, 4222A!; nr Ba'aj, *Guest* 13379!; S of Mosul, *Anders* 2778! **FKI/FPF**: Jabal Hamrin, *Sutherland* 47! 59!

Turkey (SE Anatolia), Syria, Palestine, Jordan.

Guest 13379 notes that this plant is called QUTAINA, and used by the Badawin as tinder to light cigarettes (and for other purposes) from the sparks made by striking stones together.

2. **Filago contracta** (*Boiss.*) *Chrtek & Holub* in Preslia 35: 3 (1963); Holub in Fl. Turkey 5: 111 (1975); Feinbrun-Dothan, Fl. Palaest. 3: 308 (1978); Wagenitz in Fl. Iranica 145: 24 (1980); Boulos & Hind in Fl. Egypt 3: 199 (2002); Taifour & El-Oqlah, Pl. Jordan Ann. Checkl.: 42 (2017).

Evax contracta Boiss., Diagn. Pl. Orient. ser. 1, 11: 3 (1849); Boissier, Fl. Orient. 3: 243 (1875); Dinsmore in Post, Fl. Syria, Palest. & Sinai ed. 2, 2: 38 (1933); Zohary in Dep. Agr. Iraq Bull. 31: 156 (1950); Smol'yaninova in Fl. S.S.S.R. 25: 309 (1959); Rawi in Dep. Agr. Iraq Tech. Bull. 14: 115 (1964); Meikle, Fl. Cyprus 2: 874 (1985).

Dwarf tomentose annual, 2.5–5(–7) cm, acaulescent, often simple, sometimes with a few short branches from base. Leaves crowded in a rosette, broadly to narrowly spatulate, mucronate, largest 10–30(–35) × 3–5(–7) mm, narrowed and subpetiolate below, expanded at base to a short, finally indurate sheath, floral mostly 3–4 × as long as clusters of capitula. Capitula in groups of about 6, clusters 5–15 mm wide, central larger and probably equivalent to several fused capitula, bases of all capitula indurate and ± fused; phyllaries rhomboid to obovate, acuminate, sparsely lanate dorsally, narrowed to long, setaceous point which protrude from clusters. Female and central functionally male florets about 2 mm. Achenes 1.2–1.3 mm, papillose. Fig. 69: 1–3.

HAB. Apparently very rare in the eastern foothills region of Iraq, recorded on dry barren patches of soil at alt. 220 m; fl. Apr.
DISTRIB. **FKI**: Tuz, *Guest* 1979! **FPF**: Pilkhana, 10 km E of Khanaqin, *Rawi* 12814!

Crete, Aegean Is., Cyprus, Egypt, Palestine, Jordan, Syria, Lebanon, Turkey, Iran, N Africa (Libya, Algeria).

3. **Filago anatolica** (*Boiss. & Heldr.*) *Chrtek & Holub* in Preslia 35: 3 (1963); Holub in Fl. Turkey 5: 105, f. 6 (1975); Wagenitz in Fl. Iranica 145: 23 (1980).

Evax anatolica Boiss. & Heldr. in Boiss., Diagn. Pl. Orient. ser 1, 11: 2 (1849); Boissier, Fl. Orient. 3: 243 (1875); Dinsmore in Post, Fl. Syria, Palest. & Sinai ed. 2, 2: 38 (1933); Smol'yaninova in Fl. S.S.S.R. 25: 308 (1959).

Dwarf, greyish-tomentose annual, 1–8 cm, acaulescent or with stems up to 4 cm, simple or with slender lateral branches from base and/or beneath a central capitulum. Leaves linear-lanceolate to elliptic-oblong, mucronate, largest 5–20 × 3–4.5 mm, narrowed to each end or somewhat more so but not subpetiolate, floral up to 3 × as long as clusters of capitula. Capitula in groups of about. 10, clusters 5–11 mm wide; phyllaries oblanceolate, sharply folded and keeled, dorsally tomentose, narrowed to a whitish arista which protrudes from the woolly clusters. Female florets *c.* 2 mm long, many; hermaphrodite florets 1.5–1.8 mm, bearing achenes. Achenes oblong-obovoid, 1.2–1.4 mm, papillose. Fig. 70: 1–3.

HAB. Apparently very rare in Iraq; only found once in the mountain region of Iraq; alt. ± 1360 m; fl. Jul.
DISTRIB. **MAM**: Khantur, *Rawi* 23398!

Turkey, Lebanon, Syria, Transcaucasia, NW Iran.

Fig. 69. **Filago contracta**. 1, habit × 3; 2, leaf × 3; 3, central involucre bract × 9. All from *Amdursky* 295 (Palestine material). Artist unknown.

The specimen cited by Boissier (Fl. Orient. 3: 244 (1875)) as *Evax exigua* Sibth., "Babylonia ad Mohammera (*Noë*)!, i.e. *Noë* 922! is referable to *Filago desertorum* Pomel, q.v.

4. **Filago desertorum** *Pomel*, Bull. Soc. Sci. Phys. Algérie 11: 46 (1874); Holub in Fl. Turkey 5: 106 (1975) & Fl. Europ. 4: 122 (1976); Feinbrun-Dothan, Fl. Palaest. 3: 306 (1978); Collenette, Wild Flow. Saudi Arabia: 192 (1999); Wagenitz in Fl. Pakistan 210: 134 (2003); Boulos & Hind in Fl. Egypt 3: 199 (2002); Breckle, Hedge, Podlech & Rafiqpoor, Vasc. Pl. Afghan. checklist: 173 (2013); Kilian & Ghazanfar in Fl. Oman 3: 243 (2015); Taifour & El-Oqlah, Pl. Jordan Ann. Checkl.: 42 (2017).

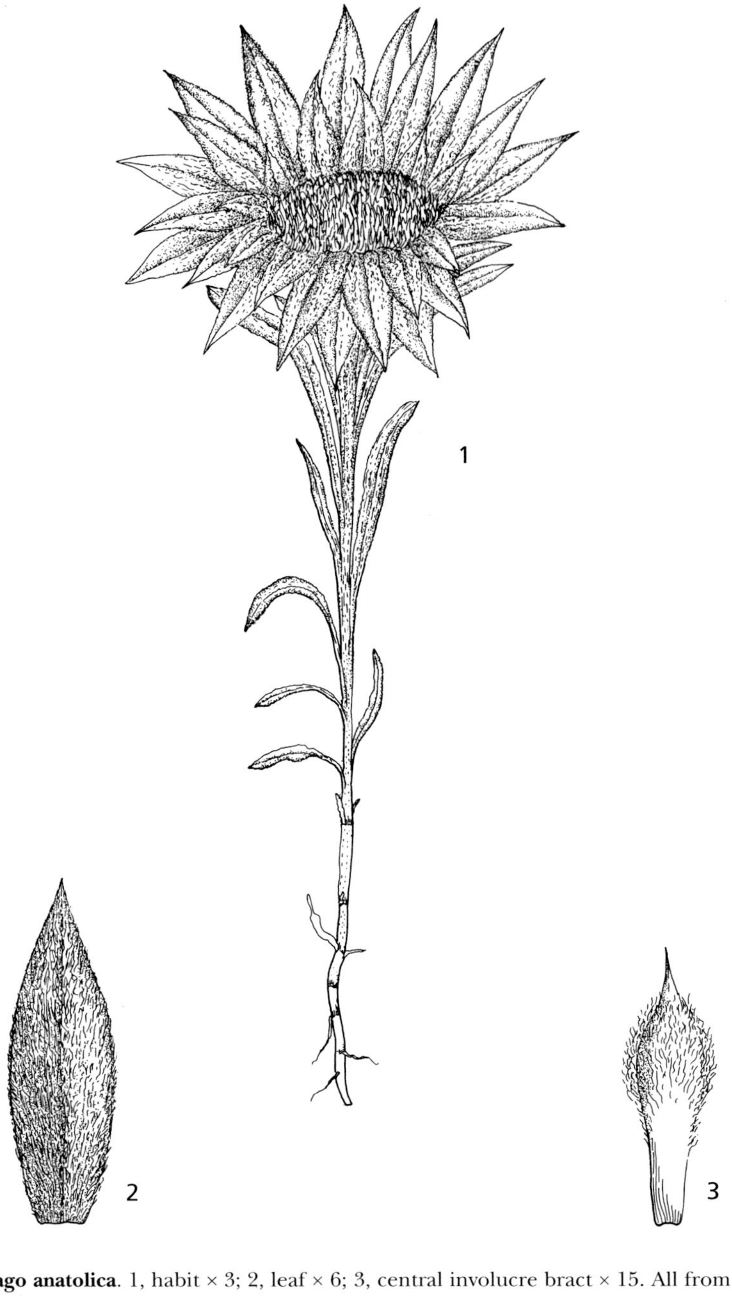

Fig. 70. **Filago anatolica**. 1, habit × 3; 2, leaf × 6; 3, central involucre bract × 15. All from *Rawi* 23398. Artist unknown.

F. spathulata C.Presl var. *desertorum* (Pomel) Batt. in Battandier & Trabut, Fl. Algérie, Dicot. 441 (1889).
F. spathulata C.Presl var. *alexandrina* Bornm., Repert. Spec. Nov. Regni Veg. 18: 41 (1922).
F. spathulata C.Presl var. *evaciformis* Bornm., Repert. Spec. Nov. Regni Veg. 18: 41 (1992) & subvar. *lancifolia* Bornm., loc. cit. 43 (1922).
Evax mauritanica Pomel var. *cyrenaica* Pamp., Nuovo Giorn. Bot. Ital. n.s., 34: 962 (1927).
F. spathulata C.Presl forma *desertorum* (Pomel) Pamp., Atti R. Ist. Ven. Sci. Lett. Arti 92: 234 (1932).
F. germanica L. subsp. *spathulata* (C. Presl) Rouy var. *desertorum* (Pomel) Jah. & Maire, Cat. Pl. Maroc 3: 745 (1934).

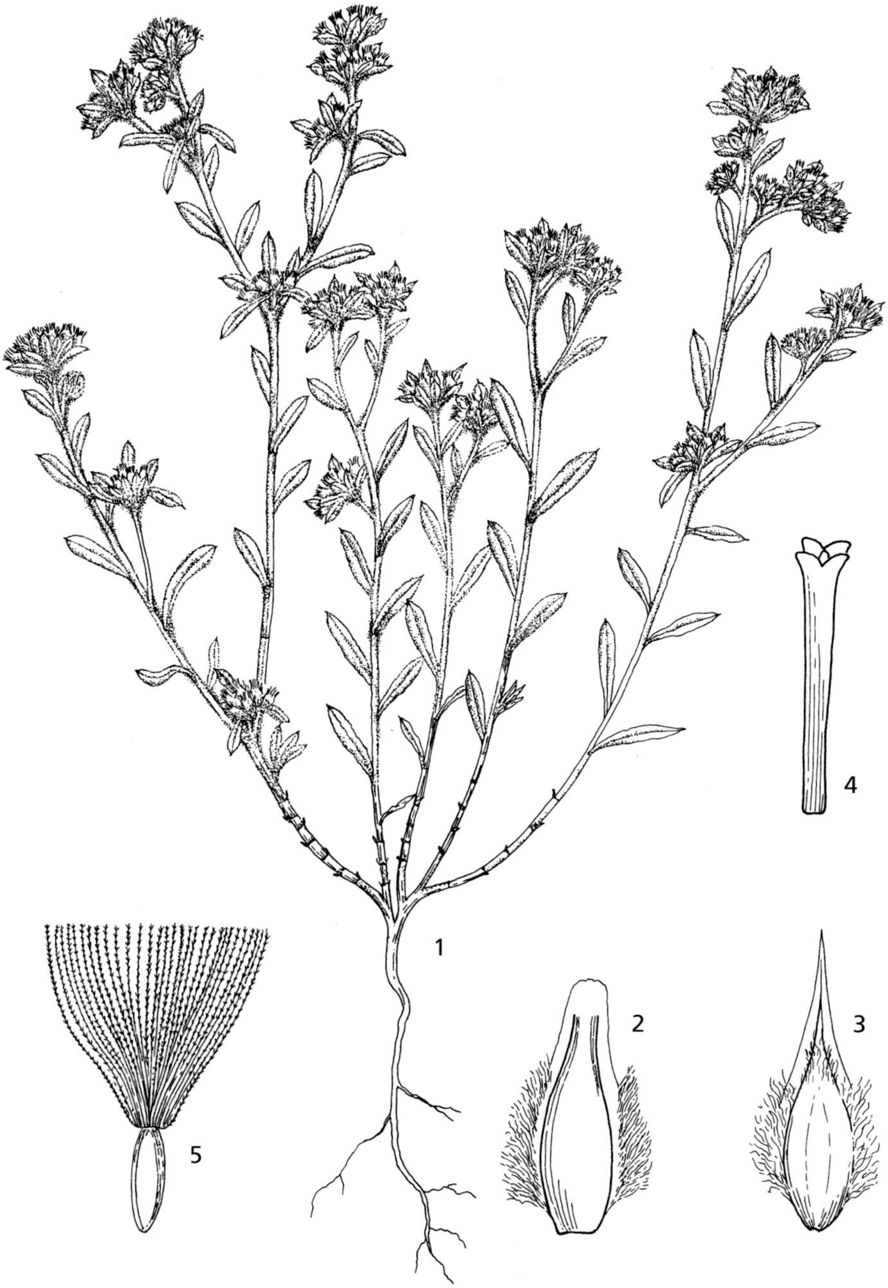

Fig. 71. **Filago desertorum**. 1, habit × 1; 2, inner phyllary × 15; 3, outer phyllary × 15; 4, floret × 18; 5, achene and pappus × 18. 1–4 from *Rawi & Haddad* 25780; 5 from *Alizzi & Husain* 33983. Artist unknown.

Small grey or brownish-grey woolly herb, 2–15 cm tall, main stem terminating in a capitulum shortly above ground level with numerous branches arising below it. Leaves obovate to oblanceolate, 1–1.5 × 0.2–0.4 mm. Capitula in spherical groups of 6–12 overtopped by leaves, each 4–5 mm; phyllaries 5-seriate, lanceolate, terminating in a 1–1.5 mm long straw-coloured scarious tip, brownish or purplish on glabrous inner surface, lanate on outer surface and margins. Female florets in concavity of phyllaries, numerous. Hermaphrodite florets 5–7, reddish-tipped. Achenes 0–8 mm long, elliptic, papillose; pappus deciduous. Fig. 71: 1–5.

HAB. Desert plains and dry steppic hills, most frequently on rather sandy soils, also as a weed in fields; up to alt. ± 650 m; fl. & fr. (Feb.-)Mar.-Apr.(-May).
DISTRIB. Occasional in the dry steppe zone and also on the alluvial plain in the Lower Mesopotamian region, very common in the Desert region of Iraq. **FKI**: 7 km W of Kirkuk, *Rawi, Noori & Kass* 27944! **FNI**: Bashiqa, *Hadač* 1549 p.p. **FPF**: nr the Iranian frontier E of Zurbatiya, *Gillett* 6714!; Jabal Muwaila, 70 km N of Amara, *Guest et al.* 17573!; Khanaqin, *Rechinger* 9055a, 9055b, p.p.; 16 km SE of Badra, *id.* 13964b; 10 km E of Mandali, *Rechinger* 12781. **FPF/LEA**: between Ba'quba and Khanaqin, *Hadač* 612 p.p. **FPF/LSM**: 24 km from the Kut-Amara turning to Badrah, *id.* 13983b & 14054a. **DWD**: H.3, 50 km W by N of Rutba, *Janan al-Mukhtar* 33377!; Rutba, *Rawi* 14651!; Qa'ra depression, *id.* 31278!; Wadi al-Ajramiya, *id.* 31270!; Hazimiya, 40 km SE of Rutba, *Chakravarty, Rawi, Khatib & al-Izzi* 31449!; 'Anaza, 125 km NE of Rutba, *id.* 31549!; 12 km N of T.1, on road to Husaiba, *id.* 31742!; H.1, 160 km ENE of Rutba, *id.* 31585!; Wadi Hauran, 12 km S of H.1, *id.* 31704!; 50 km W of Ramadi, *Rawi & Noori* 27002!; between Hit and Ramadi, *Rawi & Gillett* 6801!; 3 km S of Rahhaliya, *Chakravarty & Rawi* 29795!; Shithatha, *id.* 6422!; 15 km WNW of Karbala, *Rawi* 31506!; 5 km W of Karbala, *Rawi & Gillett* 6392! **DLJ**: Umm al-Matin, *Chakravarty, Rawi, Khatib & Alizzi* 31877!; 10 km SW of Manayif, *Alizzi & Husain* 33983! **DGA**: 4 km E of Samarra, *Rawi* 20352!; Asila, *id.* 20434!; **DSD**: 99 km NE by N of Shabicha on way to Najaf, *Rawi, Khatib & Tikriti* 29265!; Wadi al-Khirr, 15 km N of Lussuf, *Guest, Rawi & Rechinger* 19451!; 5 km NE of Ma'niya, *Khatib & Hazim* 30587!; 40 km NW of Salman, *Fauzi, Hazim & Hamid* 38687!; 10 km NE of Salman, *Rawi & Gillett* 6180!; Jal ar-Rawaq, *Rechinger* 9339!; 30 km E of Salmar, nr Golaib, *Guest, Rawi & Rechinger* 18748!; 50 km N by W of Aidaha (c. 80 km SSW of Salman), *id.* 19190!; Jumaima, 140 km SW of Salman, *id.* 19124!; nr Nukhaila station, *id.* 18312!; nr Jabal Sanam, *id.* 17041!; Zubair, *Rawi & Gillett* 6057!; 25 km SE of Zubair, *id.* 16863!; Umm Qasr, *id.* 16904!; 30 km NW of Ansab, *Kaisi, Hamad & Hamid* 48171!; 50 km from Ansab on the way to Samah, *Fauzi, Hazim & Hamid* 38868!; 10 km S by W of Busaiya, *Guest & Mahalhal* 15262!; 80 km ESE of Basra, *Guest & Rawi* 14297A! **LCA**: "Falluja desert", W of Abu Ghraib, *Haines* W.156!; Sudur, *Rawi* 20011!; Iskandariya, 50 km S of Baghdad, *Agnew & Shauqi* 189! **LEA**: Shahraban (Muqdadiya), *Rechinger* 13381; nr Kut, *Alizzi & Omar* 34898!; 48 km NE of Chabbab, nr Ali Gharbi, *anon.* 18034!; Fakka, 70 km N by E of Amara, *Rawi & Haddad* 25780!; 55 km N of Amara, *Guest, Rawi & Rechinger* 17454!

S. Europe, N Africa, Macaronesia (Canary Is., Azores), Syria, Palestine, Jordan, Arabia, Egypt, Turkey, Caucasia, Iran, Afghanistan, Pakistan.

DURAIHIMA (Ir.-Busaiya, *Guest & Mahalhal* 15620; ALAICH AL-GHAZĀL ("gazelle's fodder"), Ir.-Jabal Sanam, *Guest et al.* 17041), confirming that it is considered by local shepherds to be a plant readily grazed by sheep and other livestock – as also recorded by Muhammad Salim ar-Radhi under *F. pyramidata* in Arbil province, and by Guest (1933).

5. **Filago eriocephala** *Guss.*, Pl. Rar. Sic. 344, t.69 (1826); Holub in Fl. Turkey 5: 103 (1975); Feinbrun-Dothan, Fl. Palaest. 3: 306 (1978); Wagenitz in Fl. Iranica 145: 17 (1980); Ahmad in Feddes Repert. 124: (2013).

Filago germanica L. var. *erioceplaha* (Guss.) Boiss., Fl. Orient. (1875); Dinsmore in Post, Fl. Syria, Palest. & Sinai ed. 2, 2: 39 (1933).

Annual (5–)10–20(–35) cm, densely greyish-white lanate. Stems erect to procumbent, branched above middle. Leaves oblong to lanceolate, acute, 8–24 × 2–5(–8) mm, upper leaves not overtopping the primary cluster of capitula; floral leaves as long as or shorter than clusters of capitula. Capitula in groups of 30–50, 9–17 × 10–12 mm; phyllaries in several whorls, ovate to lanceolate, ± 3 ×1 mm, keeled at apex, shortly aristate, scarious and glossy. Inner female florets, 10–20, *c.* 2.5 mm; hermaphrodite florets 2–4, *c.* 2 mm. Achenes pappose.

HAB. Sandy soil, eroded places, roadsides; rare in the mountain region; alt. 1880–1900 m; fl. Jun.
DISTRIB. **MSU**: Lase Marf Mt., *Ahmad* 12-1065 (SUFA); above Taweela, *Ahmad* 13-117 (SUFA).

E, S and C Mediterranean region, Cyprus, W Iran, Palestine, Syria, Turkey.

6. **Filago pyramidata** (*L.*) *Huds.*, 1762: 328 (1762); Holub in Fl. Turkey 5: 104 (1975); Collenette, Wild Flow. Saudi Arabia: 192 (1999); Boulos & Hind in Fl. Egypt 3: 200 (2002); Breckle, Hedge, Podlech & Rafiqpoor, Vasc. Pl. Afghan. checklist: 173 (2013); Kilian & Ghazanfar in Fl. Oman 3: 244 (2015); Taifour & El-Oqlah, Pl. Jordan Ann. Checkl.: 42 (2017).

F. germanica L., Sp. Pl. ed. 2, 2: 1311 (1763); Boissier, Fl. Orient. 3: 245 (1875); Zohary in Dep. Agr. Iraq Bull. 31: 146 (1950); Rawi, ibid. Tech. Bull. 14 : 115 (1964); Rechinger, Fl. Lowland Iraq: 600 (1964).

Gnaphalium pyramidatum (L.) Lam., Encycl. 2 : 760 (1792).

Gnaphalium pyramidatum (L.) Willd., Sp. Pl., ed. 4 (Willdenow) 3(3): 1895 (1803), nom. illeg. comb. superfl.

Filago spathulata C.Presl, Delic. Prag. 99 (1822); Bornm. in Beih Bot. Centralbl. 60B: 189 (1940); Rechinger, Fl. Lowland Iraq: 600 (1964), incl. *F. prostrata* (Parl.) Boiss.

F. germanica L. var. *pyramidata* Gaudin, Fl. Helv. 5 : 253 (1829).

F. germanica L. var. *spathulata* (C.Presl) DC., Prodr. 6 : 247 (1838).

Gifola spathulata (as *spatulata*) (C.Presl) Rchb.f., Icon. Fl. Germ. Helv. (H.G.L. Reichenbach) 16: 26 (1853).

Filago germanica L. subsp. *spathulata* (C.Presl) Rouy, Fl. France [Rouy & Foucaud] 8 : 173 (1903).

[*F. germanica* auctt. non L.: Handel-Mazzetti. in Ann. Naturh. Mus. Wien 27: 431 (1913); Nábělek in Publ. Fac. Sci. Univ. Masaryk 52: 11 (1925); Guest in Dep. Agr. Iraq Bull. 27: 34 (1933).]

A softly woolly annual herb 5–30 cm tall. Branches mostly erect, 1-many from base, several in upper part of plant, these arising immediately below a capitulum cluster. Stems grey-lanate, visible between leaves. Leaves simple, sessile, oblanceolate, widest above middle, 10–20 × 1–3 mm, densely white-lanate. Capitula massed into spherical groups, individual capitula distinct, at ends of stem branches and usually shortly overtopped by a subtending series of leaves. Capitulum 5-angled, 5 × 1.5–2 mm bearing a few scales without florets, with much woolly indumentum at base giving way to scales with female florets in several series; phyllaries *c.* 5 mm, ovate, concave, abruptly contracted to a 1.5–2 mm long hair-tip, glabrous. Female florets *c.* 2 mm, filiform; achene *c.* 6 mm, obovoid. Inner female florets filiform, fertile, pappose; pappus longer than the achene. Hermaphrodite florets *c.* 2.2 mm, 4-lobed, pappose.

HAB. Steppic grassy plains and gypsiferous hills, in the desert on sandy soil along a limestone ridge, in clay in a depression, along irrigation channels and by roadsides on irrigated alluvial soil; alt. 30–600(–800) m; fl & fr. (Feb.-)Mar.-Apr.(-May).

DISTRIB. Common in the steppe and desert regions of Iraq and also in the irrigated alluvial plain, occasional in the mountains. **MJS**: Jabal Qilat, between Balad Sinjar and Ain Tellawi, *Field & Lazar* 419! **MAM**: 103 km N of Mosul towards Zakho, *Rechinger* 12180! **MRO**: Koi Sanjaq, *Rawi, Noori & Kass* 28138!; 5 km SW of Rania, *id.* 28424!; Rowanduz gorge, *Thesiger* 852. **MSU**: Jarmo, *Helbaek* 1289!; 1 km S of Darband-i Basian, *Rawi* 21707!; Darband-i Khan, *Omar & Karim* 38147!; E of Chemchemal, *Rechinger* 12497; 19 km W of Sulaimaniya towards Dukan, *id.* 10072; nr Dukan, *Hadač* 1216. **FUJ**: Tall Kochak, *Gillett* 10834!; 10 km N of Ba'aj, *Barkley & Haddad* 7913!; Baidha, 78 km S of Sinjar, *Chakravarty, Rawi, Khatib & Alizzi* 32085!; Tall Afar, *T. Sarsam* 9941!; 15 km N of Qaiyara, *Ani & Hasan Hadi* 9723!; Qaiyara, *Bayliss* 57!; Wadi Sefa, between Shargat and Hadhr, *Hand.-Mazz.* 1108; S of Mosul, *Anders* 2532, 2780; **FUJ/DLJ**: Hadhr, *Chakravarty, Rawi, Khatib & Alizzi* 33029! **FNI**: Nineveh, *Anders* 1262; nr 'Ain Sifni, *Field & Lazar* 738!; Bashiqa, *Hadač* 1549 p.p. **FUJ/FNI/FAR**: Makhlat, *Guest* 4246!, 4269! **FKI**: Kifri, *Rawi & Gillett* 7398!; Hawija, *Bornmüller* 1389, *Gillett* 10687!; between Tuz and Kifri, *Nábělek* 3896; NE of Sar Qal'a in valley of Narin Chai, *Poore* 526! **FKI/FPF**: Jabal Hamrin, *Bornmüller* 1388. **FPF/LEA**: Jabal Hamrin between Ba'quba and Khanaqin, *Hadač* 550. **FPF**: Pilkhana, *Rawi* 12791!; Khanaqin, *Rechinger* 9055, 9055b; 10 km N of Jalaula, *Hadač* 2981; 16 km SE of Badra, *Rechinger* 13964b; 1 km S of Badra, *anon.* 18139! **DWD**: 40 km N of Rutba, *Rawi & Khatib* 32277!; Wadi al-Ajramiya, *Rawi* 31235!; H.3 (50 km W by S of Rutba), *Janan al-Mukhtar* 33377!; Kambra Umm Rashid, *E. Palmatier & Barkley* 1046!; between Ana and Wadi Qasir, *Omar, Qaisi, Hamad & Hamid* 44384!; 20km E of Ana, on road to Qaim, *id.* 44417!; 10 km W of K.3, *id.* 44716!; K.3, *Chakravarty, Rawi, Khatib & Alizzi* 32933!; 10 km N of Kubaisa, *Hamid* 39026! **DLJ**: 5 km N of Manaif, *Qaisi & Hamad* 49009!; Rawa, *Hamid* 39105!; 8 km E of Ana, *Omar, Qaisi, Hamad & Hamid* 45100A! **DGA**: 4 km E of Samarra, *Rawi* 20357! **DSD**: nr Ansab, *Guest, Rawi & Rechinger* 18967!; 25 km from Busaiya, *Fauzi, Muhammad, Hamid & Kadhim* 40359!; 80 km NE of Zubair, *Thamir* 47397! **LCA**: Daltawa (Khalis), *Guest* 2486!; 13 km E of main road to Samarra from Baghdad & Tarmiya, *Barkley, Hikmat Abbas & Jum'a Brahim* 1604A!; nr Khan Bani Sa'd, *Alizzi & Hazim* 33680!; Baghdad, *Schafli* 46! **LCA**: Rustamiya, *Rogers* 074!; Latifiya, nr Mahmudiya, *Guest* 2388!; Hilla, *Bornmüller* 420 (f. *latifrons*). **LEA**: 60 km S of Baghdad, nr Hafariya, *Rawi* 18399!; 60 km N of Amara, nr Wadi Tib, *Guest, Rawi & Rechinger* 17479! **LEA/LSM**: between Fakka and Musaida, *Rawi* 14995! **LSM**: Majarr, *Alizzi & Omar* 34941!; Hor ach-Chibba, *Qaisi* 44032!

S Europe, Aegean Is., Syria, Palestine, Jordan, Arabia, Turkey, Caucasia, Iran, Afghanistan, Pakistan, N Africa, Macaronesia (Madeira, Canary Is., Azores).

Said to be "grazed by all animals", Arbil province, *Radhi* 3869 & Khalis, *Guest* 2486; QUTAINA "cottony" (floss), diminutive of QUTN (cotton, Ar.), "used by the Bedouin for tinder to obtain a light from a spark made by striking two stones together", nr Ba'aj, *Guest* 13377 (cf. under *Cymbolaena*, q.v.); Carter (1917) gives its local name in Kuwait as QUTTAIN, another diminutive of QUTN, and suggests comparison of the generic name *Filago* and the Latin word *filum*, a thread, which no doubt refers to the plant being covered with cottony hairs.

7. **Filago arvensis** *L.*, Sp. Pl.: 856 (1753); Dinsmore in Post, Fl. Syria, Palest. & Sinai ed. 2, 2: 40 (1933); Breckle, Hedge, Podlech & Rafiqpoor, Vasc. Pl. Afghan. checklist: 173 (2013).

Gnaphalium arvense (L.) Scop., Fl. Carn. ed. 2, 2: 153 (1772).
G. lagopus Steph. ex Willd., Sp. Pl. ed. 4, 3(3): 1897 (1803).
Oglifa arvensis (L.) Cass., Dict. Sci. Nat. ed. 2, 35: 448 (1825).
Filago arvensis var. *lagopus* DC., Prodr. 6: 249 (1838); Nábělek in Publ. Fac. Sci. Univ. Masaryk 52: 11 (1925).
Logfia arvensis (L.) Holub, Notes Roy. Bot. Gard. Edinburgh 33: 432 (1975) & in Fl. Turkey 3: 108 (1975) & in Fl. Europ. 4: 123 (1976).
F. davisii (Holub ex Grierson) Feinbrun-Dothan, Fl. Paleastina 3: 308 & Addenda (1978); Holub in Fl. Turkey 5: 109 (1975).

An erect tawny-grey woolly annual, 10–20 cm tall, with rather wiry stems. Leaves grouped towards tips of branches, spatulate to oblong, 6–20 × 1–4 mm. Capitula in groups at branching points and terminally, equaling or overtopping leaves, broadly ovoid, 2.5–6 × 2.5–4 mm; phyllaries lanceolate, 1–4 × 0.8 mm; outer female florets fertile, without pappus; outermost florets exceeding subtending scale; inner *c.* 20, filiform, fertile, broader, coriaceous in lower part but membranous above, deeply concave, with white lanate tomentum on outer surface, ± hidden by phyllaries. Hermaphrodite florets 6, tubular, possibly fertile. Achenes ellipsoidal, *c.* 1 mm long, pale brown; pappus pure white, 2–4 mm long, bristles barbellate below.

HAB. montane slopes; alt. 900–1400(–2220) m; fl. May-Jun.
DISTRIB. Occasional in the northern sector of the mountain region of Iraq. **MAM**: Aradin, *Nábělek* 3886; 'Ain Nuni, *id.* 3889. **MRO**, mt. Qandil, in gorges above Pushtashan, *Rechinger* 11802; Nawanda valley, *Hadač* 2357; Handren, *Nábělek* 3885. **MSU**, nr Penjwin, *Rechinger* 12247; mt. Malakawa, *id.* 10415 & 12292 p.p.

S Europe, Cyprus, Syria, Lebanon, Palestine, Turkey, Caucasia, Iran, Afghanistan.

70. **GAMOCHAETA** Wedd.

Chlor. And. 1: 151 (1856)
Bayer et al. in Kubitzki, Fam. Gen. Vasc. Pl. 8: 265 [2006] (2007).
Gnaphalium L., Sp. Pl.: 850 (1753)

Frances G. Davies
Revised by D. J. Nicholas Hind

Annual or perennial herbs. Leaves alternate, tomentose on both surfaces, flat, margins entire. Capitula many, in glomerule-like or spike-like clusters. Capitula heterogamous, disciform; phyllaries papery, brownish, stereome undivided; receptacle flat, epaleaceous. Outer florets female, numerous, corollas filiform, purple. Disc florets hermaphrodite, few, corollas cylindrical, (4–)5-lobed, purple; style arms truncate, apically hairy. Achenes oblong, setuliferous, setulae of globose twin-hairs; pappus setae capillary, barbellate, bases connate into a ring.

About 50–80 species, predominantly from South America, with adventive weeds in Africa, Asia, Australia, and Europe. Only one species is found in the Flora area.

Gamochaeta (from Gr. γαμο, *gamo*, marriage and χαιτε, *chaite*, mane, in reference to the hair covering).

Drury, D.G. (1971). The American spicate cudweeds adventive to New Zealand: (*Gnaphalium* Section *Gamochaeta*–Compositae). New Zealand J. Bot. 9: 157–185.
Anderberg, A.A. (1991). Taxonomy and phylogeny of the tribe Gnaphaleae (Asteraceae). Opera Bot. 104: [1–]5–195. [p. 155].

1. **Gamochaeta pensylvanica** (*Willd.*) *Cabrera*, Bol. Soc. Argent. Bot. 9: 375 (1961); Qaiser & Abid in Fl. Pakistan 210: 122, f. 28 A–D (2003).

Gnaphalium polycaulon auctt. non Pers. (1807). *Gnaphalium pensylvanicum* Willd., Enum. Hort. Berol. 867 (1809).
Gnaphalium peregrinum Fernald in Rhodora 45: 479, t. 795 (1943).
Gnaphalium purpureum auct. non L. (1753); Hook. f., Fl. Brit. India 3: 289 (1881).

Erect scarcely-branched annual herb 20–50 cm tall, vegetative parts thinly white-tomentose. Leaves spatulate, (2.5–)4–6(–8) × (0.4–)0.7–1.3(–1.8) cm, withering at anthesis, apex obtuse, lower half tapering into a pseudopetiole, surfaces concolorous, arachnoid pubescent, margins ± entire. Inflorescence of groups of capitula in spikelets subtended by leafy overtopping bracts. Capitula campanulate, 3–5 mm in diameter; outer phyllaries lanceolate, herbaceous, bright green, margins hyaline; inner phyllaries rounded or short-acuminate, straw-coloured, glabrous. Female florets filiform, c. 10; hermaphrodite fewer, tubular. Achenes minute; pappus coherent into a basal ring, deciduous.

HAB. & DISTRIB. Very rare in Iraq; only found once as a weed in the old shady neglected garden of the former British Embassy on the W bank of the R. Tigris at Karrada Mariam; alt. 35 m; fl. & fr. Apr.
LCA: Baghdad, *Agnew & Haines* W. 1926!

Native of S America, occurring as an alien in most warm parts of the world.

71. **GNAPHALIUM** L.

Sp. Pl.: 850 (1753); Gen. Pl. ed. 5: 368 (1754)
Bayer et al. in Kubitzki, Fam. Gen. Vasc. Pl. 8: 266 [2006] (2007)
Omalotheca Cass., Dict. Sci. Nat., ed. 2, 56: 218 (1828)

Frances G. Davies

Annual or perennial, soft densely grey or tawny tomentose herbs. Leaves usually spatulate or oblanceolate, margins flat. Inflorescence spicate or corymbose, individual capitula discrete, usually sessile. Capitula small, heterogamous, disciform; phyllaries few-seriate, stereome undivided; receptacle flat, epaleaceous. Female florets numerous, filiform, corollas yellow or cream; hermaphrodite florets fewer, corollas cylindrical, yellow (rarely mauve), corolla lobes apically hairy. Apical anther appendages flat; style arms truncate. Achenes small, ellipsoidal, brown, glabrous or with short subglobose duplex hairs, hairs not mucilaginous; pappus uniseriate, white, bristles free or coherent into a basal ring persistent (not in Iraq) or deciduous, sometimes absent in female florets.

A genus of ± 80 species world-wide, rather artificially separated from *Helichrysum* on the grounds of a greater number of female than hermaphrodite florets. In practice *Gnaphalium* species are drab and small-headed while those of *Helichrysum* are brighter coloured and relatively large-headed. One species in Iraq.

The English common name for this genus is Cudweed (from the soft down, – γναφάλιον, *gnaphalion*, Gr. – with which the leaves are covered; it is also one of a group of plants sometimes known as Everlasting.

Grierson, A.J.C. (1971). The identity of *Gnaphalium indicum* L. Notes Roy. Bot. Gard. Edinburgh 31: 138.
Hilliard, O.M. (1981). Some generic concepts in Compositae-Gnaphaliinae. Bot. J. Linn. Soc. 82(3): 181–232 [*Gnaphalium* 190–193].

1. **Gnaphalium supinum** *L.*, Syst. Nat. ed. 12, 3: 234 (1768); Boissier, Fl. Orient. 3: 226 (1875); Rawi in Dep. Agr. Iraq Tech. Bull. 14: 116 (1964); Grierson in Fl. Turkey 5: 99 (1975); Lack in Fl. Iranica 145: 50 (1980).

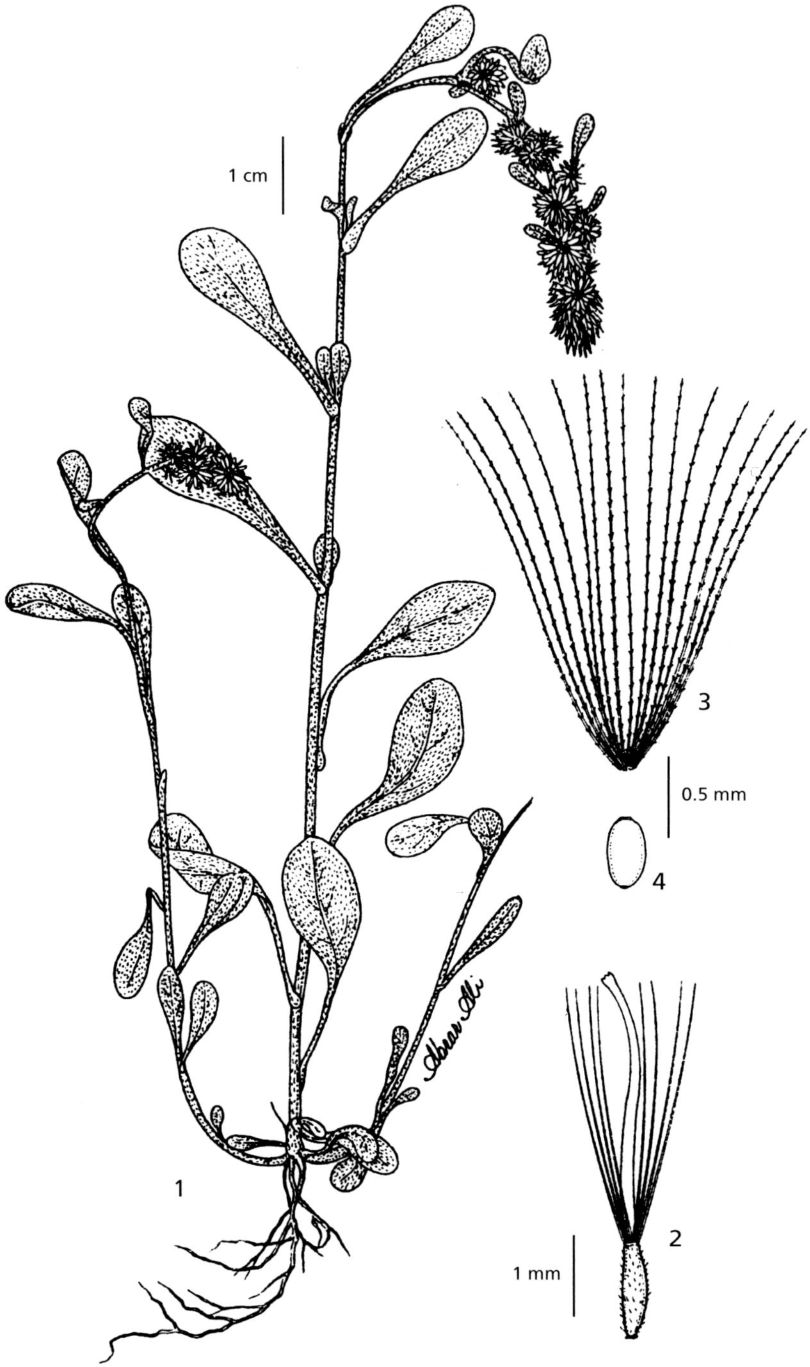

Fig. 72. **Gamochaeta pensylvanica**. 1, habit; 2, female floret; 3, pappus; 4, achene. Drawn by Abrar Ali. Reproduced with permission from Flora of Pakistan 210: fig. 28.

Omalotheca supina (L.) DC., Prodr. 6: 245 (1838); Holub in Fl. Europ. 4: 127 (1976).

Low greyish-white cushion perennial. Leaves narrowly elliptic, 10–20 × 2 mm, acute, entire. Flowering stems (0–)3–8 cm tall with a few leaves below inflorescence. Capitula campanulate, 1.2 cm in diameter, 5 mm tall, subtended in a brief spike by leaf-like bracts about equalling it; phyllaries lanceolate, 5 × 1 mm, outer ones with a narrow herbaceous centre, inner ones totally hyaline, dark brown, acute. Female florets 3-lobed, filiform, 3 mm. Hermaphrodite florets about 15, 3 mm. Achenes 0.7 mm long, pilose; pappus setae 12, free basally, eventually deciduous, 3.5 mm.

HAB. Damp alpine grassy patches, on metamorphic rock; alt. ± 3100 m; fl. & fr. Aug.
DISTRIB. Very rare in Iraq, only found once in the central sector of the alpine region. **MRO**: Ser Kurawa, *Gillett* 9729!

Turkey, Caucasus, Iran; also on high mountains in Europe, Iceland, Greenland and arctic N America.

72. **HELICHRYSUM** Mill. *nom. cons.*

Gard. Dict. ed. 4 (1754) as '*Elichrysum*' emend. Pers., Syn. Pl. [Persoon] 2: 414 (1807)
Bayer et al. in Kubitzki, Fam. Gen. Vasc. Pl. 8: 267 [2006] (2007)

J. R. Edmondson

Suffruticose herbs and subshrubs with many ± densely leafy stems, tomentose or glandular-tomentose. Resting buds often present at base of stem, stalked or sessile; basal leaves long-petiolate, ± cuneate; cauline leaves alternate, entire. Inflorescence a rather dense terminal corymb with usually many capitula; involucre scarious, yellowish or whitish, rarely reddish; phyllaries multiseriate, imbricate; receptacle flat, foveolate or areolate, with or without teeth or bristles. Florets all tubular, usually hermaphrodite, corollas 5-lobed, yellow, glandular above, or rarely outer female and corolla 3-lobed; anthers caudate; style arms truncate. Achenes smooth, cylindrical, glandular-papillose; pappus setae uniseriate, scabrid or barbellate, apices acute or claviculate.

Helichrysum (from Gr. ἥλιος, helios, sun, and χρησός, khrysos, gold, in reference to its shiny yellow bracts).

Georgiadou, E. (1982). The systematics of the genus *Helichrysum* Miller in the countries bordering the Mediterranean". Unpublished thesis, Department of Botany, University of Reading.

1. Leaf margins revolute; capitula turbinate; bracts densely imbricate 2
   Leaf margins flat; capitula broadly campanulate to hemispherical 3
2. Leaves curly with wavy margins, glandular; capitula 4–15 per inflorescence . . . . 7. *H. oligocephalum*
   Leaves plane, lanate-tomentose and ± glandular; capitula 8–30(–50) per inflorescence . . . . 6. *H. armenium*
3. Resting buds on elongated branches . . . . 2. *H. plicatum*
   Resting buds basal, sessile or on short stalks 4
4. Resting buds very thick, *c.* 10 mm in diameter; pappus apices club-shaped 1. *H. aucheri*
   Resting buds slender < 10 mm in diameter; pappus apices acute 5
5. Capitula in clusters of 10–12. . . . . 3. *H. globiferum*
   Capitula solitary or 3–8 6
6. Flowering stems ± erect; cauline leaves narrow; pedicels elongate, longer than capitula; capitula (1–)3–8 . . . . 4. *H. pallasii*
   Flowering stems deflexed or ascending; pedicels shorter than capitula; capitula 1–3(–5) . . . . 5. *H. psychrophilum*

1. **Helichrysum aucheri** *Boiss.*, Diagn. Pl. Orient., ser. 1, 11: 28 (1849); Boissier, Fl. Orient. 3: 235 (1875); Dinsmore in Post, Fl. Syria, Palest. & Sinai, ed. 2, 2: 35 (1933); Rawi in Dep. Agr. Iraq Tech. Bull. 14: 117 (1964); Georgiadou & Rechinger in Fl. Iranica 145: 59 (1980).

H. arenarium (L.) Moench subsp. *aucheri* (Boiss.) P.H.Davis & Kupicha in Notes Roy. Bot. Gard. Edinburgh 33: 239 (1974); Davis & Kupicha in Fl. Turkey 5: 95 (1975).

Perennial, scarcely caespitose at base, indumentum appressed greyish or whitish-tomentose to lanate. Stems 25–35 cm, erect, indumentum persistent, canescent. Resting buds thick, *c.*10 mm in diameter. Leaves tomentose, basal oblong-spatulate, cuneate, 4–5 × 0.7–0.9 cm; cauline narrower, subacute, leafy up to inflorescence. Capitula homogamous, in dense clusters of 30–50 capitula; pedicels 2–5 mm, whitish-tomentose. Phyllaries bright lemon-yellow, ± 50, 8–9-seriate, imbricate, opaque, outer phyllaries shorter than inner. Achenes 1–1.5 mm, papillose; pappus with rays club-shaped at apex.

HAB. Rocky mountain slopes, stony hillsides, on sandstone and limestone, in thin dry *Quercus* scrub; alt. 200–1200 m; fl. Jun.-Jul., fr. Aug.
DISTRIB. Widespread in the mountain and foothill regions of northern Iraq. **MAM**: 23 km NE of Zakho, *Rawi* 23196!; Bikhair mt., nr Zakho, *Rawi* 23055!, *Rawi* 22985!, *Field & Lazar* 776; Jabal Rubal S of Atrush, *Guest* 3617!; Zawita, *Hunting Aerosurveys* 86!; Dagh al-Razhiem to Sharanish, *Rechinger* 12115; 8 km S of Zakho, *Rechinger* 10701; Dihok (Dohuk), *Low* 293. **MRO**: Der Harir, *Nábělek* 3864, 3871; Penjwin, *Guest* 12970! **MSU**: Jarmo, *Helbaek* 1582; Sulaimaniya, *Paterson* 815! **MJS**: foot of Jabal Sinjar, *Haussknecht* s.n. **FUJ**: Mosul, Qaiyarah, *Bayliss* 129! **FAR**: Arbil to Altun Kupri, *Haussknecht* s.n. **FKI**: Kirkuk to Darband-i Bazian, *Haussknecht* s.n. **FNI**: Nineveh, *Noë* s.n.

Turkey (S Anatolia), NW Iran.

2. **Helichrysum plicatum** *DC.*, Prodr. 6: 183 (1838); Boissier, Fl. Orient. 3: 231 (1875); Dinsmore in Post, Fl. Syria, Palest. & Sinai, ed. 2, 2: 35 (1933); Rawi in Dep. Agr. Iraq Tech. Bull. 14: 117 (1964); Davis & Kupicha in Davis, Fl. Turkey 5: 90 (1975); Georgiadou & Rechinger in Fl. Iranica 145: 62 (1980).

H. anatolicum Boiss., Diagn. Pl. Orient., ser. 1, 4: 11 (1844).

Perennial with several ascending to erect sparsely glandular-pilose to glabrescent stems. Resting buds ovoid to ellipsoid, borne on short procumbent runners. Stems 20–40 cm, leafy throughout; basal leaves 3–5 × 0.3–0.6 cm, oblanceolate to linear; cauline leaves narrowly lanceolate to linear, appressed-tomentellous. Capitula in lax clusters of 7–30 heads, heterogamous or homogamous; pedicels densely white-tomentose and glandular; phyllaries 35–60, 7–10-seriate, pale yellow to cream, laxly imbricate, inner twice as long as outer, oblong-spatulate. Florets 40–60 (–90), all hermaphrodite or marginal female and inner hermaphrodite. Achenes 1–1.5 mm; pappus rays 14–20, tips acute.

Two subspecies are recognized, both present in Iraq.

Involucre 5–8 mm in diameter; phyllaries whitish-cream . . . . subsp. *pseudoplicatum*
Involucre 6.5–12 mm in diameter; phyllaries pale yellowish . . . . subsp. *plicatum*

subsp. **plicatum**

HAB. Rocky and stony mountainsides, near stream; alt. 1320–2500 m; fl. & fr. Jun.–Aug.
DISTRIB. Rare and local in the montane zone of NE Iraq. **MRO:** Chia-i-Mandau, *Guest* 2786!; Harawa Blinda mt. (NE of Haji Umran), *Kass & Noori* 27832!; Warshanka on Magar range, *Rawi & Serhang* 24338!; Shirwan Mazin, *Hadač* 5859!; Mt. Botin, *Hadač* 6137 p.p.; Tenji Saipaiyan, *Hadač* 2231.

subsp. **pseudoplicatum** (*Nábělek*) *P.H.Davis & Kupicha* in Notes Roy. Bot. Gard. Edinburgh 33: 241 (1974) & in Fl. Turkey 5: 93 (1975).

H. plicatum DC. var. *lacteum* Boiss., Fl. Orient. 3: 231 (1875).
H. pseudoplicatum Nábělek in Publ. Fac. Sci. Univ. Masaryk Brno 52: 9 (1925).

HAB. Stony ground under *Quercus*, sometimes cultivated; alt. 1100–1800(–2000) m; fl. Jun; fr. Aug.
DISTRIB. Locally common in most areas of the mountain zone of Iraq. ?Kurdistan, *Layard* s.n. (1889)!; *Brant & Strangways* s.n. (1840)! **MAM**: 1 km beyond Ispindari saddle, Swaratuka, *Chapman* 26345!; Haisi village 10 km S of Kani Mazi, *Dabbagh & Hamad* 5539! **MRO**: Merga Draiga (Mergadurezhan), *Rawi* 9123!; Baisar village, Qandil range, *Rawi & Serhang* 24165!; Sarkal at foot of mt. Helgord, *Hadač* 2117, 2688. **MSU**: Mt. Avroman, nr Tawila, *Rechinger* 10342; Penjwin, *Rawi* 22657! **FPF**: Kani Masi, *Dabbagh & Hamad* 45569!

GULARUN (*Rawi* 9123).

The two subspecies are poorly differentiated.

Distrib. of species: Turkey (SE Anatolia), Syria, Lebanon, W Iran.

3. **Helichrysum globiferum** *Boiss.*, Diagn. Pl. Orient., ser. 1, 11: 29 (1849); Boissier, Fl. Orient. 3: 232 (1875); Georgiadou & Rechinger in Fl. Iranica 145: 57 (1980).

Stems 20–35 cm., ascending to erect. Leaves shortly glandular and whitish-tomentose, basal leaves somewhat congested, oblong-spatulate, cuneate; lower cauline leaves 20–25 × 5–7 mm, linear-lanceolate. Resting buds slender, 3–5 mm in diameter. Capitula homogamous, in a terminal cluster 15–20 mm in diameter; capitula 10–12, subsessile, 6–7 × 7–9 mm; phyllaries pale lemon-yellow (dry), obtuse, lacking glands. Pappus yellowish, setae acute at apex.

HAB. not stated; alt. 1100–1500 m; fl. Jun.
DISTRIB. Very rare, found only once in the upper forest zone of NE Iraq. **MSU**: Zewiya nr Pir Omar Gudrun (Pira Magrun), *Rawi* 11521!

W Iran. Treated in Fl. Iranica 145: 57–58 as endemic to Iran; our material has leaves less densely lanate than the Iranian specimens, but its inflorescence characters match those of *H. globiferum.*

The quoted vernacular name GULA ZARDA (*Rawi* 11521) is problematic, as this is also the name of a village in the vicinity of the site where this specimen was collected (Gilazarda or Kilah Zard) on Baranand Dagh.

4. **Helichrysum pallasii** (*Spreng.*) *Ledeb.*, Fl. Ross. 2: 606 (1845); Boissier, Fl. Orient. 3: 230 (1875); Dinsmore in Post, Fl. Syria, Palest. & Sinai ed. 2, 2: 34 (1933); Davis & Kupicha in Fl. Turkey 5: 88 (1975); Georgiadou & Rechinger in Fl. Iranica 145: 64 (1980).

Gnaphalium pallasii Spreng., Syst. Veg. 3: 470 (1829).
Helichrysum callichrysum DC., Prodr. 6: 183 (1838).
H. fieldianum Rech.f. in Österr. Bot. Z. 97: 225 (1950).

Suffruticose, stems erect, several together, 20–38 cm, canescent to whitish-tomentose and somewhat glandular; resting buds fusiform, nestling in leaf rosettes, short-stalked. Basal leaves oblong-spatulate, obtuse, 2.5–3 × 5–7 mm; cauline oblanceolate to linear, subacute, canescent-tomentose. Capitula homogamous. Capitula globose, solitary or occasionally in groups of 3–4(–9) in a lax inflorescence; phyllaries 50–90, 8–12-seriate, somewhat cucullate, laxly imbricate, bright yellow, shiny, outer < ½ length of inner, scarious, elliptic, innermost oblong, slender. Florets 50–70; corollas 4–5 mm. Achenes 1.7–2 mm; pappus setae 20—30, apices acute.

HAB. Rocky mountain slopes in *Astragalus* zone; alt. 2650–3340 m; fl. Jun.-Jul., fr. Jul.-Sep.
DISTRIB. Locally frequent in the central area of the mountain zone of Iraq. **MRO**: Qandil range, *Rawi & Serhang* 24512!, *Serhang & Rawi* 26757!; by Iranian frontier, *Rawi & Serhang* 24576!; Chiya-i Mandau, *Guest* 2729!; Shaqlawa, Jabal Safin, *Omar, Sahira, Karim & Hamid* 38260!: NE of Qandil, *Rawi & Serhang* 24385!; Helgord Dagh, *Guest & Alizzi* 15826A, *Gillett* 9576!; Jabal Baradost nr Diana, *Field & Lazar* 909!; Mt. Baradost N of Rowanduz, *Thesiger* 1208.

Turkey, Syria, Lebanon, W Iran, Caucasia.

5. **Helichrysum psychrophilum** *Boiss.*, Diagn. Pl. Orient., ser. 1, 11: 28 (1849).

H. pallasii (Spreng.) Ledeb. var. *psychrophilum* (Boiss.) Boiss., Fl. Orient. 3: 231 (1875); Rawi in Dep. Agr. Iraq Tech. Bull. 14: 117 (1964).
H. chionophilum Boiss. & Balansa var. *albidum* Heimerl in Stapf, Denkschr. Akad. Wiss. Wien, Math.-Nat. Kl. 50: 39 (1885).

Dwarf, stems procumbent to ascending, 8–10(–15) cm, forming a dense clump. Resting buds numerous, fusiform, nestling within basal rosette. Basal leaves 10–22 × 3–6 mm, oblanceolate to lingulate, obtuse, eglandular; cauline leaves linear to narrowly oblong, obtuse; pedicels very short. Capitula homogamous, 1–3(–5) in lax hemispherical clusters; phyllaries 35–65, 7–9-seriate, laxly imbricate, pale golden-yellow to lemon-yellow or cream; florets 50–90; pappus scales 30–50, apices acute.

HAB. high alpine rocks and cliffs, often igneous or metamorphic; alt. (1000–)2400–3900 m; fl. Jul.-Aug.
DISTRIB. Locally frequent in the central area of the mountain zone of Iraq. **MRO**, Arl Gird (Helgord) Dagh, *Gillett* 12376; *Guest & Ludlow-Hewitt* 3069!; *Ludlow-Hewitt* 1507!; *Rechinger* 11456E, 11878, *Hadač*

2425, *Rawi & Serhang* 24917!; Ser Kurawa, *Gillett* 9754!; Qandil range, NE of Rania, *Rawi & Serhang* 18271!; *Rechinger* 11166; NE of Qandil, *Rawi & Serhang* 24427!; Shaikh wa Shaikan valley, *Rawi & Serhang* 24879!; Qandil range, *Rawi & Serhang* 26719!; Chiya-i Mandau, *Guest & Ludlow-Hewitt* 2729; mt. Helgord by Iranian frontier, *Rechinger* 11456a!

Turkey, NW & C Iran.

6. **Helichrysum armenium** *DC.*, Prodr. 6: 183 (1838); Boissier, Fl. Orient. 3: 235 (1875); Dinsmore in Post, Fl. Syria, Palest. & Sinai, ed. 2, 2: 35 (1933); Rawi in Dep. Agr. Iraq Tech. Bull. 14: 117 (1964); Davis & Kupicha in Davis, Fl. Turkey 5: 93 (1975); Georgiadou & Rechinger in Fl. Iranica 145: 66 (1980).

Gnaphalium glanduliferum Sch.Bip., Bot. Zeit. (Berlin) 3(11): 173 (1845). –Type. Iraq, Mt. Gara, *Kotschy* 310 (GOET).
Helichrysum glanduliferum Sch.Bip., Bot. Zeit. (Berlin) 3(11): 173 (1845). –Type. Iraq, Mt. Gara, *Kotschy* 310 (GOET), pro syn.
H. glanduliferum Sch.Bip. ex Boiss., Fl. Orient. 3: 236 (1875).
H. armenium Sch.Bip. var. *lacteum* Boiss., Fl. Orient. 3: 236 (1875), pro parte. –Type. Iraq, Rowanduz, Mt. Handren, *Bornmüller* 1384 (syn. JE)

Little-branched suffruticose perennial, stems ascending to erect, sparsely glandular, 28–38 cm. Resting buds broadly fusiform, born on ± elongate leafy runners. Indumentum of whole plant somewhat variable, typically laxly tomentellous and glandular, canescent. Basal leaves to 50 × 5 mm, oblanceolate-linear, margins somewhat revolute, apex acute; cauline oblanceolate to linear, acute, becoming narrower above. Inflorescence a cluster of (7–) 12–22 capitula, somewhat lax, often slightly nodding. Capitula homogamous, 7–8 mm, turbinate to campanulate; pedicels tomentose and glandular, 2–5 mm. Phyllaries 40–60, bright golden-yellow to pale yellow or whitish (var. *lacteum* Boiss. p.p.), 8–9-seriate; inner to 3 × length of outer, oblong. Florets 20–35. Achenes 1.2–1.8 mm; pappus setae 21–18, whitish or yellowish, apices acute.

HAB. dry slopes under *Quercus* scrub, *Rhus-Quercus* forest, rocky mountain slopes, stony hillsides; alt. 800–1900 (–2300) m; fl. Jun.-Jul., fr. Jul.-Aug.
DISTRIB. Common in the mountain region of N.E. Iraq. **MAM**: mt. Gara, Kotschy 310 (type of *H. glanduliferum*); Jabal Khantur, NE of Zakho, Rawi 23434!; Ser Amadia, *Guest* 4988!; N of Zakho, Jabal Khantur (on Turkish frontier), *Rechinger* 10854! *Rechinger* 10819!; Zawita, *Guest* 3733!; Aysifku 18 km NW of Serhang, *Dabbagh & Hamid* 45784!; Dori village 5 km E of Kani Masi, *Omar & Qaisi* 45428!; Kani Masi, *Hamid & Fadhil* 45619!; Bamerny 20 km NW of Sarsang, *Qaisi & Hamad* 45922!; Sarsang, *Haines* W. 519! **MRO:** Rowanduz above Zeita, *Hadač* et al. 5961; Mt. Handren, *Bornmüller* 1384; Chia-i-Mandali nr Walza, *Guest* 2702!; Mt. SE of Serva (Zirva), *Gillett* 9688!; between Karoukh Mt. and Dergala, *Rawi, Noori & Kass* 27717!; Rowanduz, Mt. Handren, *Bornmüller* 1384 (type of var. *lacteum* Bornm.); Qandil range NE of Rania, *Rawi & Serhang* 26826!; Talan region nr Dukan, *Karim* 39294!; nr Penjwin, *Rechinger* 10498!; *Noori & Hamid* 41240!; Tawila, *Rawi* 21922!; Mt. Avroman, *Gillett* 11772! (fls. white); Penjwin, *Qaisi & Hamad* 43520; *Rawi* 22620!; 5 km from Penjwin, *Noori & Hamid* 41240! **MSU**: Milakawa, *Rawi* 22485!

Turkey (S & E Anatolia), Syria, N & W Iran.

7. **Helichrysum oligocephalum** *DC.*, Prodr. 6: 184 (1838); Georgiadou & Rechinger in Fl. Iranica 145: 67 (1980).

H. kotschyi Boiss., Diagn. Pl. Orient., ser. 1, 6: 92 (1846).
H. armenium DC. var. *lacteum* Boiss., Fl. Orient. 3: 236 (1875), pro parte.

A much branched suffruticose perennial with ascending stems 25–45 cm, minutely glandular-papillose, ± lacking tomentum. Resting buds fusiform, long-stalked. Cauline leaves linear-lanceolate, subacute, narrowing above. Capitula homogamous, (2–)4–9 in a lax cluster 15–35 mm in diameter; pedicels tomentose and glandular, to 3 mm; involucre turbinate, 7–9-seriate; outer phyllaries obovate, inner linear, acute, golden-yellow. Achenes *c.* 1.5 mm. Pappus rays 20–25, acute.

HAB. Limestone and metamorphic rocks; alt. 2100–2400 m; fl. Jun.-Sep.
DISTRIB.: Rare, only one or possibly two records from alpine areas of Iraq. **MRO**: Sula Khal, 2100 m, *Rawi & Serhang* 24684. **MSU**: (or Iran), Mt. Avroman & Shahu, *Haussknecht* s.n. (1867).

N & W Iran.

73. **IFLOGA** Cass.

Bull. Sci. Soc. Philom. Paris 1819: 142 (1819)
Dict. Sci. Nat., ed. 2, 23: 13 (1822)
Bayer et al. in Kubitzki, Fam. Gen. Vasc. Pl. 8: 268 [2006] (2007)
Trichogyne Less. sect. *Ifloga* (Cass.) DC., Prodr. 6: 26 (1838)
Comptonanthus B. Nord., J. S. African Bot. 30: 54 (1964) pro parte et quoad typum

Frances G. Davies

Annual herbs with erect and ascending branches, lateral ones not overtopping main stem. Leaves alternate, sometimes opposite below, linear to linear-lanceolate, tomentose. Capitula sessile, solitary or (for the most part) groups of up to 15 distributed throughout the stems, heterogamous; receptacle of several rows, each except outermost subtending a fertile female filiform floret. Corolla of bisexual florets cylindric or campanulate, yellow. Achene ovoid to ellipsoidal, with myxogenic hairs (hairs that swell on contact with water); pappus absent in female florets, shortly plumose in upper part in bisexual florets.

Six species, most in southern Africa, extending to W Asia, with one species widespread; only one species in Iraq.

Ifloga (an anagram of *Filago*, q.v.).

Fayed, A.A. & Zareh, M. (1988). Systematic revision of Compositae in Egypt: 3. Tribe Inuleae: *Filago* and *Ifloga*. Willdenowia 17: 115–123.
Hilliard, O. (1988). A revision of *Ifloga* in southern Africa, with special comments on the northern hemisphere species. Bot. J. Linn. Soc. 83(3): 293–312.

1. **Ifloga spicata** (*Forssk.*) *Sch.Bip.* in Webb. & Berth., Phyt. Canar. 2: 310 (1845); Boissier, Fl. Orient. 3: 248 (1875); Carter in Rec. Bot. Surv. India 6, 6: 202 (1917); Zohary in Dep. Agr. Iraq Bull. 31: 147 (1950); Dickson, Wild Fls. Kuwait: 57 (1955); Rechinger, Fl. Lowland Iraq: 599 (1964); Rawi in Dep. Agr. Iraq Tech. Bull. 14: 117 (1964); Holub in Fl. Turkey 5: 100 (1975) & Fl. Europ. 4: 123 (1976); Feinbrun-Dothan, Fl. Palaest. 3: 304(1978); Wagenitz

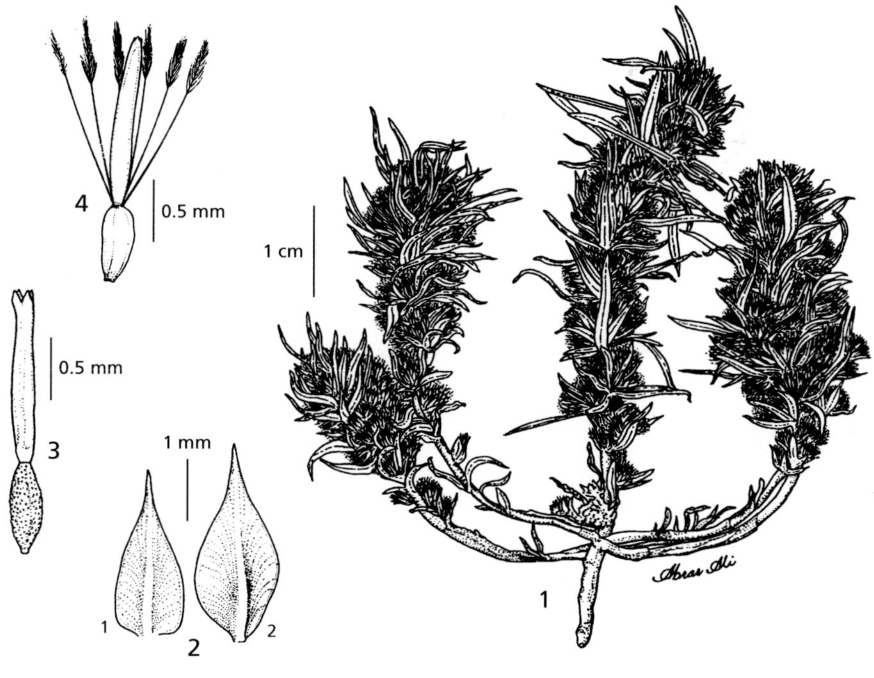

Fig. 73. **Ifloga spicata**. 1, habit; 2, phyllaries (outer (1) and inner (2)); 3, female floret; 4, achene with pappus. Drawn by Abrar Ali. Reproduced with permission from Flora of Pakistan 210: fig. 29.

in Fl. Iranica 145: 27 (1980); Boulos & Hind in Fl. Egypt 3: 197 (2002); Qaiser & Abid in Fl. Pakistan 210: 130 (2003); Kilian & Ghazanfar in Fl. Oman 3: 247 (2015); Taifour & El-Oqlah, Pl. Jordan Ann. Checkl.: 42 (2017).

Chrysocoma spicata Forssk., Fl. Aeg.-Arab. 73 (1775).

Annual herb up to 10 cm tall, with a few shorter branches from base. Leaves linear-lanceolate, entire, 10–25 × 0.5–1 mm, appearing narrower due to the inrolled margins, dull green, grey-sericeous to lanate on the (inrolled) lower surface, ± glabrous and sessile-glandular above, sheathing at base, shortly apiculate at apex. Capitula much exceeded by leaves, ovoid, 3–5 mm long including florets. Phyllaries about 19, rhomboid, 2–2.5 mm long, acuminate to acute, scarious, yellow to reddish-brown, slightly concave. Female florets 2.5 mm long without pappus; style branches rounded apically. Bisexual florets about 12, 1.7 mm long, 4–5-lobed, ?fertile, their achenes with pappus, style branches broadly rounded and penicillate. Achenes ovoid to ellipsoidal, 0.7 mm long; pappus of white hairs, plumose above. Fig. 73: 1–4.

HAB. Sandy gravel soils, sometimes gypsiferous, in the desert and occasionally the subdesert; alt. to 600(–650) m; fl. & fr. Mar.-Apr.
DISTRIB. Common in the desert region of Iraq, very rare in the steppe region and on the irrigated alluvial plain: **FUJ**: nr Ain Dibs, 2 km E of Wadi Thirthar, *Rawi & Gillett* 7145! **FPF**: Jabal Muwaila, 70 km N of Amara, *Gillett, Rawi & Rechinger* 17660! **DLJ**:5 km S. of Baiji, *Gillett* 10715! **DGA**: Asila, E of Samarra, *Rawi* 20426!; **DWD**: Wadi al-Ajramiya, 73 km N of Rutba, *Rawi & Noori* 2763!; 60 km N of Rutba, *Rawi & Khatib* 32308!; Rutba, *Rawi* 14656!; 5 km S of Rutba, *Janan al-Mukhtar* 33339!; Wadi Hauran, 12 km SE of H.2, *Chakravarty, Rawi, Khatib & Alizzi* 31619!; Nukhaib, *Guest, Rawi & Long* 14032!; 220 km W of Ramadi, *Rawi & Noori* 27048!; 15 km W of Falluja, *Chakravarty & Rawi* 29827!; between Najaf and Karbala, *Rawi* 20168!; 5 km W of Karbala, *Rawi & Gillett* 6382!; 12 km W of Ukhaidhir, *Rawi* 30873! **DSD**: 15 km SE of Ashuriya, *Guest, Rawi & Rechinger* 19387!; 99 km NE of Shabicha, along road from Najf, *Rawi, Khatib & Tikriti* 29264!; Salman, *Rawi & Gillett* 6206!; nr Jumaima, *Guest, Rawi & Rechinger* 19105!; 50 km from Ansab on road to Samah, *Fauzi, Hamid & Hazim* 38869!; Julaida, nr Samah, *Karim Hamid & Hazim* 38902!; Samah, *Alizzi & Omar* 35746!; Busaiya, *Karim et al.* 38944!; 50 km E of Busaiya on way to Khadhr al-Ma'i, *Weinert, Fauzi, Adil & Husain* 47612!; W of Zubair, *Guest et al.* 5020!; 23 km SE by S of Zubair, *Guest, Rawi & Rechinger* 16837!; Jarishan, *id.* 17122!; nr Jabal Sanam, *Qaisi, Hamad & Hamid* 48559!; Umm Qasr, *H. Ani* 39923!; al-Batin, *id.* 17058!; Chilawa, 110 km SW by S of Basra, *id.* 17195! **LCA**: nr Baghdad, *Lazar* 71! **LBA**: 10 km to Um-Qasir on road to Al- Zubair, *Al-Kaisi, Ghina & Saif* 58412!

S Europe (SE Spain), Aegean Is., Turkey, Syria, Lebanon, Palestine, Sinai, Arabia, Kuwait, Iran, Pakistan, Afghanistan, N Africa (Libya, Tunisia, Algeria, Morocco), Macaronesia (Canary Is.).

HASAJ (Ar. Ghazanfar, l.c.) HASACH (Kuw., *Carter* who noted: "grazing animals feed on it"; but no colloquial name has as yet been recorded from Iraq).

74. **LASIOPOGON** Cass.

Bull. Soc. Philom. Paris 1818: 75 (1818)
Bayer et al. in Kubitzki, Fam. Gen. Vasc. Pl. 8: 269 [2006] (2007)
Comptonanthus B.Nord., J. S. Afr. Bot. 30: 54 (1964), p.p. excl. type.

Gail Bromley

Dwarf prostrate or decumbent annuals, densely villous-lanate. Leaves small, entire, spatulate. Capitula small, campanulate, aggregated into clusters, heterogamous, disciform; involucral bracts in 2–3 series, almost equal, with membranous hyaline borders; receptacle small, flat and honeycombed, naked. Outer florets in several rows, female, corolla filiform with minute dentations at apex; central florets few, hermaphrodite, corolla tubular and 4–5–dentate; anthers caudate and having an apical appendage. Achenes obovate, slightly compressed or fusiform, glabrous or pubescent, glandular; pappus setae uniseriate, long-plumose, deciduous.

A genus of eight species, with centre of distribution in southern Africa and extending with one species to SW Asia; one species in Iraq.

Lasiopogon (from the Gr. λάσιος, lasios, hairy or woolly and πωγόν, pōgon, a beard – a reference to the woolly pubescence of the plant).

Nordenstam, B. 1964. Notes on *Lasiopogon* Cass. in South Africa. J. S. Afr. Bot. 30: 60–65.

1. **Lasiopogon muscoides** (*Desf.*) *DC.*, Prodr. 6: 246 (1838); Boissier, Fl. Orient. 3: 224 (1875); Dinsmore in Post, Fl. Syria, Palest. & Sinai ed. 2, 2: 32 (1933); Bornmüller in Beih. Bot. Centralbl. 60B: 187 (1939); Zohary in Pal. J. Bot., Jerusalem Ser. 1: 249 (1939); Zohary in Proc. Linn. Soc. 153: 104 (1941); Zohary in Dep. Agr. Iraq Bull. 31: 146 (1950); Rawi, ibid. Tech. Bull. 14: 119 (1964); Rechinger, Fl. Lowland Iraq: 602 (1964); Feinbrun-Dothan, Fl. Palaest. 3: 310 (1978); Boulos & Hind in Fl. Egypt 3: 201 (2002); Qaiser & Abid in Fl. Pakistan 210: 147 (2003); Kilian & Ghazanfar in Fl. Oman 3: 247 (2015); Taifour & El-Oqlah, Pl. Jordan Ann. Checkl.: 43 (2017).

Gnaphalium muscoides Desf., Fl. Atlant. 2: 267 (1799); Boissier, Fl. Orient. 3: 224 (1875); Dinsmore in Post, Fl. Syria, Palest. & Sinai ed. 2, 2: 32 (1933); Feinbrun-Dothan, Fl. Palaest. 3: 310 (1978).

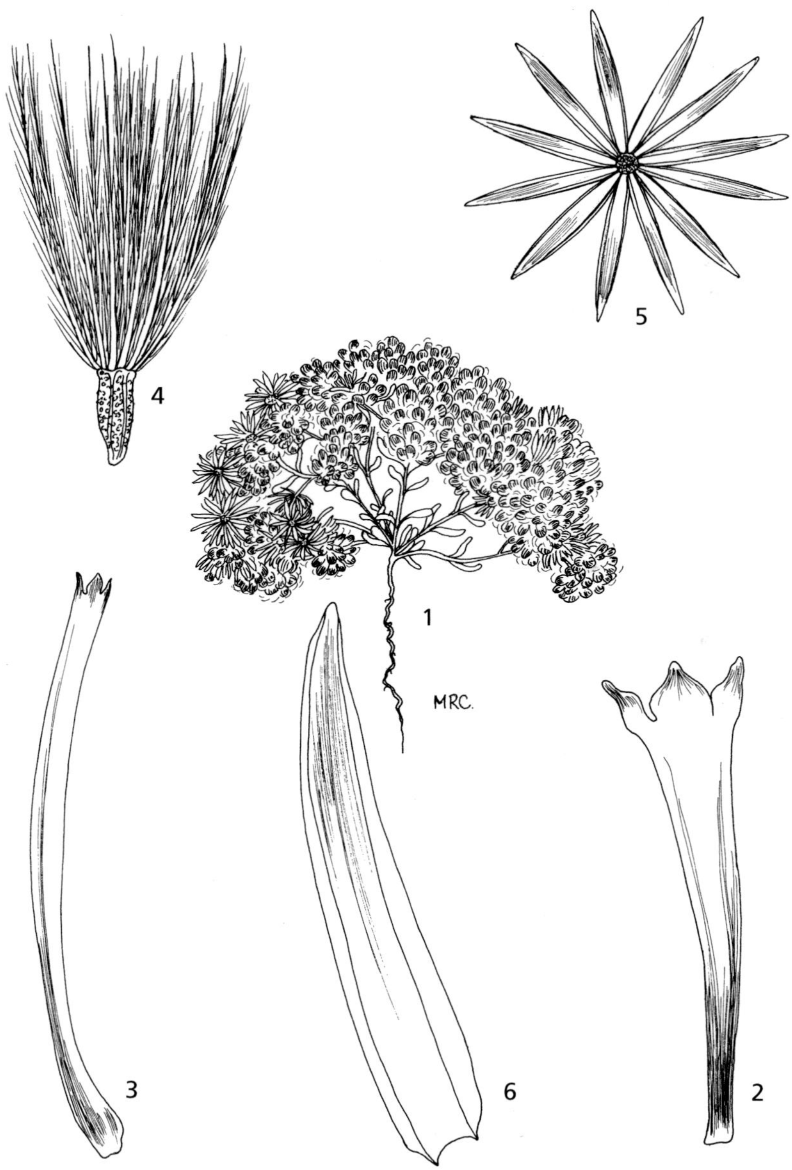

Fig. 74. **Lasiopogon muscoides**. 1, habit × 1½; 2, hermaphrodite floret × 90; female floret × 90; 4, achene and pappus × 48; 5, receptacle × 12. Drawn by Moira Rae Carter.

Lasiopogon lanatum Cass., Bull. Soc. Philom. Paris 1818: 75 (1818); Boissier, Fl. Orient. 3: 224 (1875); Dinsmore in Post, Fl. Syria, Palest. & Sinai ed. 2, 2: 32 (1933).

A dwarf annual, 2–10 cm tall, densely covered with woolly pubescence. Leaves small, 2–7 mm, obovate-spatulate, entire, those subtending the capitula about the same size as the capitula, 2–4 × 1 mm; leaves often obscured by the dense pubescence. Clusters of capitula numerous, densely woolly, composed of several capitula of discoid flowers. Phyllaries biseriate, of about equal length, 2.5–3 mm, linear with broadly hyaline margins. Female florets numerous, ± 2 mm, hermaphrodite florets few, ± 2 mm. Achenes ± 0.75 mm, glandular; pappus bristles about 10, 2 mm long, plumose. Fig. 74: 1–5.

HAB. In the desert in depressions, on low hill slopes and in wadis, on sand, loam or clay; alt. 50–650 m; fl. & fr. Mar.-Apr.

DISTRIB. Occasional in the Western Desert of Iraq. **DWD**: 95 km NE of Rutba, *Rawi & Khatib* 32348!; 40 km N of Rutba, *id.* 32286!; 18 km S of Rutba, *Rawi* 14648!; 19 km NE of Rutba, *id.* 31171!; 20 km W of Rutba, *id.* 14689!; 100 km NE of Rutba, *Chakravarty, Rawi, Khatib & Alizzi* 31520!; Ud-as-Saba, 2 km SW of H.2, *id.* 31450!; Nukhaib, *Rawi* 14803!; nr Karbala, *Agnew, Haines & Hadač* W2186!; Akashat, *Khayat & Hamad* 51559!; Wadi Hauran,15 km E of Rutba, *S. Omar, H. Abbass, Al-Kaisi & K. Hamad* 54669!; Wadi Hauran,10 km W of Rutba, *H. Abbass, Al-Kaisi & K. Hamad* 54569!

Spain, N Africa (Morocco to Egypt), Sinai, Arabia, Palestine, Jordan, Syria, Lebanon, Iran, Afghanistan, Pakistan, India (Punjab), S Africa, Namibia, Botswana.

75. **LEYSERA** L.

Sp. Pl. ed. 2: 1249 (1763)
Bayer et al. in Kubitzki, Fam. Gen. Vasc. Pl. 8: 271 [2006] (2007)

Frances G. Davies

Herbs or subshrubs with alternate, subulate leaves, glandular-hairy. Capitula radiate, heterogamous, solitary, borne on capillary leafless pedicels. Receptacle paleaceous; phyllaries many-seriate, ± unequal, imbricate. Outer florets rayed, fertile or not; inner florets hermaphrodite, tubular and fertile. Anthers truncate above, conspicuously caudate; style arms of female flowers rounded, of hermaphrodite flowers truncate and papillose. Achenes cylindrical; pappus of oblong short scales and distally plumose bristles.

A genus of 3 species; 1 species occurs in Iraq

Leysera (named in honour of the German botanist Friedrich Wilhelm von Leysser (1731–1815), the author of *Flora Halensis*).

K. Bremer, K. (1978). The genus *Leysera* (Compositae). Bot. Notiser 131: 369–383.
Anderberg, A.A. & Bremer, K. (1981). Parsimony analysis and cladistic reclassification of the *Cephaelis* generic group (Asteraceae, Gnaphalieae). Ann. Missouri Bot. Gard. 78 (4) : 1061–1072.
Fayed, A.A. (1981). Systematic revision of Compositae in Egypt. 7. Tribe Inuleae: *Phagnalon* and *Leysera*. Willdenowia 20 : 97–102.

1. **Leysera leyserioides** (*Desf.*) *Maire*, Bull. Soc. Hist. Nat. Afrique N. 20: 186 (1929); Dinsmore in Post, Fl. Syria, Palest. & Sinai ed. 2, 2: 36 (1933); Rechinger, Fl. Lowland Iraq: 604 (1964); Rechinger, Fl. Iranica 145: 73 (1980); Bramwell in Fl. Europ. 4: 133 (1976); Collenette, Wild Flow. Saudi Arabia: 203 (1999); Boulos & Hind in Fl. Egypt 3: 211 (2002); Qaiser & Abid in Fl. Pakistan 210: 206 (2003); Taifour & El-Oqlah, Pl. Jordan Ann. Checkl.: 44 (2017).

Gnaphalium leyserioides Desf., Fl. Atlant. 2 : 267 (1798).
Lonchampia capillifolia Willd. in Berlin Mat. 5 : 159 (1811); Migahid & Hamouda, F. Saudi Arabia : 331 (1974).
Leysera capillifolia (Willd.) Spreng., Novi proventus : 25 (1819); Post, Fl. Pal. ed. 1 : 36 (1896); Dickson, Wild Fls. Kuwait : 60 (1955); Rawi in Dep. Agr. Iraq Tech. Bull. 14 : 119 (1964).
Leysera leyseroides (Desf.) Dandy in F. W. Andrews, Fl. Pl. Sudan 3: 40 (1956), comb. superfl.
Asteropterus leyserioides (Desf.) Rothm., Repert. Spec. Nov. Regni Veg. 53: 4 (1944); Täckholm, Stud. Fl. Egypt, ed. 2: 559 (1974).

Delicate annual herb 10–20 cm tall with ascending branches. Leaves subulate, 5–17 × 0.3–0.5 cm, viscid with stalked glands, woolly-pubescent, dull green. Capitula solitary on

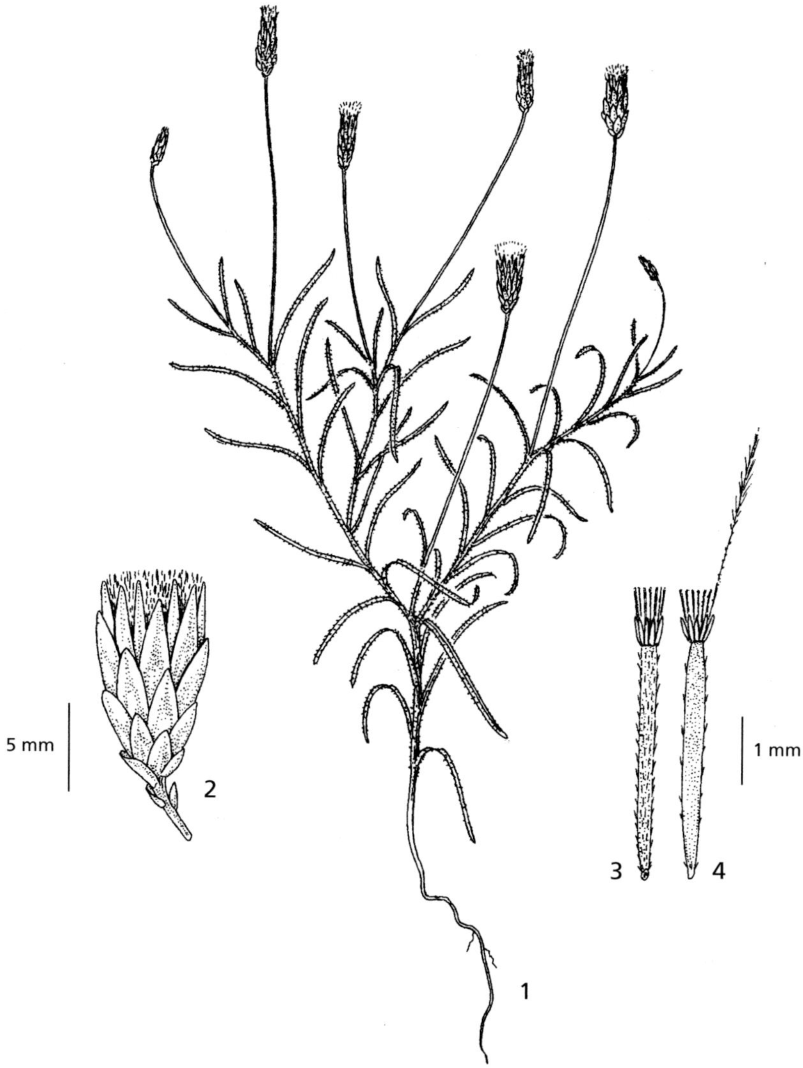

Fig. 75. **Leysera leyserioides**. 1, habit; 2, capitulum; 3, ray achene; 4, disc achene. Drawn by Margaret Tebbs. Reproduced with permission from Flora of Egypt 3: fig. 51.

3–6 cm long thread-like pedicels; involucre turbinate, *c.* 10 mm long; phyllaries graded, margins and apices scarious, centre green- or purple-tinged, glabrous, stipitate-glandular; outer receptacular paleae ovate, lacerate, uniseriate, central paleae minute. Outer florets female, 4 mm long, yellow, clasped by inner phyllaries and outer paleae, rays slightly longer than 3–4 mm long tubular florets. Achenes cylindrical, obscurely 5-ribbed, brown, sessile-glandular and with scattered white hairs; achenes of female florets with a pappus of scales, pappus of hermaphrodite florets 5–10 scales together with 5 bristles alternating with them of which upper half is plumose; mature pappus exceeding the involucre. Fig. 75: 1–4.

HAB. Desert plains, usually on rather sandy soil, sandy places in silty depressions, sometimes on gypsiferous soil; alt. 200–600 m; fl. & fr. (Mar.-)Apr.-Jun.

DISTRIB. Occasional in the western and southern districts of the Desert Region of Iraq. **DWD**: 20 km W of Rutba, *Rawi* 21132!; 230 km W of Rutba, *id.* 20927! **DSD**: by Wadi al-Khirr, 90 km NW by W of Shabicha, *Guest, Rawi & Rechinger* 19409!; 40 km WNW of Shabicha, *id.* 19279!; 13 km W by N of Shabicha, *id.* 19303!; 22 km E of Shabicha, *Chakravarty, Khatib, Rawi & Tikriti* 30107!; 38 km N of Neutral Zone, *Rawi, Khatib & Tikriti* 29181!

SW Europe (nr Almeria in SE Spain), Syria, Palestine, Jordan, Sinai, Egypt, Arabia (Saudi Arabia), Iran, N Africa (Morocco, Algeria, Tunisia, C Sahara).

76. **MICROPUS** L.

Sp. Pl.: 927 (1753); Gen. Pl. ed. 5: 398 (1754)
Bayer et al. in Kubitzki, Fam. Gen. Vasc. Pl. 8: 273 [2006] (2007)

Frances G. Davies and J. R. Edmondson

Silvery-grey annual herbs. Stems branched from base. Leaves opposite, simple, spatulate. Capitulum in leaf axils, heterogamous, disciform, few-flowered. Receptacle cylindrical; phyllaries wedge-shaped, laterally compressed, the inner phyllaries each enveloping a female floret leaving only the style visible and emergent, at maturity hardened, crested on the keel, falling with the achene. Marginal florets female, corolla adnate to ovary; hermaphrodite florets functionally male, tubular, 5-lobed, borne centrally at apex of receptacle. Achenes glabrous; pappus absent.

One species in the Old World centered on the Mediterranean and SW Asia, and a few species in N America; only 1 species in Iraq.

Micropus (from Gr. μικρο, *mikro*, small and πους, *pous*, foot).

1. **Micropus supinus** *L.*, Sp. Pl.: 927 (1753); Boissier, Fl. Orient. 3: 242 (1875); Dinsmore in Post, Fl. Syria, Palest. & Sinai ed. 2, 2: 38 (1933); Bornmüller in Beih. Bot. Centralbl. 60B: 189 (1939); Smoljaninova in Fl. S.S.S.R. 25: 282 (1959); Rawi in Dep. Agr. Iraq Tech. Bull. 14: 120 (1964); Holub in Fl. Turkey 5: 115 (1975) & Fl. Europ. 4: 126 (1976); Feinbrun-Dothan, Fl. Palaest. 3: 303 (1978); Wagenitz in Fl. Iranica 145: 10 (1980); Taifour & El-Oqlah, Pl. Jordan Ann. Checkl.: 45 (2017).

A low herb, branched from base, branches ascending; branches and leaves wholly covered in a silver sericeous tomentum. Leaves spatulate, 4–20 × 3–10 mm, very obtuse, entire, concealing stems. Inflorescence comprising the greater part of the plant, capitula paired in axils of leaves and partly concealed by them. Receptacle epaleate, bearing about 5 hard elliptic keeled pectinate woolly scales (inner phyllaries) enclosing female florets. Hermaphrodite florets 4, tubular, 2 mm long, corolla red-tinged when dry. Achenes obovoid, 2.5 × 1 mm. Fig. 76: 1–2.

HAB. In the lower mountains on limestone; alt. (250–)700–1000 m; fl. & fr. (Mar.-)May.
DISTRIB. Quite common in the central sector of the lower forest zone, occasional in the steppe region. **MRO**: below Salah ad-Din, *Barkley & Haddad* 5734!; Salah ad-Din, *Gillett* 11250!; Kuh-i Sefin, above Shaqlawa, *Bornmüller* 1390!, *Haines* W. 685!; nr Spindara, between Shaqlawa and Rowanduz, *Gillett* 8333!; 31 km N of Kirkuk on the way to Koi Sanjaq, *Rawi, Noori & Kass* 27947!; Plingan, 17 km NW of Rania, *id.* 28659!; Gali Warta, 30 km NW by N of Rania, *id.* 28816! **MSU**: 5 km SW of Chemchemal, *Rawi & Gillett* 7589! **FUJ/MAM**: nr Dohuk, *Emberger, Guest, Long, Schwan & Yusuf (Serkahia)* 15345! **FUJ**: Eski Kellek, *Rawi & Gillett* 10410!; **FAR**: Ankawa, nr Arbil, *Bornmüller* 1391!

SE Portugal and C & S Spain, Syria, Palestine, Turkey, Caucasia, Iran, N Africa (Morocco, Algeria, Libya); casual elsewhere in the Mediterranean region.

77. **PHAGNALON** Cass.

Bull. Sci. Soc. Philom. Paris 174 (1819)
Bayer et al. in Kubitzki, Fam. Gen. Vasc. Pl. 8: 276 [2006] (2007)

Frances G. Davies

Herbs or subshrubs, branched from base; stem and often at least undersides of leaves white-woolly. Capitula large, about 1 cm in diameter, long-pedicelled, solitary; phyllaries multiseriate, imbricate, at least partly scarious, often with brown markings, sterome undivided; receptacle flat, epaleaceous. Outer florets female, multiseriate, corollas filiform; style arm apices subulate. Disc florets hermaphrodite, corollas cylindrical; anther bases sagittate; style arm apices truncate. Achenes oblong, setuliferous, sericeous; pappus setae uniseriate, barbellate, basally connate, white.

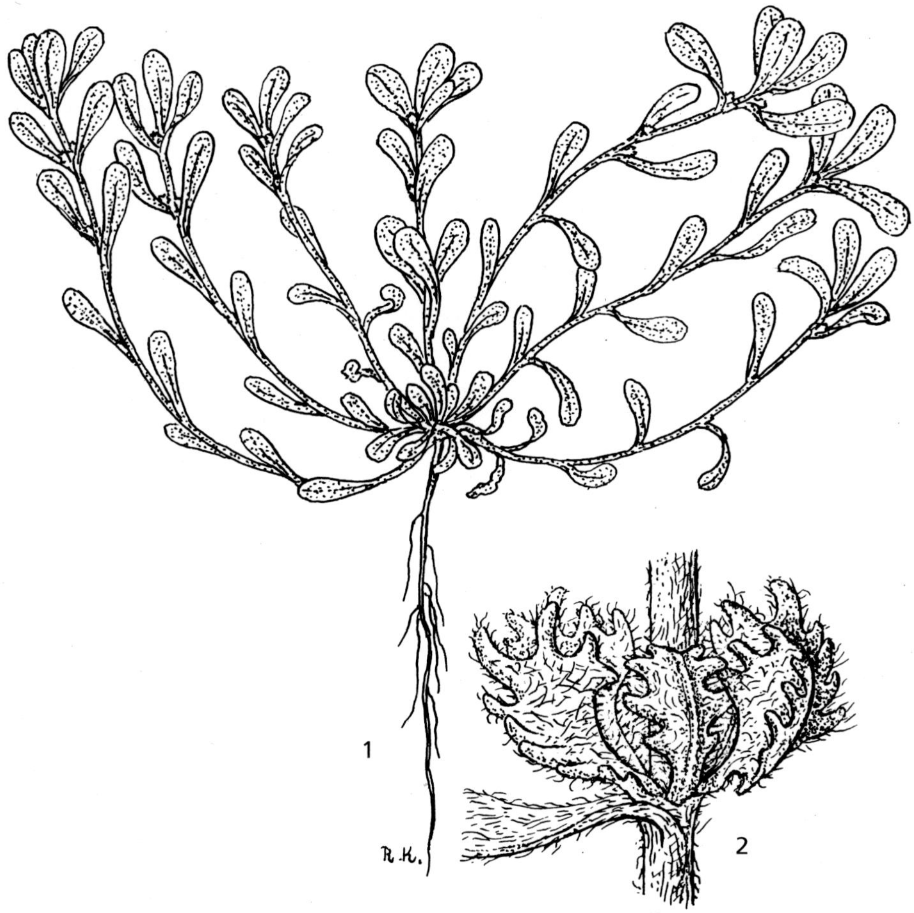

Fig. 76. **Micropus supinus**. 1, habit × 2; ; 2, capitulum showing phyllaries (enlarged). Drawn by Ruth Koppel. From Naomi Feinbrun-Dothan, *Flora Palaestina*, Part three: Ericacaceae to Compositae, Plates, Jerusalem: The Israel Academy of Sciences & Humanities, 1977, pl. 503. © The Israel Academy of Sciences and Humanities. Reproduced with permission.

Some 43 species in S. Europe, C. & S. Asia and N. Africa; only 2 species found in Iraq.

Phagnalon (of unknown derivation; perhaps, as is sometimes the case with Cassini's generic names, invented and meaningless).

Qaiser, M. & Lack, H.W. (1985). The genus *Phagnalon* (Asteraceae, Inuleae) in Arabia. Willdenowia 15(1): 3–22.

Qaiser, M. & Lack, H.W. (1985). The genus *Phagnalon* (Asteraceae, Inuleae) in Tropical Africa. Willdenowia 15(2): 437–450.

Qaiser, M. & Lack, H.W. (1987). *Phagnalon rechingeri* (Inuleae, Asteraceae), a new species from Iran, Pl. Syst. Evol. 155: 45–48.

Fayed, A.A. (1991). Systematic revision of Compositae in Egypt. 7. Tribe Inuleae: *Phagnalon* & *Leysera*. Willdenowia 20: 97–102.

Leaf margins revolute; outer phyllaries ovate, obtuse, with a submarginal brown band. . . . . 1. *P. rupestre*

Leaf margins flat; outer phyllaries lanceolate to linear, acuminate, tip and midline sometimes brown . . . . .2. *P. kotschyi*

1. **Phagnalon rupestre** (*L.*) *DC.*, Prodr. 5: 396 (1836); Boissier, Fl. Orient. 3: 220 (1875); Handel-Mazzetti in Ann. Naturh. Mus. Wien 27: 435 (1913); Dinsmore in Post, Fl. Syria,

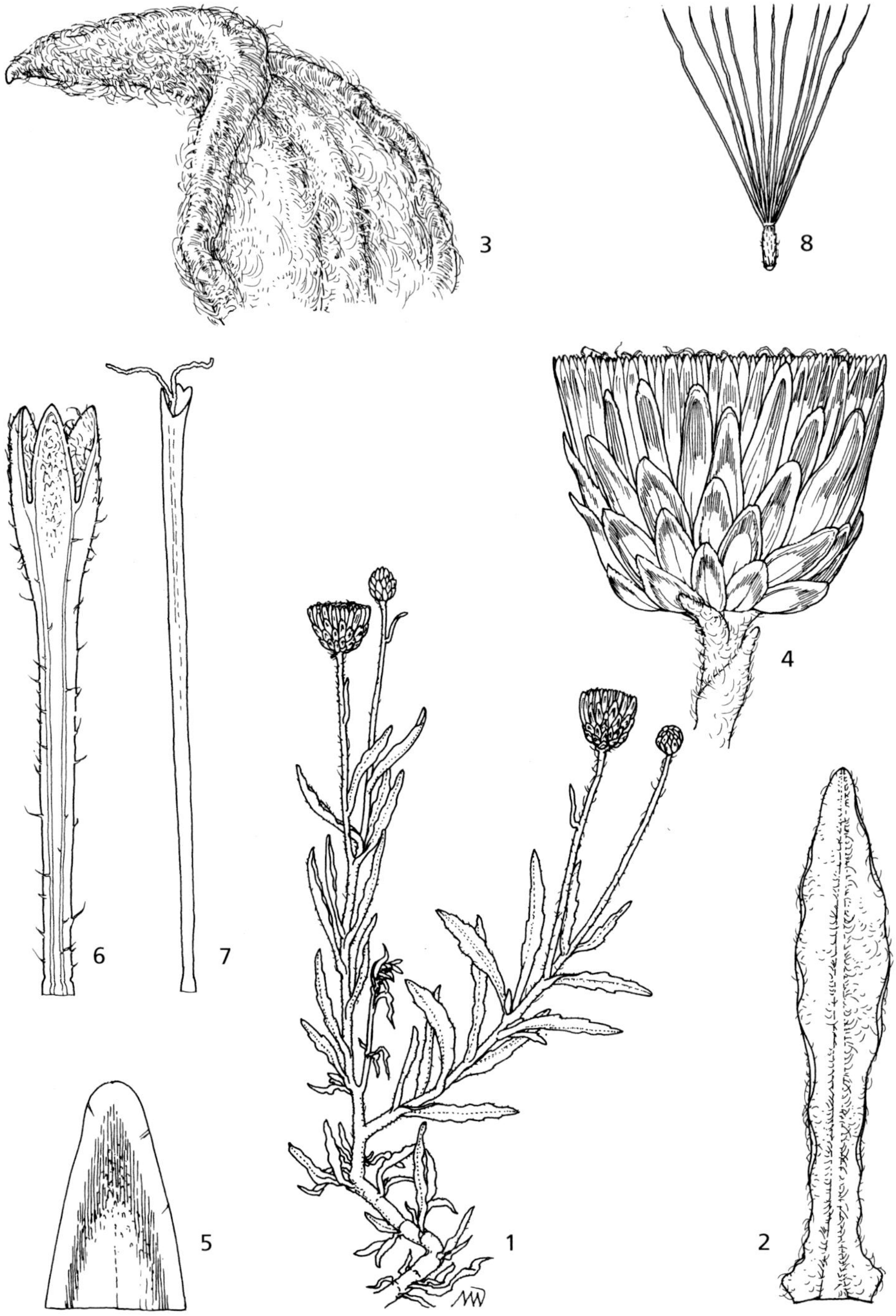

Fig. 77. **Phagnalon rupestre**. 1, habit × 1; 2, leaf × 3; 3, upper portion of leaf × 18; 4, capitulum × 6; 5, phyllary × 18; 6, female floret × 18; 7, hermaphrodite floret × 18; 8, achene and pappus × 9. All from *Omar et al.* 44408. Drawn by M. Millar Watt.

Palest. & Sinai ed. 2,2: 31 (1933); Zohary in Dep. Agr. Iraq Bull. 31: 145 (1950); Rawi, ibid. Tech. Bull. 14: 121 (1964); Täckholm, Stud. Fl. Egypt ed. 2: 558 (1974); Grierson in Fl. Turkey 5: 79 (1975); Bramwell in Fl. Europ. 4: 133 (1976); Feinbrun-Dothan, Fl. Palaest. 3: 311 (1978); Lack in Fl. Iranica 145: 40 (1980); Boulos & Hind in Fl. Egypt 3: 205 (2002). Taifour & El-Oqlah, Pl. Jordan Ann. Checkl.: 45 (2017).

Conyza rupestris L., Mant. Pl.: 113 (1767).

Woody-based herb to 40 cm with white-felted, ascending branches. Leaves sessile, half-clasping the stem, lanceolate, 1–2 × 1–3 mm, acute, margins strongly revolute, entire but sometimes undulate, upper surface green, ± pubescent, undersurface white, densely lanate. Peduncles terminal and also axillary towards shoot tips, 2–7 cm long initially, lanate but later glabrous and brown, each bearing a solitary capitulum. Involucre campanulate, 8–12 mm × 10 mm, phyllaries not or scarcely spreading, involucre remaining erecto-patent after flowering; outer phyllaries ovate-triangular, scarious, obtuse, with a dark brown submarginal marking; inner phyllaries oblong to linear, straw-coloured. Florets 6 mm, corolla bearing club-shaped multicellular hairs. Achenes 1 mm; pappus of 6 free bristles, moderately deciduous. Fig. 77: 1–8.

HAB. Rocky hillsides, on rocks and gravelly soil; alt. (200–)500–600(–800) m; fl. & fr. (Mar.-)Apr.-May.
DISTRIB. Occasional in the NW sector of the desert region of Iraq, very rare in the lower forest zone and southern desert. **MRO**: Shan-i Dar, Baradost, *Erdtman & Goedemans* 15650! **MJS**: Karsi N.W.slope, *S.Omar, Al-Khayat & Al-Kaisi* 52468, Karsiid 52630! **DWD**: Wadi al-Ajramiya, *Rawi* 31255!; Afaiyif, *Rawi & Noori* 27184!; Tal an-Nisr, *id.* 27124!; Rutba, *Field & Lazar* 137!; W Wadi Fuhaimi, between Ana and Haditha, *Handel-Mazzetti* 864; 12 km E of H.1, *Chakravarty, Rawi, Alizzi & Khatib* 31698!; between Nahiya and Qaim, *Omar, Qaisi, Hamad & Hamid* 44408!; 12 km S of Rutba, *Khatib & Hazim* 32441! **DLJ**: Rawa, *Rawi & Gillett* 6981!; 15 km E of Rawa, *Omar, Qaisi, Hamad & Hamid* 45126! **DSD**: nr Salman, *Rawi, Alizzi & Omar* 36272! & *Rawi, Agnew & Haines* W.1688!

W and C Mediterranean Europe (C & S Portugal, Spain, France, Italy, Balkans), Cyprus, Turkey (SE Anatolia), Syria, Lebanon, Palestine, Jordan, Egypt, Arabia (N Yemen), N & W Iran.

KADHA (Ir.-Rawa "whole plant pounded into a fluffy mass used for tinder", *Rawi & Gillett* 6981). The plant is used as fuel in Palestine for the same reason.

2. **Phagnalon kotschyi** *Boiss.*, Fl. Orient. 3: 221 (1875). – Type. Kurdistan: In rupestribus subhumidis m. Gara Kurdist. odor gravis et suavis, 24 July 1841, *Kotschy* 313 (FI, G, LE, M, P; syn.). Dinsmore in Post, Fl. Syria, Palest. & Sinai ed. 2,2: 32 (1933); Zohary in Dep. Agr. Iraq Bull. 31: 146 (1950); Rawi, ibid. Tech. Bull. 14: 121 (1964); Lack in Fl. Iranica 145: 43 (1980).

Conyza varthemioides Nábělek in Pub. Fac. Sci. Univ. Masaryk 52: 7 (1925).

Woody-based herb about 20 cm. Stems several, ascending, at first white-felted. Leaves sessile, broad-based, elliptic or oblong, apiculate, 10–40 × 4–10 mm, entire, bright green, both surfaces with slight lanate pubescence which in the mature leaf does not obscure the surface. Peduncles 2–5 cm, slender, arising terminally and subterminally, each bearing 1–2 capitula. Involucre campanulate, 6–8 × 10 mm, patent after flowering; outer phyllaries spreading, linear, up to 1 mm broad, gradually narrowed to tips, greenish, midline darker green or brown, arachnoid. Florets 4.5 mm long, corollas with a few multicellular hairs; styles of female florets with rounded apices, of hermaphrodite flowers truncate. Achenes 1 mm, pilose; pappus of ± 8 bristles, scarcely deciduous.

HAB. Rocky mountain slopes and moist humid places, steep wooded gorges, fissures in calcareous rocks facing S.; alt. 1200–1600 m; fl. & fr. (Jun.-)Jul.-Aug.
DISTRIB. Quite common in the NW sector of the lower forest zone of Iraq, rare in the central and SE sectors. **MAM**: Galli Zawita, on Turkish frontier N of Zakho, *Rawi* 23542!; Severa Gaurik, between Sharanish & Mt. Zawita, NE of Zakho, *Rawi* 23656!; above Basingera, nr Sharanish, *Rechinger* 11524!; Bamerne, 20 km NW of Sarsang, *Qaisi & Hamad* 45925!; above Sarsang, *Rechinger* 11676; 3 km W of Amadiya, *Rechinger* 11620; Sarsang, *E. Chapman* 26391!, *Haines* W.427!; Gara Dagh (Chiya-i Gara), *Kotschy* 313 (type)!; Dohoka village 12 km NESarsang, *Dabbagh & Hamad* 46132! **MRO**: Sakran, on the mountain 15 km SE of Choman, *Dabbagh & Hamad* 46208!, 46245! **MSU**: Avroman, *Rawi, Chakravarty, Alizzi & Noori* 19729!

Lebanon, Turkey (SE Anatolia), Iran.

78. **PSEUDOGNAPHALIUM** Kirp.

Trudy Bot. Inst. Akad. Nauk S.S.S.R., Ser 1, Fl. Sist. Vyssh. Rast. 9: 33. 1950
Bayer et al. in Kubitzki, Fam. Gen. Vasc. Pl. 8: 277 [2006] (2007).

?*Hypelichrysum* Kirp. in Trudy Bot. Inst. Akad. Nauk S.S.S.R., Ser. 1, Fl. Sist. Vyssh. Rast. fasc. 9: 33 (1950); *Gnaphalium* L. sect. *Calolepis* Kirp. in Bot. Mater. Gerb. Bot. Inst. Komarova Akad. Nauk S.S.S.R. 20: 309 (1960); *Pseudognaphalium* Kirp. subgen. *Laphangium* Hilliard & B. L. Burtt in Bot. J. Linn. Soc. 82(3): 205 (1981); *Laphangium* (Hilliard & B. L. Burtt) Tzvelev in Byull. Moskovsk. Obshch. Isp. Prir., Otd. Biol. 98(6): 105 (1993)[post 12 Jan 1994], Greuter in Willdenowia 33(2): 243 (2003).

D. J. Nicholas Hind

Annual, biennial or perennial herbs. Stems erect or ascending, simple or branched, solitary or many, tomentose or lanose, stipitate-glandular, leafy throughout, often clothed in dead leaves below. Leaves alternate, sessile or pseudopetiolate to petiolate, bases often decurrent, lamina linear, lanceolate, linear-lanceolate, oblanceolate or spatulate, discolorous or sometimes concolorous, surfaces woolly or tomentose, or glandular, margins entire, flat or undulate, apex acute to acuminate. Inflorescences terminal, usually with dense clusters of capitula, corymbose-paniculate, or cymose, capitula sessile or short-pedicellate. Capitula heterogamous, disciform, numerous; involucres broadly cylindrical to campanulate; phyllaries 3- or 4- seriate, hyaline, outer broadly elliptic, apices rounded to obtuse, tomentose at base, middle oblanceolate tomentose at base, innermost glabrous, equalling or slightly exceeding florets, white to yellow, often buff- or cream- coloured, stereome usually divided, rarely undivided; receptacle smooth or alveolate, epaleaceous. Marginal florets female, (25–) 40–130, corollas filiform or narrowly tubular, glabrous, 3- or 4- lobed, yellow. Disc florets hermaphrodite, florets 5–10 (–25), corollas tubular, glabrous, yellow, white or purplish, scarcely expanded, usually 5-lobed, lobes stipitate-glandular; apical anther appendages ovate, flat, basal anther appendages sagittate; style arms truncate, apically pubescent. Achenes ellipsoid or obovoid, brown, glabrous, rarely setuliferous, setulae when present of clavate myxogenic twin-hairs; carpopodium minute; pappus setae uniseriate, capillary, barbellate, bases free but coherent by patent cilia, white, apical cells ±clavate.

A genus of about 90 species (even recorded as high as 150 – Dillon, 2005: 139) mostly in North and South America, but with several in Africa and Asia and a number of weedy species. *Laphangium* is not recognized in this treatment, neither is the return of this species to *Helichrysum.* Only one species has been recorded in the Flora area.

Dillon M. O. (2005). *Pseudognaphalium.* In P. Sklenár et al. (eds), Fl. gen. Páramos, Mem. New York Bot. Gard. 92: 139.

Freire, S. E. Bayón, N. D. Baeza, C. M. Giuliano D. A. & Monti, C. (2014). Revision of the genus *Pseudognaphalium* (Asteraceae, Gnaphalieae) in Chile. Gayana Bot. 71(1): 68–107.

Hilliard O. M. & Burtt, B. L. (1981). [*Pseudognaphalium*] Some generic concepts in Compositae-Gnaphaliinae. Bot. J. Linn. Soc. 82(3): 202–206.

1. **Pseudognaphalium luteoalbum** (*L.*) *Hilliard & Burtt*, Bot. J. Linn. Soc. 82(3): 206 (1981); Boulos & Hind in Fl. Egypt 3: 209 (2002); Qaiser & Abid in Fl. Pakistan 210: 125 (2003).

Gnaphalium luteoalbum L., Sp. Pl.: 851 (1753); Boissier, Fl. Orient. 3: 224 (1875); Hand.-Mazz. in Ann. Naturh. Mus. Wien 27: 432 (1913); Nábělek in Publ. Fac. Sci. Univ. Masaryk 52: 9 (1925); Blatter in Journ. Ind. Bot. Soc. 11 (1): 37 (1932); Dinsmore in Post, Fl. Syria, Palest. & Sinai ed. ed. 2, 2: 33 (1933); Anthony in Notes Roy. Bot. Gard. Edinburgh 18: 290 (1935); Zohary in Dep. Agr. Iraq Bull. 31: 146 (1950); Rawi, ibid. Tech. Bull. 14: 116 (1964); Rawi & Chakravarty, ibid. 15: 48 (1964); Rechinger, Fl. Lowland Iraq: 603 (1964); Täckholm, Stud. Fl. Egypt, ed. 2: 558 (1974); Grierson in Fl. Turkey 5: 98 (1975); Feinbrun-Dothan, Fl. Palaest. 3: 311 (1978); Rechinger, Fl. Iranica 145: 47 (1980); Boulos & Hadidi, Weed Fl. Egypt ; 35 (1984); Boulos & Hind in Fl. Egypt 3: 207 (2002); Chen & Bayer in Fl. China 20-21: 816 (2011).

Gnaphalium trifidum Thunb., Prodr. Pl. Cap. 2: 150 (1800).
G. nanum Kunth in Humb., Bonpl. & Kunth, Nov. Gen. Sp. Pl. 4 (ed. folio): (1818), nom. iileg. non Willd. (1803) [= Leontonyx pusillus Drège ex Harv. & Sond.].
G. depressum Steud., Nomencl. Bot. [Steudel]: 375 (1821), nom. illeg., pro *Gnaphalium luteo-album* L.,

Fig. 78. **Pseudognaphalium luteoalbum**. 1, habit x ⅔; 2, capitulum × 6; 3, florets × 20; 4, capitulum after after flowering × 6; 5, achene and pappus × 30. 1– 3 from *Josckt* 6439; 4–5 from *Harajian* 993 (Turkish material). Drawn by Christine M. Grey-Wilson.

non Roxb. (1814) [= Homognaphalium pulvinatum (Delile) Fayed & Zareh].
Helichrysum luteo-album (L.) Rchb., Handb. Gewächsk., ed. 2, 2(2): 1460 (1829).
Gnaphalium luteo-album L. var. *incanum* A.Rich. ex Endl., Prodr. Fl. Norfolkicae: 50 (1833), nom. illeg. Pro *Gnaphalium luteo-album* L.
G. multiceps Wall. ex DC., Prodr. 6: 222 (1838).
G. dichotomum? [sic!] Blanco, Fl. Filip., ed. 2: 439 (1845), nom. illeg., non Salisb. (1796), nom. illeg. [= Filago pyramidata L.], nec Willd. ex Spreng. (1826) [= ?].
G. luteo-fuscum Webb, Niger Fl. [W. J. Hooer]: 143 (1849).
Filaginella luteo-alba (L.) Opiz, Seznam: 44 (1852).
Gnaphalium helichrysoides Wedd., Chloris Andina 1: 146 (1856).
G. luteo-album L. var. *multiceps* (Wall. ex DC.) Hook.f., Fl. Brit. India 3: 288 (1881).
G. luteo-album L. var. *pallidum* Hook.f., Fl. Brit. India 3: 288 (1881).
Dasyanthus conglobatus Bubani, Fl. Pyren. (Bubani) 2: 199 (1899), nom. illeg. pro *Gnaphalium luteo-album* L.
Gnaphalium luteo-album L. var. *compactum* Kirk, Stud. Fl. New Zealand: 298 (1899).
G. makranicum Rech.f. & Esfand., Oesterr. Bot. Zeit. 97: 224 (1950).
G. luteo-album L. f. *makranicum* (Rech.f. & Esfand.) Parsa, Fl. Iran [Parsa] 10: 44 (1980).

Erect annual herb 20–50 cm tall, stem and both surfaces of leaves grey-lanate. Leaves oblanceolate, 2–5 × 0.2–0.5 cm, acute, base slightly decurrent, without a clearly defined petiole. Inflorescence branched, without leaves, the capitula in subsessile groups of 10–25. Involucre 4–5 mm in diameter; phyllaries lanceolate, 3 mm long, greenish or hyaline, spreading in fruit. Achenes 0.5 mm long, brown; pappus ± deciduous, bristles mostly falling separately. Fig. 78: 1–5.

HAB. On lower mountain slopes in valleys, on a sandy bank in a dry river bed, on canal banks in a desert plain; alt. to c. 1000+ m; fl. & fr. Apr.-Jun. (-Aug.).
DISTRIB. Occasional in lower forest zone of Iraq, rare on alluvial plan in the desert region. **MAM**: "Pl. Mesopot. Kurdistan & Mossul" (?Gara Dagh), *Kotschy* 224!; Jabal Baykhair (Bekher), nr Zakho, *Field & Lazar* 833!; Sikreen (Sikrin), nr Sarsang, *Haines* W. 1594! **MRO**: Shewasur (Wadi Ahmar), *Omar, Karim, Hamza & Hamid* 37121! **MSU**: Biyara, *A.Hojaze* 50455 **DWD**: Nahr Husaini canal, nr Karbala, *Hand.-Mazz.* 864, *Nábělek* 3899. **DSD**: Jabal Sanam, *Whitehead* 108 (in Blatter, l.c.). **DSD/LBA**: "Mesopotamia" (probably near Shu'aiba or Basra), *Watson* s.n.

Cyprus, Lebanon, Palestine, Sinai, Egypt, Arabia (Saudi Arabia, Yemen, Oman), Turkey, Caucasia, Iran, Turkmenia, Pakistan, Afghanistan, N Africa (Morocco, Algeria, Central Sahara), Macaronesia (Azores, Madeira, Canary, Tenerife); Also widespread in South America.

Jersey Cudweed or Everlasting. Dinsmore (1933) give the Arabic name SĀBŪN AL-'ARAB ("Arabian soap") and the Egyptian local name SĀBŪN 'AFRIT ("demon soap") is recorded on the label of *Shabetai* 215 from the Baharia Oasis; both suggest that the plant is considered to have some medicinal properties. Another species of Cudweed is mentioned by Wren (1956) as an astringent – a herb well known to the old herbalists like Parkinson and also Culpeper who quotes Pliny's comments on that particular plant.

79. **RHODANTHE** Lindl.

Edwards's Bot. Reg. 20: t. 1703 (1834)
Bayer et al. in Kubitzki, Fam. Gen. Vasc. Pl. 8: 279 [2006] (2007)

C. C. Townsend
Revised by D. J. Nicholas Hind

Annual or perennial herbs or rarely subshrubs, tomentose to glabrescent, with alternate, entire, lanceolate to ovate or elliptical leaves. Capitula terminal on branches, usually homogamous and discoid with all flowers hermaphrodite and fertile, rarely heterogamous and disciform with some marginal florets female, frequently some central florets sterile; involucre campanulate or ellipsoid; phyllaries multiseriate, glumaceous, outer ovate and frequently dorsally lanate, inner with a large, usually white or coloured, lanceolate appendage; receptacle flat, sometimes alveolate, sometimes even fimbrillate or paleate. Corollas 5-lobed, tubular, ± widened above. Achenes angular, terete or ± compressed, glabrous to densely villous; pappus setae plumose.

About 40 species in Australia and southern Africa.

From the Gr. ρος, ros, pink and άνθος, anthos, flower – probably from the reddish corollas.

Wilson, P.G. (1992). The classification of Australian species currently included in *Helipterum* and related genera (Asteraceae: Gnaphlieae): Part 1. Nuytsia 8(3): 379–438.

1. **Rhodanthe chlorocephala** (*Turcz.*) *Paul G. Wilson*, Nuytsia 8(3): 386 (1992).
subsp. **rosea** (*Hook.*) *Paul G. Wilson*, Nuytsia 8(3): 387 (1992).

Acroclinium roseum Hook., Bot. Mag. 80: t. 4801 (1854).

Erect annual, 15–30 cm, with numerous slender, erect branches from base, glabrous or almost so. Leaves linear, acute or obtuse, c. 5–40 × 0.75–3 mm, lower sometimes linear-spatulate. Capitula hemispherical, c. 2–5 cm diam., outer phyllaries shining brownish, inner with a thicker brown or greenish base. Achenes densely villous; pappus of c. 10–15 plumose bristles with a club-shaped brush of denser, coloured barbs at tip.

Appendages of inner phyllaries white or pink.

HAB. & DISTRIB. In gardens only. **LBA**: Basra, *Alizzi & Omar* 35107!

A native of Australia cultivated as an ornamental and for dried flower decorations. According to Husain & Kasim this species is cultivated throughout Iraq, but only one specimen has been seen. These authors also list "Heliptrum (sic) album" with no authority, this probably referring to the horticultural name taken up by Nicholson and later, apparently independently, by Bailey; it is certainly not *Helipterum album* Ewart., a synonym of *Rhodanthe corymbosa* (F. Muell.) Paul G. Wilson.

TRIBE **ASTEREAE** CASS.

80. **LACHNOPHYLLUM** Bunge

Mem. Sav. Étr. Petersb. 7: 327 (1847)
Nesom & Robinson in Kubitzki, Fam. Gen. Vasc. Pl. 8: 292 [2006] (2007)

C. C. Townsend

Erect or ascending annual or biennial aromatic herbs clothed with eglandular and stipitate-glandular viscid hairs. Leaves petiolate to sessile, simple, entire. Inflorescence racemose or usually paniculate with short, erect branches, capitula pedicellate. Capitula heterogamous and radiate; involucre broad-campanulate; phyllaries multiseriate, gradate, recurved, margins membranaceous; receptacle concave centrally, epaleaceous. Ray florets biseriate, female, rayed, corollas pink to bluish. Disc florets hermaphrodite, fertile, cylindrical, corollas yellow; apical anther appendages lanceolate, pubescent. Achenes compressed, conspicuously narrowed at apex, body sericeous-villous setuliferous; pappus setae uniseriate, barbellate, very shortly connate at base.

A genus of two species, one occurring in Iraq and the other with a more easterly range.

Lachnophyllum (from the Gr. λαχνε, *lachne*, soft hair or down, φύλλον, *phyllon*, a leaf – referring to the foliar indumentum).

1. **Lachnophyllum noeanum** *Boiss.*, Diagn. Pl. Orient. ser. 2, 3: 17 (1856); Boissier, Fl. Orient. 3: 172 (1875); Zohary in Dep. Agr. Iraq Bull. 31: 144 (1950); Rawi, ibid. Tech. Bull. 14: 118 (1964); Grierson in Fl. Turkey 5: 130 (1975); Feinbrun-Dothan, 3: 300 (1978).

Plant (5–)10–60(–80) cm tall, usually with several stems from base, clothed throughout its vegetative parts with long, spreading, fine, whitish hairs and dense short darker glandular hairs; stems finely ridged but scarcely angular, simple or much more usually paniculately branched in upper half. Basal and lower cauline leaves oblong-spatulate, 10–35 × 3–8 mm, median narrower, sessile and somewhat amplexicaul, upper much reduced and bract-like in branches. Capitula usually solitary, on divergent or ascending pedicels; involucre campanulate; phyllaries linear-oblong, densely glandular-viscid, spreading after fruit-fall, inner 7–9 mm long, slightly exceeding pappus, often purplish-tipped. Marginal florets ± 7 mm, pink or bluish rays ± 2 mm; disc florets corollas ± 5 mm. Achenes ± 2.5 mm, setuliferous; pappus 4.5–5 mm long, whitish. Fig. 79: 1–6.

HAB. A weed of cultivation, in tobacco fields and in irrigated and inundated places on a farm; alt. 850–1500 m; fl. & fr. Sep.
DISTRIB. Only found twice (the two gatherings made within three days) in the Sulaimaniya district. **MSU**: Qaradagh, *Zohary* s.n.!; Surchinar Bakaraji farm nr Sulaimaniya, *Eig & Feinbrun* s.n.!

Palestine, Turkey, Iran.

Apparently scattered and uncommon throughout its range. It occurs elsewhere on field margins, in fallow ground etc. on heavy soil which retains moisture.

81. **BELLIS** L.

Sp. Pl.: 886 (1753); Gen. Pl.: 864 (1754)
Nesom & Robinson in Kubitzki, Fam. Gen. Vasc. Pl. 8: 303 [2006] (2007)

Frances G. Davies

Annual or perennial herbs, rosulate or with alternate, spatulate to oblanceolate leaves. Capitula heterogamous, radiate, scapose; involucre ± hemispherical; phyllaries 1–2-seriate; receptacle conical, naked. Ray florets uniseriate, female, limbs white to pinkish, conspicuous; disc florets hermaphrodite, corollas tubular, yellow, 5-lobed. Achenes obovate, compressed with a thickened marginal rib, ± pubescent; pappus absent.

About 8 species in the northern temperate regions.

Bellis (from the name of a plant in Pliny believed to be of this genus).

Annual with a short leafy stem; leaves 10–30 × 4–10 mm, with 2–4 teeth on each side; capitula including ray florets < 1.5 cm in diameter (in Iraq) . . 1. *B. annua*
Scapose, rosulate perennial; leaves 40–60 × 10–15 mm, at least some often with more than 4 teeth on each side; capitula including ray florets usually > 1.5 cm in diameter . . . . .2. *B. perennis*

1. **Bellis annua** *L.*, Sp. Pl.: 887 (1753); Boissier, Fl. Orient. 3: 175 (1875); Dinsmore in Post, Fl. Syria, Palest. & Sinai ed. 2, 2: 20 (1933); Grierson in Fl. Turkey 5: 135 (1975); Feinbrun-Dothan, Fl. Palaest. 3: 279 (1978); Meikle, Fl. Cyprus 2: 869 (1985).

Annual herb with a (1.5)2.5–15(–20) cm stem, leafy in lower part, sparingly pilose. Leaves spatulate to oblanceolate, 10–30 × 4–10 mm, tapering below to a winged petiole shorter than lamina, ciliate towards base of petiole but otherwise glabrous, acute or more usually obtuse at apex, with 2–3 angular teeth on each side. Capitula on a slender, 1–5(–10) cm peduncle; involucre 2–5 mm long; phyllaries dark-pink tipped, lanceolate–elliptic to lanceolate. Ray floret limbs 3.5–6(–8) mm, white or pink-tinged. Disc florets 1.5–2 mm. Achenes 1–1.5 mm, truncate at apex, finely pubescent. Fig. 80: 1–6.

HAB. Damp places under rocks, in a gully and in degraded *Quercus* forest; alt. 390–850 m; fl. & fr. Mar.-Apr.
DISTRIB. Apparently scattered and rare, but possibly under-collected. **MSU**: Qaradagh village, *Gillett* 7847!; Jarmo, *Helbaek* 482!; Darband-i Khan, *Wheeler Haines* W1626! **FPF**: 60 km W of Khanaqin, *Cowan & Darlington* 264! **LEA**: Ba'quba, *Colvill* 4!

Mediterranean Europe from Portugal to Turkey including the islands, Bulgaria, Cyprus, Syria, Lebanon, Palestine, Iran, N Africa (Morocco to Libya), Macaronesia (Madeira, Tenerife).

2. **Bellis perennis** *L.*, Sp. Pl.: 886 (1753); Boissier, Fl. Orient. 3: 173 (1875); Nábělek in Publ. Fac. Sci. Univ. Masaryk 52: 4 (1925); Dinsmore in Post, Fl. Syria, Palest. & Sinai ed. 2, 2: 19 (1933); Guest in Dep. Agr. Iraq Bull. 27: 14 (1933); Anthony in Notes Roy. Bot. Gard. Edinburgh 18: 290 (1935); Blakelock in Kew Bull. 4: 44 (1949); Zohary in Dep. Agr. Iraq Bull. 31: 144 (1950); Tamamshian in Fl. S.S.S.R. 25: 54 (1959); Rawi in Dep. Agr. Iraq Tech. Bull. 14: 107 (1964); Grierson in Fl. Turkey 5: 135 (1975); Feinbrun-Dothan, Fl. Palaest. 3:

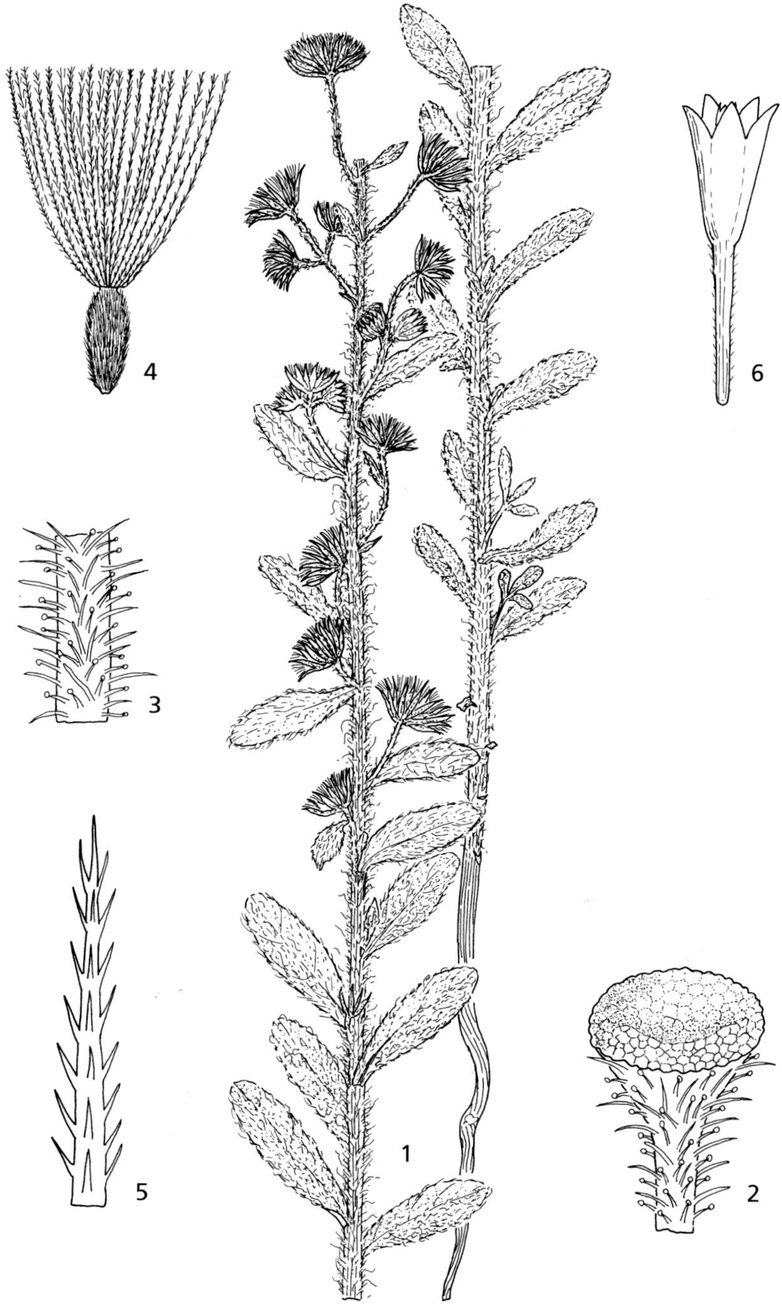

Fig. 79. **Lachnophyllum noeanum**. 1, habit × ⅔; 2, head with all flowers fallen × 1 ½; 3, portion of stem × 6; 4, achene and pappus × 6; 5, part of pappus hair × 20; 6, floret × 6.. 1, 3–6 from *Eig & Feinbrun* s.n.; 2 from *Hakererm* s.n. (Palestine material). Artist unknown.

296 (1978); Meikle, Fl. Cyprus 2: 868 (1985); Yilin Chen & Brouillet in Fl. China, 20–21: 559 (2011); Breckle, Hedge & Rafiqpoor, Vasc. Pl. Afghan: 153 (2013).

Perennial scapose herb, spreading by short stolons, mat-forming, with a root system of many fibrous rootlets; scape up to 25 cm but often much less, thinly pilose. Leaves spatulate,

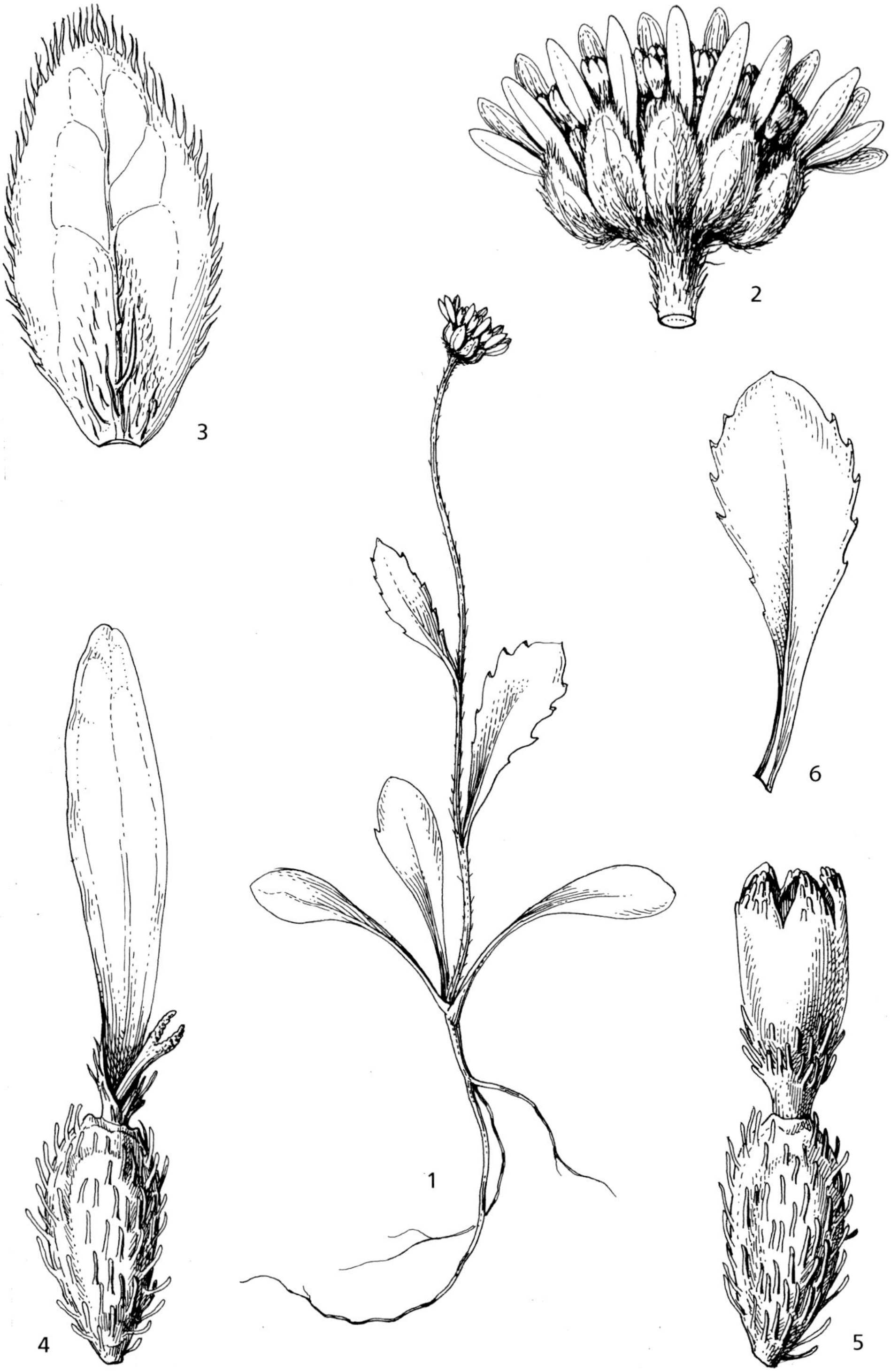

Fig. 80. **Bellis annua**. 1, habit × 2; 2, capitulum × 6; 3, phyllary × 18; 4, ray floret × 15; 5, disc floret × 20; 6, leaf × 4. All from *Wheeler Haines* W1626. Drawn by Laura (née Ripley) Mason.

20–60(–90) × (6–)10–20 mm, tapering below to a winged petiole shorter to longer than lamina, thinly hairy to glabrous on both surfaces, margin entire to shallowly serrate or dentate, some leaves usually with 4 or more teeth on each side. Involucre 3–5(–7) mm long; phyllaries dark green, obtuse or acute, elliptic to elliptic-ovate, pilose about base or rarely ± throughout, usually with an apical tuft of white hairs. Ray floret limbs 5–8 mm, white or pink-tinged beneath. Disc florets 2–2.5 mm. Achenes ± 2 mm, truncate above, thinly pubescent.

HAB. In damp places on hillsides and in valleys, in damp grassy places and orchards, in irrigated gardens and by or even in streams, on clay; alt. 800–1370 m; fl. Mar.-Jun.
DISTRIB. Widespread in the mountain regions of Iraq, rare in the foothills and plains. **MAM**: Amadiya, *Guest* 1233!; Zawita to Sawara Tuka, *Guest* 13286!; Aqra, *N. Polunin* 93!; S of Sarsang, *Anders* 1370. **MRO**: Diana, *Guest* 706!; Shaqlawa, *Rawi & Gillett* 10452!; E of Salah ad-Din, *Helbaek* 743!; Baradost Dagh, *Erdtman & Goedemans* (*Rechinger* 15622)! **MSU**: Sulaimaniyah, *Graham* 723!, *Thesiger* 339; Penjwin, *Rawi* 12160!; 10 km W of Halabja, *Barkley* 7539!; Susa, *Omar* 42610!, 42905! **FKI**: Kirkuk, *Rogers* 0489! **LCA**: Baghdad, *Rogers* 0390!

Almost throughout Europe, Cyprus, Turkey, Palestine, Iraq, Transcaucasia, Macaronesia (Azores, Madeira). Naturalized as a weed in Pakistan, India, Australia, New Zealand, N & S America.

Said by Husain & Kasim (1975: Cultivated plants of Iraq) WARD AL-DUKMAH, ZEHRAT AL-RABEE'A, ZEHRAT AL-LULU to be "wild in the north, cultivated in the middle and south of Iraq". The Baghdad specimen cited above may thus have been cultivated or an escape from cultivation – or washed down by the river from further north. No ecological notes are given.

Related to *Bellis* is the Australasian genus *Brachycome* Cass., of which one species, *B. iberidifolia* Benth., is said by Husain & Kasim (1975) to be "occasionally cultivated" in Iraq. This is a dainty, corymbosely branched annual 15–30(–50) cm tall with small, alternate leaves pinnately divided into narrow segments. The capitula have spatulate-oblong biseriate phyllaries, and are radiate, the blackish-purple disc being 4–7 mm in diameter and the blue, roseate or white ray limbs 6–11(–14) mm. Swan River Daisy (English).

82. **CALLISTEPHUS** Cass. *nom. cons.*

Dict. Sci. Nat. ed. 2, 37: 491 (1825)
Nesom & Robinson in Kubitzki, Fam. Gen. Vasc. Pl. 8: 318 [2006] (2007)
Callistemma Cass. in Bull. Soc. Philom. Paris: 1817: 32 (1817)

Frances G. Davies

Annual herb with alternate leaves. Capitula solitary on branches, showy, heterogamous; receptacle plane or slightly convex. Ray florets female, 1–2-seriate; disc-florets hermaphrodite. Achenes compressed; pappus in two rows; outer row connate into a very short, bristly corona, the inner of slender, caducous, rather long barbellate setae.

A monotypic genus.

Callistephus (from the Gr. Κάλος, kalos, beautiful and στεφος, stefos, crown).

1. **Callistephus chinensis** (*L.*) *Nees*, Gen. & Sp. Aster.: 222 (1832); Husain & Kasim, Cult. Pl. Iraq: 109 (1975); Zhu Shi et al. in Fl. China. 20-21: 569 (2011).

Aster chinensis L., Sp. Pl.: 1232 (1753).
Callistephus hortensis Cass., Dict. Sci. Nat. ed. 2, 37: 491 (1825).

Robust branched herb with scabrid stems. Leaves rhomboid to elliptic, ± sessile, up to 5 × 3.5 cm, acute, margin irregularly toothed to ¼(–½) the lamina width, the veins with short scabrid hairs, both surfaces green. Involucre 2–3 × 2–3 cm; outer phyllaries numerous, spreading, herbaceous, oblong, 2 × 0.7 cm, ciliate; inner phyllaries ovate, *c.* 1 cm long, hyaline, white or pink or purple. Ray-florets very numerous, rays up to 3 × 0.4 cm, pink or purple; disc-florets yellow, up to 1 cm. Achenes ± 4 mm long, brown.

HAB. Cultivated as a garden ornamental; alt. up to 35 m; fl. & fr. Apr.-Aug.
DISTRIB. Known only in cultivation. **MSU**: Sulaimaniya, *Thamer* C960! **DWD**: Al Habbaniyah, *Kalf, Janan, Wedad & Hana* 914; **LCA**: Zaffaraniya, *Omar & Sahira* 34810; Baghdad, *Janan al–Mokhtar* 413; Abu Ghraib, *Jenan & Sahira* 277; Baghdad, *Harthya, Fawzi* 770.

A native of northern China, widely cultivated as an ornamental. In cultivation, the rays are considerably increased in number.

83. **PSYCHROGETON** Boiss.

Fl. Orient. 3: 156 (1875)
Nesom & Robinson in Kubitzki, Fam. Gen. Vasc. Pl. 8: 319 (2007)

Frances G. Davies

Perennial or more rarely biennial herbs with a thick and woody or more slender rootstock. Leaves oblanceolate to rotund, entire to subpinnatifid, pubescent to tomentose, frequently glandular. Capitula commonly solitary or few, rarely racemose or corymbose, heterogamous. Corollas yellow, reddish when dry or withered. Marginal florets rayed or tubular, female, with fertile achenes. Disc florets hermaphrodite, functionally male. Achenes ± sericeous or pubescent and sometimes glandular; pappus setae 1–2–seriate, barbellate.

About 20 species in the mountainous regions of temperate Asia.

Psychrogeton (from the Gr. ψυχρος, *psuhros,* the cold, γειτον, *geiton,* a neighbour – a reference to the montane habitat of the plants.)

Grierson, A.J.C. (1967). The genus *Psychrogeton* (Compositae). Notes Roy. Bot. Gard. Edinburgh 27: 101–147.

Plant subscapose, leaves mostly basal, stems terminating in a single head; marginal female florets rayed .... 1. *P. amorphoglossus*
Plant with an erect leafy stem bearing numerous capitula; marginal female florets tubular ....2. *P. nigromontanus*

1. **Psychrogeton amorphoglossus** (*Boiss.*) *Novopokr.* in Not. Syst. (Leningrad) 7: 136 (1938); Grierson, Notes Roy. Bot. Gard. Edinburgh 27: 119 (1967) & in Fl. Turkey 5: 130 (1975); Breckle, Hedge, Podlech & Rafiqpoor, Vasc. Pl. Afghan. checklist: 181 (2013).

Erigeron amorphoglossum Boiss., Diagn. Pl. Orient. ser. 1, 6: 80 (1845); Boissier, Fl. Orient. 3: 166 (1875); Rawi in Dep. Agr. Iraq Tech. Bull. 14: 115 (1964).

Dwarf perennial herb, 3.5–20 cm, rootstock usually thick and woody, ± densely crowned with scaly remains of previous years' leaves. Leaves mostly basal, broadly to narrowly elliptic to obovate or oblanceolate, 1.5–5 × 0.3–1 cm, obtuse or acute, narrowed and ± petiolate below. Stems several, ± tomentose and glandular, slender, erect or ascending, lowest leaves ± similar to basal, rapidly reducing and bract-like above; all leaves densely pubescent with tangled white hairs and sessile shining glands. Capitula terminal and solitary; phyllaries ± white-lanate, ciliate-margined and glandular. Corollas yellow, red when withered; female florets 4.5–5.5 mm, ray limb 1.5–2 mm, mostly 3–toothed, not inrolled when dry, tube sparsely pilose. Fertile achenes 4–5.1 mm, sericeous, brownish, narrowly oblanceolate; pappus 3.5–4.5 mm, setae numerous. Fig. 81: 1–8.

HAB. Rocky and stony slopes and summits of the higher mountains; alt. 2900–3800 m; fl. & fr. Jul.-Sep.
DISTRIB. Confined to the NE mountain region of Iraq on and near the Turkish and Iranian frontiers. **MRO**: Algurd Dagh, *Guest & Ludlow-Hewitt* 2886!, *Rechinger* 11448!, *Haley* 164; Sia Kuh, *Ludlow-Hewitt* 1510!; Qandil range, *Rawi & Serhang* 24473!, *Rechinger* 11165; NE of Qandil, *id.* 24403A!; top of Qandil range between top of Perrish ('Persian') and Bardanas, *id.* 24562!; between the top of Persian (?Perrish) and Bardanas, on the Iranian frontier, *Rawi & Serhang* 24575!; N of Algurd range E of Bermasand lake, *id.* 24759!; Ser Kurawa, *Gillett* 9759!

Turkey, Afghanistan, C Asia.

2. **Psychrogeton nigromontanus** *(Boiss. & Buhse) Grierson,* Notes Roy. Bot. Gard. Edinburgh 27: 144 (1967); Grierson in Fl. Turkey 5: 131 (1975); Chen Yilin & Luc Brouillet in Fl. China 20–21: 570 (2011); Breckle, Hedge, Podlech & Rafiqpoor, Vasc. Pl. Afghan. checklist: 182 (2013).

Erigeron nigromontanus Boiss. & Buhse, Aufz. Reise Transkauk. Pers.: 114 (1860).

Erect perennial herb 15–75(–90) cm tall, rootstock rather slender. Stems 1-several, unbranched below, somewhat to considerably paniculate above, striate and somewhat angular to terete, paniculately branched with numerous capitula, moderately to rather densely whitish pubescent and sometimes glandular. Basal leaves mostly withered at flowering time, similar to lower cauline leaves, these with oblanceolate to elliptic lamina, acute, margins entire or shallowly dentate, narrowed below to a distinct petiole to ± 2 cm long, whole leaf 1.5–5 × 0.3–1 cm; upper cauline leaves ± whitish-pubescent and usually with shining subsessile glands. Capitula 0.8–1 cm in diameter; phyllaries linear, 3.5–4.5 mm, whitish-pilose with subsessile glands centrally, margins stramineous. Corollas yellow; female florets tubular, 1.5–3 mm, tube puberulent; disc florets almost twice as. Fertile achenes 1.5–2 mm, pubescent, brownish, narrowly oblanceolate; pappus setae numerous, 3–4 mm.

HAB. Damp patch on mountainside, mountain slope; alt. 1700–1800 m; fl. Aug.-Sep., fr. Aug.-Oct.
DISTRIB. Only found twice in the eastern mountain region of Iraq, not far from the Iranian frontier. **MRO**: Haji Umran, *Rechinger* 11308. **MSU**: Pira Magrun, *Agnew, Hadač, Haines & Walid* W1849!

China (N Xinjiang), Afghanistan, Kazakhstan, Kyrgyzstan, Iran, Turkey.

Zohary (1950) and Rawi (1964) both record *Psychrogeton persicus* (Boiss.) Grierson for Iraq, as *Erigeron persicus* Boiss., on the basis of the record of Guest (1933) – misidentification of *Erigeron uniflorus* – and Haussknecht's "Avroman et Schahu" specimens, cited in Fl Iranica 154: 50 (1982) with the collection number 522–b. Grierson (1967) lists a specimen from "Pir Omar Gudrun" (Iraq), "ad nives Mt. Schahu" (Iran). It is clear that Haussknecht's printed Pir Omar Gudrun label was used and the correct locality added in manuscript. Since the "Avroman et Schahu" specimen is labelled with exactly the same details, "ad nives 12000' [ft]" it is clear that this too came from Schahu and not from Avroman. *P. persicus* thus has no claim to be considered as an Iraqi species.

84. **TRIPOLIUM** Nees

Gen. Sp. Aster. 152 (1832)
Nesom & Robinson in Kubitzki, Fam. Gen. Vasc. Pl. 8: 319 [2006] (2007)
Aster L., Sp. Pl.: 872 (1753), pro parte.

D. J. Nicholas Hind & J. R. Edmondson

Annual, biennial or perennial herbs with fibrous, rhizomatous or ±caudiciform roots. Leaves alternate, sessile to pseudopetiolate, simple, often somewhat fleshy, oblanceolate-spatulate, entire or ± dentate. Inflorescences of solitary capitula, or many headed in a much-branched panicle or corymb. Capitula heterogamous and radiate or homogamous and discoid, usually heterochromous; involucre campanulate; phyllaries (1–)2–3(–4)-seriate, gradate; receptacle flat to convex, alveolate, epaleaceous. Ray-florets when epresent female, corollas usually blue or purple. Disc florets hermaphrodite, fertile, corollas 5-lobed, yellow, sometimes purplish-tipped. Achenes compressed, ± obovate or angled, glabrous or setuliferous, sometimes glandular; pappus setae few-seriate, barbellate, subequal, ± caducous, white, whitish or pale red, conspicuously elongating after anthesis.

A widespread monotypic genus from Eurasia, N. America and N. Africa also present in the Flora area.

Tripolium (meaning 'three cities' in Gr., the name of several places including the region now known as Libya whose capital is Tripoli).

Nesom, G.L. (1994). Review of the taxonomy of Aster sensu lato (Asteraceae: Astereae), emphasizing the New World species in Phytologia 77(3): 141–297.

Fig. 81. **Psychrogeton amorphoglossus**. 1, habit × ⅔; 2, leaf × 3; 3, capitulum × 4; 4, ray floret × 6; 5, disc floret × 6; 6, achene and pappus × 5; 7, phyllary × 9; 8, pappus hair × 60. All from *Rawi & Serhang* 24575. Drawn by M. Millar Watt.

1. **Tripolium pannonicum** *Schur*, Enum. Pl. Transsilv. 304 (1866).

Aster tripolium L., Sp. Pl.: 1: 872 (1753); Boissier, Fl. Orient. 3: 158 (1875); Dinsmore in Post, Fl. Syria, ed. 2, 2: 17 (1933): Rechinger, Fl. Lowland Iraq: 594 (1964); Hassawy, Tammimi & Alizzi in Dep. Agr. Iraq Tech. Bull. 167: 109 (1968); Grierson in Fl. Turkey 5: 120 (1975); Feinbrun-Dothan, Fl. Palaest. 3: 297 (1978); Chen Yilin & Brouillet in Fl. China 20–21: 559 (2011); Abdulridha, Taha & Wedad, Ecology and Fl. Basra: 195 (2016).
Tripolium vulgare Nees, Gen. Sp. Aster: 152 (1832); Tamamshian in Fl. S.S.S.R. 25: 182 (1959).

Short-lived perennial fleshy halophytic herb, with a short succulent rhizome. Stem 0.2–1 m tall, glabrous, erect. Leaves entire or slightly toothed, basal oblanceolate to narrowly obovate, 6–12 × 0.7–1 cm, narrowed and pseudopetiolate below, upper gradually reducing, linear-oblong to linear, obtuse to subacute. Inflorescence corymbose. Capitula heterogamous and radiate; involucre hemispherical, 6–8 mm long; phyllaries rather few, blunt, linear-oblong, membranous-margined, 1.75–2.5 mm wide, glabrous. Ray florets female, corollas 10–15 mm, purplish-blue, rarely absent. Disc floret corollas yellow. Achenes 3–4 mm, compressed with thickened angles, brownish, setuliferous or ± glabrescent; pappus 8–12 mm, dirty white. Fig. 82: 1–3.

HAB.: In an alfalfa field and along muddy ditches, always on ± saline soil; alt. s.l.; fl. & fr. Apr.-Oct.
DISTRIB.: Only found in the SE extremity of Iraq; common along canals in date palm orchards in southern Basrah and Fao. **LBA**: Duva, between Basra and Fao, *Alizzi & Omar* 35865!; 17 km NW of Fao, *Khatib & Alizzi* 33412! Basra to Fao, *Rechinger* 3612!

Throughout most of Europe along coastal and island saline areas, also in similar places in temperate SW and C Asia, China, and eastwards from Russia and Mongolia to Japan, N Africa (Algeria) and in N America.

85. **SYMPHYOTRICHUM** Nees

Gen. Sp. Aster. 135 (1832)
Nesom & Robinson in Kubitzki, Fam. Gen. Vasc. Pl. 8: 335 (2007)
Aster L., Sp. Pl.: 1: 872 (1753), pro parte

Frances G. Davies
Revised by J. R. Edmondson

Annual, biennial or perennial herbs with fibrous or rhizomatous roots. Leaves alternate, petiolate or sessile, simple, entire or crenate to dentate. Synflorescences paniculiform, sometimes racemiform or subcorymbiform or sometimes capitula solitary. Capitula heterogamous, radiate or discoid; ray-florets female, usually blue or purple; disc florets hermaphrodite, yellow, sometimes purplish-tipped, 5-lobed. Involucre campanulate or cylinric; phyllaries in (3–)4–6(–9) rows, imbricate or finally ± spreading, equal or unequal, glabrous or hairy, rarely glandular. Receptacle flat or almost so. Achenes compressed, ± obovate or angled; pappus of ± equal scabrid, bristles, in 4 series, persistent.

A large and widely distributed genus differently circumscribed in the various regions where it occurs. About 90 species, Asia, Europe N and S America; especially well represented in N America.

Symphyotrichum (from the Gr. σημ, a prefix 'joined to' and τρίχα, tricha, hair).

Rays very short, shorter than to scarcely exceeding pappus; disc-florets rather few, purplish-tipped . . . . 1. *S. subulatum*
Rays long and conspicuous, much exceeding pappus; disc-florets many, yellow throughout . . . . 2. *S. novi-belgi*

1. **Symphyotrichum subulatum** (*Michx.*) *G.L.Nesom*, Phytologia 77(3): 293 (1995).

Aster subulatus Michx., Fl. Bor. Amer. 2: 111 (1803); Rechinger, Fl. Lowland Iraq: 595 (1964); Rawi in Dep. Agr. Iraq Tech. Bull. 14: 106 (1964); Hassawy, Tammimi & Alizzi, ibid. 167: 106 (1968); Feinbrun-Dothan, Fl. Palaest. 3: 297 (1978); Yilin Chen & Brouillet in Fl. China 20–21: 652 (2011); Abdulridha, Taha & Wedad, Ecology and Fl. Basra: 195 (2016).

Annual, 30–100 cm tall; stems slender, wiry, glabrous or thinly pilose upwards, often

Fig. 82. **Tripolium pannonicum**. 1, habit × ½; 2, ray floret × 4; 3, disc floret × 4. From Flora of China, Asteraceae 20-21, fig. 484. Reproduced with permission from Missouri Botanical Garden Press, St. Louis, and Science Press, Beijing.

purplish-tinged, divergently branched above to form a usually large, many-headed panicle. Leaves sessile, entire, linear to linear-oblanceolate or lower oblanceolate-spatulate, 5–10(–15) × 0.4–1 cm, obtuse, upper always narrow, obtuse or acute, glabrous or with scattered hairs. Involucre rather narrow, narrowly campanulate to obconical, 5–8 mm; phyllaries green with pale membranous margins and pale midribs, unequal, passing into peduncular bracts, 0.5 mm wide, minutely glandular. Ray florets female, 4–5 mm, ray limbs pale lavender to purple, finally circinnately coiled, shorter than to scarcely exceeding the pappus; disc-florets narrow, rather few, purplish-tipped at least when dry. Achenes ± 2 mm, pubescent, light brown, ribbed; pappus dirty white, 5–6 mm. Involucre reflexed after fruit fall. Fig. 83: 1–8.

HAB.: In irrigated cultivation (alfalfa, corn, rice etc.), along ditches, river banks and roadsides, on loam, silty or sandy soil, often saline; alt. 10–68 m; fl. & fr. almost throughout the year.
DISTRIB. Locally common in the alluvial plains of Iraq, apparently rare in the foothills but may well spread rapidly especially along the great rivers. **FUJ**: between Mosul and Qaiyara, *Alizzi & Omar* 35262! **LCA:** Abu Ghraib, *Gillett & H. Ahmed* 12501!, *Alizzi* 32801!, *Janan al-Mukhtar* 35224!, *Chakravarty* 38216!; Baghdad, *Haines* W33!; nr R. Tigris at Namaniya, *Juma Brahim* 6184!; nr Musaiyib on road to Karbala, *Guest, Rawi & Rechinger* 16147!; 46 km SE of Musaiyib, *Thamer & Wedad* 47259! **LEA**: Musharra (=Halfaya), *Thamer* 47115!; 20 km NE of Baquba, *Botany Staff* 47136! **LSM**: Amara sugar factory, *Dabbagh & Qaisi* 41896!; 50 km from Basra towards Nasiriya, *Husham & Mohammed* 29627! **LBA**: Basra, *Omar & Thamer* 47889!; *Abul-Khasib, Alizzi & Omar* 34680!

A native of N America, where it grows in saline, brackish or fresh coastal marshes; naturalized elsewhere.

2. **Symphyotrichum novi-belgii** (*L.*) *G.L.Nesom*, Phytologia 77: 287 (1994).

Aster novi-belgii L., Sp. Pl.: 877 (1753); Bailey, Manual Cult. Pl. ed. 2: 1010 (1949).

Tall perennial, 10–80 cm with numerous slender but wiry, striate stems, glabrous below but sometimes pubescent above, leafy to the top, much-branched above with very many capitula. Leaves oblong to linear-lanceolate, entire or serrate-dentate, the lower narrowed below and ± petiolate, upper sessile and ± amplexicaul, glabrous or slightly pubescent on upper surface and along margins. Capitula in a laxer or denser panicle, 1.5 cm in diameter. Ray florets bright blue-violet, mostly 10–20 mm; disc florets yellow. Phyllaries linear, acute, somewhat divergent, glabrous or puberulent.

HAB. & DISTRIB. Cultivated in a garden and perhaps naturalized along roadside; alt. 25–45 m; fl. Jun.-Oct.: **LCA**: Hilla, *Fawzi* 39525!; Habbaniyah, cult., *Janan & Hana* C.909!

A native of N America much cultivated in Europe and elsewhere as a late-flowering ornamental. New York Aster (Am.); with other related species known in England as "Michaelmas Daisy". Widely naturalized in C & NW Europe along roadsides, railway embankments and waterways and on waste ground.

86. **CONYZA** Less. *nom. cons.*

Syn. Gen. Comp. 203 (1832)
Nesom & Robinson in Kubitzki, Fam. Gen. Vasc. Pl. 8: 340 [2006] (2007)

Frances G. Davies

Annual to perennial herbs. Leaves alternate, of varying form, entire, toothed or lobed, often with an indumentum of long or short hairs (or both). Inflorescence a branched corymbose panicle. Capitula heterogamous; phyllaries 2–4–seriate, herbaceous, imbricate,often with resinous brown midribs, involucre exceeded at flowering time by conspicuous dirty-white pappus, eventually reflexed with inner surfaces of phyllaries outermost. Florets all fertile, with several outer series of minute filiform or ligulate female florets and fewer central tubular hermaphrodite disc-florets. Corolla cream, yellow or purple. Achenes compressed, obovate or oblong, off-white to brown, ± pilose; pappus uniseriate, of minutely barbellate bristles, ± equalling florets in length and giving the characteristically fluffy appearance of the inflorescence at flowering.

Conyza (a name used by Pliny for a kind of Fleabane, possibly derived either from the Gr. κουνούπι, *kounoupi,* gnat, or from the Gr. σκόνη, *skoni,* dust, a reference to the plant's use as a flea-powder).

A genus of 60–100 species of the tropics and subtropics.

Plant bright green; stems ciliate with spreading setose hairs; fruiting capitula to 7 mm in diameter . . . . 1. *C. canadensis*
Plant grey-green, stems with spreading and appressed hairs; fruiting capitula usually 1 cm in diameter or more. . . . . 2. *C. bonariensis*

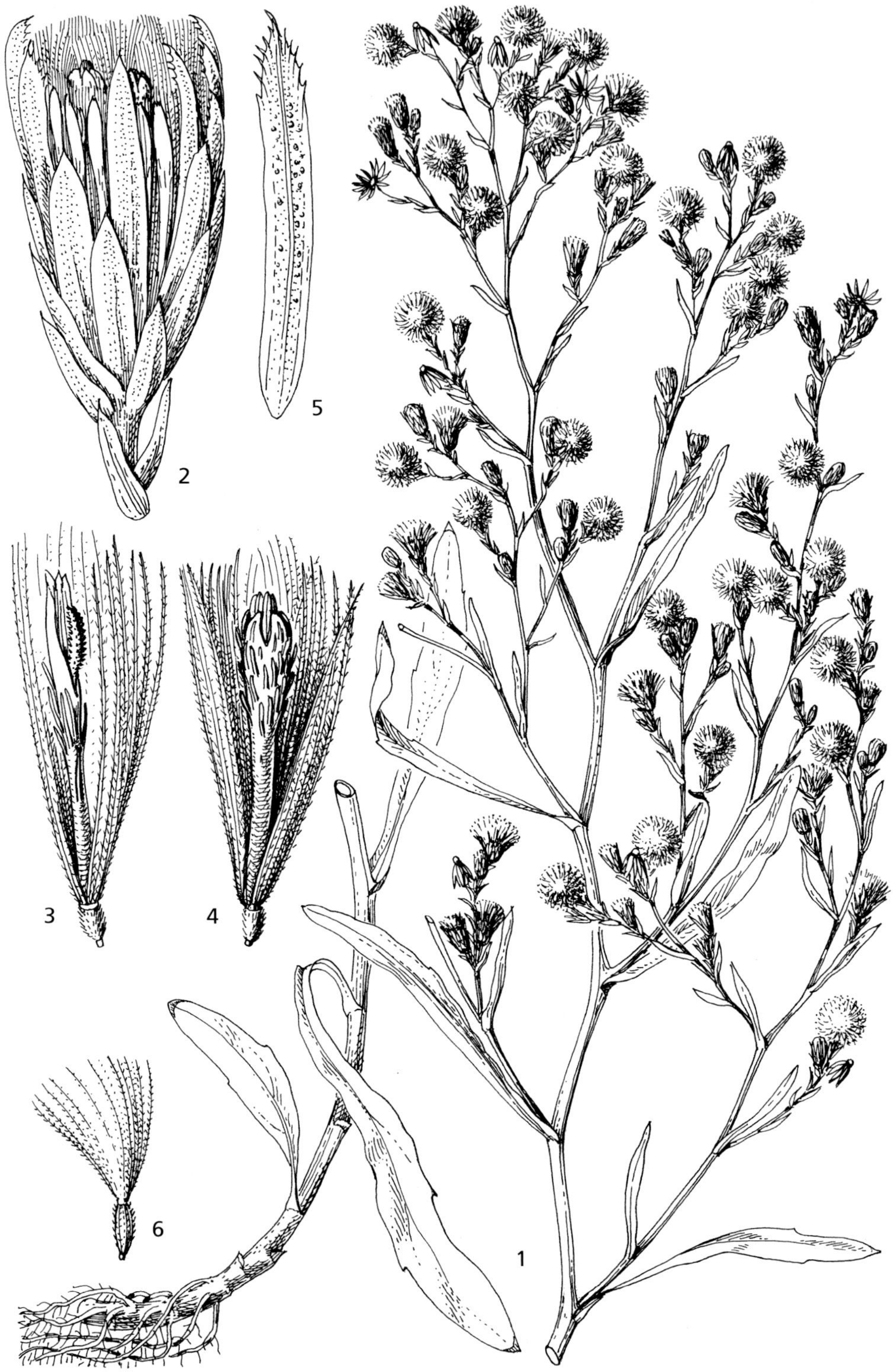

Fig. 83. **Symphyotrichum subulatum**. 1, habit × ⅔; 2, capitulum × 8; 3, ray floret × 12; 4, disc floret × 12; 5, phyllary × 12; 6, achene and pappus × 4. All from *Alizzi* 32801. Drawn by Laura (née Ripley) Mason.

1. **Conyza canadensis** (*L.*) *Cronquist* in Bull. Torrey Bot. Club 70: 632 (1943); Grierson in Fl. Turkey 5: 132 (1975); Cronquist in Fl. Europ. 4: 120 (1976); Feinbrun-Dothan, Fl. Palaest. 3: 298 (1978).

Erigeron canadensis L., Sp. Pl.: 863 (1753); Boissier, Fl. Orient. 3: 163 (1875); Nábělek in Publ. Fac. Sci. Univ. Masaryk 52: 3 (1925); Dinsmore in Post, Fl. Syria, Palest. & Sinai ed. ed. 2, 2: 17 (1933); Guest in Dep. Agr. Iraq Bull. 27: 32 (1933); Anthony in Notes Roy. Bot. Gard. Edinburgh 18: 290 (1935); Blakelock in Kew Bull. 4: 54 (1949); Zohary in Dep. Agr. Iraq Bull. 31: 144 (1950); Bochantsev in Fl. S.S.S.R. 25: 239 (1959); Rawi in Dep. Agr. Iraq Tech. Bull. 4: 115 (1964); Rawi & Chakravarty, ibid. 15: 40 (1964); Rechinger, Fl. Lowland Iraq: 595 (1964); Hassawi, Tammimi & Alizzi in Dep. Agr. Iraq Tech. Bull. 167: 121 (1968).

A weedy annual, occasionally perennating if heavily grazed, to 1 m tall, sparingly spreading-pilose. Leaves oblanceolate to linear, 1.2–5(–7) × 0.1–1.2 cm, gradually narrowed and petioloid towards base, margin with stout white hairs, generally entire or sometimes weakly toothed. Inflorescence of many to numerous capitula, in a broad corymbose panicle, in which the subsequent branches equal the primary ones. Phyllaries linear to narrowly lanceolate, 1.5–3 mm. Outer florets female, rayed, cream, 1.75–2 mm, ligule 1 mm. Disc-florets hermaphrodite, few, 4-lobed, yellow. Achenes 1–1.5 mm long, sparsely shortly hairy; pappus setae straw-coloured, ± 3 mm.

HAB. Valley bottom, by streams, riversides, and as a weed of roadsides, disturbed ground and cultivation; alt. 25–1200 m; fl. & fr. almost throughout the year.
DISTRIB. Scattered but probably increasing in the mountains and foothills, locally common in the lower alluvial plain, especially in the Baghdad district. **MAM**: S of Sharanish village, *Rawi, Noori & Tikriti* 29058! **MRO**: Qerna Qaw valley, N of Pushtashan, *Rawi & Serhang* 26619!; Rowanduz gorge, *Guest & Husham* 15920! **FUJ**: banks of R. Tigris nr Mosul, *Kotschy* 428! **FPF**: 2 km N of Sa'diya, *Qaisi* 47233! **LEA**: Kut, *Guest* 211!; Baghdad, Karrada Mariam, *Haines* W. 1173! **LCA**: Abu Ghraib, *Gillett* 12500!; Rustam, *Guest* 481!; Diwaniya, *Gillett* 9928!

Canadian Fleabane (Eng.), Horse's Tail.

A native of N America, now widely naturalized in the temperate regions of the world and in many tropical countries.

2. **Conyza bonariensis** (*L.*) *Cronquist* in Bull. Torrey Bot. Club 70: 632 (1943); Grierson in Fl. Turkey 5: 133 (1975); Cronquist in Fl. Europ. 4: 120 (1976); Meikle, Fl. Cyprus 2: 872 (1985).

Erigeron bonariensis L., Sp. Pl.: 863 (1753); Bochantsev in Fl. S.S.S.R. 25: 240 (1959); Rawi in Dep. Agr. Iraq Tech. Bull. 14: 115 (1964); Rechinger, Fl. Lowland Iraq: 596 (1964).
E. crispus Pourr. in Hist. Mém. Acad. Sci. Toulouse 3: 318 (1788); Dinsmore in Post, Fl. Syria, Palest. & Sinai ed. 2, 2: 18 (1933).
E. linifolius Willd., Sp. Pl. ed. 4, 3: 863 (1802); Boissier, Fl. Orient. 3: 169 (1875).
Conyza hispida Kunth in Humb., Bonpl. & Kunth, Nov. Gen. Sp. Pl. 4 (ed. folio): 55 (1818).
C. albida Willd. ex Spreng., Syst. Veg., ed. 16, 3: 514 (1826).
C. linearis DC., Prodr. 5: 378 (1836).
Erigeron albidum (Willd. ex Spreng.) A. Gray, Proc. Amer. Acad. Arts 5: 319 (1861-62).
Conyza plebeja Phil., Anales Univ. Chile 87: 685 (1894).
Leptilon bonariense (L.) Small, Man. S. E. Fl. [Small]: 1231 (1903).
Erigeron bonariensis L. var. *angustifolius* Cabrera, Revista Mus. La Plata, Secc. Bot. 4: 88 (1941).
E. bonariensis L. var. *microcephala* Cabrera, Revista Mus. La Plata, Secc. Bot. 4: 88 (1941).
Marsea bonariensis (L.) V. M. Badillo, Bol. Soc. Venez. Ci. Nat. 10: 256 (1946).
Conyza bonariensis (L.) Cronquist var. *angustifolia* (Cabrera) Cabrera, Man. Fl. Buenos Aires: 481 (1953).
C. bonariensis (L.) Cronquist var. *microcephala* (Cabrera) Cabrera, Man. Fl. Buenos Aires: 481 (1953).

Erect annual up to 60 cm tall, greyish, stems with a mixed indumentum of short appressed and longer patent hairs. Leaves linear, lanceolate or oblanceolate, 1–8(–10) × 0.2–2 cm, appressed-pubescent and scabridulous, upper usually entire and lower broader and irregularly jaggedly toothed or lobed. Inflorescence an elongate panicle, overtopped by subsequent lateral branches, capitula few to many. Phyllaries linear to narrowly lanceolate, 2.5–5 mm. Marginal florets female, filiform, sharply lobed or minutely ligulate at tip; disc-florets hermaphrodite, corolla yellow, 5–lobed. Achenes 1.25–1.5 mm, narrowly oblong, sparsely shortly hairy; pappus setae straw-coloured, 3–3.5(–4) mm. Fig. 84: 1–8.

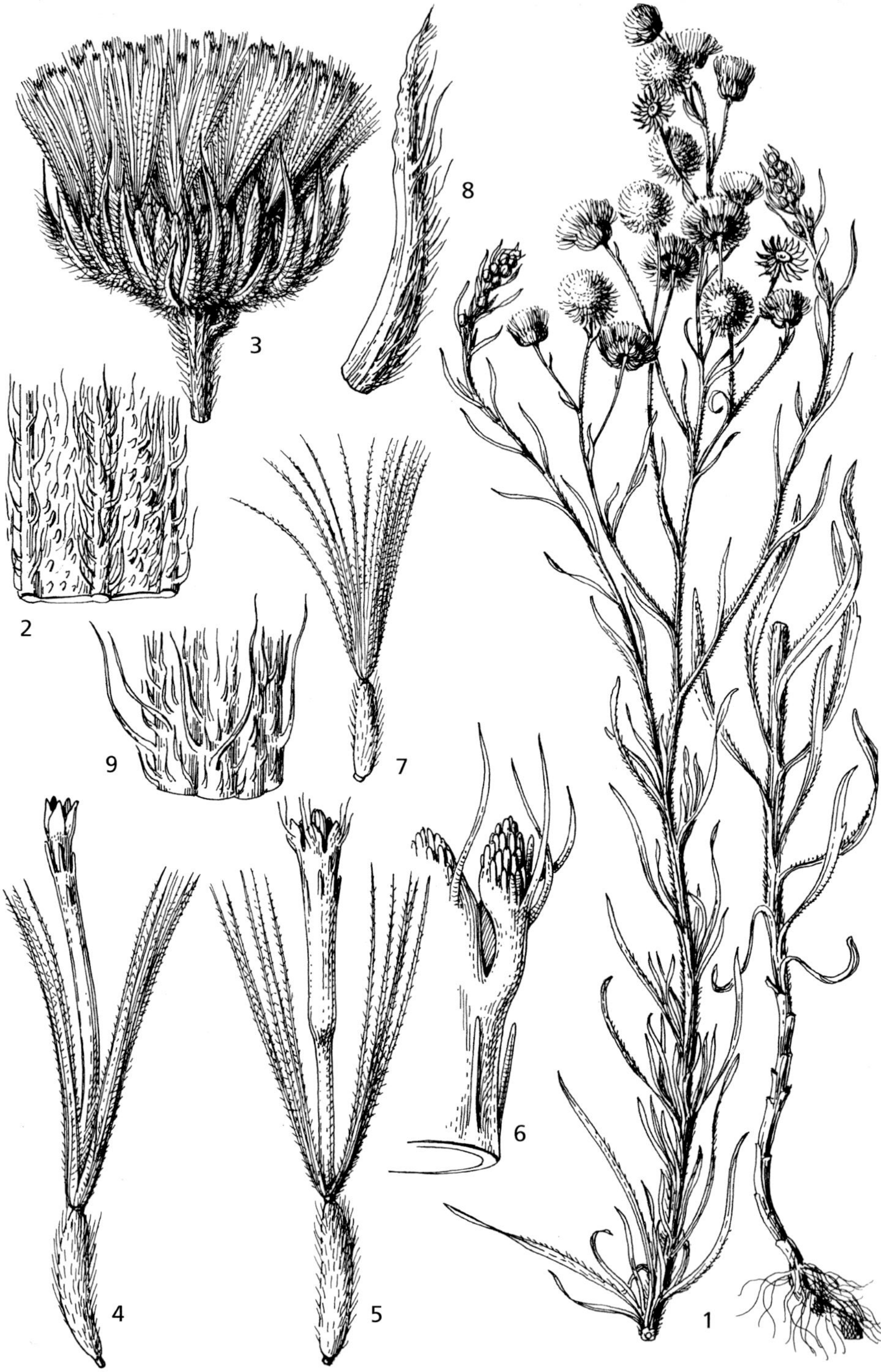

Fig. 84. **Conyza bonariensis**. 1, habit × ⅔; 2, part of leaf surface × 12; 3, capitulum × 6; 4, ray floret × 16; 5, disc floret × 16; 6, petal × 32; 7, achene and pappus × 10; 8, phyllary × 14; 9, part of stem enlarged. All from *Wheeler Haines* W30. Drawn by Laura (née Ripley) Mason.

HAB. A weed of cultivation (especially where irrigated), waysides and disturbed ground; alt. s.l.-250 m; fl. & fr. almost throughout the year.
DISTRIB. Quite common on the alluvial plain of Iraq, occasional in the foothills. **FNI**: Nineveh, *E. Chapman* 26190! **FPF**: Badra, *Qaisi & Khayat* 50614!; 2 km NW of Sa'diya, *Qaisi* 42563! **LEA**: Shatt al-Gharraf, *Karim & Hamid* 37805!; 15 km E of Amara, *Alizzi & Omar* 34715!; Amara, *Botany staff* 42488! **LCA**: Abu Ghraib, 14 km W of Baghdad, *Gillett* 5842!, 12490!; Baghdad, *Ahmed* 9833!, *Haines* W. 30!; Diwaniya, *Gillett* 9927! **LSM**: Musaida, *Rawi* 16526! **LBA**: Basra, *Gillett* 10037!

A native of S America now widespread in the warmer regions of the world as an introduced and naturalized weed.

87. **ERIGERON** L.

Sp. Pl.: 863 1753); Gen. Pl. ed. 5: 371 (1754)
Nesom & Robinson in Kubitzki, Fam. Gen. Vasc. Pl. 8: 340 [2006] (2007)

C. C. Townsend
Revised by D. J. Nicholas Hind

Annual, biennial or perennial herbs. Leaves simple, alternate, spatulate, entire to coarsely dentate; indumentum of vegetative parts of simple, white spreading cover hairs and in some species also sessile or short glandular hairs. Inflorescence of remote hemispherical capitula, capitula solitary or several in corymbs or thyrses. Capitula heterogamous, radiate; involucre turbinate to hemispherical; phyllaries 2–3(–7)-seriate, herbaceous, imbricate, gradate, sometimes subequal; receptacle flat to convex or conical, epaleaceous. Florets of 2–3 types, outer female, 1–4-seriate, rays limbs narrow, purplish to bluish, lilac or white, intermediate female or absent, corollas filiform, inner disc florets tubular, hermaphrodite, corollas yellow, 5-lobed; female florets with filiform corolla-tube and ray lacking or very reduced. Achenes obovate, compressed, 2–4-ribbed, pale, glabrous or setuliferous; pappus of few to numerous scabrid, dirty white or brownish hairs, sometimes with outer minute subsquamose hairs, accrescent in fruit to up to 1.5 × involucral length, pappus of ray-florets sometimes very rudimentary.

A world-wide genus of perhaps ± 170 to as many as ± 400 species (according to Nesom & Robinson 2006) – but on a world scale its delimitation from closely allied genera is incomplete.

Erigeron (from the Gr. ἠρι, ēri, early and γέρον, geron, old man – the pappus is soon conspicuous in the capitula).

Hind, D.J.N. (2012). Plate 729. *Erigeron karvinskianus*. Curtis's Bot. Mag. 29(1): 52–65.
Nesom, G.L. (1989). Infrageneric taxonomy of New World *Erigeron* (Compositae: Astereae). Phytologia 67(1): 67–93.
Solbrig, O.T. (1962). The South American species of *Erigeron*.Contr. Gray Herb. 191: 3–79.
Vierhapper, F. (1906). Monographie der alpinen *Erigeron*-Arten Europas und Vorderasiens. Beih. Bot. Centralblatt 19(2): 385–560.

Dwarf perennial high-alpine plant rarely exceeding 15 cm; capitula solitary; capitula without filiform female florets between hermaphrodite disc florets and marginal ray florets; plants perennial . 1. *E. uniflorus*
Taller plants, rarely as short as 15 cm; capitula very rarely solitary, usually several to numerous in a panicle; capitula with filiform female florets between hermaphrodite disc florets and marginal ray florets; plants annual, biennial or short-lived perennial . . . . 2. *E. acris*

1. **Erigeron uniflorus** *L.*, Sp. Pl.: 864 (1753); Boissier, Fl. Orient. 3: 165 (1875); Vierhapper in Beih. Bot. Centralbl. 19 (2): 495 (1906); Dinsmore in Post, Fl. Syria, Palest. & Sinai ed. 2, 2: 17 (1933); Bochantsev in Fl. S.S.S.R. 25: 227 (1959); Grierson in Fl. Turkey 5: 12 (1975); Grierson & Rechinger in Fl. Iranica 154: 31 (1982); Breckle, Hedge, Podlech & Rafiqpoor, Vasc. Pl. Afghan. checklist: 173 (2013).

E. persicum auct. non Boiss.: Guest in Dep. Agr. Iraq Bull. 27: 32 (1933).

Dwarf perennial, 2–15(–20) cm tall, stems always simple with a single capitulum, several from a rhizomatous rootstock, clothed with spreading white hairs, with or without shorter hairs and a few glands. Basal leaves linear-oblanceolate, obtuse to spatulate and often emarginate, mostly 12–35 × 2–7 mm, sparingly pilose or ciliate at least along basal margins; cauline leaves lanceolate to linear, indumentum similar, upper reducing and bract-like. Inflorescence scapiform. Capitula heterogamous and radiate, lacking intermediate female florets; involucre 10–15 mm in diameter; phyllaries narrowly to linear-lanceolate, acute, inner 4–9 mm long, all thinly to densely clothed with long white hairs (often dense and deflexed at base of capitulum), often blackish or purplish at least about tip. Female florets usually all radiate, 5–8 mm, rays 2.5–5.5 mm, narrow, purple to lilac or white; disc florets 3–3.5(–4) mm; eligulate female flowers with a 2 mm filiform tube occasionally present, at least in Iraq. Achenes 2–2.5 mm, pale stramineous, pubescent; pappus 3–4.5 mm, brownish.

HAB. On stony mountain tops, on alpine rocks on the slopes of high mountains, also near a mountain stream; alt. 2250–3600 m; fl. & fr. Aug.-Sep.
DISTRIB. Only found in a restricted area on the highest mountains of Iraq. **MRO**: Algurd (Helgord) Dagh, *Gillett* 12356!, *Rechinger* 11873, *Rawi & Serhang* 24915!; Kani Khanjar Khan, NE of Helgord Mt., *id.* 24707!; NE of Qandil, *id.* 24403A!; between the top of Persian (?Perrish) and Bardanas, on the Iranian frontier, *id.* 4257B!

Arctic and subarctic Europe, also in the high mountain ranges of C & S Europe southwards to the Pyrenees and Corsica, eastwards to Bulgaria, mountains of SW & C Asia from Turkey, Lebanon and Caucasus to Tibet; also in Arctic N America.

It is extremely interesting that the material of this species from Helgord Dagh has filiform eligulate female florets present, and in one case (*Rawi & Serhang* 24707) plentifully, while the Qandil plants have none. In all other respects the plants are identical, and it would seem absurd to regard them as different species even though this cuts across the technical character used by Vierhapper, l.c., following Cassini, to separate the two genera *Erigeron* and *Trimorpha*. Vierhapper (l.c., p. 451) makes reference to the fact that in Greenland forms of *Trimorpha borealis* Vierh. (*Erigeron borealis* (Vierh.) Simmons) occur without ligulate female flowers; also that *T. nevadensis* Vierh. may have them or not. He speculates as to whether the Greenland plants may be hybrids between *E. borealis* × *E. uniflorus*, but leaves the question of this variation, quite reasonably, open. *E. olympicus* Schott & Kotschy, which occurs on the same mountain in Turkey as *E. uniflorus* – both are also recorded from Bulgaria – may be a similar variation of *E. uniflorus* to the Helgord Dagh plants, though the leaves of the type of *E. olympicus* are much hairier than those of the Iraqi plant. Neither *E. olympicus* nor *E. nevadensis*, described from the Sierra Nevada of Spain, receive mention from Halliday in *Fl. Europaea*, but here too a note is given concerning the doubt which hangs over the presence or absence of eligulate female florets as a prime taxonomic character.

2. **Erigeron acris** *L.*, Sp. Pl.: 863 (1753); Boissier, Fl. Orient. 3: 166 (1875); Bochantsev in Fl. S.S.S.R. 25: 246 (1959); Rawi in Dep. Agr. Iraq Bull. 14: 114 (1964), Grierson in Fl. Turkey 5: 127 (1975); Grierson & Rechinger in Fl. Iranica 154: 17 (1982); Yilin Chen & Luc Brouillet in Fl. China 20–21: 646 (2011).

Annual, biennial or perennial herb 15–70 cm tall. Stems solitary or several, ridged, with patent or ± crisped white hairs, branched above. Basal leaves oblanceolate to linear-oblanceolate, 3–10 × 0.4–1.5 cm, tapering to a petioloid base, obtuse to acute, hirsute on both surfaces or glabrescent, cauline leaves numerous, lower resembling basal leaves, median linear-lanceolate or linear-oblanceolate and sessile, much reducing in inflorescence. Capitula usually several to many in a corymbose panicle; involucre 8–14 mm in diameter; phyllaries linear, 4–8 mm, ± clothed with white hairs and minutely glandular, often purplish at tip. Ray florets female, limbs 3–4 mm, purplish-pink or white. Disc florets 4–5 mm. Achenes 2–2.5 mm, pale stramineous, pubescent; pappus 4.5–5.5 mm, white to brownish.

HAB. Rocky mountain and in an orchard, in both cases near water; alt. 820–1280 m; fl. & fr. Jun.
DISTRIB. Found only twice in the extreme northern mountain region, not far from the Turkish frontier. **MAM**: Daimka village 35 km NE of Zakho, *Qaisi, Hamad & Hamid* 45357!; Dori village 5 km E of Kani Masi, *Omar & Qaisi* 45448!

Almost throughout Europe, N Africa (Morocco), Turkey, Caucasia, Iran, Afghanistan, Pakistan, C Asia, E & W Siberia, China, Korea, Japan, Canada, northern U.S.A.

Erigeron karvinskianus [as *karvinskianum*] DC., Prodr. 5: 285 (1836); Hind, Curtis's Bot. Mag. 29(1): 52–65 (2012). Cultivated for ornament and widely naturalized, widespread in

Africa, New Zealand and the west coast of the United States of America, Central America, West Indies, South America (Venezuela south to Chile), and the Pacific Islands may be cultivated in Iraq, but not mentioned by Husain & Kasim (1975). It is a herbaceous, often short-lived, perennial with radiate, heterogamous capitula, white (turning pink or lavender with age) ray florets, and disc florets with yellow corollas.

Erigeron speciosus (Lindl.) DC., a native of N America, is said by Husain & Kasim (1975: 114) to be "occasionally cultivated" in Iraq. It is an erect perennial 45–60 cm tall with petiolate, sparsely hirsute, marginally ciliate leaves, the basal spatulate, the cauline lanceolate and very acute. The disc is yellow, 12–15 mm in diameter, and the numerous narrow, violet or pinkish rays 12–15 mm. Since no specimen has been seen, the identity of this cultivated plant cannot be confirmed.

TRIBE **ANTHEMIDEAE** CASS.

88. **ARTEMISIA** L.

Sp. Pl.: 845 (1753); Gen. Pl. 5: 357 (1754)
Oberprieler, Vogt & Watson in Kubitzki, Fam. Gen. Vasc. Pl. 8: 358 [2006] (2007).

Shahina A. Ghazanfar

Annual, biennial or perennial low shrubs with a taproot and woody horizontal rootstock, much-branched, often aromatic. Leaves petiolate or upper sessile, alternate, unarmed, simple to shallowly or deeply toothed or incised or lobed to pinnatisect. Inflorescence paniculate or racemose. Capitula heterogamous; involucre ovoid to campanulate; phyllaries 2–4(–6)-seriate, imbricate, hairy or glabrous, inner with scarious margins or scarious. Receptacle convex, conical, glabrous or pilose. Ray florets pistillate, corolla narrow tubular, 2–3(–6)-toothed, ± oblique, eligulate; style exserted, 2-fid; disc florets bisexual, functionaly male, or sterile; corolla tubular to campanulate, 5-toothed; style exserted or included, 2-fid, or fimbriate. Achenes oblong-obovoid, striate or smooth; pappus absent or sometimes achenes with a minute annulus or scarious ring.

About 540 species (in its wider circumscription including, e.g., *Seriphidium,* which is supported by all recent systematic studies), distributed throughout the northern hemisphere and extending to southern Africa, South America and Pacific islands.

Malik, S., Vitales, D., Qasim Hayat, M., Korobkov, A. A., Garnatje, T., & Vallès, J. (2017). Phylogeny and biogeography of *Artemisia* subgenus *Seriphidium* (Asteraceae: Anthemideae). Taxon, 66(4), 934-952.

Torrell, M., Garcia-Jacas, N., Susanna, A., & Vallès, J. (1999). Phylogeny in *Artemisia* (Asteraceae, Anthemideae) inferred from nuclear ribosomal DNA (ITS) sequences. Taxon, 48(4), 721-736.

Watson, L. E., Bates, P. L., Evans, T. M., Unwin, M. M., & Estes, J. R. (2002). Molecular phylogeny of subtribe Artemisiinae (Asteraceae), including *Artemisia* and its allied and segregate genera. BMC Evolutionary Biology, 2(1), 17.

1. Ray florets female; disc florets bisexual 2
   Ray florets female; disc florets male or all florets bisexual 3
2. Shrub glabrous except for basal leaves; capitula heterogamous; ray florets 5–6, female, tubular, 2-toothed; disc florets 5–6, male, tubular, 5-toothed, glandular . . . . 6. *A. scoparia*
   Greyish-tomentose to glabrescent, strongly aromatic shrub; capitula homogamous; florets (flowers) 3–4(–5), all bisexual, tubular, 5-toothed . 7. *A. sieberi*
3. Cauline leaves linear, mucronate, undivided or very rarely with terminal linear lobes, sessile, mostly with 1–5 prominent veins on the abaxial side, densely covered with appressed, silky hairs, appearing glaucous; second order cauline leaves clustered at the axiles of the first order leaves, glabrous, frequently covered by yellowish transparent viscous fluid. . . . . 8. *A. jordanica*
   Basal and cauline leaves pinnatisect. 4
4. Leaves grey arachnoid tomentose on lower surface, sparsely arachnoid pubescent above; basal leaf blade broadly ovate in outline, 5–12 × 3–6 cm, 1–2-pinnatisect, with 3–4 pairs of segments . . . . 1. *A. vulgaris*

Leaves puberulent or grey-sericeous, not grey tomentose only on lower surface; basal leaves 2-more pinntisect with 4–6 pinnately lobed segments. 5
5. Tufted shrub forming hummocks; flowering stems densely pilose to sericeous; synflorescence a narrow panicle or secund raceme; capitula ± 5 mm in diameter, depressed globose; peduncule 10–15 mm. . . . . .2. *A. splendens*
Shrubs erect to stems spreading; synflorescences spreading paniciform; peduncle <10 mm 6
6. Basal leaf rosettes absent at flowering; capitula sessile to shortly pedunculate; involucre and receptacle densely tomentose to villous; corolla of disc florets densely pilose . . . . .3. *A. haussknechtii*
Basal leaves present at flowering; capitula with short peduncle to 2 mm; corolla of disc florets glabrous 7
7. Basal leaves with petiole 6–8 cm; leaf blade 8–10 × 7–9 cm in outline; lobes of pinntisect leaves lanceolate to linear, obtuse . . . . 5. *A. absinthium*
Basal leaves with petiole 3–5 cm; leaf blade 4–5 × 2–5 cm in outline; lobes of pinnatisect leaves linear, mucronulate . . . . 4. *A. campestris*

1. **Artemisia vulgaris** *L.*, Sp. Pl.: 848 (1753); Rawi, in Dep. Agr. Tech. Bull. 14: 106 (1964); Cullen in Fl. Turkey 5: 313 (1975); Podlech in Fl. Iranica 158: 170. t. 153 (1986); Poljakov in Fl. U.S.S.R. 26: 416 (2000); Ghafoor in Fl. Pak. 207: 123, f. 28 (2002); Tutin in Fl. Europ. 4: 180 (1976); Mozaffarian in Fl. Iran 59 (Anthemideae & Echinopeae): 203 (2008); Yourun Lin (Ling Yuou-ruen), Humphries & Gilbert in Fl. China 20: 700 (2011); Breckle, Hedge & Rafiqpoor, Vasc. Pl. Afghan: 152 (2013).

Large shrub 50–80 cm tall, with a thick woody rootstock. Stems many, reddish brown, erect to ascending, pubescent to glabrescent. Leaves grey arachnoid tomentose on the lower surface, sparsely arachnoid pubescent above. Basal leaves shortly petiolate, leaf blade broadly ovate in outline, 5–12 × 3–6 cm, 1–2-pinnatisect, with 3–4 pairs of segments; lobes linear-elliptic, acute; middle stem leaves ± sessile, leaf blade pinnatisect with 2–3 pairs of segments; uppermost leaves entire, lanceolate, acute. Inflorescence forming a lax, conical panicle. Capitula 3–4 mm, shortly pedunculate, erect or ± nodding; involucre 4-seriate; phyllaries densely arachnoid pubescent, all with broad scarious margins; outer phyllaries ovate, midrib green; inner elliptic-oblong, scarious. Receptacle discoid, glabrous. Florets yellow; ray florets female, 5–10, corolla 2-toothed, glandular; disc florets bisexual, 5–20, corolla 5-toothed, glandular. Achenes ± 1 mm, ovoid. Fig. 85: 1–9.

HAB. Mountains slopes, by streams in gorge; only 2 collections from Iraq; rare; alt. ± 1200 m(?); fl. & fr. August.
DISTRIB. **MAM**: Sersang, *Haines* W1226!; *Daoud & Karmani* 43774 (BAG)!

N Africa; Europe, Caucasus, Turkey; Iran, Afghanistan, Pakistan, NW India; Siberia, Mongolia, China; introduced and naturalised in N America.

2. **Artemisia splendens** *Willd.*, Sp. Pl., ed. 4 [Willdenow] 3(3): 1822 (1803); Boissier, Fl. Orient. 3: 375 (1875); Rawi, in Dep. Agr. Tech. Bull. 14: 106 (1964); Poljakov in Fl. U.S.S.R. 26: 469 (2000); Cullen in Fl. Turkey 5: 318 (1975); Podlech in Fl. Iranica 158: 189. t. 176 (1986); Mozaffarian in Fl. Iran 59 (Anthemideae & Echinopeae): 208. f. 61 (2008).

A. lasiocline Boiss., Diagn. Pl. Orient. ser.1, 6: 92 (1846).

Tufted shrub forming hummocks. Flowering stems to 20 cm, densely pilose to sericeous. Basal leaves: petiole 2 cm, 2-pinnatisect; lobes oblong to linear, acute; middle and upper leaves and leaf-like bracts similar to to basal leaves but smaller in size. Inflorescence a narrow panicle or secund raceme. Capitula erect or nodding, peduncule 10–15 mm, head ± 5 mm in diameter, depressed globose. Receptacle pilose; phyllaries imbricate; outer phyllaries ± oblong, median part thickened, with dark drown scarious margins, tometose; inner phyllaries longer than the outer, mostly scarious. Ray florets female, corolla reddish, pilose on lobes; disc florets bisexual.

HAB. Mountain slopes, amongst rocks; 2800–3350 m; fl. & fr. Jul.-Aug.
DISTRIB. **MRO**: NE of Qandil, *Rawi & Serhang* 24400!; Chiya-i Mandau, *Guest* 2699!; Halgurd Dagh,

Fig. 85. **Artemisia vulgaris**. 1, habit (upper portion); 2, habit (lower portion showing rhizome and roots); 3, capitulum; 4,5,6, phyllaries; 7, female floret; 8, bisexual floret; 9, achene. From Flora of China, Asteraceae 20-21, fig. 573. Reproduced with permission from Missouri Botanical Garden Press, St. Louis, and Science Press, Beijing.

Gillett 9592!; E side of Qandil range, *Rawi & Serhang* 24457!; E side of mountain, *Rawi & Serhang* 24503 (BAG); id, Perrish mountain 24520!; id, top of Qandil Rang Bet. top of Perrish & Bara-danas.24597!

Eastern and southern Transcaucasia, Turkey, Iran.

3. **Artemisia haussknechtii** *Boiss.*, Fl. Orient. 3: 374 (1875). –Type: Mt. Hawraman & Shahu, Dalekhan, 1867, *Haussknecht* s.n.; Rawi, in Dep. Agr. Tech. Bull. 14: 106 (1964); Podlech in Fl. Iranica 158: 178. t. 164 (1986); Cullen in Fl. Turkey 5: 319 (1975); Mozaffarian in Fl. Iran 59 (Anthemideae & Echinopeae): 214. f. 63 (2008).

Perennial shrub, grey-green, 10–30 cm, densely sericeous-tomentose, with many branches from a woody stock. Basal leaf rosettes absent at flowering; lower leaves petiolate, petiole 2–5 mm, balde 2-pinnatisect with lanceolate, ± obtuse segments; middle leaves shortly petiolate, as lower leaves but smaller, upper leaves and bracts simple, oblong to oblanceolate, entire. Inflorescence a panicle with the lower branches ascending. Capitula erect to nodding, sessile to shortly pedunculate, ± 4 mm in diameter; involucre densely tomentose; outer phyllaries ovate, with narrow, obscure scarious margins, inner phyllaries longer than outer phyllaries, oblong-ovate, obtuse with scarious margins. Receptacle densely villous. Ray florets 7–8, female, corolla 2–toothed, ovary oblique; disc florets bisexual, corolla 5-toothed, densely pilose.

HAB. Mountains, amongst rocks and increvices of rocks; alt. (650–)1700–2400 m; fl. & fr. Aug.-Oct.
DISTRIB. **MAM**: Sersang, *Haines* W 1337!; Jabal Khantur NE of Zakho, *Rawi, Noori & Tikriti* 29019! 29029! **MRO**: Rowanduz gorge, *Guest* 1597!; Germasur lake, Qandil range, *Rawi & Serhang* 26710!; Malikan vally S of Rawandoz, *Riyadh et al* 58772, 58774.

Turkey, Iran.

4. **Artemisia campestris** *L.*, Sp. Pl.: 846 (1753); Boissier, Fl. Orient. 3: 363 (1875); Rawi, in Dep. Agr. Tech. Bull. 14: 106 (1964); Rechinger, Fl. Lowland Iraq: 630 (1964); Cullen in Fl. Turkey 5: 319 (1975); Poljakov in Fl. U.S.S.R. 26: 525 (2000); Mozaffarian in Fl. Iran 59 (Anthemideae & Echinopeae): 219 (2008); Yourun Lin (Ling Yuou-ruen), Humphries & Gilbert in Fl. China 20: 725 (2011).

Oligosporus campestris (L.) Cass., Bull. Sci. Soc. Philom. Paris (1817) 33 (1826).

Shrub to 40 cm tall, with a woody rootstock, faintly aromatic. Stems many, pubescent to glabrescent. [Basal leaves: petiole 3–5 cm, leaf blade broadly ovate in outline, 4–5 × 2–5 cm, 2(–3)-pinnatisect, with 4 or 5 pairs of 3–5-lobed segments; lobes linear, mucronulate; middle stem leaves sessile, leaf blade ovate 1–2-pinnatisect, with 3–4 pairs of segments; uppermost leaves and leaf-like bracts entire or 3–5-lobed. Descrip. from material from Iran]. Synflorescence forming a lax panicle. Capitula 2–4 mm in diameter, shortly pedunculate, erect or nodding; phyllaries outer ovate, inner oblong, with broad scaious margins, shiny. Ray florets 3–6(–20), female, corolla 2-toothed; disc florets 6–10(–30) bisexual. Achenes ± 1 mm, obovoid.

HAB. Undulating depressions in "haswa" plain, in gravelly and silty soil; alt. 40–160 m; fl. & fr. Apr., Oct.-Nov.
DISTRIB. **FPF**: 80 Km W by N of Busaiyah, *Guest, Rawi & Gillett* 14132 (BAG!); **LCA**: *c.* 20 km E of Falluja, *Guest, Rawi & Noori* 13501!

China, Japan, Russia; NW Iran, C Asia, Europe, North America.

5. **Artemisia absinthium** *L.*, Sp. Pl.: 848 (1753); Boissier, Fl. Orient. 3: 373 (1875); Podlech in Fl. Iranica 158: 182. t. 167 (1986); Cullen in Fl. Turkey 5: 318 (1975); Poljakov in Fl. U.S.S.R. 26: 488 (2000); Ghafoor in Fl. Pak. 207: 114. f. 24 (2002); Mozaffarian in Fl. Iran 59 (Anthemideae & Echinopeae): 220 (2008); Gabrielian & Fragman-Sapir, Flow. Transcaucasus & Adj. Areas: 56 (2008); Yourun Lin (Ling Yuou-ruen), Humphries & Gilbert in Fl. China 20: 679 (2011); Breckle, Hedge & Rafiqpoor, Vasc. Pl. Afghan: 148 (2013).

Perennial herb 30–60 cm high, with a woody base, puberulent or grey-sericeous. Stems few. Basal and lower leaves petiolate; basal leaves: petiole 6–8 cm, leaf blade ovate in outline, 8–10 × 7–9 cm, 2(–3)-pinnatisect, with 4 or 5 pairs of pinnately lobed segments; lobes lanceolate to linear, obtuse; middle stem leaves: petiole 1.5–3 cm; leaf blade ovate or elliptic-ovate, 2-pinnatisect; lobes linear-lanceolate; uppermost leaves 4 × 2–4 cm in outline, pinnatisect or 5-lobed; leaflike bracts 3-lobed with lanceolate or linear-lanceolate lobes or entire. Inflorescence paniculate, with ascending to spreading branches. Capitula with short pedicels to 2 mm, nodding. Receptacle discoid, densely pubescent; phyllaries imbricate, outer oblong; inner oblong to orbicular, obtuse, with broadly scarious margins. Ray florets female, corolla yellow, 2-toothed; disc florets many, bisexual, corolla yellow. Achenes ± 1 mm, oblong, with apical corona or not. Fig. 86: 1–10.

HAB. In valley, in shade, near stream; alt. (only once recorded) 1450 m; fl. & fr. Aug.
DISTRIB. **MRO**: Kanirash, Qurna Qaw valley N of Pushtashan, *Rawi & Serhang* 26665!

Afghanistan, India, Japan, Kazakhstan, Kyrgyzstan, Pakistan, Russia, China; N Africa, SW Asia, Europe, North America.

6. **Artemisia scoparia** *Waldst. & Kit.*, Pl. Rar. Hung. 1: 66, t. 65 (1802); Boissier, Fl. Orient. 3: 364 (1875); Poljakov in Fl. U.S.S.R. 26: 560 (1961); Rawi, in Dep. Agr. Tech. Bull. 14: 106 (1964); Rechinger, Fl. Lowland Iraq: 631 (1964); Cullen in Fl. Turkey 5: 322 (1975); Podlech in Fl. Iranica 158: 189. t. 176 (1986); Mandaville, Fl. Eastern Saudi Arabia: 305 (1990); Ghafoor in Fl. Pak. 207: 107 (2002); Mozaffarian in Fl. Iran 59 (Anthemideae & Echinopeae): 228. f. 68 (2008); Yourun Lin (Ling Yuou-ruen), Humphries & Gilbert in Fl. China 20–21: 730 (2011); Breckle, Hedge & Rafiqpoor, Vasc. Pl. Afghan: 151 (2013).

Shrub 40–60 cm high, with a woody rootstock, glabrous except for basal leaves, with purplish brown erect to spreading branches. Basal and lower leaves petiolate; petiole 1–1.5 cm, appressed-hairy to glabrescent; leaves ovate-rounded in outline, 2(–3)-pinnatisect, with linear-lanceolate, acute to obtuse segments; upper and middle leaves subsessile to sessile, basally auriculate, hairy or glabrous; bracts linear, entire, auriculate. Inflorescence a wide panicle with ascending branches. Capitula heterogamous, oblong-globose, nodding, in paniciform inflorescences with ± erect primary and secondary branches, the whole synflorescence forming a spreading panicle. Receptacle conical, glabrous; involucre 3–4-seriate; phyllaries imbricate, outermost ovate-orbicular, with narrow scarious margins; inner oblong-ovate, with broadly scarious margins. Ray florets 5–6, female, tubular, 2-toothed; disc florets 5–6, male, tubular, 5-toothed, glandular. Achenes ± 1 mm long, obovate. Pappus absent. Fig. 87: 1–9.

HAB. In desert, by roadside, in "haswa" plain, and in sandy and rocky limestone desert; alt. 45–300 m; fl. & fr. Mar.-Nov.

DISTRIB. In the foothills, alluvial plains and the western desert. **MAM**: Sersang, *Haines* W 1196! **MSU**: Sulaimaniya, *Fawzi* 39533! **FUJ**: Hatra (Hadhr), *Guest* 3549! islands in r. Tigris near Mosul, *Kotschy* 435!; N of Balad Sinjar, *Guest* 13350! **FKI**: nr Kirkuk, *Alizzi* 15569! **FPF**: 10 km E of Sudur, *Botany Staff* 47184!; a few km NE of Mandali towards Khir-i Charmiq, *Qaisi & Hamad* 43567! **DLJ**: Wadi Thirthar, in the Jazira, *Guest* 3548! **DWD**: nr Rahhaliya, *Rawi & Joseph* 16310!; 10 km NW of Rahhaliya, *Rawi* 26926!; Shithatha, *Rawi* 26909!; Falluja, *Guest* 13172!; *c.* 17 km W by S of Ramadi, *Guest, Rawi & Rechinger* 15953!; Assufi, 80 N of Rutba, *Rawi & Noori* 26864!; 165 km W of Ramadi, *Rawi & Noori* 26845! **DSD**: nr Salaibiyat al Hamir, *c.* 35 km W of Ur, *Guest, Rawi & Rechinger* 16050!; 5 km from Ma'niya, *Hazim* 32479! 32520!; 4 km SW of Ummaiyid, *c.* 20 km SW of Samawa, *Guest, Rawi & Rechinger* 18714!; 30 km W of Ma'niya, *Hazim* 32529!; 4 km SW of Ummayid, *c.* 20 km SW of Samawa, *Guest, Rawi & Rechinger* 18714!; 4 km SW of Ummayid, 22 km SW of ?Senna, *Guest, Rawi & Rechinger* 18714!; 60 km W of Shabicha, *Chakravarty, Rawi, Khatib & Tikriti* 30083! **LCA/FKI**: between Khalis and Kirkuk, *Rawi, Chakravarty, Noori & Alizzi* 19685! **LCA**, Baghdad, *Rogers* 0493!; Baghdad, Rustam farm, *Lazar* 438!; Baghdad desert, *Haussknecht* s.n.!; between Nasiriya and Diwaniya, *Alizzi & Mohammed* 29634! **LEA**: by Baquba bridge, *Rechinger & Khudairi* 99!; Baquba, by river Diyala, *Rechinger* 8051!; Fakka, 50 km NE of Amara, *Rawi* 14920!

SALMAS (*Guest* 3549; *Guest* 3548).

Widespread across Eurasia from France to Japan, including China, India, Russia, Germany, Poland, central and southwest Asia.

7. **Artemisia sieberi** *Besser* in Bull. Soc. Imp. Naturalistes Moscou 9: 80 (1836); Poljak in Fl. U.S.S.R. 26: 712 (1961); Podlech in Fl. Iranica 158: 203 (1986); Mandaville, Fl. Eastern Saudi Arabia: 303 (1990); Mozaffarian in Fl. Iran 59 (Anthemideae & Echinopeae): 236. f. 68 (2008); Breckle, Hedge & Rafiqpoor, Vasc. Pl. Afghan: 152 (2013); Kilian & Ghazanfar in Fl. Oman 3: 252 (2015); Taifour & El-Oqlah, Pl. Jordan Ann. Checkl.: 36 (2017).

Seriphidium sieberi (Besser) K. Bremer & Humphries ex Y.R.Ling, Bull. Res. Hatbin 11(4): 18 (1991); Ghafoor & Al-Turki in Fl. Kingd. Saudi Arabia 2(3): 184. Pl. 65a (2000).

Artemisia herba-alba Asso var. *laxiflora* Boiss., Fl. Orient. 365 (1875); *A. herba-alba* auctt., non Asso., Rawi, in Dep. Agr. Tech. Bull. 14: 106 (1964); Rechinger, Fl. Lowland Iraq: 631 (1964).

Shrublet to ± 40 cm high, greyish-tomentose to glabrescent, strongly aromatic. Stems many erect with spreading branches. Basal and lower leaves petiolate, rounded in outline, bipinnatisect, 5–16 × 5–12 mm, with lanceolate, obtuse segments to ± 4 mm long; upper and middle leaves subsessile to sessile, basally auriculate, in the upper floral region palmatisect or linear, entire. Capitula numerous, discoid, sessile in densely paniciform Inflorescences; involucre oblong, 6–7-seriate, glabrous except for the base; phyllaries outermost herbaceous, inner oblong-

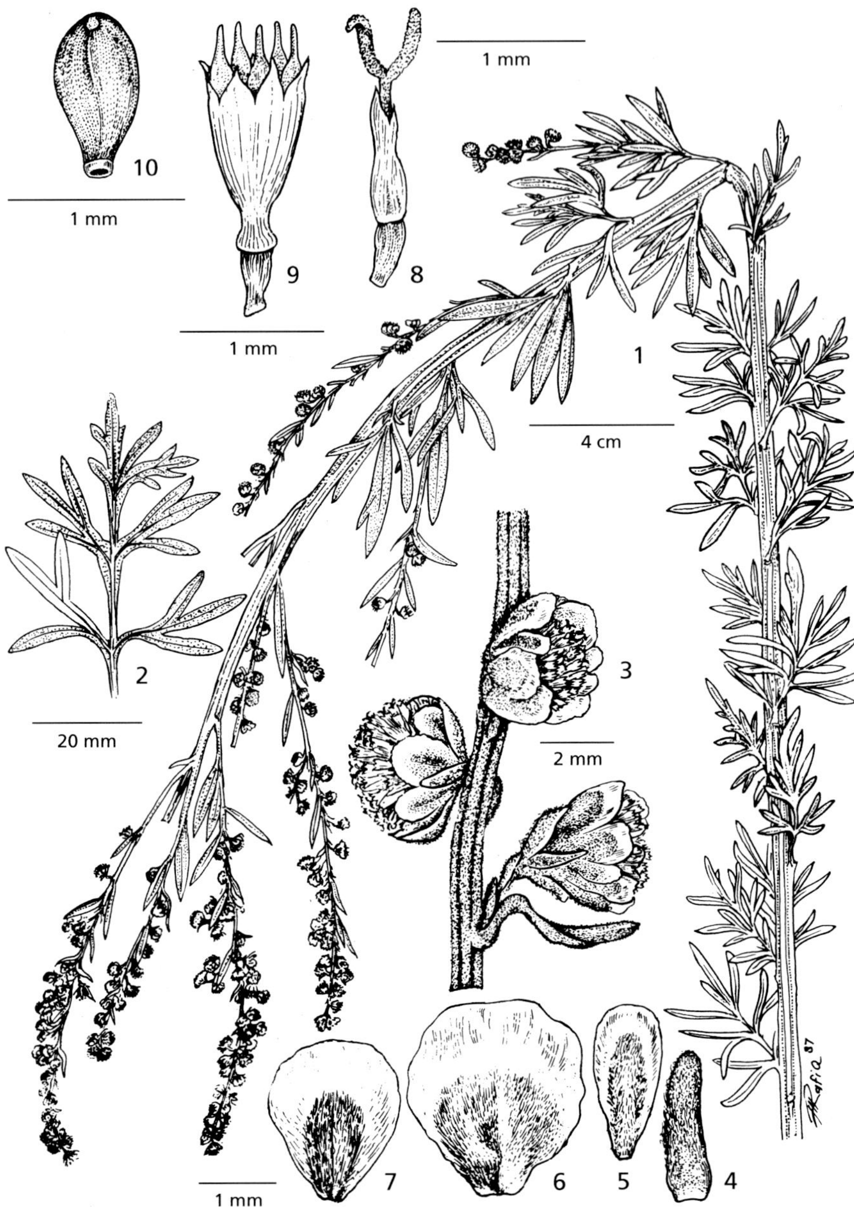

Fig. 86. **Artemisia absinthium**. 1, habit flowering branch; 2, leaf; 3, portion of flowering branch showing capitula; 4, 5, 6, 7, phyllaries; 8, marginal female floret; 9, disc floret; 10, achene. Drawn by Rafiq. Reproduced with permission from Flora of Pakistan 207, fig. 24.

oblanceolate, ± glabrous, obtuse, with broadly scarious margins. Florets 3–4(–5), all bisexual, tubular, 5-toothed, yellowish to purple. Achenes ± 1 mm long, obovate. Pappus absent.

HAB. Clay soil, stony plains, limestone & gypsum desert plateau with sandy patches, open wadi beds, dry steppe, roadside pastures; alt. 160–640 m; fl. Jun.

DISTRIB. Widespread in the desert and dry steppe regions of Iraq. **MJS/FUJ**: Jazira N of Jabal Sinjar, *Guest* 13136! **FUJ**: Balad Sinjar, *Guest* 4097!; Tal Afar to Sinjar, *Guest* 13623! **FAR**: Makhmur, *Hunting Aero Surveys* 105 & V.C.R. (R.B. 61)!; Arbil, *Guest* 13021! **FPF**: Makatu nr Mandali, *Guest* 872!; Koma Sang police station nr Mandali on Iranian frontier, *Rawi* 20620!; Mandali, *Guest* 1712!; Jabal Hamrim, *Sutherland* 109; 10 km E of Mandali, *Rechinger* 9629; **DLJ**: 6 km E of Thirthar on road to Beiji, *Rawi & Gillett* 7162! **DWD**: 10 km from Rutba to H3, *Al-Kaisi, Hamad & Hamid* 43586!; Wadi Musaad-al-Rutba,

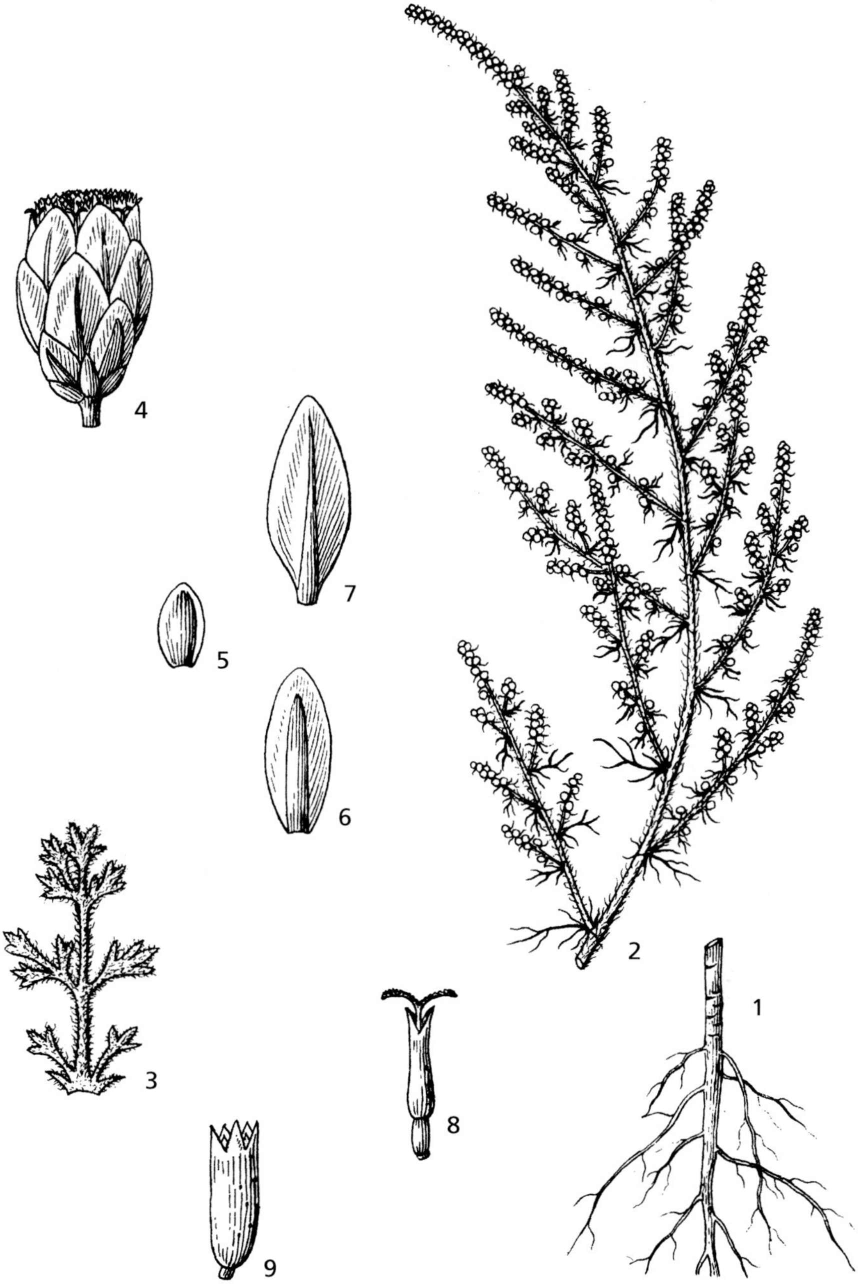

Fig. 87. **Artemisia scoparia**. 1, habit (lower portion); 2, habit (upper portion); 3, basal leaf; 4, capitulum; 5, 6, 7, phyllaries; 8, female floret; 9, bisexual floret; 10, achene. From Flora of China, Asteraceae 20-21, fig. 591. Reproduced with permission from Missouri Botanical Garden Press, St. Louis, and Science Press, Beijing.

30 km S of Rutba, *Tikrity & Hazim* 29709!; 20 km W of Rutba, *Rawi* 14701!; 270 km W of Ramadi, *Rawi* 26853!; 23 km SW of Ramadi, *Guest, Rawi & Rechinger* 15946!; 23 km SW of Ramadi, *Guest, Rawi & Rechinger* 15946!; Wadi al-Ajrumiya, 73 km N of Rutba, *Rawi & Noori* 27185!; Wadi Ghadaf, half way between Rutba and Nukhaib, *Rawi* 14731!; 25 km S of Ramadi, *Guest, Rawi & Rechinger* 16200!; 23 km SW of Ramadi, *Guest & Rechinger* 15943!; **DSD**: 5 km from Ma'niya, *Hazim* 32498!; 65 km W by N of Busaiya, *Guest, Rawi & Long* 14146!; 25 km SE of Salman, *Rawi* 14908!; 14889!; 6 km E of Shabicha,

Rawi 13841!; nr Jahama, *c.* 80 km SW of Busaiya, *Guest, Rawi & Rechinger* 16089!; Chibritiya, 60 km W of Shabicha, *Hazim* 32480!; 10 km N of Salman, R*awi & Tikrity* 24979!; 10 km W of Ma'niya, *Hazim* 32537!; nr Jahama, *c.* 80 km SW of Busaiya, *Guest, Rawi & Rechinger* 16089!; nr Ansab, 145 km S by E of Salman, *Guest, Rawi & Rechinger* 18955!; Wadi al-Khirr, 90 km NW by W of Shabicha, *Guest, Rawi & Rechinger* 19435!; Faithat um Rubain – Almaniya 95 km W of Shabicha, *Hazim* 30689A!; 18 km S by W of Busaiya, *Guest, Rawi & Rechinger* 16067!; nr Jumaima, *c.* 140 km SW of Salman, *Guest, Rawi & Rechinger* 19093!; Jir'aibiyat 38 km N of Neutral Zone, *Rawi, Khatib & Tikrity* 29180!; 5 km from Ma'niya, *Hazim* 32506!; Ma'niya, *Hazim* 32537!; 5 km from Ma'niya, *Hazim* 32487! 32512!; between Salman and Aidaha, *Fawzi, Hazim & H. Hamid* 38779!; 64 km S by W of Busaiya, *Guest & Mahallal* 15278!; 8 km N by W of Aidaha, *c.* 94 km SW of Salman, *Guest, Rawi & Rechinger* 19176!; neutral zone, 30 km N by W of Sama, *Guest, Rawi & Rechinger* 16079!; Shaib al-Batin, Al Oudja, *Weinert, Fawzi, Adel & Hussain* 47586!; Nr Adam, 50 km SSW of Busaiya, *Guest & Mahallal* 15282!; nr Jahama, *c.* 80 km SW of Busaiya, *Rechinger* 8211!

Vernacular name Kurd. BOKHOSHKAH (*Guest* 1712, Mandali) Ar. SHĪH or SHĪAH (*Guest et al.* 16089, Jahama; *Guest* 18136, nr Jabal Sinjar). Used for making "laban" or curdled milk, and is also used for medicine.

E Mediterranean across SW Asia to SW Pakistan and C Asia.

8. **Artemisia jordanica** *Danin* in Willdenowia 29: 147, f. 1, 2A (1999); Taifour & El-Oqlah, Pl. Jordan Ann. Checkl.: 36 (2017).

Ascending semishrub, 50–100 cm high, aromatic; stems ascending to erect, base of stems with appressed silky indumentum. Cauline leaves 4–45(–80) × 1–2(–8) mm, solitary, linear, mucronate, undivided or very rarely with terminal linear lobes, sessile, mostly with 1–5 prominent veins on the abaxial side, densely covered with appressed, silky hairs, appearing glaucous; second order cauline leaves clustered at the axiles of the first order leaves, glabrous, frequently covered by yellowish transparent viscous fluid. Inflorescence conical-shaped, many-capitulate; capitula heterogamous, minute in axils of glabrous bracts, sessile or pedicels to 1.5 mm; involucre glabrous; outer phyllaries orbicular, herbaceous, inner ones oblong and scarious; male florets 4–5, corolla 2.5 mm, glandular, yellowish white, 5-dentate, with a trumpet-like derivative of a style rich in papillae at its broad top; female florets 2, corolla obliquely terminating, glandular, without apical teeth. Achenes 1.5 × 0.5 mm, brown; pappus absent.

HAB. In sandy desert; alt. ± 100 m; fl. Dec.
DISTRIB. **DSD**: 4 km NE of Rahba, near the road from Najaf to Rahba, sandy desert, 5.12.1973, *Weinert & Mousawi* (GH).

S Jordan, N, E and C Saudi Arabia.

89. **ACHILLEA** L.

Sp. Pl.: 896 (1753); Gen. Pl. ed. 5: 382 (1754)
Oberprieler, Vogt & Watson in Kubitzki, Fam. Gen. Vasc. Pl. 8: 364 [2006] (2007)

A. Huber-Morath

Perennial pilose herbs, often with thick woody rootstocks, or subshrubs. Leaves alternate, entire to 3–4-pinnatipartite. Inflorescence a terminal corymb, more rarely of solitary capitula. Capitula heterogamous and radiate, rarely homogamous and discoid, pedunculate or subsessile, small or medium-sized; involucre oblong-cylindrical, ovoid, hemispherical or depressed; phyllaries multiseriate, margins ± scarious; receptacle ± flat or convex, paleaceous, paleae lanceolate or oblong, hyaline, with midrib not reaching apex of paleae. Ray florets female, fertile, rays ± dentate, white or yellow, pink or purplish. Disc florets hermaphrodite, regularly 5-toothed, corollas white, yellow or pink, corolla-tubes ± compressed, bases pouched and enveloping tops of achenes. Achenes oblong or obovate, apices rather broad, body glabrous, smooth, compressed, not winged, but with myxogenic setulae on ribs; pappus absent.

About 115 species occurring in the N hemisphere of the Old and New Worlds, with its centre of distribution in E Mediterranean and in SW Asia. Ten species in Iraq.

Achillea (commemorating the Gr. hero Ἀχιλλεύς, Achilles).

Ehrendorfer, F. & Guo, Y.P. (2006). Multidisciplinary studies on *Achillea* sensu lato (Compositae-Anthemideae): new data on systematics and phylogeography. Willdenowia 69–87.

1. Leaves undivided, serrulate, ovate or oblong .... 10. *A. fragrantissima*
   Leaves pinnatipartite, or pinnatisect 2
2. Leaves pinnatipartite, lower cauline and basal leaves oblong to ovate, 3–7 cm broad .... 8. *A. filipendulina*
   Leaves pinnatisect, lower cauline and basal leaves linear to lanceolate, 0.1–4 cm broad 3
3. Leaves 5–40 mm broad, with setaceous to linear-lanceolate, flat, neither transverse nor imbricate ultimate segments 4
   Leaves 0.8–6 mm broad, with minute, transverse, imbricate, oblong to orbicular ultimate segments; rays yellow 5
4. Ray limbs white .... 7. *A. millefolium*
   Ray limbs yellow .... 9. *A. arabica*
5. Ray limbs 5–13, 2.5–6 mm long .... 1. *A. oligocephala*
   Ray limbs (3–)4–6(–8), (0.5–)1–3(–4.5) mm long 6
6. Involucre oblong-cylindrical, ± angular, 4–7 × 2–3 mm, about twice as long as broad .... 6. *A. aleppica*
   Involucre oblong-ovoid to hemispherical and depressed, not angular, not more than 1½ × as long as broad 7
7. Stem deeply furrowed 8
   Stem terete, longitudinally striate or obtuse-angled 9
8. Leaves 1–3 mm broad, ultimate segments 0.8–1.5 mm long; capitula congested; pedicels 1–4 mm long .... 4. *A. conferta*
   Leaves 0.8–1.2 mm broad, ultimate segments 0.5–1 mm long; capitula loosely arranged; pedicels 3–8 mm long .... 5. *A. talagonica*
9. Involucre hemispherical to depressed, with short spreading hairs; phyllaries with broad pellucid margins .... 3. *A. vermicularis*
   Involucre broadly ovoid to hemispherical, ± appressed-tomentose, phyllaries without or with narrow pellucid margins .... 2. *A. wilhelmsii*

SECT. **ARTHROLEPIS** (BOISS.) BOISS.

1. **Achillea oligocephala** *DC.*, Prodr. 6: 32 (1838); Boissier, Fl. Orient. 3: 273 (1875); Handel-Mazzetti in Ann. Naturh. Mus. Wien 27: 435 (1913); Dinsmore in Post, Fl. Syria, ed. 2, 2: 48 (1933); Blakelock in Kew Bull. 4: 42 (1949); Zohary in Dep. Agr. Iraq Bull. 31: 148 (1950); Rechinger, Fl. Lowland Iraq: 626 (1964); Rawi, in Dep. Agr. Tech. Bull. 14: 104 (1964); Huber-Morath in Fl. Turkey 5: 530 (1975); Mouterde, Nouv. Fl. Liban et Syrie 3: 399 (1983); Mozaffarian, Flora Iran 59: 108 (2008).

A. gossypina Hand.-Mazz. in Ann. Naturh. Mus. Wien 27: 435 (1913).
A. membranacea auct. non Labill.: Rechinger, Fl. Lowland Iraq: 625 (1964).

Perennial, 15–40 cm tall. Basal and lower cauline leaves ± densely woolly, linear, 1.5–5 × 0.3–0.6 cm, pinnatisect, segments up to 2 mm long, 3–partite, lobes obovate to lanceolate, denticulate. Capitula 1–6(–15), corymbs 2–8 mm broad, pedicels (0.3–)1–4(–10) cm; involucre hemispherical to depressed, 4–6 × 4.5–8 mm; phyllaries broadly ovate, obtuse, margins narrow, pellucid, puberulent to glabrescent. Rays (5–)8–13, yellow, 2.5–6 mm long. Disc florets 30–50.

HAB. On steppic and desert plains; alt. 300–600 m; fl. & fr. May-Jun.
DISTRIB. Occasional in the steppe region and NW sector of the desert region of Iraq, very rare in the lower mountain region. **MJS**: Jabal Qilat, *Field & Lazar* 479! **MSU**: Chemchemal, *Agnew & Haines* 2189! **FUJ**: Ain Tellawi, *Rawi* 5704!; Mir Qasim, *Field & Lazar* 535!; Ain Al-Husan, *Guest* 4209!; Qal'a Sharqat, *Hand.-Mazz.* 1030; **FAR**: 7 km E of Jabal Makhmur, *Gillett* 11224! **FKI**: between Tuz and Tauq, *Rawi & Gillett* 7453!; Baba Gurgur, Guest 4378!; 10–15 km E of Kirkuk, *Rechinger* 9984!; between Sar Qal'a and Qal'a Shirwana, *Poore* 373! **FPF**: Jalaula, *Poore* 373! 20 km SE of Badra, Al-Kaisi Khayat 50679 (BAG!). **DWD**: Rutba, *Rawi* 21082!; 45 km W of Rutba, *Rechinger* 9958!; Tal an-Nisr, *Rawi* 31203!; 50 km N of Rutba, *Rawi & Noori* 27116!; 260 km W of Ramadi, *Rechinger* 9837!; Wadi Hauran, 12 km SE of H.1, *Rawi et al.* 31648!; 100 km W of Ramadi, *Omar & Hamid* 36736!; 30 km NE of Nukhaib, *Guest et al.* 14027!

Syria, Turkey, Iran.

SECT. **SANTOLINOIDEA** (DC.) HOFFM.

2. **Achillea wilhelmsii** *C.Koch* in Linnaea 24: 328 (1851); Afanasiev in Fl. S.S.S.R. 26: 100 (1961); Huber-Morath in Fl. Turkey 5: 232 (1975); Mouterde, Nouv. Fl. Liban et Syrie 3: 399 (1983); Huber-Morath in Fl. Iranica 158: 49 (1986); Ghafoor in Fl. Pakistan 207 (1): 38 (2002); Mozaffarian, Flora Iran 59: 70 (2008); Breckle, Hedge & Rafiqpoor, Vasc. Pl. Afghan: 146 (2013).

A. santolina auctt. non L. (1753): Handel-Mazzetti in Ann. Naturh. Mus. Wien 27: 44 (1913); Dinsmore in Post, Fl. Syria, ed. 2, 2: 46 (1933); Guest in Dep. Agr. Iraq Bull. 27: 2 (1933); Anthony in Notes Roy. Bot. Gard. Edinburgh 18: 290 (1935); Bornmüller in Beih. Bot. Centralbl. 60: 190 (1939); Blakelock in Kew Bull. 4: 42 (1949); Zohary in Dep. Agr. Iraq Bull. 31: 147 (1950); Rechinger, Fl. Lowland Iraq: 624 (1964); Rawi, Dep. Agr.Tech. Bull. 14: 104 (1964); Rawi & Chakravarty, Dep. Agr. Tech. Bull 15: 8 (1964); Täckholm, Stud. Fl. Egypt ed. 2: 577 (1974); Feinbrun-Dothan, Fl. Palaest. 3: 341 (1978).

A. teretifolia auct. non Willd.: Ledeb., Fl. Ross. 2: 536 (1845).
A. kermanica Gand. in Bull. Soc. Bot. France. 65: 37 (1918).
A. krascheninnikovii Afan. in Not. Syst. (Leningrad) 15: 395 (1953).

Perennial, 10–35 cm tall. Leaves woolly-pubescent, often glabrescent, cauline linear, ± of equal length, uppermost reaching corymbs, median cauline leaves (1–)1.5–4 × 0.1–0.3 cm, pinnatisect, segments imbricate or somewhat distant, 1 cm long, 3–partite, lobes orbicular to broadly cuneate, denticulate. Capitula 5–40, corymbs 1.5–5.5 cm, broad, pedicels 1–4(–7) mm; involucre broadly ovoid to hemispherical, 2.5–5 mm long and broad; phyllaries lanceolate to oblong, obtuse or subacute, with narrow pellucid margins, appressed-tomentose. Rays 3–5, bright yellow, 1–1.5 mm long; disc florets 15–25.

HAB. Low dry hills and plains, by roadside and in fields; alt. 50–500(–1000+) m; fl. & fr. Mar.-Jun.
DISTRIB. Rare in the SW sector of the lower forest zone of Iraq, common in the steppe region, occasional in the desert region on the irrigated alluvial plain: **MSU**: Chemchemal, *Rogers* 0233!; mountains N of Sulaimaniya, *Thesiger* s.n.!; **FUJ**: Upper Jazira, nr Sinun, *Guest* 13339!; Balad Sinjar, *Guest* 4120; Ain Ghazal, *Guest* 4087!; Ain Tellawi, *Rawi* 5711!; between Balad Sinjar and Tal Afar, *Guest* 4150!; Tal ash-Shaur, *Field & Lazar* 574! & *T. Sarsam* 9936!; Hammam, Alil, *Ari* 9731!; **FUJ/FNI**: Mosul, *Howes* 1152 p.p.!; **FAR**: Arbil, *Gillett* 8010!; **FKI**: Kirkuk province, *Ali Haideri* 3955!; Daquq, *Ahmad et al.* 50110!; Tuz, *Rogers* 0156!; **FKI/FPF**: Jabal Hamrin, *Sutherland* 105 p.p.!; Sa'diya, *Hamid* 46632!; Mandali, *Hadač* et al. 4574!; between Badra and Zurbatiya, *Hikmat Abbas al-Ani & Barucha* 7615!; **FPF**: Badra, *Gillett* 6673!; **DWD**: 20 km S of Husaiba, *Chakrawarty, Rawi, Khatib & Alizzi* 31800!; **DLJ**: 10 km NE of Qaim, *id.* 32928!; 17 km E of Haditha, *Qaisi* 44239!; **DGA**: "Musaiyif al-Ghizlan" (summer habitat of gazelles) in Daur, *Janan, Sahira, Widad & Khalaf* 49905!; Adhaim (north), *CV.C.R.* (Robertson of Hunting Aero Surveys) 89!; **LCA**: Sudur, *Omar, Fauzi (Karim), Qaisi & Noori* 44294!; Khan Bani Sa'd, *Alizzi & Hazim* 33688!, *B. Anderson* 1966!; Baghdad, *Schäfli* 34!, *Lazar* 491!; Rustamiya, *Rogers* 075!, *Lazar* 186!, 3888!; Hinaidi, (*Mu'askar al-Rashid, Lazar* 1158!; **LEA**: Ba'quba, *Paranjpye* 91!; Diyala, *F.N. Howes* 1152! p.p.

Cyprus, Syria, Lebanon, Palestine, Jordan, Turkey, Transcaucasia, Iran, Afghan.

3. **Achillea vermicularis** *Trin.* in Mém. Acad. Imp. Sci. St. Pétersbourg Hist. Acad. 6: 494 (1818); Boissier, Fl. Orient. 3: 266 (1875); Nábělek in Publ. Fac. Sci. Univ. Masaryk 52: 13 (1925); Guest in Dep. Agr. Iraq Bull. 27: 3 (1933); Bornmüller in Beih. Bot. Centralbl. 60B: 191 (1939); Blakelock in Kew Bull. 4: 42 (1949); Zohary in Dep. Agr. Iraq Bull. 31: 147 (1950); Rechinger, Fl. Lowland Iraq: 624 (1964); Rawi in Dep. Agr. Iraq Tech. Bull. 14: 104 (1964); Huber-Morath in Fl. Turkey 5: 234 (1975); Gabrielian & Fragman-Sapir, Flow. Transcaucasus & Adj. Areas: 53 (2008); Mozaffarian, Flora Iran 59: 74 (2008).

Perennial, 20–60 cm tall. Leaves with dense short spreading hairs, linear, median cauline 1–2.5 × 0.15–0.3 cm, ± incurved, pinnatisect, segments densely imbricate or somewhat distant, 0.5–1 mm long, 3–lobed or 3–partite, lobes oblong, denticulate; upper cauline leaves short, not reaching corymbs. Capitula 2–30(–45), corymbs 2–8(–10) cm broad, pedicels (3–)5–18 mm; involucre hemispherical to depressed, 2–6 × 2.5–8 mm, rounded at base; phyllaries lanceolate to ovate, obtuse, ± carinate, margins narrow, scarious (sometimes brownish) with spreading hairs or glabrescent. Rays (3–)6–8, golden yellow, 1–2.5 mm long. Disc florets (20–)40–90.

HAB. On rocky mountains and in valleys among *Quercus* forest, with *Pistacia* etc., stony hillsides, among limestone, serpentinite and other metamorphic rocks; alt. 500–3000+ m; fl. & fr. (May-)Jun.-Jul.(-Aug.).
DISTRIB. Common in the forest and lower thorn-cushion zones of Iraq, rare in the steppe region. **MAM**:

Severa Gaurik valley, NE of Zakho, *Rawi* 23659!; Mt. Zawita, NE of Zakho, nr. the Turkish frontier, *id.* 23602!; Khantur NE of Zakho, *Rawi* 23469!, *Rechinger* 10851!; Maya, 20 km W of Kani Masi, *Hamid & Fadhil* 45482!; Ser Amadiya, *Majid Mustafa* 3607!, *Guest* 4989!; Gali Mazurka, 2 km W of Amadiya, *Dabbagh & Jasim* 46895!; Mt. Gara, *Kotschy* 315!, *Rawi* 9255!; Sarsang, *Haines* W1339!; Alho, *Rawi* 8668! **MRO**: Kuh-i Sefin, *Bornmüller* 1893!; Botin, *Hadač* 6156!; Halgurd, *Guest & Ludlow-Hewitt* 2952!, 3066!, *Gillett* 12364!, *Rechinger* 11445!, *Guest & Alizzi* 15833!, *Rawi & Serhang* 24807!; Chiya i-Mandau, *Guest* 2660!; Rayat, *Guest & Alizzi* 15886!; Sakri-Sakran, *Bornmüller* 1379!; Haji Umran, *Gillett* 9467!, *G.W Chapman* 11947!, *Guest* 13062!, *Mooney* 4292!; Merga Draija (Mergadurezhan), nr. Iranian frontier, *Rawi* 9163!; Serin, *Rawi & Serhang* 24011!; Khalana, *Rawi* 13816!; between Bardanas and Qandil Dagh, *Rawi & Serhang* 24622!; at S foot of Karokh, *Alkas & Noori* 27555! **MSU**: Avroman, *Rawi & Alizzi* 19785! **FKI**: Altun Kupri, *Nábělek* 3596; Kirkuk, *Bornmüller* 1382.

Turkey, Transcaucasia, Iran.

BŪÎZHĀN (Kurd.) *Guest ex Majid Mustafa* 3607 (cf. BAURJĀN, Kurd.-Algurd, *Guest & Husham (Alizzi)* 15836 and BOWIJAN, *Rawi* 8668. According to Wahby & Edmonds' *Kurdish-English dictionary* (1966) BOJANE is the Kurdish name of a herb dried and used for keeping moths out of clothes – thus this common name of the plant might aptly be translated as "stinkwort" (Parsa 1946 gives BŪMĀDRĀN as the common Persian name for *Achillea*).

4. **Achillea conferta** *DC.*, Prodr. 6: 32 (1838); Boissier, Fl. Orient. 3: 270 (1875); Nábělek in Publ. Fac. Sci. Univ. Masaryk 52: 13 (1925); Blakelock in Kew Bull. 4: 41 (1949); Zohary in Dep. Agr. Iraq Bull. 31: 147 (1950); Rawi, in Dep. Agr. Iraq Bull. 14: 104 (1964); Mouterde, Nouv. Fl. Liban et Syrie 3: 400 (1983); Mozaffarian, Flora Iran 59: 80 (2008).

Perennial, 25–50 cm tall. Stems deeply furrowed. Leaves loosely pilose to glabrescent, linear, median cauline 10–25 × 1–3 mm, pinnatisect, segments loosely or ± densely imbricate, 1.5 mm long, obovate, denticulate. Capitula 3–20, corymbs dense, 1–5 cm broad, pedicels 1–4 mm; involucre broadly ovoid to hemispherical, 4–5 mm long and broad; phyllaries ovate to oblong, obtuse, not or very narrowly scarious-margined, canescent. Rays 4–8, pale yellow, 1 mm long. Disc florets 25–35. Fig. 88: 1–3.

HAB. Foothills and desert plains, on sandy, gravelly and sometimes silty soil; alt. 50–350(–1000) m; fl. & fr. Apr.-May(-Jun.).
DISTRIB. Quite common in the steppe region of Iraq, very common in the desert region, rare in the lower mountains and on the irrigated alluvial plain. **MJS/FUJ**: Balad Sinjar, *Rawi* 41370! **MSU**: Tasluja, *Omar & Karim* 47028! **FUJ**: "Muwasul tiatan mukzuq Nuwar", nr. Jabal Qilat, *Field & Lazar* 462!; "in deserto Singarae", *Haussknecht* s.n.; Umm Shababich, *Gillett* 11158!; Jazira, S of Balad Sinjar, *Guest* 13128!; Mir Qasim, *Field & Lazar* 539!; Qaiyara, *Bayliss* 130!; 50 km NW of Hadhr, *Ani & Brahim* 8778!; telegraph pole 90 between Baiji and Mosul, *Field & Lazar* 375; Ain Dibs, nr Baiji, *Guest* 13456; **FUJ/DLJ**: 15 km S of Hadhr, *Alizzi & Husain* 3860! **FAR**: Jabal Mahmur, *Gillett* 11209; Injana, *Guest* 4006, *Omar, Karim, Hamza & Hamid* 37061! **FAR/FKI**: between Injana and Tuz, *Rechinger* 10613; **FKI**: nr Kirkuk, *Haussknecht* s.n.; Jabal Naft, Kirkuk, *Nábělek* 3593; Kirkuk province, *Ali Haidari* 3934! **FKI/FPF**: Jabal Hamrin, *Sutherland* 105 p.p., 106 p.p., *Rawi & Gillett* 7324! **FPF/LCA**: between Khanaqin and Baghdad, *Nábělek* 3498! **FPF**: Makatu, nr Mandali, *Guest* 866!; 20 km SE of Mandali, *Rechinger* 8689!; 25 km SE of Badra, *Rawi* 20765!; Lajama, *Rawi* 20714! **DLJ**: 50 km WNW of Baiji, *Barkley* 4517!; nr Baiji, *Rawi & Hamada* 34146!; 55 km NE of Falluja, *Rawi* 20233!; **DGA**: Wadi Thirthar canal, nr Samarra, *E & C Guest & Sharif Muhammad Amin Haidar* 18328!; Ghurfa plain, nr Jabal Hamrin, *Guest* 4006; Adhaim, *Robertson* (Hunting Aero Surveys 99 (R.B. 26); *Karim, Hamid & Jasim* 40442!; 26 km E of Samarra, *Rawi* 20374! **DWD**: 10 km S of Ana, *Alizzi & Husain* 34000!; 206 km E of Rutba, *Rechinger* 12844!; 46 km N of Ramadi, *Guest, Rawi & P. Knowles* 25014!; 23 km SE of Ramadi, *Guest & Rawi* 15944!; Habbaniya, *Alizzi & Rawi* 34468A! 40 km W Ramadi to Rutba, *Al-Kiasi & K.Hamad* 48931 (BAG!). **DSD**: Wadi Khirr, S of Lussuf, *Guest, Rawi & Rechinger* 19438!; 30 km S E of Salman, *id.* 18771!; Busaiya, *Fauzi, Hazim & Hamid* 38952!; 97 km SW of Busaiya, *Khatib & Alizzi* 32763!; 50 km W of Busaiya, *Guest, Rawi & Rechinger* 10677!; **LCA**: between Falluja and Baghdad ("Falluja desert"), *Haines* W155!, *Omar & Wedad* 47519!

NE Syria, Lebanon, Turkey, Iran.
QAISŪM (Ir.-Jazira, *Guest* 13128) sometimes spelled after the Kurdish spelling GAISŪM (Kurd., *Ali-Haidari* 3934).

5. **Achillea talagonica** *Boiss.*, Diagn. Pl. Orient. ser. 1, 11: 17 (1849); Boissier, Fl. Orient. 3: 271 (1875); Mozaffarian, Flora Iran 59: 80 (2008).

A. tenuifolia Lam. var. *talagonica* (Boiss.) Bornm. in Bull. Herb. Boiss. 2 (7): 36 (1907).
A. oxylepis Boiss. & Hausskn. in Boissier, Fl. Orient. 3: 271 (1875).

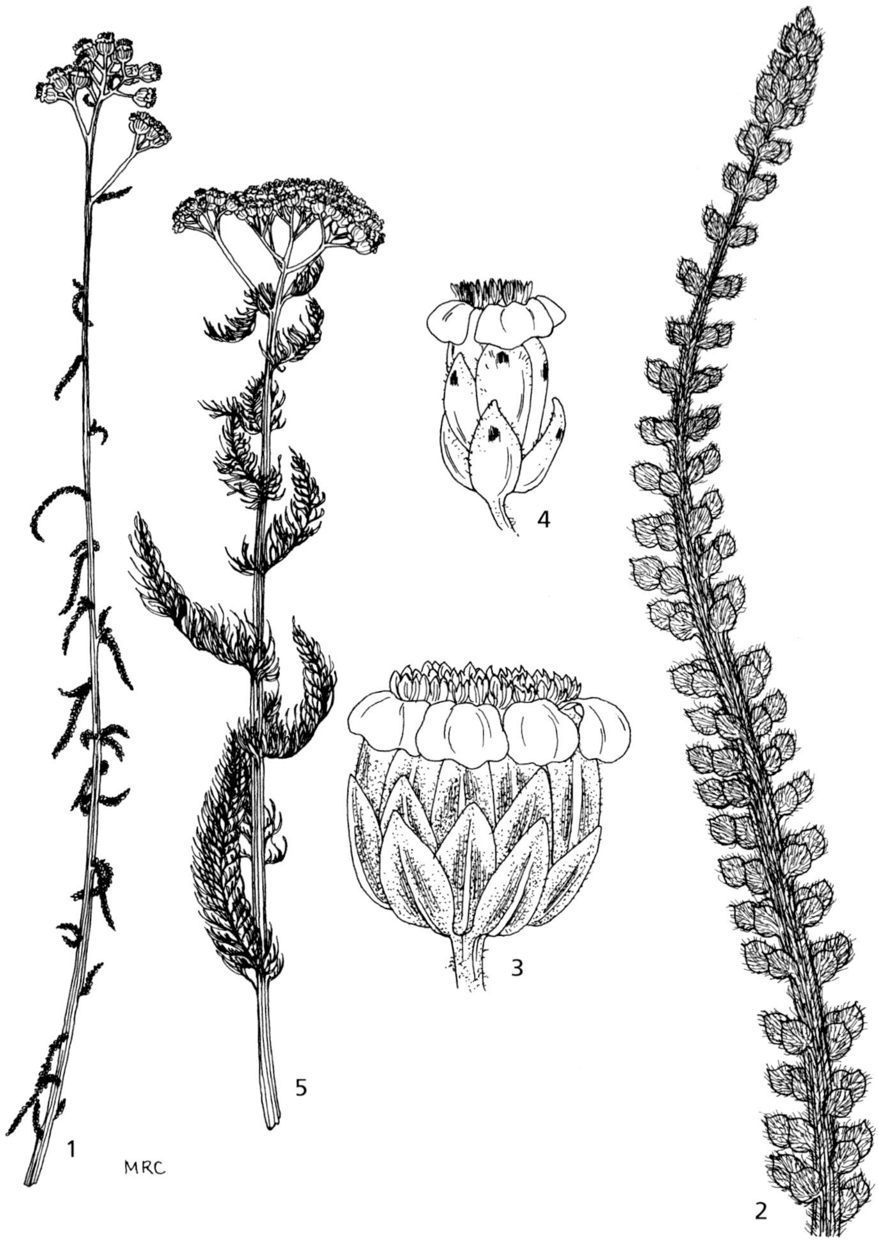

Fig. 88. **Achillea conferta**. 1, habit × ⅔; 2, single leaf × 12; 3, capitulum × 9. 2, **Achillea aleppica** subsp. **aleppica**. 4, capitulum × 9. **Achillea arabica**. 5, habit × ⅔. 1–3 from *Lazar* 539; 4 from *Gillett* 8057; 5 from *Hunting Aerosurveys* 87. Drawn by Moira Rae Carter.

Perennial, 30–50 cm tall. Stems deeply furrowed. Leaves adpressed-canescent to glabrescent, linear, median cauline 5–20(–30) × 0.8–1.2 mm, pinnatisect, segments densely imbricate, 0.5–0.7(–1) mm long, ± circular, denticulate, entire or 3-lobed. Capitula (5–)15–40, ± loosely arranged, corymbs 1.5–4 cm broad, pedicels 3–38 mm; involucre hemispherical, 3–4.5 mm long and broad; phyllaries linear-lanceolate, entire and acute to lanceolate-triangular and obtuse, margins scarious, canescent. Rays 2–4, yellow, 1.2–2 mm long. Disc florets 20–30.

HAB. Rocky mountain slopes; alt. ± 2000 m; fl. & fr. Jun.-Aug.
DISTRIB.?**MSU**, 'Avroman' (of which only the western slopes are in Iraq, the eastern in Iran), 'et Schahu' (Shahu, a more easterly mountain wholly in Iran), *Haussknecht* s.n. (the only possible collection was made over a century ago). Very doubtful if this species is to be found in Iraq.

Iran.

6. **Achillea aleppica** *DC.*, Prodr. 7: 296 (1838); Boissier, Fl. Orient. 3: 271 (1875); Handel-Mazzetti in Ann. Naturh. Mus. Wien 27: 435 (1913); Guest in Dep. Agr. Iraq Bull. 27: 2 (1933); Bornmüller in Beih. Bot. Centralbl. 60B: 91 (1940); Blakelock in Kew Bull. 4: 41 (1949); Zohary in Dep. Agr. Iraq Bull. 31: 147 (1950); Rawi in Dep. Agr. Iraq. Tech. Bull. 14: 104 (1964); Huber-Morath in Fl. Turkey 5: 239 (1975); Feinbrun-Dothan, Fl. Palaest. 3: 342 (1978); Mouterde, Nouv. Fl. Liban et Syrie 3: 401 (1983); Mozaffarian, Flora Iran 59: 86 (2008).

Perennial, 25–60 cm tall. Stems terete, longitudinally striped, adpressed-canescent, glabrescent. Leaves ± densely appressed- or subappressed-pilose, linear-filiform, median cauline (5–)10–40(–50) × 0.8–1 mm, pinnatisect, segments densely, at last ± loosely imbricate, 0.5 mm long, 3-lobed, lobes broadly ovate to orbicular, denticulate. Capitula (6–)25–120, corymbs 8–50 mm broad, pedicels absent or up to 4 mm; involucre oblong-cylindrical, ± angular, 4–7 × 2–3 mm; phyllaries triangular-ovate, ovate-oblong to oblong, obtuse, carinate, margins very narrowly scarious, appressed-canescent to glabrescent. Rays 2–4, golden or pale yellow, 1–2 mm long. Disc florets 6–20. Fig. 88: 4.

HAB. & DISTRIB. (of species). Rocky mountain slopes from lower margin of the thorn-cushion zone, the forest region in *Quercus* scrub and the moist steppe zone; alt. 150–2000; fl. & fr. (Apr.) May-Jul.

Syria, Lebanon, Palestine, Jordan, Turkey, Iran.

Involucre 4–5.5 mm long; corymbs 1.5–5 cm broad; capitula (15–)25–120, rays golden-yellow . . . . a. subsp. **aleppica**
Involucre 6–7 mm long, corymbs 0.8–1.5 cm broad, capitula 6–25, rays pale yellow . . . . b. subsp. **densa**

a. subsp. **aleppica**

DISTRIB. Common in the forest zones of Iraq, occasional in the moist steppe zone. **MJS**: Jabal Sinjar, *Handel-Mazzetti* 1372, *Gillett* 11095!; Karsi, *Gillett* 10925! **MAM**: Gali Mazurka, 2 km NW of Amadiya, *Dabbagh & Jasim* 45548; Sulaf, *Omar* 37706!; Zawita, *Guest* 4959!, *Mooney* 4315A!; Bakurman, *Rawi* 8523!; **MRO**: Kuh-i Sefin (Sefin Dagh), nr Shaqlawa, *Bornmüller* 1376!, *Gillett* 8057!; Darband, *Hadač* 2047; 27 km from Koisanjak to Erbil, *Omar, Al-Kaisi & Wedad* 49529 (BAG!); **MRO/MSU**: 31 km N of Kirkuk on the road to Koi Sanjaq, *Rawi, Noori & Kass* 27954!; **MSU**: Jarmo, *Helbaek* 1541!, *Haines* W332!; 12 km E of Chemchemal, *Rawi & Gillett* 11612!; Tasluja, *Hadač* 1960!; Haisi on Baranan Dagh, 10 km S of nr Sulaimaniya, *Rawi & Gillett* 11714!; Gilazarda, *Karim, Hamid & Jasim* 40693!; Azmir, *Karim* 39323!; Pira Magrun, *Haussknecht* s.n.; between Sulaimaniya and Qara Dagh, *Poore* 568!; Penjwin, *Guest* 12954!; Kamarspa, between Halabja and Tawila, *Rawi* 22214!; Tawila, *Rawi* 21931!, *Rechinger* 10335! **FUJ**: Tal Kochak, *Gillett* 10814!; between Tal Kochak and Mosul, *V. Dickson* 370!; nr Balad Sinjar, *Guest* 4120!; Karsi NWSlope, *Omar, Al-Khayat & Al-Kaisi* 52466!; Jabal Sinjar E slope, *Omar, Al-Kaisi & Al-Khayat* 52561 (BAG!); Jabal Khatchra, *Field & Lazar* 635!; "Muwasil Tiatan Mukzuk Nuwar", *id.* 463!; Tal Afar, *Guest* 4150!; **FUJ/FNI**: Mosul, *Noë* s.n.; **FAR**: Arbil, *Guest* 2163!; **FKI**: Baba Gurgur, *Guest* 4377!; **FPF**: Mandali, *Hadač et al.* 4574!

Throughout the range of the species.

QAISŪM (Ir.-Kursi, *Gillett* 10925).

b. subsp. **densa** (*Blakelock*) *Huber-Morath* in Ber. Schweiz. Bot. Ges 84: 146 (1975); Huber-Morath in Fl. Turkey 5: 240 (1975).

A. densa Blakelock in Kew Bull. 4: 41 (1949); Rawi in Dep. Agr. Iraq Tech. Bull. 14: 104 (1964).

DISTRIB. Rare in the NW sector of the forest zone of Iraq. **MAM**: Dohuk, *Mekki Beg* 3244!, *Rawi* 8642!; Zawita, *Guest* 4959!, *Gillett* 8351!, *Mooney* 4315A!

Endemic, but it may possibly be found in adjacent parts of Turkey (SE Anatolia).

? GULÎLK or GÎYĀ MESH (Kurd.-Dohuk, *Guest* 3244 – "not considered a useful pasture plant", *Rawi* 8642).

SECT. **MILLEFOLIUM** (DC.) W.KOCH.

7. **Achillea millefolium** *L.*, Sp. Pl.: 899 (1753); Afanasiev in Fl. S.S.S.R. 26: 78 (1961); Huber-Morath in Fl. Turkey 5: 244 (1975); Richardson in Fl. Europ. 4: 159 (1976); Huber-Morath in Fl. Iranica. 158: 62 (1986); Ghafoor in Fl. Pakistan 207 (1): 42 (2002); Mozaffarian, Flora Iran 59: 90 (2008); Zhu Shi, Humphries & Gilbert in Fl. China 20–21: 760 (2011); Breckle, Hedge & Rafiqpoor, Vasc. Pl. Afghan: 146 (2013).

A. ossica C.Koch in Linnaea 24: 323 (1851).
A. subhirsuta Gilib., Fl. Inch. 3.: 217 (1879) nom. inval., opera utique oppressa.
A. nábělekii Heimerl ex Nábělek in Publ. Fac. Sci. Univ. Masaryk 52: 12 (1925).

Perennial, 10–100 cm tall. Stems obtuse-angled, woolly-pilose. Cauline leaves similar to basal, sparingly pubescent to glabrescent, basal leaves linear to lanceolate, including petiole (up to 5 cm), (6–)10–20 × (0.5–)1–4 cm, (2–)3–pinnatisect into linear to linear-lanceolate segments, rachis 0.5–1.5 mm broad, rarely toothed, ultimate segments 0.2–0.5 mm broad, median cauline leaves 4–9 × (0.7–)1–1.5 cm. Capitula 50–100+, corymbs 4–15 cm broad, pedicels 1–5 mm; involucre oblong to ovoid, 4–5.5 × 2.5–4 mm; phyllaries oblong to lanceolate, subacute to acute, margins narrow, scarious, ± brownish, subglabrous. Rays 4–6, white, 1.5–2.5 mm long. Disc florets 10–20.

HAB. In the mountains, on metamorphic rock, and also in fields; alt. 1200–2000 m; fl. & fr. Jun.-Aug.
DISTRIB. Rare in the central sector of the lower thorn-cushion and upper forest zones of Iraq. **MAM**: Betanor (Bat Nura) 7 km E of Kani Masi, *Botany Staff* 43862! *Al-Dabagh & K.Hamad* 45575!; *K.Hamad & Fadhil* 45588! **MRO**: on E slope of Algurd Dagh (Helgord), *Gillett* 9525!, 12475!; Mergadurezhan, *Rawi* 9122!; Haj Umran, *Al-Kaisi & K. Hamad* 43548 (BAG!); Khalana district, *Rawi* 13770 (BAG!). **MRO/FAR**: 19 km SW of Salah ad-Din, *Barkley* 5631.

Most of Europe (except Mediterranean region), Turkey, Caucasia, Iran, C Asia to Siberia and Himalaya. Pink-rayed forms (the source of much of the horticultural colour range in cultivars stems from the British Isles) are present in several parts of the distribution of this species but not apparently in Turkey or Iraq.

SECT. **FILIPENDULINAE** (DC.) AFAN.

8. **Achillea filipendulina** *Lam.*, Encycl. [J. Lamarck et al.] 1: 27 (1783); Boissier, Fl. Orient 3: 259 (1875); Afansiev in Fl. S.S.S.R. 26: 90 (1961); Zohary in Dep. Agr. Iraq Bull. 31: 147 (1950); Rawi, Dep. Agr. Iraq Tech. Bull. 14: 104 (1964); Huber-Morath in Fl. Turkey 5: 249 (1975); Gabrielian & Fragman-Sapir, Flow. Transcaucasus & Adj. Areas: 52 (2008); Mozaffarian, Flora Iran 59: 98 (2008); Breckle, Hedge & Rafiqpoor, Vasc. Pl. Afghan: 146 (2013).

Tanacetum angulatum Willd., Tract. de Achilleis 52 (1789).
Achillea eupatorium M.Bieb., Tabl. Prov. Casp. 119 (1798).
A. filicifolia M.Bieb., Fl. Taur.-Cauc. 2: 338 (1808).

Perennial, 60–100 cm tall. Stems terete, obtuse-angled, loosely appressed-pilose, ± glabrescent. Leaves green, ± densely appressed-hairy, glandular-dotted, basal and lower cauline leaves oblong, 10–20 × 3–7 cm, sessile or with petiole up to 5 cm long, pinnatipartite with 10–15 pairs of oblong to lanceolate, lobed or dentate-serrate segments, upper cauline leaves smaller and less divided. Capitula (30–)50–300+, corymbs 4–10 cm broad, pedicels 2–8 mm; involucre obovoid to oblong, 3.5–5 × 2.5–3.5 mm; outer phyllaries linear-setaceous to linear-lanceolate, acute, inner lanceolate, subacute to obtuse, not marginate, sparsely hairy and gland-dotted. Rays 2–4, gold, 0.7–1 mm long. Disc florets 15–30.

HAB. In the mountains, on metamorphic rock, in fields and orchards; alt. 1400–2000 m; fl. & fr. Jun.-Aug.
DISTRIB. Quite common in the upper and middle forest zone of Iraq, occasional in the lower thorn-cushion zone: **MAM**: Maya, 25 km W of Kani Masi, *Hamid & Fadhil* 45509!; Dori, 5 km E of Kani Masi, *Omar & Qaisi* 45420!; Gali Mazurka, by Ser Amadiya, *Dabbagh, Qaisi & Hamid* 46015!, 46920!; Gara Dagh, *Rawi* 9256!; Gara village (on Gara Dagh), *Kotschy* 301!; Betanor (Bat Nura), *Botany Staff* 43857. Kani Mazi and Matina mt. summit of Barugha mt., *AL-Khayat, Jawhar & Abdulla* 57985 (BAG!); **MRO**: Gali Ali Beg, *Kass & Noori* 27203!; Sairo, *Gillett* 9673!; Algurd Dagh (Halgurd), *Gillett* 9518!; Rayat, *Guest* 13073!; Gunda Zhor, *Gillett* 12322!; **MSU**: Penjwin, *Haussknecht s.n.*!; between Benawa Suta and Penjwin, *Rawi* 22895!

Turkey, Caucasia, Iran, Aghanistan.

9. **Achillea arabica** *Kotschy*, Sitzungsber. Akad. Wien, Math.-Nat. lii. Abth. I, 255 (1866).

A. biebersteinii Afan. in Not. Syst. (Leningrad) 19: 361 (1958), nom. inval.; Fl. S.S.S.R. 26: 97 (1961); Huber-Morath in Fl. Turkey 5: 250 (1975); Richardson in Fl. Europaea 4: 165 (1976);

Mozaffarian, Flora Iran 59: 102 (2008); Collenette, Wild Flow. Saudi Arabia: 167 (1999); Boulos, Fl. Egypt Checklist: 241 (2009).

A. micrantha auct. non Willd., Tract. de Achilleis 33 (1789), Dinsmore in Post, Fl. Syria, Palest. & Sinai ed. 2, 2: 46 (1933).

A. micrantha M.Bieb., Fl. Taur.-Caucas. 2: 336 (1808); Boissier, Fl. Orient. 3: 264 (1875); Handel-Mazzetti in Ann. Naturh. Mus. Wien 27: 434 (1913); Nábělek in Publ. Fac. Sci. Univ. Masaryk 52: 13 (1925) forma *tenuisecta* Heimerl. ex Nábělek, ibid.; Guest in Dep. Agr. Iraq Bull. 27: 2 (1933); Bornmüller in Beih. Bot. Centralbl. 60B: 190 (1939); Blakelock in Kew Bull. 4: 42 (1949); Zohary in Dep. Agr. Iraq Bull. 31: 147 (1950); Rechinger, Fl. Lowland Iraq: 624 (1964).

Perennial, 10–100 cm tall. Stems terete, longitudinally striped, ± densely spreading-pubescent. Leaves laxly or densely pilose, 2–3-pinnatisect, cauline similar to basal, median cauline linear to linear-lanceolate, 2.5–12 × 0.5–4 cm, rachis entire, 0.5–1 mm broad, primary segments numerous, ultimate segments setaceous to linear (to lanceolate), acute to acuminate, 1–4(–6) × 0.2–1 mm. Capitula 30–200+, corymbs 2–10 cm broad, pedicels 0.5–4 mm; involucre oblong to broadly ovoid, 3–4 × 2–3 mm; phyllaries ovate-triangular to oblong, ± obtuse, pale, pubescent, inner phyllaries with scarious tip. Rays 4–5, golden-yellow, 1–2 mm long. Disc florets 10–30. Fig. 88: 5.

HAB. Mountain slopes of igneous and metamorphic rocks, sometimes in *Quercus* or *Pinus* woods, also by streams and in fields and on steppic plains; alt. (150–)300–1800(–2500) m; fl. & fr. (May-)Jun.-Jul. (-Aug.) in the mountains, Apr.-May(-Jun.) on the plains.

DISTRIB. Very common in the forest zone of Iraq, occasional in the lower thorn-cushion zone, quite common in the NW sector of the moist steppe zone. **MAM**: Daimka, 35 km NE of Zakho, *Qaisi, Hamid & Hamad* 4535!; between Zakho and Sharanish, *Karim, Hamid & Jasim* 41095!; Basingara, *Rechinger* 11516; nr Maya and Bidahu, 20 and 50 km W of Kani Masi respectively, *Hamid & Fadhi* 45468, 45464; Amadiya, *Anders* 2330!; Gali Mazurka, by Ser Amadiya, *Dabbagh, Qaisi & Hamid* 45967!; Gara Dagh (Chiya-i Gara), *Kotschy* 33 & 34, *Dabbagh & Kasim* 46936!; Sarsang, *Alkas* 19522!; Zawita, *Guest* 87!, *Mooney* 4315!, *V.C.R.* (Robertson, Hunting Aero surveys) 87; Sawara Tuka, *E. Chapman* 26365, *Dabbagh & Jasim* 46836; Shaikh Adi, *Guest* 3669!, *Field & Lazar* 763!; Aqra, *Rawi* 11417! **MRO**: Shaqlawa, *Rawi & Gillett* 11580!; Gali Ali Beg, *Omar, Qaisi & Widad* 49554!; Rowanduz, *Guest* 2672!; between Gunda Zhor and Darband, *Gillett* 12403!; Rayat, *Omar, Sahira, Karim & Hamid* 38453!; Haji Umran, *Omar et al.* 38462!, *Qaisi & Hamad* 43552!, *Dabbagh & Hamad* 46291!; between Darband and Haji Umran, *Hadač* 2343!; Mergadurezhan, *Rawi* 9162!; Warshanka-Magar, *Rawi & Serhang* 24437!; Berd Agha Gin, 18 km NW of Rania, *Rawi, Noori & Kass* 28683!; Kani Rash, Qurnago valley, *Rawi* 26682!; Pushtashan, *Rawi & Serhang* 26558!; Karoukh, between Saran and Kilkil, *Kass & Noori* 27369! **MSU**: 20–25 km E of Qaranjir, *Rawi* 21660!; Chemchemal, *Rawi & Gillett* 7598!, *Omar & Karim* 37914!; Jarmo, *Helbaek* 1597!, *Haines* W242!; between Kirkuk and Sulaimaniya, *Rawi* 21513!; 18 km N of Kani Sipi, *Rawi* 22426!; Penjwin, *A.S Agha* 5502!, *Rawi* 12180!; 11 km N of Penjwin, *Rawi* 22885!; 5 km 'from' (W of?) Penjwin, *Noori & K. Hamad* 41232!; Tawila, *Rechinger* 10316!, *Rawi* 22291! **MJS**: 47 km E of Karsi, *Bharucha & Ani* 8802!; Jabal Qilat, *Field & Lazar* 307! **FUJ**: Tal Kochek, *Gillett* 10812!; between Tal Kochak and Mosul, *Dickson* 370A; Tal ash-Shaur, *Field & Lazar* 573!; Tal Afar, *Sarsam* 9940!; between Balad Sinjar and Tal Afar, *Field & Lazar* 619!; Upper Jazira, N of Sinjar, *Guest* 13306! **FUJ/FKI**: between Arbil and Kirkuk, *Thesiger* 461!, 462!, 465! **FKI**: Altun Kupri, *Nábělek* 3594, *Bornm.* 13731!; Jabal Naft, nr Kirkuk, *Nábělek* 3587!, 3587A!; Kirkuk, *Haussknecht s.n.*; *Guest* 1987. **DWD**: Ana, *Omar, Al-Khayat & Adel* 55471 (BAG!).

Extreme SE Europe (Bulgaria), Syria, Lebanon, Palestine, Jordan, Turkey, Caucasia, Iran.

? BURZHĀN, *Gillett* 9504, BEJANK, *Rawi* 9162 (cfl. BŪÎZHĀN, *Guest, Majid Mustafa* et al. under *A. vermicularis*), ? BRINDA or BRINDASI, *Omar & Karim* 37914, ? TUREIK AD-DUBĀN "said to be better than (*Gillett*) 10814 for killing insects".

SECT. **BABOUNYA** (DC.) O.HOFF.

10. **Achillea fragrantissima** (*Forssk.*) *Sch.-Bip.*, Fl. (Regensb.) 38: 13 (1855); Boissier, Fl. Orient. 3: 272 (1875); Post in Dinsmore, Fl. Syria, Palest. & Sinai ed. 2, 2: 48 (1933); Zohary in Dep. Agr. Iraq Bull. 31: 147 (1950); Rechinger, Fl. Lowland Iraq: 625 (1964); Rawi in Dep. Agr. Iraq Tech. Bull. 14: 105 (1964); Feinbrun-Dothan, Fl. Palaest. 3: 343 (1978); Mouterde, Nouv. Fl. Liban et Syrie 3: 399 (1983); Collenette, Wild Flow. Saudi Arabia: 167 (1999); Boulos & Hind in Fl. Egypt 3: 249 (2002); Al-Mayah, Al-Edani & Al-Asadi, Ecol. Fl. Basra: 189 (2016).

Santolina fragrantissima Forssk., Fl. Aeg.-Arab. 147 (1775).

Fragrant perennial, suffruticose, 50–100 cm tall. Leaves oblong-linear to ovate, subobtuse, 2–10 × 1–4 mm, thick, appressed-tomentose, glabrescent, comb-like callose-serrate. Capitula discoid, numerous in lax compound corymbs 1–3 cm across, sessile or pedicels up to 6 mm. Capitula 30–60; involucre oblong to broadly ovate, 4–6 × 2.5–5 mm; phyllaries oblong, obtuse, canescent. Corollas yellow.

HAB. On desert plains, often in slight silty depressions and on sandi wadi banks, sandstone plateau and low sandstone hills, usually on sandy or gravelly places but also on silty flats encircled by limestone ridges; alt. (50–) 200–650 m; fl. & fr. (Mar.-)Apr.-May, but with dry flowers often remaining on the plants until autumn.
DISTRIB. Quite common in the Desert Region of Iraq: **DWD**: 50 km N of Rutba, *Rawi* 26856!; Sawab, NW of Rutba, *Rawi* 26871!; 10 km from Rutba on the way to H.3, *Qaisi, Hamad & Hamid* 43587!; 23 km SW of Qa'im, *E. Palmatier, Ani & Barkley* 803!; Masa'd, S of Rutba, *Alizzi & Omar* 35380!; Wadi Masa'd al-Rutba'i, 30 km S of Rutba, *Tikriti & Hazim* 29710!; 240 km W of Ramadi on road to Rutba, *id.* 29700!;180 & 90 km W of Ramadi, *Rawi* 26851! 20884!; 46 km W of Ramadi, *Guest, Rawi & P. Knowles* 25002!; **DSD**: 5 & 30 km W of Ma'niya, *Hazim* 32507!, 32514!, 32527!; Faidhat Umm (Rubaim) al-Ma'niya, *Hazim* 30690!; Cibritiya, 60 km W of Shabicha, *id.* 32478!, 32483!; nr Ashuriya well, 55 km NW of Shabicha, *id.* 14064!; 70 km SW of Samawa, *Guest, Rawi & Rechinger* 18725!; 30 & 5 km W of Ma'niya, *Hazim* 32514!, 32527!; 9–10 km N of Salman, *Rawi & H. Tikriti* 24978!, *Chakrawarty, Khatib, Rawi & Tikriti* 30010!; 8 km N by W of Aidaha, *Guest, Rawi & Rechinger* 19175!; nr Jalib Bakur, 95 km SSE of Salman, *id.* 18882!; Busaiya, *Sa'dun Yusuf Serkahia* 19514!; 18 km S by W of Busaiya, *Guest, Rawi & Rechinger* 16068!; 10 km S by W of Busaiya,
Guest & Mahalhal 15257!; between Aidaha and Ansab, *Fauzi (Karim), Hazim & Hamid* 38802!; 62 km WNW of Ansab on Darb al-Haj at Saudi frontier, *Guest, Rawi & Rechinger* 19052!; 20 km WNW of Ansab, *id.* 18997!; 50 km S of Ansab on road to Samah, *Fauzi, Hazim & Hamid* 38839! **LCA**: 10 km E of Falluja on road to Baghdad ("Falluja Desert"), *Guest, Rawi & Noori* 13518!, *Haines* W 791!

Syria, Lebanon, Palestine, Jordan, Sinai, Egypt, Arabia, Kuwait.

QAISŪM (Ir., *Guest et al.* 16068, *Guest & Mahalhal* 15257 – "sweet aromatic shrublet").

90. **ANACYCLUS** L.

Sp. Pl.: 892 (1753); Gen. Pl. ed. 5: 381 (1754)
Oberprieler, Vogt & Watson in Kubitzki, Fam. Gen. Vasc. Pl. 8: 364 [2006] (2007)

Patricia Halliday

Erect annual (occasionally perennial), glabrous or pubescent herbs. Leaves alternate, 2–3-pinnatisect, rosettiform or cauline. Capitula solitary, terminal, heterogamous, radiate, rarely discoid, or homogamous and discoid. Involucre hemispherical, turbinate or cylindrical-campanulate; phyllaries 3-seriate, inner series scarious, apices ± lacerate; receptacle flat or shortly conical, scaly; receptacular scales obovate, ± scarious and flat. Ray florets uniseriate, female fertile; corollas white, dorsiventrally compressed, tube persistent on achene at maturity. Disc florets corollas yellow, campanulate, 5–lobed, sometimes with 2 lobes longer and more erect, base expanded and capping achene on ventral side. Achenes dorsiventrally compressed, oblanceolate, with a scarious, surrounding wing, glabrous; ray achenes epappose, disc achenes with unilateral corona, ± continuous with wing.

About 12 species, distributed around the Mediterranean, in Europa, N Africa and Egypt; one species in Iraq.

Anacyclus (from Gr. *ana-*, a prefix meaning variously upon, up, upwards or above) and κυκλος, *kyklos*, cycle, perhaps with reference to the scarious involucre).

Humphries, C. J. 1979. A revision of the genus *Anacyclus* L. (Compositae: Anthemideae)." Bull. Brit. Mus. (Nat. Hist.), Bot. 7: 83–142.
Humphries, C. J. 1981. Cytogenetic and cladistic studies in *Anacyclus* (Compositae: Anthemideae). Nordic J. Bot. 1(1): 83–96 (1981).
Schweizer, D. & Ehrendorfer, F. 1976. Giemsa banded karyotypes, systematic, and evolution in *Anacyclus* (Asteraceae – Anthemideae). Pl. Syst. Evol. 126 (2): 107–148.

1. **Anacyclus nigellifolius** *Boiss.*, Diagn. Pl. Orient. ser. 1, 11: 13 (1849); Boissier, Fl. Orient. 3: 322 (1875); Dinsmore in Post, Fl. Syria, Palest. & Sinai ed. 2, 2: 58 (1933); Zohary in Dep. Agr. Iraq Bull. 31: 149 (1959); Rawi, in Dep. Agr. Iraq. Tech. Bull. 14: 105 (1964); Grierson in Fl. Turkey 5: 223 (1975); Humphries in Bull. Br. Mus. (Nat. Hist.), Bot. ser., 7 (3): 135 (1979).

A. nigellifolius subsp. *orientalis* Grierson in Notes Roy. Bot. Gard. Edinburgh 33: 411 (1975).

Erect annual herb to 30 cm tall, branched in upper part, or stem simple. Stems ribbed, 3 mm in diameter, sparsely long-hairy, especially near apex. Leaves 2-pinnatisect, oblanceolate in outline, 15–25 × 8–12 mm, rachis with 3 strong main veins, and somewhat widened at base, divisions linear-lanceolate, acute, about 6 pairs, lowest 4 pairs opposite, sparsely long-hairy, subsessile. Inflorescence with 1 capitulum or a lax corymbose cyme. Scapes up to 12 cm long, swollen immediately beneath capitula. Capitula 1.2 cm long, 1–1.5 cm in diameter. Involucre hemispherical-turbinate, 5–15 mm in diameter, ± long-hairy, phyllaries with erose, scarious margins and conspicuous midribs, outer oblanceolate, 3.5–4 × 1.5 mm, acute; inner obovate, 7 × 2–3 mm, acute to obtuse; receptacle flat, scales obovate, 6 × 2.5–3 mm, scarious, glabrous, midrib excurrent in a cusp. Ray florets few, ray oblong, 4 × 2 mm, with 2–3 apical teeth, tube ± 3 mm long, with narrow, lateral wings, sparsely papillose. Disc florets ± 4 mm long, base expanded into a ± circular appendage, 2 mm wide, covering top of achene on ventral side; anthers ± 2 mm long (including collar); style arms short, <1 mm long, apices truncate, papillose. Achenes dorsiventrally compressed, 4.5–5 × 4 mm, brown, with straw-coloured, shining wing 1.75 mm wide. Fig. 89: 1–11.

HAB. In the mountains; alt. 900–1500 m; fl. & fr. May-Jun.
DISTRIB. Very rare in Iraq, having been found only once. **MSU**: Pira Magrun, *Haussknecht s.n.*!

Syria, Lebanon, Turkey (S Anatolia).

I agree with C.J. Humphries (loc. cit.) that the characters used by Grierson (1975) to separate two subspecies are not sufficient to merit recognition in this variable species, and have accordingly reduced subsp. *orientalis* to synonymy.

91. **ANTHEMIS** s.l.

L., Sp. Pl., ed. 1, 2: 893 (1753); Gen. Pl., ed. 5: 381 (1754).
Oberprieler, Vogt & Watson in Kubitzki, Fam. Gen. Vasc. Pl. 8: 365 [2006] (2007).

Anthemis L. subgen. *Maruta* Cass., Bull. Sci. Soc. Philom. Paris 1818: 167 (1818); *Maruta* (Cass.) Gray, Nat. Arr. Brit. Pl. 2: 456 (1821); *Anthemis* L. sect. *Maruta* (Cass.) Rchb., Deutsch. Botaniker, ed. 2, 2: 139 (1844). *Cota* J. Gay, Fl. Sic. Syn. [Gusson] 2: 866 (1844); *Anthemis* sect. *Cota* (J. Gay) Rupr., Fl. Ingr. 1: 589 (1860). *Anthemis* subgen. *Cota* (J. Gay) Rouy, Fl. Franc. [Rouy & Foucaud] 8: 228 (1903); *Anthemis* L. sect. *Odontostephana* Eig, Palest. J. Bot., Jerusalem Ser. 1: 203 (1938); *Anthemis* L. subgen. *Maruta* (Cass.) Tzvelev, Fl. Europeiskoi Chasti SSSR 7: 111 (1994).

A. Ghafoor & S. I. Ali
Revised by D. J. Nicholas Hind

Annual, biennial or perennial herbs, sometimes subshrubby, variously pubescent. Leaves alternate, pinnatisect to variously incised to pinnatifid or rarely entire. Capitula heterogamous and radiate or homogamous and discoid; phyllaries (2–)3(–4)-seriate, margins narrowly to broadly scarious, entire or lacerate; receptacle convex to conical or hemispherical, paleaceous or rarely naked; paleae all over receptacle or in lower half only, subulate, linear-lanceolate or oblanceolate to spatulate, apices truncate to acute or acuminate-spinescent with excurrent midrib, usually straw-coloured and scarious, rarely with brown-coloured or purplish margins or apices (or very rarely coloured throughout). Ray florets, when present, uniseriate, female, fertile, or neuter; ray limb white or rarely yellow or reddish, base glabrous or pubescent, rarely indurate. Disc florets numerous, corollas 5-lobed, yellow or occasionally reddish-purple, lower half of tube not or inflated in fruit, rarely pubescent, lobes rarely appendaged; anther bases lacking appendages; style arms flattened, apices truncate and penicillate. Achenes terete to oblong-obpyramidal, 4–5-angled or dorsiventrally compressed, body smooth, tuberculate or sometimes pitted, sulcate and ribbed; pappus a minute dentate-lacerate rim, corona or an adaxial auricle, or absent.

A genus of between 175 and 210 species (depending upon whether *Cota* J.Gay is separated out), *Anthemis* principally circum-Mediterranean, from Europe, N and NE Africa, and SW Asia. It is widely introduced, and several naturalized species have resulted from those commonly cultivated; 34 species are known from Iraq, 7 are endemic In this account *Anthemis* is treated *sensu lato* (inclusive of *Cota* – spp. 23–27).

Fig. 89. **Anacyclus nigellifolius**. 1, habit × ⅔; 2, leaf × and part of stem (enlarged); 3, capitulum × 2; 4, ray floret × 6; 5, disc floret with achene × 6; disc floret × 6; 7, 8, receptacular scale × 6; 9, anthers × 12; 10, style × 12; 11, achene × 6. 1 from *Samuelsson* 4911. Drawn by M. Millar Watt.

Eig, A. (1938). Taxonomic studies on the Oriental species of the genus *Anthemis*. Palest. J. Bot., Jerusalem Ser. 1: 161–224. (1938).
Ghafoor, A. & Al-Turki, T.A. (1997). A synopsis of the genus *Anthemis* L. (Compositae-Anthemideae) in Saudi Arabia." Candollea 52: 457–474. (1997).

Ghafoor & S. I. Ali (2002). The genus *Anthemis* L. (Compositae) in Iraq: a synopsis. Compositae Newslett. 38: 1–41.

Ghafoor, A. (2010). The genus *Anthemis* L. (Compositae-Anthemideae) in Arabian Peninsula: A taxonomic study. Pakistan J. Bot. 42: 79–98.

Iranshahr, M. (1981). Fünf neue Arten der Gattung *Anthemis* (Compositae) aus Iraq und Afghanistan. Pl. Syst. Evol. 139(1–2): 159–162.

Iranshahr, M. 1982. Acht neue Arten der Gattung *Anthemis* aus Persien. Pl. Syst. Evol. 139(3–4): 313–317. (1982).

Lo Presti, R.M., Oppolzer, S. & Oberprieler, C. (2010). A molecular phylogeny and a revised classification of the Mediterranean genus *Anthemis* s.l. (Compositae, Anthemideae) based on three molecular markers and micromorphological characters. Taxon 59(5): 1441–1456.

Oberprieler, C. (1998). The systematics of *Anthemis* L. (Compositae, Anthemideae) in W and C North Africa." Bocconea 9: 1–328.

Key modified from Ghafoor & Ali (2002).

1. Ray limbs yellow, sometimes absent 2
   Ray limbs white or cream-coloured 5
2. Plants annual; phyllaries glossy, glabrous, pale brown, outer ovate, inner obovate with a large ± obovate 5 × 5 mm translucent appendage . 2. *A. scariosa*
   Plants perennial; phyllaries neither glossy nor glabrous, sparsely or densely white-lanate, lacking appendages 3
3. Pedicels <5 cm long; achenes with entire corona; outer phyllaries acute, inner obtuse to ±acute and apices pale or dark brown to diffuse brown . . . . 23. *A. tinctoria*
   Pedicels >5 cm long; achenes with a unilateral auricle; phyllaries narrow, apices acute 4
4. Leaves pinnatipartite, lower ±spatulate, upper cuneate, apices entire or shallowly toothed or lobed . . . . 31. *A pauciloba*
   Leaves bipinnatisect, ovate or oblong, upper leaves pinnatisect . . . . . . 32. *A. kotschyana*
5. Receptacle paleaceous all over 6
   Receptacle paleaceous in upper half only 33
6. Disc floret corolla tube hairy 7
   Disc floret corolla tube glabrous 10
7. Corolla lobes of disc florets ending in a long mucro or apiculum. . . . 5. *A. marismartui*
   Corolla lobes of disc florets acute, lacking mucro or apiculum 8
8. Paleae with broad dark brown margins; achenes *c.* 1 mm long . . . . 6. *A. melanacme*
   Paleae without dark brown margins 9
9. Phyllaries linear-lanceolate, 1 mm wide; paleae narrowly oblong-spatulate, ± 2.5 mm long, deciduous; rays 9 or 10 . . . . 7. *A. microlepis*
   Phyllaries oblong-ovate, 2–3 mm wide; paleae oblong-lanceolate, ± 3 mm long, persistent; rays 15–20 . . . . 20. *A. schizostephana*
10. Inner phyllaries with broad membranous hyaline margins and brown, shiny obtuse apical appendages nearly as long or longer than lower part . . . . 3. *A. hyalina*
    All phyllaries with narrow membranous margins and without apical appendages . . . 11
11. Capitula <1 cm in diameter. . . . . 8. *A. micrantha*
    Capitula usually >1 cm in diameter 12
12. Achenes (all or at least a few) auriculate at apex 17
    Achenes truncate, exauriculate or apices slightly marginate 13
13. Paleae purplish-brown at maturity, either all cuspidate to long-acuminate or inner short-cuneate and outer long-cuspidate . . . . . . 24. *A. melanolepis*
    Paleae scarious, straw-coloured, sometimes with darker, brown margins, all entire, ±erose or denticulate, acute, acute-acuminate, obtuse-truncate or short-mucronate 14
14. Achenes cylindrical or oblong-cylindrical, smooth; phyllaries glabrous or nearly so . 15
    Achenes obpyramidal, ± curved, tuberculate; phyllaries hairy. 16
15. Capitula 2.5–4.5 cm in diameter; paleae almost flat, oblong-spatulate, apices ±erose, obtuse-truncate, and short-mucronate . . . . 9. *A. leptophylla*
    Capitula <2 cm in diameter; paleae carinate, oblong-spatulate, margins narrowly scarious, apices obtusely denticulate. . . . . 22. *A. rascheyana*

16. Leaves sessile; pedicels 5–6 cm long; phyllaries oblong-spatulate, margins and apex broad-scarious, all obtuse . . . . 19. *A. handel-mazzettii*
Leaves petiolate; pedicels 1.5–3 cm long; phyllaries lanceolate, margins narrowly scarious, apices acute . . . . 10. *A. deserti*
17. Achenes cylindrical or nearly so 18
Achenes obconical or obpyramidal 23
18. Phyllary margins scarious, pale-brown; ray floret tubes indurate, persistent; inner achenes crown with a unilateral, erose-dentate auricle as long as or about half as long as achene . . . . 4. *A. haussknechtii*
Phyllaries margins scarious, hyaline, colourless; ray floret tubes neither indurate nor persistent; auricle less than half as long as achenes 19
19. Paleae flat . . . . 21. *A. homalolepis*
Paleae carinate 20
20. Paleae ± oblong-spatulate, ±truncate and abruptly acuminate, acumen nearly as long as basal part or slightly shorter . . . . 25. *A. coelopoda*
Paleae not as above, acumen very short 21
21. Paleae oblong-obovate; pedicels ultimately thickened upward . . . . 11. *A. gillettii*
Paleae oblong or oblong-lanceolate; pedicels not or slightly thickened 22
22. Leaves sessile; phyllaries pubescent . . . . 12. *A. hamrinensis*
Leaves with pectinate petiole; phyllaries ± glabrous . . . . 22. *A. rascheyana*
23. Fruiting pedicels thickened upwards 24
Fruiting pedicels not or harly thickened upwards. 29
24. Paleae spatulate or oblong-spatulate 25
Paleae oblong-lanceolaate to oblanceolate 26
25. Receptacle ± convex-conical; paleae 2.5–3 mm long, short-mucronate; achenes obconical-turbinate, outer nearly smooth, exauriculate, inner tubercled with auricle ¼–1/2 of achene length . . . . 13. *A. wettsteiniana*
Receptacle hemispherical; paleae *c.* 3.5–4 mm long, abruptly contracted apically and bearing equally long slender acumen; all achenes 4-angled, smooth, rhombic in cross-section, auricle minute, entire. . . . . 26. *A. altissima*
26. Leaves oblong-spatulate to cuneate-obovate, basally pectinate-dentate, entire or crenate-dentate; phyllaries ciliate, outer linear-lanceolate, inner oblong, 8–10 mm long. . . . . 1. *A. leucanthemifolia*
Leaves oblong-lanceolate or obovate-oblong, basally not pectinate-dentate; outer phyllaries triangular, acute, inner obtuse, <8 mm long 27
27. Paleae obovate-spatulate, inconspicuously mucronate . . . . 14. *A susiana*
Paleae lnceolate to oblanceolate, ± stiffly acuminate 28
28. Disc floret corollas not or very slightly inflated in fruit, deciduous; achenes smooth; paleae flat, siffly acuminate . . . . 15. *A. arvensis*
Disc floret corollas inflated and indurated in fruit; achenes tuberculate; paleae keeled, not stiffly acuminate . . . . 17. *A. melampodina*
29. Receptacle hemispherical 30
Receptacle conical. 31
30. Pedicels up to 3 cm long; paleae oblong-spatulate to subcordate-emarginate, with 1–1.5 mm long mucro; achenes with denticulate auricle . . . . 27. *A. brevicuspis*
Pedicels more than 3 cm long; paleae oblong, apices irregularly fimbriate with a short mucro; achenes truncate with a very short unilateral rim at apex . . . . 16. *A. plebeia*
31. Achenes smooth, with a thickened undulating apical rim (sometimes with a lacerate hyaline auricle) . . . . 15. *A. arvensis*
Achenes tuberculate all over or only abaxially, apex auriculate. 32
32. Leaf lobules approximate; pedicels gradually slightly thickened in fruit; achenes tuberculate all over; auricle lobes not acute . . . . 17. *A. melampodina*
Leaf lobules remote; pedicels hardly thickened in fruit; achenes only tuberculate abaxially; auricle deeply divided into acute triangular lobes . . . . 18. *A. zoharyana*
33. Achenes smaller than persistent paleae 34

Achenes as long as or longer than deciduous paleae 35
34. Phyllaries glabrous; disc floret corolla tube glandular-punctate, ± compressed. . . . . .30. *A. corymbulosa*
Phyllaries hairy; disc floret corolla tube ± inflated and eglandular 37
35. Fruiting pedicels distinctly thickened and strongly arcuate; ray florets fertile; paleae apices cuspidate; achenes ± tetragonous, distinctly 10-ribbed, occasionally tubercled 36
Fruiting pedicels neither thickened nor arcuate; ray florets sterile; paleae apices bristly; achenes ± cylindrical-turbinate, obscurely 8–10-ribbed, tuberculate . . . . 29. *A. cotula*
36. Pedicels 2–4 cm long; phyllaries persistent; paleae narrowly linear-lanceolate . . . . .33. *A. tubicina*
Pedicels >5 cm long; phyllaries deciduous; paleae setaceous-subulate. . . . . 35. *A. odontostephana*
37. Achenes auriculate 38
Achenes exauriculate 39
38. Auricle 1/3–1/2 as long as achenes, entire to obscurely lobed or divided into a few triangular lobes 40
Auricle very short, divided into a few obtuse lobes . . . . 28a.II. *A. pseudocotula* ssp. *pseudocotula* var. *radiata*
39. Achene apex oblique, slanting outwards . . . . 28b.I. *A. pseudocotula* ssp. *rotata* var. *rotata*
Achene apex rounded, not slanting outwards . . . . 28b.II. *A. pseudocotula* ssp. *rotata* var. *rotundata*
40. Auricle entire or obscurely lobed . . . . .28a.I. *A. pseudocotula* ssp. *pseudocotula* var. *pseudocotula*
Auricle divided into a few triangular lobes . . . . 28a.III. *A. pseudocotula* ssp. *pseudocotula* var. *massadensis*

1. **Anthemis leucanthemifolia** *Boiss. & C.I.Blanche*, Diagn. Pl. Orient., ser. 2, 3: 20 (1856). Boissier, Fl. Orient. 3: 296 (1875); Eig in Palest. J. Bot. Jerusalem Ser. 1: 163 (1938); Täckholm, Stud. Fl. Egypt, ed. 2. 574 (1974); Feinbrun-Dothan, Fl. Palaest. 3: 328. Pls. 547, 548 (1978); Mouterde, Nouv. Fl. Lib. Syr. 3: 408. PI. CCXXIII, n. 1 (1983).

Anthemis philistea Boiss., Fl. Orient. 3: 302 (1875).
A. leucanthemifolia Boiss. & C.I.Blanche subsp. *philistea* (Boiss.) Eig, Palestine J. Bot., Jerusalem Ser. 1: 164 (1938).
A. leucanthemifolia Boiss. & C.I.Blanche subsp. *rafaensis* Eig, Palestine J. Bot. Jerusalem Ser. 1: 164 (1938).
A. eliezrae Eig, Palestine J. Bot., Jerusalem Ser. 1: 179, & errata [as Eliezeri] (1938).

Herb, sparsely crisped hairy, up to 40 cm tall, with erect or ascending branches from base. Leaves simple, slightly fleshy or not, oblong-spatulate to cuneate-obovate, base pectinate-dentate, apices entire or crenate-dentate or lobed, or uni- to bipinnatisect, if pinnately dissected then lobes and ultimate segments narrow, cuneate. Peduncles short, distinctly thickened upwards in fruit. Capitula radiate, 2–3 cm in diameter; involucre 1–1.5 cm in diameter; phyllaries ciliate, pale green, all with white membranous apices, outer linear-lanceolate, inner oblong, *c.* 8–10 mm long; receptacle conical, chaffy all over; paleae oblong-lanceolate, 3–4 mm long, keeled, acuminate, persistent. Ray florets fertile, ray limbs oblong-elliptic, 1.2–1.4 × 4–6 mm, white, tubes compressed, glabrous. Disc florets *c.* 3–4 mm long, corollas yellow, glabrous. Achenes obpyramidate, 1.5–2 mm long, neither tuberculate nor compressed, almost ribless, outer acutely triangular, inner ± terete, beset with 3–4-lobed unilateral auricle.

HAB. Sandy or sandy-clay flood plains, wheat fields, dry steppe, stony ground; alt. 200–550; fl. & fr. Mar.-Apr.

DISTRIB. **DSD**: ± 15 km SE of Ashuriya (± 50 km WNWof Shabicha), 360 m, 1957, *Guest, Al-Rawi & Rechinger* 19370; near wadi Al-Tib Police Post nr. Kuwait (± 70 km N ofAmara), 1957, *Guest, Al-Rawi & Rechinger* 17477; Faggarat al-Shadaf, 1961, *Al-Rawi* 31151! **DLJ**: Near Tikrit, 1969, *Omar & Hamid* 36296! **DWD**: Hazimae (± 40 km SE of Rutba), 550 m, 1966, *Omar* 34215! fki: Dibs, Beiji, 200 m, 1947, *Al-Rawi & Gillett* 7173!

Subspecies *philistea* (Boiss.) Eig (based on *A. philistea* Boiss. (1875), is treated under synonymy for *A. indurata* Delile (1814) in Boulos, Fl. Egypt 3: 242 (2002). The explanation given by Boulos & Hind in Fl. Egypt suggest that *A. leucanthemifolia* is *Anthemis indurata* as treated by them. Ghafoor & Ali treat it as *Anthemis leucanthemifolia.* Lo Presti et al. (2010) suggested that *A. leucanthemifolia* and *A. indurata* were still separate species, but apparently have not sampled any material of the infraspecific taxa. We are keeping this taxon as *Anthemis leucanthemifolia* until all material for Iraq is examined critically.

Palestine, Lebanon.

2. **Anthemis scariosa** *Banks & Sol.*, Nat. Hist. Aleppo, ed. 2 [A. Russell], 2: 263 (1794); De Candolle, Prodr. 6: 4 (1838); Boissier, Fl. Orient. 3: 311 (1875); Eig in Palest. J. Bot., Jerusalem Ser. 1: 187 (1938); Rawi, in Dep. Agr. Tech. Bull. 14: 106 (1964); Grierson & Yavin in Fl. Turkey 5: 208.f. 10.19 (1975); Feinbrun-Dothan, Fl. Palaest. 3: 329. Pl. 550 (1978); Iranshahr in Fl. Iranica 158: 29 (1986); Taifour & El-Oqlah, Pl. Jordan Ann. Checkl.: 35 (2017).

Annual, erect or ascending, up to 50 cm tall, glabrous to sparsely hairy, showy herb. Leaves sessile, elliptic-oblong, 3–5 cm long, 1.5–2.5 cm wide, bipinnatisect into linear-lanceolate, 4–6 mm long, acute-mucronate ultimate segments. Peduncles slender, up to 20 cm long, unthickened in fruit. Capitula radiate, 2.5–4.5 (–5) cm in diameter. Involucre 10–20 mm in diameter; phyllaries glossy, glabrous, pale brown, scarious, outer ovate, inner obovate, basal part of margins membranous, 3–4 × 2–3 mm, bearing apically a large, ± orbicular, 5 mm long and broad, translucent appendage mostly longer than basal part; receptacle hemispherical, 4–5 × 4–5 mm, chaffy all over; paleae linear-lanceolate to oblanceolate, somewhat keeled, 3–4 mm long, with narrow, pellucid margins, acuminate, caducous. Ray florets female, fertile, limbs oblong, 8–20 × 4–8 mm, persistent, yellow, tubes pilose, swollen. Disc corolla yellow, about 3.5 mm long, pilose, inflated below in fruit, deciduous. Achenes subcylindrical, ± angular, distinctly tuberculate, inner somewhat oblique, bald, outermost tipped with persistent ray corollas.

HAB. Silty depressions in steppe and fields in the upper plains and foothills; alt. 250–350 m; fl. & fr. Mar.-May.
DISTRIB. Mesopotamia, Biredjik, in campis ad kefre, 1888, *Sintenis* 347 (F!, G!, K!). **MAM**: in deserte Chabur (Khabur), 1867, *Haussknecht* 529 (JE!). **FUJ**: Albaiyder, 78 km S of Sinjar, 260 m, 1962, *Chakravarty, Al-Rawi & Alizzi* 32079! **FKI**: Kirkuk, ± 350 m, 1929, *Rogers* 203 (G!), *Rogers* s. n. (K!); Col. Chesney's Expedition to Euphrates, 140!

Iran, Lebanon, Jordan, Syria, Turkey.

3. **Anthemis hyalina** *DC.*, Prodr. 6: 4 (1838); Boissier, Fl. Orient. 3: 307 (1875); Eig in Palest. J. Bot., Jerusalem Ser. 1: 161 (1938); Rawi, in Dep. Agr. Tech. Bull. 14: 105 (1964); Grierson & Yavin, in Davis, Fl. Turkey 5: 199 (1975); Feinbrun-Dothan, Fl. Palaest. 3: 329. Pl. 549 (1978); Mouterde, Nouv. Fl. Lib. Syr. 3: 410. PI. CCXXV, n. 4 (1983); Iranshahr in Fl. Iranica 158: 29 (1986); Ghafoor & Al-Turki in Candollea 52: 467. F. 4, L-0 (1997); Taifour & El-Oqlah, Pl. Jordan Ann. Checkl.: 35 (2017).

A. crassipes Boiss., Diagn. Pl. Orient. ser. 1, 11: 9 (1849).

Annual, up to 35 cm tall, densely to sparsely greyish hairy herb with suberect, divaricate branches from near base. Leaves short-petioled, ovate, 2–3 cm long, 2–3-pinnatisect into ovate-oblong or oblanceolate, acute lobules. Capitula 2.5–3.5 cm in diameter, on thickened, arcuate fruiting pedicels. Phyllaries 5–7 mm long, apices all obtuse, inner with broad, margins hyaline, brown, shiny obtuse appendage, nearly as long as or longer than lower hairy part; receptacle broadly conical, 5–7 mm long; paleae lanceolate to oblanceolate, 3–3.5 mm long, ± keeled, abruptly and short acuminate, persistent. Ray florets female, 14–16, fertile, limbs 7–12 × 3–5 mm, white, tube glabrous. Disc corollas terete in fruit. Achenes obpyramidate, 2–2.5 mm long, smooth, obscurely ribbed to ribless, auricle unilateral, short, opaque.

HAB. alt. 350–400 m; fl. & fr. Mar.-Apr.
DISTRIB. **FUJ**: Jazireh Derek, 1933, *Edmonds* Rustam No. 2800 (K!); Balad Singar-Tal Afor, 350 m, 1933, *Evan Guest* 4139A!; midway between Tal Afar & Sinjar, 1969, *S. Omar & H. Hamid* 36465!; Ain Ghazal (Mosul province), 400 m, *Guest* 4096!; 5 km S of Mosul, Mosul Liwa, *Fred A. Barkley* 331! **DWD**: Wadi Massad Al-Rutba, 22 km SE of Rutba, Ramadi Liwa, *Fred A. Barkley & Hikmat Abbas* 991!; Between Baghdad & Syrian desert, 1930, *F. R. Jackson* A-240/1930!; Mesopotamia, 1888, *R Sintenis* 577 (G-Boiss); Mesopotamia, 60 miles W of Euphrates, 1955, *H. Helbaek* 290!

Iran, Turkey, Lebanon, Jordan, Palestine, Syria, Saudi Arabia.

4. **Anthemis haussknechtii** *Boiss. & Reut.* in Boissier, Fl. Orient. 3: 310 (1875); Eig in Palest. J. Bot. Jerusalem Ser. 1: 170 (1938); Grierson & Yavin in Fl. Turkey 5: 204 (1975); Feinbrun-Dothan, Fl. Palaest. 3: 330. Pl. 551 (1978); Mouterde, op. cit. 410. Pl. CCXXV, n. 1 & 2 (1983); Ghafoor & Al-Turki in Candollea 52: 461. f.1, E-L (1997); Iranshahr in Fl. Iranica 158: 27 (1986); Taifour & El-Oqlah, Pl. Jordan Ann. Checkl.: 35 (2017).

Anthemis damascena Boiss. & Gaill., Fl. Orient. 3: 315 (1875).
A. grossheimii Sosn., Vestn. Tiflissk. Bot. Sada 1926-27, n.s. pt. 3–4: 176 (1927).
A. karabaghensis Mikheev, Bot. Zhurn. (Moscow & Leningrad) 70(9): 1258 (1985).

Annual, silvery to greyish, up to 25 cm tall herb with divaricate, erect or ascending branches from near the base. Leaves short-petioled, ovate, 1.5–4.5 × 1–2.5 cm, 2–3-pinnatisect into linear-oblanceolate, 1.5–3 × *c.* 0.5 mm, acute lobules. Pedicels 4–5 cm long, slightly thickened in fruit. Capitula 1.5–2.5 cm in diameter; phyllaries with pale-brown scarious margins, hairy, outer triangular-lanceolate, 4–5 mm, apex acute, inner oblong-oblanceolate, 6–7 × 1.5–2 mm, apex obtuse; receptacle conical, 6–8 mm long; paleae oblanceolate, *c.* 4 mm long, ± keeled, eroded-fimbriate to sometimes lacerate, abruptly mucronate. Ray florets female, fertile, 15–20, limbs elliptic-oblong, 5–8 × 3.5–5 mm, deciduous, corolla tube indurate, glabrous, persistent. Disc corollas glabrous, globose inflated in lower part. Achenes cylindrical, 1.75–2 mm, white, finely striate, tuberculate, straight or outer ones curved and bald, inner achenes crowned by a unilateral, erose-dentate auricle as long as or nearly half the size of achenes.

HAB. Gravel soil, mountainside steppe on conglomerate hill, fallow fields, fine gravelly hills with grey soil, decomposed limestone hills, red-clay or stony-clay hillsides, red-clay hillside, open steppe; alt. 400–1000 m; fl. & fr. March–April.
DISTRIB. **MRO**: Gali Ali Beg, *F. Karim, H. Hamid & M. Jasmin* 40892!; Sultan Dagh (mountain), N of Koi Sanjaq, *Al-Rawi & Noori* 28200! **MSU**: 1 km S of Durbandikan, Sulaimaniya Liwa, *Fred A. Barkley & Ihsan Ali Shehbaz* 7717!; Sulaimaniya fotthills, *D C. Paterson* 820! **MJS**: Jebel Sindjar, 600 m, 1933, *Eig & Zohary* s. n. (HUJ!). **FUJ**: NW Iraq, 2 km N of Abu Wajnam, *Eig & Zohary* s. n. (HUJ!); in deserto inter Chabur (Khabur) & Sindshar (Sinjar), *Haussknecht* 528 (G!; JE!); Tel Kotchek, 400 m, steppe, frequent, 1948, *Gillett* 10836!; Ain Al Hassan (near Sinjar), *Guest* 4237! **FNI**: N Iraq, 35 km S of Khanaq (N of Mosul), ± 450 m, *Eig & Zohary* s.n. (HUJ!). **FKI**: 31 km N of Kirkuk to Kori Sanjaq, near Schwan village, *Rawi, Noori & Al Kass* 27962!; Acra, Qaimqam of Acra, *Rustam* No. 3093A! **FPF**: Kani Kirmaj, *Moore* 317! Jabal Hamrin S of Al-Saadiya, *Al-Kaisi* 50527 (BAG!). **DWD**: 20 km WNW of Rutba, *Al-Rawi* 21131!

Iran, Turkey, Lebanon, Jordan, Palestine, Syria, Saudi Arabia.

5. **Anthemis maris-mortui** *Eig*, Palest. J. Bot., Jerusalem Ser. 1: 181 (1938); Feinbrun-Dothan, Fl. Palaest. 3: 331. Pl. 554 (1978); Taifour & El-Oqlah, Pl. Jordan Ann. Checkl.: 35 (2017).

Anthemis kurdica Iranshahr, Pl. Syst. Evol. 139(1–2): 160 (1981); Iranshahr in Fl. Iranica 158: 36 (1986).

Glabrescent, branched, annual herb, up to 25 cm tall, branches ascending from near base. Leaves ovate-oblong, bipinnatisect or nearly so into remote, linear, acutemucronate ultimate segments. Pedicels slender, 4–10 cm long, unthickened in fruit. Capitula radiate or discoid, 1–2.5 cm in diameter. Involucre 6–8 mm in diameter; phyllaries lanceolate, outer 3–4 mm long, membranous margined, inner *c.* 5 mm long, apices broadly hyaline, nearly glabrous; receptacle conical, 4–6 mm long, chaffy all over; paleae narrowly oblong-spatulate, 3–3.5 mm long, membranous, ± keeled, acuminate-mucronate, deciduous. Ray florets fertile, limbs oblong, 7–8 × 4–5 mm, white, tubes compressed, pilose. Disc florets *c.* 3 mm long, yellow, tube pilose, inflated below and subcompressed above, lobes somewhat patent, terminated by a mucro halfto twice size of lobes. Achenes ± obconical, 1.75–2.25 mm long, outer somewhat curved, indistinctly striate, apices shortly auriculate or with an inconspicuous unilateral rim.

HAB. Gravelly hillside, calcareous rocks; alt. 150–1200 m; fl. & fr. Apr.-May.
DISTRIB. **MAM**: Zawita mountain, between Rania & Shaqlaua, 1973, *F. Karim, H. Hamid &. M. Jasmin* 40853! **FUJ**: Jabal Khantur, in saxosis cal*c* N Zakho prope Sharanish, 1200 m, *Rechinger* 12096 (W! holo.). **FPF**: Bus Kaya (Bagsaya), *Rawi & Sh. Haddad* 25591! **LCA**: Near Baghdad, *Yusuf Lazar* 239 (F!).

Jordan, Palestine, (Syria?).

6. **Anthemis melanacme** *Boiss. & Hausskn.* in Boissier, Fl. Orient. 3: 315 (1875); Grierson & Yavin in Fl. Turkey 5: 207 (1975).

A low growing, 10–15 cm tall, greyish-pubescent, annual herb with decumbent branches from near base. Leaves narrowly oblanceolate, *c.* 2 cm long, bipinnatisect into oblong-oblanceolate, *c.* 2 mm long, acute ultimate segments. Pedicels small, apparently remaining unthickened in fruit. Capitula radiate, ± 2 cm in diameter; involucre 10 mm in diameter; phyllaries all ± triangular-lanceolate, densely hirsute outside, margins dark brown, narrow, apex acute; receptacle convex, chaffy all over; paleae lanceolate, keeled, with broad, dark brown margins, apices acuminate. Ray florets fertile, limbs oblong-spatulate, 7.5–10 mm long, white, corolla tube basally pilose. Disc floret corollas yellow, 3 mm long, tube pilose, seemingly not inflated below. Achenes immature, ± 1 mm long, apparently short-auriculate.

HAB. Fallow fields and limestone hillsides; fl. & fr. May–June.
DISTRIB. Mesopotamia, 1920, *R. J. D. Graham* s. n.!

Syria (?), Turkey.

7. **Anthemis microlepis** *Eig*, Palest. J. Bot., Jerusalem Ser. 1: 185 (1938). – Type. Iraq, 7 km W of Tel-Afar, ± 400 m, 1933, *Eig & Zohary* s.n. (W, HUG, syn.); Rechinger, Fl. Lowland Iraq: 620 (1964); Rawi, in Dep. Agr. Tech. Bull. 14: 105 (1964);

Anthemis singarensis Eig, Palest. J. Bot., Jerusalem Ser. 1: 186 (1938); Rawi, in Dep. Agr. Tech. Bull. 14: 106 (1964).

Annual, 7–20 cm tall, appressed hairy herb with ascending or erect branches. Leaves petiolate, ± patulous hairy to glabrous, ovate to oblong-ovate, (1–) 1.5–2 cm long, 8–10 mm wide, bipinnatisect into narrowly linear, ± remote, mucronulate segments. Pedicels 3–5 cm long, ± thickened in fruit. Capitula radiate, (1.5–) 2.5–3 cm in diameter; involucre 6–9 mm in diameter; phyllaries linear-lanceolate, 3.5–4 × *c.* 1 mm, greenish, hairy outside, margins narrowly scarious; receptacle conical, 5–6 × 2.5 mm, chaffy all over; paleae narrowly oblong-spatulate, 2.5 mm long, slightly keeled, apices 3-dentate and mucronate, not persistent. Ray florets female, fertile, 8–10, tubes glabrous; limbs obovate-oblong, 12–15 (–20) × 4–5 mm, white. Disc florets *c.* 3 mm long, corolla tubes basally pilose and ± inflated in fruit. Achenes obconical, somewhat curved, nearly as long as paleae, whitish, striate, obscurely tuberculate, auricles unilateral, short, denticulate.

HAB. Fallow fields, grey stoneless soil, hill with projecting rocks, sandy soil in low bushy steppe; alt. 400–500 m; Apr.May.
DISTRIB. Endemic to Iraq. **FUJ**: 7 km W of Tel-Afar, *c.* 400 m, 1933, *Eig & Zohary* s.n. (type); 53 km W of Balad Sindjar, 460 m, 1933, *Eig & Zohary* s. n. (HUJ); Jazira, N of Sinjar, field, ± 400 m, 1954, *Evan Guest* 13344!; Jazira (Nr. Tslafuq), ± 350 m, 1954, *E. Guest* 13414! **FKI**: 9 km N of Taug (between Bakuba & Kirkuk), 1933, *Eig & Zohary* s.n. (HUJ), syntype of *A. microlepis* var. *leiantha* Eig, in sched.); 14 km NW of Kirkuk, 1933, *Eig & Zohary* s.n. (HUJ).

Endemic.

8. **Anthemis micrantha** *Boiss. & Hausskn.* in Boissier, Fl. Orient. 3: 300 (1875). – Type. Pira Magrun (Pir Omar Gudrun), 1867, *Haussknecht* 531 (G!). Rawi, in Dep. Agr. Tech. Bull. 14: 105 (1964); Iranshahr in Fl. Iranica 158: 31, tab. 224, fig. 2 (1986).

Annual herb, appressed greyish hairy, low growing, basally branched, stems erect, filiform. Leaves sessile, oblong to lanceolate, 10–15 mm long, 5–10 mm broad, lower pinnatipartite into short, linear-lanceolate, acute, remote ultimate segments, uppermost simple. Pedicels filiform, 1.5–2 cm, thickened in fruit. Capitula radiate, 6–8 mm in diameter, few-flowered; involucre 5 mm wide; phyllaries acute, outer triangular-ovate, 2.5–3.5 × 1.5–2 mm, inner lanceolate, margins whitish scarious; receptacle convex, 3–4 mm long, chaffy all over; paleae lanceolate, 2.5–3 mm long, acuminate. Ray florets female, fertile, ± 6, limbs ovate, 3 × 2 mm, white, tubes compressed, glabrous. Disc florets ± 2.5 mm long, corollas yellow, glabrous. Achenes (juvenile) smooth, unilaterally narrowly marginate.

HAB. High mountain peaks; alt. above 2500 m; fl. & fr. May-Jun.
DISTRIB. Endemic to Iraq. **MSU**. Pir Omar Gudrun, 35°46’N, 45°14’E, 1867, *Haussknecht* 531 (G!).

Endemic.

9. **Anthemis leptophylla** *Eig*, Palest. J. Bot., Jerusalem Ser. 1: 162 (1938); Rechinger f, Fl. Lowland Iraq: 616 (1964); Rawi, in Dep. Agr. Tech. Bull. 14: 105 (1964); Iranshahr in Fl. Iranica 158: 29, tab. 21 (1986); Ghafoor & Al-Turki in Chaudhary, Fl. Kingd. Saudi Arabia 2(3): 174 (2000).

Annual herb, erect, up to 40 cm tall, appressed hairy, unbranched or basally few-branched. Lower leaves with pectinate-dentate short petiole, median and upper sessile, broadly lanceolate-oblong to oblong-ovate, 1–4 cm long, 0.8–1.8 (–2) cm wide, finely bi- or ± tripinnatisect, lobes remote, ultimate segments linear, shortly acuminate. Pedicels 5–8 cm long, gradually thickened upwards in fruit. Capitula radiate, 2.5–4.5 cm in diameter; involucre 10–12 mm in diameter; phyllaries smooth or nearly so, outermost narrowly triangular, ± 3 × 2 mm, hyaline margined, inner oblong-spatulate, 5–6 × 3–4 mm, margins broadly hyaline and ± lacerate, apex obtuse; receptacle hemispherical, 3–4 × 4–5 mm, chaffy all over; paleae oblong-spatulate, 3–3.5 × 1.5–2 mm, almost flat, membranous with a whitish vein running through and forming a short mucro at ± erose obtuse-truncate apices. Ray florets female, 15–20, tubes glabrous, base ± winged; limbs oblong-obovate to oblanceolate, 10–16 × 5–8 mm, white. Disc corollas glabrous, basally not swollen in fruit. Achenes oblong-cylindrical, 1.5–2 mm long, 10-ribbed, smooth, straw-coloured, apices unilaterally short marginate.

HAB. Sandy clay soil, gravelly silty plain, disturbed ground, sandy washes on stony plain; alt. 200–370 m; fl. & fr. Apr.-May.

DISTRIB. **FUJ**: Qaiyarah, Mosul, 215 m, *Baylin* 93a! **DLJ**: 20 km E of Ana to Al Qaim, 1976, *S. Omar, Al0Kaisi, K. Hamad & H. Hamed* 44422!; Between Ana & Al Qaim, 1972, *H. Hamid* 39178!; Rawa, 34°28'N, 41°51'E, 1972, *H. Hamid* 39120!; 50 km from Haditha to Boji, 1974, *S. Omar, M. Noori, H. Hamid, KH. Muhammad & M. Jasim* 41478! **DWD/DLJ:** Nr. Haditha, 1972, *H. Hamid* 39149! **DSD**: 50 km N by W of Aidaha (Al-Aida: ± 80 km SSW of Sahnan, ± 370 m, 1957, *Guest, Al-Rawi & Rechinger* 19192!

Iran, Saudi Arabia.

10. **Anthemis deserti** *Boiss*., Fl. Orient. 3: 305 (1875); Dinsmore in Post, Fl. Syria, Palest. & Sinai ed. 2, 2: 53 (1933); Rechinger, Fl. Lowland Iraq: 618 (1964); Rawi, in Dep. Agr. Tech. Bull. 14: 105 (1964); Al-Rawi, Fl. Kuwait 2: 240 (1987); Ghafoor & Al-Turki in Candollea 52: 467, f. 4, 1-K (1997); Collenette, Wild Flow. Saudi Arabia: 168 (1999); Ghafoor in Fl. Kingd. Saudi Arabia 2(3): 172, Pl. Comp. 52, 1-K (2000).

Anthemis melampodina Delile subsp. *deserti* (Boiss.) Eig, Palest. J. Bot., Jerusalem Ser. 1: 177 (1938); Boulos & Hind in Fl. Egypt 3: 245 (2002).

Annual, greyish-tomentose herb with ascending branches from near the base. Leaves petiolate, petiole 3–6 (–8) mm long, lamina oblong or linear-oblong, 2–3 × 1–1.5 cm, pinnatipartite into linear or oblong, entire or 3 -fid ultimate segments. Pedicels slender, unthickened in fruit. Capitula radiate, 2–2.5 (–3) cm in diameter; involucre 8–10 mm in diameter; phyllaries lanceolate, margins narrowly scarious, apex acute; receptacle hemispherical, 2.5–3 mm long, chaffy all over; paleae oblong-lanceolate, 3 mm long, carinate, acute. Ray florets female, 10–15, fertile, limbs 4–6 (–7) × 2.5–3.5 mm, white, tubes glabrous. Disc corollas glabrous, about 3 mm long and inflated at base in fruit. Achenes ± obpyramidate, slightly curved or not, 1.5–1.75 mm long, longitudinally tubercled on ridges, exauriculate.

HAB. Sandy gravelly soil, depauperate 'haswa' plains and subdesert, 'haswa' ridge with large pebbles, plains with low flat dunes and sand ridges, sandy open wadis; alt. 10–410 m; fl. & fr. Mar.-May.

DISTRIB. **DSD**: Faihat Urn al Minaikhar Nr. Sabnan, 1973, *F. Karm, M. Noori, H. Hamid & H. Khadim* 40216!; Al-Maaniya, ± 95 km W of Shabicha, 1961, *Khaitb & Hazm*, 30561!; Shabicha al-Salman, 1953, *Rami & A. Al-Jab* 21430!; Al-Samah, 29°45'N, 45°50'E, 1968, *Alizzi & S. Omer* 35748!; Khadar al-mai, 29°45'N, 46°50'E, ± 190 m, 1960, *Al-Rawi, M. Al-Khatib & H. Tikrity* 29109!; 10 km SE of Jiraibiyat, ± 125 km S of Basra, ± 175 m, 1957, *Guest, Al-Rawi & Rechinger* 17155!; Khadar Almai, 180 m, 1963, *Khatib & Alizzii* 32686 & 32728!; ± 30 km S by E of Rumailla station (*c.* 55 km W by S of Basra), 1957, *Guest, Al-Rawi & Rechinger* 17332!; 20 km NW of Sahnan, 1972, *Fauzi, Hazim & H. Hamid* 38697!; 23 km SE by S of Zubair, ± 10 m, 1957, *Guest, Al-Rawi & Rechinger* 16853!; Near Al-Baniya, 30°12'N, 47°15'E, ± 65 km WSW of Basra, 70 m. *Guest, Al-Rawi & Rechinger* 17311!; Southern desert, Al-Ishrishi (*c.* 35 km E by N Bwaiya), 115 m, 1955, *Guest & Al-Rawi* 14187!; Southern Desert, 60 km SW by W of Zubair, ± 70 m, 1955, *Guest & Al-Rawi* 14307!; Nr. Al-Aidha (Al-Aida) (± 110 km SW of Salman, 35 km NW of Jumaima), 1957, *Guest, Al-Rawi & Rechinger* 19130!; Southern desert, ± 80 km ESE Buraiqa, *c.* 140 m, 1955, *Guest & Al-Rawi* s.n.!; Near Safai al-Maghif (NE of Ghazlani), ± 100 km WSW of Basra, ± 130 m, *Guest, Al-Rawi & Rechinger*, 17259! **LEA**: Rustam Farm, near Baghdad, 1933, *Yusuf Lazar* 36 (F!, K!). **LCA**: 16 km W of

Falluja, 1961, *Chakravarty & Al-Rawi* 30282!; **LBA**: Al-Batin Nr. (WNW) Jarishan, ± 85 km SW of Basra, 80 m, *Guest, Al-Rawi & Rechinger* 17092!;

E Egypt (rare), Jordan, Lebanon, Palestine, Saudi Arabia, Syria(?).

QAHWIYAN, REBYAN.

11. **Anthemis gillettii** *Iranshahr*, Pl. Syst. Evol. 139(1–2): 159 (1982); Iranshahr in Fl. Iranica 158(4): 33, t. 32 (1986).

Annual, up to 1 cm tall, hairy, simple or branched herb. Leaves petiolate, ovate, 1.5–2.5 cm long, 5–10 mm wide, bipinnatisect into linear, acute segments. Pedicels short, 2–4 cm long, thickened upwards in fruit. Capitula radiate, 1 .5–2 cm in diameter; involucre 8–10 mm in diameter; phyllaries hairy outside, outer oblong, *c.* 3 mm long, acute, inner obovate, 5–5.5 × ± 3 mm, margins broadly hyaline, apex obtuse; receptacle ovoid-hemispherical, ± 5 mm long, chaffy all over; paleae oblong, 3–3.5 mm long, carinate, apically short mucronate. Ray florets female, fertile, 8–10, limbs oblong, white tubes glabrous. Disc florets yellow, as long as paleae, not inflated in fruit. Achenes cylindrical, 1.5–1.8 mm long, faintly striate, smooth, almost bald or inconspicuously denticulate at apices.

HAB. Dry to moist steppe on gypsiferous soils and eroded slopes; 200–400 m; fl. & fr. Mar.-May.
DISTRIB. **FPF**: W of Khanaqin, 1929, *Cowan & Darlington* 164! **FKI**: 12 km E of Kirkuk, 1948, *Al-Rawi & Gillett* 10579! **LEA**: Badra, 1947, *Gillett* 6657!; Mountains on Persian border, E of Badra, 1947, *Gillett* 6706 (K! iso).

Iran.

12. **Anthemis hamrinensis** *Iranshahr*, Pl. Syst. Evol. 139(1–2): 161 (1982). –Type. Baghdad, inter Tigris flumen et Monte Jabal Hamrin, 24 km ab oppido Samarra Orientem versus, ca. 34°15'N, 44°E, 1957, *Rechinger* 9580 (W!, holo.); Iranshahr in Fl. Iranica 158: 43, t. 45 (1986).

Annual, 10–20 cm tall, whitish-tomentose, much branched herb with patent to suberect branches. Leaves sessile, ovate-oblong, 1–2 cm long, 5–10 mm wide, bipinnatisect into linear-oblanceolate ultimate segments. Pedicels slender, 4–7 cm long, unthickened in fruit. Capitula radiate, 1.5–2 cm in diameter; involucre 7–10 mm in diameter; phyllaries dimorphic, outer ovate-oblong, 1.5–2 mm long, ± obtuse, inner narrowly obovate, 3–4 × 1.5–2 mm, margins scarious, apex obtuse; receptacle conical, chaffy all over; paleae oblong, flat, *c.* 2.5 mm long, mucronate. Ray florets female, fertile, 12–15, tubes glabrous; limbs oblong, *c.* 6 mm long, white. Disc corollas yellow above, purplish below, *c.* 2–2.5 mm long, ± compressed. Achenes cylindrical, *c.* 1.5 mm long, striate, smooth, very short-auriculed.

HAB. Gravel mixed sandy soil; alt. 50–400 m; fl. & fr. May–Jun.
DISTRIB. Endemic to Iraq. **DGA**: Baghdad, inter Tigris flumen et Monte Jabal Hamrin, 24 km ab oppido Samarra Orientem versus, ca. 34°15'N, 44°E, 1957, *Rechinger* 9580 (W!); 26 km E Samarra, ± 60 m, gravel soil, 1957, *Al-Rawi* 20376!

Endemic to Iraq.

13. **Anthemis wettsteiniana** *Hand.-Mazz.* in Ann. Nat. Hofsmus. Wien 27: 433, tab. 16, fig. 7 (1913). Rechinger, Fl. Lowland Iraq: 617 (1964); Rawi, in Dep. Agr. Tech. Bull. 14: 106 (1964); Mouterde, Nouv. Fl. Lib. Syr. 3: 409. Pl. CCXXIV, n. 3 (1983); Iranshahr in Fl. Iranica 158: 30, t. 27 (1986); Haloob & Al-Kaisi, Illustr. Fl. Lowland Iraq 1(1): 29 (2016); Abdulridha, Taha & Wedad, Ecol. & Fl. Basra: 193 (2016).

Anthemis deltawensis Eig, Palest. J. Bot., Jerusalem Ser. 1: 172 (1938); Rawi, in Dep. Agr. Tech. Bull. 14: 105 (1964).
A. desertisyriaci Eig, Palest. J. Bot., Jerusalem Ser. 1: 174 (1938).

Annual, greyish-green, appressed-lanate, up to 25 cm tall herb with ascending branches from the base. Leaves petiolate, broadly ovate to ovate-oblong, 2.5–4.5 × 1.5–2.5 cm, bipinnatisect into linear-lanceolate, remote, obtuse, mucronate ultimate segments. Pedicels (3–) 4–6 cm long, gradually thickened upwards in fruit. Capitula radiate, (1.5–) 2–2.5 (–3) cm in diameter; involucre 8–10 mm in diameter; phyllaries densely hairy outside, dimorphic, margins white membranous, outer narrowly triangular-lanceolate to lanceolate, 4–5 × 1–1.5 mm, inner oblong-lanceolate, 6–7 × 1.5–2 mm, apices white, membranous; receptacle ±

convex-conical, chaffy all over; paleae spatulate or oblong-spatulate, 2.5–3 mm long, slightly carinate, membranous, shortly mucronate, deciduous. Ray florets female, fertile, 12–15, limbs oblong, 7–12 × 4–5 mm, white, tubes glabrous, unindurated. Disc florets pale-yellow, as long as or shorter than paleae, glabrous. Achenes ± curved, obconical-turbinate, ± 2 mm long, tuberculate; inner with 0.5–1 mm long auricle, outer nearly smooth, exauriculate.

HAB. Foothill plains, sandy or clay soil or stony desert, cultivated fields; alt. 200–550 m; fl. & fr. Apr.-May.
DISTRIB. **DLJ**: Jezira, 20 km W of Sumaicha-Belad line, Baghdad Liwa, 1960, *Agnew* 272 (E!); Falluja desert, by cement works, 1960, *Agnew* 345 (E!). **DGA**: N Iraq, foothills of Jebel Hamrin, 1933, *Eig & Zohary* s. n. (HUJ!); Sammara, Baghdad Liwa, 1961, *Agnew* 5240 (E!). **DWD**: 60 km from Qaim, 1974, *Botany Staff* 41713!; Rutba, 1955, *Al-Rawi* 640!; E of Shitatha, 200 m, 1947, *Al-Rawi & Gillett* 6432!; 30 km S of Shanana, 200 m, 1955, *Al-Rawi* 15603!; nr. Rutba wells, 1933, *H. R. R Dickson* iv! **LCA**: 3 km N of Garma, bridge on Falluja-Baghdad canal, 1960, *Agnew* 278 (E!); Nh-aq, 35 km NE of Deltawah (N of Baghdad), 1933, *Eig & Zohary* s. n. (HUJ!).

A. wettsteiniana, and first of the synonyms, *A. desertisyriaci*, were kept separate under the analyses of Lo Presti et al. 2010: A molecular phylogeny and a revised classification of the Mediterranean genus Anthemis s.l. (Compositae, Anthemideae) based on three molecular markers and micromorphological characters. Taxon, 59(5): 1441–1456.

Iran, Syria.

14. **Anthemis susiana** *Nábelĕk*, Spisy Prír. Fac. Masarykovy Univ. [= Publ. Fac. Sci. Univ. Masaryk, Brno.] 52: 161 (1925); Iranshahr in Fl. Iranica 158: 34, t. 31 (1986).

Annual, 5–15 cm tall, appressed whitish-pubescent herb with ascending branches from near the base. Leaves petiolate, oblong-lanceolate, 15–25 × 5–10 mm, bipinnatisect into remote, linear, callose mucronulate segments. Pedicels up to 5 cm long, upwardly thickened in fruit. Capitula radiate, 1.8–2 cm in diameter; involucre 6–9 mm in diameter; phyllaries dimorphic, outer triangular-ovate, 2.5–3 mm long, inner oblong-spatulate, 4–5 × *c.* 2 mm, margins hyaline, apex obtuse; receptacle conical, ± 5 mm long, chaffy all over; paleae obovate-spatulate, ± 2.5 mm long, membranous, ± carinate, inconspicuously mucronate. Ray florets female, fertile, 8–10, limbs ovate-oblong, 8–10 mm long, white, tube compressed, glabrous. Disc floret corollas yellow, slightly longer than paleae, terete in fruit. Achenes cylindrical, oblong-turbinate, 1 .5–2 mm long, striate, smooth, unilaterally minutely auriculate.

HAB. Alluvial plain; alt. 10–20 m; fl. & fr. Apr.-May.
DISTRIB. No specimens from Iraq have been seen by the authors and it is included here based on Iranshahr (1986), who cited a specimen (Shatt at-Tib, 60–70 km N of Amarah).

Iran.

15. **Anthemis arvensis** *L.*, Sp. Pl.: 894 (1753); De Candolle, Prodr. 6: 10 (1838); Boissier, Fl. Orient. 3: 301 (1875); Fedorov in Scraschkin & Bobrov, Fl. SSSR 26: 36 (1961); Täckholm, Stud. Fl. Egypt, ed. 2. 574 (1974); Grierson & Yavin Fl. Turkey 5: 198. f. 9 (1975); Fernandes in Fl. Europ. 4: 153 (1976); Ghafoor & Al-Turki in Candollea 52: 465. f. 2, G-K (1997).

Annual or biennial herb, sometimes perennial, up to 50 cm tall, sparsely to densely hairy. Leaves obovate-oblong, 1–3-pinnatisect into linear-oblong or linear-lanceolate to elliptic, acute-mucronate ultimate lobules. Pedicels slender, not or slightly to distinctly thickened in fruit. Capitula 1–4 cm in diameter; involucre 5–10 mm in diameter; phyllaries oblong or oblong-obovate, outer acute, inner subacute to subobtuse, margins and apex hyaline to pale-brown and scarious; receptacle convex, becoming shortly to elongate conical in fruit, 2–10 mm long; paleae lanceolate-oblanceolate, broadening toward ± stiffly acuminate apex, ± persistent. Ray florets fertile, limbs elliptic-oblong, 4–8 × 3–4 mm, white, deciduous, tube glabrous. Disc corolla basally not or slightly inflated, glabrous. Achenes obconical, 1.5–2.5 mm long, smooth, with an undulate rim or lacerate hyaline auricle.

HAB. Clay soil, roadsides in upper plains and foothills; alt. 200–300 m; fl. & fr. Mar.-Apr.
DISTRIB. **FUJ**: Disintegrating marble inside of inner wall of old Hatra, Al-Jezira desert, 23.3.1964, *E. D. Barkley & F. A. Barkley* 6574!; Midway between Tal-Afar & Sinjar, 1969, *S. Omer & H. Hamid* 36478! **FPF**: 3 km N of Saadya, 1975, *Al-Kaisi* 42544! **FKI**: Tur Shermati dist. Kirkuk, 1910, *Nábělek* 3632 (SAV!). **LCA**: Babylon, 32°33'N 44°25'E, *Reuter* 3631 (SAV!).

Cultivated and waste ground, and widely naturalized. Egypt, Iran, most of Europe, Saudi Arabia, Turkey; introduced into the New World, Australia, New Zealand.

16. **Anthemis plebeia** *Boiss. & Noë* in Boissier, Fl. Orient. 3: 313 (1875). –Type. Kurdistan, *Noë* 228 (G-Boiss.!). Iranshahr in Fl. Iranica 158(4): 41, t. 42 (1986).

Annual herb, 5–10 (–15) cm tall, densely appressed greyish-hairy, unbranched. Leaves subsessile, narrowly oblong, 1–1.5 cm long, 5–10 mm wide, bipinnatisect into linear-oblong, 1.5 × 0.5 mm, acute ultimate segments. Pedicels 2.5–3 cm long, unthickened in fruit. Capitula radiate, 1.5–2 cm in diameter; involucre 8–10 mm in diameter; phyllaries ovate-oblong, outer small, inner 3–4 × 1–1.5 mm, all membranous, margins fimbriate; receptacle hemispherical, *c.* 3 × 2 mm, chaffy all over; paleae oblong, 3–3.5 × *c.* 1 mm, transparent, slightly keeled, irregularly fimbriate, shortly mucronate. Ray florets female, fertile, 10–12, limbs oblong-obovate, 6–8 × 3–4 mm, white, tubes compressed, pilose. Disc corollas yellow, glabrous, inflated below in fruit. Achenes obpyramidate, smooth, sulcate-striate, truncate or with a very short unilateral rim.

HAB. Foothills, gravel wadis, fine silt depressions and sandy desert; alt. 100–150 m; fl. & fr. Mar.-Apr.
DISTRIB. Endemic to Iraq. **FPF**: 18 km SE of Baksaya, 100 m, 1958, *Al-Rawi & Sh. Haddad* 25627! **DSD**: 20 km W of Subaicha, 1961, *Haines, E. Hadač, Waleed-El-Hashmi & Agnew* 5549 (E!). **DWD**: North End of Habbaniya Lake, Ramadi Liwa, 1960, *A. D. Q. & S. Agnew* 237 (E!).

17. **Anthemis melampodina** *Delile*, Descr. Egypte Hist. Nat.: 268, Planche 45, fig. 11 (1814). Eig, Palest. J. Bot., Jerusalem Ser. 1: 176 (1938), pro parte; Rechinger, Fl. Lowland Iraq: 618 (1964); Rawi, in Dep. Agr. Tech. Bull. 14: 105 (1964); Feinbrun-Dothan, Fl. Paalest. 3: 332. Pl. 557 (1978), pro parte; Ghafoor & Al-Turki in Candollea 52: 469. Fig. 4, A-D (1997); Boulos & Hind in Fl. Egypt 3: 243 (2002); Collenette, Wild Flow. Saudi Arabia: 169 (1999); Taifour & El-Oqlah, Pl. Jordan Ann. Checkl.: 35 (2017).

Annual herb, 1 5–25 cm tall, densely appressed canescent, branches ascending from near base. Leaves shortly petiolate, oblong to ovate-oblong, 1.5–3 cm long, 1(–2)-pinnatisect into linear-oblong, mucronate lobules. Pedicels gradually slightly thickened in fruit. Capitula 2–3 cm in diameter; involucre 5–8 mm in diameter; phyllaries lanceolate to ovate-oblong, hairy outside, scarious on margins and apex, subobtuse to obtuse; receptacle ovate-conical; paleae oblong-oblanceolate, *c.* 3 mm long, ± keeled, usually mucronate. Ray florets female, 9–15, fertile, limbs elliptic, 8–12 × 4–6 mm, white, deciduous. Disc corollas inflated and dark brown in lower half. Achenes obpyramidate, *c.* 1.5–1.75 mm long, auricle scarious, unilateral, 1/6 to as long or longer than achenes.

HAB. Sub-desert; alt. ±100 m. fl.
DISTRIB. **DWD**: Iraq, 15 km W of Ramadi, ± 100 m, *Al-Rawi & Gillett* 6783!

Egypt, Palestine, Jordan, Saudi Arabia.

A very rare species in Iraq where it is replaced by the more common *A. deserti* which is recognised by the acute outer phyllaries and exauriculate achenes.

18. **Anthemis zoharyana** *Eig*, Palest. J. Bot., Jerusalem Ser. 1: 178 (1938); Täckholm, Stud. Fl. Egypt, ed. 2. 575 (1974); Feinbrun-Dothan, Fl. Palaest. 333. Pl. 558 (1978); Collenette, Wild Flow. Saudi Arabia: 170 (1999); Ghafoor & Al-Turki, Candollea 52: 465, f. 2, L-P (1997); Boulos & Hind in Fl. Egypt 3: 245 (2002); Taifour & El-Oqlah, Pl. Jordan Ann. Checkl.: 36 (2017).

Annual herb, up to 20 cm tall, sparsely tomentose to almost glabrous, greyish-green with rigid, ± upright branches from near base. Leaves short-petiolate, oblong, 3–5 × 1–2 cm, 1–2-pinnatisect into remote, linear, 3–8 × *c.* 1 mm, mucronulate ultimate segments. Pedicels slender, hardly thickened in fruit. Capitula radiate, 2–3 cm across; involucre *c.* 1 cm in diameter; phyllaries sparsely hairy, dimorphic, outer lanceolate-oblanceolate, *c.* 3 × 1.5 mm, narrowly scarious margined, inner ones oblong-spatulate, almost twice as long as outer ones, broadly scarious margined towards obtuse apices; receptacle conical, 4–5 mm long, chaffy all over; paleae narrowly oblong-spatulate, 3–3.5 × 1–1.5 mm, ± carinate, membranous, nerve greenish, mucronate. Ray florets female, fertile, 10–15, limbs elliptic-oblong, 8–12 × 3–5 mm, white tube glabrous, compressed. Disc floret corollas yellow, as long as paleae, tubes glabrous, inflated below in fruit. Achenes obpyramidate, 1.5–2 × *c.* 0.75 mm, slightly compressed and curved, conspicuously tuberculate on dorsal side, ventrally ± ribbed, unilateral auricle deeply lacerate into acute, triangular lobes, as long as to slightly longer or shorter than achenes.

HAB. Depressions with silty soil or near water; alt. 100–550 m; fl. & fr. Apr.-Jun.
DISTRIB. **FKI**: Injana, 34°29'N, 44°38'E, (Jabal Hamrin), 1971, *S. Omar & F. Karim* 37835! **LEA**: Uod al-Sada (2 km SW of Hz.) 550 m, 1962, *Chakravarty, Al-Rawi, Alizzi & Khatib* 31471!

Egypt, Palestine, Saudi Arabia.

19. **Anthemis handel-mazzettii** *Eig*, Palest. J. Bot., Jerusalem Ser. 1: 162 (1938). Rechinger, Fl. Lowland Iraq: 616 (1964); Rawi, in Dep. Agr. Tech. Bull. 14: 105 (1964).

Anthemis melampodina Delile var. *truncata* Eig, Palest. J. Bot., Jerusalem Ser. 1: 176 (1938). – Type. ± 92 km W of Basra, 40 m, 1933, *Eig & Zohary* s.n. (HUJ!, syn.).

Annual herb, appressed pubescent, erect, 10–15 cm tall with ascending branches. Leaves sessile, oblong, 2–3.5 cm long, 1.5–2 cm wide, 1-pinnatisect to ± bipinnatisect into linear-oblanceolate or ± spatulate, entire or shallowly dissected, up to 5 × 2 mm, obtuse, spinulose ultimate segments. Pedicels 5–6 cm long, unthickened in fruit. Capitula radiate, 2.5–3 cm in diameter; involucre 6–8 mm in diameter; phyllaries appressed hairy, obtuse, outer oblong, *c.* 3 × 1.5 mm, margins scarious, inner oblong-spatulate, *c.* 5 × 2 mm, with margins and apices broadly membranous; receptacle hemispherical, *c.* 4–5 mm in diameter, chaffy all over; paleae narrowly ovate, 2.5–3 × *c.* 1 mm, apices acute-acuminate, hyaline. Ray florets 8–12, female, fertile, limbs oblong, 8–10 × 4–5 mm, white, tubes glabrous. Disc florets yellow, *c.* 4 mm long, tube glabrous, basally uninflated in fruit. Achenes obpyramidate, mostly curved, *c.* 1.25 mm long, tuberculate, apices truncate and exauriculate.

HAB. Sandy gravelly soil, stony desert with sand, hard sandy plain; alt. 40–400 m; fl. & fr. Mar.-Apr.
DISTRIB. Endemic to Iraq. **DWD**: 5 km W of Ramadi, on road to Hit, 60 m, 1947, *Al-Rawi & Gillett* 6791! **DSD**: Between Aidaha and Ansab, 1968, *Alizzi & S. Omar* 35655!; Al-Batin, ± 65 km SW of Basra, ± 70 m. *Guest, Al-Rawi & Rechinger* 17086!; Between Kuwaibda and Ar Rumail (40 km SW of Basra), 1933, *Eig & Zohary* s.n. (HUJ!); ± 92 km W of Basra, 40 m, 1933, *Eig & Zohary* s.n. (HUJ!), syntype of *A. melampodina* var. *truncata* Eig); Nr. Jumaima (± 140 km SW of Salman), ± 390 m, 1957, *Guest, Al-Rawi & Rechinger* 19122!; 65 km W of Basra (Petrol Resource), 1975, *Botany Staff* 42403!; Tumela, Toba Railway Station, 20 km W of Ghubaishiya Basra, Basra Liwa, 1965, *Shariff Y Haddad* 9537!; Khaidar el Mai, Southern desert near Kuwait, 1960, *Agnew & Haines* 259 (E!); E of Ghazlani, ± 105 km SW of Basra, ± 120 m, 1957, *Guest, Al-Rawi & Rechinger* 17239!; 65 km NW of Subaicha, *Haines, Waleed Al-Hashmi & Agnew* 5560 (E!).

Endemic.

20. **Anthemis schizostephana** *Boiss. & Hausskn.* in Boissier, Fl. Orient. 3: 314 (1875); Rawi, in Dep. Agr. Tech. Bull. 14: 106 (1964); Iranshahr in Fl. Iranica 158(4): 40, t. 39 (1986).

Annual herb, up to 10 cm tall, low growing, appressed hairy, branches spreading from base. Leaves petiolate, oblong, 15–25 × 8–15 mm, bipinnatisect into linear, 1–3 mm long, mucronulate ultimate segments. Pedicels 1.5–2.5 (–3.5) cm long, slightly thickened in fruit. Capitula radiate, medium, 1.5–2 cm in diameter; involucre 8–10 mm in diameter; phyllaries hairy outside, oblong-ovate, 4–5 × 2–3 mm, green in middle, with scarious whitish margins and apices; receptacle conical, 5–6 mm long, chaffy all over; paleae oblong-lanceolate, *c.* 3.5 mm long, ± keeled, abruptly short mucronate, persistent. Ray florets fertile, 15–20, limbs elliptic, 8–10 × 5–6 mm, white, tubes hairy, inflated, purplish. Disc florets yellow, as long as to slightly longer than paleae, corolla tubes pilose, inflated and purplish below in fruit. Achenes finely cylindrical, 2–2.5 mm long, whitish, sulcate, smooth, with a unilateral, acutely fimbriate, *c.* 1 mm long auricle.

HAB. Hard sandy gravel desert and hard silty plains; alt. 250–350 m. fl. Mar.-Apr.
DISTRIB. **MRO**: Galzarda mountain, 1973, *F. Karim, H. Hamid & M. Jasmin* 40676!; ± 40 km WNW of Shabicha, 1957, *Guest, Al-Rawi & Rechinger* 19297! **DWD**: Lake Tharthar, Ramadi Liwa, 1964, *Hashim & Matti* 384!; 55 km W of Ramadi (Ramadi-Rutba highway), 1957, *Al-Rawi* 20863! **DSD**: Nageib, Northern desert province, 25 m, 1955, *Guest, Al-Rawi & Long* 14038!; ± 95 km SSE of Salman (nr. Bakur well or Jalib Bakur), *c.* 31 m, *Guest, Al-Rawi & Rechinger* 18890!

Iran.

21. **Anthemis homalolepis** *Eig*, Palest. J. Bot., Jerusalem Ser. 1: 175 (1938). – Type. Syrian desert, wadi Muhammadi, 70 km W of Ramadi (190 km W of Baghdad), 220 m, 1933, *Eig & M. Zohary* s.n. (E, HUJ!); Rechinger, Fl. Lowland Iraq: 619 (1964); Rawi, in Dep. Agr. Tech. Bull. 14: 105 (1964).

Erect annual herb, appressed tomentose, 6–25 cm tall with diffuse, divaricate, ascending branches. Leaves sessile, ovate-oblong, 1–2.5 cm long, (5–)8–12 mm wide, finely and deeply bipinnatisect into linear-oblong, 1–1.5 × 0.5 mm, apiculate ultimate segments. Pedicels 6–9 cm long, slender or filiform, remaining unthickened in fruit. Capitula radiate, 1–2.5 cm in diameter; involucre 8–10 mm in diameter; outer phyllaries oblong-lanceolate, 3.5–4.5 × *c.* 2 mm, with broad hyaline margins, inner oblong-spatulate, 4.5–5.5 × *c.* 2.5 mm, margins and apices broadly scarious and erose lacerate, all hairy outside, apices obtuse; receptacle convex-hemispherical, *c.* 5–6 mm across, chaffy all over; paleae flat, oblong-lanceolate, 3–3.5 × *c.* 1.5 mm, abruptly shortly mucronate, persistent. Ray florets 10–12, female, fertile, limbs oblong, 6–10 × *c.* 4.5 mm, white, tubes glabrous. Disc floret corollas yellow, tubes glabrous, slightly longer than paleae. Achenes thin, cylindrical, ± curved, *c.* 2 mm long, sulcate, smooth, unilateral auricle very small.

HAB. Wadi banks, sandy gravelly or loamy clay soils, dry gypsiferous soil with small gravel; alt. 75–440 m; fl. & fr. Feb.-May.
DISTRIB. Endemic to Iraq. **DLJ**: 40 km NW of Falluja, 75 m, 1961, *Al-Rawi* 30191! **DWD:** 140 km W of Karbala, 200 m, 18.4.61, *Chakravarty & Al-Rawi* 30973!; Syrian desert, 20 km W of Ramadi on the way to Baghdad, 170 m, 1933, *Eig & Zohary* s. n. (HUJ!); Syrian desert, wadi Muhammadi, 32°22'N, 42°40'E, 70 km W of Ramadi (190 km W of Baghdad), 220 m, 1933, *Eig & M. Zohary* s. n. (E, HUJ!, type); 180 km W of Ramadi to Rutba, *c.* 450 m, 1962, *Chakravarty, Al-Rawi, Khatib & Alizzi* 31416!; 20 km W of Nukhaib, 1961, *Al-Rawi* 30999!; Swab, N of Rutba, 1967, *Alizzi* 35210!; 70 km W of Nukhaib, 1961, *Al-Rawi* 31019!; ± 160 km E of Rutba, 440 m, *Chakravarty, Al-Rawi, Khatib & Alizzi* 3 1591!; Syrian desert (± 15 km W of Ramadi), ± 40 m, 1955, *Guest & Al-Rawi* 13917!; 80 km NW of Nukhaib, 1954, *Al-Rawi* 23707!; 20 km S of Nukhaib, *Alizzi & S. Omar* 35522!

Endemic.

22. **Anthemis rascheyana** *Boiss.*, Diag. Pl. Orient. ser. 1, 11: 8 (1849); Dinsmore in Post, Fl. Syria, Palest. & Sinai ed. 2, 2: 53 (1933); Eig in Palest. J. Bot., Jerusalem Ser. 1: 166 (1938); Feinbrun-Dothan, Fl. Palaest. 3: 334. Pl. 562 (1978); Mouterde, Nouv. Fl. Lib. Syr. 3: 412. PI. CCXXIV, n. 1 (1986); Ghafoor & Al-Turki in Candollea 52: 469, f. 4, E-H (1997); Chaudhary, Fl. Kingd. Saudi Arabia 2(3): 173. Pl. Comp. 52, E-H (2000); Taifour & El-Oqlah, Pl. Jordan Ann. Checkl.: 36 (2017).

Annual herb, up to 15 cm tall, ash-grey, appressed hairy with upright or ascending branches from near base. Leaves oblong with pectinate petiole, somewhat bipinnatisect into linear-oblong, acute ultimate segments. Pedicels slender, not or only slightly thickened in fruit. Capitula small, radiate, 10–15 mm in diameter; involucre *c.* 6–8 mm in diameter; phyllaries ovate-oblong, almost glabrous, obtuse with broad, scarious, ± lacerate margins; receptacle hemispherical, 3–4 mm in in diameter, chaffy all over; paleae oblong-lanceolate, carinate, persistent, apex obtusely denticulate and mucronate-acuminate. Ray florets female, tube ± compressed, glabrous and basally inflated, limbs, ovate-oblong, white. Disc floret corollas yellow, with glabrous, basally inflated, ± compressed corolla tube. Achenes thin, cylindrical, 1.5–2 mm long, dorsiventrally slightly compressed, striate, smooth, apex truncate to minutely auricled.

HAB. Slight depressions on open stony plain, compact silty soil; alt. up to 450 m; fl. & fr. Apr.-May.
DISTRIB. **DWD**: 7 miles south of T-1 station, Iraq, 1935, *R. I. Attiyah* s. n. (F!); Iraq Pefroleum Co., Station T-4, collector unknown (probably *Yousuf Lazar*) 33 (F!). **DSD**: 8 km N by W of Aidaha (Al-Aida) (± 95 km W of Sahnan), ± 410 m, 1957, *Guest, Al-Rawi & Rechinger* 19153!

Palestine, Saudi Arabia, Syria.

23. **Anthemis tinctoria** *L.*, Sp. Pl.: 896 (1753). Boissier, Fl. Orient. 3: 280 (1875) – pro parte; Fedorov in Schischkin & Bobrov, Fl. SSSR 26: 41 (1961); Rawi, in Dep. Agr. Tech. Bull. 14: 106 (1964); Grierson & Yavin in Fl. Turkey 5: 211, f. 10 (1975); Fernandes in Fl. Eur. 4: 155 (1976); Feinbrun-Dothan, Fl. Palest: 335. Pl. 565 (1978); Mouterde, Nouv. Fl. Lib. Syr. 3: 404. PI. CCXX, n. 1 (1983); Iranshahr in Fl. Iranica 158(4): 18. Tab. 9 (1986); Taifour & El-Oqlah, Pl. Jordan Ann. Checkl.: 36 (2017).

Chamaemelum tinctorium (L.) Schreb., Spic. Fl. Lips.: 145 (1771).
Buphthalmum bipinnatum Gilib., Fl. Lit. Inch. 1: 213 (1782), nom. inval., opera utique supressa
Chamaemelum discoideum All., Fl. Pedem. 1: 188 (1785).
Anthemis discoidea (All.) Willd., Sp. Pl., ed. 4 [Willdenow], 3(3): 2188 (1803).
A. coarctata Sm., Fl. Graec. Prodr. 2(1): 191 (1813).
A. tinctoria L. var. *discoidea* (All.) DC., Prodr. 6: 11 (1838).
A. tinctoria L. var. *pallida* DC., Prodr. 6: 11 (1838).
Cota tinctoria (L.) J. Gay ex Guss., Fl. Sicul. Syn. 2(2): 867 (1845).
Anthemis chrysantha Schur, Enum. Pl. Transsilv. 1: 331 (1866), nom. illeg. non J.Gay (1848/49)
A. brachyglossa C.Koch, Linnaea 24(3): 318 (1851).
A. tinctoria L. var. *fussii* Griseb. & Schenk, Arch. Maturgesch. 18: 338 (1852).
Cota parnassica Boiss. & Heldr., Diagn. Pl. Orient., ser. 2, 3: 20 (1856).
Anthemis euxina Boissier, Fl. Orient. [Boissier] 3: 282 (1875).
A. pallescens Heldr. ex Nyman, Consp. Fl. Eur. (fasc. 2): 359 (1879), nom. nud. et nom. inval.
A. parnassica (Boiss. & Heldr.) Nyman, Consp. Fl. Eur. (fasc. 2): 359 (1879).
A. tinctoria L. ssp. β *discoidea* (All.) Arcang., Compend. Fl. Ital.: 357 (1882).
A. gaudium-solis Velen., Fl. Bulg. Suppl. 1: 152 (1898).
A. kelwayi Hort. ex L. H. Bailey & N. Taylor, Stand. Cycl. Hort. 1: 299 (1914), nom. illeg. sub. syn.
A. tinctoria L. var. *virescens* Bornm., Notizbl. Bot. Gart. Berlin-Dahlem 7: 19 (1917).
Anacyclus tinctorius (L.) Samp., Ann. Acad. Polyt. Porto 14: 162 (1921), in obs.
Anthemis sancti-johannis Stoj., Stef. & Turrill, Gard. Chron., ser. 3, 80: 270 (1926).
A. saguramica Sosn., Vestn. Tiflissk. Bot. Sada n.s. (3–4): 150 (1927).
A. tinctoria L. var. *briquetii* Fiori, Nuov. Fl. Italia 2: 649 (1927); *A. tinctorius* L. var. *orientalis* Eig, Palest. J .Bot., Jerusalem Ser. 1: 211 (1938); *A. tinctorius* L. var. *parviflora* Eig, Palest J. Bot., Jerusalem Ser. 1: 211 (1938).
A. meinkeana Rech. f., Ann. Nat. Hofmus. Wien 57: 90 (1949).
A. maris-nigri Fed., Fl. U.R.S.S. 26: 41, 866 (1961).
A. debilis Fed., Fl. U.R.S.S. 26: 44, 867 (1961).
A. subtinctoria Dobrocz., Ukrayins'k. Bot. Zhurn. 18(2): 69 (1961).
A. zephyrovii Dobrocz., Ukrayins'k. Bot. Zhurn. 18(2): 79, fig. 1 (1961).
A. tinctoria L. subsp. *subtinctoria* (Dobrocz.) Soó, Acta Bot. Acad. Sci. Hung. 12(3–4): 366 (1966); *A. tinctoria* L. subsp. *fussii* (Griseb. & Schenk) Beldie, Fl. Veg. Munt. Bucegi: 270 (1967).
A. tinctoria L. subsp. *subtinctoria* (Dobrocz.) Soó ex Smejkal, Folia Fac. Sci. Natur. Univ. Purkyn. Brun., Biol. 26, 11(3): 114 (1970); *A. tinctoria* L. subsp. *subtinctoria* (Dobrocz.) Soó, Acta Bot. Acad. Sci. Hung. 17(1–2): 124 (1972).
Cota euxina (Boiss.) Holub, Folia Geobot. Phytotax. 9(3): 270 (1974).
C. gaudium-solis (Velen.) Holub, Folia Geobot. Phytotax. 9(3): 270 (1974).
C. tinctoria (L.) J. Gay ssp. *subtinctoria* (Dobrocz.) Holub, Folia Geobot. Phytotax. 9(3): 270 (1974).
C. sancti-johannis (Stoj., Stef. & Turrill) Holub, Folia Geobot. Phytotax. 9(3): 270 (1974).
Anthemis tinctoria L. subsp. *australis* R.Fern., Bot. J. Linn. Soc. 70(1): 14 (1975)
A. tinctoria L. var. *euxina* (Boiss.) Grierson, Notes Roy. Bot. Gard. Edinburgh 33: 428 (1975)
A. tinctoria L. f. *halophila* Thin, Fitologiya 22: 87 (1983)
A. tinctoria L. var. *intermedia* Thin, Fitologiya 22: 87 (1983)
A. tinctoria L. var. *pectinata* Thin, Fitologiya 22: 87 (1983)
A. tinctoria L. var. *ramulosa* Thin, Fitologiya 87 (1983)
A. tinctoria L. subsp. *parnassica* (Boiss. & Heldr.) Franzén, Mountain Fl. Greece [Strid] 2: 430 (1991)
A. tinctoria L. var. *australis* (R.Fern.) Govaerts, World Checkl. Seed Pl. 1(1): 7 (1995)
A. tinctoria L. var. *gaudim-solis* (Velen.) Govaerts, World Checkl. Seed Pl. 1(1): 8 (1995)
A. tinctoria L. var. *sancti-johannis* (Stoj., Stef. & Turrill) Govaerts, World Checkl. Seed Pl. 1(1): 8 (1995)
A. tinctoria L. var. *subtinctoria* (Dobrocz.) Govaerts, World Checkl. Seed Pl. 1(1): 8 (1995)
Cota tinctoria (L.) J.Gay subsp. *australis* (R. Fern.) Oberpr. & Greuter, Willdenowia 33(1): 41 (2003)
C. tinctoria (L.) J.Gay subsp. *euxina* (Boiss.) Oberpr. & Greuter, Willdenowia 33(1): 41 (2003)
C. tinctoria (L.) J.Gay subsp. *fussi* (Griseb. & Schenk) Oberpr. & Greuter, Willdenowia 33(1): 41 (2003)
C. tinctoria (L.) J.Gay subsp. *gaudium-solis* (Velen.) Oberpr. & Greuter, Willdenowia 33(1): 41 (2003)
C. tinctoria (L.) J.Gay subsp. *parnassica* (Boiss. & Heldr.) Oberpr. & Greuter, Willdenowia 33(1): 41 (2003)
C. tinctoria (L.) J.Gay subsp. *sancto-johannis* (Stoj., Stef. & Turrill) Oberpr. & Greuter, Willdenowia 33(1): 41 (2003)
C. tinctoria (L.) J.Gay subsp. *virescens* (Bornm.) Oberpr. & Greuter, Willdenowia 33(1): 41 (2003)
C. tinctoria (L. J.Gay var. *discoidea* (All.) Kuzmanov & Gussev, Fl. Reipubl. Popularis Bulg. 11: 320, 478 (2012), *nom. inval.*
C. tinctoria (L.) J.Gay var. *discoidea* (All.) Özbek & Vural, Türk. Bitkileri List.: 147 (2012)
C. tinctoria (L.) J.Gay var. *pallida* (DC.) Özbek & Vural, Türk. Bitkileri List.: 147 (2012)
C. tinctoria (L.) J.Gay var. *gaudium-solis* (Velen.) Kuzmanov & Gussev, Fl. Reipubl. Popularis Bulg. 11: 321, 478 (2012)

Perennial herb, erect, 30–50 (–70) cm tall, stiff, sparsely to densely tomentose-canescent, rarely almost glabrous, greyish-green with branches from near base. Leaves obovate-oblanceolate, 1–5 cm long, 0.6–2.5 (–3) cm wide, 2 (–3) pinnatisect into triangular-lanceolate, 1.5–2.5 × *c.* 1 mm, cuspidate segments. Pedicels 2.5–5 cm long, unthickened in fruit. Capitula radiate or discoid, (2–) 2.5–4.5 cm in diameter; involucre 1–1.5 cm in diameter; outer phyllaries triangular-lanceolate, smaller than inner, acute, inner phyllaries oblong to oblong-lanceolate, 3–4 × 2–3 mm, apex obtuse to subacute, ciliate, pale or dark brown to diffuse brown, all sparsely or densely whitish-lanate in lower part; receptacle hemispherical, 4–5 mm long, 10–15 mm in diameter, chaffy all over; paleae oblong-lanceolate, 4–5 × 1–1.5 mm, ± keeled, apices sharply cuspidate or attenuate acuminate. Ray florets when present fertile, 18–20, limbs oblong, 4–8 (–10) × *c.* 2 mm, white or creamy to golden yellow, tube compressed. Disc floret corollas yellow, 3–4 mm long, tube basally not inflated at maturity. Achenes ± dorsally compressed, 4-angled, rhombic in cross section, 1.5–2.5 mm long, faintly ribbed on each side, smooth, auricle less than 0.5 mm long, membranous, entire.

HAB. Stony or muddy soil, mountain slopes, stream banks; stony soil; alt. 1250–2000 m; fl. & fr. Apr.-Jul. DISTRIB. **MAM**: Daimka village, 35 km NE of Zakho, 1976, *S. Omar & K. Al-Dabbagh* 45373-A! 10 km S of Sarsng, *Al-Kaisi, Al-Khayat & F. Karim* 51013! **MRO**: Iraqi Kurdistan: Env. of Rayat, 1280 m, near the bridge, 1933, *Zohary* s.n. (HUJ); Hoz (Kurdistaniae) Turc, alt. ca. 1900 m, in graminosis, 1910, *Fr. Nábělek* 3602 (SAV!); Al-Gird Dagh (nr. Nawauda) by a stream, 2000 m, 1932, *Evan Guest* Rustam No. 2868! (F!); Kurdistan, *J. Brandt & W. H. F. Strangeways* 1840 (herb. Benthamianum), K-140!; Hawara Blinder Mountain, NE of Haji Omran, 1700–1900 m, 1959, *Al-Rawi, Noori & Al-Kass* 27797!; Haji Omran, 1650 m, 1959, *Noori & Al-Kass* 27779!; Between Kujar & Kani Grawl region, E. Karoukh, 1959, 1500–1700 m, *Al-Kass, Noori Sarhank* 27649!; Haji Omran, 1971, *S. Omar, Sahira, F. Karm & H. Hamid* 38459!; **MSU**: Azmir, on mountain, 1972, *F. Karim* 39342! **FUJ**: Pl. Mesopotameae & Mosul, 1841, *Kotschy* 177!; Kurdistan (herb. Hookerianum 1867), K-141!

Throughout Europe, Iran, Lebanon, Palestine, Syria, Turkey and Central Asia.

24. **Anthemis melanolepis** *Boiss.*, Diagn. Pl. Orient. Suppl.: 297 (1888). Grerson & Yavin in Fl. Turkey 5: 219 (1975); Mouterde, Nouv. Fl. Lib. Syr. 3. 405. Pl. CCXX, n. 3 (1983); Iranshahr in Fl. Iranica 158: 21 (1986).

Cota palestina Reut., Inger & Kotschy, Ins. Cypern: 240 (1865), p.p.
Anthemis palestina Reut., Fl. Orient. [Boissier] 3: 283 (1875), p.p.
A. syriaca Bornm., Repert. Spec. Nov. Regni Veg. 10: 470 (1912).
A. melanolepis Boiss. var. *macrocarpa* Eig, Palest. J. Bot., Jerusalem Ser. 1: 207 (1939).

Annual herb, ± glabrous, up to 30 cm tall with procumbent, divaricate branches. Leaves subsessile, oblong-obovate, 2–3-pinnatisect into triangular, callose-spinulose ultimate segments. Pedicels thickened upwardly. Capitula radiate, 2–2.5 cm in diameter; involucre 10–15 mm wide, outer phyllaries ovate, apices ± scarious, inner ones oblong, acute; receptacle convex, chaffy all over; paleae thick, leathery, brownish purple, apically cuneate, mostly shorter than achenes, outer long cuspidate, occasionally all cuspidate and shorter than disc florets. Ray florets female, *c.* 15, limbs oblong, *c.* 7.5 mm long, white, tube glabrous. Disc florets yellow, *c.* 4 mm long, base hardly inflated in fruit. Achenes tetragonous, compressed, *c.* 2.5 mm long, 8–10-ribbed on each side, ± winged on margins, auricle undulate, minute.

HAB. Cultivated areas, fields and river margins; fl. & fr. Apr.-May.
DISTRIB. Included here on a report from Iraq (Diyala: ad confines Persiae in collibus conglom., 10 km E Mandali, ad ripam dexfram fluvi, *Rechinger* 12789) by Iranshahr (1986). The authors have not seen any material from the flora of Iraq area.

Cyprus, Turkey, Palestine.

25. **Anthemis coelopoda** *Boiss.*, Diagn. Pl. Orient., ser. 1: 11: 12 (1849). Boissier, Fl. Orient. 3: 283 (1875); Fedorov in Schischkin & Bobrov, Fl. SSSR 26: 64 (1961); Grierson & Yavin in Fl. Turkey 5: 217, f. 10 (1975); Fernandes in Fl. Eur. 4: 157 (1976); Iranshahr in Fl. Iranica 158: 25 (1986).

Cota coelopoda (Boiss.) Boiss., Diagn. Pl. Orient., ser. 2, 3: 21 (1865).
Anthemis coelopoda (Boiss.) var. *bourgaei* Boissier, Fl. Orient. 3: 284 (1875); Breckle, Hedge & Rafiqpoor, Vasc. Pl. Afghan. Checkl.: 148 (2013).

Anthemis rigida Náb̆elek, Publ. Fac. Sci. Univ. Masaryk 52: 14, t. 2, fig. 2 (1925). –Type. Kurdistan Turcicae districtu Serijur, Hondran Dar supra pagum Rewanduj dit Erbil (Arbela), in graminosis alt. ca. 900 m, 23.5.1910, *Nábělek* 3625 (SAV!): Rawi, in Dep. Agr. Tech. Bull. 14: 106 (1964).
Cota bourgaei Boiss., Fl. Orient. [Boissier] 3: 284 (1875).
C. coelopoda (Boiss.) Boiss. var. *bourgaei* (Boiss.) Özbek & C. Vural, Türk. Bitkileri List.: 146 (2012);
C. coelopoda (Boiss.) Boiss. var. *longiloba* (Grierson) Özbek & C. Vural, Türk. Bitkileri List.: 147 (2012).

Erect annual herb, 20–60 cm tall, sparsely pubescent to glabrescent, with divaricate suberect branches. Leaves elliptic-oblong, 2.5–7 cm long, 1 .5–3 cm broad, bi- or tripinnatisect into oblong or oblanceolate, 2–8 mm long, 0.5–1.5 mm wide, acute-mucronulate ultimate segments. Pedicels 5–8 cm long, hollow and upwardly thickened in fruit. Capitula radiate, 2–3 cm in diameter; involucre 1–2 cm in diameter; outer phyllaries ovate to oblong-linear, ± acute, inner oblong, 4–6 × 2–2.5 mm, apices ± obtuse, scarious; receptacle hemispherical, *c.* 5–6 mm across, chaffy all over; paleae ± oblong-spatulate, 3.5–4.5 × *c.* 2 mm, gradually narrowed above, persistent, somewhat truncate and abruptly acuminate, acumen nearly as long as basal part or slightly shorter. Ray florets female, *c.* 20, fertile, limb 10–15 × 3–4 mm, white, tube glabrous. Disc florets yellow, *c.* 4 mm long, tube glabrous, basally not inflated in fruit. Achenes oblong, *c.* 2 mm long, compressed, faintly (2–) 3-ribbed on each side, with minute, entire or crenulate auricle.

HAB. Mountainside, bank of irrigation ditch, clayey soil, fields; alt. 800–1000 m; fl. & fr. Mar.-May.
DISTRIB. **MAM**: Sarsang to Hillside, 1973, *F. Karim, H. Hamid & M. Jasim* 41012!; Surdor, nr. Dohuk, 1960, coll. ignot., W-187!; Sakho (Zakho) to Sharanish, 1973, *F. Karim, H. Hamid & M. Jasim* 41135! **MRO**: Kurdistan Turcicae districtu Serijur, Hondran Dar supra pagum Rewanduj dit Erbil (Arbela), *Nábělek* 3625 (SAV! Type of *A. rigida* Nábělek). **MSU**: Sulaimania, 1930, *D. W. Peterson* 820!; Bakraja, 1971, *S. Omar & F. Karim* 37994!

Bulgaria, Yugoslavia, Greece, Turkey, Iran, Afghanistan.

26. **Anthemis altissima** *L.*, Sp. Pl.: 893 (1753) emend. Sprengel, Syst. Veg. ed. 16, 3: 594 (1826); Boissier, Fl. Orient. 3: 282 (1875); Fedorov in Fl. SSSR 26: 64 (1961); Grierson & Yavin in Fl. Turkey 5: 216, f. 10 (1975); Fernandes in Fl. Europ. 4: 157 (1975); Mouterde, Nouv. Fl. Lib. Syr. 3: 405. PI. CCXX, n. 2 (1983); Iranshahr in Fl. Iranica 158: 23, t. 14 (1986); Breckle, Hedge & Rafiqpoor, Vasc. Pl. Afghan. Checkl.: 148 (2013); Taifour & El-Oqlah, Pl. Jordan Ann. Checkl.: 35 (2017).

Cota altissima (L.) J. Gay in Guss., Fl. Sic. Syn. 2: 867 (1844).
Anthemis cota L., Sp. Pl.: 893 (1753); Rawi, in Dep. Agr. Tech. Bull. 14: 105 (1964); Ghafoor & Al-Turki in Fl. Kingd. Saudi Arabia 2(2): 174, pl. 53 p–s (2000); Ghafoor in Fl. Pakistan 207: 20 (2002).
A. cota L. var. *brevicuspidata* Eig, Palest. J. Bot., Jerusalem Ser. 1: 204 (1938).
A. cota L. var. *longicuspidata* Eig, Palest. J. Bot., Jerusalem Ser. 1: 204: (1938).

Annual herb, erect, up to 120 cm tall, hairy to nearly glabrous, robust with branches from above base. Leaves ovate-oblong, 2–3-pinnatisect into linear-lanceolate, 3–6 mm long, less than 1 mm wide, spinulose-mucronate ultimate segments. Pedicels upwardly thickened in fruit, up to 7.5 cm long. Capitula radiate, (2–) 2.5–4 (–5) cm in diameter; involucre 1–2 cm in diameter; outer phyllaries ovate, acute, margins not scarious, inner oblong-lanceolate, to ± elliptic-oblong, obtuse, with hyaline or scarious margins and apices; receptacle hemispherical, becoming umbonate, *c.* 10–15 mm in diameter, chaffy all over; paleae oblong-spatulate, ± keeled, 3.5–4 × 1.5 mm, straw-coloured, abruptly contracted at the apices with 3.5–4 mm long, slender acumen. Ray florets female, 18–20, mostly fertile, limbs elliptic-linear, 7.5–12.5 × *c.* 3 mm, white, tube glabrous. Disc floret corollas yellow, as long as paleae, tube glabrous, not inflated in fruit. Achenes ± compressed, obpyramidate, 2–2.5 mm long, 4-angled and rhombic in cross section, usually 10-ribbed on each face, smooth, apices with a very short, entire auricle.

HAB. Limestone soil, cultivated fields; alt. 300–1100 m; fl. & fr. May-Jul.
DISTRIB. **MAM**: Ain Nuni, inter Aradan et Hasitha, 1910, *Nábělek* 3621 (SAV!). **MRO**: Kurdistan, Herb. Hookerianum, Kew No. 204! **MSU**: Sulaimaniya, 1934, Field Museum Near East Expedition, *Henry Field & Yousuf Lazar* 952 (F!). **MJS**: Jebel Sinjar, above Kursi, 1100 m, 1948, *Gillett* 11055! **DJL**: Jazira (south of Sinjar near Baiji), *c.* 300 m, 1954, *Guest* 13390!

Central and SW Asia (Afghanistan, Pakistan) westwards to Algeria, Spain, Italy, Germany; Balkans, Turkey, Palestine, Syria, Iran,

27. **Anthemis brevicuspis** *Bornm.*, Beih. Bot. Centralbl. 32(2): 397 (1914); Iranshahr in Fl. Iranica 158: 22, t. 13 (1986).

Anthemis feinbruniae Eig, Palest. J. Bot., Jerusalem Ser. 1: 204 (1938). – Type. Northern Iraq, Jebel Sindjar, 800 m, 1933, *Eig & Zohary* s. n. (HUJ! holo.; S iso.); Rawi, in Dep. Agr. Tech. Bull. 14: 105 (1964).

A. rayatensis Eig, Palest. J. Bot., Jerusalem Ser. 1: 205 (1938). – Type. Iraqi Kurdistan, Env. of Rayat, 1280 m, 1933, *Eig & Zohary* s. n. (HUJ!, iso.); Rawi, in Dep. Agr. Tech. Bull. 14: 106 (1964).

Erect herb, up to 35 cm tall, sparsely patulous hairy with divaricate, ascending branches. Lower leaves with dentate petiole, upper sessile, oblong to ovate-oblong, bipinnatisect to ± tripinnatisect into remote, linear-oblong to narrowly lanceolate, acute, mucronate ultimate segments. Pedicels up to 10 (–12) cm long, remaining unthickened in fruit. Capitula radiate, *c.* 2.5 cm in diameter; involucre 8–10 mm in diameter; phyllaries ovate to oblong-lanceolate, hairy, brown membranous margined, lacerate-fimbriate in upper half and at acute apex; receptacle hemispherical, 5–6 mm long, chaffy all over; paleae oblong-spatulate, 3–3.5 mm long, *c.* 1.5 mm wide, slightly carinate, truncate to subcordate-emarginate, with 1–1.5 mm long mucro. Ray florets female, fertile, limbs oblong, 9–12 mm long, 4–5 mm broad, white tube glabrous. Disc florets slightly longer than paleae, corollas yellow, tube glabrous. Achenes tetragonal, compressed, *c.* 3 × 1.25 mm, smooth, inconspicuously 10-striate, with 0.25–0.5 mm long, denticulate auricle at apex.

HAB. Near moist cultivated land in mountainous regions; alt. 800–1280 m; fl. Apr.-Oct.
DISTRIB. **MRO**: Iraqi Kurdistan, Env. of Rayat, 1280 m, 1933, *Eig & Zohary* s. n. (HUJ!, isotype of *A. rayatensis* Eig). **MSU**: Northern Iraq, Jebel Sindjar, 800 m, 1933, *Eig & Zohary* s. n. (type of A. *feinbruniae* Eig; HUJ! holo.; S iso.);

Iran.

28. **Anthemis pseudocotula** *Boiss.*, Diagn. Pl. Orient., ser. 1, 6: 86 (1846), emend. Eig, Palest. J. Bot., Jerusalem Ser. 1: 196 (1938); Rawi, in Dep. Agr. Tech. Bull. 14: 106 (1964); Rechinger f., Fl. Lowland Iraq 621 (1964); Täckholm, Stud. Fl. Egypt, ed. 2: 574 (1974); Grierson & Yavin in Fl. Turkey 5: 209, f. 8 & 10 (1975); Feinbrun-Dothan, Fl. Palaest. 3: 338, pl. 570 (1978); Iranshahr in Fl. Iranica 158(4): 16. Tab. 8 (1986); Al-Rawi, Fl. Kuwait 2: 240 (1987); Ghafoor & Al-Turki in Candollea 52: 472, f. 6, A-G (1997); Chaudhary, Fl. Kingd. Saudi Arabia 2 (3): 174, pl. A-G. (2000); Boulos & Hind in Fl. Egypt 3: 242 (2002); Taifour & El-Oqlah, Pl. Jordan Ann. Checkl.: 35 (2017).

Annual herb, (10–)15–50 cm tall, appressed hairy or glabrescent with erect or ascending branches from near the base. Leaves short-petiolate, ovate-oblong, 1.5–3 × 1–1.5 cm, 2(–3)-pinnatisect into linear-oblanceolate, 1–3 × *c.* 0.5 mm, acute ultimate segments. Pedicels 3–6 cm long, gradually thickened upwards and strongly arcuate in fruit. Capitula radiate, 1.5–2.5 cm in diameter; involucre 5–8 mm in diameter; phyllaries ovate-oblong, 3–4 × 1–2 mm, pubescent, scarious margined, subacute; receptacle elongate-conical, 6–8 × 3–4 mm, chaffy in upper half only; paleae linear-subulate, 2.5–3 mm long, cuspidate, persistent. Ray florets female, 10–12, fertile, limbs oblong, 3–7 × 2.5–3.5 mm, white, tube glabrous, compressed. Disc florets yellow, as long as paleae, tube glabrous, somewhat swollen at base in fruit. Achenes turbinate-obpyramidate, ± tetragonous, 1–1.5 mm long, distinctly 10-ribbed, occasionally ± tuberculate, oblique, truncate or rounded at the apices, auricle absent or nearly up to half as long as achenes.

HAB. (of sp.). Fine sandy-clay to light clay gravelly soils or silty soil or moist black loam in depressions, sometimes in heavy silt; alt. 30–750 m; fl. & fr. March–July.
DISTRIB (of sp.). **MAM**: Dohuk, 1932, *Mieke Bee* 3281!; Dohuk, 500 m, 1951, *Guest* 1309! **MRO**: 2 km N of Darband, 650 m, 1975, *Botany Staff* 43207! **MSU**: Between Sasar & Dokan, 510 m, 1975, *Botany Staff* 43059!; just E of Durbendikan, Sulaimaniya Liwa, 1964, *F A. Barkely & Shariff Y. Haddad* 7750! Qaradagh Moutain, Timara village, 1973, *F Karim, H. Hamid & M. Jasim* 40590! **FUJ:** 40 km E of Sinjar, *Chakravarty, Al-Rawi, Khatib & Alizzi* 33137!; roadside weed near Hamman Al-Alil, *c.* 30 km S of Mosul, 1974, *M. Hussain* 511!; road to Sinjar, 1969, *S. Omar & H. Hamid* 36421!; between Baghdad and Syrian desert, *F. K. Jackson* H 240/1930!; **FAR**: Arbil Province, 1933, *Al-Radhi* 3865!; **FPF**: Qizil Robat, 1932, *Guest* 1768!; E of Badra, 1949, *Al-Rawi & Gillett* 7253!; 25 km SE of Buksaya, Khatima Police Station, *Al-Rawi & Sh. Haddad* 25639!; Pilkana, near Khaniqin Chehale, near Amara, 1928, *H. Field & Yousuf Lazar* H. 35! (F!). **DLJ**: Balad, *Al-Rawi & Gillett* 6508!; Telegraph pole M30, between Baiji & Mosul, 1925, *Henry Field & Yousuf Lazar* R 389!; Rawah-Sinjar, 115 km SW of Sinjar, 250 m, 1962, *Chakravarty, Al-Rawi, Khatib &*

Alizzi 31989!; Tharthar Dam, 1965, *Hikmat Abbas Al-Ani & Haddad* 9659! **DGA**: Khorqin, 1957, valleys of Jebel Hamrin, *Haines* W93!; Baghdad, 12 km, inter Tigris flumen et Monte Jabal Hamrin, ad oppido Samara Orientem versus, *Rechinger* 9558! (G!); **DWD**: Al-Masad, 15 km S of Rutba, 650 m, *Chakravarty & Al-Rawi et al.* 32839 & 32875!; 45 km NW of Rutba, Nr. H3, ± 710 m, 1957, *Al-Rawi* 21161!; Jadida, W of Nukhaib, 350 m, 1961, *Al-Rawi* 31047!; 15 km S of Rutba, 620 m, *Khaitb & Hazim* 32453!; 5 km S of H3 (Nr. Rutba), ± 710 m, 1957, *Al-Rawi* 21207!; 75 km W of Rutba (Wadi Al-Walaj), ± 710 m, 1957, *Al-Rawi* 21190!; in arenosis 10 km N of Sheikh Mohammad, 1957, *Rechinger* 9593 (G!); Western desert, ± 700 m, 1957, *Rechinger* 9944! (G!); Shabicha, Southem desert, ± 300 m, 1955, *Guest, Al-Rawi & G Long* 14060!; Wadi Al-Khirr, 15 km N of Lussuf (± 100 km NNW of Shabicha), 1957, *Guest, Al-Rawi & Rechinger* 19442!; 35 km W of Rutba, ± 600 m, 1957, *Rechinger* 9920! (G!); Western desert, Wadi Al-Walaj, 75 km SW of Rutba, ± 700 m, 1957, *Rechinger* 9931 (G!). **DSD**: Southern desert, 30 km W by N, Butiaya, 1955, 150 m, *Guest, Al-Rawi & G. Long* 14153!; Takhadid, 330 m, 1978, *Al-Kaisi, K. Hamid & H. Hamid* 48084!; Southern desert, Shabicha, 1947, *Al-Rawi & Gillett* 6283!; Southem desert, 52 km E by Salman, 1955, *Guest, Al-Rawi & G Long* 14126! **LEA**: Shimal (Sinjar), 1932, near Baghdad, *Yousuf Lazar* 129!; Fakkia, 70 km E of Amara, ± 40 m, 1958, *Al-Rawi* 25790!; Tarmiya, Baghdad, 1955, Haines W113!; Baghdad, 1920, *R. J. D. Graham* s. n.!; near Baghdad, 1933, *Yousuf Lazar* 198 (F!); Mosul, *R. G. Bayliss* 93! **LCA/LEA**: Al-Sudor, Diyala Liwa, 1964, *W. Suleiman* 215! **LCA**: Sudur, 1976, *S. Omar, Fauzi, Al-Kaisi & Noori* 44298! **LSM:** Chalat Police Station, near Persian border, ± 150 km, 1958, *Al-Rawi & Sh. Hadad* 25676! **LSM**: Diwaniya, Southern desert, Wadi Al-Khirr, 15 km W of Lusuf (± 100 km NNW of Shabicha), 32 m, 1957, *Rechinger* 9489! (G!).

Cyprus, Turkey, Palestine, Syria, Egypt, Saudi Arabia, Iran.

The species is divided into several varieties represented in Iraq:
subsp. **pseudocotula** var. **pseudocotula**
Auricle 1/3 to ½ as long as achenes, triangular, entire or obscurely lobed.

HAB. Silty depressions; alt. 150–400 m; fl. & fr. Apr.-Jun.
DISTRIB. **MSU**: Derbandi Khan, mountain, 1971, *S. Omar & F. Karim* 38138! **FUJ**: Tell Kotchek, ± 400 m, 1948, *Gillett* 10836! **DSD**: Desert Province, *c.* 55 km NW of Shabicha, 210 m, 1955, *Guest, Al-Rawi & H. Long* 14050! **LCA**: Near Baghdad, 1933, *Y. Lazar* 197 (F!); Baghdad, 1933, *Y. Lazar* 3897!

Egypt (Sinai), Turkey, Lebanon, Palestine, Syria, Iran.

subsp. **pseudocotula** var. **radiata** *Eig*, Palest. J. Bot., Jerusalem Ser. 1: 197 (1938).
Auricle very short, divided into few lobes.

HAB. Clayey sand or sandy silty soils in low depressions; alt. 100–950 m; fl. & fr. Apr.-Jun.
DISTRIB. **MAM**: 8 km S of Dohuke, 590 m, 1975, *Botany Staff* 43370!; **MRO**: Gali Warta, ± 30 km NW by N of Rania, 950 m, 1959, *Al-Rawi, Noori & Al-Kass* 28861!; Koi Sanjaq, 590 m, 1959, *Al-Rawi, Al-Kass & Noori* 281 17!; Plingan village, ± 17 km NW of Rania, 900 m, 1959, *Al-Rawi, Noori & Al-Kass* 28663!; Kalala Junction, 1958, *Al-Kass* 18584!; Qara village, on road to Koi Sanjag, ± 440 m, 1959, *Al-Rawi, Noori & Al-Kass* 128034! **MSU**: Qara Chulam, between Sula & Chuwarta, 1971, *S. Omar & R Karim* 38094! **FAR**: Arbil Province, 1933, *Al-Radhi* 3865! **FKI**: 30 km from Kirkuk to Sulaimaniya, hillside, 1973, *F. Karim, H. Hamid & M. Jasmin* 40514! **FPF**: 38 km SE of Mandali, ± 127 m, 1957, *Al-Rawi* 20698A! **DLJ**: Telegraph pole M 90, between Baiji & Mosul, May 1925, *H. Field & Y. Lazar* 389! **DWD**: 35 km NW of Rutba, ± 600 m, 1957, *Al-Rawi* 21145! 75 km W of Rutba (Wadi Al-Walaj), 1957, *Al-Rawi* 21190!

Jordan, Syria.

subsp. **pseudocotula** Boiss., var. **massadensis** *Yavin*, Israel J. Bot. 19: 146 (1970).
Auricle 1/3 to ½ as long as achenes, divided into a few triangular lobes.

HAB. Loamy clay soils in depressions; alt. 50–400 m; fl. & fr. Mar.-Apr.
DISTRIB. **FUJ:** Near Baaj, roadside, 1969, *S. Omar & H. Hamid* 36529!: Balad Sinjar-Tal Afar, 350 m, 1933, *Guest* 4139! **DLJ**: Falluja-Ain-Ehwrait Etal, 1930, *Guest* 1120! **DGA**: 13 km E of Samarra, *c.* 65 m, *Al-Rawi* 20359! **DWD:** 70 km W of Nukheib, 300 m, 1961, *Al-Rawi* 31394! (?) Inside the castle of Hodos, 230 m, 1964, *Chakravarty & Al-Rawi* 33043!;

Palestine.
subsp. **rotata** (Boiss.) *Eig* var. **rotata** *Eig*, Palest. J. Bot., Jerusalem Ser. 1: 197 (1938).
Apex of achenes oblique, slanting outwards.

HAB. Gravelly soils by roadsides and hillside; alt. 800–1160 m; fl. & fr. Mar.-Jul.
DISTRIB. **MSU**: Sulaimaniya Arbat, 1971, *S. Omar & F. Karm* 38115!; 5 km from Ranjawin, 1160 m, 1973, *M. Noori & K. Hamid* 41244! **FPF**: 10 km N of Khaniqin, hillside, 1977, *H. Hamid* 46605! **DGA**: between Samara & Tekrit, 1974, *Sahira, Raja & Khalida* 41807!

Egypt (Sinai), Jordan, Palestine.

subsp. **rotata** (Boiss.) *Eig.* var. **rotundata** *Eig*, Palest. J. Bot., Jerusalem Ser. 1: 197 (1938). Achenes rounded at apex, not slanting outwards.

HAB. Sandy soil near irrigation ditches; alt. 50–100 m; fl. & fr. Mar.-May.
DISTRIB. **MSU**: Tainal, 1971, *S. Omar & F. Karm* 37925! **DGA**: 50 km E of Samara, Asila village (10 km N Shaikh Mohammad), ± 50 m, 1957, *Al-Rawi* 20439! **LEA**: 13 km E of Baquba, 1929, *Cowan & Darlington* 184! **LCA**: work-ground, near irrigation ditch, Baghdad, 1955, *Haines* 113!

Palestine.

29. **Anthemis cotula** *L.*, Sp. Pl.: 894 (1753); Boissier, Fl. Orient. 3: 3 15. (1875); Fedorov in Schischkin & Bobrov, Fl. SSSR 26: 67. PI. 4, fig. 1 (1961); Rawi, in Dep. Agr. Tech. Bull. 14: 105 (1964); Rechinger f., Fl. Lowland Iraq: 621 (1964); Grierson & Yavin in Davis, Fl. Turkey 5: 208. Fig. 10 (1975); Fernandes in Tutin et al., Fl. Eur. 4: 155 (1976); Feinbrun-Dothan, Fl. Palaest. [M. Zohary & N. Feinbrun-Dothan] 3: 337. Pl. 569 (1978); Mouterde, Nouv. Fl. Lib. Syr. 3: 413. PI. CCXXX, n. 1 (1983); Ghafoor & Al-Turki, Candollea 52: 473. Fig. 6, P-S (1997); Boulos & Hind in Fl. Egypt 3: 239 (2002).

Cotula foetida S.G.Gmel., Reise Russl. 1: 137 (1774), non *Cotula foetida* Poepp. ex DC., Prodr. 6: 139 (1838).
Anthemis foetida Lam., Fl. Franç. (Lamarck) 2: 164 (1779), nom. illeg. superfl. pro *Anthemis cotula* L.
Chamaemelum cotula (L.) All., Fl. Pedem. 1: 186 (1785).
Maruta foetida (Cass.) Gray, Nat. Arr. Brit. Pl. 2: 456 (1821), nom. illeg. superfl. pro *Anthemis cotula* L.
Maruta foetida Cass., Dict. Sci Nat., ed. 2, 29: 174 (1823), nom. illeg. superfl. pro *Anthemis cotula* L.
Anthemis ramosa Link ex Spreng., Syst. Veg., ed. 16 [Sprengel], 3: 593 (1826).
Anthemis psorosperma Ten., Syll. Pl. Fl. Neapol., App.: 555 (1831); Relaz. Viaggio Abruz. [Ind. Pl. Aprutio Coll. 1831]: 128 (1832).
Maruta cotula (L.) DC., Prodr. 6: 13 (1838).
Anthemis cotula L. subsp. β *psorosperma* (Ten.) Arcang., Comp. Fl. Ital.: 358 (1882).

Foetid annual herb, up to 60 (–70) cm tall, sparsely hairy to almost glabrous, with erect or ascending basal branches and corymbosely branched above. Leaves short-petiolate or subsessile, sometimes fleshy, obovate-oblong or ovate-oblong, 1.5–6.5 cm long, 0.5–3 cm wide, finely 2–3-pinnatisect into narrowly linear, 1–3 × *c.* 1 mm, entire or 2–3-dentate, acute-mucronate ultimate segments. Pedicels slender, up to 15 cm long, unthickened in fruit. Capitula radiate, occasionally discoid, 1.2–3 cm in diameter; involucre 6–10 mm in diameter; phyllaries ovate-oblong, 3–4 mm long, whitish hairy, margins narrowly scarious, apices ±obtuse; receptacle oblong-conical, 5–8 × 3–4 mm, chaffy in upper half only; paleae linear-subulate, 2.5–3 mm long, bristly, persistent. Ray florets female, 14–16, sterile, limbs oblong-elliptic, 5–14 × 3–6 mm, white, tube glabrous. Disc floret corollas yellow, as long as to slightly longer than paleae, tube terete, glabrous, ± inflated below in fruit. Achenes subcylindrical-turbinate, 1–1.5 mm long, obscurely 8–10-ribbed, tuberculate, light brown, exauriculate to rarely crenulate-rimmed. Fig. 90: 1–4.

HAB. Rare in Iraq in clay and alluvial damp soil in upper plains and foothills; alt. 250–300 m; fl. & fr. Mar.-May.
DISTRIB. **FUJ**: Mosul, in silvulis salicum ad ripam Tigridis, 10.5.1910, *Fr. Nábělek* 3639 (SAV!).

SW Asia, common in almost all European countries and North Africa; introduced into N & S America, Australia and New Zealand.

A very variable species which can be easily confused with *A. pseudocotula* and *A. arvensis.*

30. **Anthemis corymbulosa** *Boiss. & Hausskn.* in Boissier, Fl. Orient. 3: 316 (1875). –Type. Iraq, In deserte ad fluviam Chabur Mesopotamiae, *Haussknecht* 527 (G-BOISS, holo.; JE!, iso.); Rechinger, Fl. Lowland Iraq: 621 (1964); Rawi, in Dep. Agr. Tech. Bull. 14: 106 (1964).

Annual herb, erect or ascending, up to 45 cm tall, unbranched below, somewhat corymbosely-branched above upper half, glabrous or glabrescent. Leaves short-petiolate or somewhat sessile, linear-oblong, 1.5–3 cm long, 1–1.5 cm wide, uni- or bipinnatisect into small, remote, finely filiform, ± obtuse ultimate segments. Pedicels slender, 2.5–4 cm long, unthickened in fruit. Capitula radiate, 1.2–1.6 cm in diameter; involucre 4–6 mm in diameter; phyllaries ovate, 2–2.5 × 1.5 mm, glabrous, obtuse, apices and margins broadly membranous; receptacle conical, 3–4 mm long, pitted in lower and chaffy in upper half;

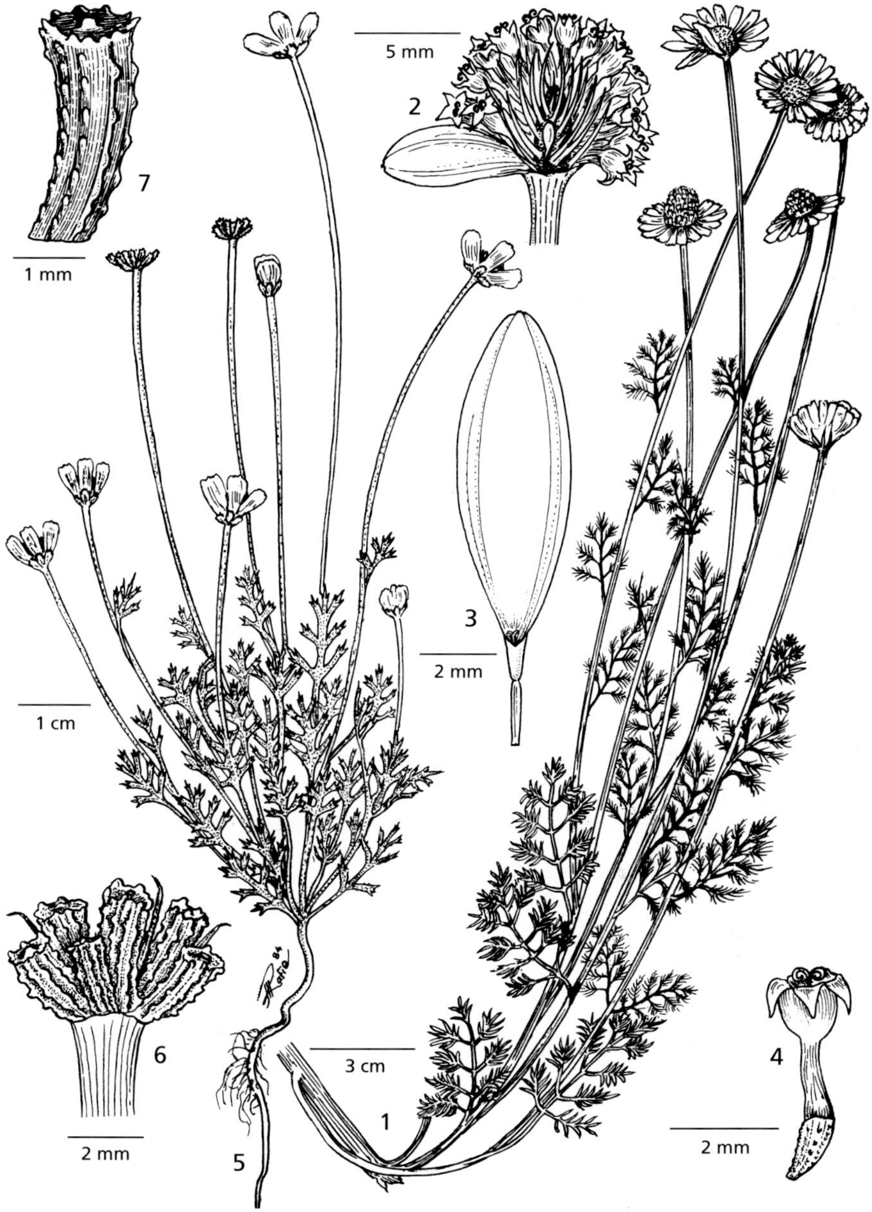

Fig. 90. **Anthemis cotula**. 1, habit flowering branch; 2, capitulum; 3, ray floret; 4, disc floret with partly mature achene. **Anthemis odontostephana**. 5, habit; 6, fruiting capitulum; 7, achene. Drawn by Rafiq. Reproduced with permission from Flora of Pakistan 207, fig. 1.

paleae narrowly lanceolate, *c.* 2 mm long, keeled. Ray florets female, sterile, limbs narrowly oblong-ovate, 6–7 × 2.5– 3 mm, white. Disc florets yellow, with glabrous, punctate-glandular, ± compressed corolla tube. Achenes deciduous, turbinate, *c.* 1 .5 mm long, somewhat tuberculate by sessile glands, with nearly half as long or shorter unilateral auricles.

HAB. In shallow depressions in the vicinity of irrigated land; alt. 100–200 m; fl. & fr. May–Sep.
DISTRIB. Endemic to Iraq. **FUJ**: In deserte ad fluviam Chabur Mesopotamiae, *Haussknecht* 527 (G-BOISS;

JE!); LEA: P.4. NE of Baquba, (Shatvaban road), 1957, *Hunting Technical Services Limited* 8!; lower Diyala, 1957, *V. C. R.* 161!

Endemic.

31. **Anthemis pauciloba** *Boiss.*, Diagn. Pl. Orient., ser. 1, 6: 83 (1845). Grierson & Yavin in Fl. Turkey 5: 195. Fig. 9 (1975); Mouterde, Fl. Nouv. Lib. Syr. 3: 408. Pl. CCXXII, n. 3 (1983); Iranshahr in Fl. Iran. 158: 12, t. 3 (1986).

Anthemis argentea Eig, Palest. J. Bot., Jerusalem Ser. 1: 192 (1938).
Anthemis pauciloba Boiss. var. *brachysperma* Eig, Palest. J. Bot., Jerusalem Ser. 1: 190 (1938).

Perennial herb, upright, up to 45 cm tall, basally ± woody, appressedly greyish or pale pubescent with branches from near base. Leaves long-petiolate, lower leaves ovate or oblong-spatulate, up to 10 cm long (including petiole), pinnatipartite or incompletely 2-pinnatisect into linear-oblanceolate or cuneate segments, upper leaves cuneate-spatulate, entire or shallowly 2 or 3-toothed. Pedicels up to 30 cm long, unthickened in fruit. Capitula radiate or discoid, 1 .5–2.2 cm in diameter; involucre 10–15 mm in diameter; phyllaries ovate-lanceolate, 3–4 mm long, sparsely to densely lanate-tomentose, acute; receptacle depressed hemispherical, *c.* 5 mm long, chaffy all over; paleae oblanceolate, *c.* 2.5 mm long, keeled, acuminate. Ray florets 10–30 when present, limbs 2.5–7 mm long, yellow, tube glabrous. Achenes angular, 2–3 mm long, incurved, obscurely ribbed, whitish, with unilateral, 0.5–1 mm long, ± dentate auricle.

HAB. Limestone steppes; alt. 1300–2000 m; fl. & fr. Apr.-Jul.
DISTRIB. Included here based on a report from Iraq by Iranshahr (1986) who cited *Rechinger* 10968 (W) and *Rechinger* 10874 (W) from Sharanish area of Iraqi Kurdistan. The authors, however, have not examined any material.

Lebanon, Syria, Turkey.

32. **Anthemis kotschyana** *Boiss.*, Diagn. Pl. Orient. n.s. 2, 3: 23 (1856); Grierson & Yavin in Fl. Turkey 5: 196, f. 9 (1975).

Anthemis kotschyana Boiss. var. *discoidea* (Bornm.) Grierson, Notes Roy. Bot. Gard. Edinburgh 33: 428 (1975).
A. kotschyana Boiss. f. *discoidea* Bornm., Repert. Spec. Nov. Regni Veg. Beih. 89: 331 (1944).
A. warburgii Eig, Palest. J. Bot., Jerusalem Ser. 1: 189 (1938).
A. bulgardaghica Eig, Palest. J. Bot., Jerusalem Ser. 1: 192 (1938).
A. rotundifolia Eig, Palest. J. Bot., Jerusalem Ser. 1: 192 (1938).

Perennial herb, appressed greyish hairy with suberect, up to 25 (–40) cm tall, basally woody branches. Leaves long-petiolate, lamina ovate or oblong, 0.8–2 cm long (excl. equally long petiole), bipinnatisect into linear-oblanceolate, 2–5 mm long ultimate segments, upper leaves becoming 1-pinnatisect. Pedicels up to 15 cm long, leafless. Capitula discoid, 1–1.5 cm in diameter; involucre 7–12 mm wide; phyllaries oblong to ovate-lanceolate, 2–4 mm long, sparsely appressed pubescent, apices slightly scarious, margins pale or dark, acute or ± acute; receptacle hemispherical; paleae linear-oblong to ± spatulate, 3–4 mm long, membranous, mucronate. Disc florets yellow, 2.5–3.5 mm long, inflated basally in fruit. Achenes obconical, 1.5–2.5 mm long, pale-whitish, obscurely striate, exauriculate or unilaterally short-auriculate.

HAB. *Quercus* woods, on steppe and cultivated slopes; alt. 1800–2060 m; fl. & fr. May-Aug.
DISTRIB. Of this highly polymorphic species only var. *discoidea* (Bornm.) Grierson has been reported from Iraq by Iranshahr (1986) who cited *Al-Rawi* 23604 from Iraqi Kurdistan.

Iran, Turkey.

It is very difficult to separate this variety from discoid forms of *A. cretica.*

33. **Anthemis tubicina** *Boiss. & Hausskn.*, in Boissier, Fl. Orient. 3: 319 (1875); Rawi, in Dep. Agr. Tech. Bull. 14: 106 (1964); Grierson & Yavin in Fl. Turkey 5: 211 (1975);

Anthemis odontostephana Boiss. var. *tubicina* (Boiss. & Hausskn.) Bornm., Beih. Bot. Centralbl. 28(2): 247 (1911).

Annual herb, 10–25 cm tall, greyish hairy with decumbent branches from near base. Leaves small, petiolate, ovate-oblong, 1.5–2.5 cm long, 1–1.5 cm wide, bipinnatisect into linear-oblong, 1–3 mm long, less than 1 mm wide, acute ultimate segments. Pedicels filiform, 2–4 cm long, becoming much (3–4 mm in diameter) thickened in fruit. Capitula small, radiate, 8–12 mm in diameter; involucre 4–5 mm wide. Phyllaries oblong-ovate or inner obovate, 3–4 mm long, margins broad-scarious, hairy in midrib region, obtuse, deciduous; receptacle convex, 3–4 mm in in diameter, chaffy in upper part; paleae narrowly linear-lanceolate, shorter than to equalling achenes, hyaline, deciduous. Ray florets 5–6, sterile, limbs ± obovate, 4–5 mm long, white. Disc florets yellow, tube glabrous, not inflated at base in fruit. Achenes cylindrical, 3–4 mm long, slightly dilated to base, 10-ribbed, all prickly tuberculate, persistent and ± patent on receptacle at maturity, truncate with acutely 10-lobed or dentate radiating crown at apex.

HAB. Calcareous rocks; alt. 650–1300 m; fl. & fr. Mar.-Jun.
DISTRIB. **MRO**: Gali Warta, 30km NW by N of Rania, *Rawi, Nuri & Kass* 28814!; Shakh-i- Harir (nr. Harir: Rawanduze), *Emberger, Guest, Shwani & Serkahin* 15439! **MSU**: Pir Omar Gudrum, in rupestris calc, 3–4000 ft., June 1867, *Haussknecht* 534 (G-BOISS!, JE!). **MJS**: Sinjar, 650–700 m, 1948, *Gillett* 10883!; Jabal Sinjar, *Al-Kaisi & Wedad* 52074!; Kursi,id 10886A!; Mesopotamia, 1867, calcareous rocks, Mt. Singare, *Haussknecht* 539 (G!, JE!). Daratri (on the road bet. Halabja & Tawaila, *Rawi* 22031!

Turkey, Syria (?), Iran.

34. **Anthemis odontostephana** *Boiss.*, Diagn. Pl. Orient., ser. 1, 6: 85 (1846). Boissier, Fl. Orient. 3: 319 (1875); Fedorov in Schischkin & Bobrov, Fl. SSSR 26: 70 (1961); Rawi, in Dep. Agr. Tech. Bull. 14: 105 (1964); Iranshahr in Fl. Iranica 158: 10, t. 1 (1986) excl. var. *tubicina* (Boiss. & Hausskn.) Bornm.; Ghafoor in Fl. Pakistan 207: 18, f. 1 A-C (2002); Jongbloed, Wild Flow of United Arab Emir.: 139 (2003); Breckle, Hedge & Rafiqpoor, Vasc. Pl. Afghan. Checkl.: 148 (2013); Kilian & Ghazanfar in Fl. Oman 3: 252 (2015).

Dwarf annual herbs, up to 15 cm tall, appressed greyish hairy, simple or branched and leafy in lower half only, branches upright, leafy only up to middle. Leaves shortly petiolate, ovate, 1.5–2 cm long, 6–10 mm wide, uni- or bi- pinnatisect into 2–3 × *c.* 1.5, acute ultimate segments. Pedicels filiform, 8–10 cm long, gradually thickened in fruit. Capitula radiate, 1–1.2 cm in diameter; involucre 6–7 mm in diameter; phyllaries oblong, 2–3 mm long, appressed-hairy, broadly membranous margined, obtuse, deciduous; receptacle conical, 2 mm long, chaffy in upper half; paleae few, setaceous, subulate, 2.5–3 mm long, very narrow, deciduous. Ray florets few, female, sterile, limbs elliptic, *c.* 5 × 4 mm, white. Disc floret corollas yellow, with a glabrous, terete corolla tube. Achenes cylindrical, ± curved, 2.5–3 mm long, dilated at bases and apices, 10–15-ribbed, outer tuberculate-muricate, truncate, with a 10–15-dentate rim, inner with a unilateral, oblique, ± lacerate auricle. Fig. 90: 5–7.

HAB. Red soil; alt. 50–1500 m; fl. & fr. Mar.-May.
DISTRIB. **MRO**: Deibrun ad fontem naphtae in rad. mont. Siah Kuh (Suriana), 1909-10, *Nábělek* 3624 (SAV!). **FPF**: Kuh-Beijat inter Deibrun et Amara, 1909-10, *Nábělek* 3634 (SAV!). **DGA**: Jabal Hamrin, 1917, *Sutherland* 111 (BM!). **LBA**: Basrah, 1957, *Haines* s.n. (E!).

Afghanistan, Pakistan, Iran, United Arab Emirates, Oman.

92. **TANACETUM** L.

Sp. Pl.: 843 (1753)
Oberprielor, Vogt & Watson in Kubitzki, Fam. Gen. Vasc. Pl. 8: 366 [2006] (2007)
Chrysanthemum L., Sp.Pl.; 887 (1753) p.p.; *Balsamita* Mill., Gard. Dict. 4 (1754); *Pyrethrum* Zinn, Cat. Pl. Goetting 414 (1757)

Shahina A. Ghazanfar

Perennial herbs, rarely subshrubs. Indumentum absent or hairs basifixed or sometimes stellate (not in Iraq plants). Leaves alternate, 1–3-pinntipartite, rarely entire. Capitula solitary or in corymbs, heterogamous and radiate or disciform, or homogamous and discoid. Involucre campanulate; phyllaries in 3–5 rows, margins broad or narrow, scarious, usually brown. Receptacle flat to hemispherical, epaleate, usually pilose. Marginal florets female, fertile or neuter; ray limb white, yellow (in Iraq plants) or pink; disk florets bisexual;

corolla yellow, tubular, apex 5-lobed or 5-toothed. Achenes terete or obscurely 3-angled, 5–10(–15)-ribbed; apex marginally rounded or with corona present, rarely with an adaxial auricle; pericarp without myxogenic cells.

A genus of about 160 species in N Africa, C Asia, Europe; some species widely cultivated. 9 species in Iraq.

1. Capitula solitary 2
   Capitula in dense or loose corymbs 5
2. Capitula homogamous; all florets tubular 3
   Capitula heterogamous; marginal florets ligulate, disc florets tubular 4
3. Plant sparsely to densely pubescent, glandular; achene with corona oblique, divided into 6–8 linear-oblong scales . . . . . 8 *T. gillettii*
   Plant densely whitish- silvery-tomentose; achenes with corona unilateral, developed on posterior side of achene apex ± denticulate . . . . . 9. *T. argenteum* subsp. *canum*
4. Stems and leaves glabrous; corona of achene divided to base into 6–8 lanceolate scales . . . . . 5. *T. nivale*
   Stems and leaves grey-green pubescent; corona developed on posterior side of achene, irregularly notched at apex . . . . . 4. *T. kotschyi*
5. Capitula in dense corymbs 6
   Capitula in loose corymbs, each capitulum on long floral axis 8
6. Phyllaries 1-seriate; ligules white; achene 5-ribbed, ribs white; corona with irregular margins . . . . . 1. *T. cilicium*
   Phyllaries more than 1-seriate; corolla limb yellowish; achene with entire coronas . . . 7
7. Perennial to 20 cm; stems densely grey-pubescent and glandular; phyllaries several-seriate, imbricate; marginal florets yellowish, longer than the disc florets . . . . . 7. *T. aureum*
   Perennial 25–60 cm; stems densely pubescent; phyllaries 1–2-seriate; marginal florets scarcely longer than the disc florets; disc florets yellow . . . . . 6. *T. polycephalum* s.l.
8. Leaves ovate-oblong to ovate-elliptic, base cuneate, margins crenate to serrate; corolla limb 4–5 mm . . . . . 2. *T. balsamita* subsp. *balsamitoides*
   Leaves pinnatisect, segments obtuse; corolla limb 6–7 mm . . . . . 3. *T. parthenium*

1. **Tanacetum cilicium** (*Boiss.*) *Grierson* in Notes Roy. Bot. Gard. Edinburgh 33: 434 (1975); Grierson in Fl. Turkey 5: 261 (1975); Podlech in Fl. Iranica 158: 96 (1986).

Pyrethrum cilicium Boiss., Diagn. Pl. Orient. ser. 2, 3: 28 (1856); Dinsmore in Post, Fl. Syria, Palest. & Sinai ed. 2, 2: 62 (1933).
Chrysanthemum cilicium (Boiss.) Bornm., Repert. Spec. Nov. Regni Veg. Beih. 89: 342 (1944).

Perennial herb to 70 cm. Stems unbranched, erect, grooved, pubescent to glabrescent. Leaves pubescent, to 20 cm; basal leaves with long petioles, lamina pinnatisect, lobes obtuse. Capitula numerous, in dense terminal corymbs; floral axis to 25 cm. Involucre campanulate, 4–5 mm broad; phyllaries 1-seriate, linear to oblanceolate, 4–4.5 mm, pubescent, apex scarious-lacerate. Marginal florets: ligules white, ± 2 mm; disc florets: corollas 2 mm, tubular, 5-toothed above. Achene ± 2 mm, 5-ribbed, ribs white; corona with irregular margins.

HAB. In oak forest, in shade, near stream; known from a single collection; alt. not known; fl. Aug.
DISTRIB. **MAM**: Dohok nr Sersang, *Agnew & Haines* W.2055.

Lebanon, Turkey.

2. **Tanacetum balsamita** (*L.*) *Sch.Bip.*, Tanaceteen: 55 (1844)
subsp. **balsamitoides** (*Sch.Bip.*) *Grierson* in Notes Roy. Bot. Gard. Edinburgh 33: 434 (1975); Grierson in Fl. Turkey 5: 266 (1975); Podlech in Fl. Iranica 158: 98 (1986).

Chrysanthemum balsamita L., Sp. Pl. ed. 2: 1252 (1763) non *Tanacetum balsamita* L.
Matricaria balsamita (L.) Desr. in Lam., Encycl. 3:733 (1792).
Pyrethrum balsamita (L.) Willd., Sp. Pl. 3:2153 (1803); Rawi, in Dep. Agr. Tech. Bull. 14: 122 (1964).
Tanacetum balsamitoides Sch.Bip., Tanaceteen: 51 (1844).

Chrysanthemum balsamitoides Nábělek in Publ. Fac. Sci. Univ. Masaryk Bmo 52:22 (1925).
Pyrethrum balsamitoides (Nábělek) Tzvel. in Fl. URSS 26:196 (1961). Ic: Nábělek in Publ. Fac. Sci. Univ. Masaryk Bmo 52:'t. 5 f. 2 (1925) as *Chrysanthemum balsamitoides*.

Perennial rhizomatous herb, aromatic, 20–70 cm tall. Stems several arising from the woody rhizome, ± erest, branched above, sparsely pubescent. Leaves yellowish-green, glabrous to adpressed-pubescent, punctate glandular; petiole slender 2–4 cm; lamina ovate-oblong to ovate-elliptic, 2–6 × 0.8–3 cm, base cuneate, margins crenate to serrate, pubescent; cauline leaves similar, reduced in size, sessile. Capitula radiate, heterogamous, in loose corymbs on long floral axes. Involucre hemispherical; phyllaries 2–3-seriate, outer lanceolate, 2 mm; apex blackish-brown; inner oblong-lanceolate, 3–4 mm, ± pubescent, scarious-lacerate on apices. Receptacle flat to convex. Marginal florets female, ligules white, 4–5 mm, shallow3-lobed at apex; disc florets hermaphrodite, corollas yellow, tubular, 2.5–3 mm, tubular, 5-toothed at apex. Achenes ± 2 mm, ribs 6–8, white, with sessile glands; apex with irregularly lobed corona.

HAB. Mountains slopes, bedside road in *Astragalus* zone, in orchard under trees and near water; alt. 1450–2800 m.; fl. Jul.-Sep.
DISTRIB. **MAM**: Ser Amadiya, *Haines* W. 2056. **MRO**: Mergan to near Bardanes, *Rawi* 24363!; between Haji Umran and Iranian frontier, *Rawi* 22463!; Road between Bardanas and Qandil range, *Rawi* 224633!; Road between Bardanas and Haji Umran, *Rawi* 24641!; Kani Sawaran range, nr Alana, *Rawi* 24670!; Halgurd range, Goum Tawerah, *Rawi* 24725!; Sula Khal, *Rawi* 24691!; Kani Rash, Qerna Qaw valley, *Rawi & Serhang* 26674!; mt. Serin on road to Qandil, *Rawi & Serhang* 24025!; Pushtashan, 15 km NE of Rania, *Rawi & Serhang* 24208!; mt. Sakran, SE of Chuman, *Dabbagh & Hamad* 46228, 46238!; Halgurd Dagh, *Guest* 2824, 2869 & 2917, *Gillett* 9484!; Lalan to Halgurd Dagh, *Gillett* 12467!; Haji Umran, Dabbagh & Hamad 46280, *Qaisi & Hamad* 43550.

Armenia, Turkey, Iran.

GULIKASPI (Kurd. *Gillett* 9484).

Grierson (l.c.) maintains that the two subsp. are very similar vegetatively, with the main difference is the presence of ray florets (subsp. *balsamitoides*) and their absence (subsp. *balsamita*). Subsp. *balsamita* is more widespread.

3. **Tanacetum parthenium** (*L.*) *Sch.Bip.*, Tanaceteen: 55 (1844); Grierson in Fl. Turkey 5: 268 (1975); Heywood in Fl. Eupop. 4: 171 (1976); Meikle, Fl. Cyprus 2: 923 (1985); Podlech in Fl. Iranica 158: 98 (1986); Ghafoor & Ronaq in Fl. Pakistan 207: 54, f. 9 (2002); Breckle, Hedge & Rafiqpoor, Vasc. Pl. Afghan. Checkl.: 190 (2013).

Matricaria parthenium L., Sp. Pl.: 890 (1753).
Pyrethrum parthenium (L.) Sm., Fl. Brit. 2: 900 (1800).
Chrysanthemum parthenium (L.) Bernh., Syst. Verz. Erfut: 145 (1800).
Pyrethrum parthenifolium Willd., Sp. Pl. 3: 2156 (1803); Rawi, in Dep. Agr. Tech. Bull. 14: 122 (1964). Dinsmore in Post, Fl. Syria, Palest. & Sinai ed. 2, 2: 62 (1933).
Tanacetum parthenifolium (Willd.) Sch.-Bip., Tanaceteen: 56 (1844); Grierson in Fl. Turkey 5: 268 (1975).
Pyrethrum persicum Boiss., Diagn. Pl. Orient. ser.1, 6: 89 (1846).
Chrysanthemum persicum (Boiss.) Nábělek, Publ. Fac. Sci. Univ. Masaryk 52: 21 (1925).

Annual or perennial, strongly aromatic herbs, 13–30 cm tall. Stems leafy. Leaves glabrous to pubescent, punctate glandular; petiole slender 2–4 cm, expanding at base; lamina ovate oblong in outline, 2–7 × 1.5–5 cm, pinnatisect into obtuse segments; cauline leaves similar to basal leaves but reduced in size, sessile. Capitula radiate, heterogamous in loose corymbs; floral axis long. Involucre hemispherical; phyllaries 2–3-seriate, pubescent, outer lanceolate, ± 3 mm, keeled; inner oblanceolate, 3–4 mm, ± pubescent, scarious-lacerate on apices. Receptacle flat to convex. Marginal florets female, ligules white, 6–7 mm, shallow3-lobed at apex; disc florets hermaphrodite, corollas yellow 2.5–3 mm, tubular, 5-toothed at apex. Achenes ± 1 mm, 6–8 white ribs and sessile glands; apex with irregularly lobed corona. Fig.

HAB. Mountain sides, in stony soil, near streams; alt. 600–2500 m; fl. May-Aug.
DISTRIB. **MRO**: Mergan to near Bardanes, *Rawi* 24363!; Hadren Dagh, Rowandaz, *Gillett* 8287!; Hawara Blinda mt. near NE Haji Umran, *Kass & Noori* 27812! **LCA**: Zafaraniya, *Sahira* C-973 (cult.)!

Africa, SE & E Europe, SW Asia (Turkey to Caucasus), Iran, Afghanistan, Pakistan, C Asia; naturalized in many countries.

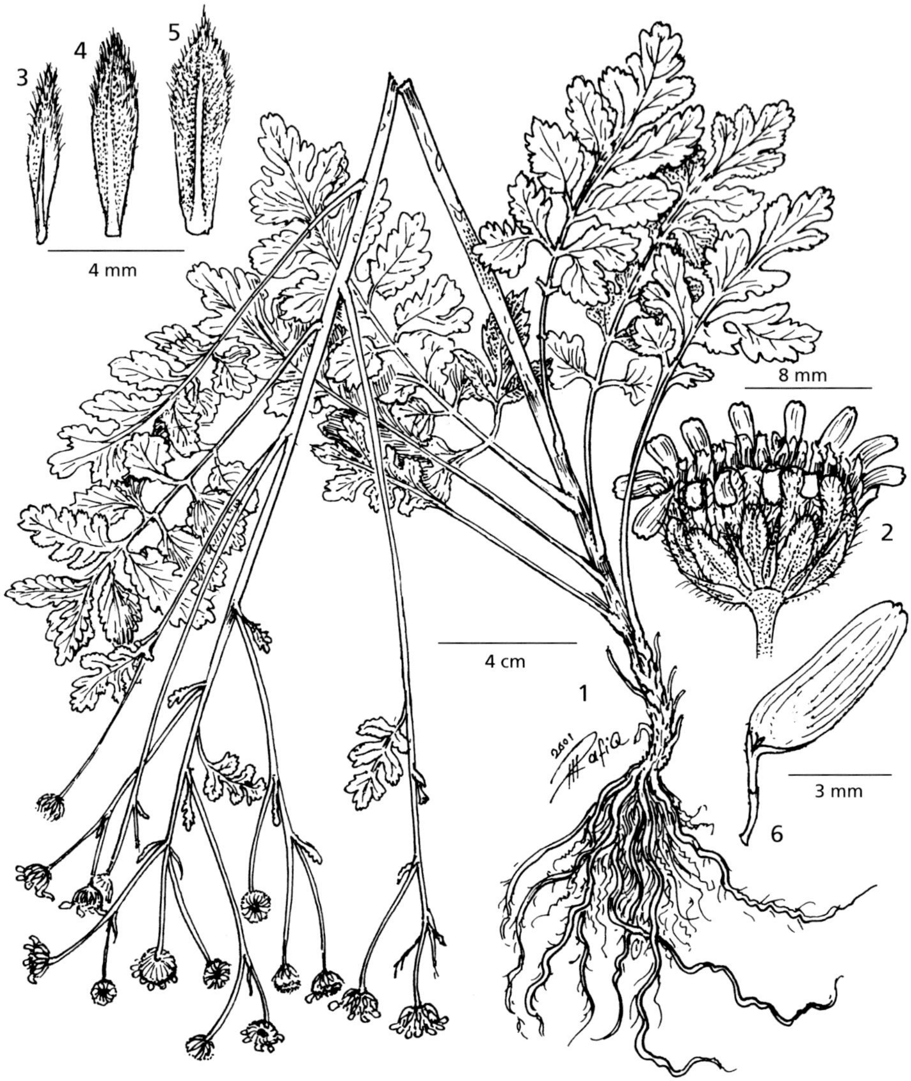

Fig. 91. **Tanacetum parthenium**. 1, habit; 2, capitulum; 3, 4, 5, phyllaries; 6, ray floret. Drawn by Rafiq. Reproduced with permission from Flora of Pakistan 207, fig. 9.

4. **Tanacetum kotschyi** (*Boiss.*) *Grierson* in Notes Roy. Bot. Gard. Edinburgh 33(3): 435 (1975); Grierson in Fl. Turkey 5: 276 (1975); Podlech in Fl. Iranica 158: 102 (1986).

Pyrethrum kotschyi Boiss., in sched. in Kotschy, Pl. Pers. Austr. Nr. 770 (1845); Boissier in Diagn. Pl. Orient. ser 1, 6: 88 (1846); Nábělek, Publ. Fac. Sci. Uni. Masaryk (Brno) 52: 19 (1925); Rawi, in Dep. Agr. Tech. Bull. 14: 122 (1964).

Aromatic perennial with several stems arising from a woody rootstock. Stems 15–25 cm, erect, densely grey-green pubescent. Basal leaves broadly oblanceolate in outline, grey-green pubescent with sessile glands, 1–3 cm, palmatisect with linear segments; petiole to 10 cm expanded at base; cauline leaves with fewer segments, reduced in size, sessile. Capitula solitary; peduncles 5–12 cm, leafless. Involucre hemispherical, 1–1.5 cm broad; phyllaries in several serieswith blacjish-brown scarious lacerate margins; outer phyllaries lanceolate 3–4 mm; inner oblanceolate, 5–6 mm. Marginal florets: ligules white, 7–11 mm, apex shallowly 3-toothed at apex; disc florets yellow, corollas tubular, 3–4 mm. Achene ± 2 mm, brown 4–5-ribbed; corona developed on posterior side of achene, irregularly notched at apex.

HAB. Stony mountain slopes, alpine rocks, along roadside; not uncommon; alt. 1600–3500 m; fl. Jul.-Sep.
DISTRIB. **MRO**: Chia Mandali, Rowandaz, *Guest & Ludlow-Hewit* 2700!; southern slope of Karoukh mt., Kass & *Noori* 27528!; 27403!; Ser Kurawa, *Gillett* 9737!; rd to Bandanas & Haji Omran, *Rawi* 24636!; Qandil range, NE of Rania, *Rawi & Serhang* 18255!, 24458!, 26717!, *Rawi* 24097!; distr. Erbil, *Rechinger* 11134!; Erbil, Helgurd mts., *Rechinger* 11432!; Algird Dag, *Guest* 3064!; Helgurd range, *Rawi* 24745!, 24829!; Perrish mt., *Rawi* 24539!

Caucasus, Turkey, Iran.

5. **Tanacetum nivale** *Sch.Bip.*, Tanaceteen: 60 (1844), non *Pyrethrum nivale* Adam ex Sprengel (1826). –Type. Iraq, in monte Kurdistanico Gara ad terminum nivalem, *Kotschy* 605 (P, holo. [P00706096!]). Rawi, in Dep. Agr. Tech. Bull. 14: 122 (1964); Grierson in Fl. Turkey 5: 269 (1975); Podlech in Fl. Iranica 158: 104 (1986).

Perennial glabrous herb with several stem arising from a horizontal woody rhizome. Stems ± erect, unbranched, Basal leaves petiolate, pinnatisect, glabrous; upper leaves similar but smaller in size, sessile. Capitula solitary. Peduncle leafless, 3–5 cm. Involucre 1 cm broad; phyllaries lanceolate, 4–5 mm, margins pale brown, scarious; inner phyllaries with scarious lacerate apex. Marginal florets: ligules 7–10 mm, white; disc florets ± 3 mm. Achenes ± 2mm, 5-ribbed; corona divided to base into 6–8 lanceolate scales.

HAB. Not given; alt. ± 2000m; fl. Sep.
DISTRIB. **MAM**: Gara Dag, 1841, *Kotschy*, 605 type; 405; Th. Kotschy inter cilicico-kurdicum 1859 *Kotschy*, 460!

Iran, Turkey.

6. **Tanacetum polycephalum** *Sch.Bip.*, Tanaceteen: 47 (1844); Podlech in Fl. Iranica 158: 116 (1986); Rawi, in Dep. Agr. Tech. Bull. 14: 123 (1964).

(Including subsp. *argyrophyllum* (C.Koch) Podlech in Rech.f., Fl. Iranica 158: 117 (1986) and subsp. *azerbaidjanicum* Podlech in Rech.f., Fl. Iranica 158: 119 (1986)).

subsp. **polycephalum**

Pyrethrum polycephalum (Sch.Bip.) Boiss., Fl. Orient. 3: 351 (1875). – Type. Iraq, mt. Gara, *Kotschy* 341 (G, K!, W).
Tanacetum argyrophyllum (C.Koch) Tzvel. var. *polycephalum* (Sch.Bip.) Grierson, Notes Roy. Bot. Gard. Edinburgh 33: 434 (1975), comb. illegit.

Grey-green perennial 25–60 cm tall, with several stems arising from a woody rootstock. Stems grooved, densely grey-pubescent and glandular. Basal leaves to 19 cm with along petiole to 8 cm; lamina bipinnatisect with ± linear obtuse segments; cauline leaves similar but sessile. Capitula many in terminal corymbs. Involucre ± 5 mm broad, hemispherical; phyllaries 1–2-seriate; outer phyllaries oblong-lanceolate, ± 3 mm, pubescent with brown scarious, lacerate apex; inner phyllaries oblanceolate, 2.5–3 mm, apex brown scarious, lacerate. Marginal florets: ligules scarcely longer than the disc florets; disc florets yellow, corollas tubular, ± 2 mm, 5-lobed at apex. Achene ± 2.5 mm, 4–5-ribbed; corona ± entire.

HAB. Stony mountain slopes, limestone screes and amongst oak; alt. 1000–2600 m; fl. May-Aug.
DISTRIB. Recorded to be locally common. **MAM**: Jabal Khantur, *Rechinger* 10814!, *Rawi* 23436!; Gali Zawita, NE of Zakho near Turkish frontier, *Rawi* 23548!; Sersang, slopes of Gara Dagh, *E. Chapman* 26379!; mt. Maya 25 km W of Kani Masi, *Hamid & Fadhil* 45509!; Alho, *Rawi* 8668!; N of Bijara, *Gillett* 11794!; mt. Gara, *Kotschy* 341! (type of *Pyrethrum polycephalum* C.H. Schultz Bip.); Ser Amadiya, Gali Mazurka, *Dabbagh, Qaisi & Hamid* 46015!, *Dabbagh & Jasin* 46920!; mt. Maya 25 km W of Kani Masi, *Hamid & Fadhil* 45509!; Dori (Dudi), 5 km E of Kani Masi, *Omar & Qaisi* 45420! Kopi Qara Dagh, *Haines* 1554 (subsp. *argyrophyllum* (C.Koch) Podlech). **MRO**: Qandil range, *Rawi & Serhang* 26727!; Pushtashan, 15 km NE of Rania, *Rawi* 24189!; mt. Sakri-Sakran, *Bornmüller* 1396!; Qandil, *Rawi* 24464 (subsp. *azerbaidjanicum* Podlech). **MSU**: Tawila by Mt. Avroman (Hawraman) near Iranian frontier, *Rechinger* 10375!; mt. Zalam, *Rawi, Alizzi & Noori* 29380!; Kamarspa, on road from Halabja to Tawila, *Rawi* 22209!; mt. Hawraman, Hawara Birza, *Rawi, Alizzi & Noori* 29344!

Iran.

subsp. **heterophyllum** (*Boiss.*) *Podlech* in Fl. Iranica 158: 117 (1986).

Tanacetum heterophyllum, Boiss., Diagn. Pl. Orient. ser. 1, 6: 90 (1845).
Pyrethrum szovitsii Boiss, Diagn. Pl. Orient. ser. 1, 11: 25 (1849) p.p.
P. myriophyllum C.A.Mey. var. *variegatum* Boiss., Fl. Orient. 3: 351 (1875).
P. myriophyllum auctt. non C.A.Mey.: Rawi, in Dep. Agr. Tech. Bull. 14: 122 (1964).

Similar to subsp. *polyphyllum* and growing in the same habitats, sometimes difficult to tell apart. Differing mainly in the outer phyllaries narrow lanceolate, distinctly keeled dorsally and inner phyllaries with very narrow scarious margins.

HAB. Stony mountain slopes; alt. 1350–2600 m; fl. May-Aug.
DISTRIB. **MAM**: Jabal Khantur, NE Zakho, *Rawi* 23378, *Rawi et al.* 29061!; Ser Amadiyah *Evans & Guest* 4990! **MRO**: Arbil, Haji Omran *Rechinger* 11321; Algurd Dag, *Rechinger* 11355; Kani Mazi Shirin nr mt. Botin, *Agnew & Hadač* 5977; Qandil nr. Iran, *Rechinger* 11148; above Pushtashan *Rechinger* 11092; Kermasur, *Rawi* 24130!; Malakh, *Rawi* 24031!; Chiya-I Mandau, *Guest* 2798! **MSU**: Avroman, *Rawi et al.* 19732!; Pira Magrun, *Rawi* 12050!; Kopi Kara Dagh *Haines* 1086!

Iran.

7. **Tanacetum aureum** (*Lam.*) *Greuter, Agab. & Wagenitz* in Taxon 54: 156 (2005).

Achillea aurea Lam., Encycl. 1: 26 (1783).
Pyrethrum chiliophyllum Fisch. & C.A.Mey. ex DC., Prodr. [A. P. de Candolle] 6: 59 (1838).
Tanacetum chiliophyllum (Fisch. & C.A.Mey. ex DC.) Sch.Bip., Tanaceteen: 47 (1844); Podlech in Fl. Iranica 158: 113 (1986).
Pyrethrum millefoliatum auctt. non (L.) Willd. (1803): Rawi, in Dep. Agr. Tech. Bull. 14: 122 (1964).

Grey-green perennial to 20 cm with several stems arising from a woody rootstock. Stems grooved, densely pubescent. Basal leaves oblong to oblanceolate in outline, 2–3 pinnatisect, segments oblong, obtuse, sedsely pubescent; petiole 1–2 cm, expanded at base and decurrent; cauline leaves similar but reduced in size and sessile. Capitula heterogamous, in terminal corymbs. Involucre ± 5 mm broad; phyllaries several-seriate, imbricate, glabrous to pubescent; outer phyllaries ovate-lanceolate, 2–2.5 mm, nargins brown, apex scarious, lacerate; inner 3–4 mm, mostly glabrous, apex scarious lacerate. Marginal florets female; ligules yellowish, ± 2 mm, apex 3-dentate; disc florets hermaphrodite, corollas ± 2mm, tubular. Achenes ± 2 mm, 4–5-ribbed; corona short, entire.

HAB. On side of mountain; alt. 2850–3100 m; fl. Aug.
DISTRIB. **MAM**: NE Qandil, *Rawi* 24426.

Iran, Caucasus, NE Anatolia, Armenia, Azerbaijan.

8. **Tanacetum gillettii** *Podlech* in Rech.f. Fl. Iranica 158: 140 (1986). –Type. Iraq, Erbil: Ser Kurawa, *Gillett* 9784 (K!, holo.).

Perennial herb, sparsely to densely pubescent, glandular. Stems many, to 12 cm, ascending. Basal and lower cauline leaves 10–18 mm, petiolate; petiole 5–10 mm; lamina ovate to almost orbicular, 2-pinnatisect, segments ovate-obling, acute; cauline leaves similar, reducing in size, sessile. Capitula solitary, homogamous. Involucre napiform, 8–10 mm broad; Phyllaries 3-seriate, hirsute; outer phyllaries lanceolate, acute; inner phyllaries longer twice longer than the inner, oblong, apex scarious, lacerate. Receptacle convex. Florets, yellow, tubular, ± 4 mm, apex 5-dentate. Achene cylindrical, 3.5–4 mm, 5–6-ribbed, glandular; corona oblique, divided into 6–8 linear-oblong scales. (Descript. from in Fl. Iranica (1986)).

HAB. On mountain, amongst rocks; alt. ± 2990 m; fl. Aug.
DISTRIB. **MRO**: Ser Kurawa, *Gillett* 9784 (type, not located).

Endemic.

9. **Tanacetum argenteum** (*Lam.*) *Willd.*, Tract. De Achilleis 51, t. 2, f. 4 (1789); Dinsmore in Post, Fl. Syria, Palest. & Sinai ed. 2, 2: 63 (1933).
subsp. **canum** (*C.Koch*) *Grierson*, Notes Roy. Bot. Gard. Edinburgh 33: 427 (1975); Grierson in Fl. Turkey 5: (1975); Podleck in Fl. Iranica 158: 135 (1986).

Whitish- silvery-tomentose perennial with a woody rootstock covered with remnants of old leaves. Stems to 20 cm, almost leafless above. Basal leaves 2–3-pinnatisect with lanceolate segments; petiole 1–3 cm; cauline leaves similar to basal but reducing in size, fewer divisions and sessile. Capitula homogamous, in dense terminal corymbs. Involucre 3-4 mm broad, campanulate. phyllaries ± densely sericeous; outer phyllaries linear-oblong, ± 3 mm, margins and apex whitish scarious; inner oblanceolate, 5mm. Florets yellowish, all tubular, hermaphrodite, glandular-punctate. Achenes 5-ribbed, glandular; corona unilateral, developed on posterior side of achene apex ± denticulate.

HAB. Mountains; known from a single collection from Iraq; alt. & fl. not given.
DISTRIB. **MAM**: Ser Amadiyah, *Haines* 2061.

Turkey, Lebanon, Caucasus.

93. **TRIPLEUROSPERMUM** Sch.Bip.

Tanaceteen: 31 (1844)
Oberprielor, Vogt & Watson in Kubitzki, Fam. Gen. Vasc. Pl. 8: 342 [2006] (2007)

Shahina A. Ghazanfar

Annual or perennial herbs, ± glabrous. Leaves alternate, 1–3-pinntisect. Capitula solitary or in corymbs, pedunculate, discoid, disciform or radiate. Involucre ± hemispherical; phyllaries in 2–5 rows, margins scarious, pale brown to black. Receptacle convex to conical, epaleate. Ray florets, female, fertile, limb white or pale pink; disc florets bisexual, fertile; corolla 5-lobed, yellowish or greenish, lobes with a resin sac at apex. Achenes triquetrous with one adaxial and 2 lateral ribs or 1–2 abaxial ribs, rugulose to tuberculate abaxially and between ribs. Pappus absent or coronate, adaxially auricled or a few scales or sometimes rounded; pericarp with myxogenic cells with (1–)2(–3–5) apical resin sacs.

Forty species in Europe, northern Africa, temperate Asia and N America; 4 species in Iraq.

1. Capitula in lax corymbs, disciform; marginal florets eligulate; pappus absent . . . . 4. *T. disciforme*
   Capitula solitary, radiate; marginal florets ligulate; pappus coronate 2
2. Phyllaries glabrous, margins white scarious . . . . 1. *T. auriculatum*
   Phyllaries sparsely hirsute, margins reddish- blackish-brown, scarious 3
3. Perennial rhizomatous herb; phyllaries with brownish-black margins; receptacle hemispherical. . . . . 2. *T. caucasicum* var. *melanolepis*
   Annual herb; phyllaries with reddish-brown margins; receptacle conical. . . . . 3. *T. parviflorum*

1. **Tripleurospermum auriculatum** (*Boiss.*) *Rech.f.*, Fl. Lowland Iraq: 629 (1964); Täckholm, Stud. Fl. Egypt ed. 2: 577 (1974); Batanouny, Fl. Ecol. Qatar: 171 (1981); Collenette, Wild Flow. Saudi Arabia: 222 (1999); Boulos & Hind in Fl. Egypt 3: 251 (2002); Taifour & El-Oqlah, Pl. Jordan Ann. Checkl.: 48 (2017).

Chamaemelum auriculatum Boiss., Diagn. Pl. Orient. ser. 1, 11: 23 (1849); Dinsmore in Post, Fl. Syria, Palest. & Sinai ed. 2, 2: 59 (1933).
Matricaria auriculata (Boiss.) Ball, J. Linn. Soc., Bot. 16: 512 (1878).
M. auriculata auct. non Benth. & Hook.f.: Gen. Pl. [Bentham & Hooker f.] 2(1): 428 (1873), nom. inval.; Rawi, in Dep. Agr. Tech. Bull. 14: 120 (1964).
M. auriculata Muschl., Man. Fl. Egypt ii. 1011 (1912).

Annual herb, 10–25(–30) cm tall, branching from base with stems ascending, glabrous. Leaves 10–30 mm (including petiole), deeply 1–2-pinnatisect with linear-lanceolate to linear-filiform lobes, sparsely hairy; cauline leaves similar and ± sessile. Capitula solitary, 10–15 mm in diameter, radiate; peduncle 6–11 cm, slender, leafless. Involucre hemispherical; phyllaries oblong, 2–3 mm including the broadly scarious margins, obtuse at apex, glabrous. Receptacle ovate. Ray limbs 2–4 mm, white, apex minutely dentate; disc florets yellowish, corollas 2–3 mm, eglandular at apices. Achenes oblong, 1–1.5 mm, abaxially 3-ribbed, corona as long as the achene. Fig. 93: 8–10.

HAB. In sandy and gravelly and clayey and rocky soil, sandy depressions, edges of wheat fields; alt. 150–400 m; fl. Mar.-May.
DISTRIB. **DWD**: 10 km from Rutba to H3, *Botany Staff* 41749!; Nukhaib, *Rawi* 13861!; Al Sabz, 2 km SW of H2, *Chakravarty, Rawi, Alizzi & Khatib* 31464!; Just W of H2, *Barkley* 5360!; 18 km W of T1 to Al Qaim, *Omar, Qaisi, Hamad & Hamid* 45083!; Al Massad S of Rutba, *Alizzi* 35195! *Omar & Hamid* 36686!; Fathat Imam 70 km SW of Nukhaib, *Rawi* 31072!; Faithat Al Masad, *Jenan Al Mokhtar* 33325!; 95 km NW of Rutba, *Rawi & Khatib* 32330!; 120 km NW of Rutba, *Omar* 34248!; Jadida, W of Nukhaib, *Rawi* 31076!; 5 km above Rawa, *Rawi & Gillett* 7024!; ?Between Baghdad and Syrian desert, *Jackson* H.2401!

Egypt, Jordan, Palestine, Lebanon, Syria, Arabia.

2. **Tripleurospermum caucasicum** (*Willd.*) *Hayek*, Denkschr. Akad. Wiss. Math.-Nat. Kl. 99: 204 (1924) var. **melanolepis** (*Boiss. & Buhse*) *Podlech* in Fl. Iranica 158: 75 (1986).

Chamaemelum melanolepis Boiss. & Buhse in Mem. Soc. Nat. Mosc. 12: 118 (1860).
C. kochii Boiss., Fl. Orient. 3: 329 (1875).
Matricaria melanophylla Nábělek in Publ. Fac. Sci. Univ. Masaryk Brno 52:17, t. 3 (no. 3), f. 4, (1925).
M. arlgirdensis Blakelock, Kew Bull. 4(1): 58, f. 3 (1949). –Type. Arl Gird Dagh [Halgurd Dagh], 3300 m., 22 July 1932, *Guest* 3065 (K!, holo.); Rawi, in Dep. Agr. Tech. Bull. 14: 120 (1964).
Tripleurospermum melanolepis Pobed., Bot. Mater. Gerb. Bot. Inst. Komarova Akad. Nauk S.S.S.R. 20: 490 (1960); Hossain in Fl. Turkey 5: 298 (1975).

Perennial rhizomatous herb 6–15 cm with erect to ascending stems, sparsely hirsute. Basal leaves 10–30 mm, (including petiole of 3–4 mm), 1–2-pinnatisect with linear-lanceolate to linear-filiform lobes, sparsely hairy; cauline leaves similar but sessile and smaller in size. Capitula solitary, 1–1.5 cm in diameter, radiate; phyllaries oblong-elliptic, *c.* 4 × 2–5 mm, sparsely hirsute, margins brownish-black, scarious. Receptacle hemispherical. Ligules 7–10 mm, white, apex minutely 3-dentate; disc florets yellowish, corolla 3–4 mm. Achenes (immature) angular-oblong, up to 1.5 mm, with 3 prominent ribs; corona 2–3-lobed.

HAB. In the mountain region, on mountain slopes, in sandy, gravelly soil; alt. 2800–3000 m; fl. Jul.
DISTRIB. **MRO**: Qandil range, *Rawi & Serhang* 24493! 24741! 26730! 26759! 26779! 26784! 24572!; Ser Kurawa, *Gillett* 9776!; Halgurd Dagh, *Gillett* 9608! *Guest & Alizzi* 15822! *Guest* 3065 (type of *M. arlgirdensis* Blakelock)

Iran, Turkey, Georgia, Armenia.

3. **Tripleurospermum parviflorum** (*Willd.*) *Pobed.*, Bot. Mater. Gerb. Bot. Inst. Komarova Akad. Nauk S.S.S.R. 21: 354. 1961; Hossain in Fl. Turkey 5: 302 (1975); Podlech in Fl. Iranica 158: 75 (1986); Ghafoor in Fl. Pakistan 207: 34 (2002).

Pyrethrum parviflorum Willd., Sp. Pl. ed. 4, 3(3): 2158 (1803).
P. praecox M.Bieb., Fl. Taur.-Cauc. 2: 324 (1808).
Marticaria parviflora (Willd.) Poir. in Lam., Encycl. Method. Bot. Suppl. 3: 608 (1814); Collenette, Wild Flow. Saudi Arabia: 204 (1999).
Chamaemelum praecox (M.Bieb.) Vis, Osserv. Matric. 13 (1845); Dinsmore in Post, Fl. Syria, Palest. & Sinai ed. 2, 2: 59 (1933).
Tripoleurospermum praecox (M.Bieb.) Bornm. in Beih. Bot.Centralbl. 60: 192 (1940); Rechinger, Fl. Lowland Iraq: 628 (1964).

Annual herb, strongly aromatic, up to 35 cm tall with ascending stems, glabrous to sparsely pubescent. Leaves bright green, 30–50 (including petiole), deeply 1–2-pinnatisect with linear-lanceolate to linear-filiform lobes, sparsely hairy; cauline leaves similar and sessile to very shortly petiolate. Capitula solitary, ± 15 mm in diameter, radiate; peduncle 3–10 cm, slender, leafless. Involucre hemispherical; phyllaries narrow-oblong, 2–3 mm, sparsely hirsute, margins reddish-brown, scarious. Receptacle conical. Ligules 3–5 mm, white, apex minutely 3-dentate; disc florets yellowish-orange, corolla 2–3 mm, 4-toothed, eglandular at apices. Achenes oblong, up to 1–1.5 mm, abaxially 3-ribbed and adaxially 2-ribbed, smooth, pericarp with myxogenic cells; corona 2–3-lobed.

HAB. In the Upper Jazira District, in open grassland in the foothills, western and southern desert and alluvial plain; locally common; alt. 250–400 m; fl. Jul.
DISTRIB. **FUJ**: Mosul experimental farm, *Rawi & Gillett* 10361!; Zummar, Tal Afar, *Mudur of Zummar* 3143!; Jezira N of Balad Sinjar, *Guest* 13303!; Hammam Ali, ± 30 km S of Mosul, *Hossain* 511; Qa'la

Shargat, *Guest* 405!; Shargat to A-Hadhr, *Ani* 9442! **FNI**: Mosul, *Shahwani* 25175! **FAR**: 27 NE of Arbil, on road to Shaqlawa, *Shahwani* 25134!; Arbil province, *Radhi* 3848!; Wadi Bastura between Arbil and Salahuddin, *Barkley & Abbas* 7061! **FKI**: Kirkuk, *Rogers* 0458!; **FPF**: Qaraghan (Jalaula), *Guest* 1879A!; 37 km W of Khanaqin, *Cowan & Darlington* 194. **DWD/DSD**: Gaiyara road, *Rogers* 50. **LEA**: Nahrwan, *Robertson* 91!; 10 km SW of Baquba, *Anderson* 1971!

BEIBUN (*Rawi & Gillett* 10361, but not the true plant of that name which grows further south), BAIBŪN (*Mudur of Zummar* 3143, *Guest* 13303), RUBYĀN (*Radhi* 3848). Recorded to be grazed by animals and used as a fodder when dry. Said to be used to dye grey hair in Syria (*Rawi & Guest* 10361).

Russia, Georgia, C Asia, Pakistan, Iran, Turkey, Saudi Arabia.

4. **Tripleurospermum disciforme** (*C.A.Mey.*) *Sch.Bip.*, Tanaceteen: 34 (1844); Hossain in Fl. Turkey 5: 310 (1975); Podlech in Fl. Iranica 158: 75 (1986); Pobed in Fl. U.S.S.R. 26: 174 (2000, Engl. transl.); Ghafoor in Fl. Pakistan 207: 34, f. 5 (2002); Breckle, Hedge, Podlech & Rafiqpoor, Vasc. Pl. Afghan. checklist: 196 (2013).

Chrysanthemum disciforme C.A.Mey., Verz. Pl. Cauc.: 75 (1831).
Matricaria disciformis (C.A.Mey.) DC., Prodr. 6: 51 (1838).

Perennial or biennial herb, aromatic, up to 50 cm tall, with erect branched, often woody stems, glabrous, ± grooved. Leaves 30–60 mm, lower leaves with a long petiole 3–5 mm, expanding at base, lamina deeply 1–2-pinnatisect with linear-lanceolate to linear-filiform lobes, ultimate lobes aristate-acute at apex; cauline leaves smaller in size, sessile. Capitula in lax corymbs, 8–12 mm in diameter, heterogamous; peduncle 4–10 cm. Involucre hemispherical; phyllaries in 2 rows, outer broadly obovate, inner oblanceolate, obtuse, margins broad to narrow, white-membranous. Receptacle ovoid-oblong. Marginal florets female, eligulate, 2-toothed; disc florets yellowish, corolla 3 mm, 5-toothed, with red glands at apices. Achenes triquetrous, up to 1.5 mm, abaxially 3-ribbed, smooth or obscurely tuberculed, with 2 glands dorsally. Pappus absent. Fig. 92: 1–9.

HAB. Mountain sides, in rocky and clayey soils; alt. 1000–2200 m; fl. Jul.
DISTRIB. In the Upper Jazira District, foothills, western and southern desert and alluvial plain. **MRO**: Haji Umran, *Qaisi & Hamad* 43557! *Dabbagh & Hamad* 46300! *Guest & Alizzi* 15907! *Rawi* 24285! *Rawi* 24309!; N of Haji Umran to Iranian frontier, *Rawi* 24962!; Kholan, *Omar, Sahira, Karim & Hamid* 13707! Jabal Sakran 15 km SE of *Chuman, Dabbagh & Hamad* 46210!; Rayat, *Omar, Sahira, Karim & Hamid* 38443!; Halgurd Dagh, *Gillett* 12480! **MRO/MSU**: NE of and descent to Darband, *Gillett* 9535!

Turkey, Armenia, C Asia, Iran, Afghanistan, Pakistan.

94. **MATRICARIA** L.

Sp. Pl.: 890 (1753); Gen. Pl. ed. 5: (1754)
Oberprielor, Vogt & Watson in Kubitzki, Fam. Gen. Vasc. Pl. 8: 368 [2006] (2007)

J. R. Edmondson

Annual fragrant herbs; indumentum absent or of simple, basifixed hairs. Leaves 1- or 2-pinnatisect, with numerous linear segments. Inflorescences ± flat-topped. Capitula pedunculated, heterogamous or homogamous, radiate or discoid; phyllaries in 2 or 3 rows, margins scarious. Receptacle conical to subulate, lacking scales. Ray florets (if present) in 1 row, female, white, fertile. Disk florets yellow or pale green, bisexual; corolla tubular, apex 4- or 5-lobed. Anther bases obtuse; apical appendage triangular. Style branches linear, apex truncate. Achenes small, terete, slightly flattened and with 3–5 thin ribs; pappus reduced to a corona or absent.

Matricaria, from L. *matrix*, womb, so called because a decoction was used to treat menstrual disorders.

Applequist, W. L. (2003). A reassessment of the nomenclature of *Matricaria* L. and *Tripleurospermum* Sch.Bip. (Asteraceae). Taxon 51: 757–761.

Capitula discoid; phyllaries with conspicuous membranous brown margins; all florets tubular, yellow, 4-lobed . . . . 1. *M. aurea*

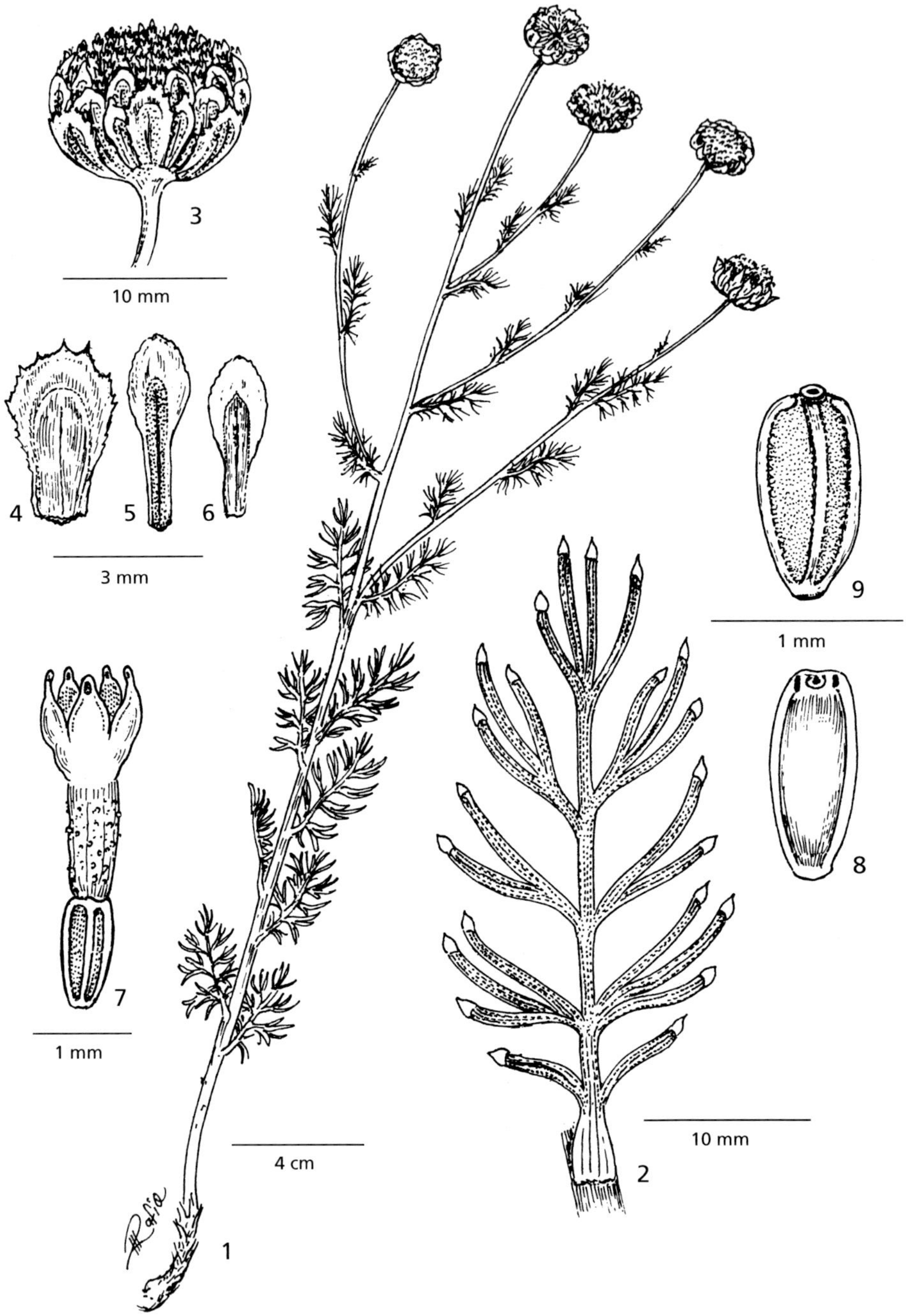

Fig. 92. **Tripleurospermum disciforme**. 1, habit, 2, leaf; 3, capitulum; 4, 5, 6, phyllaries; 7, disc floret; 8, achene (dorsal view); 9, achene (ventral view). Drawn by Rafiq. Reproduced with permission from Flora of Pakistan 207, fig. 5.

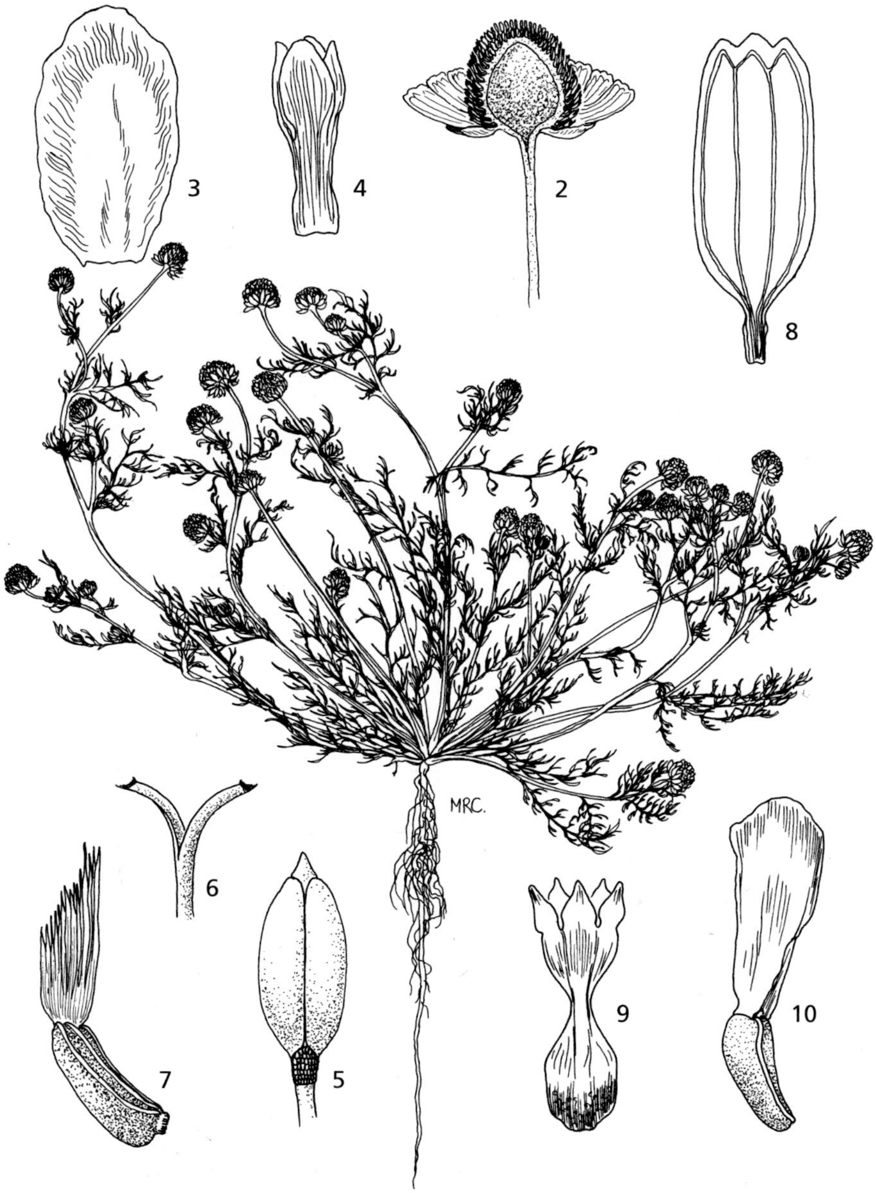

Fig. 93. **Matricaria aurea**. 1, habit × ½; 2, capitulum dissected × 2; 3, phyllary × 10; 4, floret × 20; 5, anther (enlarged); 6, style (enlarged); 7, achene with corona × 20. Tripleurospermum auriculatum. 8, ray floret × 6; disc floret × 8; 10, achene and pappus × 8 from *Ani & Haddi* 9452; 10 from *Alizzi* 35195. Drawn by Moira Rae Carter.

Capitula radiate; phyllaries with pale green membranous margins; disc florets white, 5-lobed . . . . 2. *M. chamomilla*

1. **Matricaria aurea** (*Loefl.*) *Sch.Bip.*, Bonplandia 8: 369 (1860); Boissier, Fl. Orient. 3: 324 (1875); Rawi, in Dep. Agr. Tech. Bull. 14: 120 (1964); Pobed in Fl. U.R.S.S. 26: 147 (2000); Grierson in Fl. Turkey 5: 295 (1975); Feinbrun-Dothan, Fl. Palaest. 3: 345 (1978); Podlech in Fl. Iranica 158: 81. t. 79 (1986); Jongbloed & al., Ann. Checkl.: Pl. UAE: 32 (2000); Chaudhary, Fl. Kingd. Saudi Arabia 2(3): 178 (2000); Boulos & Hind in Fl. Egypt 3: 253

(2002); Ghafoor in Fl. Pakistan 207: 28, f. 4A–G (2002); Kilian & Ghazanfar in Fl. Oman 3: 253 (2015); Abdulridha, Taha & Wedad, Ecology and Fl. Basra: 219 (2016); Taifour & El-Oqlah, Pl. Jordan Ann. Checkl.: 44 (2017).

Cotula aurea Loefl., Iter Hispan. 163 (1758).
Chamomilla aurea (Loefl.) Gay ex Cosson & Kralik, Cat. Pl. Syrie Palaest. 10 (1854); Kay in Fl. Europ. 4: 167 (1976).
Anacyclus aureus L., Mantissa Alt. 287 (1771).

Annual herbs, stems branched from base, often flexuous. Leaves alternate, 2–3-pinnatipartite, with narrowly linear acute segments to 0.5 mm wide, mucronulate. Capitula 4–5 mm in diameter, discoid. Phyllaries glabrous, oblong, obtuse. Corollas golden-yellow, with 4 teeth 1.5–2.5 mm. Achenes 0.8 × 0.2 mm (excluding corona), brown, ribbed on ventral face; pappus a dentate corona 0.6-0.8 mm. Fig. 93: 1–7.

HAB. In sandy and silty soils, semi-desert, fields, waste places, weed in gardens; alt. 140–1300; fl. Mar.-Jun.
DISTR. Largely in the Mesopotamian plains and desert areas of Iraq. **MRO**: Chinaruk, *Omar, Karim, Hamza & Hamid* 37215! **MSU**: Penjwin, *Rawi* 22751!; Qarachitan, *Gillett* 7708!, Mosul, Qaiydrah, *Bayliss* 90! **FUJ**: Bayder 78 km S of Balad Sinjar, *Chakravarty, Rawi, Khatib & Alizzi* 32090!; Tal Afar, *Qaisi & Hamad* 49233!; Jazira S of Balad Sinjar nr Ba'aj, *Guest* 13398! **FKI**: Kirkuk, *Rogers* 0488! **FPF**: Qaraghan (Jalaula), *Guest* 1879!; Wadi Gifgir nr Tarsaq, *Rawi* 12588!; Qizil Robat (Sa'diya), *Guest* 1772!; 5 km W of Rshaida PO near Iranian frontier, *Ani* 39817!; **DLJ**: Wadi Thirthar E of Hadra, *Rawi & Gillett* 7251!; midway between Rawa and Haditha, *Omar, Qaisi, Hamad & Hamid* 44623!; **DWD**: 15 km E of Baghdadi, *Omar, Qaisi, Hamad & Hamid* 44546!; bottom of Wadi Hawran 12 km SE of H2, *Chakravarty, Rawi, Khatib & Alizzi* 31617!; 8 km E of Ana, *Omar, Qaisi, Hamid & Hamad* 44348!; near K3, *Hamid* 39064!; 95 km NE of *Rutba, Rawi & Khatib* 32316! *Aftan Rawi* 23737!; 70 km W of Nukhaib, *Rawi* 31392!; 30 km W of Rahalliya, 180 km W of Karbala, *Chakravarty & Rawi* 29813!; **DWD/DLJ**: nr Haditha, *H. Hamid* 39154! ; 17 km E of Haditha, *Qaisi* 44238!; **DSD**: 55 km NW of Shabicha, *Guest, Rawi & Long* 14051!; depression S of Haidar (Khadhar) al-Mai, *Weinert, Fawzi, Adel & Hussain* 47595!; Ath-Thulaima nr Jumaima, 145 km SW of Salman, by Saudi frontier, *Guest, Rawi & Rechinger* 19087!; Takhaded 60 km S of Salman, *Botany Staff* 42217!; Salman, Al 'Ain al Hilwa, *Kasim, Haberi, Hamid & Kadhim* 40138!; Jiraibiyat 30 km W of Tib range fence, *Ani* 39809! **LEA**: 10 km SW of Ba'quba, *Anderson* 1963!; Ba'quba, *Paranjpye* (comm. R.J.D. Graham) s.n.!; Nahrwan, *V.C. Robertson* (R.A. 17)!; Sudur, nr Muqdadiya, *Rawi* 20009!; Nahrwan South, *Hunting Aerosurveys* 93! **LCA**: Baghdad, *Colvill* s.n.! *Rogers* 018!; Baghdad airport meteorological station, *Ani & Haddi* 9452!; Hillah, *anon.* (comm. R.J.D. Graham) s.n.!; Baghdad, Rustam, *Lazar* 3893! 10 km from Nasiriya to Chabaish, *Dabbagh & Qaisi* 41887!; Abu Ghraib, *Matti & Jumaily* 304!; Karradat Mariam, Baghdad, *Haines* W. 594! *Graham* s.n.!; **LSM**: 70 km E of Amara nr Rshaida well, *Ani* 39725!; 20 km N of Musharrah, *Ani* 39718!; 15 km N of Amara, *Guest, Rawi & Rechinger* 17416!; nr Azair, between Hor and R. Tigris, *Guest, Rawi & Rechinger* 17385! **LBA**: Basra to Fao, *Rechinger* 8615.

BABON (Kurd.-Qarachitan), *Gillett* 7708; BAIBŪN, Balad Sinjar nr Baaj (*Guest* 13398), BĀBUNNAJ, E of Hadra (*Rawi & Gillett* 7251); flowers used to treat stomach ache.

2. **Matricaria chamomilla** *L.*, Sp. Pl.: 891 (1753); Boissier, Fl. Orient. 3: 323 (1875); Grierson in Fl. Turkey 5: 293 (1975); Rawi, in Dep. Agr. Tech. Bull. 14: 120 (1964).

Matricaria recutita L., Sp. Pl.: 891 (1753); Pobed in Fl. U.R.S.S. 26: 147 (2000); Feinbrun-Dothan, Fl. Palaest. 3: 344 (1978); Podlech in Fl. Iranica 158: 82. t. 80 (1986); Meikle, Fl. Cyprus 2: 924 (1985); Boulos & Hind in Fl. Egypt 3: 251 (2002); Ghafoor in Fl. Pakistan 207: 30, f. 4H–Q (2002).
Chrysanthemum chamomilla (L.) Bernh., Syst. Verz. (Bernhardi) 145 (1800).
Chamomilla vulgaris Gray, Nat. Arr. Brit. Pl. 2: 454 (1821).
Matricaria chamomilla L. var. *recutita* (L.) Fiori, Nuov. Fl. Italia 2: 620 (1927); Grierson, Notes Roy. Bot. Gard. Edinburgh 33(2): 253 (1974).
Chamomilla recutita (L.) Rauschert, Folia Geob. Phytotax. 9: 255 (1974); Kay in Fl. Europ. 4: 167 (1976).

Glabrous annual herb 10–30 m, branched from base. Leaves 2-pinnatisect, segments linear, setaceous, flat, mucronate. Capitula 15–20 cm in diameter, long-stalked. Phyllaries oblong, obtuse, margins scarious. Ray florets 6–10 mm, white, reflexed; disc florets golden-yellow. Achenes ± terete, tapering towards base, apex truncate, 5-ribbed on ventral face; pappus reduced to a corona.

HAB. Waste places near cultivation; alt. 50–250 m; fl. Apr.-Jun.
DISTR. Scattered in Mesopotamian alluvial plain. **FNI**: Nenawa (Nineveh) farm, *Hikmat Abbas Al-Ani* 9772! Mosul to Seiramun, *Handel-Mazzetti* 1208; bank of r. Tigris nr Mosul, *Nábělek* 3638!

Europe, Mediterranean area, Palestine, Syria, Turkey, Iran. Often cultivated for its medicinal uses, and probably not native in Iran.

95. **AARONSOHNIA** Warb. et Eig

Bull. Zi. Org. Inst. Agr. Nat. Hist. 6: 40 (1927)
Oberprielor, Vogt & Watson in Kubitzki, Fam. Gen. Vasc. Pl. 8: 368 [2006] (2007)

Gail Bromley

Annual herbs with alternate, pinnatisect leaves. Capitula small, terminal on long scapiform pedicels, solitary, radiate, disciform or discoid; phyllaries 2–3-seriate, margins broadly scarious; receptacle naked, hemispherical to shortly conical. Marginal florets female, few, corollas urceolate, yellow. Disc florets numerous, hermaphrodite, corollas tubular, 4–5-dentate. Achenes dimorphic; marginal achenes broader, finely multistriate and rounded dorsally, angular ventrally; inner achenes smaller, narrower and more curved, not rounded or multistriate dorsally; all achenes with a ± developed (rarely absent) apical auricle, mucilaginous when moist.

A genus of two species, one of which occurs in Iraq.

Aaronsohnia (in honour of the Jewish botanist and agronomist Aaron Aaronsohn, 1886–1919).

1. **Aaronsohnia factorovskyi** *Warb. & Eig* in Bull. Zi. Org. Inst. Agr. Nat. Hist. 6: 40 (1927); Dinsmore in Post, Fl. Syria, ed. 2, 2: 60 (1933); Zohary in Dep. Agr. Iraq Bull. 31: 150 (1950); Rawi in Dep. Agr. Iraq Tech. Bull. 14: 104 (1964); Rechinger, Fl. Lowland Iraq: 629 (1964); Feinbrun-Dothan, Fl. Palaest. 3: 346, pl. 585 (1978); Dickson, Wild. Fls. Kuwait: 11 (1955); Collenette, Wild Flow. Saudi Arabia: 166 (1999); Al-Mayah, Al-Edani & Al-Asadi, Ecology & Fl. Basra: 187 (2016).

Annual, 6–20(–30) cm tall, glabrous to slightly white-pilose. Leaves arising from lower part of plant, somewhat fleshy, pinnatisect into linear, entire to pinnately lobulate lobes *c.* 1–9 × 0.5–2 mm. Capitula 6–12 mm in diameter, on slender pedicels 3–13 cm long according to size of plant; outer phyllaries *c.* 1–2 mm, deltoid-ovate, inner phyllaries oblong to ovate-oblong, obtuse, *c.* 3–4 mm; all phyllaries scarious-margined, inner more broadly so. Ray florets female, few, corollas urceolate with 3 minute apical teeth. Disc florets hermaphrodite, numerous, corollas tubular with 4–5 teeth, occasionally tinged with green and showing darker resinous striations. Outer achenes *c.* 1.5–2 × 1 mm; inner not exceeding 1.5 mm long and narrower, with a larger auricle mostly exceeding achene body. Fig. 94: 1–7.

HAB. In desert, in depressions and wadis, chiefly on sand, sandy gravel, loam and clay, also on rocky hills and rises and on silty plains; alt. 10–410 m; fl. & fr. Jan.-Apr.
DISTRIB. Common in the Southern Desert of Iraq, apparently only occasional in the Western Desert: **DWD**: 20 km S of Nukhaib, *Alizzi & Omar* 35523!; **DWD**/ **DSD**: 95 km W of Shabicha, *Khatib & Hazim* 30534!; Wadi al-Khirr, 15 km N of Lussuf, *Guest, Rawi & Rechinger* 19453!; **DSD**: Khidhr al-Mai, *Karim, Sharif, Hamad & Hamid* 48062!, *Khatib & Alizzi* 32687!; 10 km SE of Busaiya, *Rawi* 26003!; 75 km ESE of Busaiya, *Guest & Mahalhal* 15233D!; Al-Taamin, midway between Jumaima and Ansab (locality appears to be within Saudi Arabia, formerly Neutral Zone), *Qaisi, Hamad & Hamid* 48319!; near As-Salman, *Rawi & Gillett* 6177!; 25 km SE of As-Salman, *Rawi* 14901!; 20 km from Salman on the way to Takhadid, *Qaisi, Hamad & Hamid* 48110!; between Salman and Aidaha, *Fawzi, Hazim & Hamid* 38783!; near Golaib, *Guest, Rawi & Rechinger* 18737!; Jabal Sanam, 20 km W by S of Safwan, *id.* 17016!; Jabal Sanam, *Qaisi, Hamad & Hamid* 48638!, id., *Guest, Rawi & Schwan* 14368!, *Rawi & Gillett* 6141!; Batin, WNW of Jarishan, 85 km SW of Basra, *Guest, Rawi & Rechinger* 17102!; 10 km S.E. by S of Jiraibiyat, id. 17156!; nr Safai al-Maghif, NE of Ghazlani, 100 km WSW of Basra, id. 17257!; 60 km SW by W of Zubair, *Guest & Rawi* 14311!; **DSD/LBA**: near Um Qasr, *Khatib & Alizzi* 33450!; **LBA**: 80 km SE of Basra, Chakrawarty, *Khatib, Rawi & Tikriti* 29928!

Palestine, Jordan, Syria, Arabian Peninsula, Libya.

The following colloquial names have been noted for this plant by various collectors: DRIHMA (Salman to Samawa, *Qaisi et al.* 48229); SAFARA (nr. Jabal Sanam, "many sheep grazing", *Guest et al.* 17016); HUNKRAIS (Batin, *Guest, Rawi & Rechinger* 17102); GIRRAIS or HINGIRRAIS (Zubair, *Abdal Mustafa* 3166); HODHAN (E of Busaiya, *Guest & Mahalhal* 15233D); and RUBIAN (E. of Salman, *Guest, Rawi & Rechinger* 18737).

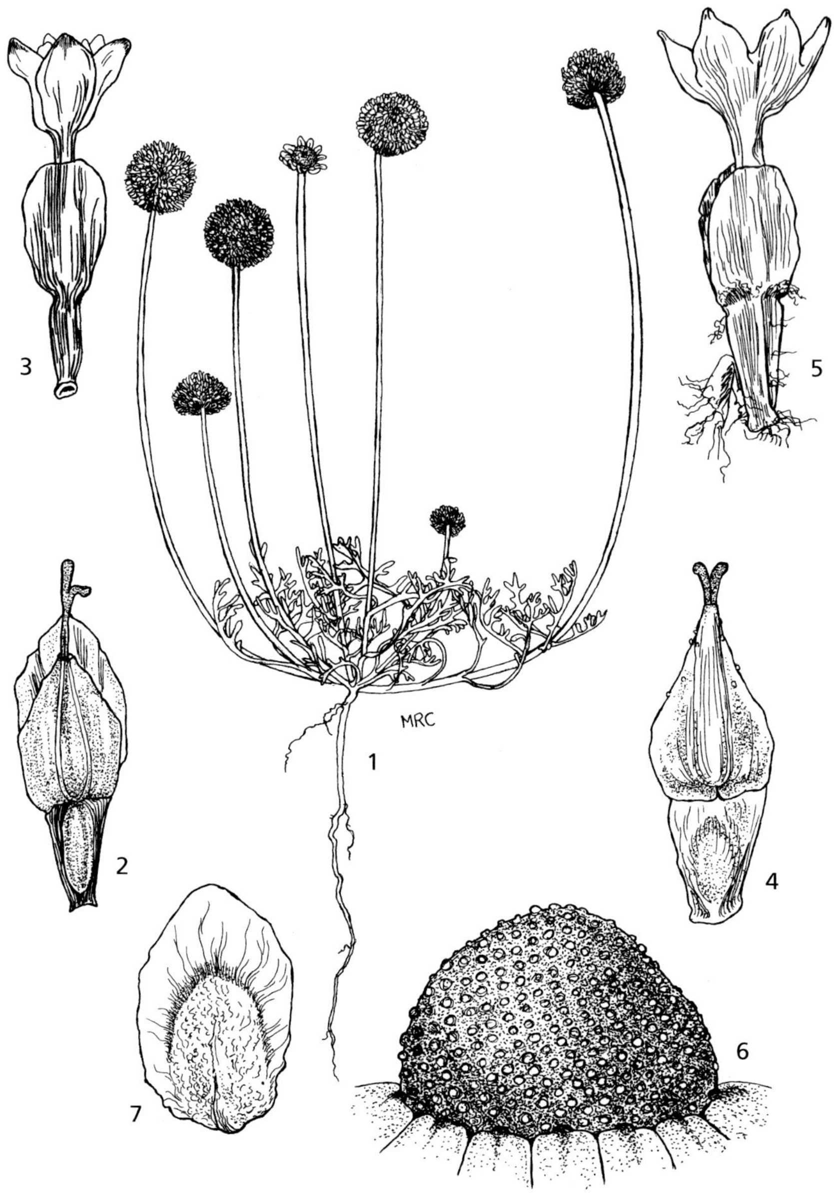

Fig. 94. **Aaronsohnia factorovskyi**. 1, habit × 1½; 2, ray floret (female) × 24; 3, disc floret (bisexual) × 24; 4, mature achene (ray floret) × 24; 5, mature achene (disc floret) × 24; 6, receptacle × 15; 7, phyllary × 15. Drawn by Moira Rae Carter.

96. **GLEBIONIS** Cass.

Dict. Sci. Nat., ed. 2, 41: 41 (1826)
Oberprielor, Vogt & Watson in Kubitzki, Fam. Gen. Vasc. Pl. 8: 369 [2006] (2007)

Patricia Halliday
Revised by D. J. Nicholas Hind

Erect, annual leafy herbs. Leaves alternate, dentate or 1–3-pinnatisect. Synflorescences cymose, lax, or reduced to single terminal capitulum. Capitulum pedunculate, heterogamous, many-flowered, radiate. Involucre hemispherical; phyllaries 4-seriate, with conspicuous scarious margins. Receptacle convex, naked. Ray florets female, fertile, white to yellow; disc-florets bisexual, yellow, tubular, 5-fid. Achenes dimorphic, those of rays 3-angled, winged laterally, those of disc-florets turbinate or compressed, ribbed, not or slightly angular; pappus absent.

Three species in the Mediterranean region and temperate Eurasia.

Glebionis (apparently a made-up name, Cassini being rather fond of creating meaningless epithets).

1. **Glebionis coronaria** (*L.*) *Spach*, Hist. Nat. Vég. 10: 181 (1841); Boulos & Hind in Fl. Egypt 3: 255 (2002).

Chrysanthemum coronarium L., Sp. Pl. 1: 890 (1753); Boissier, Fl. Orient. 3: 336 (1875); Dinsmore in Post, Fl. Syria, ed. 2, 2 : 61 (1933); Tsvelev in Fl. S.S.S.R. 26: 134 (1961); Rawi in Dep. Agr. Iraq Bull. 14: 111 (1964); Täckholm, Stud. Fl. Egypt ed. 2: 579 (1974); Grierson in Fl. Turkey 5: 254 (1975); Feinbrun-Dothan, Fl. Palaest. 3: 349 (1978); Meikle, Fl. Cyprus 2: 921 (1985); Zhu Shi, Humphries & Gilbert in Fl. China 20-21: 771 (2011).
Buphthalmum oleraceum Lour., Fl. Conchinch. 2: 506 (1790).
Matricaria coronaria (L.) Desr., Encycl. [J. Lamarck et al.] 3: 737 (1791).
Chrysanthemum coronatum Dum. Cours., Bot. Cult. 2: 446 (1802), nom. illeg. pro *Chrysanthemum coronarium* L.
C. speciosum Brouss. ex Pers., Syn. Pl. [Persoon] 2(2): 463 (1807), nom. nud. pro syn. sub *Chrysanthemum coronarium* L.
Pyrethrum indicum Roxb. ex Sims, Curtis's Bot. Mag.: t. 1521 (1813), non *Chrysanthemum indicum* L.
Chrysanthemum roxburghii Desf., Tabl. École Bot., ed. 2: 119 (1815), nom. nov. pro *Pyrethrum indicum* Roxb. ex Sims
C. coronarium (L.) Spach var. *discolor* d'Urv., Mém. Soc. Linn. Paris 1: 368 (1822).
Pinardia coronaria (L.) Less., Syn. Gen. Compos.: 255 (1832).
Chrysanthemum senecioides Dunal ex DC., Prodr. 6: 64 (1838), nom. nud. pro syn.
C. coronarium (L.) Spach var. *concolor* Batt. in Fl. Algérie (Battandier et al.) 1(3): 462 (1889).
Chamaemelum coronarium (L.) E.H.L.Krause, Deutschl. Fl. (Sturm), ed. 2, 13: 203 (1905).
Chrysanthemum coronarium (L.) Spach var. *spatiosum* L. H. Bailey, Gentes Herb. 1, fasc. 1, Art. 1: 47 (1920).
C. coronarium (L.) Spach subsp. *discolor* (d'Urv.) Rech.f., Beih. Bot. Centralbl. 54(2): 634 (1936).
Glebionis coronaria (L.) Tzvelev, Bot. Zhurn. (Moscow & Leningrad) 84(7): 117 (1999), comb. superfl.
G. roxburghii (Cass.) Tzvelev, Bot. Zhurn. (Moscow & Leningrad) 84(7): 117 (1999), nom. illeg. (as Cassini had not made the valid combination in the work to which Tzvelev referred, based on '*Chrysanthemum roxburghii* Cass.']
Xanthophthalmum [sic!] *coronarium* (L.) Trehane ex Cullen, Eur. Gard. Fl. 6: 615 (2000), nom. illeg.
Dendranthema coronarium (L.) M. R. Almeida, Fl. Maharashtra 3A: 97 (2001).
Glebionis coronaria (L.) Spach var. *discolor* (d'Urv.) Turland, Taxon 53(4): 1073 (2004).
Xanthophthalmum [sic!] *coronarium* (L.) P.D.Sell, Fl. Gr. Brit. Ireland 4: 556 (2006).
X. [sic!] *coronarium* (L.) P.D.Sell var. *discolor* (d'Urv.) P.D.Sell, Fl. Gr. Brit. Ireland 4: 556 (2006).

Branched, glabrous herb up to 38(–90) cm tall. Leaves usually bipinnatisect, sometimes tripinnatisect, oblanceolate-obovate in outline, 3–7 × 1.5–6 cm, upper leaves 1–2-pinnatisect and smaller; divisions variable in size and shape, always narrow, acute, ultimate segments 1 mm wide. Capitula 4 cm in diameter, solitary, terminal, on an ebracteate scape and borne well above the leaves; involucre up to 2–2.5 cm in diameter; phyllaries numerous, 4-seriate, outermost series ovate, 5–6 × 4 mm, middle series to 9 × 4–5 mm; innermost ± oblong, thinner in texture, 8 × 4–5 mm; receptacle 8–10 mm in diameter. Ray florets 14–19, 13–14 × 5 mm, ray oblong, emarginate, tube narrowly cylindrical, ± 3 mm long, somewhat papillose externally. Disc florets 4–5 mm. Achenes of ray florets 3 × 3–4 mm, 3-angled, ± prismatic; those of disc florets 2.5 × 1 mm, angled, ± terete, 10-ribbed. Fig. 95: 1–11.

HAB. In wadi and sandy depressions, and as a weed of cultivation; alt. 70–670 m; fl. Apr.-May.
DISTRIB. Rare in the Western Desert of Iraq and the eastern alluvial plain, also as a ruderal in Baghdad. **DWD**: 112 km W of Karbala, *Rawi* 30491!; Wadi al-Ajrumiya, *Rawi* 30507!; 30 km N of Rutba, *Chakravarty, Rawi, Alizzi & Khatib* 32888! **LEA**: 45 km E. of Tib, *Thamer* 47710!; **LCA**: Baghdad, *Haussknecht* s.n.!

Mediterranean Europe from Spain to Turkey including the islands, Portugal, Syria,

Fig. 95. **Glebionis coronaria**. 1, habit × ⅔; 2, leaf × 2; 3, capitulum (dissected) × 1.; 4, outer phyllary × 4; 5, inner phyllary × 4; 6, ray floret × 4; 7, disc floret × 4; 8, anthers × 12; 9, style × 12; 10, achene from ray floret × 6; 10, achene from disc floret × 6. 1–5 from *Thamer* 47710; 6–11 from *Thamer & Abdul Jahar* C. 975. Drawn by M. Millar Watt.

Lebanon, Palestine, Jordan, N Africa (Libya to Morocco), Macaronesia (Canary Is., Madeira); elsewhere as an escape from cultivation, or as an adventive weed.

According to Husain & Kasim (1975), in addition to its wild occurrence, this plant is cultivated throughout Iraq. Three cultivated specimens have been seen: **LCA**, Abu Ghraib, *Alizzi* 34867!; Baghdad, *Paranjpye* (comm. Graham) s.n.!; **LBA**: Basra, *Thamer & Abdul Jahar* C. 975!

The above authors quote the name DAWOODI (Ar.) for *C. coronarium.* Crown Daisy (Eng.).

Recent taxonomic work has resulted in the reduction in the number of species in this genus to three: the present species, *G. segetum* Fourr. and the Moroccan *G. carinatum* (Schousb.) Tzvelev. The latter has been recorded by Husain & Kasim (1975) under *Chrysanthemum carinatum* L. as "commonly cultivated" in Iraq, but no specimens have been seen. It has very narrow leaf-segments and purple disc-florets contrasting with rays with two bands of different colours, giving a tricoloured effect from above.

Other species formerly included within the genus *Chrysanthemum* s.l. which have been recorded as cultivated in Iraq are: *C. parthenium* (L.) Bernh. (now *Tanacetum parthenium* (L.) Sch.Bip.), *C. cinerariifolium* (Trev.) Trev. (now *Tanacetum cinerariifolium* (Trev.) Sch. Bip.), and a "C. fruticans" recorded by Husain & Kasim – presumably *C. frutescens* L. (now *Argyranthemum frutescens* (L.) Sch.Bip.). No specimens of any of these have been seen.

TRIBE **INULEAE** CASS.

97. **RHANTERIOPSIS** Rauschert

Taxon 31 (3): 557 (1982)
Anderberg & Eldens in Kubitzki, Fam. Gen. Vasc. Pl. 8: 342 [2006] (2007)
Postia Boiss. & C.I.Blanche, Fl. Orient. [Boissier] 3: 182 (1875), nom. illeg. (non Fries 1874)

C. C. Townsend

Shrubby perennial herbs with many erect stems arising from a woody rootstock. Leaves entire, ± leathery, lanceolate. Capitula many-flowered, heterogamous and radiate or homogamous and discoid; involucre hemispherical to subglobose; phyllaries 2–3-seriate, adpressed, imbricate; receptacle small, convex, paleaceous; paleae scarious, narrowly oblong to linear, midrib yellowish-brown, margins often serrulate to fimbriate towards apices, apices acute. Marginal florets few, female, rayed or filiform; disc florets hermaphrodite; anthers shortly caudate. Achenes ± compressed, tetragonal, apex setuliferous; pappus of 2–3(–4) unequal, linear, caducous bristles.

Genus of four species with a markedly disjunct distribution: two species found in Lebanon and southwestern Syria, and two species in the mountains of western Iran and eastern Iraq.

Rauschert, S. (1982). Nomina nova generica et combinationes novae spermatophytorum et Pteridophytorum. Taxon 31(3): 554–563.
Wiklund, A. (1987). The genus *Rhanteriopsis* (Asteraceae-Inuleae). Bot. J. Linn. Soc. 95(1): 27–42.
De, A. B. (1988). *Takhtajanianthus* De, gen. nov.' Curr. Sci. 57(11): 615.

1. **Rhanteriopsis puberula** (*Boiss. & Hausskn.*) *Rauschert* in Taxon 31 (3): 558 (1982); Wicklund in Bot. J. Linn. Soc. 95(1): 27 (1987).

Postia puberula Boiss. & Hausskn. in Boissier, Fl. Orient. 3: 183 (1875).
Chrysophthalmum leptocladum Rech. f. in Anz. Math.-Nat. Kl. Österr. Akad. Wiss. 88(10): 258 (1951); Rechinger, Fl. Iranica 145: 133, t.108 (1980).
Takhtajanianthus puberula (Boiss. & Hausskn.) A.B.De, Curr. Sci. 57(11): 615 (1988), nom. illeg. superfl.

Erect, to 37(–40) cm tall, "neck" of rootstock lanate at ground level. Stems slender, smooth, often flexuous, leafy throughout and branched in upper ¼. Young growth lanate, older growth glandular and shortly pubescent. Leaves all cauline, simple, alternate, coriaceous,

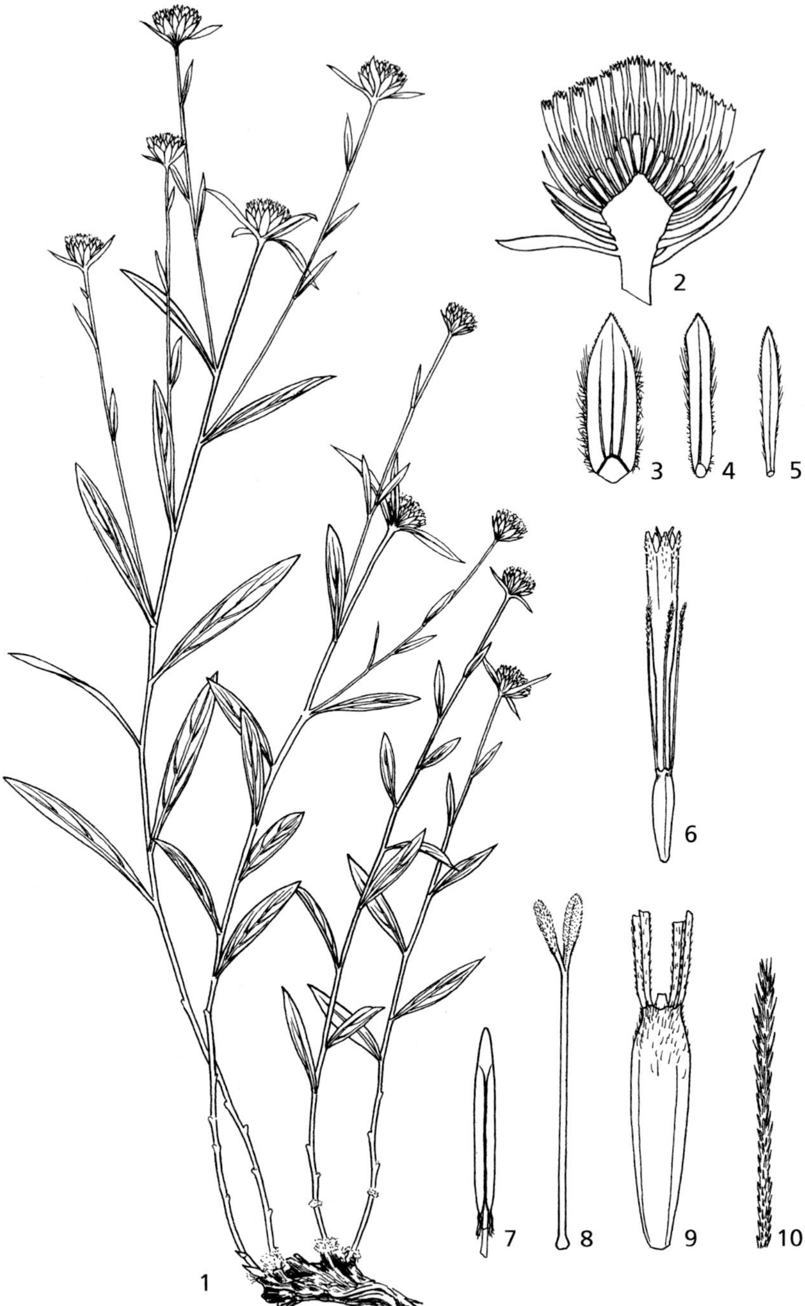

Fig. 96. **Rhanteriopsis puberula**. 1, habit × $^{2}/_{3}$; 2, capitulum × 2½; 3, outer phyllary × 4; 4, inner phyllary × 4; 5, receptacle scale × 4; 6, floret × 6; 7, anther × 10; 8, style × 10; 9, achene × 16; 10, tip of pappus hair × 20. 1 from *Rawi, Husham (Alizzi) & Noori* 29416; 2–10 from *Wheeler-Haines* W.2025.

stiffly spreading, shortly petiolate or subsessile, attenuate, acute, upper linear-lanceolate, 2–4.2(–5.5) × 5–10(–11) mm, decreasing in size upwards, entire or with several white callus-tipped teeth arising along centre section of margins, shortly pubescent and ± lanate, with a substratum of yellow, sessile glands on both surfaces, venation subparallel, projecting from lower surface, with conspicuous reticulation, especially beneath. Inflorescence branched, with 1–3(–7) capitula, each branch single-headed. Capitula hemispherical or subglobose, small, (5–)8–11 mm in diameter, (6–)8–10 mm long, homogamous, discoid, solitary, terminal; phyllaries subequal, villous and glandular, often with 2–3 leaf-like bracts subtending the capitulum, unequal, spreading, to 23 × 4 mm; outer phyllaries linear-lanceolate to oblanceolate, acuminate, ± 5 × 1.5 mm, 3-veined; inner narrowly linear-lanceolate to oblanceolate, acuminate, 6 × 1 mm; paleae pallid, shining, linear, slightly widened above and scabrid, acute, 4–5 × 0.5 mm. Florets hermaphrodite, exceeding involucre by ⅓, regular, tubular, papillose in upper half, 5–lobed, 5 × 1 mm; corollas yellow. Style forked, slender with compressed clavate to spatulate style arms. Anthers 2.5 mm long, apical appendage narrow, lanceolate, base sagittate, caudate. Achenes (immature) dark brown, glossy, usually 4–angled, at least 2 × 0.5 mm, white-hairy at tip; pappus white, of scabrid bristles to 4 mm long, exceeding involucre. Fig. 96: 1–10.

HAB. Rocky clefts in cliff-side in gorge, on a stony hillside; alt. 700–750 m; fl. & fr. Jun.-Jul.
DISTRIB. Rare in Iraq; only found twice in one subdistrict of the SE sector of the lower forest zone. **MSU**: Ahmad Awa, nr Halabja, *Wheeler-Haines* W.2025!; Zalam, nr Khurmal, *Rawi, Husham (Alizzi) & Noori* 29416!

SW Iran (Khuzistan and Luristan).

98. **RHANTERIUM** Desf.

Fl. Atlant. 2: 291 (1799); Bentham & Hooker f., Gen. Pl. 2 : 338 (1873).
Anderberg & Eldenäs in Kubitzki, Fam. Gen. Vasc. Pl. 8: 378 [2006] (2007)

Patricia Halliday

Perennial, canescent subshrubs with woody rootstock, much-branched and sparsely leafy. Leaves alternate, lanceolate, pubescent, entire or toothed. Capitula heterogamous and radiate, many-flowered, terminal, solitary; involucre hemispherical; phyllaries multiseriate, inner enclosing ray florets, coriaceous; receptacle flat or slightly convex, paleaceous; paleae conduplicate. Ray florets female, uniseriate, rays yellow. Disc florets hermaphrodite, fertile, cylindrical; anthers caudate, tails short; style arms compressed, papillose in upper part. Achenes angled or ribbed, glabrous; pappus uniseriate, of broad flattened brtistles, marginal achenes usually epappose, or all achenes epappose.

A genus of 2 or 3 species, one in Asia, the others in N. Africa; only one species in Iraq.

Rhanterium (from Gr. ραντεριον, *ranterion, Aspergillus* according to its original author René Desfontaines, l.c.).

Wicklund, A. (1986). The genus *Rhanterium* (Asteraceae: Inuleae). Bot. J. Linn. Soc. 93(2): 213–246.

1. **Rhanterium epapposum** *Oliv.* in Hooker's Icon. Pl. 14: 50, t. 1367 (1881); Guest in Dep. Agr. Iraq Bull. 27: 9 (1933); Blakelock in Kew Bull. 4: 61 (1949); Zohary in Dep. Agr. Iraq Bull. 31: 145 (1950); Dickson, Wild Fls. Kuwait & Bahrain: 80 (1955); Rechinger in Bull. Coll. Arts & Sciences, Baghdad 2: 7 (1957) & Bericht. Deutsch. Bot. Ges. 72, 7: 279 (1959) & Fl. Lowland Iraq: 609 (1964); Rawi in Dep. Agr. Iraq Tech. Bull. 14: 123 (1964); Rechinger, Fl. Iranica 145: 128 (1980); Collenette, Wild Flow. Saudi Arabia: 213 (1999); Jongbloed, Wildflow. UAE: 185 (2003); Bari, Fl. Qatar 1: 215 (2012); Kilian & Ghazanfar in Fl. Oman 3: 269 (2015); Abdulridha, Taha & Wedad, Ecology and Fl. Basra: 223 (2016).

Aromatic, somewhat arachnoid perennial subshrub to 50(–55) cm tall, with thick, fibrous rootstock and spreading, much-branched, ribbed, slender stems, sparsely leafy throughout, occasionally leaves crowded on young growth, decreasing in size upwards, young growth white-woolly, older growth less woolly or glabrous. Leaves olive- or blue-green, alternate, almost fleshy, linear, 1.0–2.3(–3.7) × 0.2–0.4 cm, base truncate, often with 2 or 3 small teeth either side of leaf at junction with stem, acute, entire or with remote, irregular teeth to 2 mm

Fig 97. **Rhanterium epapposum**. 1, habit × ⅔; 2, capitulum (dissected) × 4; 3, phyllary × 4; 4, receptacle scale (paleae) × 4; 5, ray floret × 4; 6, disc floret × 4; 7, anther × 8; 8, style × 4; 9, receptacle after achene abscission × 4 All from *Guest, Ali & Rawi* 14531. Drawn by Christine M. Grey-Wilson.

long, both surfaces somewhat arachnose. Capitula pedunculate, 7–10 × 8–14(–18) mm; involucre yellow-green, hemispherical; phyllaries ± 5-seriate, free, lanceolate, imbricate, rigid, erect to spreading, to 6 × 2 mm, midrib strongly marked, impressed on outer surface, acute to acuminate, margins minutely erose in upper half, glabrous or very sparsely arachnose at base externally; receptacle ± flat, paleaceous; paleae 5 mm long, subulate, hairy at tip. Ray florets female, uniseriate, ray limb strap-shaped, 5–8 × 1.5–1.75 mm, yellow, glabrous, with 2–3(–4) apical teeth and usually 4 veins, tube of floret 1 mm long; style slender, forked, style arms compressed, lanceolate, 1–2 mm long, papillose. Disc florets hermaphrodite, regular, corollas tubular, yellow, glabrous, 5–5.5 × 1 mm, 5-lobed, lobes erect, ovate, acute; anther cylinder much exceeding corolla, 2.5–3.5 mm long, appendage ovate, acute, base sagittate with long slender adpressed tails and well developed anther-collars; style arms short, lanceolate, c. 1 mm. Achenes (immature) yellow, those of female florets obconical, smooth, somewhat curved, 2.5–3 × c. 1 mm, glabrous, those of hermaphrodite florets very slender, filiform, slightly ribbed, 1.5 mm; pappus absent. Fig. 97: 1–9.

HAB. In the desert on compact sandy soil, sandy gravel, sometimes in silty depressions, or in sandy patches over gypsiferous subsoil or with a pebbly substrate, in calcareous sandy pockets on a rocky limestone plateau, sandy patches in pebbly desert ("haswa") etc.; alt. up to ± 350 m; fl. & fr. Mar.-Apr.
DISTRIB. Very common in the Southern Desert of Iraq, particularly the S sector, occasional in the Western Desert. **DWD**: 5 km S of Shanana, nr Nukhaib, *Rawi* 14813!; 20 km SW of Falluja, *Rechinger* 8299; 13 km W of Ukhaidhir, *Rawi* 30839! **DWD/DSD**: Wadi al-Khirr, 40 km NE of Lussuf, *Guest, Rawi & Rechinger* 19459! **DSD**: 120 km N of Shabicha, *Rawi & Gillett* 6325!; 15 km SE of Ashuriya, *Rechinger* 9495!; Faidhat Umm Ribian, 95 km W of Shabicha, *Hazim* 30693!; 30 km E by S of Shabicha, *Guest, Rawi & Long* 14063!; 30 km E of Salman, *Haddad* 9605!; 25 km from Maniya, *Hazim* 32474!; 45 km SW of Samawa on road to Ansab, *Qaisi, Hamad & Hamid* 48352!; 14 km SSW of Busaiya, *Guest & Mahalhal* 15229!; Khasm ibn Hallaf, *id.* 15268!; nr Adan, 50 km SSW of Busaiya, *id.* 15283!; Ichrishi, *Guest & Rawi* 14203!; 75 km ESE of Busaiya, *Guest & Mahalhal* 15232!; 46 km WNW of Ansab, *Guest, Rawi & Rechinger* 19045!; 15 km SE of Ansab, along road to Samah, *Fauzi, Hazim & Hamid* 38853!; Khadr al-Ma'i, *Khatib & Alizzi* 32711!; Chilawa, *Guest, Rawi & Rechinger* 17192!; nr Zubair, *E.S. Brown* s.n.!; 44 km S of Zubair, *Thamir* 47432!; Samah, Qaisi, *Hamad & Hamid* 48435!; Jabal Sanam, *Watson* s.n. comm. *Sharples*; between Safwan and Umm Qasr, *H. Hamid* 38551!; Umm Qasr, *Khatib & Alizzi* 32660! **LCA**: Rustamiya, *Lazar* 26! (see note below).

Arabia, S Iran, Pakistan.

'ARFAJ, sometimes pronounced by the Badawin as ARAFAJ; Guest (1933) noted that this was an important grazing plant for camels and also, when young, for sheep; he mentioned that it was one of the principal fuel plants brought into Basra by the donkey load from the desert S of Zubair. Dickson (1955) reported that this plant was also sold in Kuwait for firewood, that it is one of the best foods for camels and that sheep also graze on it at night in early summer; she added that, since the Burzan Oilfield had been opened in Kuwait, the country had been denuded of ARFAJ by the workmen for firewood and a veritable dust bowl had been formed. Where it is growing in very sandy soil, hummocks of wind-blown sand accumulate around each bush and the weary traveller cannot find a more fragrant and comfortable pillow; for the plants have a most pleasant aromatic smell – see also Fl. Iraq vol. 1, fig. 19B. It is reported to have been chosen as Kuwait's national flower.

As regards *Lazar* 26, cited above, the collector states that the plant was growing as a weed on the Agricultural Experiment Station at Rustamiya, nr Baghdad; but it seems most unlikely that this characteristic desert species, normally found on sandy desert soils, should have grown there – and the possibility of a mixed label springs to mind.

99. **PULICARIA** Gaertn.

Fruct. Sem. Pl. 2: 461 (1791)
Anderberg & Eldenäs in Kubitzki, Fam. Gen. Vasc. Pl. 8: 379 [2006] (2007)
Francoeuria Cass., Dict. Sci. Nat., ed. 2, 34: 44 (1825)
Platychaete Boiss., Diagn. Pl. Orient., ser. 1, 11: 5 (1849)
Sclerostephane Chiov., Fl. Somala 1: 200 (1929)

Frances G. Davies

Annual or perennial herbs or low shrubs, often grey-felted or glandular or both. Leaves petiolate or more usually sessile, often semi-amplexicaul, margins plane or undulate. Inflorescence a cymose arrangement of rather distant capitula. Capitula usually heterogamous and radiate with outer florets 1–2-seriate, female and rayed, ray limb 3-toothed, inner florets numerous, tubular and hermaphrodite, or capitula homogamous and discoid; phyllaries in few to several rows, imbricate, herbaceous, narrow, outermost occasionally much broader than others; receptacle naked, achene-pits sometimes with fimbrillate or glandular margins. Florets very numerous, outer 1–2 rows female and ligulate, ray florets 3-toothed, inner florets tubular, hermaphrodite, or sometimes discoid, with all florets tubular and hermaphrodite; corollas yellow or cream; style-arms subulate, not appressed, tips rounded. Achenes cylindrical or slightly compressed, finely ribbed, glabrous or pilose, often with an annulus of glands below pappus; pappus markedly biseriate, outer ring of short scales ± fused into a cup, inner of barbellate to subplumose setae falling either as a ring or separately.

A genus of about 75 species, of which 6 occur in Iraq.

Pulicaria (from the Lat. *pulex*, a flea, the strong smell of certain species when burnt being said to repel fleas). Fleabane (Eng.).

Gamal-Eldin, E. (1981). Revision der Gattung *Pulicaria* (Compositae – Inuleae) für Afrika, Makaronesia und Arabien. Phanerog. Monogr. 14: 1–311.

1. Rays of outer florets conspicuous, exceeding phyllaries by about 2.5–3.5 mm . . . . 1. *P. dysenterica*
   Rays of outer florets inconspicuous, not or only slightly exceeding phyllaries 2
2. Tube and dorsal surface of ray limb of marginal florets pilose; pedicels elongate, to *c.* 10 cm, thickening in fruit . . . . 2. *P. guestii*
   Tube and dorsal surface of ray limb of marginal florets glabrous; pedicels short, rarely to 6 cm, slender, not thickening in fruit. 3
3. Achenes glabrous; pappus falling in its entirety, setae with a short, narrow tuft of slightly longer hairs at tip . . . . 6. *P. crispa*
   Achenes ± hairy; pappus hairs falling separately, setae not narrowly tufted at tip 4
4. Phyllaries broader, oblong-elliptic, in 2(–3) series; leaves and stem (especially about base) ± densely white-woolly . . . . 5. *P. gnaphalodes*
   Phyllaries narrower, lanceolate to linear, multiseriate; leaves and stem not densely white-woolly 5
5. Outer phyllaries lanceolate, acute; pappus shorter than or equalling achene. . . . . 4. *P. arabica*
   Outer phyllaries linear, acuminate; pappus at least twice as long as achene. . . . . 3. *P. vulgaris*

SECT. **PULICARIA.**

1. **Pulicaria dysenterica** (*L.*) *Bernh.*, Syst. Verz. Pfl.: 153 (1800); Boissier, Fl. Orient. 3: 201 (1875); Dinsmore in Post, Fl. Syria, Palest. & Sinai ed. 2, 3: 25 (1933); Zohary in Dep. Agr. Iraq Bull. 31: 145 (1950); Rawi, ibid. Tech. Bull. 14: 122 (1964); Rawi & Chakravarty, ibid. 15: 79 (1964); Rechinger, Fl. Lowland Iraq: 607 (1964); Grierson in Fl. Turkey 5: 75 (1975); Feinbrun-Dothan, Fl. Palaest. 3: 317 (1978); Lack in Fl. Iranica 145: 113 (1980); Gamal-Eldin, Phanerog. Monogr. 14: 122 (1981); Qaiser & Abid in Fl. Pakistan 210: 60 (2003); Breckle, Hedge & Rafiqpoor, Vasc. Pl. Afghan. Checkl.: 183 (2013); Abdulridha, Taha & Wedad, Ecol. & Fl. Basra: 223 (2016); Taifour & El-Oqlah, Pl. Jordan Ann. Checkl.: 46 (2017).

Inula dysenterica L., Sp. Pl.: 882 (1753).
Pulicaria dysenterica var. *microcephala* Boiss., Fl. Orient. 3: 202 (1875); Nábělek in Publ. Fac. Sci. Univ. Masaryk 52: 7 (1925); Guest in Dep. Agr. Iraq Bull. 27: 79 (1933); Blakelock in Kew Bull. 4: 60 (1949); Rawi in Dep. Agr. Iraq Tech. Bull. 14: 122 (1964).
P. dysenterica var. *stenophylla* Boiss., Fl. Orient. 3: 202 (1875).

Perennial herb up to 70 cm tall, usually much-branched. Stems ascending, greyish-green and softly pubescent throughout. Leaves 2–6 × 0.5–1.5 cm, oblong to lanceolate, semi-amplexicaul to auriculate, margin entire or toothed, pubescent and glandular on both surfaces. Capitulum hemispherical, 5–16 × 5–10 mm; phyllaries lanceolate to linear, 2–3.5 mm long, herbaceous, long-hairy and sessile-glandular, apices ciliate; Receptacle smooth or with fimbriate margins to achene pits. Female florets uniseriate, 7–10 mm, ray limb 5–7 mm, bright to pale yellow; hermaphrodite florets numerous, 3 mm long, lobes papillose and glandular. Achenes cylindrical, 1–1.25 mm long, finely ribbed, pilose; pappus of short lacerate scales and 15–20 setae, ± 3 mm long.

HAB. Damp places near springs and streams, in shade of *Quercus* forest, orchards, fields etc.; alt. 500–1700(–1900) m in the mountains, 5–200 m in the steppe region and on the irrigated alluvial plain; fl. & fr. (Jul.-)Aug.-Oct in the mountains and steppe region, Apr.-May on the irrigated alluvial plain.
DISTRIB. Quite common in the mountain region of Iraq, occasional in the steppe region, rare on the irrigated alluvial plain. **MAM**: *Khantur, Rawi, Noori & Tikriti* 28981!; Zawita, *Guest* 1584!; Sarsang, *Haines* W.470!; Dairaluk (Dereluk), 16 km E of Amadiya, *Dabbagh & Hamid* 4580! **MRO**: Shaqlawa, *Sahira* 37560!; Haruna, *Gillett* 9642!; Gali Ali Beg (Rowanduz gorge), *Guest* 488!; Bekhal falls, nr Rowanduz, *Guest & Husham* 5915!; Rayat, *Guest* 13040!; Qandil range, *Rawi & Serhang* 26497! **MSU**: between Chemchemal and Tasluja, *Hadač* 2787; Sarchinar, *Radhi* 527!; Sulaimaniya, *Qardar* 5395!; 5 km N of Saiyid Sadiq, *Khayat, Qaisi & Thamir* 4631! **FUJ/FNI**: on an island in the R. Tigris near Mosul, *Kotschy* 438! **FNI**: near Tal Kaif, *Alizzi & Omar* 35269! **FPF**: 10 km E of Mandali, *Barkley et al.* 9022!; Badra, *Qaisi & Yahya* 45287! **LEA**: between Muqdadiya and Baquba, *Rechinger* 8037. **LBA**: Abul-Khasib, *Thamir & Abdul-Jabbar* 50007!; Siba, *Hamad, Hamid & Kadhim* 46357!

Almost throughout Europe except the more northerly regions, Mediterranean Islands, Cyprus, Syria, Palestine, Lebanon, Turkey, Iran, Caucasia, Afghanistan, C Asia (Turkmenia to Tajikistan), Pakistan, NW India, W China, Oman, N Africa (Morocco, Algeria), tropical Africa (Sudan, Ethiopia).

ZERD SHIRE (Kurd.-Sul., *Radhi* 527), ZARBANG (Kurd.-Sul., *Qadr* 5395).

Boissier's var. *microcephala* cannot be maintained as there exist all intermediates of capitulum size from 1–3 cm. However, it is notable that in Iraq almost all plants are diminutive in habit and have capitula less than 1.5 cm in diameter.

2. **Pulicaria guestii** *Rech.f. & Rawi* in Bull. Coll. Sci. Baghdad 2: 6 (1957); Rechinger in Ber. Deutsch. Bot. Ges. 72: 278 (1959); Collenette, Wild Flow. Saudi Arabia: 210 (1999); Gamal-Eldin, Phanerog. Monogr. 14: 140 (1981); Abdulridha, Taha & Wedad, Ecol. & Fl. Basra: 221 (2016).

Perennial tap-rooted herb with many spreading to erect branches from the base upwards, vegetative parts covered with both a cottony soft white pubescence and dense sessile glands. Leaves oblanceolate to narrowly spatulate, 3–5 × 0.4–0.7 cm, base widened where it joins stem, apex obtuse, margins entire and (especially in lower leaves) undulate. Inflorescence a loose cyme, each capitulum on pedicels up to 10 cm, which thicken in fruit. Involucre 1–1.6 × 0.4–0.5 cm; phyllaries linear to linear-lanceolate, herbaceous, densely glandular and pubescent, extreme tip membranous and glabrous. Receptacle naked. Female florets 5–6 mm long, ray limb golden yellow, 3 mm, dorsal surface and tube hairy. Disc florets 5–6 mm long, lobes gradually tapered, papillose and glandular. Achenes obscurely ribbed, those of ray florets brown, of disc florets grey or brown, all pilose, ± 1.5 mm; outer pappus a stalked, toothed cup; inner of 8–12 white subplumose setae, ± equalling disc florets. Fig. 98: 1–4.

HAB. In the desert on hard silty soil, in flat silty depressions on rocky limestone, silty plain with scattered stones, on sandy gravel desert; alt. 200–350 m; fl. & fr. Apr.
DISTRIB. Quite common in the S part of the Southern Desert of Iraq. **DSD**: Salman, *Guest, Rawi & Long* 14092!; 40 km SSE of Salman, *Guest, Rawi & Rechinger* 18861!, *Rechinger* 9319; 95 km SE of Salman, *Rechinger* 9335; Takhadid, 60 km S by W of Salman, *Botany Staff* 42220!; between Salman and Shawila, *Karim, Nuri, Hamid & Kadhim* 40165!; 65 km N by W of Busaiya, *Guest & Rawi* 14147!; 12 km WNW of Ansab, ± 143 km S of Salman, *Guest, Rawi & Rechinger* 18966!, *Rechinger* 9357.

Saudi Arabia, Kuwait.

According to Dickson in *Wild Fls. of Kuwait,* this species is known as GARAIS in Saudi Arabia.

3. **Pulicaria vulgaris** *Gaertn.*, Fruct. Sem. Pl. 2: 461, t. 173 f. 7 (1791); Boissier, Fl. Orient. 3: 304 (1875); Dinsmore in Post, Fl. Syria, Palest. & Sinai ed. 2, 2: 27 (1933); Bornmüller in Beih. Bot. Centralbl. 60B: 183 (1939); Täckholm, Stud. Fl. Egypt ed. 2: 562 (1974); Gamal-

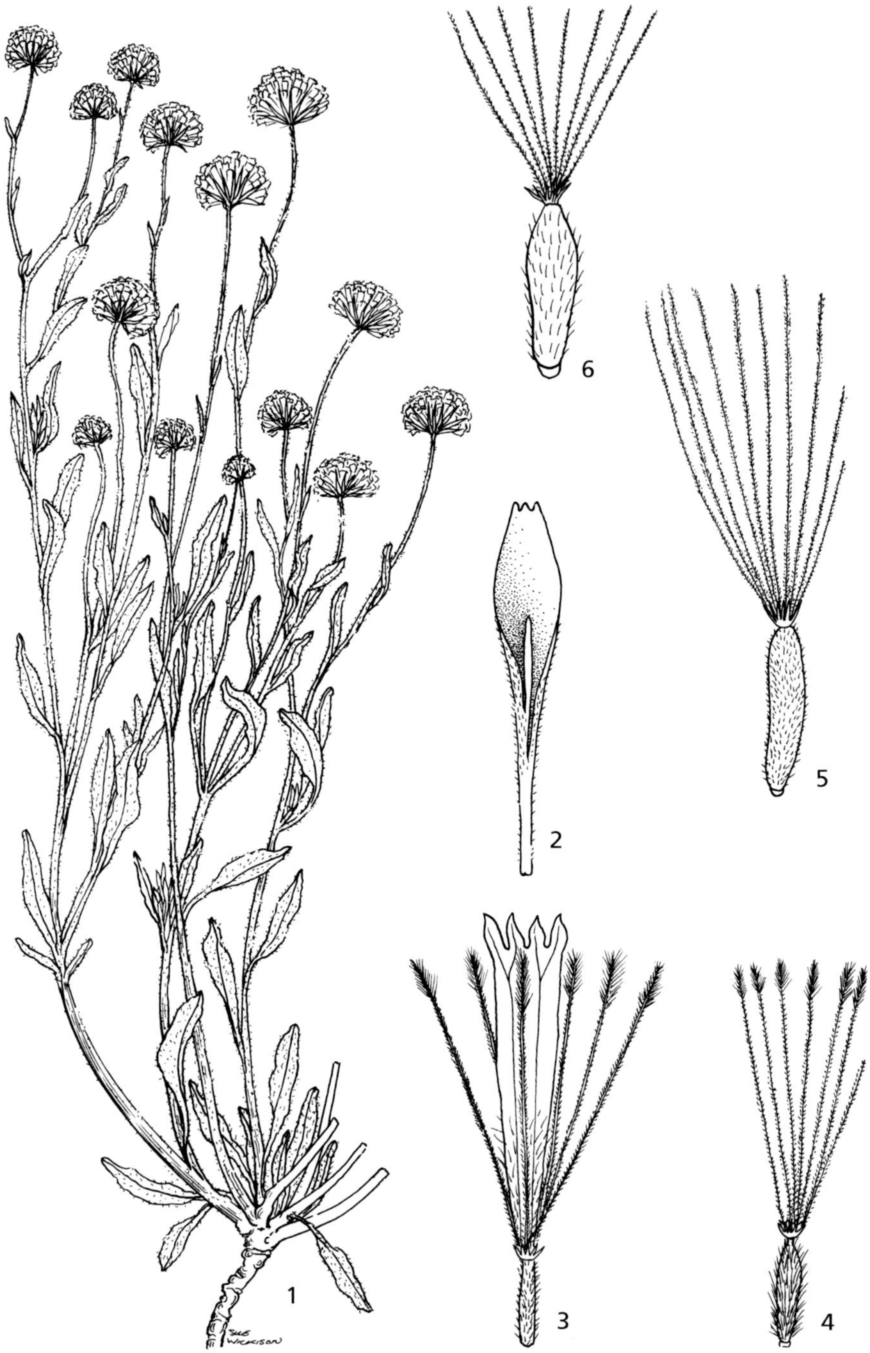

Fig. 98. **Pulicaria guestii**. 1, habit × ⅔; 2, ray floret × 10; 3, central disc floret × 10; 4, achene and pappus × 10. **Pulicaria arabica**. 5, achene and pappus × 16. **Pulicaria vulgaris**. 6, achene and pappus × 16. 1–3 from *Guest, Rawi & Rechinger* 18966; 4 from *Karim et al.* 40165; 5 from *E. Chapman* 25397; 6 from *Bornmüller* 1371. Drawn by Sue Wickison.

Eldin, Phanerog. Monogr. 14: 79 (1981); Boulos & Hind in Fl. Egypt 3: 220 (2002); Qaiser & Abid in Fl. Pakistan 210: 62 (2003).

Annual herb, with many spreading or ascending branches, to 30 cm tall, vegetative parts sublanate-pubescent and glandular. Leaves oblong-elliptic to oblong-ovate, sessile, 1.2–5 × 0.3–1.2 cm, acute, entire, sometimes undulate-margined, lamina pilose and glandular on both surfaces. Capitula borne in irregular groups on pedicels 1–4 cm long often with small bract-like leaves; involucre 8–10 × 6–7 mm, apices of phyllaries somewhat spreading. Phyllaries with outer row clearly broader, all densely hairy and glandular. Female florets in one row, 2–2.5 mm, ray limb not exserted. Disc florets 2.5–3 mm. Achenes 1.25–1.5 mm, pilose; outer pappus united into a very short cup; inner pappus of 7–10 white hairs 1–1.5 mm. Fig. 98: 6.

HAB. Among ancient ruins, on mountain slope and by a spring; alt. [only once recorded] ± 225 m.
DISTRIB. Rare in Iraq, only found on three occasions, once in the NW part of the moist steppe zone, once in the E mountain zone, the third locality uncertain but probably in the mountains. **MRO**: Ain Sattan Zimmar, *Albert Mamarjan & Kheirah Jbarah* 5182 (BAG!). *anon.* (Nat. Herb. Iraq 5182)! **MSU**: Penjwin, *Haussknecht* s.n.! **FNI**: Nineveh, *Bornmüller* 1371!

Most of Europe from England, S Scandinavia and C Russia southwards, Turkey, Lebanon, Egypt (very rare), Sudan, Iran, Caucasia, C Asia, Siberia, Mongolia, China, Kashmir, N Africa (Algeria, Morocco), Macaronesia (Tenerife).

4. **Pulicaria arabica** (*L.*) *Cass.*, Dict. Sci. Nat. 44: 94 (1826); Boissier, Fl. Orient. 3: 205 (1875); Dinsmore in Post, Fl. Syria, Palest. & Sinai ed. 2, 2: 27 (1933); Guest in Dep. Agr. Iraq Bull. 27: 79 (1933); Anthony in Notes Roy. Bot. Gard. Edinburgh 18: 290 (1935); Bornmüller. in Beih. Bot. Centralbl. 60B: 185 (1939); Blakelock in Kew Bull. 4: 60 (1949); Zohary in Dep. Agr. Iraq Bull. 31: 145 (1950); Dickson, Wild Fls. Kuwait: 77 (1955); Rawi in Dep. Agr. Iraq Tech. Bull. 14: 122 (1964); Rechinger, Fl. Lowland Iraq: 608 (1964); Täckholm, Stud. Fl. Egypt ed. 2: 563 (1974); Grierson in Fl. Turkey 5: 76 (1975); Feinbrun-Dothan, Fl. Palaest. 3: 318 (1978); Gamal-Eldin, Phanerog. Monogr 14: 87 (1981); Boulos & Hadidi, Weed Fl. Egypt: 38 (1984); Collenette, Wild Flow. Saudi Arabia: 209 (1999); Boulos & Hind in Fl. Egypt 3: 221 (2002); Kilian & Ghazanfar in Fl. Oman 3: 263 (2015); Abdulridha, Taha & Wedad, Ecol. & Fl. Basra: 221 (2016); Qaiser & Abid in Fl. Pakistan 210: 63 (2003);Taifour & El-Oqlah, Pl. Jordan Ann. Checkl.: 46 (2017).

Inula arabica L., Mant. : 114 (1767).
Pulicaria trichocephala DC., Prodr. 5: 478 (1836).
P. elata Boiss., Diagn. Pl. Orient., ser. 1, 2: 6 (1849).
P. demissa Boiss., Fl. Orient. [Boissier] 3: 205 (1875).
P. arabica var. *angustifolia* Post, Fl. Pal. ed. 1 : 423 (1882).
P. lanceps Bornm., Notizbl. Bot. Gart. Berlin-Dahlem 7: 152 (1917).
P. foliolosa auctt. non DC.: Blakelock in Kew Bull. 4: 61 (1949); Rawi in Dep. Agr. Iraq Tech. Bull. 14: 122 (1964); Rechinger, Fl. Lowland Iraq: 608 (1964).

Annual or perennial herb 30–50 cm tall with a much-branched stem, side branches making an angle of 45° to main stem and often overtopping it, green to reddish, pilose and glandular with hairs finally evanescent. Leaves oblong to oblanceolate, sessile or with a petiolate base, 1.25–6 × 0.2–4 cm, entire, pilose and sessile-glandular on both surfaces. Inflorescence consisting of remote capitula on 1–3.5 cm long pedicels. Involucre 5–7 × 9–12 mm; phyllaries linear, herbaceous, glandular with long spreading hairs. Female florets 2–4 mm long, disc florets 2–3 mm long, lobes glandular and papillose. Achenes ellipsoid, 1.25–1.5 mm long, pilose; outer pappus a toothed cup, inner of 6–10 slender setae 2.5–3.5 mm. Fig. 98: 5.

HAB. In damp places, near streams and wells, marshy fields and sides of ditches, on silty, sandy and clay soils; alt. 3–250(–1450) m; fl. & fr. (Apr.-)Jun.-Aug.(-Sep.).
DISTRIB. Occasional in the mountain region of Iraq, more frequent in the steppe region and also on the irrigated alluvial plain in the desert region. **MRO**: Shaikh wa Shaikhan valley, *Rawi* 24886!; Saruchawa, ± 10 km from Rania, *Omar, Sahira, Karim & Hamid* 38234! **MSU**: Jarmo, *Helbaek* 2015!; Penjwin, *Qaisi & Hamad* 43528! **FUJ**: between Mosul and Qaiyara, *Alizzi & Omar* 35260! **FUJ/FNI**: nr Mosul, *Kotschy* 449! **FNI**: Nineveh, *Bornmüller* 1307, *E. Chapman* 25397! **FKI**: Zain al-Abdin, 40 km S of Kirkuk, *Juma'a Brahim* 6113! **FPF**: Koma Sang, NE of Mandali, *Rawi* 20593!; Mandali, *Haddad* 9994!; 16 km S of Zurbatiya, *Qaisi*

& Yahya 45269!; 4 km E of Badra, *Haddad* 9945! **LCA**: Daltawa (Khalis), *Guest* 2428!; Khan Bani Sa'd, 25 km N of Baghdad, *Anderson* 2002!; Baghdad, *Aucher-Eloy* 3095!; Abu Ghraib, *Janan al-Mukhtar* 36980!; Hilla, *Paranjpye* s.n. comm. Graham!; Suwaira, *Alizzi & Omar* 34658!; Shatra, *Omar & Karim* 36768!; on Shatt al-Gharaf, nr Kut, *Karim & Hamad* 37806!; Garma Bani Sa'd, *Rawi* 26033! **LSM**: Shahala, nr Amara, *Field & Lazar* 41! **LBA**: Abul-Khasib, *Alizzi & Omar* 34687!

Aegean Is., Cyprus, Palestine, Jordan, Sinai, Egypt, Syria, Lebanon, Turkey, Iran, Afghanistan, Pakistan, Saudi Arabia, Oman, Saudi Arabia, C Asia (Turkmenia), N Africa (Algeria, Morocco), Mauritania.

Specimens which have been referred to *P. foliolosa* DC. either have short but distinct rays and are clearly referable to *P. arabica* (*Guest* 290, 1619, 10036) or are too old and poor for the rays to be present (*Calder* 2122, 2123, 2124, 2126) but otherwise so identical to the remainder that one cannot record the Indian *P. foliolosa* from Iraq on such a basis; they are almost certainly *P. arabica* also.

SECT. **FRANCOEURIA** (CASS.) HOOK.F.

5. **Pulicaria gnaphalodes** (*Vent.*) *Boiss.*, Diagn. Pl. Orient. ser. 1, 6: 76 (1845); Boissier, Fl. Orient. 3: 203 (1875); Carter in Rec. Bot. Surv. India 6, 6: 203 (1917); Dinsmore in Post, Fl. Syria, Palest. & Sinai ed. 2, 2: 26 (1933); Rawi in Dep. Agr. Iraq Tech. Bull. 14: 122 (1964); Lack in Fl. Iranica 145: 116 (1980); Gamal-Eldin, Phanerog. Monogr. 14: 286 (1981); Qaiser & Abid in Fl. Pakistan 210: 70 (2003); Breckle, Hedge & Rafiqpoor, Vasc. Pl. Afghan. Checkl.: 183 (2013).

Inula gnaphalodes Vent., Descr. Pl. Jard. Cels.: 75 (1802).

Perennial herb, woody at base, erect or sometimes decumbent or forming dwarf cushions up to 70 cm tall. Stems and leaves very conspicuously white-lanate and brown-glandular. Leaves obovate, 0.2–2 × 0.1–1 cm, sessile, ± amplexicaul, obtuse or acute, entire or dentate, undulate-margined. Inflorescence a ± dichotomously branching cyme, pedicels 1–3 cm long, densely yellowish- or brown-glandular. Involucre variable in size, 6–9 × 4–5 mm, obconical to hemispherical; phyllaries few, 2(–3)-seriate, elliptic-oblong, 4 × 0.8 mm, herbaceous, acute, somewhat lacerate, glandular and lanate, not reflexing in fruit. Female florets 4–5 mm, rays 2–3 mm, falling early, few. Disc florets 4–5 mm. Achenes 2.2–2.5 mm, stramineous to brown, ± densely pilose; outer pappus a ring of scales, inner pappus of about 18 straw-coloured subplumose bristles, fused with outer only at extreme base.

HAB. On hillsides, open sandy plains, by river, streams and irrigation channels; alt. 65–700 m; fl. & fr. May-Nov.
DISTRIB. Frequent, locally common, in the lower mountain and foothill regions of Iraq. **MRO**: road from Chinarok to Qandil, *Rawi & Serhang* 26472!; Diana, *Guest & Husham* 15873! **MRO/FAR**: 5 km from Taqtaq on road to Koi Sanjaq, *Rawi & Serhang* 25362! **FNI**: Mindan bridge, on Khatir river between Mosul and Aqra, *Alizzi & Omar* 35284! **FKI**: 50 km NNE of Kirkuk, *Rawi* 26462! **FPF**: Khanaqin, *Rogers* 0400!; Sa'diya, *Qaisi* 46706!, *Omar & Thamer* 47129!; Sudur, *Botany Staff* 47133!

Iran, Afghanistan, Pakistan, C Asia (Turkmenia), Tibet.

According to H.G. Carter (1917), this species is known as NAFÎJ in Kuwait. *Guest & Husham* 15873 cites the Kurdish name BŪNŌSHAKAR from the Rowanduz area, noting that the plant is eaten by all animals and said to be used for flavouring food.

6. **Pulicaria crispa** (*Forssk.*) *Benth.* ex *Oliv.* in Trans. Linn. Soc. 29: 96 (1873); Gamal-Eldin, Phanerog. Monogr. 14: 258 (1981).

Inula undulata L., Mant. 115 (1767), nom. ambig. rejic. (see Jeffrey, Lack, Gamal Eldin & Wagenitz in Taxon 29: 5/6 694–695. 1980).
Aster crispus Forssk., Fl. Aeg.-Arab.: 150 (1775).
Francoeuria crispa (Forssk.) Cass., Dict. Nat. Sci. 34: 44 (1825); Boissier, Fl. Orient. 3: 207 (1875); Handel-Mazzetti in Ann. Naturh. Mus. Wien 27: 433 (1913); Nábělek in Publ. Fac. Sci. Univ. Masaryk 52: 7 (1925); Guest in Dep. Agr. Iraq Bull. 27: 79 (1933); Bornmüller in Beih. Bot. Centralbl. 50: 186 (1940); Blakelock in Kew Bull. 4: 55 (1949); Zohary in Dep. Agr. Iraq Bull. 31: 145 (1950); Rawi, ibid. Tech. Bull. 14: 22 (1964); Rechinger, Fl. Lowland Iraq: 609 (1964); Täckholm, Stud. Fl. Egypt ed. 2: 563 (1974).

Pulicaria undulata (L.) C.A.Mey., Verz. Pfl. Cauc.: 79 (1831); Boulos & Hind in Fl. Egypt 3: 221 (2002); Qaiser & Abid in Fl. Pakistan 210: 72 (2003); Breckle, Hedge & Rafiqpoor, Vasc. Pl. Afghan. Checkl.: 183 (2013); Kilian & Ghazanfar in Fl. Oman 3: 268 (2015); Taifour & El-Oqlah, Pl. Jordan Ann. Checkl.: 46 (2017).
Francoeuria undulata (L.) Lack in Fl. Iranica 145: 120 (1980).

Perennial twiggy subshrub up to 0.7 m tall, much-branched from base upwards, forming extensive cushions. Stems grey or white-lanate. Leaves 3–25 × 1–5 mm, lanceolate, semi-amplexicaul and decurrent on one side of base, acute, margins strongly crispate-undulate, lanate, or glabrescent on upper surface. Inflorescence of rather distant capitula in irregular cymes. Involucre 5–8 × 3–4 mm when dry; phyllaries numerous, linear, green or brown with a dark central stripe, long-acuminate, glandular, glabrous to sparingly pilose; receptacle densely golden-glandular, otherwise naked. Female florets 2.5–3 mm long, ray 1 mm, whitish-yellow or yellow. Disc florets 2.5–3.5 mm. Achenes 0.8 mm long, glabrous, brown, 5-ribbed; outer pappus a ring of scales, inner pappus of setae joined at base into a tube, each seta with a narrow apical tuft of slightly longer hairs.

HAB. In thickets of *Salix* and *Populus euphratica* by river, on sandy loam in a vegetable garden, in a *Sorghum* field on the irrigated alluvial plain, clay soil with large stones; alt 650 m; fl. & fr. Apr.-Jun.
DISTRIB. Occasional in the mountain region and the moist steppe zone of Iraq; also on the irrigated alluvial plain in the desert region. "Assyria", *Aucher-Eloy* 3100. **MJS**: Montafah (?Mantufa), *Field & Lazar* 89! **FUJ**: bank of r. Tigris below Mosul, *Nábělek* 3915. **FUJ/FNI**: on island in r. Tigris nr Mosul, *Kotschy* 439! **FUJ/FNI**: Mosul, *Anders* 2415. **FNI**: Nineveh, *Bornmüller* 1372. **FPF**: 17 km ES of Zurbatiya, *Al Kaisi & Al Khayat* 50654! **LCA**: Bada, N of Baghdad, *Guest* 2509!; Baghdad, *Guest* 179!; Rustamiya, *Lazar* 1182! **LCA/LEA**: Oja canal, nr Ctesiphon, *Guest* 1036! **LEA**: Aziziya, *Guest* 3568! **DWD**: Rawa, *Al Izzi & S. Omar* 35358!; Rutba Range manegment station, *S. Omar, Al-Kaisi & K.Hamad* 43942!; 18 km from Rutba, *S. Omar, H. Abass & Al-Kaisi* 55922! **DSD**: 18km S by W of Busaya, *Guest, Rawi & Rechinger* 16065!

?JIFJAF (Baghdad, *Guest* 2509).

Handel-Mazzetti (Ann. Naturh. Mus. Wien 27: 433 (1913)) recorded *Pulicaria odora* (L.) Reichb. with a query from Iraq, on the basis of a specimen collected by himself "ad ripam dextram Tigridis inter Baghdad et Mosul in desertis sabulosis inter Sumedscha et Beled, c. 80 m, 5.5.1910", *Hand.-Mazz.* 971. This is a small sterile specimen, little more than a seedling either of an annual plant or of a perennial in its first year. It is not determinable and certainly cannot be used as the basis for establishing *P. odora* as an Iraqi plant.

100. **DITTRICHIA** Greuter

Exsicc. Genav. Conserv. Bot. Distrib. Fasc. 4: 71 (1973)
Anderberg & Eldenäs in Kubitzki, Fam. Gen. Vasc. Pl. 8: 380 [2006] (2007)

Shahina A. Ghazanfar

Annual or perennial herbs, branched, densely glandular-pilose. Leaves alternate, sessile, linear-lanceolate, glandular, viscid. Capitula heterogamous, radiate, many in lax racemes. Receptacle epaleate. Phyllaries 3–5-seriate, imbricate. Outer florests female, yellow, rayed, 3-toothed; disc florets tubular, yellow, 5-lobed. Achene obovoid, shorter than corolla, without ribs. Pappus basally connate.

Native to Southwestern Europe, Southeastern Europe and Western Asia. Two species; one in Iraq.

1. **Dittrichia graveolens** (*L.*) *Greuter* in Exsicc. Genav. Conserv. Bot. Distrib. Fasc. 4: 71 (1973); Ball in Fl. Europ. 4: 135 (1976); Rechinger, Fl. Iranica 145: 106 (1980); Boulos & Hind in Fl. Egypt 3: 212 (2002); Qaiser & Abid in Fl. Pakistan 210: 44 (2003).

Inula graveolens (L.) Desf., Fl. Atlant. 2: 275 (1799); Boissier, Fl. Orient. 3: 199 (1875); Dinsmore in Post, Fl. Syria, Palest. & Sinai ed. 2, 2: 25 (1933); Guest in Dep. Agr. Iraq Bull. 27: 49 (1933); Zohary in Dep. Agr. Iraq Bull. 31: 145 (1950); Rawi in Dep. Agr. Iraq Tech. Bull. 14: 117 (1964); Rechinger, Fl. Lowland Iraq: 605 (1964); Grierson in Fl. Turkey 5: 72 (1975); Feinbrun-Dothan, Fl. Palaest. 3: 314 (1978); Meikle, Fl. Cyprus 2: 892 (1985).
Erigeron graveolens L., Cent. Pl. 1: 30 (1755).
Inula brahuica Boiss., Fl. Orient. 3: 199 (1875).

Fig. 99. **Dittrichia graveolens**. 1, habit; 2, capitulum, 3, achene with pappus; 4, top of achene with pappus. Drawn by Ruth Koppel. From Naomi Feinbrun-Dothan, *Flora Palaestina*, Part three: Ericacaceae to Compositae, Plates, Jerusalem: The Israel Academy of Sciences & Humanities, 1977, pl. 524. © The Israel Academy of Sciences and Humanities. Reproduced with permission.

Strong-smelling annual herb with many ascending slender branches from about mid-stem, becoming woody, (0.1–)0.3–1 m tall, densely clothed with short glandular and long simple hairs. Leaves numerous; median cauline leaves sessile, narrowly oblanceolate to linear, 1.5–5 × 1–8 mm, entire or with distant small teeth; upper cauline and branch leaves rapidly reducing. Inflorescence forming the branched upper half of the plant, its primary branches spreading out at 30–60° from the main stem with secondary branches arise from these in robust plants. Capitula in racemes on stem and branches, borne singly on short peduncles or sessile, in either case subtended by one or more leaf-like bracts; involucre funnel-shaped, 7–12 mm in diameter; phyllaries green and glandular centrally, increasing in length and in width of the scarious margin from the outer to the inner; inner phyllaries linear-lanceolate, 6–7 mm, hyaline apart from a narrow green median strip; all phyllaries reflexed after fruit fall. Female florets 6–8, ray *c.* 2 mm long, shortly emergent; central florets hermaphrodite, 3 mm long; all florets frequently drying purplish. Achenes obovoid, grey, *c.* 2 × 0.6 mm, pubescent; pappus bristles *c.* 3.5 mm, whitish to pinkish. Fig. 99: 1–4.

HAB. In moist, at least sometimes saline clay soil; alt. ±10 m; fl. & fr. Oct.-Dec.
DISTRIB. Occasional, sometimes abundant, in the southern alluvial plain of Iraq and on the adjacent desert. **DWD**: 10 km NW of Qalat Salih, *Rawi & Tikriti* 24989! **DSD**: 120 km S of Basra, near the frontier, *Juma'a Brahim* 6189! **LBA**: Basra, *Guest* 311!, 1619A!, *Omar & Thamer* 47980!

S Europe from Portugal to Turkey, Aegean Is., Cyprus, Palestine, Jordan, Syria, Lebanon, Iran, Afghanistan, Pakistan, N Africa (Tunisia to Morocco).

Said by Guest (1933) to be used locally in the treatment of eye diseases.

101. **ANVILLEA** DC.

Prodr. 5: 487 (1836); Benth. & Hook. f., Gen. Pl. 2: 338 (1873)
Anderberg & Eldenäs in Kubitzki, Fam. Gen. Vasc. Pl. 8: 381 [2006] (2007).
Sycodium Pomel, Nouv. Fl. Atlant. 1: 39 (1874); *Anvilleina* Maire, Bull. Soc. Hist. Nat. Afrique N. 30: 346 (1940)

C. C. Townsend

Perennial, canescent subshrubs with many short, rigid, spreading branches, only young stems leafy. Leaves small, obovate or spatulate. Capitula large, hemispherical, many-flowered, heterogamous, discoid and radiate, or homogamous and florets all tubular, shortly and rigidly pedunculate; involucre becoming indurate after flowering, concrescent into a cupule and persisting on plant for more than one season; outer phyllaries few, leaf-like, ± spreading, inner 2 series imbricate, lanceolate, sometimes connate in lower part; receptacle concave, scaly, scales intermixed with florets. Corollas yellow, sometimes a single row of rayed female marginal florets, more often all florets regular, tubular and hermaphrodite; anthers caudate; style slender, forked, arms compressed, papillose. Achenes tetragonal, hairy, apex umbilicate; pappus absent.

A genus of 1 or 2 species, 1 species in the Orient (from Syria, Arabia and S Iran) and possibly another species in N Africa (Libya).

Anvillea (possibly to commemorate the mapmaker Jean Bourguignon d'Anville, 1697–1782).

Anderberg, A. A. (1982). The genus *Anvillea* (Compositae). Nordic J. Bot. 2: 297–305.

1. **Anvillea garcini** (*Burm. f.*) *DC.*, Prodr. 5: 487 (1836); Boissier, Fl. Orient. 3: 181 (1875); Handel-Mazzetti in Ann. Naturh. Mus Wien 27: 433 (1913); Dinsmore in Post, Fl. Syria, Palest. & Sinai ed. 2, 2: 22 (1933); Bornm. in Beih. Bot. Centralbl. 60B: 184 (1939); Zohary in Dep. Agr. Iraq Bull. 31: 145 (1950); Dickson, Wild Fls. Kuwait: 19 (1955); Rechinger, Fl. Lowland Iraq: 610 (1964); Rawi in Dep. Agr. Iraq Tech. Bull. 14: 106 (1964); Feinbrun-Dothan, Fl. Palaest. 3: 320 (1978); Rechinger, Fl. Iranica 145: 129 (1980); Collenette, Wild Flow. Saudi Arabia: 171 (1999); Boulos & Hind in Fl. Egypt 3: 222 (2002); Kilian & Ghazanfar in Fl. Oman 3: 278 (2015).

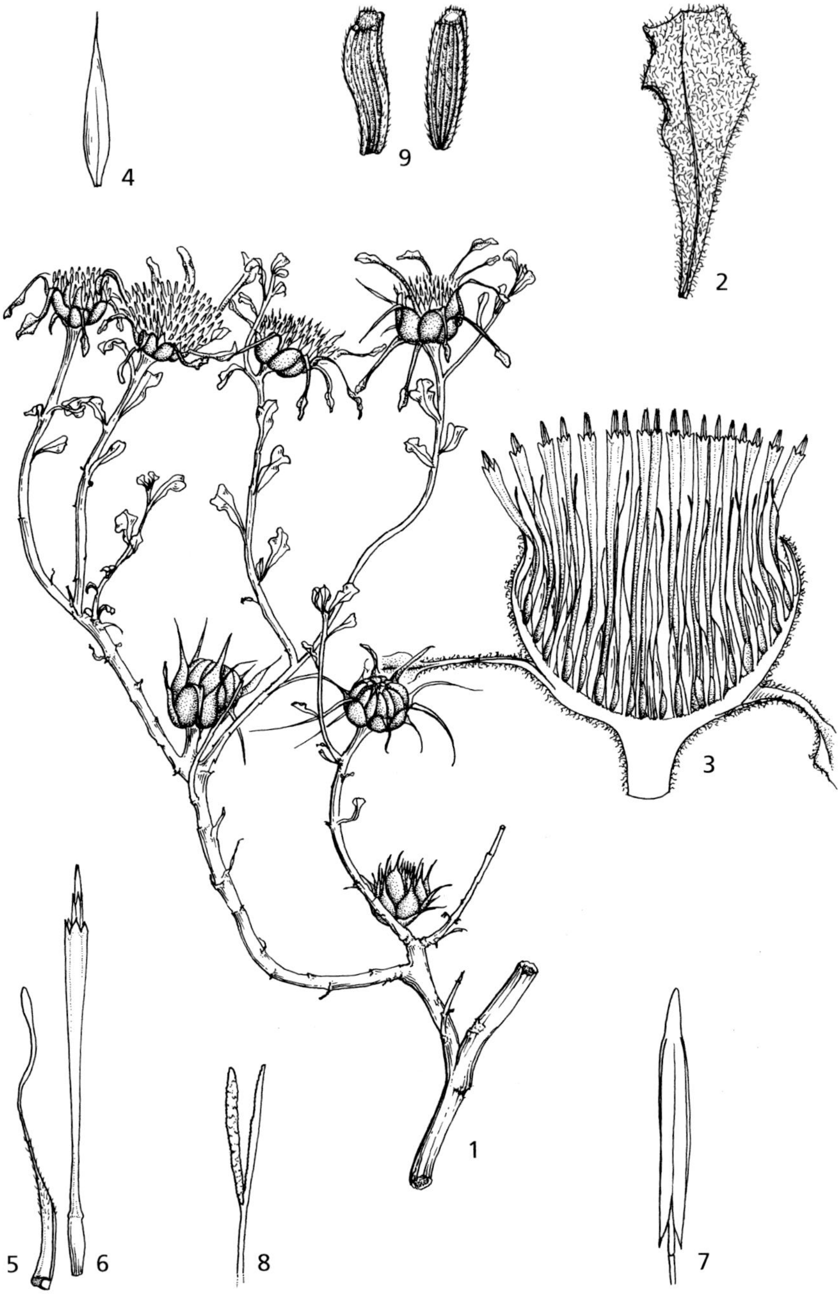

Fig. 100. **Anvillea garcinia**. 1, habit × ⅔; 2, leaf × 2; 3, capitulum (l.s.) × 3; 4, inner phyllary × 4; 5, receptacle scale × 4; 6, disc floret × 4; 7, anther × 8; 8, style × 8; 9, achenes × 4. All from *Guest, Rawi & Long* 14044. Drawn by Christine M. Grey-Wilson.

Anthemis garcini Burm. f., Fl. Indica: 183, t. 60 f. 1 (1768), also citing *Buphthalmum garcini* as a synonym.
Buphthalmum flosculosum Vent., Pl. Nouv. Jard. Cels t. 25 (1801).
B. arabicum Del., Fragm. Fl. Arab. Petr. 14, t. 4 (1833).

Perennial, stiff-branched shrublet, up to 30 cm tall, with a long, woody tap-root and dense leafy crown, developing into low, twiggy bushes. Stem smooth or ribbed, with white peeling bark. Leaves alternate, simple, obovate to spatulate, 2–2.7 × 0.8–1 cm, usually attenuate, but sometimes abruptly narrowed into a very short petiole, obtuse or truncate, often with one large apical tooth, margins entire to dentate, with few, comparatively large teeth in upper part, canescent and glandular-punctate. Capitula homogamous, discoid, solitary, 2–2.3 cm diam., 1.7 cm tall; phyllaries ± 3-seriate, outer phyllaries foliaceous, 15–20 × 5 mm, inner 2 series connate at base, lanceolate, acuminate, up to 7 × 2.5 mm, enlarging to 10 × 4 mm in fruit, minutely erose in upper part; receptacle concave, scaly, scales rigid, each subtending a floret, persisting after flowering and exceeding achene, 3.5 mm long, emarginate, midrib excurrent in a flattened bristle, 5 mm, widened slightly at tip, top of scale and lower half of bristle glandular-hairy. Corollas yellow to deep yellow, regular, tubular, funnel-shaped, 5–8 × c. 2 mm, lobes 5, erect; anthers narrowly cylindrical, 5 mm long, sagittate, caudate, tails appressed, appendage narrowly ovate to lanceolate, anther-collars well-developed; style slender, forked, arms compressed, papillose with rounded tips. Achenes ± oblong in outline or widening slightly at base, compressed, 4-angled, 3 × 1 mm, glandular-hairy, persisting on receptacle after maturing; pappus absent. Fig. 100: 1–9.

HAB. On desert soils, sandy gravel and silt, sometimes overlying gypsum, often stony or sometimes in pockets between limestone or other rocks, in silty depressions etc.; up to 600+ m; fl. & fr. (Mar.-)Apr.-Jun. and Oct.(-Nov.).

DISTRIB. Occasional (locally common to rare) in the western sector of the Western Desert of Iraq, abundant to frequent (in a few places locally dominant) in the Southern Desert, rare (only found in one limited locality) in the Northern Sector. **DLJ:** nr Tikrit, *Guest, Rawi & Rechinger* 15945!; **DWD:** nr Rutba, *Alizzi & Husain* 34093!; 80–100 km E of Rutba, *Alizzi* 35158!, *Botany Staff* 41826!; 230 km W of Ramadi, *Guest & Rawi* 13958!, *Rawi* 14716!, *K. Hamad* 48880!; 60 km S of Rutba, *Rawi* 14716!; Wadi at-Tibul, 110 km N of Nukhaib, *Rawi* 31093!; 25 km E of Nukhaib, *Guest, Rawi & Long* 14044!; 5 km S of Shanana, *Rawi* 14809!; 114 km W of Karbala, *Rawi* 30945!; Faidhat al-Qaiyara, 65 km W by S of Najaf, *Guest, Rawi & Long* 19467!; **DWD/DSD:** between Nukhaib and Busaiya (precise details missing), *Serkahia & H. Tikriti* 15127!; **DSD:** 50–60 km W of Shabicha, *Hazim* 30684!, 32485!; 25 km 'from' (E of?) Ma'niya, *id.* 32487!; 40 km WNW of Shabicha, *Guest, Rawi & Rechinger* 19292!; Firch, 40 km NW of Salman, *Fauzi, Hazim & Hamid* 38682!; nr Golaib, 30 km E of Salman, *Fauzi, Rawi & Rechinger* 18762!; 40 km SSE of Salman, *id.* 18862!; Jumaima, *Qaisi, K. Kadhim & H. Hamad* 48261!; 8 km N by W of Aidaha, *Guest, Rawi & Rechinger* 19174!; Khadhar al Ma'i, *Rawi, Khatib & Tikriti* 29199! (and many others); 14 km ESE of Khadhar al-Mai, *Guest & Mahalhal* 15238!; 75 km S of Khadhar al-Ma'i, *Alizzi* 34415!; 50 km W of Busaiya, *Haddad* 9585!; 62 km WNW of Ansab, *Guest, Rawi & Rechinger* 19058!; 46 km WNW of Ansab, *id.* 19046!; Faidhat ach-Chabd, 125 km N of the neutral zone, *Rawi, Khatib & Tikriti* 2200!; 79 km W of Basra, on road to Busaiya, *Botany Staff* 42394!; 46 km WNW of Ansab, on Saudi frontier, *Guest, Rawi & Rechinger* 19046!; Wishih, 103 km S (? for SE) of Ansab, *id.* 42281!; 24 km W of *Busaiya, Guest & Mahalhal* 15263!; nr Adan, *id.* 15281!; nr northern apex of neutral zone, *c.* 50 km N of Samah, *Guest, Rawi & Rechinger* 16080!; nr Samah, *Alizzi* 34399!

Syria, Palestine, Jordan, Egypt, Arabia, S Iran.

A radiate form of the species is also sometimes found in Arabia which is considered to be a subspecies of *A. garcini* (subsp. *garcini*) by Arne Anderberg (Opera Bot. 104: 1–19 1981), and may perhaps be the same plant as the radiate *Anvillea* found in N Africa.

The following vernacular names have been noted for this plant, the first two in Iraq, the third in Kuwait: NAKHD (Ir.-DWD, Shanana, S of Nukhaib, *Rawi* 14809; Nuaqib, *Guest, Rawi & Long* 14044 "eaten by all animals; camels especially relish it, though its taste is rather pungent"; HARR, Ir.-northern sector of DSD, Salman district, *Guest, Rawi & Rechinger* 19046, 19058; Samah, *id.* 16080; Busaiya, *Guest & Mahalhal* 15263 "eaten by sheep and camels"; Adan, *id.* 15281. NIQT, Kuwait, *Dickson* 121.

These are apparently local variants of the same colloquial Arabic name for the plant which is generally regarded as a useful grazing plant for animals and also, no doubt when dry, as suitable fuel for the camp fire. Precisely how this name should be spelled, or its original derivation, remain obscure.

102. **ASTERISCUS** Miller

Gard. Dict. Abr., ed. 1, 1 (1754)
Anderberg & Eldenäs in Kubitzki, Fam. Gen. Vasc. Pl. 8: 382 [2006] (2007)
Bubonium J. Hill, Veg. Syst. 2: 74 (1761)
Nauplius Cass., Dict. Sci. Nat., ed. 2, 23: 566 (1822).
Odontospermum Necker ex Sch.Bip. in Webb & Berth., Phyt. Canar. 2: 232 (1844).

Patricia Halliday

Annual or perennial herbs; leaves simple, alternate. Capitula many-flowered, heterogamous, radiate, subsessile, axillary and terminal; involucre hemispherical; phyllaries 2–3-seriate, imbricate, outermost spreading, foliaceous, appendiciform, bases indurate and apices foliaceous; paleae concave or shallowly boat-shaped. Corollas yellow; ray florets uniseriate, 3-toothed; disc florets regular, tubular, 5-toothed. Achenes turbinate, ribbed, marginal 3-angled, inner 4-angled or ± terete.

A genus comprising about 8 species, mainly eastern Mediterranean in distribution, but also extending to other parts of Europe, Macaronesia (Canary Is., Cape Verde Is.), and N Africa (S to central Sahara). A single species in Iraq.

Asteriscus (from Gr. ἀσΤήρ, star, and the diminutive suffix -iscus).

Greuter, W. (1997). Save *Asteriscus*, sink *Nauplius* (Compositae). Fl. Med. 7: 41–48.
Halvorsen, T. & Borgen, L. (1986). The perennial Macaronesian species of *Bubonium* (Compositae-Inuleae). Sommerfeltia 3: 1–103.
Wicklund, A. (1985). The genus *Asteriscus* (Asteraceae-Inuleae). Nordic J. Bot. 5: 299–314.
Wiklund, A. (1987). The genus *Nauplius* (Asteraceae-Inuleae). Nordic J. Bot. 7(1): 1–23.

Annual, subacaulescent herb to 5(–7) cm; paleae 2–3 mm, boat-shaped, hardly keeled; disc florets 2–3 mm long; anthers *c.* 1 mm, bases truncate; pappus *c.* 1 mm, subulate . . . . 1. *A. pygmaeus*
Perennial, much-branched shrub to 20 cm; paleae 5 mm, keel winged; disc florets *c.* 5 mm long; anthers 3–3.5 mm, bases long-caudate; pappus 1.5–2 mm, lanceolate . . . . 2. *A. graveolens*

1. **Asteriscus pygmaeus** (*DC.*) *Coss. & Dur.*, Ann. Sci. Nat., ser. 4 (Bot.), 4: 284 (1855) & in Bull. Soc. Bot. France 4: 471 (1857); Boissier, Fl. Orient. 3: 179 (1875); Nábělek in Publ. Fac. Sci. Univ. Masaryk 52: 4 (1925); Dinsmore in Post, Fl. Syria, Palest. & Sinai. ed, 2, 2 : 21 (1933); Anthony in Notes Roy. Bot. Gard. Edinburgh 18: 290 (1935); Zohary in Dep. Agr. Iraq Bull. 31: 144 (1950); Rechinger, Fl. Lowland Iraq: 611 (1964); Rawi in Dep. Agr. Iraq Tech. Bull. 14 : 107 (1964); Täckholm, Stud. Fl. Egypt, ed. 2: 565 (1974); Feinbrun-Dothan, Fl. Palaest. 3: 321 (1978); Rechinger, Fl. Iranica 145: 131 (1980).

Asteriscus aquaticus (L.) Less. var. *pygmaeus* DC., Prodr. 7: 287 (1838) pro parte (Arabian plants)
Odontospermum pygmaeum (Coss. & Dur.) Hoffm. in Engl. & Prantl, Pflanzenfam. 4, 5: 209 (1890); Handel-Mazzetti in Ann. Naturh. Mus. Wien 27: 433 (1913); Dickson, Wild Fls. Kuwait: 68 (1955).

Annual sub-acaulescent herb to 5(–7) cm tall, glandular and long-hairy throughout and with a long tap-root; branched at ground level beneath central sessile or subsessile capitulum, branches short, rigid, plant sometimes unbranched and rosettiform with as solitary capitulum, or stems simple and erect. Leaves basal and cauline, upper much exceeding the involucre, obovate, spatulate, long attenuate, 2.1–5 gland–1.5 cm, obtuse, entire. Capitula solitary, short pedunculate, terminal or axillary, 1.2–2 cm in diameter, persisting on dead plant for more than one season. Phyllaries 2–3-seriate, outer much exceeding capitulum; paleae 2–3 mm long, boat-shaped, apices wider and hairy outside, obtuse. Ray florets female, limb oblong, *c.* 5 m long, with 3 apical teeth, 6-veined, tube *c.* ⅓ of total floret length, glandular and long-hairy, indumentum extending along central section of limb on outer surface. Disc florets hermaphrodite, bottle-shaped, with a narrow neck; style bifid, style arms clavate to spatulate. Achenes (immature) turbinate, sericeous, 2 along central section of limb on outer surface. Fig. 101: 1–11.

HAB. Desert "haswa" plains and rocky hills, on sandy and silty soil, gravel, clay, gypsum etc., also in depressions and along wadis; alt. up to 600 m; fl. & fr. Mar.-Apr.(-Jun.).

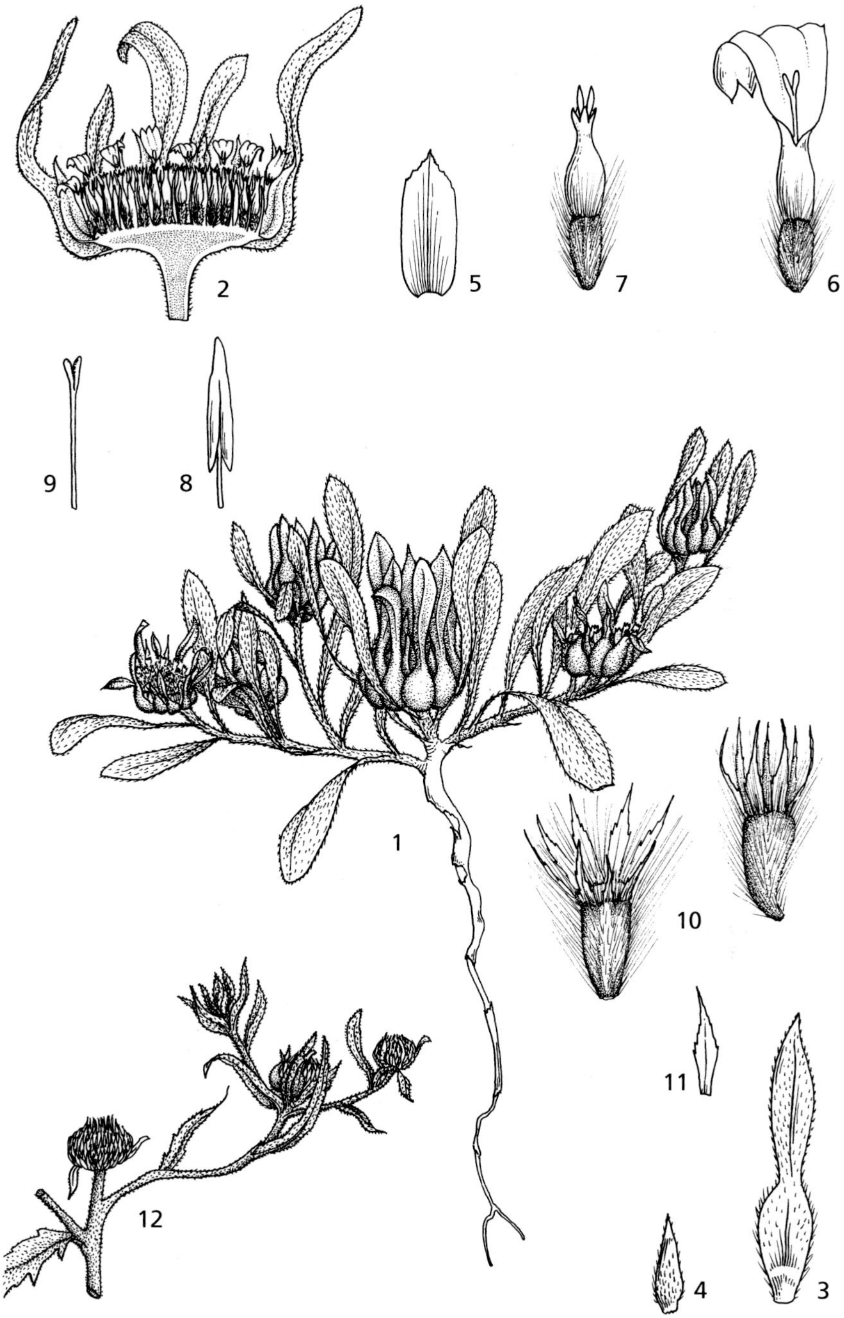

Fig. 101. **Asteriscus pygmaeus**. 1, habit × ⅔; 2, capitulum (dissected) × 2; outer phyllary × 2; 4, inner phyllary × 2; 5, receptacular scale × 6; 6, ray floret × 6; 7, disc floret × 6; 8, anther × 12; 9, style × 6; 10, achenes × 8; 11, pappus scale × 8; **Asteriscus graveolens**. 12, habit × 2/3. 1–11 from *Rawi* 31060; 12 from *Guest, Rawi & Rechinger* 19434. Drawn by Christine M. Grey-Wilson.

DISTRIB. Occasional in the NW sector of the Desert region of Iraq: **DWD**: 320 km W of Baghdad, *Cowan & Darlington* s.n.! **DWD/DLJ**: nr Haditha, *Handel-Mazzetti* 784; nr R. Euphrates, below Hit, *id.* 832. **DWD/LCA**: Falluja, *Nábělek* 3751.

Syria, Palestine, Jordan, Sinai, Egypt, Arabia, Kuwait, S Iran, Pakistan, Afghanistan, Macaronesia (Canary Is.).

Dickson (1955) noted the name BURKAT for this plant in Kuwait and recorded that it is "sometimes used by the Badawin to put into sacks of rice to keep it sweet". Shabetai has recorded the local name NUGUD in Sinai.

2. **Asteriscus graveolens** (*Forssk.*) *Less.*, Syn. Gen. Comp. 210 (1832); Dinsmore in Post, Fl. Syria, ed. 2, 2: 21 (1933); Zohary in Dep. Agr. Iraq Bull.: 31: 144 (1950); Rechinger, Fl. Lowland Iraq: 611 (1964); Raw in Dep. Agr. Iraq Tech. Bull. 16: 107 (1964); Täckholm, Stud. Fl. Egypt, ed. 2: 565 (1974); Feinbrun-Dothan, Fl. Palaest. 3: 321 (1978).

Buphthalmum graveolens Forssk., Fl. Aegypt.-Arab. 151 (1775).
Nauplius graveolens (Forssk.) Wiklund, Nord. J. Bot. 7: 16 (1987); Boulos, Fl. Egypt Checklist: 236 (2009).
Ceruana schimperi Boiss., Diagn. Pl. Orient., ser. 1, 11: 4 (1849).
Odontospermum graveolens (Forssk.) Sch.Bip. in Webb & Berth., Phyt. Canar. 2: 232 (1844).
Asteriscus schimperi (Boiss.) Boiss., Fl. Orient. [Boissier] 3: 180 (1875).

Perennial much-branched shrub to 20 cm tall in Iraq, leafy, glandular and hairy throughout. Branches short, straight, rigid, the older wood grey, with peeling bark. Leaves all cauline, alternate, mostly oblanceolate in outline, 3–4 × 0.6–0.8 cm, attenuate, widening at subamplexicaul, ± auricled base, obtuse to acute, irregularly toothed, but uppermost linear-lanceolate and entire. Capitula 1–1.5 cm diam., very shortly pedunculate or at ends of branches, persisting on the plant for more than one season. Phyllaries ± 3-seriate, outermost phyllaries up to 2 cm long, supporting the capitulum, innermost linear-lanceolate, acute, c. 6 × 2 mm; receptacle convex, hemispherical, paleae 5 mm long, boat-shaped, keeled, with membranous wing on keel, stiff, upper part hairy on outer surface. Corollas golden-yellow. Ray florets female, staminodes present, rays oblong, 7 × 1.5 mm, with 3 deeply-cut apical teeth and densely papillose on outer surface. Disc florets hermaphrodite, corollas sub-cylindrical to funnel-shaped, c. 5 × 1 mm, with 5 deeply cut teeth; anthers cylindrical, 3–3.5 mm long, appendages lanceolate, base sagittate, long-caudate, filaments stout; styles bifid, style arms compressed, with triangular, acute, papillose appendages. Achenes turbinate, ribbed, 2 × 1 mm, marginals 3-angled, inner ± terete, ± densely sericeous; pappus scales lanceolate, scarious, lacerate, 1.5–2 mm. Fig. 101: 12.

HAB. In sandy patches on fine gravel desert plain; alt. ± 330 m. fl. & fr. (Feb.-)Mar.-Apr.
DISTRIB. Apparently very rare in Iraq; only found once some 60 years ago in the northern sector of the southern desert region. **DSD**: Wadi al-Khirr, 90 km NW by N of Shabicha, *Guest, Rawi & Rechinger* 19434!

Chad, Mali, Mauritania, Algeria, Tunisia, Libya, Egypt, Sinai, Jordan, Palestine, Lebanon, Syria, Saudi Arabia.

103. **PALLENIS** Cass.

Dict. Sci. Nat., ed. 2, 37: 275 (1825)
Benth. & Hook. f., Gen. Pl. 2: 340 (1873); Anderberg & Eldenäs in Kubitzki, Fam. Gen. Vasc. Pl. 8: 382 [2006] (2007)

Patricia Halliday

Erect, stiff, ± hairy, annual herbs with simple, alternate leaves. Capitula solitary, heterogamous, ligulate, in angles of branches or terminal; involucre ± hemispherical, 2–3-seriate; outermost phyllaries foliaceous, much exceeding capitulum, spreading, spine-tipped, inner ± erect; receptacle convex, with shallowly boat-shaped, keeled scales. Florets all fertile; rays female, 1–3-seriate, with 3 apical teeth; disc florets hermaphrodite, regular, tubular, with a single wing along inner face and 5 deeply-cut apical teeth; corollas yellow. Achenes of ray florets dorsiventrally compressed, with 2 lateral wings, those of disc florets laterally compressed with a single narrow wing along inner face; pappus scales minute, numerous, free, scarious, lacerate.

A monotypic genus of the E Mediterranean region, extending into SW Asia and N, and also westwards to Macaronesia.

Pallenis (perhaps from Gr. *palea*, chaff, due to its somewhat fluffy appearance).

Wicklund, A. (1985). The genus *Asteriscus* (Asteraceae-Inuleae). Nordic J. Bot. 5: 299–314.

1. **Pallenis spinosa** (*L.*) *Cass.*, Dict. Sci. Nat. ed. 2, 37: 276 (1825); DC., Prodr. 5: 487 (1836); Boissier, Fl. Orient. 3: 180 (1875); Dinsmore in Post, Fl. Syria, Palest. & Sinai ed. 2, 2: 22 (1933); Zohary in Dep. Agr. Iraq Bull. 31: 144 (1950); Blakelock in Kew Bull. 4: 60 (1953); Rawi in Dep. Agr. Iraq Tech. Bull. 14: 120 (1964): Täckholm, Stud. Fl. Egypt ed. 2: 565 (1974); Grierson in Fl. Turkey 5: 51 (1975); Feinbrun-Dothan, Palaest. 3: 320 (1978); Rechinger, Fl. Iranica 145: 130 (1980); Taifour & El-Oqlah, Pl. Jordan Ann. Checkl.: 45 (2017).

Buphthalmum spinosum L., Sp. Pl. ed. 1: 903 (1753); Reichb., Ic. Fl. Germ. 16: t. 938, 939 (1853).
B. asteroideum Viv., Fl. Lybia Spec. 57: t. 25, f. 2 (1824).
Asteriscus spinosus Sch.Bip. in Webb & Bertol., Phyt. Canar. 2: 230 (1835–60); Fragman, Levy-Yamamori & Christodoulou, Fl. Eastern Med. 404 (2001); Boulos & Hind in Fl. Egypt 3: 225 (2002).

Erect annual herb to 72 cm tall, long-hairy. Stems rigid, simple or branched, either from base or only in upper half, leafy throughout. Leaves 3–7 × 0.8–2 cm, basal oblanceolate, attenuate, subpetiolate, entire or often shallowly, irregularly serrate, especially in lower half, withering early; cauline leaves entire or with marginal callouses, semi-amplexicaul, auricled; all leaves spine-tipped. Capitula, excluding outermost bracts, 2–2.5 cm diam.; phyllaries 2–3-seriate, outermost spreading, up to 4.2 × 9 mm, with conspicuous veins, forming a star around capitulum, abruptly contracted near base into a short stalk, spine-tipped, inner phyllaries stiff, ± erect, oblong-lanceolate, 9 × 2–3 mm, cuspidate; receptacular scales oblong or oblanceolate, ± keeled, 5 × 1 mm, acute, cuspidate, ciliate at apex. Ray florets 6–7 × 0.5–1 mm, rays narrow, oblong with expanded, winged, ciliate tube, 3 apical teeth, 3 veins, yellow; disc florets 4 × <1 mm, cylindrical, somewhat thickened at base glabrous; anthers 2 mm, cylindrical, appendages lanceolate, base sagittate, caudate, tails divergent; style bifid, arms compressed, rounded at apex. Achenes of ray-florets oblong-obovate in outline, 2.4 × 2 mm, those of disc florets turbinate, 1.5 × 1 mm; pappus scales ± lanceolate, 0.5 mm. Fig. 102: 1–14.

HAB. On a mountainside, waste land near a village; alt. 300–600 m; fl. & fr. Jun.-Jul.
DISTRIB. Apparently rare in Iraq; only found a few times in the NW sector of the lower margin of the forest zone and upper plains. **MAM**: Atrush, *Guest* 3626!, *Anders* 1591; Aqra, *Rawi* 11380! **FNI**: Bashiqa, col. ingot. 9064!; 1 km NE of Mindan bridge, *E. Chapman* 26197!

S Europe (Portugal & Spain to Italy, the Balkans, Greece, Crete, Crimea), Cyprus, Syria, Lebanon, Turkey, Egypt, Iran, Transcaucasia, Turkmenia, N Africa (Libya, Tunisia, Algeria, Morocco), Macaronesia (Canary Is.).

104. **INULA** L.

Sp. Pl.: 881 (1753); Gen. Pl. ed. 5: 375 (1754)
Anderberg & Eldenäs in Kubitzki, Fam. Gen. Vasc. Pl. 8: 382 [2006] (2007)
Codonocephalum Fenzl in Flora 26: 397 (1843); *Dittrichia* Greuter in Exsicc. Genev. 4: 71 (1973)

Frances G. Davies
Revised by C. C. Townsend

Annual to biennial or perennial leafy herbs or (not in Iraq) shrubs. Leaves simple, alternate, entire, dentate or serrate, glabrous or canescent or viscid-glandular. Inflorescence varying from a single terminal capitulum to corymbose or racemose, capitula ranging in size from small to very large, heterogamous and radiate or sometimes homogamous and discoid; phyllaries multiseriate, herbaceous or cartilaginous, inner sometimes scarious. Receptacle naked. Outer florets female, rayed, 2–3-dentate, occasionally truncate, sometimes absent. Disc-florets bisexual, equally 5-lobed, fertile. Corollas yellow; style arms subulate, tips rounded, not conspicuously hairy. Achenes small, brown, cylindrical or angular, ± pilose; pappus uniseriate, barbellate, mostly free.

About 100 species in Europe, Asia and Africa; 6 species in Iraq.

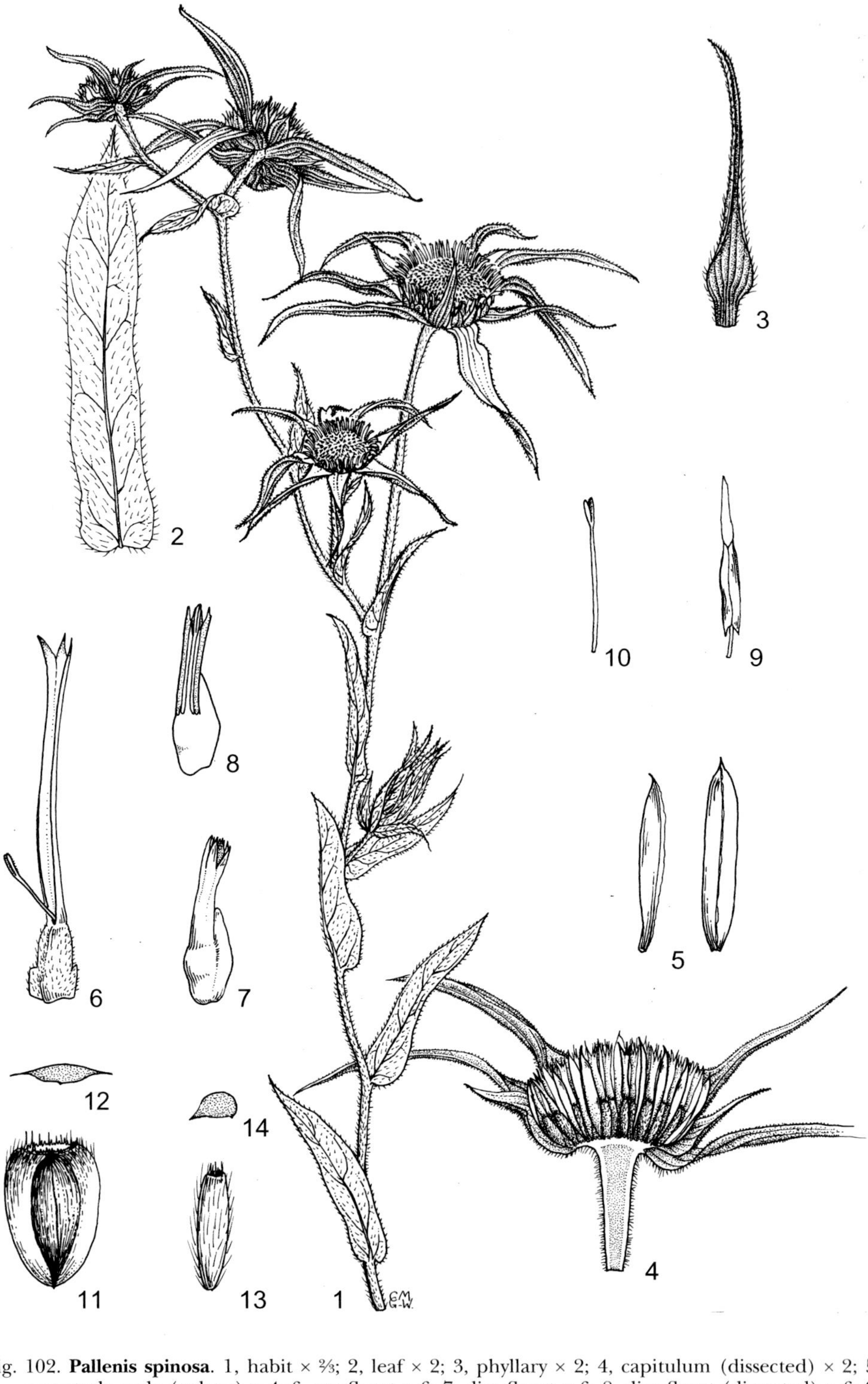

Fig. 102. **Pallenis spinosa**. 1, habit × ⅔; 2, leaf × 2; 3, phyllary × 2; 4, capitulum (dissected) × 2; 5, receptacle scale (paleae) × 4; 6, ray floret × 6; 7, disc floret × 6; 8, disc floret (dissected) × 6; 9, anther × 12; 10, style × 6; 11, marginal achene × 6; 12, T.S marginal achene × 6; 13, inner achene × 6; 14, T.S. inner achene × 6. All from *Guest* 3626. Drawn by Christine M. Grey-Wilson.

Inula (from the name of a plant in Pliny, perhaps *I. helenium*).

Inula contains species superficially resembling some yellow-rayed members of the tribe Astereae, but the style arms of the disc florets of the latter are always tipped by hairy, triangular, appressed appendages. *Inula* is distinguished from the closely-related genus *Pulicaria*, and especially the long-rayed *P. dysenterica*, by the lack of a scaly or cupular outer pappus series characteristic of *Pulicaria*.

1. Capitula conspicuously radiate, marginal florets; rays 6–25 mm 2
   Capitula discoid and florets all tubular and hermaphrodite, or inconspicuously radiate and marginal florets female; rays at most 2 mm long and only shortly exserted 4
2. Leaves densely tomentose on lower surface . . . . 2. *I. helenium*
   Leaves not tomentose below, green, sparsely pilose and/or minutely glandular or glabrous 3
3. Inner phyllaries densely softly pilose at tip; outer phyllaries oblong to panduriform . . . . 3. *I. salicina*
   Inner phyllaries with a scabrid-hairy indumentum similar to that of median ones; outer phyllaries linear . . . . 1. *I. viscidula*
4. Florets all tubular and hermaphrodite; inner phyllaries 15–18 mm long, outer at least 4 mm wide . . . . 5. *I. inuloides*
   Outer florets female, usually truncate or with a very short ray limb; inner phyllaries 8–10 mm long, outer at most 0.5 mm wide . . . . 4. *I. thapsoides*

1. **Inula viscidula** *Boiss. & Kotschy* in Boissier, Fl. Orient. 3: 188 (1875); Grierson in Fl. Turkey 5: 61 (1975); Rechinger, Fl. Iranica 145: 84 (1980).

Rhizomatous perennial herb 50–100 cm tall, generally much-branched with divergent branches, sessile-glandular above. Basal leaves not seen. Median cauline leaves sessile, elliptic, 6–14 × 2–4 cm, acute, margin serrate and bordered by minute cartilaginous teeth, lamina with dense sessile glands beneath, glabrous or shortly pilose beneath on veins and margins; upper cauline and branch leaves reducing. Capitula solitary or 2–4 together at ends of branches; involucre hemispherical, 1.75–2.5 cm in diameter; outer phyllaries loosely erect, ± reflexed after fruit fall, linear, acuminate, base cartilaginous, rest green, herbaceous except for double row of cartilaginous marginal teeth; inner phyllaries 8–15 mm, linear, yellow, ciliate at tip. Female florets 8–10 mm, ray 6–8 mm, much exceeding the style, apex 3-toothed; disc-florets *c.* 5 mm. Achenes cylindrical, 10-ribbed, 2.5–3 mm long, pubescent; pappus 4.5–6.5 mm, dirty white.

HAB. Lower slopes of high mountains, usually in shade of *Juglans* or *Quercus libani* and frequently by water; alt. 1100–1700 m; fl. & fr. Jul.-Aug.
DISTRIB. Frequent, locally abundant, in the central mountain region of Iraq not far from the Iranian frontier. **MRO**: between Gunda Zhor and Darband, *Gillett* 12397!; NE of Sairo, *Gillett* 9679!; Pushtashan, 15 km NE of Rania, lower slopes of Qandil range, *Rawi* 23918!; lower slopes of Qandil range NE of Rania, *Rawi & Serhang* 18238!; Qerna Qaw (Qurnago) valley N of Pushtashan, *id.* 26628!, 26693!

Turkey, Iran.

2. **Inula helenium** *L.*, Sp. Pl. 1: 881 (1753); Boissier, Fl. Orient. 3: 186 (1875); Zohary in Dep. Agr. Iraq Bull. 14: 145 (1950); Gorshkova in Fl. S.S.S.R. 25: 440 (1959); Rawi in Dep. Agr. Iraq Tech. Bull. 14: 117 (1964); Grierson in Fl. Turkey 5: 56 (1975); Rechinger, Fl. Iranica 145: 81 (1980); Yousheng Chen & Arne A. Anderberg in Fl. China 20-21: 839, f. 665 (2011).

Robust perennial herb from a thick subtuberous rhizome. Stems several, erect 1–2 m tall, angular below, ± densely tawny-pubescent and glandular, branched above. Basal leaves lanceolate-elliptic, up to 50 × 15 cm, tapering to a distinct petiole, margins crenulate to serrate; median cauline leaves sessile, amplexicaul, ± auriculate, oblong to ovate, crenulate; upper cauline and branch leaves reducing, narrower; all leaves shortly pilose and papillate-glandular above, ± tomentose on lower surface. Inflorescence corymbose with few to many capitula. Involucre hemispherical, 2.5–4 cm in diameter, frequently subtended by small bract-like leaves; phyllaries ± densely velvety-tomentose dorsally, outer

Fig. 103. **Inula helenium**. 1, habit flowering branch. Reproduced with permission from Flora of China Vol. 21, fig 665.

ovate and subacute, median oblong and obtuse to lanceolate and acute, 12–15 mm long, innermost subglabrous and membranous. Female florets mostly 15–35 mm, ray 10–25 mm; disc florets 7–9 mm. Achenes 4-angled, finely ribbed, 2.5–4 mm; pappus 7–10 mm, brownish. Fig. 103: 1.

HAB. Lower slopes of high mountains and in a hillside valley; alt. 1000–1200+ m; fl. & fr. Aug.-Sep.
DISTRIB. Scarce and scattered in the mountain region of Iraq. **MAM**: Shaikhan valley, *Rawi & Serhang* 24881! **MRO**: about Zirva, N of Kani Rash, *Thesiger* 1230! **MSU**: Penjwin, *Haussknecht* s.n.!

Widespread in Europe and probably native in the southeast., but widely naturalized from cultivation as a medicinal plant and its exact native area obscure.

Turkey, Iran, Caucasia, W Siberia, C Asia, China, Mongolia; naturalized in N America.

Elecampane (Eng.), Aunée (Fr.). Medicinally the root was credited with an astonishing range of virtues in olden times, from curing shortness of breath and loose teeth to snakebite. It was also candied with sugar as a sweetmeat. As with *Helianthus tuberosus*, the root contains a good deal of inulin. Pratt (1891) observes that it is "no doubt a good pectoral medicine, and it is certainly a useful remedy for the diseases of sheep", while it appears also to have a genuine effectiveness as a sedative, antispasmodic and tonic. It is mildly diuretic and used to improve appetite by stimulating gastric secretion.

3. **Inula salicina** *L.*, Sp. Pl.: 882 (1753); Boissier, Fl. Orient. 3: 187 (1875); Nábělek in Publ. Fac. Sci. Univ. Masaryk 52: 5 (1925); Dinsmore in Post, Fl. Syria, Palest. & Sinai ed. 2, 2: 23 (1933); Zohary in Dep. Agr. Iraq Bull. 14: 145 (1950); Gorshkova in Fl. S.S.S.R. 25: 454 (1959); Rawi in Dep. Agr. Iraq Tech. Bull. 14: 118 (1964); Grierson in Fl. Turkey 5: 60 (1975); Rechinger, Fl. Iranica 145: 85 (1980); Yousheng Chen & Anderberg in Fl. China 20-21: 840 (2011).

I. aspera Poir., Encycl. Suppl. 3: 154 (1813); Gorshkova in Fl. S.S.S.R. 25: 452 (1959).
I. cordata Boiss., Diagn. Pl. Orient. ser. 1, 4: 3 (1844; Boissier, Fl. Orient. 3: 188 (1875).
I. salicina var. *aspera* (Poir.) Beck. in Denkschr. Akad. Wiss. Wien, Math.-Nat. Kl. 44: 302 (1881).

Perennial herb with a slender rhizome. Stem slender, brittle, 0.3–0.8(–1.1) m tall, leafy throughout, ridged, ± pilose below, glabrous or almost so above. Median cauline leaves largest, 4–7(–11) × 0.7–3 cm, sessile, semi-amplexicaul, oblong-lanceolate, acute and apiculate, apparently entire but with a border of minute double cartilaginous teeth, lamina glabrous and bright green on both surfaces, or minutely hairy on veins beneath. Inflorescence a corymb of 1–4 (rarely up to *c.* 12) capitula; pedicels 1–1.5 cm. Involucre shallowly hemispherical, 1.25–2 cm in diameter; outer and median phyllaries oblong to panduriform, ciliate, with an apical leafy erect to reflexed appendage, median to 6–7 mm long; innermost phyllaries linear, without appendages, densely softly hairy at tip. Female florets rayed, 15–20 mm, rays 10–15 mm; disc florets 7–9 mm. Achenes cylindrical, *c.* 1.75 mm, ribbed, glabrous; pappus 5–8 mm, dirty white, very minutely barbellate.

HAB. On the lower slopes of mountains, mostly under trees (*Pinus* or *Quercus*); alt. 900–1180 m; fl. & fr. Jun.-Aug.
DISTRIB. Rather rare in the mountain region of Iraq. **MAM**: Jabal Bekher, nr Zakho, *Field & Lazar* 821!; Zawita, *Omar, Dabbagh & Qaisi* 45305!; Kani Masi (Ain Nuni), *Hamad & Fadhil* 45648!; Sarsang, *Haines* 1338!; **MRO**: Mergasur, *Haines* s.n.!

Almost throughout Europe, Turkey, Lebanon, Caucasia, Siberia eastwards to C. Asia, China, Japan.

4. **Inula thapsoides** (*M.Bieb. ex Willd.*) *Spreng.*, Pugill.: 58 (1813); Boissier, Fl. Orient. 3: 190 (1875); Gorshkova in Fl. S.S.S.R. 25: 474 (1959); Grierson in Fl. Turkey 5: 68 (1975); Rechinger, Fl. Iranica 145: 93 (1980).

Conyza thapsoides M.Bieb. ex Willd., Sp. Pl. ed. 4, 3: 1949 (1803).

Perennial herb with 1-few leafy erect branched stems from a woody rhizomatous stock bearing many dormant buds. Stems erect, 0.3–1 m tall, leafy to the apex, brownish- or whitish-tomentose throughout and with scattered glands. Basal leaves elliptic or lanceolate, 10–35 × 3–6 cm including a petiole of up to 4 cm, apex shortly acute or mucronate, base narrowed down the petiole. Cauline leaves lanceolate to oblong, very variably decurrent down the stem (hardly at all in Iraq), margins minutely serrulate, lamina densely velvety or sericeous especially beneath, veins frequently prominent beneath. Inflorescence corymbose, of congested groups of 3–9 subsessile capitula, subtended by bract-like leaves. Involucre obconical, 6–8 mm in diameter; phyllaries imbricate, outer and median herbaceous, linear-lanceolate, densely sericeous, apically hair-tufted, inner linear, with a sparser tomentum, 9–11 mm. Outer female florets 4–5 mm long, corolla truncate or occasionally very shortly rayed, ray not exserted; disc florets 5–7 mm, very numerous, narrowly tubular, lobes often ± pubescent. Achenes 2.5–2.9 × 0.4 mm, angular, ± pilose; pappus brownish to off-white, bristles barbellate, 5–7 mm.

HAB. In shaded or sheltered places on the lower slopes of high mountains, in two of the four localities recorded in *Quercus* forest; alt. 1200–1500 m; fl. & fr. Jun.-Jul.
DISTRIB. Rare in the mountain region of Iraq. **MAM**: Gali Zawita, *Rawi* 23538!; N of Zakho, *Rechinger* 11511; **MRO**: Kani Mazu Shirin, *Agnew, Hadač & Haines* 2045! **MSU**: nr Penjwin, *Rechinger* 10523.

Crimea, Turkey, Caucasia, Iran.

The Iraqi plants are referable to subsp. *australis* Grierson in Notes Roy. Bot. Gard. Edinburgh 33: 425 (1975) in having the leaves scarcely at all decurrent and with the venation prominent on the lower surface; but as Rechinger observes in *Fl. Iranica*, these characters are by no means always related.

5. **Inula inuloides** (*Fenzl*) *Grierson* in Notes Roy. Bot. Gard. Edinburgh 33: 250 (1975); Grierson in Fl. Turkey 5: 58 (1975).

Codonocephalum inuloides Fenzl in Flora 26: 145 (1950). –Type. Gara Dagh, *Kotschy* 376 (HAL, M, S, WAG, syn)!; Rawi in Dep. Agr. Iraq Tech. Bull. 14: 112 (1964); Rechinger, Fl. Iranica 145: 75 (1980
Sprunera inuloides Sch.Bip. in Walpers, Rep. Bot. Syst. 2: 954 (1843).

Robust perennial 0.5–1 m tall. Stem terete or ± angled above, striate, brownish, densely clothed with small sessile glands and with or without fine simple hairs, ± branched above with long, suberect branches. Basal leaves petiolate, petiole 3–15 cm, sometimes longer than lamina; lamina 7–16 × 3–8 cm, elliptic or oblong, coriaceous, entire, finely reticulate-veined. Lower cauline leaves very similar, upper gradually diminishing in size and length of petiole; all leaves with an indumentum similar to that of stem. Capitula solitary or 2 together, axillary in narrow racemes on stem and branches, subtended and often overtopped by bract-like leaves; involucre campanulate, 12–20 mm in diameter; phyllaries coriaceous, ciliate, abruptly contracted to a bluntly acuminate point which decreases in length inwards, inner less firm and 15–18 mm. Corollas yellowish, 8–9 mm, all tubular, hermaphrodite and fertile. Achenes cylindrical, 4-ridged, dull brown, 4–5 mm; pappus 6–8 mm, shortly plumose, creamy-white.

HAB. Frequent in *Pinus* or *Quercus* forests, also on stony and rocky hill slopes an in an orchard, on clay or marl; alt. 800–1400 m; fl. & fr. (Jun.-)Jul.-Aug.
DISTRIB. Quite frequent in the forest zone of the NE mountain region of Iraq. **MAM**: Zawita, *Guest* 4436!, 4883!; 10 km from Zawita towards Atrush, *Qaisi, Hamad & Hamid* 46084!; Atrush, *Guest* 3630!; 12 km W of Dohuk towards Amadiya, *Rechinger* 11496; Gara Dagh, *Kotschy* 376 (type)!; Sarsang, *Haines* 407!; 1 km beyond Ispindari pass, Sawara Tuka, *Chapman* 26357!; Sandur, *Qaisi & Khalid* 46111! 5 km from Bamerny to Aradin, *Al-Kaisi & K.Hamad* 45954 (BAG!); **MRO**: Kani Linja, *Gillett* 9632!; Qurnago valley, N of Pushtashan, *Serhang & Rawi* 26592!; lower slopes of Qandil range NE of Rania, *Rawi & Serhang* 18249!

Turkey (SE Anatolia).

105. **CHRYSOPHTHALMUM** Sch.Bip.

Repert. Bot. Syst. (Walpers) 2: 955 (1843)
Anderberg & Eldenäs in Kubitzki, Fam. Gen. Vasc. Pl. 8: 383 (2007)
[as *Chrysothphalmum*] Grierson in Davis, Fl. Turkey 5: 52 (1975); Rechinger, Fl. Iranica 145: 131 (1980).

C. C. Townsend

Subshrubs. Stems lacking resin canals. Leaves simple, alternate, pubescent. Capitula heterogamous and disciform, or homogamous and discoid; pedicels thickened or not thickened at maturity; involucre campanulate to hemispherical; phyllaries multiseriate, subequal, outer herbaceous, inner scarious; receptacle smooth, paleaceous; paleae linear, caniculate. All florets hermaphrodite, or marginal florets female, short-radiate, rays yellow and disc florets hermaphrodite, corollas yellow; anthers ecalcarate, tails branched; style with acute sweeping hairs. Achenes ±angled, shorter than corollas, epidermis with elongated crystals; pappus of scales, or absent.

A genus of three species from Turkey, Syria and Iraq; only one species is present in Iraq, and throughout the distribution of the genus.

Aytaç, Z & Anderberg, A. A. (2001). A new species of *Chrysophthalmum* Schultz Bip. (Asteraceae-Inuleae) from Turkey, Bot. J. Linn. Soc. 137(2): 211–214.

1. **Chrysophthalmum montanum** (*DC.*) *Boiss.*, Diagn. Pl. Orient., ser. 1, 2(11): 4 (1849).

Ceruana montana Aucher ex DC., Prodr. 7: 287 (1838).
Chrysophthalmum (as Chrysophtalmum) *sternutatorium* Sch.Bip. ex Walpers, Repert. Bot. Syst. (Walpers) 2: 955 (1843); Boissier, Fl. Orient. 3: 182 (1875).

Erect, 20–40 cm, ±densely silky pubescent. Leaves elliptic or oblanceolate, 4–10 × 0.4–1.8 cm. base pseudopetiolate, lamina prominently veined, densely sericeous at first, minutely glandular, margins entire, apices acute to acuminate. Inflorescence 1–5-headed; pedicels not thickened at maturity, few-bracteolate. Involucres 0.75–1 cm in diameter; phyllaries linear-lanceolate, 3–4.5 × 0.75–1 mm, appressed-pubescent and glandular; paleae 4 × 0.25 mm. Female florets few, corollas 4 mm long. Disc floret corollas 4.5–5 mm long. Achenes 2 mm, dark brown, glabrous, shiny; pappus of scale-like setae, *c.* 0.4 mm, white. Fig. 104: 1–10.

HAB. In rock crevices and limestone cliffs; alt. between 750 and 2200 m; fl. & fr. Jul.–Sep.
DISTRIB. Mountain areas of NE Iraq. MAM: Sarsang, *Rechinger* 11697 (E!, W); Jabal Khantur, nr Shahranish, *Rechinger* 10745 (W); Zawita, *Rechinger* 10897 (W!). Mt. Gara, *Kotschy* 314 (type of Ch. *sternutatorium*).

Turkey, Syria and Iraq.

106. **PENTANEMA** Cass.

Bull. Soc. Philom. Par. 1818: 74 (1818)
Anderberg & Eldenäs in Kubitzki, Fam. Gen. Vasc. Pl. 8: 383 [2006] (2007)
Inula sect. *Pentanema* (Cass.) Boiss., Fl. Orient. 3: 199 (1875)

Frances G. Davies
Revised by D. J. Nicholas Hind

Dwarf shrubs, annual or perennial herbs, generally rhizomatous, glabrous- glabrescent to tomentose. Leaves petiolate to amplexicaul or decurrent. Inflorescence of rather remote capitula on slender pedicels. Capitula generally many in terminal corymbs, generally heterogamous radiate to minutely radiate or heterogamous disciform or homogamous discoid. Phyllaries imbricate, many-seriate. Receptacle naked. Ray florets female, radiate, corollas yellow. Disc florets hermaphrodite, corolla 5-toothed, bright yellow; anthers ecalcarate, tails branched. Achenes ellipsoid or cylindrical, ribbed, glabrous or setuliferous; pappus setae uniseriate, barbellate, free or rarely connate at base.

About 40 species distributed in SW, S, C Asia, Africa and Europe.

Pentanema (from the Gr. πέντε, *pente,* five and νῆμα, *nima,* a thread – from the 5 pappus-hairs of the type species, *P. divaricatum.*

Gutiérrez-Larruscain, D., Santos-Vicente, M., Anderberg, A.A., Rico, E. & Martínez-Ortega, M.M. (2018). Phylogeny of the *Inula* group (Asteraceae: Inuleae): Evidence from nuclear and plastid genomes and a recircumscription of *Pentanema.* Taxon 67(1): 149–164.
Gutiérrez-Larruscain, D., Andrés-Sánchez, S., Rico, E. & Martínez-Ortega, M. M. (2019). Advances in the systematics and evolution of western Mediterranean representative speceis of the *Pentanema conyza* clade through genetic fingerprinting. J. Syst. Evol. 57 (1): 42–54.
Gutiérrez-Larruscain, D., Santos-Vicente, M., Martínez-Ortega, M. M. (2019a). Typification of 25 names in *Inula* (Inuleae, Asteraceae), and a new combinations in *Pentanema.* Phytotaxa 395 (1): 17–26.

Fig. 104. **Chrysophthalmum montanum**. 1, habit × ⅔; 2, leaf × 1; 3, capitulum (dissected) × 4; 4, phyllary × 4; 5, receptacle scale (paleae) × 4; 6, hermaphrodite disc floret × 6; 7, female floret × 6; 8, anther × 8; 9, style × 8; 10, achene with pappus × 8. 1–2 from *Rechinger* 10897; 3–10 from *Field & Lazar* 657. Drawn by Christine M. Grey-Wilson.

Annual; stems divaricately branched from base upwards, blackish-purple; involucre 6–8 mm in diameter; phyllaries linear to lanceolate, densely furnished with long, white, spreading hairs and ± glandular; female florets 4.5–5 mm . . . . 1. *P. divaricatum*
Perrenial; stems only branched above, greenish-brown; involucre 12–22 mm in diameter; outer and median phyllaries linear, sericeous, glandular; inner phyllaries narrowed to a very long subulate point, ciliate above; female florets 12–17 mm. . . . . 2. *P. oculus-christi*

1. **Pentanema divaricatum** *Cass.*, Dict. Sci. Nat. ed. 2, 38: 374 (1825); Gorshkova in Fl. U.R.S.S. 25: 480 (1959); Rechinger, Fl. Iranica 145: 99 (1980); Qaiser & Abid in Fl. Pakistan 210: 33, f. 9A–E (2003); Breckle, Hedge, Podlech & Rafiqpoor, Vasc. Pl. Afghan. checklist: 179 (2013); Kilian & Ghazanfar in Fl. Oman 3: 259 (2015); Abdulridha, Taha & Wedad, Ecology & Fl. Basra: 219 (2016).

Vicoa divaricata (Cass.) O.Fedtsch. & B.Fedtsch., Consp. Fl. Turkestan. 4 : 174 (1911), comb. illeg., non Oliv. & Hiern (1877) [= *Pulicaria petiolaris* Jaub. & Spach]; Guest in Dep. Agr. Iraq Tech. Bull. 27 : 106 (1933); Blakelock in Kew Bull. 4(1) : 65 (1949).

Inula divaricata (Cass.) Boiss., Fl. Orient. [Boissier] 3 : 200 (1875); Handel-Mazzetti in Ann. Naturh. Mus. Wien 27 : 433 (1913); Nábělek in Publ. Fac. Sci. Univ. Masaryj 52 : 6 (1925); Anthony in Notes Roy. Bot. Gard. Edinburgh 18 : 290 (1935); Zohary in Dep. Agr. Iraq Bull. 31 : 145 (1950); Rawi, ibid. Tech. Bull. 14 : 117 (1964).

Vicoa pentanema Aitch. & Hemsl. in Trans. Linn. Soc. ser. 2, 3: 75 (1886); Rechinger, Fl. Lowland Iraq: 605 (1964).

Annual, 5–30 cm, divaricately branched from base upwards. Stem and branches slender, wiry, usually blackish-purple, furnished with long, spreading white hairs and shortly glandular. Basal leaves obovate, narrowed to base, apex blunt, usually withered by flowering time; cauline leaves oblong to elliptic or obovate, c. 1.5–4 × 0.4–1.4 cm, sessile and broadly amplexicaul, obtuse or acute; leaf indumentum similar to that of stem. Capitula on slender pedicels 0.5–3(–4) cm; involucre hemispherical, 6–8 mm across; phyllaries linear to lanceolate, inner c. 4–6 mm, densely furnished with long, white, spreading hairs and ± glandular. Female florets yellow, 4.5–5 mm, ray limb c. 1.5–2 mm. Disc florets hermaphrodite, c. 4 mm. Achenes oblong-elliptic, c. 1 mm, sparsely upwardly pilose; pappus c. 3 mm, of 5 slender hairs. Fig. 105: 1–6.

HAB. Mountains in open *Quercus* forests and valleys, grassy steppe, wadis, fallow fields, orchards and other cultivated areas, roadsides and plains, even in slightly saline areas, on clay, gravel, loam and limestone; alt. 400-1200 m; fl. & fr. Apr.-Aug.
DISTRIB. Occasional in the foothills and desert areas, frequent in the steppe region and on the alluvial plain. **MJS**: Jabal Golat between Ain Tellawi and Balad Sinjar, *Field & Lazar* 478! **MAM**: above Sarsang, *Rechinger* 11700!; **MRO**: Pushtashan, on lower slopes of Qandil, *Rawi & Serhang* 23831!; Rowanduz, *Bornm.* 1366. **MSU**: Jarmo, *Helbaek* 1715! 1813! **FUJ**: Balad Sinjar, *Omar* 37598!; Qaiyara, *Bayliss* 137! **FNI**: Nineveh, *Bornm.* 1369! **FAR**: Mahmur, *Gillett* 11025!; Arbil, *Guest* 13025! **FAR/FKI**: between Arbil and Kirkuk, *Haussknecht* s.n.! **FKI**: Hawija, *Rawi* 1517!; Fatha, *Tikriti & Khatib* 29717!; Qara Lu, *Burhan ad-Din* 5514! **FPF**: Zurbatiya, *Haddad & Noori Muhamad* 9990!; Mandali, *Qaisi & Yahya* 45194!; Badra, *id.* 45294!; Lajama, *Rawi* 20705! **DLJ**: 70 km W of Tikrit, *Alizzi & Husain* 33724!; Hatra, *Barucha* 8656!; 34 km NW of Falluja, *Rawi, Shahwani et al.* 25477!; 11 km SW of Injana, *Rawi* 22874! **DGA**: 4 km W of Samarra, *Rawi* 20325! **DWD**: Shabicha, *Guest, Rawi & Long* 14060B! **DSD**: 52 km E by S of Salman, *id.* 14125!; 30 km W by N of Busaiya, *id.* 14158! **LEA**: 4 km E of Badra, *Haddad* 9949!; Nahrwan canal, *Hunting Aerosurveys* 100 R.A. 99! **LCA**: Baghdad, *Aucher-Eloy* 3846!; *Paranjpye* s.n.!; Khan Bani Sa'd, *B. Anderson* 1993!; Rustamiya, *Guest* 157!, *Lazar* 1183!; 20 km W of Abu Ghraib, *Ani & Haddad* 9921! **LSM**: 20 km S of Amara, *Rawi* 15044! **LEA/LSM**: between Fakka and Musaida, *Rawi* 14986! **LSM**:, 40 km E of Amara, *Rawi, Bharucha & Tikriti* 29309! **LSM/LBA**: Nahr al-Qarma, *Rawi & Shahwani* 25493!

Rechinger, in Fl. Iranica 145: 102 (1980) recorded *P. flexuosum* (Boiss. & Hausskn.) Rech. f. from Iraq on the basis of Haussknecht material from Mt. Avroman, which however he did not apparently see. Haussknecht's specimens from W Iran often bear printed "Montes Avroman et Schahu" labels, as does the type of *P. flexuosum*. Additional details are added in handwriting in the material in Haussknecht's own herbarium and Herb. Boissier at G, but not always in distributed isotypes. A specimen without the extra details may be the source of the above record from Mt. Avroman. Haussknecht's material was gathered from above Juanro in the Kermanshah district of Iran, and there is no evidence that *P. flexuosum* is an Iraqi species.

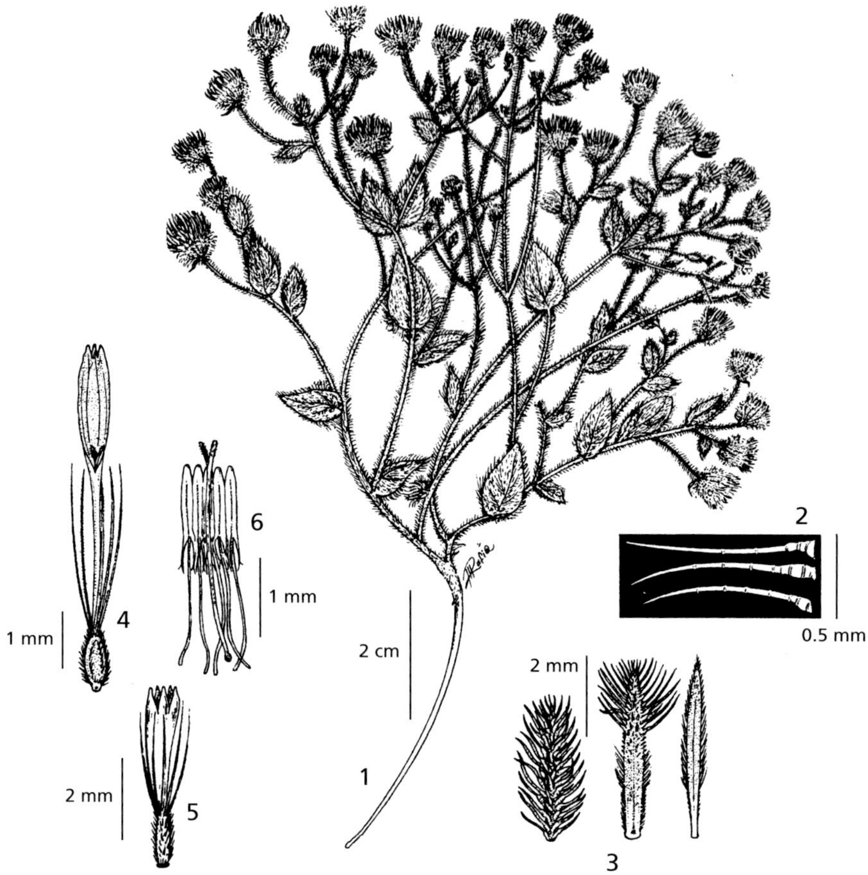

Fig. 105. **Pentanema divaricatum**. 1, habit; 2, stem and leaf indumentum; 3, phyllaries; 4, ray floret; 5, disc floret; 6, stamens, stigma and styles. Drawn by Rafiq. Reproduced with permission from Flora of Pakistan 207, fig. 9.

Lebanon, Turkey, Iran, Afghanistan, Pakistan, India, C Asia (Turkmenia to Tadjikistan).

2. **Pentanema oculus-christi** (*L.*) *D.Gut.Larr., Santos-Vicente, Anderb., E.Rico & M.M.Mart. Ort.* in Taxon 67(1): 159 (2018).

Inula oculus-christi L., Sp. Pl.: 881 (1753); Boissier., Fl. Orient. 3: 192 (1875); Gorshkova in Fl. S.S.S.R. 25: 461 (1959); Rawi in Dep. Agr. Iraq Tech. Bull. 14: 118 (1964); Grierson in Fl. Turkey 5: 64 (1975); Rechinger, Fl. Iranica 145: 88 (1980).
Aster oculus-christi (L.) Cav., Elench. Pl. Hort. Matr.: 5 (1803).

Rhizomatous perennial herb 13–50 cm tall. Stems branched only above, rarely simple and single-headed, lanate especially above. Lower leaves 7–21 × 1.4–4 cm, oblanceolate to elliptic, apex acute and often apiculate, base gradually narrowed, petiolate, petiole up to 7 cm long, margins entire, surfaces softly pubescent to lanate throughout; cauline leaves ovate to oblong, sessile, slightly to auriculate-amplexicaul, diminishing upwards. Inflorescence a terminal corymb of 1–10 capitula on densely villous pedicels. Involucre hemispherical, 12–22 mm broad; outer and median phyllaries herbaceous, linear, sericeous, glandular, median 5–8 mm long; inner phyllaries narrowed to a very long subulate point, ciliate above. Female florets 12–17 mm, ray limb 7–10 mm; Disc florets 7–8 mm. Achenes ribbed, setuliferous; pappus 6–7 mm, white.

HAB. On mountain slopes and cultivated foot of hill near stream; alt. 1300–2200 m; fl. & fr. Jun.-Jul.

DISTRIB. Scattered in the mountain region of Iraq. **MRO**: Algurd Dagh (Helgord), Rawi 13746!; Nawanda valley, Hadač 2325!; Sakri-Sakran, *Bornmüller* 1365; Kew-ar Rash, E side of Karoukh, *Kass & Nuri* 27669!; bet. the slope of Karoukh & Dargala village, *Rawi,Nuri& Kass* 27731! **MSU**: Penjwin, *Rawi* 22656!; 11 km N of Penjwin, *Rawi* 22900!; "Pl. Mesopot., Kurdistan & Mossul", *Kotschy* 319!

SE & C Europe, Turkey, Caucasia, Iran, C Asia (Turkmenia).

107. **PLUCHEA** Cass.

Bull. Soc. Philom. Paris 1817: 31 (1818)
Anderberg & Eldenäs in Kubitzki, Fam. Gen. Vasc. Pl. 8: 387 [2006] (2007)
Gynema Raf., Fl. Ludov.: 63 (1817)
Gymnostylis Raf., Amer. Month. Mag. Crit. Rev. 2: 268 (1818), nom. nud.
Stylimnus Raf., J. Phys. Chim. Hist. Nat. Nat. Arts 89: 100 (1819).
Tecmarsis DC., Prodr. 5: 93 (1836).
Berthelotia DC., Prodr. 5: 375 (1836).
Eyrea F. Muell., Linnaea 25: 403 (1852).
Spiropodium F. Muell., Fragm. [Mueller] 1(2): 33 (1858).

Frances G. Davies

Shrubs or occasionally herbs, usually pubescent and/or glandular. Leaves alternate or subopposite, simple (in Iraq) or compound, sessile and auriculate or decurrent, or petiolate. Capitula usually heterogamous and disciform, scattered or aggregated into corymbs, rarely solitary, medium-sized; involucres campanulate; phyllaries multiseriate, coriaceous; receptacle epaleaceous. Marginal florets female and filiform. Disc florets hermaphrodite, functionally male. Achenes terete or angular, all or only three filiform florets fertile; pappus setae uniseriate, free, persistent, minutely barbellate.

About 80 species, pantropical; only one species in Iraq.

Pluchea (named in honour of the French naturalist Abbé Noël-Antoine Pluche (1688–1761), author of Spectacle de la nature (1732).

King-Jones, S. 2001. Revision of *Pluchea* Cass. in the Old World. Englera 23: 1–236 (2001).

1. **Pluchea ovalis** (*Pers.*) *DC.*, Prodr. 5: 450 (1836); Collenette, Wild Flow. Saudi Arabia: 209 (1999); King-Jones in Willdenowia 23: 91 (2001); Qaiser & Abid in Fl. Pakistan 210: 107 (2003); Kilian & Ghazanfar in Fl. Oman 3: 261 (2015).

Baccharis ovalis Pers., Syn. Pl. 2: 424 (1807).
Pluchea tomentosa DC. in Wight, Contr. Bot. India: 16 (1834); Blakelock in Kew Bull. 4: 60 (1949); Rechinger, Fl. Lowland Iraq: 597 (1964); Rawi in Dep. Agr. Iraq Tech. Bull. 14: 122 (1964); Abdulridha, Taha & Wedad, Ecology and Fl. Basra: 221 (2016).
P. mollis Baker in Bull. Misc. Inform., Kew 1895: 182 (1895).
P. bravae Bolle, Bonplandia 8: 130 (1860).
P. dioscoridis (L.) DC. var. *pseudovalis* Cufod., Nuovo Giorn. Bot. Ital., ser. 2, 50: 105 (1943).

Shrub 1–2 m tall with brown, finely pubescent stems. Leaves sessile, broadly to narrowly ovate, up to 5 × 4.5 cm, acute to obtuse, dentate, the lamina on both surfaces olive- or grey-green and softly hairy. Inflorescence of corymbs terminating main and secondary branches, stems leafless above; involucres 6–7 × 0.8–1 cm; phyllaries pale green, margins hyaline, minutely glandular, innermost with fimbriate margin, outermost ovate, 2 mm, grading into linear innermost. Corollas pink to purple; female florets ± 40, 8 mm; hermaphrodite ± 12, 8–9 mm. Achenes 0.5–1 mm, cylindrical, pilose; pappus of ± 14 off-white hairs.

HAB. Damp places near water: on mud banks in marshes, on a semi-disturbed silt island in a tidal stretch of river, a weed in tidally irrigated date gardens; alt. s.l.–12 m; fl. & fr. (Dec.-)Mar.-Apr.
DISTRIB. Common in the southern marsh and Basra districts of the Desert Zone in SE Iraq. **LSM**: Amara marshes, *Guest* 1628!; Hawr as-Sikhaim, S of Majarr al-Kabir, *Thamir* 46651!; Haur al-Hawiza, 16 & 18 km E of Qalat Salih, *id.* 46672!, *Thamir, Widad & Hana* 46788!; Qurna, *Chakravarty, Rawi, Khatib & Tikrity* 29867! **LBA**: Basra, *Thamir* 47721!; on islet nr Ma'qil, Basra, *Guest, Rawi & Rechinger* 16729!; Kut as-Saiyid, *Rawi & Gillett* 5955!; Abul-Khasib, *Thamir* 47399!; Hur ash-Shibaish, *Qaisi* 44032!; Siba, 50 km ESE of Basra, *Karim, Nuri, Hamid & Kadhim* 39914!; along the Shatt al-Arab, between Basra and Fao, *Haines* W.881!, *Rawi* 25908!; 10 km S of Fao, *Hamad, Hamid & Kadhim* 46354!; mouth of Shatt al-Arab nr Fao, *Rechinger* 8619.

?SHABĀBAQ (Ir.-Amara, *Guest* 1628), KHARFASH (Ir.-Basra, *Guest* 16729 – "an erect slightly aromatic shrublet").

S Iran, Afghanistan, Pakistan, Africa.

108. SPHAERANTHUS L.

Sp. Pl.: 927 (1753); Gen. Pl. ed. 5: 998 (1754)
Anderberg & Eldenäs in Kubitzki, Fam. Gen. Vasc. Pl. 8: 389 [2006] (2007)

Frances G. Davies

Herbs of moist ground. Stems decumbent, winged by decurrent leaf-bases. Leaves simple, narrow, sessile, grass-green. Synflorescences cone-like, each cone comprising many capitula, each subtended by a bract (often reduced to a scale) on condensed secondary receptacle. Capitula heterogamous. Florets few; outer florets female, filiform, corollas purple, becoming corky at base when mature, minutely 3-lobed; central florets hermaphrodite, corolla tube cylindrical, lobes 5; anthers sagittate with short tails, apical anther appendages small, ovate; style undivided or minutely bifid. Achenes small, angular, brown, densely glandular-setuliferous; pappus absent.

40 species in Asia, tropical Africa and Australia; only one species in Iraq.

Sphaeranthus, from Gr. σφαίρα, sphaira, a globe and ἄνθος, anthos, a flower.

Ross-Craig, S. (1955). A revision of the genus *Sphaeranthus*. Bentham-Moxon Trustees.

1. **Sphaeranthus strobiliferus** *Boiss. & Noë* in Boissier, Diagn. Pl. Orient. ser. 2, 3: 6 (1856). –Type. Baghdad, *Noë* 329. Boissier, Fl. Orient. 3: 216 (1875); Nábělek in Publ. Fac. Sci. Univ. Masaryk 52: 7 (1925); Zohary in Dep. Agr. Iraq Bull. 31: 145 (1950); Rawi in Dep. Agr. Iraq Tech. Bull. 14: 125 (1964); Rechinger, Fl. Lowland Iraq: 597 (1964).

Decumbent herbs 10–30 cm tall, with a bunch of thick smooth roots at base. Stems several, leafy up to and beyond the capitula, narrowly green-winged. Leaves sessile, decurrent almost to the previous nodes, lanceolate, 2.5–5 × 0.5–0.9 cm, apiculate, margin entire or with distant sharp teeth, both surfaces almost glabrous. Synflorescences shortly (0.5–1 cm) stalked, ovate, the compound receptacle lanceolate. Scales of capitulum ovate, 1.1 × 1.1 mm, acuminate to a subulate point; primary capitula 6-flowered, outer 4 florets female, fertile, 2 outer subtended by 2 curved scales, and 2 placed centrally, inner 2 apparently hermaphrodite but functionally male and becoming very corky. Achenes densely glandular.

HAB. In fields, along ditches and at the edge of marshes in the alluvial plain; s.l.–35 m; fl. & fr. Aug.
DISTRIB. Uncommon in lowland Iraq. **LCA**: Baghdad, 1868, *Haussknecht* s.n.; Baghdad, *Noë* 329 (type-Kew Bull. 1924: 5: 187 (1924); Za'faraniya, nr Baghdad, *Aftan ar-Rawi* 19936!; Babylon, *Reuter* 3704 (SAV). **LSM**: Garma district, Nasiriya province, *Rawi* 16589!; 70 km E of Amara, *Alizzi & Omar* 34377!

Endemic? Recorded for Iran (Davies, ms.) but not mentioned in Fl. Iranica.

109. DOELLIA Sch. Bip.

in Walpers, Repert. Bot. Syst. 2: 953 (1843)
Bremer, Asteraceae Clad. & Class. 299. 1994; emend. in Willdenowia 25: 21. 1995;
Anderberg & Eldenäs in Kubitzki, Fam. Gen. Vasc. Pl. 8: 390 [2006] (2007)

Frances G. Davies
Revised by Shahina A. Ghazanfar

Perennial herbs. Stem with resin canals. Leaves alternate, linear-oblong to linear-lanceolate, dentate, serrate or almost entire, hairy or subglabrous, not decurrent along the stem. Capitula heterogamous, disciform, terminal, solitary or few together in a lax corymb; phyllaries chartaceous, narrowly linear, imbricate in several series. Receptacle flat, epaleate. Ray florets female, filiform, corolla yellowish or purple, 3-lobed. Disc florets bisexual, yellowish or purplish, 5-lobed; anthers tailed; style bifid, style branches with obtuse sweeping hairs reaching below the furcation. Achenes oblong, rather stout and somewhat angular, sparsely hairy with conspicuous red resin ducts; pappus of free barbellate capillary bristles in one row, each bristle with appressed or erecto-patent teeth and an extremely small inconspicuous coroniform ridge (Description adapted from Fl. Pakistan 102: 100. 2003).

Doellia is distinguished from *Blumea* s. str. by its style with obtuse sweeping hairs reaching far below the bifurcation, radial endothecium and cypselas without elongated epidermis crystals. The genus differs from all species of *Blumea* s.l. by its cypselas with conspicuous longitudinal red resin ducts.

A small genus with 2 species, distributed in S Africa, Madagascar, E Africa to Sinai, Iran and Pakistan.

1. **Doellia bovei** (*DC.*) *Anderb.* in Willdenowia. 25: 21. 1995; Boulos & Hind in Fl. Egypt 3: 189 (2002); Qaiser & Abid in Fl. Pakistan 210: 102, f. 24 (2003); Kilian & Ghazanfar in Fl. Oman 3: 255 (2015).

Conyza bovei DC., Ann. Sci. Nat. Bot. ser. 2, 1: 261 (1834).
Blumea abyssinica Sch.Bip. ex A.Rich, Tent. Fl. Abyss. 1: 392 (1847).
Blumea bovei (DC.) Vatke in Linnaea 39: 485 (1875); Randeria in Blumea 10: 285 (1960); Feinbrun-Dothan, Fl. Palaest. 3: 301 (1978); Rechinger, Fl. Iranica 145: 6 (1980); Collenette, Wild Flow. Saudi Arabia: 174 (1999).
Erigeron bovei (DC.) Boiss., Fl. Orient. 3: 168 (1875); Dinsmore in Post, Fl. Syria, Palest. & Sinai ed. 2, 2: 18 (1933).

Perennial herb 25–60 cm tall, with a woody base and taproot. Stems erect or ascending, glabrous to saprsely pubescent, branched from base or a little above, leafy throughout. Leaves sessile, ± clasping stem with rounded auricles, 1.5–6(–8) × 0.2–1 cm, margins shallowly denticulate to subentire, sometimes undulate, sparsely pubescent. Involucre of many linear-lanceolate phyllaries 5–6 mm long. Capitula in loose corymbs or solitary, on slender pedicels 1.3–4.5(–6) cm. Involucre hemispherical; phyllaries linear, tapering and acute with narrow scarious margins and a prominent brown midrib, dorsally pubescent; inner phyllaries 5–6 mm long; receptacle glabrous. Corollas blue to purple; female florets female, ± 3 mm; disc florets 3–4 mm. Achenes dark brown, oblong, 0.8 mm, pubescent, with conspicuous red, resin ducts; pappus setae white, 3–4 mm. Fig. 106: 1–6.

HAB. & DISTRIB. A note on a cover of this species in the Kew herbarium states that it is to be found in Iraq. No record has been traced, nor any specimen seen. It might well occur in the south of the country, and is included since it could be overlooked for other non-descript composites.

NW Africa (Somalia, Ethiopia), Sinai, Palestine, Saudi Arabia, Yemen, Oman, Iran, Afghanistan, Pakistan.

TRIBE **HELIANTHEAE** CASS.

110. **GAILLARDIA** Foug.

in Mem. Acad. Sci. Paris 1786: 6 (1788)
Panero in Kubitzki, Fam. Gen. Vasc. Pl. 8: 401 [2006] (2007)

Patricia Halliday

Annual, biennial or perennial herbs. Leaves alternate, entire, toothed or pinnatifid. Capitula heterogamous and radiate or homogamous and discoid, pedunculate. Phyllaries 2–3-seriate, ± reflexed in fruit. Receptacle convex to subglobose, alveolate, usually ± setose. Ray florets, if present, yellow, crimson-purple, or yellow and crimson-purple, ligulate, neuter, occasionally female and fertile; rays broad, cuneate or fan-shaped, 3–4-toothed. Disc florets hermaphrodite, fertile, regular, tubular, 5-toothed. Florets usually externally glandular-pubescent, hairs often moniliform. Anther-bases auriculate. Style-arms with short, glabrous or filiform, hispidulous appendages. Achenes turbinate, ± twice as long as broad, wholly or partly covered by long, stiff hairs.

A North American genus of about 20 species, of which several are cultivated in other parts of the world. One (or possibly 2) species cultivated in Iraq.

Fig. 106. **Doellia bovei**. 1, habit × ⅔; 2, immature achene with pappus × 10; 3, capitulum with achenes dispersed × 4; 4, outer phyllary × 10; 5, ray floret × 10; 6, disc floret × 10. 1, 5–6 from *Miller* 67 (Yemen); 2–4 from *Burton* Mar.-Apr. 1878 (Jordan). Drawn by Moira Rae Carter.

Marlowe, K. & Hufford, L. (2007). Taxonomy and biogeography of *Gaillardia* (Asteraceae): a phylogenetic analysis. Syst. Bot. 32: 208–226.

Gaillardia (named in honour of Antoine René Gaillard (fl. 1720–80), later styled Gaillard de Charentonneau, a magistrate, patron of botany and member of the Académie des Sciences. His property at Charentonneau, now located in a suburb of Paris near the Parc de Vincennes, once had a fine garden).

1. **Gaillardia pulchella** *Foug.* in Mém. Acad. Sci. Par. 1786: 6 (1788); Stout in Gardening for Egypt: 83 (1933); Husain & Kasim, Cult. Pl. Iraq: 501 (1975); Chakravarty in Plant Wealth Iraq 1: 251 (1976).

Annual, erect, simple or more usually branched, shortly bristly and variably long-multicellular-hairy, to 47(–60) cm tall, with ribbed stems. Leaves alternate, lower cauline ± petiolate, oblanceolate, 2–9 × 0.7–2 cm, lyrate-pinnatifid to ± toothed, teeth often rounded; upper cauline leaves oblong-lanceolate, sessile, ± clasping, often entire but sometimes toothed or lobed, usually smaller than lower leaves. All leaves variably multicellular-hairy on both surfaces, but usually more sparsely so above. Capitula solitary, terminal on stem and branches, long-pedunculate, about 5 cm in diameter; peduncle slender, striate, up to 18 cm long, multicellular-hairy; phyllaries herbaceous, subequal, 12 ×2–2.5 mm, connate at base; outer series lanceolate, acuminate, densely multicellular-hairy; inner series broad at base, abruptly narrowed into a long, bristly acumen; receptacular setae subulate. Ray florets corollas crimson-purple, with or without yellow tips; ray 1.8–2 cm long, externally hairy. Disc florets *c.* 7 mm long, tube yellow and glabrous, lobes crimson-purple and externally long-hairy; style bifid, arms compressed, appendage tapering upwards, long-papillose. Achene 2–3 × 1.5–2 mm, with erect, long, stiff hairs arising from base and ± concealing achene; pappus of 5–7 chartaceous, lanceolate scales, each 4–5 mm, tapering into an awn equalling half length of scale.

HAB. & DISTRIB. Cultivated as an ornamental in Iraq; alt. up to 35 m; fl. & fr. Apr.-Aug. **LCA**: Abu Ghraib, *Janan* 34874!, *Sahira & Janan* C. 261! C. 264! C. 269!; Baghdad, Sahira 35806!, *Janan* C. 411! **LBA**: Siba, 50 km S.E. of Basra, *Karim, Nuri, Hamid & Kadhma* C. 803!

A native of N America, this species is cultivated in many other countries.

Gaillaria aristata Pursh is also said by Husain & Kasim, l.c., to be "commonly cultivated" in Iraq (Lazar p. 430), but no specimens of this species have been seen from the Flora area. It differs from *G. pulchella* in being perennial and usually having entirely yellow rays, or occasionally rays purplish at the base; its receptacle-setae exceed the achenes, whereas in *G. pulchella* they subequal the achenes.

111. **BIDENS** L.

Sp. Pl.: 831 (1753); Gen. Pl.: 840 (1754)
Panero in Kubitzki, Fam. Gen. Vasc. Pl. 8: 411 [2006] (2007)

Patricia Halliday

Erect, procumbent or scandent, annual or perennial, glabrous or pilose herbs or shrubs. Stems terete or angled. Leaves opposite, simple, toothed or pinnatisect, rarely ternate. Inflorescence corymbose-paniculate, white, yellow or rarely red-flowered. Capitula homogamous, discoid or heterogamous and radiate; involucre campanulate or hemispherical, 2–(3–4)-seriate, outer phyllaries unequal, foliaceous; inner series smaller and membranous; receptacle flat, scaly; scales ± flat, subtending disc florets. Ray florets 1-seriate, ligulate, neuter, rarely pistillate; disc florets tubular, bisexual, fertile; style arms with shortly conical appendage. Achenes oblong-cuneate, ± 3–4-angular, apex truncate; pappus of 2–4, rigid, persistent, retrorsely or occasionally antrorsely barbed awns.

A genus of about 200 species, distributed throughout temperate Eurasia, N Africa and N & S America; one species in Iraq.

Bidens (from Lat. *bi*, two, *dens*, tooth).

1. **Bidens tripartita** *L.*, Sp. Pl.: 831 (1753); Boissier, Fl. Orient. 3: 251 (1875); Dinsmore in

Post, Fl. Syria, Palest. & Sinai ed. 2, 2: 42 (1933); Sherrff in Field Mus. Nat. Hist., Bot. ser. 16 (1): 265, tab. 67 (1937); Zohary in Dep. Agr. Iraq Bull. 31: 641 (1950); Rawi in Dep. Agr. Iraq Tech. Bull. 14: 107 (1964); Chakravarty & Rawi in Dep. Agr. Iraq Tech. Bull. 15: 19 (1964); Kupicha in Fl. Turkey 5: 46 (1975); Brummitt in Fl. Europ. 4: 140 (1975); Feinbrun-Dothan, Fl. Palaest. 3: 325 (1978); Collenette, Wild Flow. Saudi Arabia: 174 (1999); Yousheng Chen & Hind in Fl. China, 20–21: 858 (2011); Breckle, Hedge & Rafiqpoor, Vasc. Pl. Afghan: 153 (2013); Kilian & Ghazanfar in Fl. Oman 3: 274 (2015).

Erect, branched, annual herb (30–)45–74(–100) cm tall, with ± multicellular hairy stems. Leaves all cauline, lanceolate-ovate in outline, 2(–5)-lobed, up to (4–)8(–15) × 2 cm, segments lanceolate, dentate, or leaves consistently small (3 × 1 cm), petiolate, cuneate, acute, glabrescent above, sparsely multicellular-hairy beneath, with minutely hispid, cartilaginous margins. Inflorescence solitary, pedunculate, terminal and axillary. Capitula erect, 1–2 cm in diameter, subtended by about 6 foliaceous bracts; peduncle ± sparsely multicellular-hairy. Involucre 1–2 cm in diameter; outer phyllaries 10–40 × 2–10 mm, margins cartilaginous and hispid; inner phyllaries 9 × 3 mm, linear-lanceolate with narrow, scarious, entire margins. Paleae linear-lanceolate, 8–9 mm long, striate with wide, scarious, minutely irregularly-toothed margins. Disc florets ± cylindrical, 4 mm long, lobes 0.5 mm long; stamens included, anthers 1 mm long, sagittate; style slender. Achenes 8 mm long excluding awns, 2 mm wide, margins retrorsely hairy; pappus of 2(–4) erect, retrorsely barbed awns 2 mm long. Fig. 107. 1–10.

HAB. Usually near water by streams, ditches and river banks, in an irrigated *Populus* plantation; alt. 50–1150 m; fl. & fr. Jun.-Sept.
DISTRIB. Occasional (rather rare) in the lower forest zone, steppe and desert regions of Iraq. **MAM**: 25 km from Zakho on road to Kani Masi, *Botany Staff* 43811!, 43823! **MSU**: Sulaimaniya bridge over R. Tinjaro, *Haines* W2043!; Qaranjir, 22 km E of Kirkuk, *Qaisi & Hamad* 43506! **FKI**: between Khalis and Kirkuk, *Rawi, Chakraverty, Alizzi & Noori* 19704! **FPF**: Kai Baz, *Rawi* 24859! **DSD**: 30 km SE of Kufa, *Thamir* 47651!

Syria, Palestine, Turkey, Iran, Caucasia, Afghanistan, Egypt, Arabia, India, China & Japan, throughout temperate Europe (including Macaronesia, Balearic Is. & Crimea), N Africa, SE Australia and N America.

The medicinal uses of this plant are described by Rawi & Chakravarty (1964); but the colloquial names QUNNAB MĀ ("water hemp") and TÎL MĀ appear to be suggestions rather than records of names actually used in Iraq.

TRIBE **COREOPSIDEAE** LINDL.

112. **COREOPSIS** L.

Sp. Pl.: 907 (1753); Gen. Pl. ed. 5: 388 (1754)
Panero in Kubitzki, Fam. Gen. Vasc. Pl. 8: 411 [2006] (2007)

Patricia Halliday
Revised by D. J. Nicholas Hind

Erect or decumbent, glabrous or pilose annual or perennial herbs, subshrubs or shrubs. Stems poorly to densely branched. Leaves opposite (rarely alternate), sessile to petiolate or pseudopetiolate, simple and entire to ternately or pinnately dissected. Inflorescences of slitary capitula or corymbose-paniculate. Capitula heterogamous and radiate; involucre campanulate or hemispherical; phyllaries 2–(rarely 3–4)-seriate, series equal or unequal, outer herbaceous to submembranaceous, appressed to spreading, inner usually larger, scarious; receptacle flat to strongly convex, paleaceous, paleae flat or concave, subtending disc florets, lanceolate to linear, and rapidly deciduous. Ray florets uniseriate, rayed, sterile or rarely pistillate and fertile, corollas usually yellow. Disc florets tubular, hermaphrodite, all fertile or inner sterile; anther thecase and apical anther appendages blackish. Achenes roundish to linear-oblong, ±obcompressed, usually 2-winged; pappus absent or of 2 antrosely barbed bristles or scales.

A genus of ± 120 species in N & S America and Africa.

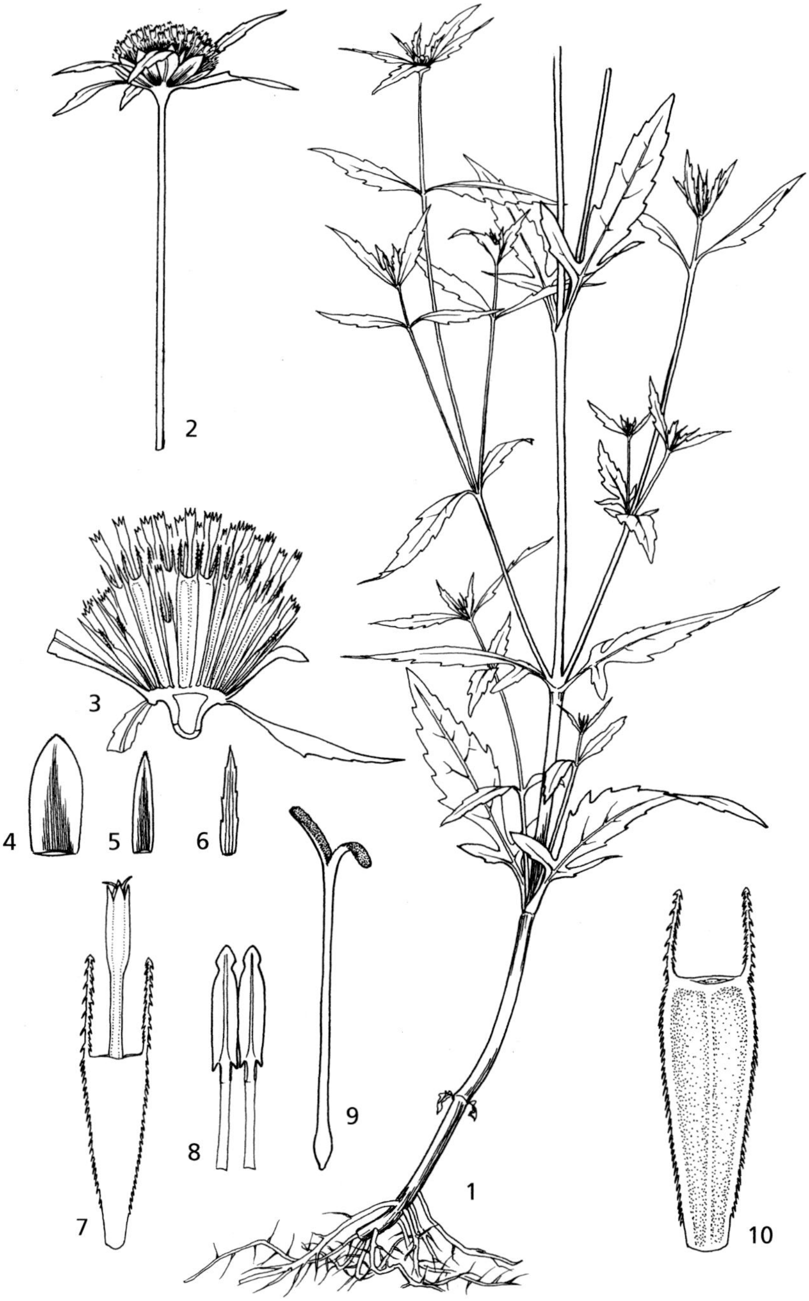

Fig. 107. **Bidens tripartita**. 1, habit × 2/3; 2, capitulum × 2/3; 3, capitulum dissected × 2; 4, outer phyllary × 2; 5, inner phyllary × 2; 6, receptacular scale × 2; 7, floret × 4; 8, anther × 15; 9, style × 15; 10, achene × 6. All from *Wheeler Haines* W2043. Drawn by Pat Halliday.

Coreopsis (from the Gr. κοριός, korios, a bug, and the suffix -οψις, -opsis, resembling; from the form of the achene.)

Sherff, E.E. 1936. Revision of the genus *Coreopsis*. Publ. Field Mus. Nat. Hist. Bot. Ser. 11 : 279–475.

1. Lobes of disc-florets yellow; ray florets fertile . . . . *C. stillmanii*
   Lobes of disc-florets purplish-brown; ray florets sterile. 2
2. Inner phyllaries 2–3 × longer than usually deltoid-ovate outer phyllaries; achenes thin, almost flat . . . . *C. tinctoria*
   Inner phyllaries ± equalling subulate or linear-lanceolate outer phyllaries; achenes thick, convex . . . . *C. basalis*

1. **Coreopsis tinctoria** *Nutt.* in Acad. Nat. Sc. Phil. 2: 114 (1821); Sherff, op. cit. 417 (1936); Vassilchenko in Fl. S.S.S.R. 25: 547 (1959); Husain & Kasim in Cult. Pl. Iraq: 113 (1975).

Diplosastera tinctoria (Nutt.) Tausch, Hort. Canal.: t. 4 (1823).
Calliopsis tinctoria (Nutt.) DC., Prodr. 5: 568 (1836).
Bidens tinctoria (Nutt.) Baill., Hist. Pl. (Baillon) 8: 305 (1882).

Erect, glabrous annual herbs to about 1.2 m, divaricately branched above with slender branches. Leaves opposite, lowest shortly petiolate, median and upper sessile; median and lower leaves 1–2-pinnate with narrowly linear to linear-lanceolate segments up to 9 cm. Capitula numerous, on slender, elongate pedicels; involucre hemispherical, 0.5–1 cm in diameter; outer phyllaries deltoid-ovate, 1–2 mm long; inner phyllaries broadly oblong or oblong-ovate, 5–6 mm. Ray limbs broadly obovate, 8–15 mm long, usually 3-lobed at apex, yellow with a purplish-brown base. Disc-florets lobes purplish-brown. Achenes flat, very compressed, blackish, unwinged, 1.25–4 mm.

HAB. & DISTRIB. Cultivated in gardens and on Agricultural Research Stations. **LCA**: Rustam Farm, *Lazar* 437!; Abu Ghraib, *Samir Monim* 16498! **LBA**: Basra, *Alizzi & Omar* 35106!

A native of central and western USA and Canada, much cultivated and escaping in other parts of N America and elsewhere.

A common garden decorative, popular because both showy and easy of cultivation. According to Husain & Kasim (1975) it is cultivated throughout Iraq. A yellow dye can be extracted from the flowers.

C. stillmanii (A. Gray) Blake, a native of California, and *C. basalis* (Dietr.) Blake, from Texas, are both recorded by Husain & Kasim (1975) as commonly cultivated in Iraq, and are included in the key. However, no specimens have been seen.

113. **COSMOS** Cav.

Icon. 1: 9 (1791)
Panero in Kubitzki, Fam. Gen. Vasc. Pl. 8: 412 [2006] (2007)

Shahina A. Ghazanfar

Annuals or perennials with an erect or ascending stem, ± branched. Leaves opposite, mostly cauline, usually pinnately 1–3-lobed or undivided, petiolate or sessile, usually glabrous, sometimes glabrate, hispid, puberulent, or scabridulous. Capitula solitary or corymbose, radiate. Involucre hemispherical; phyllaries ± 2-seriate, ± equal, membranous or herbaceous, margin ± scarious, persistent; receptacles flat, paleate; scales linear, flat or slightly concave-convex, scarious or entire. Ray florets sterile; corollas white to pink or purple, or yellow to red-orange. Disk florets bisexual, fertile; corollas yellow or orange, 5-lobed. Achenes quadrangular-cylindric or -fusiform, sometimes slightly arcuate, beaked, wingless or winged; pappus of 2–4(–8) retrorsely barbed awns, sometimes absent.

About 26 species, tropical and subtropical America, especially Mexico; widely introduced elsewhere; two species cultivated in Iraq, both native to Mexico and SW USA.

Ray limbs pink, reddish pink, purple or white . . . . 1. *C. bipinnatus*
Ray limbs yellow to reddish. . . . . 2. *C. sulphureus*

1. **Cosmos bipinnatus** *Cav.*, Icon. 1: 9 (1791); Hussain & Kassim, Cult. Pl. Iraq 113 (1975)

Annual to 2 m. Leaves ± sessile, pinnately lobed. Capitula solitary, 3–6 cm in diameter; peduncles 10–20 cm; Involucre 7–15 mm in diameter; phyllaries lanceolate to ovate-lanceolate, erect. Ray limbs pink, reddish pink, purple or white. Achenes glabrous, papillose; pappus absent, or of 2 or 3 ascending to erect awns.

HAB & DISTRIBUTION. A garden ornamental, native to Mexico and SW USA, cultivated throughout Iraq; sometimes found as an escape; fl. Aug.
MRO: Haji Omran by Iranian frontier, on mountain slope, Aug. *Rawi* 24937.

2. **Cosmos sulphureus** *Cav.*, Icon. [Cavanilles] 1: 56 (1791); Hussain & Kassim, Cult. Pl. Iraq 113 (1975).

Annual to 2 m, glabrous or sparsely pilose to hispid. Leaves petiolate; lamina lobed with margin sparsely ciliate, Capitula solitary; peduncles 10–20 cm. Involucre 6–10 mm in diameter; phyllaries, oblong-lanceolate, erect. Ray limbs bright yellow to red-orange. Achenes usually hispidulous; pappus absent, or of 2 or 3 widely divergent awns.

HAB & DISTRIBUTION. A popular garden ornamental, native to Mexico, cultivated throughout Iraq; fl. Feb., Aug.-Nov.
LCA: Zaffaraniya, Aug., cultivated, *S. Omar & Sahira* C-581; Abu Ghraib, cultivated, Feb., 40 m, *H. Ahmed* 9870, Nov., *Alizzi* 34813; Al-Hilla, roadside, Oct., *Fawzi* C-807.

114. **DAHLIA** Cav.

Icon. 1: 56 (1791)
Panero in Kubitzki, Fam. Gen. Vasc. Pl. 8: 412 [2006] (2007)

C. C. Townsend

According to Husain & Kasim (1975: Cultivated plants of Iraq) *Dahlia pinnata* Cav., the Common or Garden Dahlia (Eng.) is cultivated throughout Iraq. This is a glabrous or subglabrous plant mostly up to 1.4 m tall, with lower leaves bipinnate with lanceolate to ovate, dentate leaflets and main stem leaves usually simply pinnate. The phyllaries are leafy and ± reflexed, and the capitula radiate, ± nodding. The cultivated forms of this well-known plant are legion, and vary from small double "button" capitula with a dwarfer habit to coarse plants with very large single or double capitula. Ray limbs are red in wild forms, but wide range of colours are available in cultivation. No Iraqi specimen has been seen, and thus the identity of the plant grown in Iraq cannot be confirmed. Abdulridha, Taha & Wedad, Ecology and Fl. Basra: 209 (2016), record *D. variabilis* (Willd.) Desf. cultivated in the Basra area.

TRIBE **TAGETEAE** CASS.

115. **FLAVERIA** Juss.

Gen. Plant. 186 (1789); Benth. & Hook.f., Gen. Plant. 2: 407 (1873)
Panero in Kubitzki, Fam. Gen. Vasc. Pl. 8: 422 [2006] (2007)
Vermifuga Ruiz & Pav., Prodr. Fl. Per.: 114 (1794); *Brotera* Spreng., J. Bot. (Schrader) (1800) 2: 186, tab. 5 (1801), non Cav. (1799), nec Willdenow (1803), nec Vellozo (1829); *Nauenbergia* Willd., Sp. Pl., ed. 4, 3(3): 1489, 2393 (1803); *Dilepis* Suess. & Merxm., Mitt. Bot. Staatssamml. München, Heft 1: 14 (1950)

Patricia Halliday
Revised by D. J. Nicholas Hind

Annual or perennial, glabrous or pubescent herbs with opposite, entire or toothed leaves. Inflorescence often of densely clustered and glomerulate, or loosely aggregated capitula, cymose, corymbose-paniculate or sometimes solitary at branch-tips. Capitula small, narrow, heterogamous and radiate and peripheral to glomerules or homogamous and discoid and found throughout glomerules, sessile or shortly pedicellate; phyllaries 2–8, subequal, sometimes with 2 outer, smaller bracts; receptacle naked or scaly. Corollas yellow; ray florets fertile and fertile, ray limb smaller to larger than achene, entire, emarginate or tridentate, erect or reflexed; disc florets 1–15, hermaphrodite, actinomorphic, tubular, 5-toothed; anthers with apical appendage; styles bifid, reflexed, style arms truncate or rounded. Achenes ± oblong, black, glabrous, 8–10-ribbed; pappus usually absent (one species with pappus of scales).

A genus of 22 species largely native to North America, but with native species in Central America, and two widespread weedy species found in both Old and New Worlds, extending into Australia; one of the weedy species is found in Iraq.

Flaveria (from the Lat. *flavus*, golden-yellow, as the plants were formerly used to produce a yellow dye).

Powell, A. M. (1978). Systematics of *Flaveria* (Flaveriinae-Asteraceae). Ann. Missouri Bot. Gard. 65(2): 590–636.

1. **Flaveria trinervia** (*Spreng.*) *Mohr* in Contr. U.S. Nat. Herb. 6 : 810 (1901).

Oedera trinervia Spreng., Nachr. i. Bot. Gart. Halle : 63 (1800).
Brotera contrayerba Spreng., J. Bot. (Schrader) (1800) 2: pl. 5 (1801), non *Flaveria contrayerba* Pers. (1807). Note: Sprengel's validating description (for both genus and species) appears under the paper, '5.' on "*Millera contrayerba* Cavan., neu untersucht und bestimmt." pp. 186–189; Tab. V is a foldout plate with dissections named *Brotera contrayerba*. Sprengel's generic name is a later homonym of Cavanilles, in the Sterculiaceae).
Nauenbergia trinervata Willd., Sp. Pl., ed. 4, 3(3): 2393 (1803), nom. illeg. et nom. nov. pro. *Brotera contrayerba* Spreng.
Brotera trinervata Pers., Syn. Pl. 2: 498 (1807), nom. illegit. pro *Brotera contrayerba* Spreng.
Flaveria repanda Lag., Gen. Sp. Pl.: 33 (1816).
Brotera sprengelii Cass., Dict. Sci. Nat., ed. 2, 34: 306 (1825), nom. illegit. pro *Brotera contrayerba* Spreng., etc., etc.
Flaveria trinervata (Pers.) Baillon, Hist. Pl.: 55 (1882), comb. illeg. (based on Persoon's nom. illeg.).

Erect, dichotomously branched, glabrous or puberulent annual herb to 76(–>200) cm tall, with ribbed, often purple stems. Leaves all cauline, lanceolate or elliptic, 1–7.5 × 0.4–2.5 cm, denticulate with 3 conspicuous veins from base, cuneate, obtuse-rounded, lower leaves sometimes purple, petiolate, upper leaves grey-green, sessile or subsessile. Inflorescence much-branched, cymose. Heterogamous and radiate capitula peripheral to glomerules, homogamous and discoid capitula throughout glomerules, capitula few-flowered, terminal, in dense glomerules 1–2.5 cm in diameter, supported by 2–4 (occasionally more) ± connate foliaceous bracts up to *c.* 2 cm; receptacle scaly, scales linear, flat, somewhat expanded at tip, 3.5 mm. Florets all fertile, each floret sheathed by 2 unequal, concave, distinctly keeled, obtuse bracts 3.5–4 × 1–1.25 mm, with upper part of margins papillose; styles slender, arms short. Ray florets female, up to 2 mm long, flask-shaped, base of tube swollen and externally multicellular-hairy, ray ± equalling tube. Disc florets hermaphrodite, campanulate, 2.5 mm long, teeth reflexed, tube externally multicellular-hairy; anthers sagittate, *c.* 1 mm long, collar linear, oblong, appendage lanceolate. Achene oblong-obovate, somewhat compressed, 1–2 × 0.75 mm, with small, light-coloured callus at base; pappus absent. Fig. 108: 1–11.

HAB. Damp ground, waste places and roadside, or as a weed of cultivation; alt. 60–2300 m; fl. & fr. Mar.-Dec.
DISTRIB. **LCA**: Baghdad, Jadriyah, *Agnew* 1867!

Native to the New World, extending from the USA to Chile and Argentina; occurs as an adventive in many countries of the Eastern Hemisphere.

116. **TAGETES** L.

Sp. Pl.: 887 (1753); Gen. Pl. ed. 5: 378 (1754)
Panero in Kubitzki, Fam. Gen. Vasc. Pl. 8: 429 [2006] (2007)

Patricia Halliday

Erect or spreading, annual or perennial, branched, glabrous or glabrescent herbs. Leaves opposite, usually pinnate, rarely simple, toothed, with scattered oil glands.

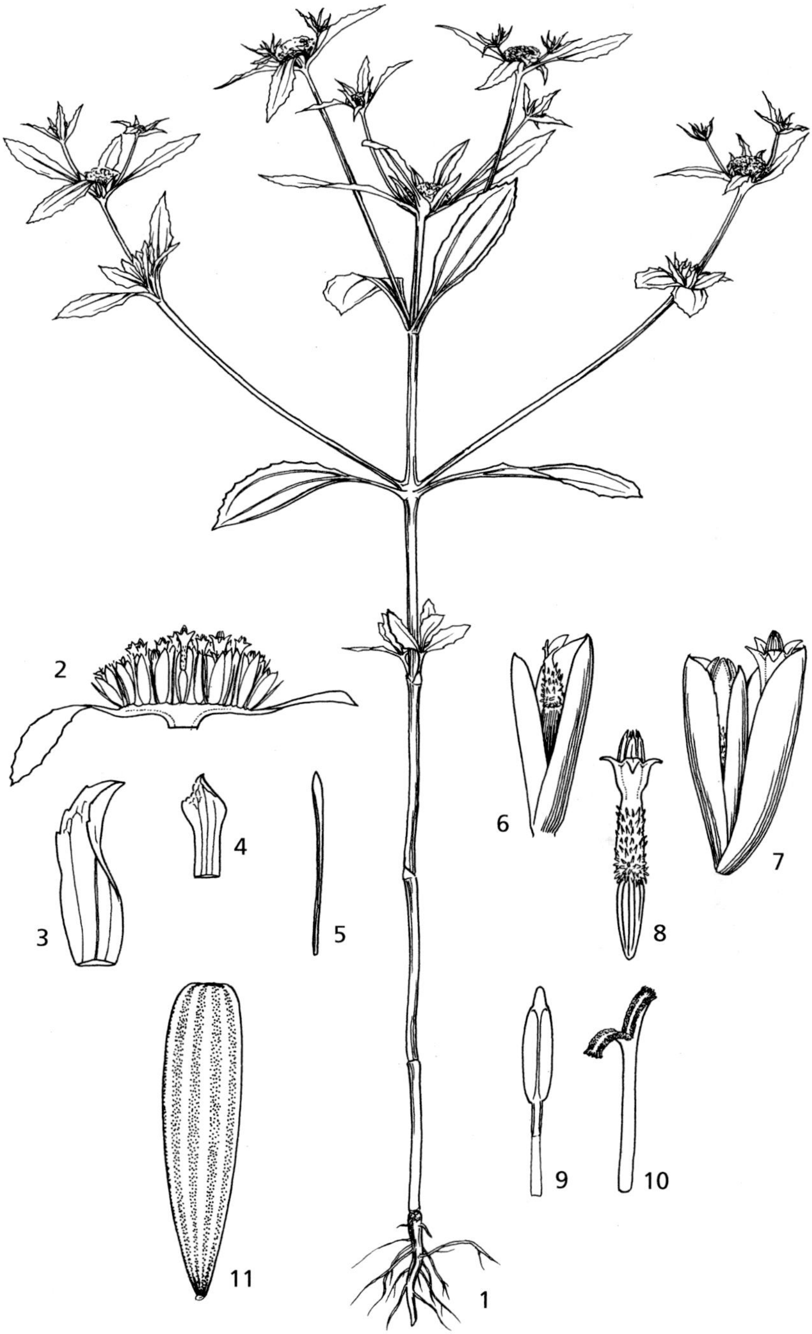

Fig. 108. **Flaveria trinervia**. 1, habit × ⅔; 2, capitulum × 2½; 3, outer phyllary × 3; 4, inner phyllary × 3; 5, paleae (receptacle scale)× 4; 6, ray floret with bracts × 8; 7, disc floret with bracts × 8; 8, disc floret (bracts removed) × 8; 9, anther × 16;10, style × 20; 11, achene × 11. 1–10 from *Miller* 228; 11 from *Agnew* 1867. Drawn by Pat Halliday.

Inflorescence cymose, corymbose or solitary, long-pedunculate. Capitula small to large, heterogamous and radiate, rarely homogamous and discoid; involucre cylindrical; phyllaries connate, uniseriate, equal, with *c.* 2 rows of oil glands; receptacle flat, naked, or with minutely fringed pits. Corollas yellow or orange, rarely white; ray florets female, fertile, ray limb entire to 3-lobed, spreading; disc florets hermaphrodite, fertile, regular, tubular, 5-lobed; style-arms truncate, penicillate or shortly appendiculate. Achenes linear, oblong-oblanceolate to clavate, compressed and ± angular or terete, glabrous or pubescent, with a callus at base; pappus of 3–10 scales, unequal, variably connate, sometimes 1–2(–5) long-aristate and much exceeding rest, which are short and truncate or rounded.

A genus of about 50 species, distributed throughout the warm regions of the New World; 2 or 3 species are widely cultivated.

Tagetes (possibly from *Tages*, the name of an Etruscan or Roman God).

1. **Tagetes erecta** *L.*, Sp. Pl.: 887 (1753); Dinsmore in Post, Fl. Syria, Palest. & Sinai ed. 2, 2: 43 (1933); Guest in Dep. Agr. Iraq Bull. 27: 60 (1933); Ba'li, Gard. Fls. 1: 381 (1946); Husain & Kasim, Cult. Pl. Iraq: 117 (1975).

Tagetes patula L., Sp. Pl.: 887 (1753); Ba'li, Gard. Fls. 1: 381 (1946); Husain & Kasim, Cult. Pl. Iraq: 117 (1975).

Strongly aromatic, erect, annual herb to 1.5 m tall, with open growth and ribbed, ± scabrid stems. Leaves pinnate, to 12 × 6 cm, with elliptic to lanceolate serrate divisions. Peduncles ± leafy, much swollen (to 1 cm in diameter) immediately below capitulum. Capitula yellow, solitary, terminal, 3–5 cm in diameter, 2–3 cm long. Ray florets numerous; rays oblong-obovate, entire or 2-lobed, up to 3.9 × 1.2 cm, tube ± 1.3 cm. Lobes of disc florets narrow, linear, 1.4–1.5 × 0.5 mm, one more deeply divided than the rest, all densely hairy on inner face; style bifid, slender, arms papillose. Achenes black, ± 1 cm; pappus scales ± 5, scarious, usually 2 aristate, barbellate, (8–)11–12 mm, others obtuse to rounded, ± connate, glabrous, 3–4(–6) mm.

HAB. & DISTRIB. Grown as an ornamental (commonly so, according to Husain & Kasim, l.c.); fl. & fr. Jul.-Dec. **LCA**: Baghdad, *Sahira* C. 630!, C. 631!

A native of Mexico, widely cultivated elsewhere.

African Marigold (Eng.).

T. tenuifolia Cav. is stated to be "occasionally cultivated" by Husain & Kasim (1975), but no specimen has been seen.

TRIBE **HELIANTHEAE** CASS.

117. **XANTHIUM** L.

Gen. Pl. ed. 6: 1056 (1764)
Panero in Kubitzki, Fam. Gen. Vasc. Pl. 8: 445 [2006] (2007)

Patricia Halliday
Revised by D. J. Nicholas Hind

Annual herbs. Leaves alternate, entire or lobed. Inflorescence terminal and axillary. Capitula unisexual, monoecious, solitary or in clusters. Male capitula terminal, many-flowered; phyllaries free, uniseriate; receptacle scaly. Corolla greenish, tubular, regular, 5-dentate; filaments monadelphous; anthers free, exserted, sagittate, minutely appendiculate; gynoecium imperfect, style undivided. Female capitula few at base of inflorescence, ovoid, 2-flowered, 2-locular, densely armed with hooked or straight spines and bearing 2 apical beaks; corolla absent; stamens absent; style persistent, bifid, exerted from hole on inner face of beak. Fruit an indehiscent burr; seed glabrous with styles persisting at apex; pappus absent.

A genus of possibly 3 species, probably originating in America, 2 species now of almost world-wide distribution; 2 species in Iraq.

Xanthium (from Gr. ξανθος *(xanthos)*, yellow or fair-haired, doubtless a reference to the colour of the spines. Two goddesses were named Xanthe, 'the golden one', in Gr. mythology (one was a Naiad, daughter of Oceanus, the other an Amazon).

Löve, D. & Dansereau, P. (1959). Biosystematic studies on *Xanthium*: taxonomic appraisal and ecological status. Canad. J. Bot. 37: 173–208.

Widder, F.J. (1923). Die Arten der Gattung *Xanthium*". Feddes Repert. Spec. Nov. Regni Veg. Beih. 20: 1–221.

Widder, F.J. (1964). Die Veränderlichkeit von *Xanthium spinosum*. Phyton (Horn) 11(1 & 2): 69–82 (1964).

Stems armed with branched yellow spines; leaves lanceolate in outline, ± 3 × as long as broad; apical lobe long and narrow; grey-white canescent beneath, green above; female phyllaries absent. . . . . 1. *X. spinosum*

Stems unarmed; leaves ovate in outline, ± 1.5 × as long as broad (often broader than long), apical lobe not as above; female phyllaries numerous, free. . . . . 2. *X. strumarium*

1. **Xanthium spinosum** *L.*, Sp. Pl.: 897 (1753); Boissier, Fl. Orient. 3: 252 (1875); Dinsmore in Post, Fl. Syria, ed. 2, 2: 43 (1933); Täckholm, Stud. Fl. Egypt, ed. 2: 568 (1974); Kupicha in Fl. Turkey 5: 48 (1975); D. Löve in Fl. Europ. 4: 143 (1976); Feinbrun-Dothan, Fl. Palaest. 3: 324 (1978); Boulos & Hadidi, Weed Fl. Egypt: 45 (1984); Collenette, Wild Flow. Saudi Arabia: 225 (1999); Boulos & Hind in Fl. Egypt 3: 227 (2002); Taifour & El-Oqlah, Pl. Jordan Ann. Checkl.: 49 (2017).

Much-branched, spiny, scabrid herb, with slightly ribbed and angled stems to 30 cm (–100 cm outside Iraq), usually (?always) without purple mottling on stems. Leaves lanceolate in outline, ± 6 × 1.6 cm, decurrent, acute, with 3(–5) entire lobes, upper leaves often entire, green above, densely grey-white canescent beneath and on veins above, subsessile or shortly petiolate, *c.* 1 cm long, canescent, on short, lateral shoots which are subtended by 1 or 2, usually trifid yellow spines, each spine-arm *c.* 2.3 cm. Capitula solitary or more usually in clusters, shortly pedunculate, male capitula *c.* 5 mm in diameter; phyllaries oblanceolate, *c.* 2 × 1 mm, hairy; receptacular scales 2.5 mm. Florets *c.* 3 mm, funnel-shaped, externally glandular in upper ½; anthers *c.* 1 mm. Female capitula sub-globose to ovoid, erect, 2–3 mm long; phyllaries absent; beaks 2, straight, slender, usually one much more developed than other. Fruits oblong-ellipsoid, 12(–14) × 7 mm (excluding spines), spines *c.* 3 mm. Seed (immature) ± cylindrical, slightly flattened on inner face. Fig. 109: 12–13.

HAB. Cultivated and diturbed land, clay and gypsum soils in field; alt.; fl. Oct.; May-Nov.
DISTRIB. Very rare in Iraq; only found once. **DWD**: Qaim, on cultivated land, *A. Sharif, K. Hamad & H. Hamid* 47315; *S. Omar & K. Hamad* 50420!; 20 km to Ana, *S. Omar,H. Abbas.Al-Kaisi & K. Hamad* 55830!; 25 km from Rawa on left, *S. Omar,H. Abass, Al-Kaisi & K.Hamad* 55854!; **MRO**: Haj Omran, *Riyadh, Hydar, Noor, Adel & Shymaa* 58733 (BAG!).

S Europe (S France, S Austria, Hungary to S Russia), Syria, Lebanon, Palestine, Egypt, Turkey: also occurring perhaps in N Africa, Ethiopia, parts of India and Siberia.

Spiny Burweed (Engl.).

2. **Xanthium strumarium** *L.*, Sp. Pl.: 987 (1753); Boissier, Fl. Orient. 3: 251 (1875); Dinsmore in Post, Fl. Syria, Palest. & Sinai ed. 2, 2: 43 (1933); Guest in Dep. Agr. Iraq Bull. 27: 108 (1933); Zohary, ibid. 31: 147 (1950); Rawi, ibid. Tech. Bull. 14: 126 (1964); Rechinger, Fl. Lowland Iraq: 612 (1964); Rawi & Chakravarty in Dep. Agr. Iraq Tech. Bull. 15: 98 (1964); Rawi & Bunni, ibid. ed. 2, 145: 129 ("Poisonous plants of Iraq") (1973); Kupicha in Fl. Turkey 5: 47 (1975); D. Löve in Fl. Europ. 4: 143 (1976); Feinbrun-Dothan, Fl. Palaest. 3: 323 (1978); Boulos & Hadidi, Weed Fl. Egypt: 45 (1984); Collenette, Wild Flow. Saudi Arabia: 225 (1999); Boulos & Hind in Fl. Egypt 3: 229 (2002); Abdulridha, Taha & Wedad, Ecol. & Fl. Basra: 229 (2016); Haloob & Al- Kaisi, Illustr. Fl. Lowland Iraq 1(1): 61 (2016); Taifour & El-Oqlah, Pl. Jordan Ann. Checkl.: 49 (2017).

X. brasilicum Vell., Fl. Flum., Ic. 10, t. 23 (1825); Blakelock in Kew Bull. 4: 65 (1949); Rawi in Dep. Agr. Iraq Tech. Bull. 14: 126 (1964); Rechinger, Fl. Lowland Iraq: 612 (1964); Täckholm, Stud. Fl. Egypt ed. 2: 568 (1974).

X. pungens Wallr., Beitr. zur Bot. 2: 231 (1844); Rechinger, Fl. Lowland Iraq: 612 (1964).

X. saccharatum Wallr., Beitr. zur Bot. 2: 238 (1844); Rechinger, Fl. Lowland Iraq: 613 (1964).

Erect, scabrid, glandular, little-branched herb, 20–68(–200) cm tall, often usually (?always) with purple-mottled stems. Leaves petiolate, lamina ovate, broadly lanceolate or triangular in outline, (1.5–)5–11 × (1.5–)5–8(–14) cm, cordate, cuneate, acute to obtuse variously ± distinctly 3–5(–7)-lobed, lobes irregularly crenulate or more sharply toothed, both surfaces clad with white bristles and sessile yellow glands, petiole (0.8–)7(–12.3) cm. Capitula solitary or more usually in clusters, sub-pedunculate, aromatic when bruised[14]; male capitula subglobose to ovoid ± 7 mm in diameter; phyllaries lanceolate-hairy and glandular, 3 × 0.5–1 mm; receptacular scales obovate, 3 mm. Florets 3.5–4 mm long, funnel-shaped, externally glandular in upper ½. Anthers ± 1.5 mm. Female capitula with uniseriate involucre of numerous, free, lanceolate, hairy and glandular phyllaries, 3 × 1 mm, beaks incurved, erect or divergent, styles filiform, papillose. Fruit an oblong-ellipsoid or ovoid burr, 12–32 mm × 6–9 mm (excluding spines), spines up to 7 mm long, often pubescent at base and between spines. Seed compressed, obovate in outline, with thin, black epidermis, longitudinally ribbed, 9–16 × 4 mm, hemispherical in transverse section, with or without 2 narrow longitudinal wings, styles persisting as a bristle-like appendage ± 5 mm. Fig. 109: 1–11.

HAB. Rocky mountain slopes and steppe hills near springs, on sandy soil or gravel, also on cultivated land; on the irrigated alluvial plain on river banks and as a weed in fields; alt. up to ± 1650 m; fl. & fr. (Jun.-)Sept.-Oct.(-Feb.).

DISTRIB. Occasional in the mountains and foothills, quite common on the alluvial plain. **MAM**: Sharanish, *Rawi, Noori & Tikriti* 29052!; Kani Mazi, NE of Zakho, *Botany Staff* 43825!, 43886!; Sarsang, *Barkley & Ani* 9063! **MRO**: Haji Umran, *Rawi* 24289! Bani Baradost, *Thesiger* 327. 19 km S. Erbil, *Botan izaddin* 48896 (BAG!). **MSU**: Penjwin, *Haussknecht* s.n.!; Avroman, *Chakravarty, Rawi, Alizzi & Noori* 19800! **FUJ/FNI**: on a sandy island in R. Tigris, nr Mosul, *Kotschy* s.n.! **FUJ**: Hadhr, *Alizzi & Omar* 35240!; between Baiji and Shergat, *id.* 35233! **FNI**: Nineveh, *Abdul-Karim al-Khadhairi* 13617! **FKI**: between Kirkuk and Khalis, *Rawi et al.* 19695!; Zain al-'Abdin, *Juma Brahim* 6120!; **FPF**: Sa'diya, *Qaisi* 42899!; Mandali, *Haddad* 9995! Galal Badra river near Zurbatia border port, *A. Haloob & Saif* 58683 (BAG!). **DWD**: [no loc.], *Qaisi, Sharif, Hamid & Hamad* 47319!; Habbaniya, *Botany Staff* 47196! **LCA**: Abu Ghraib, *Sahira* 35827!; on riverside, near Baghdad, *Chakravarty* 29682!; bank of R. Diyala at Rustamiya, *Lazar* 3467!; Daura, *Haines* W.1614!; 46 km SE of Musaiyib, *Thamir & Widdad* 47281!; bank of R. Tigris at Na'maniya, *Juma Brahim* 6192! **LEA**: Dabouni, *Guest* 219! **LSM**: Hor ach-Chiba, 10 km SE of Musharra, *Rawi & Alizzi* 34203! **LBA**: Basra, *Guest* 308!; muddy bank of Shatt al-Arab at Fao, *Haines* W.883!

Syria, Palestine, Jordan, Egypt, Turkey, N & S America, Europe, Africa etc.; distribution almost world-wide.

Burweed, Clotbur (Am.). HASACH (a general name for spiny plants) and LIZZAIJ (also a general name for "sticky" plants which cling) are given by Guest (1933). According to Rawi (1973) the germinating seeds and young seedlings (in the two-leaf stage) contain xanthostrumarin and are poisonous to sheep, cattle and pigs. On the other hand Rawi & Chakravarty (1964) draw attention to the medicinal value of the plant.

118. **ECLIPTA** L.

Mant. Pl. 2: 157 (1771), *nom. cons.*
Panero in Kubitzki, Fam. Gen. Vasc. Pl. 8: 450 [2006] (2007)
Eupatoriophalacrom Mill., Gard. Dict. Abr. ed. 4 (1754), nom. rej.; *Micrelium* Forssk., Fl. Aegypt.-Arab. : 152 (1775), nom. inval.; ?*Abasoloa* La Llave & Lex., Nov. Veg. Descr. 1: 11 (1824); *Paleista* Raf., New. Fl. 2: 43 (1837) [1836]; *Polygyne* Phil., Linnaea 33(2): 170 (1864/65); *Ecliptica* Kuntze, Revis. Gen. Pl. 1: 334 (1891), orth. var.

D. J. Nicholas Hind

Annual or perennial herbs; rootstock short, creeping, rooting at nodes, strigose. Stems decumbent to erect, terete, pith narrowly fistulose or solid. Leaves opposite scarcely connate or not at nodes, petiolate or pseudopetiolate, lamina elliptic or narrowly oblong to narrowly ovate, lanceolate or linear, base acute or acuminate, margins entire to serrate, apices narrowly rounded to narrowly acuminate, surfaces strigose, without glandular dots, weakly 3-veined to

[14] According to Wheeler-Haines (nos. 1613 & 1614, K) only the large fruits are aromatic when cut, small mature fruits being non-aromatic.

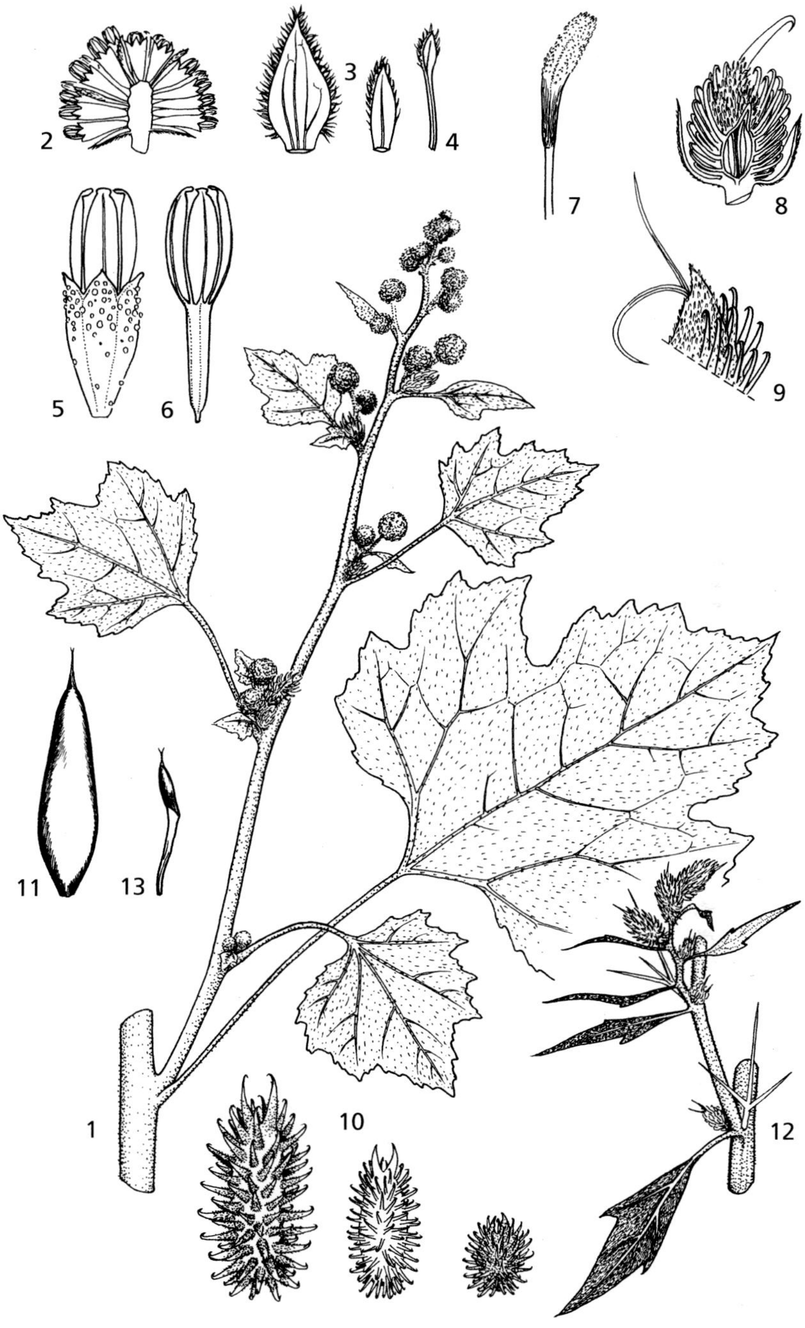

Fig. 109. **Xanthium strumarium**. 1, habit × 2/3; 2, capitulum male (dissected) × 3; 3, phyllary from male capitulum × 6; 4, receptacle scale × 6; 5, male floret × 10; 6, staminal column × 10; 7, style × 16; 8, capitulum female (dissected) × 4½; 9, female floret × 8; 10, fruiting bodies × 1; 11, achene × 2. **Xanthium spinosum**. 12, habit × 2/3; 13, achene × 2. 1 from *Hooker* 349; 2–9 from *Juma Brahim* 6120; 10 from *Hooker* 349, *Juma Brahim* 6120; 11 from *Hooker* 349; 12–13 from *Sharif, K. Hamad & H. Hamid* 47315. Drawn by Pat Halliday.

±pinnate, secondary veins strongly ascending. Inflorescences terminal or axillary, 1–5 capitula per node, pedicels thin. Capitula broadly campanulate to hemispherical, heterogamous, radiate, corollas and achenes rapidly deciduous (mature capitula rarely intact in dry material); phyllaries 8–11, herbaceous, *c.* 2-seriate, subequal or inner phyllaries somewhat shorter, elliptical to obovate, acute, strigose; receptacle ± convex, paleaceous; paleae linear to filiform, apices slightly broadened. Ray florets 8–*c.* 50, 1–3-seriate, female and mostly fertile, corollas whitish to yellow, ray limb narrowly elliptical to linear, short-bilobed, smooth to mamillose on upper surface. Disc florets 8–*c.* 40, hermaphrodite, corollas greenish-white to yellowish, mostly glabrous outside, sometimes with few hairs on lobes, throat narrowly campanulate, usually 4-lobed, lobes broadly oblong-ovate, papillose; anther thecae usually slightly blackened, apical anther appendages broader than wide, with or without a single gland outside; style base not enlarged; style arm apices with tuft of hairs forming acute or truncate tip. Achenes maturing and falling quickly, obpyramidal, tetragonal or outer ray achenes triangular with spiculiform setulae near top, surface usually becoming tuberculate with age, winged or wingless; pappus absent, scarcely coroniform or with 2 or 3 awns.

A genus of about 5 species, one of which is very widespread throughout the tropics and present in Iraq.

Orchard, A. E. & Cross, E. W. (2013). A revision of the Australian species of *Eclipta* (Asteraceae: Ecliptinae), with discussion of extra-Australian taxa. Nuytsia 23(1): 43–62.

1. **Eclipta prostrata** (*L.*) *L.*, Mant. Alt.: 286 (1771). Exell, Cat. Vasc. Pl. S. Tomé: 225 (1944); Andrews, Fl. Pl. Sudan 3: 26, t. 4 (1956); Adams in Hepper, Fl. W. Trop. Afr. ed. 2, 2: 241 (1963); McVaugh in Fl. Novo-Galiciana 12: 315–317 (1984); Hind in Fl. Mascareignes 109. Composées: 179–181 (1993); Boulos & Hind in Fl. Egypt 3: 230 (2002); Mesfin Tadesse, Fl. Ethiop. Eritr. 4(2): 286 (2004); Beentje & Hind in Fl. Trop. E. Afr. Compositae pt. 3: 728–730 (2005); Chen Yousheng & Hind in Fl. China 20-21: 869 (2011).

Verbesina alba L., Sp. Pl.: 902 (1753).
V. prostrata L., Sp. Pl., ed. 1, 2: 902 (1753).
Eclipta erecta L., Mant. Pl. 2: 286 (1771), nom. illeg. based on *Verbesina alba* L.
E. strumosa Salisb., Prodr. : 205 (1796), based on *Eclipta prostrata* (L.) L.
Cotula alba (L.) L., Syst. Nat. 2: 564 (1767).
Micrelium asteroides Forssk., Fl. Aegypt. Arab. : 152 (1775), nom. inval.
M. tolak Forssk., Fl. Aegypt. Arab. : 152 (1775), nom. inval.
Eclipta undulata Willd., Sp. Pl., ed. 4, 3(3): 2219 (1803)
E. thermalis Bunge, Enum. Pl. Chin. Bor. : 39 (1833)
E. erecta L. [var.] *β diffusa* DC., Prodr. 5: 490 (1836)
E. prostrata (L.) L. [var.] β *undulata* (Willd.) DC., Prodr. 5: 490 (1836).
E. parviflora Wall. ex DC., Prodr. 5: 490 (1836).
E. arabica Steud., Nomencl. Bot., ed. 2, 1: 542 (1840), nom. illeg., superfl. pro *E. erecta* L.
E. marginata Steud., Nomencl. Bot., ed. 2, 1: 542 (1840), nom. illeg., superfl. pro *E. erecta* L.
Eleutheranthera prostrata (L.) Sch.Bip., Bot. Zeit. (Berlin) 24: 239 (1866).
Eclipta erecta L. var. *prostrata* (L.) Baker, Fl. Maurit. Seych. : 169 (1877).

Hispid annual herbs, 20–60 cm tall. Stems basally branched, sulcate, erect, ascending or prostrate. Leaves opposite, sessile, 2–12 × 0.3–1.8 cm, oblong-elliptic, lanceolate or oblanceolate, green when fresh turning blackish on drying, apices subacute to acute. Capitula axillary, solitary or few, heterogamous and radiate; pedicels 1–8 cm; involucres broadly campanulate to hemispherical, 3–4 mm in diameter in flower expanding to 7–9 mm in diameter in fruit and appearing almost cupulate; outer phyllaries 4 × 2 mm, broad-lanceolate, subacute, hispid-pubescent, inner phyllaries similar but with narrow scarious margins; receptacle at first flat becoming slightly convex in fruit, paleaceous; paleae persistent, 2–3 mm long, outer linear, inner filiform, apices dilate, laciniate. Ray florets female, fertile, multiseriate, *c.* 3 mm, ray limb 1–2 mm long, white, apex bidentate. Disc florets hermaphrodite, fertile, numerous, corollas *c.* 2 mm long, falling rapidly after anthesis. Achenes 2.5–3 × 1 mm, at first conspicuously green and becoming blackish at maturity, obconical, truncate, tuberculate, setuliferous in upper half when immature; pappus absent or visible as a puberulent rim when immature. Fig. 110: 1–10.

HAB. A widespread weedy species, preferring damp or swampy areas, often in irrigation schemes or in cultivated areas, found on pond margins, seepage areas and in flood plains; from near sea level to c. 2000 m; flowering throughout the year in ideal conditions.

DISTRIB. Mostly restricted to the east alongside rivers and floodplains, especially alongside the Tigris and its tributaries. FPF: 3 km N of Saadiya, *Qaisi* 50493. DWD: 5 km E of Ana, *Omar & Hamad* 50387! Al Gaim, *Omar & Hamad* 50425! LEA: Ba'quba, *Paranjpye* 6!; Aziziya, *Guest* 3473! LCA, Diyala river at Rustamiya, *Guest* 3509!; Umm al Bunni, Kut, *Guest* 3578!; Al Sudur, *Omar, Mokhtar, Sahira, Abdilly & F. Karim* 36974!; ?*Noë* 820! Karradah nr Baghdad, *Guest* 1038!; by R. Tigris, Baghdad, *Guest* 13464! *Haines* s.n.! & *S. & A. Agnew* 191 (33297) (K!), *Chakravarty* 29678!; Shatra farm, *Omar & Karim* 36764!; Hilla, *Paranjpye* (comm. Graham s.n.); Kut, Shatt al Gharraf, *Karim & Hamid* 37807!; 5 km W of Shamiya, *Alizzi & Mohammed* 32146!, 32151! Kut, *Guest* 202!

BARKAIJAH (*Guest* 3473) and SANAISLAH (*Guest* 3578).

119. **HELIANTHUS** L.

Sp. Pl.: 904 (1753); Gen. Pl. ed. 5: 386 (1754)
Panero in Kubitzki, Fam. Gen. Vasc. Pl. 8: 466 [2006] (2007)

C. C. Townsend

Tall, frequently coarse, ± scabrid annual or perennial herbs wtih leaves opposite or spirally arranged, frequently opposite below and alternate above. Capitula showy, sometimes very large, heterogamous, solitary or in corymbs. Phyllaries multiseriate, herbaceous; receptacle ± convex to almost flat, paleaceous, paleae persistent, entire or tricuspidate, partially embracing mature achenes. Ray florets sterile with yellow rays; disc florets hermaphrodite, tubular, lobes yellow or brown. Achenes ± 4-angled, compressed; pappus early-caducous, of 2 thin chaffy scales and sometimes smaller intermediate scales.

50 species in the USA, extending north to Canada and south to Mexico; 3 or 4 species cultivated in Iraq.

Helianthus (from the Gr. ἥλιος (helios), sun and ἄνθος (anthos), flower, from the appearance of the capitula and the fact that they turn to face the sun.

Heiser, C.B. et al., 1969. The North American Sunflowers (Helianthus). Mem. Torr. Bot. Club 22 (3): 1–218.

1. Annual plants with fibrous taproots; lower leaves cordate at base; disc florets with brownish–purple lobes 2
   Rhizomatous perennials; lower leaves cuneate or at most subtruncate, not cordate; disc florets with yellow lobes 3
2. Slender plant; disc rarely 2 cm in diameter; phyllaries rarely 3 mm wide, glabrous to sparsely hairy. . . . . *H. debilis* (see note)
   Coarse plant, disc rarely, and only in depauperate plants, as little as 2 cm wide; phyllaries rarely as narrow as 3 mm, ± densely ciliate and dorsally pubescent . . . . 1. *H. annuus*
3. Rhizomes tuberiferous at tips; lower leaves shortly cuneate at base, then decurrent (broadly so about base of lamina) along petiole . . . . 2. *H. tuberosus*
   Rhizomes not tuberiferous at the tips; lower leaves more longly cuneate, gradually narrowed into narrowly winged petiole . . . . 3. *H.* × *laetiflorus*

1. **Helianthus annuus** *L.*, Sp. Pl.: 904 (1753); Dinsmore in Post, Fl. Syria, Palest. & Sinai ed. 2, 2: 42 (1933); Guest in Dep. Agr. Iraq Bull. 27: 43 (1933); Vassilchenko in Fl. S.S.S.R. 25: 542 (1959); Heiser et al. in Mem. Torr. Bot. Club 22 (3): 61 (1969); Husain & Kasim, Cult. Pl. Iraq: 115 (1975); Chakravarty, Plant Wealth of Iraq 1: 271 (1976); Kupicha in Fl. Turkey 5: 44 (1977).

Annual herb, simple or branched, (0.3–)0.7–3 m; stem stout, scabrid. Leaves opposite below, alternate above. Lower leaves broadly cordate-ovate to ovate-elliptic, scabrid on both surfaces, dentate-serrate, long-petiolate, lamina 10–40 × 8–35 cm, 3-veined at base. Capitula mostly solitary, often very large, nodding; phyllaries long-ciliate and dorsally pubescent, lanceolate-oblong to broadly ovate; ray limbs 3–6 mm, yellow; disc (2–)5–10+ cm broad, disc floret lobes brownish-purple. Achenes large, 8–15 × 4–9 mm, whitish or greyish with dark lines, obovate to ± round, compressed.

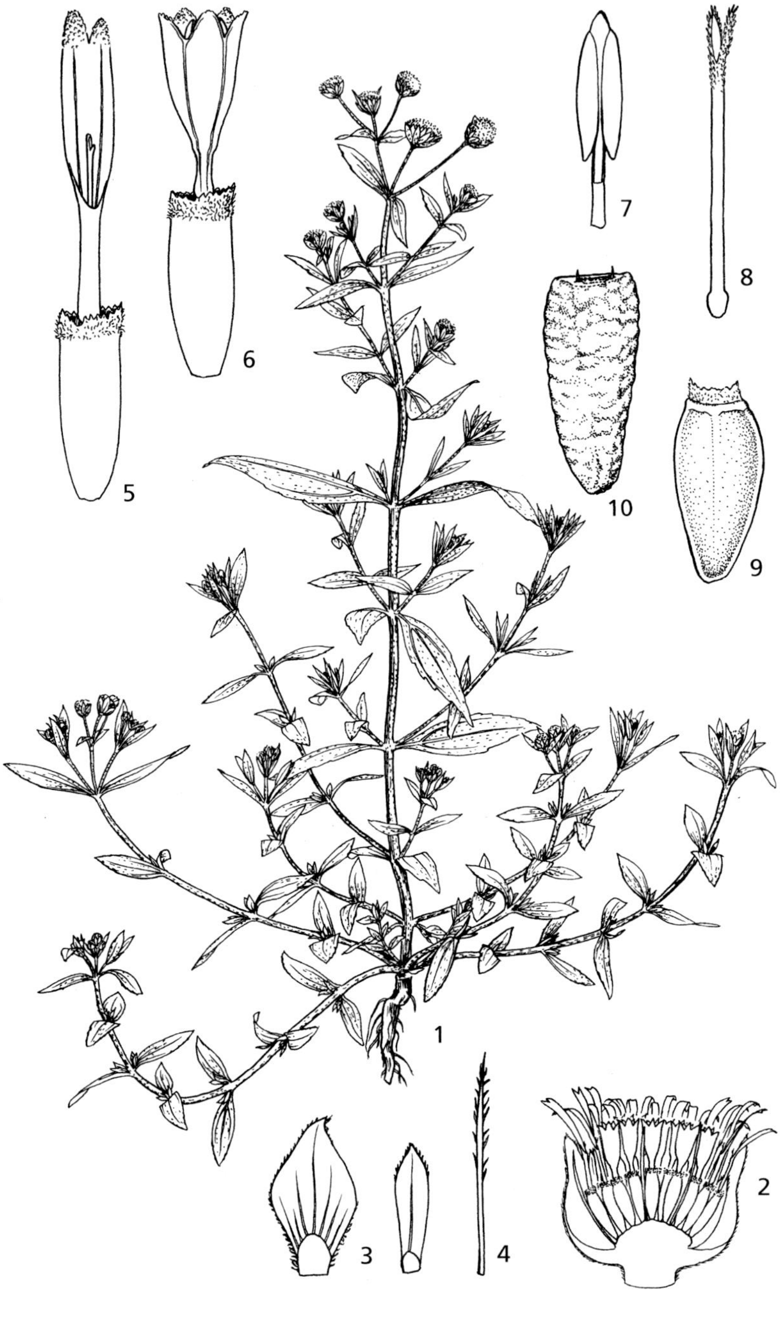

Fig. 110. **Eclipta prostrata**. 1, habit × 2/3; 2, capitulum (dissected) × 5; phyllaries × 5; 4, paleae (receptacle scale)× 10; 5, ray floret × 16; 6, disc floret × 16; 7, anther × 24; 8, style × 24; 9 achene (immature) × 10; 10, achene (mature) × 10 . 1 from *Rawi* 16590; 2–10 from *Alizzi* 33174. Drawn by Pat Halliday.

HAB. & DISTRIB. Cultivated at an Agricultural Research Station. **LCA**: Abu Ghraib, *Ahmed* 9867!, *Rawi* 10739!; *Gillett* 12505!

A native of N America grown as a crop and sometimes spontaneous in other parts of the world.

Sunflower (Eng.), Girasole (It.). Guest gives the Ar. names WARD ASH SHAMS, 'AIN ASH SHAMS, 'ABAD ASH SHAMS, SHAMS WA QAMAR; QŪNAH BĀQĀN (Turk.) and GŪLAH BARRŪZHAH (Kurd.) are also given. He states that Sunflower is sometimes cultivated in Iraq as a garden plant, chiefly for ornament, seldom as a crop, though it could be grown in lower Iraq as a summer oil-seed crop. He also observes that there is a good market for the seed (achenes), which can be used to express an oil used in soap and candle-making, as well as for salad oil and for lighting. The seeds can be used as poultry food, for mixing with bird seed and for human consumption. The crop is widely grown in Russia and C Europe, where large quantities of oil and oilcake are produced, oilcake being the residue of the achene mass after pressing, and a valuable cattle food. In the USA it is grown chiefly as a silage crop. Recent information on the cultivation of the crop in Iraq is given by Chakravarty, l.c.

2. **Helianthus tuberosus** *L.*, Sp. Pl.: 905 (1753); Dinsmore in Post, Fl. Syria, Palest. & Sinai ed. 2, 2: 42 (1933); Guest in Dep. Agr. Iraq Bull. 27: 44 (1933); Vassilchenko in Fl. S.S.S.R. 25: 544 (11959); Heiser et al. in Mem. Torr. Bot. Club 22(3): 126 (1969); Chakravarty, Plant Wealth of Iraq 1: 275 (1976); Meikle, Fl. Cyprus 2: 901 (1985).

Coarse perennial rhizomatous herb, rhizomes slender, finally enlarged at tips into fusiform, oblong-ovoid or subglobose, fleshy edible tubers. Stems erect, 1–3 m, scabrid or hirtellous, mostly branched above. Leaves opposite below, alternate above. Larger cauline leaves broadly ovate or broadly lanceolate-ovate, scabrid on upper surface, ± densely whitish-puberulent below, serrate, petiolate, lamina 3-veined at base, 9–26 × 3.5–13 cm, shortly cuneate at base and then decurrent (broadly so above) along petiole. Capitula mostly numerous, rays 2.5–4 mm, yellow; disc 1.5–2.5 cm in diameter, disc floret lobes yellow; phyllaries ciliate and dorsally pubescent, lanceolate. Achenes 5–7 mm.

HAB. & DISTRIB. Cultivated at a former Agricultural Experimental Farm; alt. 40 m; fl. Nov. **MRO**: Malikan valley S. of Rawanduz, *Riyadh, Hydar, Noor, Adel & Shaymaa* 58770 (BAG!). **LCA**: Rustam Farm, *Lazar* 433!; Baghdad, Abu-Ghraib, Botany Dep.,*Al-Khayat* C1129 (BAG!).

Jerusalem Artichoke (Eng.), Topinambour (Fr.). Guest (1933) gives the Ar. Name TARTŪF and states that it is "occasionally grown as a vegetable in kitchen gardens, but not much known". Husain & Kasim (1975) state that it is "cultivated in the northern regions" of Iraq and give the additional names TARTŪF AL ARDH and ALMAZA. Chakravarty (1976) states that it is raised to a very limited extent in C Iraq and gives nutritive information.

The tubers of this plant are a commercial source of inulin and a good carbohydrate food for consumption by diabetics.

3. **Helianthus × laetiflorus** *Pers.*, Syn. Pl. 2: 476 (1807), pro sp.; Heiser et al. in Mem. Torr. Bot. Club 22 (3): 137 (1969); Husain & Kasim, Cult. Pl. Iraq: 188 (1975).

Very similar to *H. tuberosus*, differing in the characters given in the key. The reasons for considering it a hybrid are given by Clevenger & Heiser [Rhodora 65: 121–133 (1963)], the two putative parent species being *H. tuberosus* L. and *H. rigidus* (Cass.) Desf. The scabrid hairs on the surface of the leaves of *H. rigidus* have a distinct white base almost absent in those of *H. tuberosus*, with *H.* × *laetiflorus* ± intermediate; but this is highly comparative.

HAB. & DISTRIB. Cultivated at an Agricultural Research Station near Baghdad; alt. 40 m; fl. Oct.–Nov. **LCA**: Abu Ghraib, *Rawi* 26071!, *Alizzi* 34814!

Stated on both sheets to be grown as an ornamental, as is usual.

Helianthus debilis Nutt. subsp. *cucumerifolius* (Torr. & Gray) Heiser is reported by Husain & Kasim (1975, as *H. cucumerifolius* Torr. & Gray) as being cultivated throughout Iraq, and is therefore included in the key. No specimens have been seen and its occurrence cannot thus be confirmed.

Related to *Helianthus* is the C American genus *Tithonia*, comprising mainly tall herbs or shrubs with mostly alternate, lanceolate, ovate or cordate-ovate leaves which are in some species 3–5-lobed. The capitula are large, with yellow ligules, long pedicels ± inflated below the capitula and the pappus is persistent on the achenes. Two species are recorded by Husain & Kasim (1975) as occasionally cultivated in Iraq, but no specimens have been seen to confirm their identity. They may be distinguished by the following key:

Phyllary apices rounded, 4-seriate, outer occasionally more pointed; ray limbs 10–25 mm; capitula resembling a large *Tagetes* . . *T. rotundifolia* (Mill.)S.F.Blake
Phyllaries acute or acuminate, 2-seriate; ray limbs 30–70 mm; capitula resembling a *Helianthus*. . . . . *T. diversifolia* (Hemsl.)A.Gray

Both species are widely naturalized in the tropics.

120. **RUDBECKIA** L.

Sp. Pl.: 906 (1753); Gen. Pl. ed. 5: 387 (1754)
Panero in Kubitzki, Fam. Gen. Vasc. Pl. 8: 471 [2006] (2007)
Obeliscaria Cass. subgen. *Dracopis* Cass., Dict. Sci. Nat., ed. 2, 35: 273 (1825); *Dracopis* (Cass.) Cass., Dict. Sci. Nat., ed. 2, 38: 17 (1825).

C. C. Townsend

Annual or perennial herbs; leaves entire to trilobed or pinnatifid, subglabrous to hispid. Capitula solitary to numerous, commonly ± long-pedunculate, heterogamous, radiate; phyllaries 2–3-seriate, mostly spreading or reflexed, herbaceous; receptacle conical to columnar, paleaceous, paleae partly embracing achenes. Ray florets neuter and sterile, the ray limbs yellow, orange or partly purple, rarely absent; disc florets hermaphrodite, fertile, dark purplish to brown or yellow. Achenes glabrous, quadrangular or slightly compressed; pappus a small corona or absent.

About 17 species in N America.

Rudbeckia (named in honour of Olof Rudbeck, 1630–1702, the Swedish naturalist and anatomist, who discovered the lymphatic vessels). Coneflower (Am.).

Urbatsch, L.E., Baldwin, B.G. & Donoghue, M.J. (2000). Phylogeny of the Coneflowers and relatives (Heliantheae: Asteraceae) based on Nuclear rDNA Inernal Transcribed Spacer (ITS) sequences and Chloroplast DNA restriction site data. Syst. Bot. 25(3): 539–565.

Majority of leaves 1–2-pinnatifid or pinnately compound; disc floret corollas yellow to yellowish-green; pappus coroniform or of 4 scales . . 1. *R. laciniata*
All leaves simple and unlobed; disc floret corollas proximally yellowish-green and distally brown-purple; pappus absent . . . . 2. *R. hirta*

1. **Rudbeckia laciniata** *L.*, Sp. Pl.: 906 (1753).

Rudbeckia digitata Mill., Gard. Dict., ed. 8, n. 6 (1768), non Willd. (1803).
Rudbeckia laciniata L. var. *digitata* (Mill.) Fiori, Fl. Ital. [Fiori et al.] 3: 300 (1904).
Helianthus laciniatus (L.) E. H. L. Krause, Deutschl. Fl. (Sturm), ed. 2, 13: 164 (1905).
Rudbeckia laciniata L. var. *heterophylla* (Torr. & A. Gray) Fernald & B. G. Schub., Rhodora 50: 172 (1948).
R. laciniata L. var. *ampla* (A. Nelson) Cronquist, Vasc. Pl. Pacif. N. W. [C. L. Hitchcock et al.] 5: 280 (1955).
R. laciniata subsp. *ampla* (A. Nelson) W. A. Weber, Phytologia 51(6): 375 (1982).

Tall perennial up to 3 m; stem glabrous, striate, with long ascending branches above. Basal leaves usually pinnatisect with 2–3 pairs of dentate segments, at least the lower pair bi- to trifid or bi- to trisect; cauline leaves becoming trifid or trisect or the upper simple, all dentate. Capitula few to numerous, on long slender peduncles Phyllaries lax, spreading or reflexed, glabrous or sparsely strigose, green. Rays yellow, 2–5 cm. Disc 1.25–2.25 cm in diameter; disc florets yellow or greyish-yellow; style-appendages short and blunt; Achenes tetragonous; pappus a short, toothed corona.

HAB. & DISTRIB. Cultivated; in a garden. **MRO**: Shaqlawa, *Dabbah & Hamad* C.856!

A native of N America.

The specimen cited above is the full double cultivar *Rudbeckia laciniata* cv. 'Hortensia', "Golden Glow" (Engl.). This is common in cultivation but is weak-stemmed and apt to sprawl if subjected to heavy winds or rain.

2. **Rudbeckia hirta** *L.*, Sp. Pl.: 907 (1753); Vassilchenko in Fl. S.S.S.R. 25: 539 (1959).

Centrocarpha hirta (L.) D. Don ex G. Don, Hort. Brit. [Sweet] ed. 3: 357 (1839).
Rudbeckia strigosa Nutt., Trans. Amer. Phil. Soc., 7: 354 (1840).
R. brittonii Small ex Small, Mem. Torr. Bot. Club 4: 130 (1894).
R. monticola Small, Torreya 1: 67 (1901).
Helianthus hirtus (L.) E. H. L. Krause, Deutschl. Fl. (Sturm), ed. 2, 13: 166 (1905).
Rudbeckia hirta L. var. *brittonii* (Small) Fernald, Rhodora 39: 457 (1937).
R. hirta L. var. *monticola* (Small) Fernald, Rhodora 39: 457 (1937).
R. hirta L. var. *serotina* (Nutt.) Core, Castanea 27: 62 (1962).

Biennial or short-lived perennial occasionally flowering in the first year, (30–)40–100 cm, vegetative parts hispid throughout with hairs frequently tuberculate at base. Leaves very variable in shape and size, basal lanceolate to elliptic, long-petiolate, median cauline leaves linear to ovate or ovate-lanceolate, sessile and ± auriculate at base, upper narrower and more tapering below. Capitula several or solitary, long-pedunculate; involucre open; phyllaries hispid or hirsute, herbaceous, irregularly spreading or reflexed. Disc hemispherical or ovoid, purplish or brown, 1.25–4.5 cm, sometimes with a basal dark blotch. Achenes tetragonous; pappus absent.

A native of N America well known in cultivation.

Black-eyed Susan (Am.).

var. HIRTA, with cauline leaves ovate-lanceolate or ovate, basal leaves broadly elliptical, and all leaves coarsely dentate, is not recorded from Iraq. The following var. is the Iraqi plant:

var. **pulcherrima** *Farwell* in Rep. Mich. Acad. Sci. 6: 209 (1904).
Rudbeckia serotina Nutt., J. Acad. Nat. Sci. Phil. 7: 80 (1834).
R. bicolor Nuttall in Acad. Nat. Sc. Phil. 7 : 80 (1834); Husain & Kasim, Cult. Pl. Iraq : 116 (1975).
R. strigosa Nutt., Trans. Amer. Phil. Soc., 7: 354 (1840).
R. flexuosa T. V. Moore, Pittonia 4: 180 (1900).

Cauline leaves spatulate, oblanceolate or broadly linear; basal leaves narrowly to broadly oblanceolate or lanceolate; all leaves entire or finely serrate.

HAB. & DISTRIB. Cultivated (on clay soil). **LCA**: between Baghdad and Karrada Sharqiya, *Janan al-Mukhtar* 35770!

N America from British Columbia to N Mexico.

According to Husain & Kasim (1975) this plant is "occasionally cultivated" in Iraq. Only this single specimen has been seen.

Once considered closely related to *Rudbeckia* is the N American genus *Echinacea* Moench, which differs in having the paleae rigid and spinescent at tips and exceeding the disc florets, so that the disc is prickly in appearance. According to Husain & Kasim (1975), *E. purpurea* (L.) Moench is also cultivated in Iraq. This is a perennial 0.5–1.5 m tall with broadly lanceolate to ovate, usually serrate leaves and purple (rarely pale) rays.

121. **ZINNIA** L. *nom. cons.*

Syst. Nat. ed. 10: 1221 (1759); Benth. & Hook. f., Gen. Plant. 2: 357 (1873)
Panero in Kubitzki, Fam. Gen. Vasc. Pl. 8: 476 [2006] (2007).

Patricia Halliday
Revised by D. J. Nicholas Hind

Erect, glabrous or pubescent, annual or perennial herbs or subshrubs. Leaves opposite, entire. Capitula solitary, terminal, medium to large, heterogamous, radiate, fertile; involucre hemispherical-cylindrical; phyllaries multiseriate, imbricate, outer series short, inner series progressively longer; receptacle paleaceous. Ray florets female, few to numerous, rays persistent, spreading, entire or emarginate, tube short. Disc florets few to numerous, hermaphrodite, tubular, regular. Achenes compressed, ± flat; pappus of 1–3 erect aristae, produced from angles of achene, or absent.

A genus of ± 18 species, mainly New World in origin (especially Mexico) many of which are cultivated as ornamentals in Europe, Asia, Africa etc.; only one species cultivated in Iraq.

Torres A.M. (1963). Taxonomy of *Zinnia*. Brittonia 15 (1): 1–25.

1. **Zinnia violacea** *Cav.,* Icon. 1: 57, t. 81 (1791).

Z. elegans[15] Jacq., Ic. Pl. Rar. 3 (fasc. 11): 589 (late 1792 or early 1793) and Coll. 5: 152 (1796); Curtis, Bot. Mag. t. 527 (1801); Guest in Dep. Agr. Iraq Bull. 27: 109 (1933); Dinsmore in Post, Fl. Syria, ed. 2, 2: 42 (1933); Stout, Gardening for Egypt: 83 (1935); Bai'li, Gard. Fls. in Arabia 1: 384 (1946); Husain & Kasim, Cult. Pl. Iraq: 118 (1975).
Z. australis F. Bailey, Queensland Dept. Agr. Bull. 9, Bot. [Bull.] 3: 14 (1891).

Glabrescent to densely pubescent annual herb to 30(–45) cm tall in Iraq (to 1 m in country of origin). Stems ± stout, somewhat ribbed, with multiseriate, acicular and a few branched hairs, the latter usually towards base of stem. Leaves sessile ovate-oblong or lanceolate, 5–11 × 1.6–7.2 cm, with 3–5 conspicuous main veins, amplexicaul, acute, hispid and glandular, glands sessile, yellow. Inflorescence unbranched, scape swollen and hollow beneath capitulum. Capitula 2.5–9 cm in diameter (including rays); phyllaries 4–5-seriate, oblong, obtuse with dark apical border, ± hairy and/or ciliate; receptacle becoming conical, hollow, paleae ± oblong, usually coloured at maturity. Ray floret limbs red to violet, undersurface pale, 1- to multiseriate, obovate-spatulate, 0.9–3.5 × 3–10 mm, emarginate or entire, minutely pubescent and glandular, tube long-hairy; style bifid, hairy. Disc florets 5-lobed, lobes 5–13 × 2–6 mm, spreading, bearded; anthers 35 mm (excluding collars), sagittate. Achenes of ray florets obovate in outline, ± triquetrous, 7 × 3 mm, with ± pronounced shoulders, hairy, margins shallowly toothed and ciliate; achenes of disc florets similar but narrower, 10 × 2–3 mm; pappus absent.

HAB. & DISTRIB. Cultivated as an ornamental plant for its brightly coloured flowers in Baghdad and other cities where ornamental plants are grown. **LCA**: Baghdad, *Paranjpye* 100 !, *Fauzi* C-771 !, *Sahira* C.847 !, *Janan al-Mukhtar* 35768 !; fl. & fr. (Apr.-)Jul.-Sep.(-Nov.).

Sometimes known as YANGI DUNYĀ in Iraq.

Related to *Zinnia* is the small American genus *Sanvitalia* Lam., of which one Mexican species, *S. procumbens* Lam., is well-known in cultivation. According to Husain & Kasim (1975), it is "occasionally cultivated" in Iraq. It is a low-growing procumbent annual with branched, leafy stems and ovate, opposite, entire 3-veined leaves. The capitula are subtended by bract-like leaves and radiate, the purplish disc 8–14 mm in diameter and the yellow rays 4–7 mm, and pappus usually with 1 or 2 awns.

TRIBE **EUPATORIEAE** CASS.

122. **AGERATUM** L.

Sp. Pl. ed. 1: 839 (1753); Gen. Pl. ed. 5: 363 (1754)
Hind & Robinson in Kubitzki, Fam. Gen. Vasc. Pl. 8: 522 [2006] (2007)

D. J. Nicholas Hind

Annual to perennial herbs or subshrubs, few- to many- branched, base often decumbent with numerous adventitious roots. Stems terete, striate, sparsely puberulous to sparsely

[15] Several authors have misstated the date of publication of Cavanilles *Icones* … vol. 1 as 1797; it is clear it was published in 1791 and therefore predates the publication of *Z. elegans* in *Icones Plantarum Rariorum.* McVaugh (Fl. Nova Galiciana 12, 1984) cited the publication date of Jacquin's tab. 589 as 1793 but may well have been late 1792 or early 1793. Any proposal to conserve the later *Z. elegans* against *Z. violacea,* based on 'long standing use', is a 'use in ignorance'.

hirsute. Leaves opposite or sometimes alternate, short- to long-petiolate, lamina elliptical or lanceolate to deltoid or ovate, margins entire to dentate, apex not or scarcely acuminate, lower surface usually with large sessile or partially sunken glandular-punctate, usually 3-veined from or near base. Inflorescence cymose to subcymose, sometimes subumbellate, pedicels short to moderately long. Capitula homogamous and discoid; phyllaries 30–40, distant, 2–3-seriate, equal or subequal, lanceolate, markedly indurate and often with scarious lateral margins; receptacle conical, glabrous or paleaceous. Florets 20–125; corollas white, blue, or lavender, funnelform or with distinct basal tube, lobes triangular, about as long as wide, papillose on inner surface, partially papillose and sometimes hispidulous on outer surface anther collar cylindrical; anther appendage large, somewhat longer than wide; style base not enlarged, glabrous; style arms linear, usually strongly and densely papillose. Achenes prismatic, 4–5-ribbed, glabrous or ribs setuliferous; carpopodium distinct, usually large and asymmetrical; pappus lacking, or coroniform, or of 5 or 6 free, flattened, sometimes awn-like scales.

Native of C America. Several species are now cultivated as garden plants in Europe, N Africa and SW Asia; among them Lindley & Moore (1870) mention *A. mexicanum* Sims (= *A. houstonianum* Mill.) resembled by *A. conyzoides* L. (which is sometimes called *Phalacraea coelestina* by horticulturists).

A genus of some 40 tropical species, all but one American; the exception is *A. conyzoides* L., Goatweed, a common weed now naturalized in Sri Lanka and elsewhere in the tropics; only one species is found in Iraq, widely cultivate throughout the world.

Johnson, M.F. (1971). A monograph of the genus *Ageratum* L. (*Compositae-Eupatorieae*). Ann. Missouri Bot. Gard. 58(1): 6–88 (1971).

Reveal, F.R. & Jarvis, C.E. (2009). Proposal to conserve the name *Ageratum conyzoides* (Asteraceae) with a conserved type." Taxon 58(3): 1011. (2009).

Sharma, V.H. (1987). Comments on the identity of *Ageratum conyzoides* L., and *A. houstonianum* Mill. – two naturalized weeds in India. Feddes Repert. 98(11–12): 557–560.

1. **Ageratum houstonianum** *Mill.*, Gard. Dict. ed. 8, no. 2 (1768); Husain & Kasim, Cult. Pl. Iraq: 108 (1975); Chakravarty, Pl. Wealth Iraq: 7 (1976).

A. mexicanum Sims, Bot. Mag. t. 2524 (1825).

Annual, 20–50 cm tall. Leaves opposite, base cordate, broadly ovate, crenate, long-petiolate. Inflorescence of 10–50 capitula. Involucre 4–6 mm; phyllaries green, linear, acuminate. Corollas blue, exceeded by exserted blue style arms. Achenes angular, truncate, black; pappus of 5 scales, 2–3 mm.

HAB. & DISTRIB. Cultivated under irrigation at horticultural and agricultural experiment stations in the alluvial plain in the desert region of Iraq. LCA: Baghdad, *Janan* C.481!, *Sahira* C.763; alt. ± 35 m; fl. & fr. May & Oct.

Though Husain & Kasim (1975) state that this species is "grown all over the country", no specimens from Iraq other than the above have been seen.

Stout (1935) recommended *Ageratum* (without specifying which species, but probably *A. conyzoides* L.) for cultivation in gardens in Cairo and remarks "sows itself, seldom out of flowers, acclimatized seed produces taller plants". She suggested sowing in August for the flowering season which begins in November in Lower Egypt.

Dinsmore in Fl. Syria, ed. 2, 2: 16 (1933) stated that the Mexican species *A. coelestinum* Sims (= *A. corymbosum* Zucc.) as well as *A. conyzoides* L. are sometimes cultivated in the Levant, whilst Täckholm & Drar (Stud. Fl. Egypt ed. 2: 547, 1974) said that the latter species is common along canal banks in the Delta of the Lower Nile in Egypt and that its var. *mexicanum* Sims is naturalized in Dakhla oasis in the Libyan Desert of Egypt, although *A. conyzoides* is the more widespread species in Egypt (Boulos & Hind in Fl. Egypt vol. 3: 185–186, 2002). Tutin (in Tutin et al., Fl. Eur. 4: 109, 1976) stated that *A. houstonianum* Mill. is frequently cultivated as a garden ornamental and found as a casual in Europe, while [two] species of *Ageratum* are also cultivated as decoratives in Turkey.

123. **EUPATORIUM L.**

Sp. Pl.: 836 (1753); Gen. Pl. ed. 5 : 935 (1754)
Hind & Robinson in Kubitzki, Fam. Gen. Vasc. Pl. 8: 528 [2006] (2007)

Traganthes Wallr., Sched. Crit. 1: 456 (1822); *Eutrochium* Raf., New Fl. 4: 78 (1838); *Chone* Dulac, Fl. Hautes-Pyrénées: 512 (1867), nom. illeg. pro *Eupatorium* L.; *Cunigunda* Bubani, Fl. Pyren. 2: 273 (1899), nom. illeg. pro *Eupatorium* L.; *Uncasia* E. Green, Leafl. Bot. Observ. 1: 13 (1903); *Eupatoriadelphus* R.M.King & H.Rob., Phytologia 19(7): 431 (1970)

D. J. Nicholas Hind

Herbs, annual to perennial. Leaves opposite or verticillate, upper leaves subopposite to alternate, lamina linear to ovate, deltate, or 3-lobed, serrate to subentire. Inflorescence a corymbose or pyramidal panicle. Capitula discoid; phyllaries 10–22, weakly to strongly subimbricate, 2–5-seriate, sometimes inner phyllaries deciduous; receptacle flat or weakly convex, epleaceous. Florets 3–23; corollas white to purple, lavender, or pink, narrowly funnelform or with constricted basal tube and narrowly to broadly campanulate limb, outer surface with glands often concentrated at base of throat and on outer surfaces of lobes, rarely with a few hairs, lobes 5, usually slightly longer than wide; anther collar cylindrical; anther appendages large, ovate-triangular, c. 1.5 × as long as wide; style base puberulent or rarely glabrous, with or without node; style branches filiform to slightly broadened or flattened distally, papillose. Achenes prismatic, 5-ribbed; carpopodium not or slightly differentiated; pappus setae 25–40, barbellate, persistent, apical cells with rounded to shortly acute tips.

32 species in Europe, Asia and eastern N. America.

1. **Eupatorium cannabinum** *L.*, Sp. Pl.: 838 (1753); Dinsmore in Post, Fl. Syria, Palest. & Sinai ed. 2, 2: 16 (1933); Grierson in Fl. Turkey 5: 173 (1975); Taifour & El-Oqlah, Pl. Jordan Ann. Checkl.: 41 (2017).

Eupatorium syriacum Jacq., Icon. Pl. Rar. [Jacquin] 1: 17, t. 170 (1781–1786), non Steven (1856).
E. soleirolii Loisel., Mém. Soc. Linn. Paris 6: 428 (1827).
E. cannabinum L. var. *indivisum* DC., Prodr. 5: 180 (1836).
E. heterophyllum DC., Prodr. 5: 180 (1836).
E. caucasicum Steven, Bull.Soc. Imp. Naturalistes Mscou 29(2): 371 (1856).
Chone heterophylla Dulac, Fl. Hautes-Pyrénées: 512 (1867), nom. illeg. pro *Eupatorium cannabinum* L.
Eupatorium cannabinum L. var. *syriacum* (Jacq.) Boissier, Fl. Orient. (Boissier) 3: 154 (1875).
Cunigunda vulgaris Bubani, Fl. Pyren. (Bubani) 2: 273 (1899), nom. illeg. pro *Eupatorium cannabinum* L.
Eupatorium cannabinum L. subsp. *syriacum* (Jacq.) Nyman, Consp. Fl. Eur. pt. 2: 396 (1879)
E. cannabinum L. subsp. *syriacum* (Jacq.) H. Lindb., Iter Cyprium: 35 (1946), nom. illeg. & comb. superfl.
E. cannabinum L. var. *heterophyllum* (DC.) Kitam., Acta Phytotax. Geobot. 32(5–6): 139 (1981).

Erect perennial herb with many herbaceous stems 0.5–1.5 m tall. Stems serect, purplish-red, usually simple. Leaves opposite, petiolate, petiole *c.* 5 mm, lamina 7–20 × 3–5 cm, tripinnatifid, leaflets ovate to lanceolate, puberulent, bright green, narrowed or rounded at base, margins crenate or incised-dentate, apices acute. Inflorescences densely compound-corymbose. Capitula (3–)5(–7)-flowered; involucres campanulate, after seed dispersal patent to spreading; phyllaries 2–3-seriate, imbricate, elliptical, 1–3 mm, often red-tipped. Corollas 3–4 mm, pink, purplish-red, or white. Achenes cylindrical, 2.5–3.5 × 0.4 mm, 5-ribbed, black; pappus spreading, 3.5–5 mm, straw-coloured or whitish. Fig. 111: 1–6.

HAB. By perennial mountain streams, on damp rocky cliffs, by a mill-stream, in an irrigated *Populus* plantation; alt. 600–1200 m; fl. & fr. Jul.-Oct.
DISTRIB. Occasional in the lower forest zone of Iraq. **MAM**: Sulaf, nr Amadiya, *Guest* 2767!; 25 km from Zakho on way to Kani Masi (Ain Nuni), *Janan* et al. 43828!; Sersang, *Haines* W.1336!, *Barkley & Ani* 9065. **MRO**: Rowanduz gorge, *Guest* 453! 486! 2980!; Shaikh wa Shaikhan valley, *Rawi & Serhang* 24888!

Europe, W & C Asia (Cyprus, Syria, Lebanon, Palestine, Jordan, Turkey, Iran, Caucasia); introduced and probably naturalized in China.

E. syriacum Jacq. was described with entire leaves, but it is not considered distinct as it is easy to find all leaf forms from entire to incised within the same clump of stems.

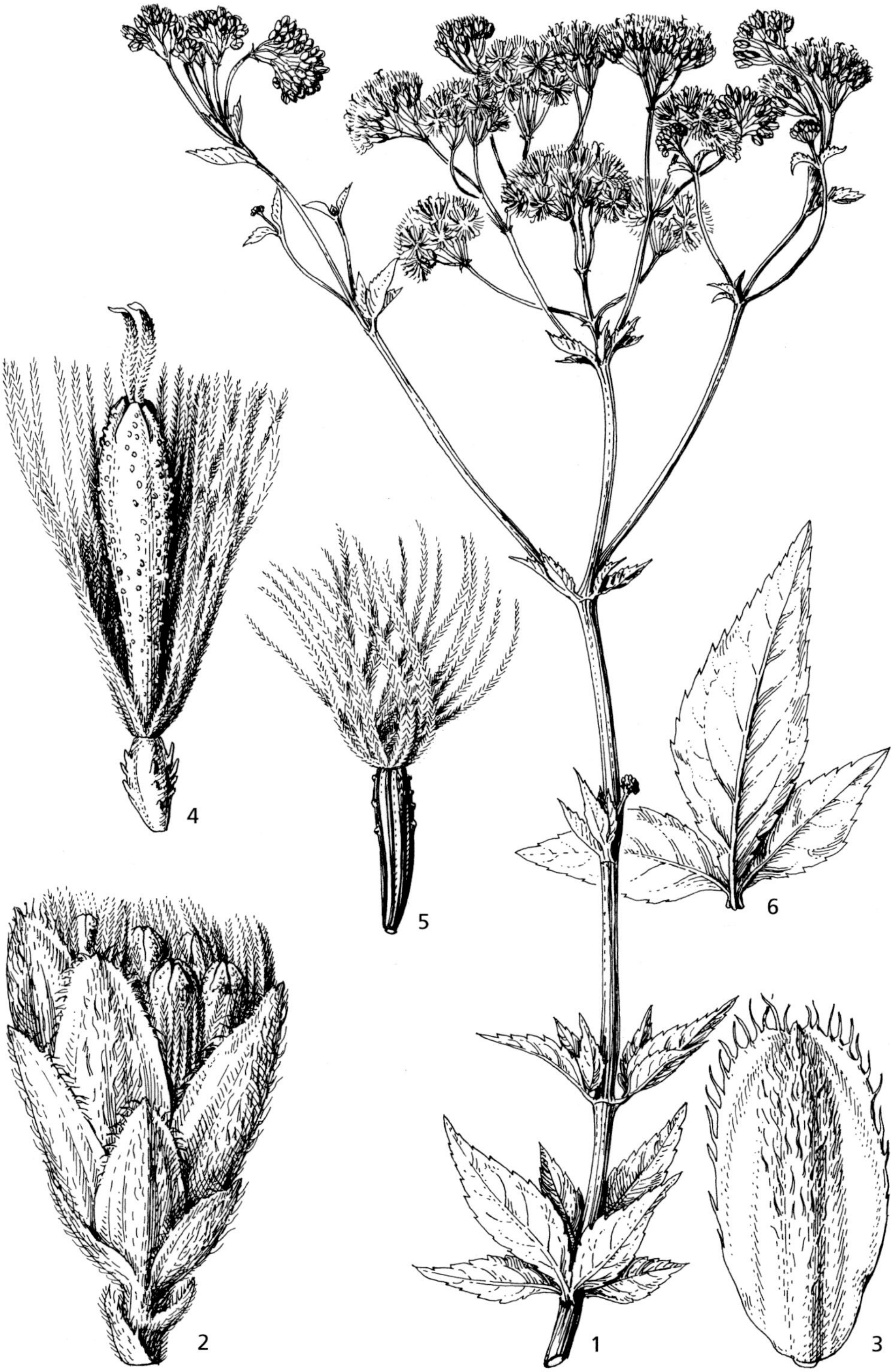

Fig. 111. **Eupatorium cannabinum**. 1, habit × ⅔; 2, capitulum × 16; 3, phyllary × 12; 5, achene and pappus × 10; 6, leaf (undersurface) × 10. All from Wheeler Haines W1336. Drawn by Laura (née Ripley) Mason.

INDEX TO VERNACULAR NAMES

Vernacular names are cited in the text along with their source, and in some cases are translated into English. Arabic and Kurdish names are capitalized in this list; English names are given in lower case letters.

INDEX TO FAMILIES, GENERA AND SPECIES

Scientific names of accepted species and infraspecific taxa are in bold, synonyms and misapplied names are in *italics* and subgeneric and sectional names are in SMALL CAPITALS. Where a name appears on more than one page, the principal page reference is given in **bold.**